Biology and Conservation of Sea Turtles

Revised Edition

Proceedings of the World Conference on Sea Turtle Conservation
Washington, D.C., 26–30 November 1979

with contributions on
Recent Advances in Sea Turtle Biology and Conservation, 1995

Biology and Conservation of Sea Turtles

Revised Edition

**Edited by
Karen A. Bjorndal**

University of Florida

Smithsonian Institution Press

Washington and London

First edition published 1982 by the Smithsonian Institution Press.

Library of Congress Cataloging-in-Publication Data

World Conference on Sea Turtle Conservation (1979 :
Washington, D.C.)
Biology and conservation of sea turtles / edited by Karen A.
Bjorndal. — rev. ed.
        p.   cm.
"Proceedings of the World Conference on Sea Turtle
Conservation, Washington, D.C. 26–30 November 1979 with
contributions on Recent advances in sea turtle biology and
conservation, 1995."
Includes bibliographical references.
ISBN 1-56098-619-0 (alk. paper)
1. Sea turtles—Congresses.   2. Wildlife conservation—Congresses.
I. Bjorndal, Karen A.   II. Title.
QL666.C536W65   1979
597.92—dc20   95-18872

British Library Cataloguing-in-Publication Data is available.

♾ The paper in this publication meets the minimum requirements
of the American National Standard for Permanence of Paper for
Printed Library Materials Z39.48-1984.

Manufactured in the United States of America.

02 01 00 99 98 97 96   5 4 3

Front cover: Adult female green turtle, *Chelonia mydas,* at French
Frigate Shoals, the major migratory breeding site for this species in
the Hawaiian Islands. Photo by G. H. Balazs.

Contents

Recent Advances in Sea Turtle Biology and Conservation, 1995

Preface to the Revised Edition

Since the publication of *Biology and Conservation of Sea Turtles* in 1982, significant advances have been made in our understanding of sea turtle biology and in the conservation of sea turtles. Of course, critical aspects of sea turtle biology still elude us and serious threats continue to jeopardize sea turtle populations. As is natural, those involved in sea turtle biology and conservation tend to dwell on the deficiencies rather than on our accomplishments. I hope that the publication of this revised edition can serve as a celebration of the progress we have made.

The original contributions to the 1982 edition have not been changed. To document the progress made since then, as well as the deficiencies that remain, eighteen authors have written reviews for fourteen topics. The topics were selected to represent those areas in which major advances have been made, not to provide an overview of all major aspects of sea turtle biology and conservation. The authors were given the difficult task of summarizing their fields in very few words and listing the significant new literature, so that readers could quickly grasp new developments and gain access to the primary literature that has appeared in each area since this volume was first published.

The progress has been fueled by improved communication among individuals involved in all phases of sea turtle work. Foremost in this area has been the *Marine Turtle Newsletter*. Initiated in 1976 by Nicholas Mrosovsky, who served as its first editor, the MTN has continued to flourish under the successive editorships of Nat Frazer and, currently, Karen and Scott Eckert.

The Annual Symposia on Sea Turtle Biology and Conservation have played a crucial role in keeping everyone abreast of research and conservation developments. These symposia have grown from a few people gathered in Jacksonville, Florida, in 1981 to gatherings in recent years of nearly 600 people from all over the world. This growth would not have been possible without the careful nurturing provided by Jim and Thelma Richardson, Barbara Schroeder, Sally Murphy, and the scores of other selfless volunteers who undertake the huge task of hosting this event each year. The Proceedings of these symposia—published since 1988 as Technical Memoranda of the National Marine Fisheries Service Southeast Fisheries Science Center—provide rapid communication of the results of the symposia.

The Marine Turtle Specialist Group of the International Union for the Conservation of Nature—a group of over 200 individuals from more than 50 countries—has increased its effectiveness as a network for communication within the international community of sea turtle conservationists. The two West Atlantic Turtle Symposia (WATS I, Costa Rica, 1983, and WATS II, Puerto Rico, 1987), convened under the auspices of IOCARIBE and, guided by Fred Berry, provided important impetus for communication and planning on a regional basis, which have been continued in other areas.

The many publications that have resulted from the decades of research on sea turtles are retrievable, thanks to the Sea Turtle On-line Bibliography developed by Alan Bolten and maintained by the Archie Carr Center for Sea Turtle Research at the University of Florida. This system is available worldwide at no charge on the Internet. Also on Internet, CTURTLE is a listserver discussion group that provides rapid communication among individuals involved in sea turtle biology and conservation.

I want to express my gratitude to Karen Eckert and George Zug, who stimulated interest and organized the funding effort for this revised edition, and to the Atherton Seidell Endowment Fund for providing financial support. I thank the authors of the new sections for writing three-page reviews when they wanted to write thirty pages, and Peter Cannell, science acquisitions editor for Smithsonian Institution Press, for his help and encouragement.

Finally, in the celebration of our progress, we should remember those early workers who made our advances possible: the pioneers who first brought the plight of sea turtles to the attention of the world, who solved many of the early problems, and who served as our mentors. Since we gathered at the World Conference on Sea Turtle Conservation in Washington, D.C., in November 1979, we have lost valued colleagues. Archie Carr passed away in May 1987; Doug Robinson, in June 1991; Leo Brongersma, in July 1994. We see far because we stand on the shoulders of giants. To those giants, I respectfully dedicate this revised edition.

Karen A. Bjorndal
January 1995

Introduction

When the World Conference on Sea Turtle Conservation convened in Washington, D.C. on 26–30 November 1979, more than three hundred participants had come together from forty different nations. Our purpose was threefold: to increase communication between areas of the world where lack of information flow has impeded conservation efforts, to take stock of our knowledge of the biology of sea turtles, and to develop a Conservation Strategy for the world's sea turtles to serve as a philosophical basis and general model for future conservation efforts.

All these endeavors were successful. The information exchange, begun at the conference, has continued unabated and already has had positive results. This volume presents the summary of sea turtle biology made at the meetings. The Conservation Strategy (including a list of specific Action Projects), which appears at the end of this volume, speaks for itself. It is not an integrated methodology for conserving sea turtles. There are too many variables among geographic areas—and too many gaps in our knowledge. Although some specific points in the strategy may become dated, the underlying principles are enduring. In his paper in this book, David Ehrenfeld stated the case succinctly: ". . . a combination of our incomplete knowledge about sea turtles and the numerous constraints imposed by their biology dictates a very conservative conservation strategy. I conclude that the best we can do is to concentrate on the protection of existing wild populations, using the simplest and least risky techniques of conservation."

The papers that make up this book are grouped into three sections, roughly in the order in which they were presented at the conference. Those in the first section deal with different aspects of the basic biology of sea turtles as these apply to conservation problems and possible solutions. The large number of papers on the topic of reproduction shows where the major research efforts have been in the past. Recent research in areas away from the nesting beaches is beginning to round out our understanding of sea turtle ecology. It is in this section that our ignorance seems greatest and most frustrating because our science really has few sure answers to the problems of sea turtle conservation. Many of the papers (for example, the population models) represent only first attempts at deciphering some facet of sea turtle biology and are not intended to be final, accurate representations.

The second section includes papers that report the status of sea turtle populations or that discuss subsistence hunting in different regions. The authors have accomplished an impressive task in placing before us summaries of what is known of the distribution and density of all the sea turtle populations of the world. The coverage is, by necessity, uneven. In some areas, every meter of beach has been walked and all nesting turtles tagged. In others, and unfortunately these are in the vast majority, we have only historical references, or a report from a single plane flight over the region, on which to base our information. Still other areas represent total blanks. In the years ahead, these poorly known areas will certainly grow smaller. It is hoped that the summaries presented here will help focus survey efforts and funds on the places where they are needed most.

The third section comprises papers that deal with conservation theory, techniques and law as well as with general conservation problems that are not restricted to one geographic region. Few papers in the areas of conservation theory and law, as they apply to sea turtles, have been published. Those that appear here are a major contribution to this important field and will be a valuable reference source for years to come. This section also includes the most controversial articles—those treating the long-debated questions of turtle farming and head-starting.

Five of the papers presented at the conference are not included here. Two were committed elsewhere: Carr's West Atlantic Survey is to be published in full by the National Marine Fisheries Service under contract NA 80–GA–C–00071, and Bjorndal's paper on marine turtle life tables appeared in *Copeia* 1980, number 3. The other three papers were not submitted for publication in this volume.

For the participants in the conference, and all sea turtle conservationists, I would like to thank the fol-

lowing organizations whose donations were a major factor in the success of the conference: World Wildlife Fund/U.S.; National Marine Fisheries Service, National Oceanic and Atmospheric Administration, U.S. Department of Commerce; U.S. Fish and Wildlife Service, U.S. Department of the Interior; U.S. Agency for International Development through the Man and the Biosphere Program; Center for Environmental Education; Chicago Zoological Society; New York Zoological Society; U.S. National Park Service, U.S. Department of the Interior; Arabian American Oil Company; Defenders of Wildlife; Truland Foundation; Fauna Preservation Society; Chelonia Institute; U.S. National Shrimp Congress; and the U.S. Navy and U.S. Marine Corps. The U.S. Department of State kindly allowed us to hold the conference in the State Department Building. Other contributors were too many to be listed, but this in no way lessens our gratitude for the generosity of them all. Vivian Silverstein and Patty Shaver worked hard, long hours to organize the conference with great success, and special thanks are due them.

Karen A. Bjorndal
15 January 1981

Contributors

The most recent mailing addresses of contributors to this volume are given below in alphabetical order. Their addresses at the time the research results were presented appear at the head of each article. The addresses of some authors (marked with *) could not be verified for the revised edition.

George H. Balazs
National Marine Fisheries Service
2570 Dole Street
Honolulu, Hawaii 96822-2396 USA

Süleyman Balik
Department of Biological Oceanography
Faculty of Science
Ege University
Bornova-Ismir, Turkey

Mohammed Amour Al Barwani
Ministry of Agriculture and Fisheries
P.O. Box 467
Muscat, Sultanate of Oman

Clark R. Bavin (deceased)
Division of Law Enforcement
U.S. Fish & Wildlife Service
Department of the Interior
Washington, D.C. 20240 USA

Satish Bhaskar
Students Sea Turtle Conservation Body
2/23 A Veteran Lines
Pallavaram
Madras 600043
India

Karen A. Bjorndal
Center for Sea Turtle Research
Department of Zoology
University of Florida
Gainesville, Florida 32611 USA

Alan B. Bolten
Center for Sea Turtle Research
Department of Wildlife Ecology & Conservation
Bartram Hall
University of Florida
Gainesville, Florida 32611 USA

Brian W. Bowen
BEECS Genetic Analysis Core
P.O. Box 110699
University of Florida
Gainesville, Florida 32611 USA

L. D. Brongersma (deceased)
Rijksmuseum van Natuurlijke Historie
Postbus 9517
2300 RA Leiden
The Netherlands

William M. Brown
Division of Education
Piedmont College
P.O. Box 10
Demorest, Georgia 30535 USA

Archie Carr (deceased)
Department of Zoology
University of Florida
Gainesville, Florida 32611 USA

Kim Cliffton
Drylands Institute
2509 N. Campbell Avenue, #405
Tucson, Arizona 85719 USA

Dennis Oscar Cornejo
5402 Avenue F
Austin, Texas 78751 USA

Stephen E. Cornelius
World Wildlife Fund - US
1250 24th Street, N.W.
Washington, D.C. 20037 USA

Deborah T. Crouse
Center for Marine Conservation
1725 DeSales Street, N.W.
Washington, D.C. 20036 USA

*Nelma Carrascal de Celis
Forest Research Institute
College, Laguna 3720
Republic of the Philippines

G. S. de Silva
10 Baldoon Road
Scarborough
Ontario M1B 1V7
Canada

Juan Díaz F.
Departamento de Pesca
Instituto Nacional de la Pesca
Mexico 7, D.F.

C. Kenneth Dodd, Jr.
National Biological Service
7920 N.W. 71 Street
Gainesville, Florida 32653 USA

Nicole Duplaix
27 Boulevard Lannes
75116 Paris, France

Karen L. Eckert
Wider Caribbean Sea Turtle Conservation Network
17218 Libertad Drive
San Diego, California 92127 USA

Scott A. Eckert
Hubbs-Sea World Research Institute
2595 Ingraham Street
San Diego, California 92109 USA

David Ehrenfeld
Department of Natural Resources
Cook College
Rutgers University
P.O. Box 231
New Brunswick, New Jersey 08903 USA

L. M. Ehrhart
Department of Biological Sciences
University of Central Florida
P.O. Box 25000
Orlando, Florida 32816 USA

Richard S. Felger
Drylands Institute
2509 N. Campbell Avenue, #405
Tucson, Arizona 85719 USA

Nat B. Frazer
Savannah River Ecology Laboratory
P.O. Drawer E
Aiken, South Carolina 29802 USA

Jack Frazier
CINVESTAV
Apartado Postal 73 "Cordemex"
Merida 97310
Mexico

Remzi Geldiay (deceased)
Department of Biological Oceanography
Faculty of Science
Ege University
Bornova-Ismir, Turkey

Edgardo D. Gomez
Marine Sciences Center
University of the Philippines
Diliman, Quezon City
Philippines

Derek Green
ESPEY, Huston & Associates, Inc.
P.O. Box 519
Austin, Texas 78767-0519 USA

Coppelia Hays
3493 Cemetery Road
Bowling Green, Kentucky 42101 USA

J. R. Hendrickson
4917 North Camino Arenosa
Tucson, Arizona 85718 USA

Lawrence H. Herbst
College of Veterinary Medicine
University of Florida
Gainesville, Florida 32610 USA

Henry H. Hildebrand
413 Millbrook
Corpus Christi, Texas 78418 USA

H. O. Hillestad
Law Environmental, Inc.
114 Townpark Drive, Suite 250
Kennesaw, Georgia 30144 USA

Huang Chu-Chien
Institute of Zoology
Academia Sinica
7 Zhongguancun Lu
Haidian
Beijing 100080
China

George R. Hughes
Natal Parks Board
P.O. Box 662
3200 Pietermaritzburg
Kwazulu-Natal
South Africa

Elliott R. Jacobson
College of Veterinary Medicine
University of Florida
Gainesville, Florida 32610 USA

C. S. Kar
Forest & Environment Department (Wildlife Wing)
Mangrove Forest Division (Wildlife)
At. & Post. Rajnagar
Dist. Kendrapara, Orissa 754225
India

F. Wayne King
Florida Museum of Natural History
Department of Natural Sciences
268 Museum Road
Gainesville, Florida 32611 USA

Edward F. Klima
92 High Point Road
Skamania, Washington 98648 USA

Tufan Koray
Department of Biological Oceanography
Faculty of Science
Ege University
Bornova-Ismir, Turkey

John Kowarsky
Department of Biology
W.A.I.T., Hayman Road
South Bentley, Western Australia 6102
Australia

Colin J. Limpus
Conservation Strategy Branch
Queensland Department of Environment
and Heritage
P.O. Box 541
Capalaba 4157 Australia

David Mack
425 Feather Rock Drive
Rockville, Maryland 20850 USA

René Márquez M.
Instituto Nacional de la Pesca
INP/CRIP - Manzanillo
AP 591, Playa Ventanas S/N
Manzanillo, Colima, CP 28200
Mexico

Mike A. McCoy
73-1091 Ahikawa Street
Kailua-Kona, Hawaii 96740 USA

Charles McVea, Jr.
Colle Towing Company, Inc.
P.O. Box 340
Pascagoula, Mississippi 39567 USA

James P. McVey
National Marine Fisheries Service
1335 East-West Highway, Room 5492
Silver Spring, Maryland 20910-3225 USA

Anne Meylan
Florida Marine Research Institute
Florida Department of Environmental Protection
100 8th Avenue S.E.
St. Petersburg, Florida 33701-5095 USA

J. D. Miller
Queensland Department of Environment
and Heritage
P.O. Box 5391
Townsville
Queensland 4810 Australia

Edward O. Moll
Biology Department
Southwest Missouri State University
Springfield, Missouri 65804 USA

Jeanne A. Mortimer
Department of Zoology
P.O. Box 118525
University of Florida
Gainesville, Florida 32611-8525 USA

N. Mrosovsky
Department of Zoology
University of Toronto
Toronto, Ontario, Canada M5S 1A1

Daniel Navid
Avenue General Guisan 90
CH-1180 Rolle
Switzerland

Bernard Nietschmann
Department of Geography
University of California
Berkeley, California 94720 USA

Masaharu Nishiwaki (deceased)
University of Ryukyus
Okinawa, Japan

Larry Ogren
6725 Broward Street
Panama City, Florida 32408 USA

Fernando Ortiz-Crespo
FUNDACYT
Ap. 17-12-792
Quito, Ecuador

David Wm. Owens
Department of Biology
Texas A&M University
College Station, Texas 77843-3258 USA

Cuauhtémoc Peñaflores S.
Secretaria de Pesca
Instituto Nacional de la Pesca
CRIP Salina Cruz
Tte. Azueta s/n
Salina Cruz, Oaxaca, C.P. 70690 Mexico

Nicholas Polunin
7 Chemin Taverney
1218 Grand-Saconnex
Geneva, Switzerland

Peter C. H. Pritchard
Florida Audubon Society
460 Highway 436, Suite 200
Casselberry, Florida 32707 USA

Henri A. Reichart
Suriname Forest Service
P.O. Box 436
Parimaribo, Suriname

James I. Richardson
Institute of Ecology
University of Georgia
Athens, Georgia 30602-2202 USA

Thelma H. Richardson
Institute of Ecology
University of Georgia
Athens, Georgia 30602-2202 USA

James Perran Ross
Florida Museum of Natural History
University of Florida
Gainesville, Florida 32611 USA

Manuel Sanchez Perez
Secretaria de Pesca
Instituto Nacional de la Pesca
Chilpancingo 70
Mexico, D.F.C.P. 06170 Mexico

Joop P. Schulz
Worp 3
7419 AB Deventer
The Netherlands

Wilber R. Seidel
National Marine Fisheries Service
Southeast Fisheries Center
Pascagoula Laboratory
P.O. Drawer 1207
Pascagoula, Mississippi 39567-1207 USA

Igal Sella
81 Derekh Ha'yam
Haifa, Israel

Stephen V. Shabica
National Biological Service
1342 USS Simon Bolivar Road
Kings Bay, Georgia 31547-2529 USA

Siow Kuan Tow
17, Jalan Hijau Mida 4
80400 Johor Bahru
Malaysia

*Iman Soetrisno
Faculty of Fisheries
Bogor Agricultural University
Bogor, Indonesia

James R. Spotila
Department of Bioscience and Biotechnology
Drexel University
Philadelphia, Pennsylvania 19104 USA

C. Silvia Spring
Marine and Coastal Section
Department Arts, Sport, Environment, Tourism &
Territories
P.O. Box 737
Canberra ACT 2601
Australia

Stephen E. Stancyk
Marine Science Program
University of South Carolina
Columbia, South Carolina 29208 USA

Njoman Sumertha Nuitja
Faculty of Fisheries
Bogor Agricultural University
Dramaga Campus
Bogor, Indonesia

Ismu Sutanto Suwelo
Directorate General of Forest Protection and Nature
Conservation
9 Jalan H. Juanda
Bogor 16122
Indonesia

Itaru Uchida
Port of Nagoya Public Aquarium
1-3 Minato-Machi
Minato-ku
Nagoya City 455
Japan

Aristóteles Villanueva O. (deceased)
Departamento de Pesca
Instituto Nacional de la Pesca
Mexico 7, D.F.

J. M. Watson, Jr.
National Marine Fisheries Service
Pascagoula Laboratory
P.O. Drawer 1207
Pascagoula, Mississippi 39567 USA

Susan Wells
56 Oxford Road
Cambridge, CB4 3PW
United Kingdom

Ross Witham
1457 N.W. Lake Point
Stuart, Florida 34994 USA

Blair E. Witherington
Florida Marine Research Institute
Tequesta Field Laboratory
19100 Southeast Federal Highway
Tequesta, Florida 33469 USA

C. L. Yntema (deceased)
Department of Anatomy
Upstate Medical Center
Syracuse, New York 13210 USA

George R. Zug
Division of Amphibians and Reptiles
National Museum of Natural History
Washington, D.C. 20560 USA

Sea Turtle Biology

Overview

Archie Carr
Department of Zoology
University of Florida
Gainesville, Florida 32611

Notes on the Behavioral Ecology of Sea Turtles

ABSTRACT

Knowledge of the behavioral ecology of sea turtles, though still riddled with gaps, has progressed to the point where a short review is bound to be inadequate. This summary calls attention to features of sea turtle behavior that seem especially noteworthy, and to the major imperfections in knowledge of the group. As an example of the degree to which innate patterns of response govern sea turtle behavior, the peculiarly stereotyped process by which the egg-cavity is dug is described. This one behavioral trait distinguishes all sea turtles from all the terrestrial and freshwater species. Stages in the behavioral development of sea turtles are rapidly reviewed, as follows: ascent of hatchlings from nest to surface; traversal of the beach; traversal of the surf; orientation behavior after passing the breakers. The bearing of the last on the lost year puzzle and on the sargassum raft theory of hatchling refuging is discussed and the status of our knowledge of orientated movements within developmental habitats and in reproductive travel is assessed. The probable existence of nonmigratory populations or population segments is pointed out; and basking and hibernation as possible alternatives to migration are discussed.

The natural history of marine turtles has received growing attention during the last 20 years, and much has been learned. Much remains to be learned, however, and the gaps are both an intellectual challenge and an obstacle to conservation. My aim in the present paper is to point out what seem to me high points in the behavioral ecology of the group, and to call attention to conspicuous gaps in our knowledge.

The ecology of an air-breathing vertebrate in the ocean is bound to be complicated. The Cetacea have solved the problems through cerebration, sociality, and parental care. The sea turtles show little of these—none of the last—and instead have evolved a remarkably successful repertoire of purely inherent responses to the demands of a changing suite of environments.

A trait that epitomizes the role of machinelike instinctive patterns in sea turtle behavior is the technique used in digging the cavity in which the eggs are laid. For a long time I have found this hard to explain and impossible to ignore. Many years ago, in an effort to make sense out of the confused taxonomy of sea turtles, I tabulated some osteological characters of the four thecophoran (hardshelled) genera and concluded that, while it was probably best to keep them all in a single family (Cheloniidae), a case could be made for giving separate familial recognition to *Chelonia* and placing *Lepidochelys, Eretmochelys* and *Caretta* together in another family (Carr 1942). Dr. Rainer Zangerl pointed out that my assessment gave undue weight to certain characters that were all just expressions of the same structural adaptation (Zangerl and Turnbull 1955). In his view, *Chelonia* and *Eretmochelys* seemed clearly allied, on the one hand, and *Caretta* and *Lepidochelys* on the other; although he saw no reason to recognize separate families for any of the group. Dr. Zangerl is a distinguished paleontologist, and I bowed to his pronouncement; but only insofar as his criticism of my skull characters was concerned. I reserved judgment on the relationships of *Chelonia*.

Later on I turned to behavioral characters that I hoped might provide criteria for judging relationships among the genera, including those of the aberrant *Dermochelys*. One aspect of sea turtle behavior which, because of its stereotyped complexity, might be expected to show relationships, is the activity of the nesting female. During the first years of our sea turtle program at the University of Florida, we accumulated films and photographic records of nesting behavior, as material for such comparisons. Almost at once it became clear that nesting behavior was a somewhat unreliable source of taxonomic criteria.

There are differences among the genera, to be sure. John Hendrickson has listed some of them in a paper in this volume. For example, a ridley compresses the sand over her finished nest by rocking laterally and slapping the sand with alternate sides of the plastron. With much diminished emphasis, this can be seen in *Eretmochelys,* and to even less degree in *Caretta*, so it might be interpreted as evidence of affinity among the three. On the other hand, the fact that *Chelonia* and the leatherback do no pounding may simply reflect their greater body weight—as may also the marked difference in gait between the same two sets of genera. Other variations in the behavior of sea turtles on shore may be merely phenotypical responses to environmental differences. For example, most Tortuguero green turtles make no trial egg-holes before completing one; on Ascension Island this task may be begun and abandoned many times during one emergence (Jeanne Mortimer, personal communication). Also, a Tortuguero green turtle almost always turns quickly back into the sea if a light is shined on her when she emerges from the surf; Ascension Island turtles seem usually unperturbed by lights. Detailed attention to each stage of the nesting process reveals minor differences that may be genetically controlled, and thus of taxonomic utility. The salient feature, however, is an astonishing conservatism.

This impression reaches a peak in the process by which a sea turtle—any sea turtle, anywhere on earth—uses the hind feet to form the flask-shaped cavity in which her eggs will be deposited. In this operation the digging is done by the back flippers. Their work is inflexibly stereotyped. The feet work alternately. A foot is brought in beneath the hind edge of the shell, and its edge is pressed against the ground and curled to pick up a small amount of sand. The cupped foot is then lifted and swung laterally, and the sand falls; and instantly, the other hind foot, which until then has rested on the sand by the rim of the egg cavity, snaps sharply forward, throwing sand from beside the hole to the front and side. This sequence is then precisely repeated in reverse. The nest-digging process is an unvarying series of these reciprocating actions of the hind feet. It continues until the nest grows to a depth equal to the reach of the hind leg. If this ritual were confined to a single species it would be arresting, because the rigidly stereotyped behavior seems to go beyond ecologic utility. When one finds it in almost unmodified form in each of the five genera of sea turtles, one longs for a logical explanation. At least I do.

No other kind of turtle appears to be bound by such discipline. I have watched the nesting of soft-shells, snapping turtles, box turtles and several species of *Pseudemys*, and can clear them of any such hidebound behavior. According to Eglis (1963) a parrot-beaked tortoise that he observed "scooped small amounts of earth 8 to 12 times with the near side of the left heel and deposited them near the right foot . . ." sometimes resting a few seconds before completing the sequence. That side-neck turtles practice no such strict manual of arms as sea turtles is clear from Vanzolini's observation (1967) of the nest digging of *Podocnemis expansa*. A female that he watched "started scooping vigorously with the (left) hind foot . . . throwing the sand toward the opposite side of the body. After 4 to 10 scoops she would stop for 15 to 30 seconds, then start another, usually shorter, series of movements with the same foot."

There is no such laxity in a sea turtle. She scoops and drops sand with the left foot, then kicks sand forward with the right foot, then invariably puts the right foot in the nest, then repeats the process in reverse. If this were the only way, or even clearly the best way,

ARCHIE CARR

to dig a hole in sand, the machinelike rigidity of the drill would seem reasonable. When one finds that all the other kinds of sea turtles make precisely the same ritualistic approach to nesting, puzzlement is compounded. Sea turtles dig nests in material that ranges from fine siliceous particles to spherical pellets of calcareous algae. The density, friability, and angle of repose of the nesting media vary widely, and suggest nothing that explains why sea turtles, as compared with turtles of land or fresh water, have so consistently stereotyped their technique of shaping an egg chamber.

When I watch film clips showing, in succession, nest digging by each of the five genera only one, very slight divergence is revealed. This occurs in the forward sand-kick of the back flipper of the leatherback. In all the others the fin that kicks returns in one smooth motion to take its turn in digging. In *Dermochelys*, immediately after slapping forward the foot swings part way back, then very weakly repeats the kick before returning to its place in the egg-hole.

Minimum chronologic separation of modern sea-turtle lines is probably no less than 30 million years. That leading to *Dermochelys* is probably at least 50 million years old. It seems almost irrational to believe that the slavish concordance of the five genera in their identical technique of digging a hole is convergent. The only alternative is that it has been inherited from a common ancestor. Zangerl (1980) believes that *Dermochelys* belongs in the same family as the other sea turtles. There is evidence from both immunological criteria (Frair, 1979) and chromosome morphology (Bickham, 1979) that supports this view. However, the two lines may have been separate since at least the Late Cretaceous; and this seems an unconscionable length of time to retain a pattern of behavior the stereotyped detail of which appears to go so far beyond adaptive demand.

Nevertheless, a person asked what a sea turtle is, might truthfully reply: "A sea turtle is a kind of turtle that never puts the same back foot into its egg-hole twice in succession."

Two fundamental features of the life cycles of sea turtles complicate their ecology. One is their adherence to the ancestral habit of laying eggs in holes in the ground. This obligation draws them to land, often from distant foraging grounds, and introduces severe ecologic problems for both the female turtle and her offspring. The other complicating factor is their large size. The advantages of being big are obvious; but in sea turtles the change from hatchling-size to ponderous maturity is accompanied by repeated shifts in foraging requirements, and thus in habitat.

The burden of being born on land rests heavily upon the hatchlings; and they have responded with dramatic adaptations. They begin coping with their environment a meter down in the sand of the sea beach. Any overview of sea turtle ecology, however cursory, should

signalize the protocooperative activity by which newly hatched young turtles, imprisoned in a hat-sized chamber far down under settled sand, move up *en masse* to the surface. The process is apparently not mediated by geotactic stimuli. The hatchlings are not striving toward the surface as a goal, but merely reacting to local stimuli in ways that take them toward the surface. They do not dig upward; they merely thrash about.

The process has been watched and photographed in nests of *Chelonia, Dermochelys, Eretmochelys,* and *Caretta,* behind glass panes, both in the laboratory and in natural nests in the field. In all cases, the rise to the surface has been achieved by sporadic outbursts of thrashing, usually triggered by one turtle and quickly spreading through the clutch. This activity automatically dislodges sand from the walls and ceiling, builds up the floor of the chamber, and carries the group to within a few inches of the surface. There the sand often sinks in a circular area above the hatchlings. They remain quiet there for a while, evidently awaiting the propitious temperature change that usually occurs near dawn (Bustard 1967; Mrosovsky 1968).

This example of group facilitation (Carr and Hirth 1961) is particularly noteworthy because from there on, the social organization of sea turtles is weak. There is fleeting social intercourse during courtship and mating, and in *Lepidochelys* the fantastic reproductive aggregations variously known as *arribadas, arribazones, morriñas,* and *flotas* must be adaptive. Otherwise, little sociality has been observed.

As soon as the hatchlings are out of the nest and on the surface of the beach, each separately faces the problem of finding the sea. A voluminous literature, summarized by Ehrenfeld (1968) and Mrosovsky (1978), describes and seeks to explain their remarkable sea-finding ability. A mechanism that will account for its full versatility has not yet been identified. Hatchlings of all the genera are able to take accurate seaward headings from most nest sites even when the surf is not in view. They can do this even when the emergence site is experimentally moved to a shore with diametrically opposite exposure (Carr and Ogren 1960). That means that the guiding sense is not a combined, genetically represented, compass sense and regional map sense. Light is clearly important in the guidance process, but the exact way in which light guides the turtles is not clear. Recent results of a series of experiments with green turtle hatchlings, in which hoods were used to interfere with vision, lead Van Rhijn (1979) to the conclusion that a mechanism drawing the turtle toward maximum brightness is inadequate. He suggested that a multiple input system, as defined by Schone (1975), must be involved, perhaps with some reference to silhouette patterns and to substrate inclination as well. His conclusions seem sound, but they still do not provide a complete model for the versatile sea-finding

once extended northward along the coast to the U.S. frontier, and a few individuals have lately been turning up in San Diego Bay (Stinson, in litt.). It may be a straw in the wind that these regularly take refuge in warm water discharged from a power plant. Margie Stinson is studying the behavior of this interesting colony.

The other alternative to emigration is hibernation. It is sad for science that depletion of sea turtles will prevent adequate study of both basking and hibernation. Both are significant adaptations, investigation of which would enhance understanding of sea turtle ecology and the physiology of large reptiles. The distribution of both has been so reduced by the decline of temperate zone turtles, however, that the original prevalence and geography of the traits will never be determined.

In the case of hibernation this is particularly distressing to me, because in Florida many years ago I was party to neglecting what now seems to have been an opportunity to study the hibernation ecology of two species—the green turtle and Kemp's ridley. Carr and Caldwell (1956) recorded fishermen's reports that ridleys and green turtles went into winter dormancy in the mud off the west coast of the Florida peninsula. They gave the reports inadequate attention, however. After Felger, Cliffton and Regal (1976) reported hibernation by *Chelonia agassizi* in the Gulf of California, I belatedly reinterviewed Florida fishermen old enough to remember times before the loss of the West Coast ridleys and green turtles. It now seems to me probable that the immature turtles which, each April in Florida in the Cedar Key-Wacassassa area, came partly in from the south as migrants, also came partly "up out of the mud," as informants consistently believed. I have recently learned from the manager of the University of Florida Marine Laboratory on Seahorse Key that as a boy he used to dive up turtles that he located from a boat, during winter months, as humps in the mud of the bottom. During the winter of 1978 we went out and searched for such humps and did exploratory trawling in the area. We found nothing. Both of these once populous colonies are nearly gone. Felger et al. (1976) gave their torpid black turtles a proper presentation in *Science*. At the time they discovered the colony, however, Mexican scuba divers were beginning to exploit the torpid turtles commercially. That drain, combined with incidental catches by trawlers is now thinning out the hibernating contingent dangerously (Cliffton, personal communication).

More recently Carr, McVea, and Ogren (1980) described an aggregation of loggerheads in a 6-mile, man-made channel leading into Port Canaveral on the eastern coast of Florida. At all seasons the concentration of turtles in the channel is extraordinary. When the colony was discovered in the winter of 1977 the water off Florida had reached exceptionally low temperatures. Loggerheads taken by trawling in the channel during March of that year were mostly immature, and nearly all were torpid. Many showed unmistakable signs of having been dislodged by the trawl from the clay walls and bottom-mud of the cut. It has since been learned that loggerheads assemble in the channel at all seasons and that, during the winter, the colony is mainly composed of subadults while in summer it includes mature turtles some of which are females bearing tags that have been put on at nearby nesting beaches.

The ecologic attraction of the Canaveral channel has not been identified. It is not known whether its use as a hibernaculum is a wholly unique occurrence, evoked by some special feature of the artificial channel, or whether it reflects a prevalent habit of North American sea turtles that has just been overlooked (Ogren and McVea, this volume).

One of the most striking attributes of marine turtles is their ability to cross open ocean and make scheduled landfalls at ecologically necessary places, or to go back to such places if experimentally displaced. Like much of what we know about the group, this homing ability was widely known to seaside people before any zoologist ever put a tag on a turtle. When tagging projects began to develop, it was quickly corroborated, and eventually the ability of turtles to find little islands in the open sea was clearly revealed. This appears to require highly evolved guidance adaptations, comparable to those of terns, albatrosses, and other birds—and just as poorly understood.

Although usually thought of as a reproductive adaptation, homing orientation is not confined to breeding migrations. It is also clearly involved in the maintenance of home-range boundaries by nonbreeding individuals. It has long been known to professional turtle fishermen that when green turtles escape after being displaced great distances away from resident foraging or developmental range, they are capable of making quick, accurate returns. Carr and Caldwell (1956) recorded such returns by young Florida green turtles that had been displaced as far as 30 miles; and more recently Burnett-Herkes (1974) and Ireland (1979; 1980) reported an impressive series of homing performances at Bermuda. Balazs (1976, 1980) found strong feeding-site fidelity in Hawaiian green turtles, and Limpus (in litt.) has observed the same tendency in Australian loggerheads. It is widely believed by Florida fishermen that some loggerheads have home rock patches to which they return year after year. Striking corroboration of this is Norine Rouse's 6-year record (personal communication) of a male loggerhead that returns each fall to the same rock 15 meters down on a reef off Palm Beach. Nietschsmann (in press) records repeated territorial returns by displaced tagged hawksbills on Miskito Bank.

Animal orientation has received increasing attention in recent years. The proceedings of a recent symposium at Tubingen, Germany, (Schmidt-Koenig and Keeton 1978) reveal dramatic advances in defining homing capacities and revealing hitherto unknown ramifications of sensory physiology. One useful upshot of all this research is the evidence it provides that a broad palette of orienting cues is available to migrating animals. In any complex pathfinding feat, the migrant almost surely uses a number of different guidesigns. Despite the brilliant progress in both laboratory and field studies, however, we still have no idea how animals navigate. One reason for this may be that most of the field research has been done with birds traveling overland, where they are probably in touch with a number of different cues and where interpretation of field experiments is accordingly complicated. The ultimate refinement of animal navigation would seem to be the capacity of many species to make accurate landfalls after long journeys in the open sea. Such travel takes place in what must be tbe least cluttered theater for tracking experiments, and, if migratory paths could be accurately traced, carefully designed experiments would yield a wealth of circumstantial evidence by which navigation theories could be compared and assessed. In the case of the marine turtles, there seems no doubt that the experiments, though logistically difficult, through satellite telemetry are within the grasp of investigators. They are long overdue. The island-finding urge and ability rank among the most imposing behavioral adaptations that natural selection has produced. The lack of a satisfactory theory to explain the guidance mechanism is an embarrassment to science.

Acknowledgments

The writer's sea turtle research has had the long-time support of the National Science Foundation and the Caribbean Conservation Corporation.

Literature Cited

Balazs, G.
1976. Green turtle migrations in the Hawaiian Archipelago. *Biological Conservation* 9:125–40.
1980. Synopsis of biological data on marine turtles in the Hawaiian Islands. National Oceanic and Atmospheric Administration Technical Memorandum NOAA-TM-NMFS-SWFC–7, 141 pp.

Bickham, J.
1979. Karyotypes of sea turtles and the karyological relationships of the Cheloniidae. *American Zoologist* 19:983 (abstract).

Burnett-Herkes, J.
1974. Returns of green sea turtles (*Chelonia mydas* Linnaeus) tagged at Bermuda. *Biological Conservation* 6:307–8.

Bustard, H. R.
1967. Mechanism of nocturnal emergence from the nest in green turtle hatchlings. *Nature* 214:317.

Carr, A.
1942. Notes on sea turtles. *Proceedings of the New England Zoological Club* 21:1–16.
1963. Orientation problems in the high seas travel and terrestrial movements of marine turtles. In *Bio-Telemetry* pp. 179–93. Oxford: Pergamon Press.
1972. The case for long range chemoreceptive piloting in *Chelonia*. NASA Publication SP 260:179–93.

Carr, A., and D. K. Caldwell
1956. The ecology and migrations of sea turtles, 1. Results of field work in Florida, 1955. *American Museum Novitates* 1793:1–23.

Carr, A., and H. Hirth
1961. Social facilitation in green turtle siblings. *Animal Behavior* 9:68–70.

Carr, A., and A. B. Meylan
1980. Evidence of passive migration of green turtle hatchlings in sargassum. *Copeia* 1980(2):366–368.

Carr, A., and L. Ogren
1960. The ecology and migrations of sea turtles, 4. The green turtle in the Caribbean Sea. *Bulletin of the American Museum of Natural History* 121:1–48.

Carr, A.; L. Ogren; and C. McVea
1980. Apparent hibernation by the Atlantic loggerhead turtle off Cape Canaveral, Florida. *Biological Conservation* 19:7–14.

Dampier, W.
1906. *Dampier's Voyages* (1679–1701), John Masefield, ed. New York:

Eglis, A.
1963. Nesting of a parrot-beaked tortoise. *Herpetologica* 19:66–68.

Ehrenfeld, D.
1968. The role of vision in the sea finding orientation of the green sea turtle (Chelonia mydas). 2. Orientation mechanism and range of spectral sensitivity. *Animal Behavior* 16:281–87.

Felger, R. S.; K. Cliffton; and P. J. Regal
1976. Winter dormancy in sea turtles: Independent discovery and exploitation in the Gulf of California by two local cultures. *Science* 191:283–85,

Frair, W.
1979. Contributions of serology to sea turtle classification. *American Zoologist* 19:983 (abstract).

Frick, J.
1976. Orientation and behavior of hatchling green turtles (*Chelonia mydas*) in the sea. *Animal Behavior* 24:849–57.

Green, D.
1979. Double tagging of green turtles in the Galapagos Islands. *IUCN/SSC Marine Turtle Newsletter* 13:4–9.

Ireland, L. C.
1979. Homing behavior of immature green turtles (*Chelonia mydas*). *American Zoologist* 19:952 (abstract).
1980. Homing behaviour of juvenile green turtles (*Chelonia mydas*). In *A Handbook on Biotelemetry and Radio Tracking,* eds. C. J. Amlaner, Jr. and D. W. Macdonald. Oxford: Pergamon Press.

Ireland, L. C., J. Frick, and D. B. Wingate

1978.   Nighttime orientation of hatchling green turtles (*Chelonia mydas*) in open ocean. In *Animal Migration, Navigation and Homing,* eds. K. Schmidt-Koenig and W. T. Keeton, pp. 420–29. Berlin: Springer-Verlag.

Mrosovsky, N.

1968.   Nocturnal emergence of hatchling sea turtles: control by thermal inhibition of activity. *Nature* 220:1338–39.

1978.   Orientation mechanisms of marine turtles. In *Animal Migration, Navigation and Homing,* eds. K. Schmidt-Koenig and W. T. Keeton. Berlin: Springer-Verlag.

Nietschmann, B.

In press.   Following the underwater trail of a vanishing species: the hawksbill. *National Geographic Society Research Reports,* 1972.

Schmidt-Koenig, K., and W. T. Keeton, eds,

1978.   *Animal Migration, Navigation and Homing.* Berlin: Springer-Verlag.

Schone, H.

1975.   Orientation in space: animals. In *Marine Ecology* vol. 2, part 2, ed. O. Kinne, pp. 499–553. London: Wiley.

Van Rhijn, F. A.

1979.   Optic orientation in hatchlings of the sea turtle, *Chelonia mydas,* 1. Brightness: not the only optic cue in sea-finding orientation. *Marine Behavior and Physiology,* 6:105–71.

Vanzolini, P. E.

1967.   Notes on the nesting behavior of *Podocnemus expansa* in the Amazon Valley (Testudines, Pelomedusidae). *Papeis Avulsos Zool. Sao Paulo,* 20:191–215.

Whittow, G. C., and G. H. Balazs

1979.   The thermal biology of Hawaiian basking green turtles (*Chelonia mydas*). *American Zoologist,* 19:981 (abstract).

Zangerl, R.

1980.   Patterns of phylogenetic differentiation in the Toxochelyid and Cheloniid sea turtles. *American Zoologist,* 20:585–596.

Zangerl, R., and W. D. Turnbull

1955.   *Procolpochelys grandaeva* (Leidy), an early carettine sea turtle. *Fieldiana Zoologica,* 37:345–82.

Reproduction, Nesting, and Migration

L. M. Ehrhart
University of Central Florida
Department of Biological Sciences
P. O. Box 25000
Orlando, Florida 32816

A Review of Sea Turtle Reproduction

ABSTRACT

Selected behavioral aspects and factors contributing to the reproductive potential of marine turtles are reviewed. Courtship and mating have been described only for *Chelonia mydas*. Interpretation of the adaptive functions of courtship and mating awaits the development of quantified description for all species. The temporal relation of a season's mating to the laying of fertilized eggs has provoked much discussion but received little research attention for 25 years. Although there is a wealth of nesting-behavior description available, most accounts lack quantification and no comparative ethologic synthesis has been attempted. There is much information concerning clutch size and egg size in the literature. Sea turtles do not conform to the direct, positive relationship of clutch size and body size, seen in turtles as a group. They do, however, adhere to the inverse relationships between egg size and clutch size and between relative egg size and adult body size. The most commonly observed multi-annual cycles are 2 and 3 years, but these are often based on small percentages of multi-annual recoveries. Within season internesting intervals average 12 to 15 days; in some forms they are 9 to 10 days; and they vary with environmental conditions in others. The number of clutches laid per nesting year varies from 1 to 11 and the frequency of 1 and 2 layings a year may be greater than previously thought. Fertility rates have been examined by only a few researchers; they seem fairly stable at 80 to 90 percent in loggerheads and leatherbacks.

Introduction

In his classic chapter entitled "A Hundred Turtle Eggs," Carr (1967) argued that "the whole race and destiny of the creature are probably balanced at the edge of limbo by the delicate weight of that magic number of eggs." He recognized that virtually all other aspects of sea turtle life history are ". . . to some degree reflected in the number of eggs the female drops into the hole she digs in the sand." Indeed, sea turtle biologists the world over have recognized the fundamental nature of clutch size and, almost to a man, have dutifully counted and reported the numbers of eggs deposited by individuals of the races with which they were concerned. The actual adaptive strategy, however, that results from the relationship of reproductive factors to demographic features is difficult to define. Placement of sea turtles on the r-K selection continuum of MacArthur and Wilson (1967) is obscured by at least two conditions. First, there is little information concerning selection correlates for most species. Second, where those factors are known, they suggest placement at different, even opposite, points on the continuum. For those attributes tabulated by Pianka (1978), for example, patterns of mortality, climate, stability, body size, and age at maturity suggest the "K-strategy," while those of survivorship and fecundity favor "r selection." It is not my intent to deal with the latter problem, but rather to address the former, with a review of what is and is not known about marine turtle reproduction. Constraints of space require that I be selective in this endeavor.

Behavior

Courtship and Mating

The literature contains few detailed descriptions of courtship and mating behavior for species other than *Chelonia mydas* (Table 1). For green turtles, however, the works of Booth and Peters (1972), Bustard (1972), and Hendrickson (1958) provide a wealth of detail upon which the comparative ethology of sea turtle mating can be built. The other works cited in Table 1 are generally shorter and less detailed but also provide

useful observations. Careful reading of these accounts reveals, however, certain differences in behavioral detail among them. Bustard (1972), for example, describes the beginning of courtship as follows: "When a male green turtle first approaches a female it swims round to face the female and nuzzles her head, rather like rubbing noses." Continued nuzzling of the neck and shoulders accompanied by nonaggressive "bites" leads to "biting actions at one of the rear flippers." If the female remains in place, attempts at mounting and copulation follow. Booth and Peters (1972), on the other hand, describe head-to-head posturing only in the context of female rejection (Hendrickson [1958] also mentions this) and note that "biting" (gentle nipping) plays a role in courtship only occasionally. Aspects of mating described by Booth and Peters (1972) but not mentioned by others include a "refusal" position ("she will turn toward him and assume a vertical position in the water, with the plastron facing the male, and all limbs wide-spread") and the existence of a "female reserve." In the "reserve area" females rest on the bottom sand, apparently immune to the blandishments of males, which appear to avoid the area completely.

The references to mating behavior in other sea turtle species (Table 1) contain accounts that are basically anecdotal. They involve no underwater observations and present no photographs, as Booth and Peters (1972) and Bustard (1972) do. Without a more extensive comparative base, it is impossible to surmise the functional significance of precopulatory activities. Surely certain aspects of green turtle courtship serve to arouse the female, to bring her into a receptive condition, but the extent to which these activities occur in other species is unknown. The ethologically more important question of the usefulness of these traits as species-specific communication signals and agents of reproductive iso-

Table 1. Descriptive accounts of courtship and mating in marine turtles

Species	Locality	Reference
Chelonia mydas	Australia	Booth and Peters (1972)
	Australia	Bustard (1972)
	Costa Rica	Carr and Giovannoli (1957)
	Aldabra	Frazier (1971)
	Borneo	Harrisson (1954)
	Malaya-Sarawak	Hendrickson (1958)
	Surinam	Schulz (1975)
	Cayman Is. (captive)	Simon et al. (1975)
	Cayman Is. (captive)	Ulrich and Owens (1974)
	Florida (captive)	Witham (1970)
Eretmochelys imbricata	Seychelles Is.	Hornell (1927)
Caretta caretta	Georgia	Caldwell et al. (1959)
	South Carolina	Caldwell (1959)
	Florida (captive)	Wood (1953)
Lepidochelys kempi	Mexico	Chavez et al. (1968)

lation is even further from solution. The fact that sexually aroused male green turtles attempt to mount almost any appropriately sized object in the water (including skin divers and roughly-fashioned decoys) suggests that complex species-typical courtship behavior may be lacking in this species and others as well. At this writing, then, we simply are not ready to construct a comparative ethological synthesis of sea turtle courtship and mating behavior.

Delayed Fertilization

The problem of the temporal relationship of copulation, fertilization and egg-laying has puzzled marine-turtle biologists for at least 25 years. It has two major aspects. The first involves the question of long-term sperm storage: are eggs laid in a given season fertilized by spermatozoa from a mating 2 or 3 years in the past?" The second involves the need for repeated matings during the nesting season: are repeated matings necessary to insure the fertility of multiple clutches? The problem emerged during the mid-1950s when Harrisson (1954), speaking of copulation in green turtles, said, "It looks to me as if it mainly occurs after the female has laid." Then in 1957, Carr and Giovannoli noted that female green turtles are sometimes pursued by males just after laying and questioned whether mating may take place before or after nesting, or both. Hendrickson (1958) also mentioned the possibility of postnesting mating and so did Carr and Ogren (1960). None of these, however, seemed even moderately convinced that such a condition necessarily prevailed. Nevertheless, the idea began to take hold in the literature of the 1960s as Carr and Hirth (1962) remarked that, "since mating brings about fertilization of eggs that will be laid two or three years later and has nothing to do with the eggs of the season, it seems likely that copulation could take place equally well before or after nesting, and that it may occur at both times." Carr (1965) attempted to clarify the issue by arguing that it seemed unlikely that any of the eggs of the current season were fertilized as a result of current mating because, even when mating precedes the first nesting, it would have to take place after at least some of the female's eggs had formed shells. He concluded that "the encounter probably serves to fertilize eggs for the next nesting season, two or three years ahead."

The idea of delayed fertilization was called into question by Frazier (1971) when he posed the following problems: 1) the assumption that nesting females without claw marks on the carapace have not mated and that eggs laid by such females are fertile, lack evidence; 2) the necessity for virgin females to make their first migration over thousands of kilometers to mate but not nest seems nonadaptive; 3) the migration of males to the breeding ground in order to deposit spermatozoa for 3 years hence instead of mating at the feeding ground a few months earlier or later seems, likewise, nonadaptive; 4) the mechanisms by which females could maintain viable sperm in the reproductive tract are unknown.

The last contention is met, at least in part, by observations of fertile eggs laid by female *Terrapene* (Ewing 1943), *Malaclemys* (Barney 1922) and *Chelydra* (Smith 1956) after several years in isolation. Perhaps sea turtles are capable of similar feats. Solomon and Baird (1979) have recently indicated the presence of living spermatozoa in the lower reproductive tract of female green turtles. Schulz (1975), however, has also expressed doubt about the delayed fertilization hypothesis.

Recently, Carr et al. (1978), noting the suggestion in Schulz (1975) and Cornelius (1975) that non-nesting migrations to the breeding grounds occur, have concluded that the temporal relation of a season's mating to the laying of fertilized eggs is still an open question. Furthermore, they argued persuasively that the ability to store sperm could be a critical preadaptation in the ecological evolution of the species, greatly increasing the chances that a single inseminated female could save a colony otherwise destroyed by a natural disaster.

Clearly, this is a problem that has begged for the attention of directed research since Harrisson's statement of 25 years ago. Until more solid information is forthcoming, it shall remain an enigma.

The problem of within-season fertilization schedules becomes moot if multiannual delayed fertilization is ever proven. Meanwhile, however, the observations of Booth and Peters (1972) and Simon et al. (1975) on green turtles and those of Caldwell, Carr, and Ogren (1959) in loggerheads suggest that mating may begin prior to the nesting season and that most females mate only once a season.

Nesting

The literature is replete with descriptive accounts of the nesting behavior of most species (Table 2). References in Table 2 (Bustard 1972, Hendrickson 1958, Schulz 1975) will show that many authors have contributed considerable detail. As long ago as 1960, Carr and Ogren noted that careful observation of each of the phases of nesting would provide considerable opportunity for ethologic comparison. The problem is that most of the information is qualitative rather than quantitative.

Nevertheless, Carr and Ogren (1960) provided a good beginning by partitioning emergence and nesting into 11 stages. The pattern has appeared in the reviews of Hirth (1971), Rebel (1974), and Ehrenfeld (1979), among others, and is as follows: 1) stranding, testing of stranding site, and emerging from wave wash; 2)

Table 2. Descriptive accounts of marine turtle nesting behavior

Species	Locality	Reference
Chelonia mydas	Costa Rica	Carr and Giovannoli (1957), Carr and Ogren (1960)
	Costa Rica and Ascension Island	Carr and Hirth (1962)
	Sarawak	Hendrickson (1958)
	Australia	Bustard and Greenham (1969)
	Gulf of Aden and Seychelles Islands	Hirth and Carr (1970)
	Surinam	Schulz (1975)
Chelonia depressa	Australia	Bustard (1972)
Caretta caretta	Georgia	Caldwell et al. (1959)
	Colombia	Kaufman (1968)
	Ceylon	Deraniyagala (1939)
	Australia	Bustard (1972)
	South Carolina	Caldwell (1959)
Lepidochelys kempi	Mexico	Chavez et al. (1968), Pritchard (1969a), Hildebrand (1963)
Lepidochelys olivacea	Surinam	Pritchard (1969a), Schulz (1975)
Eretmochelys imbricata	Costa Rica	Carr et al. (1966)
	Seychelles Islands	Frazier (in press)
Dermochelys coriacea	Trinidad	Bacon (1970)
	Ceylon	Deraniyagala (1939)
	Surinam	Pritchard (1969b)
	Costa Rica	Carr and Ogren (1959)
	Surinam	Schulz (1975)

selecting of course and crawling from surf to nest site; 3) selecting of nest site; 4) clearing of nest premises; 5) excavating of body pit; 6) excavating of nest hole; 7) oviposition; 8) filling, covering, and packing of nest hole; 9) filling of body pit and concealing of site of nesting; 10) selecting of course and locomotion back to the sea; 11) re-entering of wave wash and traversal of the surf. Kaufman (1968) and Schulz (1975) adhered to a 7-stage sequence that combined several of the steps above.

Ehrenfeld (1979) carried the process a step further by comparing the nesting behavior of *Eretmochelys* and *C. mydas* in a tabular format. I have borrowed heavily from that work and have attempted to extract from the literature information relevant to the nesting behavior of the other five forms, in order to further the synthesis. A brief summary of selected traits follows.

GAIT

When moving on land, *Eretmochelys, Caretta* and both species of *Lepidochelys* employ an alternating sequence of footfalls: diagonal flippers move together. In *Dermochelys* and both species of *Chelonia,* however, movement of the paired appendages is apparently simultaneous.

OVIPOSITION

In *Eretmochelys* and *Caretta*, and apparently in *L. kempi* and *L. olivacea*, the hind flippers are spread beside the nest during laying and their medial edges curl as eggs are extruded. In *Dermochelys, C. mydas* and *C. depressa* the hind flippers generally cover the nest cavity (one often protrudes into it behind the tail) and remain at rest as eggs are extruded.

CHARACTER OF BODY PIT

C. mydas characteristically excavates a deep body pit. Interestingly, *C. depressa* apparently does not. *Dermochelys*, nesting in Costa Rica, also prepares a substantial body pit; but those nesting in Surinam do not. *Eretmochelys, Caretta* and *L. olivacea* prepare only shallow pits, and *L. kempi* makes none at all.

TIME OF DAY

L. kempi is the only marine turtle that is exclusively diurnal in its nesting habits. Some populations of *C. depressa* also nest in the daytime. *Eretmochelys* nests during the day in the Seychelles (Diamond 1976). All of the others nest nocturnally, although occasional diurnal nestings are known for *Dermochelys, L. olivacea, C. mydas* and *Caretta*.

Table 3. Clutch size, egg size and carapace length for marine turtle species.

Species	Locality	Clutch size $\bar{x}$	Egg size $\bar{x}$ (mm)	Carapace length $\bar{x}$ (cm)	Reference
Lepidochelys kempi	Mexico	110	39	65	Chavez et al. (1968)
	Mexico	116	——	64[a]	Pritchard (1969a)
Lepidochelys olivacea	Surinam	116	40[a]	68[a]	Pritchard (1969b)
	Australia	108	39	——	Cogger and Lindner (1969)
Eretmochelys imbricata	Guyana	168	38	84	Pritchard (1969b)
	Costa Rica	161	38	——	Carr et al. (1966)
	Surinam	146	——	——	Schulz (1975)
	Seychelles	172	40	90	Frazier (in press)
Chelonia depressa	Australia	50	49	89[a]	Bustard and Limpus (1969)
Caretta caretta	Florida	100	——	96	Davis and Whiting (1977)
	Natal	117	——	94	Hughes (1971)
	Natal	118	——	93	Hughes (1970)
	Natal	118	——	93	Hughes and Mentis (1967)
	Florida	120	——	90	Worth and Smith (1976)
	Florida	125	——	93	Gallagher et al. (1972)
	Florida	110	41	92	Ehrhart and Yoder (1978), Ehrhart (1979a)
	Georgia	126	42[a]	——	Caldwell et al. (1959)
	Colombia	107	——	88	Kaufman (1975)
Chelonia mydas	Surinam	142	——	112	Pritchard (1969b)
	Surinam	138	——	109	Schulz (1975)
	Ascension	116	55	108	Carr and Hirth (1962)
	Guyana	122	48	107	Pritchard (1969a)
	Aldabra	89[b]	46	103	Frazier (1971)
	Florida	128	46	101	Ehrhart (1979b)
	Costa Rica	110	46	100	Carr and Hirth (1962)
	Yemen	160	42	96	Hirth (1971)
	Sarawak	105	40	95	Hendrickson (1958)
Dermochelys coriacea	Trinidad	98	55	158	Bacon (1970)
	Natal	106	——	164	Hughes et al. (1967)
	Natal	97	——	165[a]	Hughes and Mentis (1967)
	Natal	104	——	157	Hughes (1970)
	Natal	——	——	103	Hughes (1971)
	Surinam	92	——	160[a]	Pritchard (1969b)
	Surinam	95	53	——	Schulz (1975)
	Ceylon	110	54[a]	152	Deraniyagala (1939)

—— No data.

a. Approximately.

b. In a postscript to his paper, Frazier noted that: "Average clutch size is larger than stated."

Reproductive Potential

Clutch Size

There is an abundance of references to clutch size (Table 3). The provision, by many authors, of carapace lengths of adult females allows one to examine the extent to which marine turtles as a group conform to the rather direct relationship between clutch size and body size, seen generally in turtles (Moll 1979). A visual analysis of available data is provided by the graph of the clutch size-body size relationship in Figure 1. It is clear that sea turtles fail to adhere to the generalization. In fact, if the two ridley species and the very aberrant flatback are ignored, the relationship is essentially an inverse one. The adaptive significance of this pattern undoubtedly involves a host of factors too complex to treat here. It seems reasonable, for example, that the exceedingly small clutch size of flatbacks is an accomodation to shell shape that must be compensated for in other life history features. In the same vein, the selective advantage of the relatively small clutches of *Dermochelys* is quite obscure, at least to this writer.

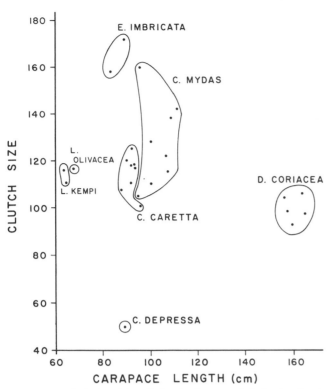

Figure 1. Relationship of clutch size to adult body size in marine turtles.

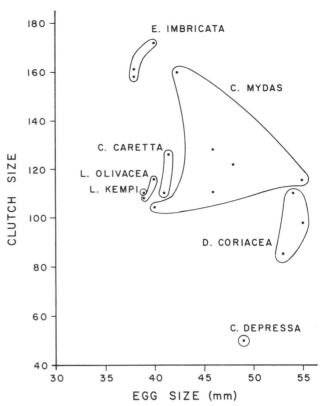

Figure 2. Relationship of clutch size to egg size in marine turtles.

Intraspecific trends in clutch size have been examined sparingly. It seems clear that clutch size varies positively with body size of the female parent in green turtles (Carr and Hirth 1962; Pritchard 1969b; Frazier, in press; Hirth 1971). We have found the same positive correlation in loggerheads nesting at Merritt Island, Florida (Ehrhart 1979b). Also, trends in clutch size over the course of a nesting season seem to vary among species and populations. Carr (1967) reported that *C. mydas* lays fewer eggs in the first and last clutches and Carr and Hirth (1962) and Pritchard (1971) observed decreasing clutch sizes as the season progressed. Caldwell (1959) and LeBuff and Beatty (1971) have reported the same trend in loggerheads. We have analyzed data gathered over five nesting seasons at Merritt Island (Florida) and are convinced that no such trend exists for the loggerheads nesting there.

Egg Size

Fewer workers have reported egg sizes (Table 3). There are enough data, however, to conclude tentatively that marine turtles generally adhere to interspecific trends in this factor seen in turtles as a group. Egg size typically correlates inversely with clutch size, and larger turtles lay relatively smaller eggs (Moll 1979). Inspection of the data in Figures 2 and 3 confirms that these relationships prevail among sea turtles. Note that in Figure 3 the y-axis is relative egg size, calculated by dividing mean egg diameter (or length) by mean adult carapace length.

Few authors have addressed intraspecific trends in egg size. Caldwell (1959) reported that larger loggerheads laid relatively smaller eggs. Regression analysis of our data for Merritt Island loggerheads, however, reveals no inverse relationship. The correlation was weak, at best (r = 0.29), and positive (Ehrhart 1976).

Cycles, Internesting Periods, and Clutch Numbers

An examination of the relevant literature reveals that the most commonly observed multiannual renesting intervals are 2 and 3 years. This is apparently true for *Dermochelys* (Hughes 1971; Pritchard 1969b; Schulz 1975), *Caretta* (Caldwell et al. 1959; Davis and Whiting 1977; Kaufman 1975; Bustard 1972), *Eretmochelys* (Carr et al. 1966; Frazier, in press) and for most populations of *C. mydas* (Hendrickson 1958; Carr and Ogren 1960; Schulz 1975; Frazier, in press). The eastern Australian population of *C. mydas* is reported to exhibit a 4-year cycle (Bustard and Tognetti 1969). *Lepidochelys*, of course, is peculiar among marine turtle genera in that the majority of individuals (of both species) actually nests annually (Pritchard 1969b; Schulz 1975). The latter author found that fully two-thirds of the Surinam *L. olivacea* population nested at 1-year intervals, one-

L. M. EHRHART

fourth at 2-year intervals, and about 8 percent at 3 years.

The most thorough examinations of this phenomenon, in light of long-term tag and recapture results, were published by Carr and Carr (1970) for *C. mydas* and by Hughes (1976) for *C. caretta.* The former account mentions the possibility of green turtles on 4-year cycles in Costa Rica. More importantly, the authors showed that, although a nesting female usually maintains a constant cycle, modulation resulting in shifts from 3- to 2-year cycles, and vice versa, may occur. Hughes (1976), however, has concluded from 12 years of tagging and recovering Tongaland loggerheads, that they exhibit no regular reproductive cycle and that irregularity is characteristic of that population.

Within-season internesting intervals, for chelonians in general, was the subject of a recent review by Moll (1979). Examination of the data for marine turtles compiled there and in several recent works (Schulz 1975; Frazier, in press) reveals that the most commonly observed intervals for *Caretta, Chelonia* and *Eretmochelys* are 12 to 15 days. *Dermochelys* departs from this pattern by nesting at 9- or 10-day intervals (Hughes and Mentis 1967; Pritchard 1969b) and Hendrickson (1958) reported 10-day intervals for green turtles at Sarawak. For *L. kempi* and *L. olivacea* the intervals are often much longer and more variable because the timing of arribadas seems to be governed by environmental factors, such as wind, tide, and surf conditions (Pritchard 1969b; Schulz 1975).

The difficulty in characterizing sea turtle reproductive potential that is caused by variations in multiannual cycles is compounded further by variation in the number of clutches a nesting year. Moll (1979) has compiled a table of data and references relating to the maximum number of clutches a year in turtles as a group. The older literature indicates that most species apparently lay between 3 and 5 clutches a year but it is becoming apparent for green turtles (Schulz 1975; Carr et al. 1978) and Florida loggerheads (Ehrhart, 1979b), at least, that many females nest only once or twice. On the other hand, Hendrickson (1958) reported as many as 11 clutches a year in Sarawak green turtles, and there are now reports of nearly that many for a few Florida loggerheads (F. Lund and C. LeBuff, personal communications). It is difficult, therefore, to make any unifying statement about the number of clutches deposited a year for sea turtles as a group, despite the considerable volume of information available.

Fertility

Not all marine turtle eggs deposited on the beach are fertile. Only a few workers have examined this reproductive feature in detail, however, perhaps because of

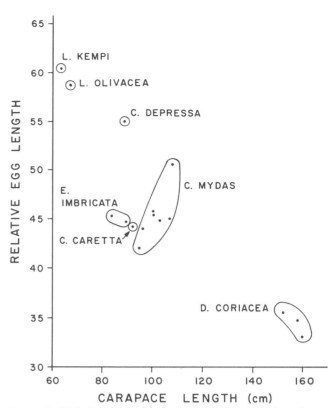

Figure 3. Relationship of relative egg size (egg length or diameter to adult carapace length) to adult body size (carapace length) in marine turtles.

the undesirability of cutting open 80- to 90-day-old eggs for examination. Also, there seems to be some confusion in the literature between hatch rates and fertility rates. Hughes and his coworkers in Natal have looked at this phenomenon most thoroughly, in loggerheads and leatherbacks (Hughes et al. 1967; Hughes 1970; Hughes and Mentis 1967). We have examined this factor in loggerheads at Merritt Island and found substantial agreement with those from Natal, in that fertility rates generally hover between 80 percent and 90 percent annually. Fowler (1979) reported a 91 percent fertility in eggs of *C. mydas* at Tortuguero, Costa Rica. These rates may be biased somewhat by the fact that visual examination of egg contents undoubtedly fails to identify embryos that expire at very early stages; and so these are counted as infertile. A fast, efficient technique for separating infertile eggs from those suffering developmental arrest at very early stages is needed. Further analysis of our data reveals that fertility rates are not correlated with size of female parent, clutch size, egg weight, or point-in-season.

Summary and Conclusions

Detailed descriptions of courtship and mating are available for only one species of sea turtle, *Chelonia mydas;* these accounts can serve as a base upon which a comparative synthesis can be built. Interpretation of the adaptive functions of courtship and mating awaits the development of quantified description for all species.

The temporal relation of a season's mating to the laying of fertilized eggs is a problem that remains unresolved after 25 years. Little new evidence has accumulated in that time. Although I do not condone the practice, adult females are legally killed far from the nesting beaches and outside the nesting season in many places. Zoologists should obtain, dissect, and examine the reproductive tracts of such females, freshly killed, for the presence of viable spermatozoa.

There is a wealth of description of marine turtle nesting behavior in the literature. Most of the accounts lack quantification, however, and a systematic comparative analysis has not been attempted. Some of us need to exchange our tagging pliers and calipers for stop watches, tape recorders, cameras, and event recorders.

There is an abundance of references to clutch size in the literature. It appears that sea turtles do not adhere to the chelonian generalization of a direct, positive relationship between clutch size and body size. Egg size has also been reported by many workers. In general, sea turtles seem to conform to interspecific trends seen in turtles as a group. Egg size is inversely related to clutch size, and larger turtles lay relatively smaller eggs.

Multiannual reproductive cycle lengths have been reported often for most species. Many of these, however, are based on a very small percentage of multiannual recoveries on the nesting beach. Where this is the case, workers should be aware of the conclusions reached by Hughes (1976), in the only study where 50 percent of the turtles tagged were recaptured at least once. He concluded that there was no regular reproductive cycle.

Within-season internesting intervals vary from 9 to 10 days in *Dermochelys* and some *C. mydas* populations to 12 to 15 days in *Caretta, Eretmochelys,* and most *Chelonia.* The interval is longer and more variable in *Lepidochelys,* apparently because of its dependence upon environmental conditions. The number of clutches deposited a nesting year varies from 1 to 11 in sea turtles. Many females lay 3 to 5 times, but the frequency of 1 and 2 clutch depositions a season may be greater than previously thought.

Fertility rates in marine turtle clutches have been examined only sparingly in the wild. They seem to be stable at 80 percent to 90 percent in loggerheads, leatherbacks, and green turtles. These rates may be somewhat higher, however, because an effective technique for identifying embryos that die in very early stages is lacking.

Literature Cited

Bacon, P. R.
1970.   Studies on the leatherback turtle, *Dermochelys coriacea* (L.), in Trinidad, West Indies. *Biological Conservation* 2:213–17.

Barney, R. O.
1922.   Further notes on the natural history and artificial propagation of the diamond-back terrapin. *Bulletin of the U. S. Bureau of Fisheries* 38:91–111.

Booth, J., and J. A. Peters
1972.   Behavioral studies on the green turtle (*Chelonia mydas*) in the sea. *Animal Behavior* 20:808–12.

Bustard, H. R.
1972.   *Sea Turtles: Their Natural History and Conservation.* London: Collins.

Bustard, H. R., and P. M. Greenham
1969.   The nesting behavior of the green sea turtle (*Chelonia mydas* L.) on a Great Barrier Reef island. *Herpetologica* 25:93–102.

Bustard, H. R., and C. Limpus
1969.   Observations on the flatback turtle, *Chelonia depressa* Garman. *Herpetologica* 25:29–34.

Bustard, H. R., and K. P. Tognetti
1969.   Green sea turtles: a discrete simulation of density-dependent population regulation. *Science* 163:939–41.

Caldwell, D. K.
1959.   The loggerhead turtles of Cape Romain, South Carolina. *Bulletin of the Florida State Museum* 4:319–48.

Caldwell, D. K.; A. Carr; and L. H. Ogren
1959.   Nesting and migration of the Atlantic loggerhead turtle. *Bulletin of the Florida State Museum* 4:295–308.

Carr, A.
1965.   The navigation of the green turtle. *Scientific American* 212:78–86.
1967.   *So Excellent a Fishe.* New York: Natural History Press.

Carr, A., and M. H. Carr
1970.   Modulated reproductive periodicity in *Chelonia. Ecology* 51:335–37.

Carr, A.; M. H. Carr; and A. B. Meylan
1978.   The ecology and migrations of sea turtles, 7. The West Caribbean green turtle colony. *Bulletin of the American Museum of Natural History* 162:1–46.

Carr, A., and L. Giovannoli
1957.   The ecology and migrations of sea turtles, 2. Results of field work in Costa Rica, 1955. *American Museum Novitates* 1835:1–32.

Carr, A., and H. Hirth
1962.   The ecology and migrations of sea turtles, 5. Comparative features of isolated green turtle colonies. *American Museum Novitates* 2091:1–42.

Carr, A.; H. Hirth; and L. Ogren
1966.   The ecology and migrations of sea turtles, 6. The hawksbill turtle in the Caribbean Sea. *American Museum Novitates,* 2248:1–29.

Carr, A., and L. Ogren
1959.   The ecology and migrations of sea turtles, 3. *Dermochelys* in Costa Rica. *American Museum Novitates* 1958:1–29.
1960.   The ecology and migrations of sea turtles, 4. The green turtle in the Caribbean Sea. *Bulletin of the American Museum of Natural History* 121:1–48.

Chavez, H.; M. Contreras G.; and T. P. E. Hernandez D.
1968. On the coast of Tamaulipas, Parts 1 and 2. *International Turtle and Tortoise Society Journal* 2(4):20–29: 37; (5):16–19, 27–34.

Cogger, H. G., and D. A. Lindner
1969. Marine turtles in northern Australia. *Aust. Zool.* 15:150–59.

Cornelius, S. E.
1975. Marine turtle mortalities along the Pacific coast of Costa Rica. *Copeia* 1975:186–87.

Davis, G. E., and M. C. Whiting
1977. Loggerhead sea turtle nesting in Everglades National Park, Florida, U.S.A. *Herpetologica* 33:18–28.

Deraniyagala, P. E. P.
1939. *The Tetrapod Reptiles of Ceylon*, vol. 1, *Testudinates and Crocodilians*. London: Dulau.

Diamond, A. W.
1976. Breeding biology and conservation of hawksbill turtles, *Eretmochelys imbricata* L., on Cousin Island, Seychelles. *Biological Conservation* 9:199–215.

Ehrenfeld, D. W.
1979. Behavior associated with nesting. In *Turtles: Perspectives and Research*, eds. M. Harless and H. Morlock, pp. 417–34 New York: Wiley.

Ehrhart, L. M.
1976. Studies of marine turtles at Kennedy Space Center and an annotated list of amphibians and reptiles of Merritt Island. In Final report to NASA/KSC: A study of a diverse coastal ecosystem on the Atlantic coast of Florida. NASA grant NGR 10–019–004.

1979a. Reproductive characteristics and management potential of the sea turtle rookery at Canaveral National Seashore, Florida. In Linn, R. M. ed., Proceedings of the First Conference on Scientific Research in the National Parks, 9–12 November 1976, New Orleans, *National Park Service Transactions and Proceedings* Series 5.

1979b. Threatened and endangered species of the Kennedy Space Center, part 1: Marine turtle studies. In Final report to NASA/KSC: A continuation of baseline studies for environmentally monitoring STS at John F. Kennedy Space Center. NASA contract NAS 10–8986.

Ehrhart, L. M., and R. G. Yoder
1978. Marine turtles of Merritt Island National Wildlife Refuge, Kennedy Space Center, Florida. In Henderson, G. E. ed., Proceedings of the Florida and Interregional Conference on Sea Turtles, 24–25 July 1976, Jensen Beach, Florida. *Florida Marine Research Publication* 33.

Ewing, H. E.
1943. Continued fertility in female box turtle following mating. *Copeia* 1943:112–14.

Fowler, L. E.
1979. Hatching success and nest predation in the green sea turtle, *Chelonia mydas*, at Tortuguero, Costa Rica. *Ecology* 60:946–55.

Frazier, J.
1971. Observations on sea turtles at Aldabra Atoll. *Philosophical Transactions of the Royal Society of London*, B, 260:373–410.

In press. Marine turtles in the Seychelles and adjacent territories. In D. R. Stoddart, editor, *Biogeography and Ecology of the Seychelles Islands*.

Gallagher, R. M.; M. L. Hollinger; R. M. Ingle; and C. R. Futch
1972. Marine turtle nesting on Hutchinson Island, Florida in 1971. *Florida Department of Natural Resources, Marine Research Laboratory Special Scientific Report* 37:1–11.

Harrisson, T.
1954. The edible turtle (*Chelonia mydas*) in Borneo. 2. Copulation. *Sarawak Museum Journal* 6:126–28.

Hendrickson, J. R.
1958. The green sea turtles, *Chelonia mydas* (Linn.) in Malaya and Sarawak. *Proceedings of the Zoological Society of London* 130:455–535.

Hildebrand, H. H.
1963. Hallazgo del area de anidación de la tortuga marina "lora," *Lepidochelys kempi* (Garman) en la costa occidental del Golfo de Mexico. *Ciencia* 22:105–12.

Hirth, H. F.
1971. Synopsis of biological data on the green turtle *Chelonia mydas* (Linnaeus)1758. *FAO Fisheries Synopsis*, 85:1:1–8:19.

Hirth, H. F., and A. Carr
1970. The green turtle in the Gulf of Aden and Seychelles Islands. *Verh. K. Ned. Akad. Wet.* 58:1–44.

Hornell, J.
1927. *The Turtle Fisheries of the Seychelles Islands*. London: H. M. Stationery Office.

Hughes, G. R.
1970. Further studies on marine turtles in Tongaland, 3. *Lammergeyer* 12:7–25.

1971. Sea turtle research and conservation in south east Africa. *IUCN Publications, New Series* 31:57–67.

1976. Irregular reproductive cycles in the Tongaland loggerhead sea turtle, *Caretta caretta* L. (Cryptodira: Cheloniidae). *Zoologica Africana* 11:285–91.

Hughes, G. R.; A. J. Bass; and M. T. Mentis
1967. Further studies on marine turtles in Tongaland, 1. *Lammergeyer* 3:5–54.

Hughes, G. R, and M. T. Mentis
1967. Further studies on marine turtles in Tongaland, 2. *Lammergeyer*, 3:55–72.

Kaufman, R.
1968. Zur brutbiologie der meeresschildkrote *Caretta caretta* L. *Mitt. Inst. Colombo-Aleman Invest. Cient.* 1:65–72.

1975. Studies on the loggerhead sea turtle, *Caretta caretta* (Linne) in Colombia, South America. *Herpetologica* 31:323–26.

LeBuff, C. R., and R. W. Beatty
1971. Some aspects of nesting of the loggerhead turtle, *Caretta caretta caretta* (Linne) on the Gulf coast of Florida. *Herpetologica* 27:153–56.

MacArthur, R. H., and E. O. Wilson
1967. *The Theory of Island Biogeography*. Princeton: Princeton Univ. Press.

Moll, E. O.
1979. Reproductive cycles and adaptations In *Turtles: Perspectives and Research* eds. M. Harless and H. Mor-

lock, pp. 305–31. New York: Wiley.

Pianka, E. R.
1978. *Evolutionary Ecology.* New York: Harper and Row.

Pritchard, P. C. H.
1969a Sea turtles of the Guianas *Bulletin of the Florida State Museum* 13:85–140.

1969b. Studies of the systematics and reproductive cycles of the genus *Lepidochelys.* Ph. D. dissertation. University of Florida.

1971. Galapagos sea turtles——preliminary findings. *Journal of Herpetology* 5:1–9.

Rebel, T. P., ed.
1974. *Sea Turtles and the Turtle Industry of the West Indies, Florida and the Gulf of Mexico.* Coral Gables: Miami Univ. Press.

Schulz, J. P.
1975. Sea turtles nesting in Surinam. *Zoologische Verhandelingen, nitgegeven door het rijksmuseum van Natuurlijka Historie te Leiden,* 143:1–144.

Simon, M. H., G. F. Ulrich, and A. S. Parkes
1975. The green sea turtle (*Chelonia mydas*): mating, nesting and hatching on a farm. *Journal of Zoology* 177:411–23.

Smith, H. M.
1956. Handbook of amphibians and reptiles of Kansas. *University of Kansas Museum of Natural History Miscellaneous Publications,* 9:1–356.

Solomon, S. E., and T. Baird
1979. Aspects of the biology of *Chelonia mydas* L. *Oceanography and Marine Biology Annual Reviews* 17:347–61.

Ulrich, G. F., and D. W. Owens
1974. Preliminary note on reproduction of *Chelonia mydas* under farm conditions. *Proceedings of the World Mariculture Society,* 5:205–14.

Witham, R.
1970. Breeding of a pair of pen-reared green turtles. *Quarterly Journal of the Florida Academy of Sciences* 33:288–90.

Wood, F. G.
1953. Mating behavior of captive loggerhead turtles. *Copeia* 1953:184–86.

Worth, D. F., and J. B. Smith
1976. Marine turtle nesting on Hutchinson Island, Florida, in 1973. *Florida Marine Research Publication* 18:1–17.

David Wm. Owens
Biology Department
Texas A & M University
College Station, Texas 77843

The Role of Reproductive Physiology in the Conservation of Sea Turtles

ABSTRACT

Three overlapping areas are discussed in defining the role of reproductive physiology in sea turtle conservation. These are: 1) identifying critical and possibly unique reproductive processes of major concern to species survival; 2) developing improved techniques for accomplishing high priority applied and basic research; and 3) moving vigorously ahead in basic reproductive physiology research, especially where critical areas have been identified. A technique for determining the sex of immature turtles is described, based on the observation that juvenile males have higher circulating titers of testosterone than do females of the same age. Because of the difficulties inherent in working with wild turtles in their natural habitats, the study of captive adult colonies in large "naturalistic" ponds is recommended. Numerous physiological techniques are discussed which may be of use in conservation. These include radioimmunoassay of hormones, blood sampling, X-ray photography, laparotomy, hormone manipulation, and electroejaculation. It is emphasized that these techniques can be accomplished without doing harm to the turtles.

Introduction

It may not be obvious that there is an important role for reproductive physiology in the conservation of beleaguered species. One could, with legitimate cause, argue to leave that sort of research out as nonessential. On the other hand, our understanding of basic chelonian reproductive biology is far too narrow to act as a base for wise management decisions. When it comes to the best techniques for handling captive individuals, incubating eggs, and rearing hatchlings, guessing replaces facts. Arguments for and examples of how the scientist and conservationist can use modern techniques to gain valuable insight into several important physiological questions are given below. Such techniques can continue to improve our knowledge without sacrificing adult animals as was once the common

Figure 1. Method of taking blood sample. In this photo the animal is restrained in a small slanting table. The technique works well on animals larger than 100 g.

practice in the study of reproductive physiology.

There are at least three primary areas where, in the case of sea turtles, the reproductive physiologist should be interacting closely with conservationists:

- identifying critical and possibly unique reproductive processes of major concern to species survival;
- developing improved techniques for accomplishing high priority applied and basic research; and
- moving vigorously ahead in basic reproductive physiology research, especially where critical areas have been identified.

For the most part these three areas overlap considerably; however, we will consider each in turn.

Identifying Critical Reproductive Processes

Critical points such as the need to protect the female while nesting are not the primary interest here since they are obvious to all concerned workers. Rather, there are more subtle phenomena which, as they are discovered, should be brought to the attention of conversationists. The best example of this process in action is the recent verification of the role of temperature in embryonic sex differentiation, which Mrosovsky and Yntema (this volume) have described.

A further area where very little work has been done, but which could prove important in the future, is in fertility studies. Great variations in hatching rates, possibly caused by differences in fertility levels, have been

seen in the wild (Hirth, 1971). Even individual females have shown striking differences in fertility in captive wild adults (Wood and Wood 1977). Still, a thorough and systematic study comparing eggs for their fertility, as distinguished from the possibility of early embryonic death, has not been conducted on wild populations. It would be very valuable to know if one could predict and assess the status of the male population based on changes in the fertility level seen in the eggs deposited by the more accessible females. In general, it can be stated that we know dangerously little about the wild male sea turtle.

Improving Conservation and Research Techniques

Blood and Cerebrospinal Fluid Sampling

A technique that has applications in several types of physiological studies is a simple method to take a blood sample from a sea turtle (Owens and Ruiz 1980). Drilling the plastron for heart puncture was the technique most commonly used until recently. Needless to say, this was a very difficult, traumatic, and laborious method. The dorsal cervical sinus sampling system, which is far simpler, was apparently developed by Edward Scura and associates while working with captive turtles on Grand Cayman Island. By inserting a syringe needle perpendicular to the neck on either side of the midline, a rapid clean sample can be drawn (Figure 1). While the technique is relatively simple, one must take great care not to strike the spinal nerve when inserting the needle. I have sampled the same turtle, without difficulty, up to six times in 24 hours. This capability of taking multiple samples is uniquely important to reptilian physiology.

In a similar manner, cerebrospinal fluid samples can be drawn directly from the fourth ventricle of the brain by passing a needle through the foramen magnum (Owens and Ruiz, 1980). This technique has been attempted fewer than 60 times; the turtles did not appear to be harmed by the process. The technique is not recommended for field use.

Methods of Determining Sex

The sex of immature sea turtles must be determined for two reasons that may be important to the conservationist. The first is to be able to determine the sex ratio in immature wild populations. Currently this can only be accomplished by sacrificing large numbers of turtles, a practice which must now be avoided. The second reason is to be able to set aside enough potential breeders of each sex in captive culture practice. Since sex ratios do not appear to be equal either in wild (Hirth 1971) or in captive-hatched populations

DAVID WM. OWENS

(Owens and Hendrickson 1978) it is essential to be able to make sex determinations as early as possible. Most turtle fishermen claim to be able to determine the sex of immature turtles. However, in my experience, this has not been verified. On rare occasion the sex is obvious in farm-reared male turtles at 4 years of age. More commonly however, at least 6 years are required for clear external differentiation in males (personal observation). Females, on the other hand, have no obvious external secondary sexual characteristics and thus large immature males can be mistaken for females (Owens et al. 1978). Our preliminary attempts at determining sex by external morphology were unsuccessful. We were successful at inserting a proctoscope into the cloacal region of a group of 4-year-olds, but could see very little differences in the 6 animals. All seemed to have similarly developed penile structure. Only later did we realize that the population may have been strongly skewed toward one sex. Thus we may have been looking at several individuals of the same sex. Therefore, this technique needs further evaluation. A much more laborious method is to do a laparotomy on each animal and then carefully seal the hole. We have done this and the turtles appeared to recover satisfactorily (Figure 2). Laparotomy is not of value in very young animals where the gonads can not be visually differentiated.

Karyotyping has also been suggested as a method to distinguish the sexes based on heterogametic chromosomes (Makino 1952). Recent studies however indicate that there are no obvious morphologically distinctive male and female chromosomes (Bickham et al. 1980).

The sex determination method which has proven most dependable is based on the fact that immature male turtles have higher circulating levels of androgens than females, even though secondary sexual characteristics are not yet obvious (Owens et al. 1978). In this radioimmunological technique, a small amount of antibody which has been made to testosterone is placed in a tube containing the turtle's serum plus a known quantity of radioactive testosterone. The testosterone in the turtle's serum then competes with the radioactive testosterone for binding sites on the antibody. When the antibodies are precipitated out, they are bound to the radioactive testosterone in an inverse proportion to the amount of competing testosterone that was in the animal's serum. The excess radioactive testosterone is then poured out of the tube leaving the precipitate behind to be quantified by a radioactivity counter. This technique is now routinely used in hospitals and endocrinology laboratories. It has proven very successful in determining sex in iguanid lizards (Judd et al. 1976) and is currently being evaluated in wild sea turtles (Mary Mendonca, personal communication). Kits for conducting the assays can be purchased from various

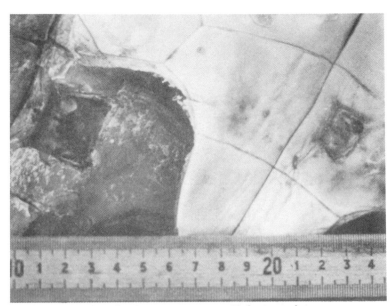

Figure 2. Plastrons from two immature *Chelonia mydas* on which laparotomy was performed.

companies. The materials provided contain most of the necessary reagents. Unfortunately, we can not recommend this procedure for most labs since extensive equipment is needed. Cayman Turtle Farm now routinely determines the sex of all turtles being set aside as future breeders (Jim Wood, personal communication). The technique works well on animals as young as 4 years of age. It is doubtful, however, that it will be useful on younger animals.

Experimental evidence indicates that mammalian follicle-stimulating hormone (FSH) will stimulate the secretion of testosterone in immature males. Since it does not cause an elevation of testosterone in females, such hormone challenging might be useful in determining the sex of animals younger than the 4-year-olds that we used. The FSH will cause the testis to produce just enough testosterone to allow clear differentiation. There is one possible problem with administering foreign (non-sea turtle) protein hormones. It has recently been found that turtles have such a sensitive immune system that they will eventually develop antibodies to foreign protein hormones (Owens, Hendrickson, and Endres 1979). Although it is doubtful that a short exposure to gonadotropin would produce sufficient antibodies, or that the antibodies (if present) would affect the animal's own gonadotropin system, it would be unwise to experiment with such a technique on depleted wild populations.

Continuing Basic Research

Much of the important basic research on sea turtles was begun about 25 years ago by a handful of dedicated scientists. These international programs have been invaluable in developing current understanding of re-

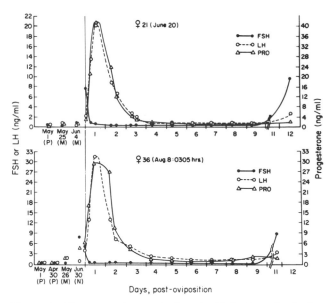

Figure 3. Hormone changes in female *Chelonia mydas* during the course of the reproductive season. P = prenesting, M = mating, N = oviposition, O = oviposition. Reprinted from Licht et al. (1979)

productive migrations, nesting ecology, and developmental biology. Although much work remains to be done, particularly with species other than *Chelonia mydas,* it is safe to say that these earlier basic research projects have provided a foundation for understanding the life histories of marine turtles. In fact, the conservation of these species would be impossible without this knowledge.

Studying reproductive physiology could be considered a relatively new frontier in sea turtle research, much as the migration studies were in the 1950s. Although the few examples which have been given above may prove useful in conservation, as with most basic research, it is difficult to predict just where the utility will materialize.

In about 1973 a collaborative research project was initiated to study the reproductive endocrinology of the green turtle at the turtle farm on Grand Cayman Island (Owens 1976, 1980). Because of their size, sea turtles are exceptionally good models for modern endocrinology. The first important breakthrough came when the turtle farm collected thousands of pituitaries from the animals being marketed. With this material Licht et al. (1977) were able to separate and verify that there are actually two distinct gonadotropins (FSH and LH type hormones) in reptiles. Prior to this time, numerous physiological studies, including our work with sea turtles (Owens 1976), had indicated that there might only be a single gonadotropin in the Reptilia. Furthermore, Licht had sufficient extra gonadotropin material to produce antibodies to the molecules. With those antibodies to sea turtle gonadotropins he was able to develop the first specific reptilian radioim-

munoassays. These radioimmunoassays have now been used to describe the hormone cycles of the green turtle (Licht et al. 1979). As can be seen in Figure 3, sea turtles have a very dynamic endocrine system with FSH peaking during nesting and LH and the steroid progesterone peaking about 1 day later (Figure 4) when presumably ovulation of the clutch is occurring. This is the first such description for any reptilian species. It certainly, once and for all, refutes the "sluggish physiology" stereotype that is often given to reptiles. These data should warn the field conservationist that the day after nesting is probably the most critical (possibly sensitive to environmental disturbance) of the entire internesting interval. Since all of these data were generated over a 4-year period at the turtle farm, one wonders if a similar picture might be present in the wild. Verification of similar patterns has recently been presented by Licht, Rainey, and Cliffton (1980) for animals sampled in Mexico, Surinam, and Aves Island. Choosing the critical times to take a few important samples in the field was made possible by the more exhaustive studies in the captive breeding pond on Grand Cayman. This pond made it possible to catch a female, take a quick sample and have her back in the water in less than 5 minutes. Samples were thus obtained for and correlated with all of the important behavioral categories that have been defined to date. This research pattern, consisting of a complete study in captive wild animals, followed by a careful but concise verification in a natural population, is an excellent way to improve the understanding of sea turtle physiology. It should be noted that not a single adult turtle has been sacrificed in these studies.

The study of captive-reared or wild-caught turtles in large "naturalistic" ponds with beaches would also be a useful alternative for other endangered sea turtle species. Such a confined system is the best approach to understanding reproduction and behavior because of the difficulty scientists have in locating and working with individuals in the marine environment. A byproduct of such studies would be the development of a captive breeding colony which could be a final line of defense against eventual extinction. Only the pelagic leatherback turtle appears to have behavioral restrictions that would exclude the possibility of maintaining such a captive colony.

Many new techniques for studying reproductive physiology would lend themselves well to a captive pond system, yet never take the life of a single turtle. One such technique, which would be interesting to correlate with the endocrine work, is the use of X-ray photography on nesting females as has recently been done with fresh water turtles (Gibbons and Greene 1979). With this technique it might be possible to verify ovulation in relation to the observed gonadotropin peaks.

DAVID WM. OWENS

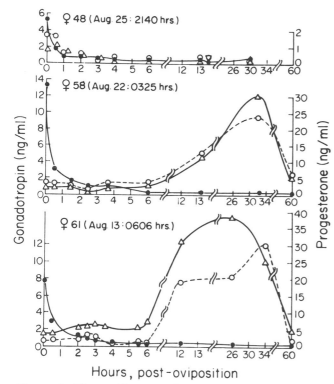

Figure 4. The rapid onset of an apparent ovulatory surge during the hours after oviposition. Symbols are the same as in Figure 3. Reprinted from Licht et al. (1979). Note that animal 48 did not demonstrate the ovulation surge because she had completed her nesting for the year.

Another technique that has produced excellent results in captured fresh water turtles is the use of oxytocic hormones to induce a turtle to deposit a clutch of eggs (Ewert and Legler 1978). Finally, if fertility is low, as has been the case on occasion at Cayman Turtle Farm, recent studies indicate that semen can be collected from males by the use of electroejaculation (Platz, Mengden, and Quinn, personal communication).

Despite the very serious threat to the survival of marine turtles, scientists and conservationists are in an excellent position to learn more about the reproductive physiology of these species than any other ectothermic vertebrate. Recent methodological advances, as well as the large size of sea turtles, render them uniquely suited for studies in reproductive physiology. We should immediately intensify our studies of the basic biology of marine turtles not only in the hope of saving these important species but also in the hope of saving many additional species on which such research is not possible.

Acknowledgments

Original data reported here were obtained with the assistance of grants from Cayman Turtle Farm, Ltd., the Janss Foundation and the National Science Foundation (PCM 75–22628). We gratefully acknowledge this support for our work. I thank Mimi Owens and J.R. Hendrickson for encouragement during all phases of this work. We would also like to thank Sandy Gibbs for typing this manuscript and Mark Grassman for commenting on early drafts.

Literature Cited

Bickham, J. W.; K. A. Bjorndal; M. W. Haiduk; and W. E. Rainey
1980.   The karyotype and chromosomal banding patterns of the green turtle (*Chelonia mydas*).*Copeia* 1980:540–43.

Ewert, M. A., and J. M. Legler
1978.   Hormonal induction of oviposition in turtles. *Herpetologica* 34:314–18.

Gibbons, J. W., and J. L. Green
1979.   X-ray photography: a technique to determine reproductive patterns of freshwater turtles. *Herpetologica* 35:86–89.

Hirth, H. F.
1971.   Synopsis of biological data on the green turtle *Chelonia mydas* (Linnaeus) 1758. *FAO Fisheries Synopsis,* 85:1:1–8:19.

Judd, H. L.; A. Laughlin; J. P. Bacon; and K. Benirschke
1976.   Circulating androgen and estrogen concentrations in lizards (*Iguana iguana*). *General Comparative Endocrinology* 30:391–95.

Licht, P.; H. Papkoff; S. W. Farmer; C. H. Muller; H. W. Tsui; and D. Crews
1977.   Evolution of gonadotropin structure and function. *Recent Progress in Hormone Research* 33:169–248.

Licht, P.; W. Rainey; and K. Cliffton
1980.   Serum gonadotropins and steroids associated with breeding activities in the green sea turtle *Chelonia mydas,* II. Mating and nesting in natural populations. *General Comparative Endocrinology* 40:116–22.

Licht, P.; J. Wood; D. W. Owens; and F. Wood
1979.   Serum gonadotropins and steroids associated with breeding activities in the green sea turtle *Chelonia mydas,* I. Captive animals. *General Comparative Endocrinology* 39:274–89.

Makino, S.
1952.   The chromosomes of the sea turtle, *Chelonia japonica,* with evidence of female heterogamety. *Annot. Zool. Jap.* 25:250–57.

Owens, D. W.
1976.   Endocrine control of reproduction and growth in the green sea turtle *Chelonia mydas.* Ph. D. Dissertation, University of Arizona, Tucson.
1980.   The comparative reproductive physiology of sea turtles. *American Zoologist* 20:549–63.

Owens, D. W., and J. R. Hendrickson
1978.   Endocrine studies and sex ratios of the green sea turtle, *Chelonia mydas. Florida Marine Research Publication* 33:12–14.

Owens, D. W.; J. R. Hendrickson; and D. B. Endres
1979.   Somatic and immune responses to bovine growth hormone, bovine prolactin and diethylstilbestrol in the green sea turtle. *General Comparative Endocrinology* 38:53–61.

Owens, D. W.; J. R. Hendrickson; V. Lance; and I. P. Callard
1978.   A technique for determining sex of immature *Chelonia mydas* using a radioimmunoassay. *Herpetologica*, 34:270–73.
Owens, D. W., and G. J. Ruiz
1980.   Obtaining blood and cerebrospinal fluid from marine turtles. *Herpetologica* 36:17–20.
Wood, J., and F. Wood
1977.   Captive breeding of the green sea turtle (*Chelonia mydas*). *Proceedings of the World Mariculture Society* 8:533–41.

Jeanne A. Mortimer
Zoology Department
Bartram Hall
University of Florida
Gainesville, Florida 32611

Factors Influencing Beach Selection by Nesting Sea Turtles

ABSTRACT

Sea turtles nest on a variety of beach types. This paper discusses how edaphic and biotic factors influence beach selection, citing evidence in the literature and data gathered at Ascension Island where the nesting beach is a discontinuous series of some 32 sandy crescents varying in shape, accessibility and substrate composition. Ascension turtles seem to prefer unlighted beaches with open sandy offshore approaches, and foreshores relatively free of rock clutter. Nesting density on the beaches is not correlated with average percent hatchling emergence. Although hatching success is influenced by characteristics of the beach sand, the turtles nest in all types of sand. This agrees with evidence, gathered elsewhere, that grain size is not important to a turtle in her choice of nesting beach. In determining the worldwide nesting patterns of different populations and species of sea turtles, biotic factors such as predation on eggs and hatchlings, and interspecific competition among nesting females, have probably been more important than purely geological characteristics of the beaches.

Introduction

Sea turtles nest on a variety of beach types, and it is not usually obvious why they choose one beach over another. In some instances, discontinuities occur because populations have become extinct. Others most probably can be explained by characteristics of the beaches themselves.

Among the basic requirements for a good nesting beach is easy accessibility from the sea. The beach platform must also be high enough that it is not inundated by spring tides or flooded by the water table below. The beach sand should facilitate gas diffusion but be moist enough and fine enough to prevent excessive slippage while the nest is being constructed.

Each of the 5 genera of sea turtles probably has slightly different beach requirements. Some of the variables that have been considered are the nature of the offshore approach, the slope of the beach, the beach-front vegetation, and the texture of the sand. Different populations of turtles, even within the same species, nest on a wide range of shores.

Ascension Island is an excellent place to investigate how physical attributes of a beach influence a gravid female in her choice of nest site. Ascension is only 8 km in diameter. It is isolated in the equatorial Atlantic Ocean and serves as the sole nesting ground for a large population of green turtles (*Chelonia mydas*). Its nesting beach is a discontinuous series of some 32 sandy crescents scattered along the western and northern sides of the island. These vary in overall size, shape, accessibility, and substrate composition.

During my 16 months on the island, I surveyed the beaches regularly and determined nesting density at each. I correlated density of nesting with various physical beach characteristics, including: beach length, rock cluttering the foreshore, rock obstructing the offshore approach, and artificial light visible on the shore. From data on hatching success I was able to determine whether clutch survival and nesting density were correlated.

Methods

The entire coastline of Ascension Island was surveyed and each patch of sandy shoreline was evaluated as to its suitability for turtle nesting. Ascension beaches show varying degrees of approachability by turtles. Some are completely unobstructed above and below the water line. At others horizontal beds of rock or strewn boulders, either submerged or on the foreshore, partially obstruct turtle emergence. The percentage of the total length of each beach along which the approach of a turtle would be inhibited by each type of obstacle was measured.

Nesting density on each beach was determined by counting turtle tracks. Counts were made just after dawn while the tracks were still damp; counts made later in the day were not reliable. At most of the 26 beaches monitored, counts were made at least once a week. For each beach, I estimated the total number of tracks during the season by plotting the morning track counts against time and measuring the area below the curve with a planimeter.

Relative emergence success was determined by examining the contents of nests from which young had hatched and emerged through the sand. Initially, an attempt was made to mark clutches when they were laid and subsequently to excavate them at the end of their incubation period; but the heavy nesting density on most of the beaches, coupled with the propensity of Ascension turtles to dig multiple nest pits, resulted in so much shifting of sand that it was difficult to locate

marked nests at a later date. Nests from which young had emerged were found either by back-tracking the seaward paths of hatchlings or by looking for the saucer-shaped depressions in the sand indicative of hatchling activity below the surface.

To calculate percentage hatchling emergence, I first determined what proportion of the eggs in a nest had successfully hatched. From that number I subtracted the number of hatchlings that had died or seemed likely to die in the sand during their ascent to the surface.

Results

Figure 1 shows the relationship between beach length and the average estimated number of clutches laid at each beach during the 1976–77 and 1977–78 seasons. The symbols differentiate among beaches with respect to the presence or absence of adverse factors such as submerged rocks offshore, artificial lighting, and obstruction by exposed slabs of rock on the foreshore. Linear regressions were performed on the 9 beaches lacking obstacles to nesting and on the 17 remaining beaches. The slopes of the two regression lines were nearly equal (0.90 without obstacles, and 0.84 with obstacles).

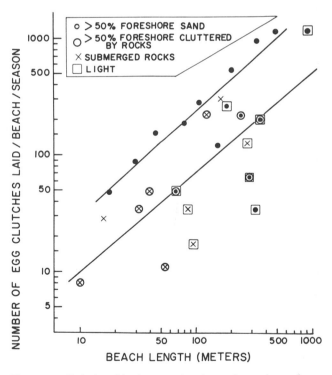

Figure 1. Relationship between estimated number of egg clutches laid on each beach during a season and beach length for 26 beaches at Ascension Island. The three symbols (X, O, and □) indicating different obstacles to nesting are used in combination for some beaches. The upper line is a regression based on the 9 beaches without obstacles (r = 0.92; p < 0.001). The lower line is from the remaining 17 beaches where one or more obstacles occur (r = 0.76; p < 0.001).

JEANNE A. MORTIMER

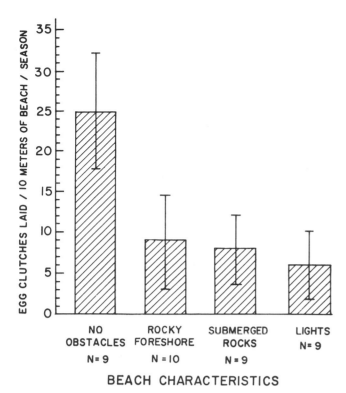

Figure 2. Relationship between physical characteristics of Ascension beaches and nesting density. Some beaches have more than one of the characteristics and thus are included in more than one category. Brackets indicate standard deviation.

Figure 2 illustrates the relationship between each of three sets of adverse shore conditions (beach rock on the foreshore, submerged rocks, and lights) and track density. Some beaches are considered more than once because they fall into more than one category. Submerged rocks and artificial lighting seem to be the greatest hindrances to nesting, followed by the presence of low slabs of beach rock along the foreshore of the beach. Each beach is considered separately in Figure 3, which shows the relationship between track density and various combinations of obstacles.

The relationship between the distribution of nesting activity on each of the three major Ascension Island beaches and offshore contour lines is shown in Figure 4. The heaviest nesting occurs on stretches of beach where the offshore approach is deepest.

Figure 5 shows the relationship between percent hatchling emergence from each nest excavated, and track density on the beach where the nest was located. Paradoxically, the turtles do not seem to prefer beaches on which my data show the higher levels of clutch viability. In fact, a Spearman rank test (Siegel 1956) showed a negative correlation between hatchling emergence and the average nesting density recorded for the beach at which the nest was located ($r = -0.2582$; $p = 0.02$; $N = 77$).

Discussion

Offshore Approach and Shoreline Composition

Data gathered at Ascension Island show that the heaviest nesting occurs on unlighted beaches with open offshore approaches and foreshores relatively free of rock clutter (Figures 1–3). For such beaches the correlation between beach length and the estimated numbers of clutches laid per beach per season is highly significant (Figure 1). On beaches characterized by the presence of obstacles to nesting, there is an overall decrease in the number of clutches laid. Within the boundaries of a given beach, turtles also seem to prefer the deepest approach to the beach (Figure 4).

There are probably two main reasons why nesting females at Ascension avoid beaches with rock strewn approaches. The most obvious is that Ascension's heavy surf makes coming ashore over rocks dangerous. Frequently Ascension females have cracks and gashes in their shells (Carr and Hirth 1962). The second reason is the heightened predation pressure on hatchlings along rocky shores. More predators occur in the vicinity of rock-strewn bottom than over sand (Mortimer 1981).

For other species of turtle in other localities, the condition of the offshore approaches seems to be important in their choice of a nesting beach. For example, the heavy body and soft skin of the leatherback turtle (*Dermochelys coriacea*) make it particularly vulnerable to mechanical injury. Probably for this reason it almost invariably nests on beaches with obstruction-free approaches (Pritchard 1971, Hughes 1974).

In contrast, loggerheads (*Caretta caretta*) in Tongaland (Hughes 1974) and at Mon Repos, in Queensland, Australia, (Bustard 1968) seem to prefer beaches adjacent to outcrops of rocks or subtidal reefs. Hughes (1974) suggests that this peculiar choice indicates the use of rocky approaches to orient to the site of emergence. He discounts the possibility that feeding is involved.

The gradient of the beach will determine the distance a turtle must crawl overland in order to reach a nest site. Nesting beaches of the leatherback often slope steeply, thus reducing the distance between the water line and the nest sites (Pritchard 1971, Schulz 1975). In Surinam, leatherbacks as well as green turtles avoid beaches behind mud banks that become partially exposed during low tide, while those are the beaches actually preferred by the small olive ridley (*Lepidochelys olivacea*)(Schulz 1975).

Beach Vegetation

In some parts of the world, the presence of beach vegetation seems to affect the choice of a nesting site. On the Great Barrier Reef, green turtles reportedly nest where there is substantial beach-front vegeta-

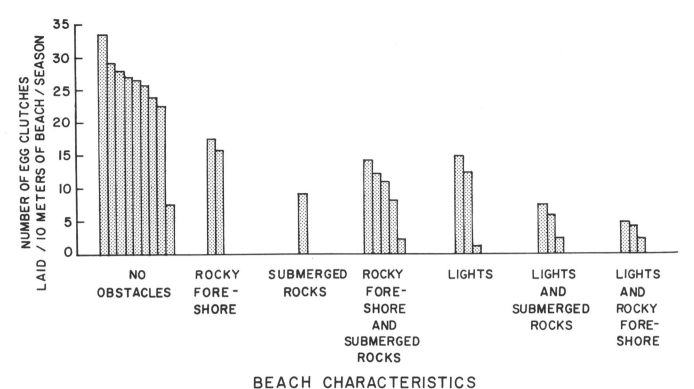

Figure 3. Relationship between combinations of beach characteristics and nesting density at Ascension Island. Each bar represents 1 beach.

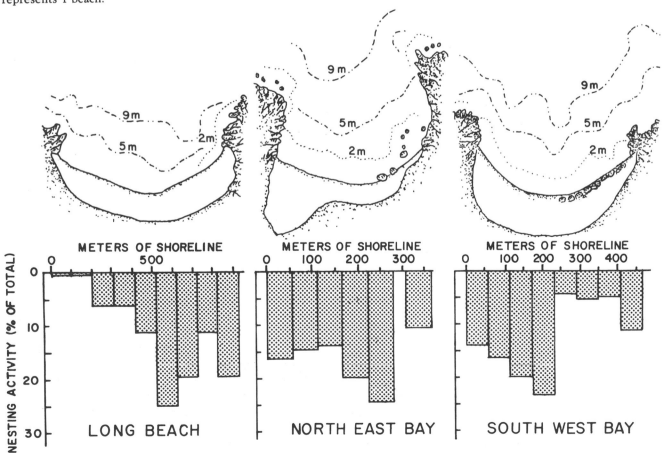

Figure 4. Relationship between nesting density along the shoreline and the position of the offshore contour lines at the 3 major Ascension beaches. Data were gathered by counting tracks during the 1977–78 nesting season.

　　　　　　　　　　　　　　　　JEANNE A. MORTIMER

tion—large bushes or even trees (Bustard 1972). Bustard and Greenham (1968) suggest that turtles nesting in these locations are more successful in digging egg chambers because plant rootlets help bind the sand grains and thus prevent slippage. Hawksbills commonly dig their nests far back on the nesting beach, amongst vegetation (Garnett, undated report; Frazier 1975; A. Carr, personal communication). In other places, however, roots in the substrate appear to constitute an important obstacle to digging. Vegetation is not a factor in beach selection at Ascension because there are no native plants on the beaches.

Beach Sand

An important property of any nesting beach is the quality of its sand. One would assume that sand type would have a bearing on two vital aspects of the nesting biology of the turtle: beach selection by the females, and survival of eggs laid. The influence of sand type, especially particle size distribution, on choice of nesting beach has been discussed by a number of authors. Little attention, however, has been paid to the manner in which the physical parameters of beach sand affect hatching success.

Surprisingly, not only does the particle size of the nesting medium vary from one nesting shore to another (Hirth and Carr 1970, Hirth 1971) but also a wide range of sand types may be utilized by the same colony. Hughes (1974) observed that on Europa Island, the beaches range in composition from fine sand to coral pebbles, and all are used by green turtles. At Ascension, Stancyk and Ross (1978) collected sand samples from 16 of the major nesting beaches and analyzed them for organic matter, water and calcium carbonate content, pH, color, and particle size distribution. They found no correlation between any of these parameters and nesting frequency, estimated by a brief nesting survey, and from the Ascension nesting records collected by Mariculture, Ltd.

Hendrickson and Balasingam (1966) suggested that in Malaysia, the texture of the sand might account for the selection of separate beaches by green turtles and leatherbacks because the beach chosen by *Chelonia* was composed of finer sand. However, the coarse sand of the *Dermochelys* beach is probably caused by an onshore current that strikes the beach perpendicularly, and produces a steep slope and a rapidly shelving bottom at the water line (Hendrickson and Balasingam 1966).In as much as leatherbacks nest successfully in fine sand in other regions (Carr and Ogren 1959, Hirth 1963, Pritchard 1971) particle size is probably less important than the slope and offshore configuration of the beach.

Nesting media of hawksbills range from fine siliceous sand to coarse shell and coral fragments (Carr et al. 1966, Hirth 1963, Hirth and Carr 1970, Bustard

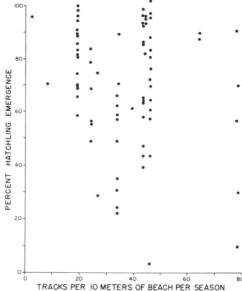

Figure 5. Relationship between percentage hatchling emergence from nests and track density at the beaches where the nests were located.

1974, and Ross, in press). At Ascension, although juvenile hawksbills are occasionally observed in the sea, adults have never been recorded. I doubt that hawksbills could nest successfully at Ascension. They are probably not large enough to dig to the depth necessary to encounter sand that is sufficiently moist for egg chamber formation and to prevent desiccation of the eggs (Mortimer 1981).

Throughout most of its range, the loggerhead shows a tendency to nest in silica sand (Hughes 1974, Caldwell et al. 1959; Cooke and Mossom 1929). However, this apparent preference may simply reflect the fact that much of the range of the turtle is in temperate regions too cold for the formation of coral sand. Some colonies of loggerheads nest on coral beaches such as at Heron Island, Australia, (Bustard and Greenham 1968). In Tongaland, sand particle size seems to be of negligible importance to loggerheads in their choice of nesting beaches (Hughes 1974).

During my study at Ascension I tried to learn what influence particle size distribution has on both a female's choice of nesting beach and on clutch viability. I found that there is an optimum range of grain sizes for hatching success (Mortimer 1981). Nests can fail in sand that is either too fine or too coarse. One would thus expect turtles to evolve an ability to select beaches at which their eggs would have the best chances of survival. Contrary to my expectations, I found no correlation between the average percentage hatchling emergence at Ascension beaches and nesting density on those beaches. Ascension turtles lay eggs in media ranging in texture from that of dust to that of gravel. This provides further and quite anomalous evidence that grain size is not of over-riding importance to a turtle in her choice of a nesting beach. Characteristics of the offshore approach may be of greater importance.

Biotic Influences on Use of Nesting Beaches

The nesting sites for many major sea turtle populations are islands, which are usually characterized by relative freedom from mammalian predators, at least prior to man's arrival. Even mainland-nesting populations often utilize stretches of coastline that are in effect islands in that they are partly cut off from optimal habitats of mammalian predators by barriers such as rivers and lagoons or by environments hostile to predators. For example, the 32 km of coastline that serve as the nesting beach of the Tortuguero green turtle population is bounded at each end and behind by rivers.

Competition from other species of sea turtles may also influence choice of nesting beach. Larger species, such as the leatherback, can easily destroy the nests of smaller sea turtles while digging their own egg chambers. When leatherbacks and other species of sea turtles nest in the same vicinity, they tend to segregate either spatially, as in Malaysia (Hendrickson and Balasingam 1966), or temporally, as at Tortuguero, Costa Rica, (Carr and Ogren 1959). It is probably no coincidence that the most important nesting beach of the small olive ridley in the western Atlantic is a stretch of shoreline in Surinam, fronted by an extensive offshore mud bank (Schulz 1975). The mud bank seems to hinder the approach of the larger green and leatherback turtles which could destroy the shallow nests of the ridleys.

Probably in the course of the evolutionary history of sea turtles, such biotic factors as predation and competition have been more important than purely geological characteristics in determining which beaches were used for nesting.

Human Alteration of Beaches

At Ascension, I found that nesting females tend to shun beaches where there is artificial lighting nearby, even if the light does not shine directly on the beach. Stancyk and Ross (1978) also found evidence of less-frequent nesting on Ascension beaches near civilization. Artificial lighting has long been recognized as a disruptive agent at turtle nesting beaches (Carr and Ogren 1959), but its effect has not been examined in sufficient depth.

Light has been shown to disorient newly-emerged sea turtle hatchlings of all species for which there are data—leatherbacks (Carr and Ogren 1959), green turtles (Carr and Ogren 1960), loggerheads (McFarlane 1963, Mann 1977, Fletemeyer 1979), and hawksbills (Philibosian 1976). However, the effect of artificial lights on nesting females seems to vary somewhat between species. Green turtles nesting at Merritt Island, Florida, (Ehrhart 1979), at Tortuguero (Carr et al. 1978) and at Ascension Island avoid beaches where artificial lights are visible. Although moving lights will frighten nesting females of all species, there is evidence that stationary artificial light has little effect on nesting female loggerhead turtles (Mann 1977, Ehrhart 1979). To establish controls that will limit human impact on nesting beaches, tolerable levels of light must first be determined.

In the past, not enough consideration has been given to the possible damage to incubating nests caused by vehicular and foot traffic on beaches. Recently, however, Mann (1977) and Fletemeyer (1979) have independently demonstrated that compaction of sand by these two factors can preclude the successful emergence of hatchlings. Mining of beach sand can also have devastating and long term effects on nesting beaches (Sella, this volume; Kar and Bhaskar, this volume).

Unfortunately, some encroachment of human activities and development on beaches is unavoidable. Information that will enable us to develop coastlines in a manner compatible with sea turtle nesting requirements is badly needed.

Acknowledgments

I would like to thank Dr. Archie Carr for his continued support and encouragement. Thanks also go to Drs. Karen Bjorndal and Jack Ewel for helpful suggestions regarding the manuscript.

My work at Ascension was primarily supported by a grant to Archie Carr from the National Geographic Society. Partial funding was also received from grants to Archie Carr from the following sources: the U.S. National Marine Fisheries Service (contract 03–78–D08–0025), the Caribbean Conservation Corporation, and the National Science Foundation (contract DES 73 06453). The work would not have been possible without the approval and assistance of the Island Administrators who served in that office during my stay—Mr. G. Guy, Brig. G. MacDonald, Mr. S. Gillett, and Mr. B. Kendall—and the governmental agencies on the island which provided me with continuous logistical support—the Government of St. Helena, Property Services Agency, the U.S. Air Force Base, the NASA Station, British Broadcasting Corporation, and Cable and Wireless.

Literature Cited

Bustard, H. R.
1968.   Protection for a rookery: Bundaberg sea turtles. *Wildlife in Australia* 5:43–44.
1972.   *Australian Sea Turtles.* London: Collins and Sons.
1974.   Barrier reef sea turtle populations. In *Proceedings of the Second International Coral Reef Symposium,* 1, pp. 227–34. Brisbane: Great Barrier Reef Committee.

Bustard, H. R., and P. Greenham
1968. Physical and chemical factors affecting hatching in the green sea turtle, *Chelonia mydas. Ecology* 49:269–76.

Caldwell, D. K.; A. Carr; and L. Ogren
1959. Nesting and migration of the Atlantic loggerhead sea turtle. *Bulletin of the Florida State Museum* 4:295–308.

Carr, A.; M. H. Carr; and A. B. Meylan
1978. The ecology and migrations of sea turtles, 7. The West Caribbean green turtle colony. *Bulletin of the American Museum of Natural History* 162:1–46.

Carr, A., and H. Hirth
1962. The ecology and migrations of the sea turtles, 5. Comparative features of isolated green turtle colonies. *American Museum Novitates* 2091:1–42.

Carr, A.; H. Hirth; and L. Ogren
1966. The ecology and migrations of sea turtles, 6. The hawksbill in the Caribbean Sea. *American Museum Novitates* 2248:1–22.

Carr, A., and L. Ogren
1959. The ecology and migrations of sea turtles, 3. *Dermochelys* in Costa Rica. *American Museum Novitates* 1958:1–29.
1960. The ecology and migrations of sea turtles, 4. The green turtle in the Caribbean Sea. *Bulletin of the American Museum of Natural History* 121.1–48.

Cooke, C. W., and S. Mossom
1929. Florida state geological survey twentieth annual report 1927–1928.

Ehrhart, L. M.
1979. A survey of marine turtles nesting at the Kennedy Space Center, Cape Canaveral Air Force Station, North Brevard County, Florida Department of Natural Resources. Report, 122 pp.

Fletemeyer, J.
1979. Sea turtle monitoring project. Submitted to Broward County Environmental Quality Board. Manuscript, 64 pp.

Frazier, J.
1975. Maziwi Island. Interim report, mimeo, 2 pp.

Garnett, M. C.
Undated. The breeding biology of hawksbill turtles (*Eretmochelys imbricata*) on Cousin Island, Seychelles. The International Council for Bird Preservation (British Section). Research Report, mimeo, 32 pp.

Hendrickson, J. R., and E. Balasingam
1966. Nesting beach preferences of Malayan sea turtles. *Bulletin of the Natural History Museum of Singapore* 33:69–76.

Hirth, H. F.
1963. The ecology of two lizards on a tropical beach. *Ecological Monographs* 33:83–112.
1971. Synopsis of biological data on the green turtle, *Chelonia mydas* (Linnaeus) 1758. *FAO Fisheries Synopsis* 85:1:1–8:19.

Hirth, H. F., and A. Carr
1970. The green turtle in the Gulf of Aden and the Seychelles Islands. *Verhandelingen der Koninklijke Nederlandse Akademie van Wetenschappen* 58:1–44.

Hughes, G. R.

1974. The sea turtles of southeast Africa. 1. Status, morphology and distributions. *Oceanographic Research Institute Investigational Report, Durban,* 35:1–144

Mann, T. M.
1977. Impact of developed coastline on nesting and hatching sea turtles in southeastern Florida. Master's thesis, Florida Atlantic University.

McFarlane, R. W.
1963. Disorientation of loggerhead hatchlings by artificial road lighting. *Copeia* 1963:153.

Mortimer, J. A.
1981. Reproductive ecology of the green turtle, *Chelonia mydas,* at Ascension Island. Ph. D. dissertation, University of Florida.

Philibosian, R.
1976. Disorientation of hawksbill turtle hatchlings, *Eretmochelys imbricata,* by stadium lights. *Copeia* 1976:824.

Pritchard, P.
1971. The leatherback or leathery turtle, *Dermochelys coriacea. IUCN Monograph, Marine Turtle Series* 1:1–39.

Ross, J. P.
In press. Hawksbill turtle, *Eretmochelys imbricata,* in the Sultanate of Oman. *Biological Conservation.*

Schulz, J. P.
1975. Sea turtles nesting in Surinam. *Stichting Natuurbehoud Suriname (STINASU)* 3:1–143.

Siegel, S.
1956. *Nonparametric Statistics for the Behavioral Sciences.* New York: McGraw Hill.

Stancyk, S. E., and J. P. Ross
1978. An analysis of sand from green turtle nesting beaches on Ascension Island. *Copeia* 1978:93–99.

J. R. Hendrickson
Department of Ecology and Evolutionary Biology
University of Arizona
Tucson, Arizona 85721

Nesting Behavior of Sea Turtles with Emphasis on Physical and Behavioral Determinants of Nesting Success or Failure

ABSTRACT

Owing to major constraints on form and function resulting from their common evolutionary history at ordinal level, sea turtles show remarkably little variation between species with respect to the *major aspects* of their nesting behavior. The 7 living species do show variation and overlap with respect to the *minor details* of their nesting behavior; this is ascribed to the enhanced selective value of the remnant capability for coping with varying nesting environments, the specializations fundamental to species formation being effected in other niche dimensions. A broad examination of the niche characteristics of the different species permits identification of cause-effect relationships which relate back to the nesting process and indicate guidelines for species-specific conservation policy in management of nesting beaches; some of these are discussed.

There is little information on the effects of perturbations on nesting success, particularly with respect to human-produced perturbations of the environment. Some poorly understood, potentially important examples are discussed.

Introduction

There is a considerable body of literature on the nesting of sea turtles. Despite their uneven coverage, the extensive reports on the 7 recognized species of sea turtles do permit several general remarks.

First, there is remarkable similarity among the species with respect to the places where they nest, the conditions under which they nest, and the behavioral steps by which the process is accomplished. This has frequently been noted (Carr and Ogren 1960; Schulz 1975; and Ehrenfeld 1979, among others). One concludes that this sameness in major features of the nesting pattern is due to anatomical and habitat constraints in the common evolutionary history of all sea turtles—a sort of macro-evolutionary characteristic of the testudine line in a marine environment. They all characteristically nest on exposed marine beaches, in deep, clean, relatively loose sand above high-tide level, usually at night, and follow approximately the same stereotyped sequence which Carr and Ogren (1960) divided into 11 discrete stages (Table 1).

Second, when searching the literature for interspecific differences in the finer details of the nesting pattern, one is impressed by the fact that there is almost as much variation in reports on the same species as can be found between species. Either we have to contend with an unsettling amount of unreliable reporting or there is, in fact, a great deal of intraspecific variation among populations, individuals, and even between one time and another in the life of the same individual.

I am persuaded that the intraspecific variation is real and better explains the divergent accounts in the lit-

erature than does faulty reporting. I conclude that these animals, under heavy evolutionary constraints with respect to the major behavioral and ecological parameters of the nesting process, are preserving all the "fitness" they can for coping with a variable environment by exploiting fully the remnant genetic and behavioral adaptability available at intraspecific level, reserving for other niche dimensions the specialization which is fundamental to species formation and maintenance. It is impossible to define any one environmental attribute or detail of behavior which is absolutely distinctive to any one species. At this "fine-grained" level within the strictly codified major pattern mentioned above, the details of action conform with immediate circumstance, and distinctions between species become so blurred as to lose most of their meaning.

Table 1. Sequential stages of sea turtle nesting, with notation of interspecific differences

Hendrickson (1958)	Carr & Ogren (1960)	Species differences[1]
A. Approach to the beach.	1. Stranding, testing of stranding site, & emergence from wave wash.	Shorter duration and less overt action in Dc and in Lo & Lk when in arribada formations
B. Ascent of the beach.	2. Selecting of course and crawling from surf to nest site.	(no consistent, major differences)
C. Wandering on the high beach.	3. Selecting of nest site.	(no consistent, major differences)
	4. Clearing of nest site premises.	Most marked in Ei and Cac, in response to surface litter and vegetation at site
D. Digging the body pit.	5. Excavating of body pit.	Most pronounced in Cm
E. Digging the egg hole.	6. Excavating of nest hole.	(no consistent, major differences)
F. Egg laying.	7. Oviposition.	(no consistent, major differences)
G. Covering the nest.	8. Filling, covering, and packing of nest hole.	Ei "ladles" sand over eggs, others scrape it; Lo & Lk rock body to "pound" sand compact, Dc pivots with weight on rear body, and other spp. "knead" sand compact
	9. Filling body pit and concealing of site of nesting.	(no consistent, major differences)
H. Return to the sea.	10. Selecting of course and locomotion back to sea.	(no consistent, major differences)
	11. Reentering of wave wash and traversal of surf.	(no consistent, major differences)

[1] Abbreviations: Cac = C. caretta; Cm = C. mydas; Dc = D. coriacea; Ei = E. imbricata; Lk = L. kempi; Lo = L. olivacea.

By way of example, the tabular comparison of behavioral differences between *Eretmochelys* and *Chelonia* on Caribbean beaches produced by Carr, Hirth, and Ogren (1966) is correctly identified by Ehrenfeld (1979) as a rare example of interspecies comparison sufficiently detailed to be of serious interest. Yet inspection of the items in this table, in the light of worldwide reports on the same species, reveals that the differences noted are so minor, so relative and nonquantifiable, and so subject to additional variation in other populations that this "best case" loses most of its meaning in the present context.

Finally, there is remarkably little documentation of the effects of perturbation on nesting success. What little information one can find on this subject is mainly on local perturbation through natural processes (such as predation pressure, space competition on the high beach, plant succession in nesting areas, climatic and ocean current variations). There are almost no objective analytical or experimental treatments of man-produced perturbations—clearly the most serious single element in problems of sea turtle management and survival at the present time. There is much speculation, but little hard data.

We must thus deal in value judgements, and the predictive powers of parametric statistics are not available to us. We are reduced to making lists of things to worry about.

For the reasons outlined above, I see little point in repeating here detailed descriptions of the various stages of the nesting process or attempting an exhaustive review and cross-comparison of variations in nesting behavior between species of living sea turtles.

The motive for this conference is conservation, which implies human action programs to make better or keep good the survival prospects of sea turtle populations. There are few specific action programs directed toward particular aspects of nesting behavior which can be suggested—the turtles can manage better without our intervention! What may be productive is to consider niche characteristics of the different species which influence nesting behavior and which have an ultimate relationship to nesting success or failure, and to derive from this some ideas of species-specific conservation-policy guidelines that may enhance probabilities of success or diminish risks of failure. In attempting the derivations from niche characteristics which follow, I am painfully aware that inadequate biological information and imperfect comprehension may cause me to make statements that individual readers (including myself at a later date!) may find outrageous or silly.

Chelonia mydas (Green Turtle)

1. Strongly directed migration and isolated breeding circumstances contributing toward reduced gene flow between groups have apparently resulted in the existence of an unknown number of genetically-distinct population entities which, although still imperfectly recognized, have the potential for important differences in behavior and physiology. Therefore,

A. The creation of a large finite number of different nesting preserves and attendant conservation programs in different parts of the world may be more important for green turtles than for any other species, and

B. What seems to work well in one preserve (for example, levels of artificial light near breeding beaches, types of physical handling of nesting females that seem to be tolerated) will be less automatically transferrable to other preserves than in the case of other sea turtle species.

2. Because individual *Ch. mydas* tend to have longer cycle times between nesting seasons (and may take longer to mature), there will be a longer lag between changes in nesting success and the effects on annual nesting figures than may be expected for other sea turtles with the possible exception of leathery turtles.

3. Management of the total life history is much more likely to be a binational problem, or a multinational problem than with other species because the migration circuits of *Ch. mydas* often cross international boundaries.

4. Land acquisition for protection of nesting sites has more promise in the case of this strongly site-fixed species than in the case of the others.

Chelonia depressa (Flatback Turtle)

1. Conservation is a one-country problem (apparently with no crisis threatening in the near future).

2. Careful information-gathering to learn more about the species without disturbing its (apparently) satisfactory position is a top priority (beware of too-enthusiastic tagging programs!).

Eretmochelys imbricata (Hawksbill Turtle)

1. Nesting preserves, with a few possible exceptions, will not work effectively because of diffuse nesting habits.

2. This species is the most likely to be inhibited from nesting by any form of human disturbance of the beach (for example, by tourists, investigating scientists, habitations, lights).

3. Lag time for observing effects of suspected disturbance should be minimal.

4. Probability of new appearances on previously non-utilized nesting beaches is high; constant wide-area surveillance is indicated as a standard part of all management programs.

5. This species, despite the severe predation pressure from humans which it is now experiencing because of

its desirable shell plates, is the most likely of all species to survive in this man-altered world. Rarity becomes a refuge, and a return to higher population levels is predictable with relief from predation pressures. Continued support of conservation programs is warranted, even after many years of apparent absence.

Caretta caretta (Loggerhead Turtle)

1. In contrast to *Ch. mydas,* the transfer of successful techniques of nesting beach management may be possible between widely separated parts of the world.
2. In the special case of beaches in the eastern United States, control of nest predators will likely be the most effective single measure possible within the scope of beach management. Such action may be considered a direct readjustment of disturbed faunal balance in the proximity of inhabited areas.
3. Also in the special case of the eastern U.S. seaboard, there may be a case for promoting beach activity by carefully organized and controlled groups of enthusiasts from the populated areas near the nesting grounds, thus tapping the very phenomenon ("civilization") which is indirectly responsible for the increased rate of nesting failures through unbalanced predation.

Lepidochelys olivacea (Olive Ridley)

1. The need for differentiation between two different program types should be considered. Programs to deal with *arribada*-forming populations may be quite different from those developed to deal with other areas where nesting is presently (or has always been) more individual and diffuse in space.
2. Where saturation nesting by large *arribadas* still occurs, nest removal and transplantation to safe hatchery areas is indicated. Doomed nests may be in two categories: those made at or below high tide or predictable storm levels (mostly at the bases of eroded cliff banks hindering ascent to the high beach), and first-*arribada* eggs in those particular areas where experience has shown high probability of later *arribadas* landing at the same section of beach.
3. Predator control during the incubation season seems to be called for in this relatively shallow-nesting species.

Lepidochelys kempi (Kemp's Ridley)

This is clearly the most gravely endangered of all sea turtle species, a situation made more acute by the annual aggregation of practically the entire reproductive resource of the species. Heroic measures are called for, and until and unless the annual loss to fisheries bycatch can be moderated, there are virtually no limits

to the actions which may be justified in the case of this species. This includes present attempts to "headstart" cultured juveniles, to establish new subcolonies, and to create captive gene pools.

Dermochelys coriacea (Leathery Turtle)

1. At present one should assume that the leathery (leatherback) turtle's lack of reaction to disturbing stimuli while nesting is due to its extraordinary level of stereotypy and that, anthropomorphic interpretations notwithstanding, disturbance may well be producing levels of traumatic alarm sufficient to cause permanent abandonment of the nesting site. More rigid beach protection should be provided than for any other species except the hawksbill.
2. Preseason removal of tree trunks and other large objects from nesting beaches should be carried out where possible.
3. It seems likely that this widely ranging, powerful swimmer has the highest level of gene flow around the world of all sea turtle species, and that, in contrast with *Ch. mydas,* any measures proved effective in increasing nest success in one place may be applied with the same results elsewhere.

Things to Worry About

As mankind continues to alter the various environments of this planet, the number of extrinsically derived, density-independent factors unfavorable to sea turtles continues to increase. The subtlety of their effects makes them difficult to detect, and the irreversibility of their presence in the environment has frightening portents. We can all readily understand the relevance of levels of artificial light on nesting beaches and its probable relation to levels of nest failure. I worry also about levels of high amplitude, low frequency vibration in the vicinity of nesting beaches. Although there is an almost total absence of objective evidence one way or the other, the structure of the sea turtle ear and the resemblance of these vibrations to those which surf could produce, providing a beach "signature" important to homing turtles, cause me to wonder how much a supposedly inoffensive diesel generator on the back-beach could interfere with normal nesting behavior. If the still unproven hypothesis of beach imprinting has reality in fact, what will be the effects of extraction of ground water and input of sewage near the beaches as tourist hotels and other "developments" encroach upon turtle nesting grounds? Indoctrinated in the effects of pesticides on the peregrine falcon, I was caused to do a good deal of thinking when I recently read an article suggesting that the plankton food-chain structure of the sea has a major dichotomy, with the really small-sized plankton leading

J. R. HENDRICKSON

to a sort of "cnidarian sink." Evidence was also presented indicating that PCB contamination may be leading to progressive shunting of the sea's productivity along this route. Now, leathery turtles are specialist feeders on jellyfish. . . .

Literature Cited

Carr, A., and L. Ogren
1960. The ecology and migrations of sea turtles. 4. The green turtle in the Caribbean Sea. *Bulletin of the American Museum of Natural History* 121:1–48.

Carr, A.; H. Hirth; and L. Ogren
1966. The ecology and migrations of sea turtles. 6. The hawksbill turtle in the Caribbean Sea. *American Museum Novitates* 2248:1–29.

Ehrenfeld, D. W.
1979. Behavior associated with nesting. In *Turtles: Perspectives and Research,* ed. M. Harless and H. Morlock, pp. 417–34. New York: Wiley and Sons.

Hendrickson, J. R.
1958. The green sea turtle, *Chelonia mydas* (Linn.) in Malaya and Sarawak. *Proceedings of the Zoological Society of London* 130:455–535.

Schulz, J. P.
1975. Sea turtles nesting in Suriname. *Nederlandsche Commissie Voor Internationale Natuurbescherming, Mededelingen* 23:1–143.

N. Mrosovsky
Departments of Zoology and Psychology
University of Toronto
Toronto M5S 1A1, Canada

C. L. Yntema
Department of Anatomy
Upstate Medical Center
Syracuse, New York 13210

Temperature Dependence of Sexual Differentiation in Sea Turtles: Implications for Conservation Practices

ABSTRACT

Data on the effect of incubation temperature of the eggs on sexual differentiation in turtles are briefly reivewed. Even a change of 1 to 2° C can make a considerable difference to the sex ratio of the hatchlings. Current conservation methods include incubation of eggs in styrofoam boxes above ground, establishment of central hatcheries, incubation in reduced clutch sizes, and egg harvesting only during certain seasons. The thermal aspects of these practices are analyzed in turn. It is concluded that incubation of eggs in styrofoam boxes runs the risk of masculinizing turtle populations and that other practices may be affecting sex ratio in ways that cannot yet be specified. More work on this problem is urgently needed before unevaluated methods become accepted procedures.

Introduction

Sexual differentiation of a number of turtle species is affected by the incubation temperature of the eggs (Figure 1). At higher temperatures there are more females, at lower temperatures more males and, at least in the freshwater snapping turtle, *Chelydra serpentina,* at still lower temperatures more females again. The temperatures at which ratios between the sexes change rapidly we will refer to as pivotal temperatures. For the loggerhead sea turtle, *Caretta caretta,* the pivotal temperature is about 30° C (Figure 1). The method of sexing hatchlings of this species has been given by Yntema and Mrosovsky (1980) and speculations about the ultimate cause of having sexual differentiation dependent on temperature, and some theoretical ramifications of this phenomenon, are being advanced elsewhere (Mrosovsky, 1980). In this paper we are concerned with the practical implications for those involved in the conservation of sea turtles. Protecting eggs during incubation sometimes entails temperatures different from those prevailing in natural conditions. We will consider in turn the thermal aspects and possible effects on sex ratio of the following: 1) incubation

Reprinted by permission from Biological Conservation *(1980), vol. 18 pp. 271–80.*

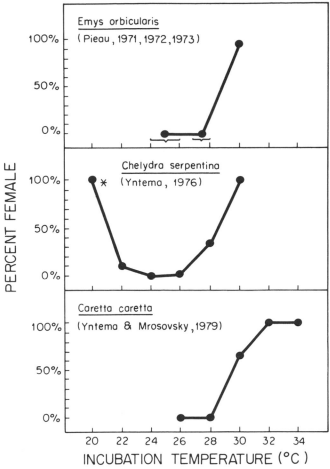

Figure 1. Sex ratio in 2 species of freshwater turtle (European pond terrapin, *Emys orbicularis,* and snapping turtle, *Chelydra serpentina*) and 1 species of sea turtle (loggerhead, *Caretta caretta*) from eggs incubated at different temperatures. Brackets show range of temperature prevailing. Star marks a group transferred to 26°C after 83 to 88 days at 20°C.

in styrofoam boxes above ground, 2) establishment of central protected hatcheries, 3) clutch size, and, 4) problems that might arise if eggs are harvested during close seasons.

Styrofoam Boxes and Incubation Duration

Styrofoam boxes, or similar forms of incubating eggs above ground in containers, have been widely used, for example, in Surinam, Cyprus, Mexico, the Caribbean and the United States. Incubation in styrofoam boxes generally takes longer than normal, presumably on account of lower temperatures (Schulz 1975, Marquez 1978). The two examples of this just cited will be considered quantitatively, but first it is necessary to be able to calculate temperature differences from the lengthened incubation periods. For this reference is made to Figure 2, showing the duration of incubation as a function of temperature for 2 species of marine

turtle (green turtle, *Chelonia mydas,* and loggerhead). The sample sizes are quite small at some points but the data are consistent enough to show that, as a general working rule over the ranges depicted, a 1° C lowering of temperature will be reflected in a 5-day increase in incubation time. This rule can of course be refined in the light of further data and can be made more complex to take into account the curvilinear relationship between temperature and incubation period that is obvious with wider temperature ranges (Yntema 1978). However, a simple rule may have value in field applications and is sufficient for the present argument.

With this rule in mind we can interpret the lengthening of incubation in styrofoam boxes. Table 1 shows data from Surinam. It seems likely that the differences between styrofoam boxes and the various procedures involving leaving the eggs in the sand, or replanting them, have been underestimated because, with buried eggs, incubation time includes time to emergence while, with animals in boxes only covered with a thin layer of sand (see Schulz, 1975), the time of hatching becomes apparent sooner. It therefore is likely, at least for leatherback (*Dermochelys coriacea*) and green turtles, that incubation temperatures in the styrofoam boxes were often 2° C lower than they would have been in the sand.

How would a 2° C drop influence the sex ratio? In Surinam the temperature of the sand at 80 cm depth, about the depth below ground level of the bottom of green turtle nests there (Schulz 1975), was 29 ± .5° C in a year that was not untypical as regards weather (Mrosovsky 1968). If the curve relating temperature and sex ratio for leatherbacks and green turtles is similar to the one shown for loggerheads in Figure 1, a 2° C drop could result in almost 100 percent of the hatchlings being male. However, to be concerned about a 2° C difference, it is not necessary even to assume that the curves for leatherbacks and green turtles in Surinam will be the same as those for the loggerhead clutch from Little Cumberland Island in the United States. All one has to assume is that the pivotal temperature for the Surinam turtles is close to temperatures commonly prevailing in natural nests, and that the shape of the curves relating incubation temperature to sex ratio are steep. They are steep in all the turtle species studied so far (Figure 1). A small temperature difference makes a considerable difference to sex ratio. The masculinizing effect of a 2° C drop could therefore be considerable.

The second example with styrofoam boxes concerns Atlantic ridley turtles, *Lepidochelys kempi,* a species perilously close to extinction. Attempts to boost the population of Atlantic ridleys in Mexico have included incubation of eggs in styrofoam boxes. Marquez (1978) reports that incubation took about 5 days longer in the boxes than in the sand (details on whether incubation

N. MROSOVSKY

includes time for emergence not given). Using our simple rule for converting time to temperature, this would mean the boxes were 1° C cooler on average than the sand. However, Marquez suggests they were 2 to 3° C cooler. The discrepancy can possibly be resolved: the average temperature difference might have been nearer 1° C but at times differences of 2 to 3° C, or even more, occurred. More details about incubation duration and prevailing temperatures are needed to assess this matter properly. But if the curves relating sex ratio and incubation temperature are steep for the Atlantic ridley, as for other species studied (Figure 1), then even 1° C could result in an appreciable increase in the percentage of males.

In both these examples we have not asserted that there must have been changes in sex ratio, only that it is a possibility that should be very seriously investigated. There are two reasons at least for this caution. First, when eggs are exposed above ground in styrofoam boxes, the temperatures will fluctuate much more than when eggs are in the sand. At 80 cm depth, the sand can be an almost perfect constant-temperature incubator (Mrosovsky 1968). We do not know, therefore, whether a brief spell of warm temperature in a styrofoam box might protect the eggs against the masculinizing effects of generally lower temperatures. This question is not easy to answer with field studies involving uncontrollable fluctuations in ambient temperature levels. Laboratory work is needed. Second, if temperature in styrofoam boxes is considerably colder than in sand, it is possible that females will be produced. For the freshwater snapping turtle, *Chelydra serpentina,* the curve relating sex ratio to temperature is U-shaped (see Figure 1). The possibility that there are 2 pivotal temperatures in sea turtles needs to be investigated.

Because of these unknowns, there is a need for actual data on sex ratios from eggs incubated artificially. To date, the following information is available. At the Cayman Turtle Farm, Grand Cayman Island, where many

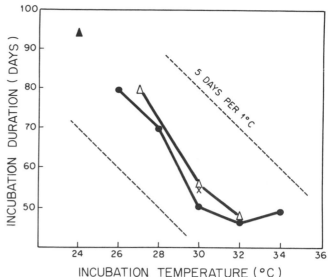

Figure 2. Incubation duration of sea turtle eggs as a function of incubation temperature. Dotted lines show slopes for a 5-day change in incubation duration per 1°C change in incubation temperature.

Open Triangle	*Chelonia mydas* (Bustard and Greenham 1968)
Closed Triangle	*Chelonia mydas* (Ackerman and Prange 1972)
Cross	*Caretta caretta* (Dimond 1965)
Circles	*Caretta caretta* (Yntema and Mrosovsky 1979)

In the last case, a variable 1 to 3 day correction has been added to the incubation times in the laboratory to cover what would probably have happened if the animals had also been at the various different temperatures during the 2 days spent in transit.

eggs have been incubated in a variety of ways, the sex ratio of green turtles varies greatly. Owens and Hendrickson (1978) give the following figures for the percentage females in different batches: 59, 83, 33, 17, 74, 97 and circa 100 percent. They suspect that incubation temperature varied in different seasons, hatcheries and containers.

Table 1. Mean incubation duration (days) of large samples of nests incubated in different ways in Surinam[1] and estimates of corresponding temperature differences

Species	In sand	In styrofoam boxes	Lengthening of incubation in styrofoam boxes	Estimated average degrees C cooler, styrofoam boxes
Leatherback (*Dermochelys coriacea*)	61–66	73	7–12	1.4–2.4°C
Green turtle (*Chelonia mydas*)	54–61	64–65	3–11	0.6–2.2°C
Olive ridley (*Lepidochelys olivacea*)	49–53	54	1–5	0.2–1.0°C

[1.] From Schulz 1975, Tables 21 and 22.

Table 2. Onset of metabolic heating of eggs in green turtles (temperature within clutch exceeding sand temperature by >1°C)

Incubation duration in days	Days when metabolic heating begins	Percentage incubation period elapsed at start of metabolic heating	Reference
61	42	62	Carr and Hirth (1961)
68	34	50	Bustard (1972)

Location of Hatcheries and Nest-Site Selection

On Heron Island, Australia, incubation time for green turtle nests in shady areas exposed to the wind is about twice as long as that for nests incubating simultaneously in sunny protected areas (Bustard 1972). The period varies from 42 to 77 days, averaging 56 days; in seasons with much rain, nests have taken as long as 91 days to hatch. No details on sample sizes and methods were given. In his work on hawksbill turtles, *Eretmochelys imbricata,* Garnett (1978) failed to find any relationship between incubation duration and position of the nest in relation to shade.

Until nest-site selection is understood better and information has been obtained about sex ratios of hatchlings in undisturbed turtle populations, it may be difficult to design the ideal hatchery. On beaches where turtles nest in a variety of places, affecting incubation duration differentially, this factor should be considered in choosing a protected location for a central hatchery that may have relatively uniform thermal characteristics. A further reason for taking this matter seriously is that studies of freshwater turtle eggs incubated outdoors in natural conditions have shown that whether the eggs are in the sun or the shade and ambient temperatures do indeed affect sex ratio (Yntema, unpublished; Pieau 1975).

Clutch Size, Metabolic Heat and Critical Periods

In Malaysia, eggs of leatherback turtles are often buried in the sand in batches of 50 instead of in their natural clutches which average 84 eggs (Balasingam 1965, 1966). Apparently, hatching rates are better with 50 eggs. Because eggs produce metabolic heat, one should ask whether small clutches are cooler and so affect sex ratio.

Schulz (1975) reported that incubation times for green turtle eggs in styrofoam boxes with either 76 to 180 eggs per box or with 56 eggs per box were 65 days and 64 days, respectively. However, with natural nests, clutch size can affect incubation duration. Garnett (1978) found for hawksbill turtles in the Seychelles, that incubation (plus emergence) time was negatively correlated with clutch size (n = 44, r = 0.4, p < .01). However, in quantitative terms the effect was small, with 10 extra eggs in the clutch shortening incubation

by less than half a day on average. Perhaps effects of clutch size are seen only in conditions where heat loss from the egg mass is relatively small. In cases where clutch size influences incubation duration (presumably through attending metabolic heating, although the relationship is not necessarily causal) would clutch size also affect the sex ratio?

Metabolic heating occurs mainly later on in incubation. Although heating up of the egg mass is gradual, it is convenient to be able to specify some day in incubation when this heating becomes important. We arbitrarily define the onset of metabolic heat as occurring when the egg mass becomes more than 1° C warmer than a control site in the sand at similar depth. Using this definition, it can be calculated that metabolic heating is not important until more than halfway through incubation (Table 2).

The critical period for sexual differentiation of the gonads in sea turtles has not yet been worked out, but this has been studied in snapping turtles (Yntema 1979). The exact critical period depends on the incubation temperature and details of the thermal schedules used. However, with all temperature regimes employed so far, the critical period does not start before stage 14 or continue after stage 20 of embryonic development is reached (for a description of embryonic stages see Yntema 1968). How many days it takes to reach stage 20 depends, of course, on the incubation temperature. Nevertheless, even at different temperatures, this stage is reached before halfway through the total incubation period (Table 3). If sea turtles are like snapping turtles, then the critical period for sexual differentiation occurs before metabolic heating becomes important. Clutch size should not, therefore, make much difference to sex ratio.

Before accepting this, it is necessary to study critical periods in sea turtles, especially in a unique species like the leatherback. Knowledge of the critical periods also is obviously of value in managing artificial hatcheries. It should also be pointed out that if critical periods for sea turtles turn out to be relatively early in incubation, before metabolic heating becomes important, it means that temperatures of sand adjacent to nests, rather than those within the actual egg mass itself, can be used to discover what temperature eggs in a given place or season are undergoing at the time

N. MROSOVSKY

Table 3. Critical periods for sexual differentiation in snapping turtles (stages 14–20 of embryonic development)

Incubation temperature °C	Days to reach stage 14 (Yntema 1979)	Days to reach stage 20 (Yntema 1979)	Incubation period (Yntema 1978)	Percentage incubation period elapsed at end of critical period
26°	16	31	70	49%
30°	11	24	62	39%

of gonadal differentiation. This would simplify the application to field situations of information obtained in laboratory studies, with separated eggs incubated at a constant temperature.

Egg Harvesting and Close Seasons

Because sea turtles sometimes dig up the nests of turtles that have laid before, there are attractions in harvesting eggs that are laid early in a season. For instance, Pritchard (1978) has suggested that the logical method of exploiting olive ridleys, *Lepidochelys olivacea,* in Mexico would be to permit egg collection from the first *arribada* [synchronized mass nesting] or up to a certain date, because these eggs would be most liable to destruction by turtles laying later on. While this suggestion has obvious merit, especially when large arribadas are involved, it should nevertheless be evaluated in light of information about sand temperatures at different times in the nesting season and knowledge of the pivotal temperatures for these turtle populations.

Especially interesting problems arise with places where turtles nest all year round. For instance, in the Sarawak turtle islands (Talang Talang islands and Satang Besar) incubation plus emergence times of green turtles average 54 days during much of the summer: in February, during the monsoons, they average as high as 71 days (Hendrickson, 1958). This 17-day difference suggests a 3.4° C temperature difference if our simple rule is used. A more conservative estimate of 2° C can be made by referring to Bustard and Greenham's (1968) curve for green turtles in Figure 2 over the particular range of incubation periods involved. But even 2° C is enough to make a large difference to the sex ratio (Figure 1).

Green turtles also nest all year round in the Mozambique Channel on the island of Europa (Servan 1976). Incubation times range from 58 days in February, during the austral summer, to 85 to 99 days in June during the austral winter. Although sample sizes were small, it is clear that incubation can take at least 30 days longer at certain times of the year. Presumably temperature is largely responsible because not only is the sun weaker during the winter but the beaches are in shade during the morning (Servan 1976). Even if allowance is made for imperfection in our simple rule (which for 30 days

would mean a 6° C difference) when applied over large ranges, it does not seem at all unreasonable to suppose that there is a 4° C difference in incubation temperatures (Figure 2).

How do turtle populations evolve in such situations? Three possibilities will be mentioned. First, there might be a second pivotal temperature, as occurs in the snapping turtle (Figure 1). If this was fairly close to the upper one, it could permit some female differentiation, even at cooler times of year. Or, as suggested by Servan (1976) for Europa, different populations of turtles might use the island at different times of year. These populations might then have different pivotal temperatures. Perhaps this could even be used as a distinguishing characteristic. It is also possible that, although fewer turtles nest on Europa and the Sarawak turtle islands during the cooler seasons (Servan 1976, Hendrickson 1958) a high percentage of hatchlings produced then might be males. Should this be the case, there would be implications for egg harvesting schemes. For instance, eggs have been collected from Europa for turtle ranching (Fretey 1978). If this practice is to continue, it might be wise to consider spreading out egg collection over different seasons.

Furthermore, if clutch size turns out to be a significant factor in determining sex ratio, contrary to what we suspect at the moment, on the basis of the meagre information available ("Clutch Size. . .," above), then any seasonal changes in clutch size would also have to be taken into account.

Summary and Conclusions

A number of current conservation practices such as use of styrofoam boxes, close seasons, and establishment of central hatcheries, are likely to affect the sex ratio of sea turtles. The temperature changes involved, though slight in absolute terms, are probably large enough to affect sexual differentiation. But before this can be asserted with confidence, further work is needed. In particular it is important to:

• Learn about the effects of fluctuating temperatures such as are experienced by eggs above ground in styrofoam boxes.
• Discover what the pivotal temperature or temperatures are, and how much they differ in different species and populations.

• Locate the critical periods for sexual differentiation. There is a need to obtain at least some further information on the questions listed above now—before unevaluated methods become accepted procedures.

The possibility should also be considered that incubation in sand on a natural beach may provide the eggs with important benefits in addition to the correct temperature regime. Absorption of minerals and stimuli for imprinting are two possibilities that have already been raised (Simkiss 1962, Mrosovsky 1978). But there may be other facets of natural incubation that have not even been thought of yet. Certainly more research on the thermal aspects of incubation, as listed above, is urgently needed. But, in case we are not sufficiently in tune with the natural processes involved in incubation, we also advocate allowing at least some of the eggs in any conservation program to develop in the ground where the turtle laid them. In this context it is worth recalling the sentiments of Henry David Thoreau (1967) on the incubation of turtle eggs:

I am affected by the thought that the earth nurses these eggs. They are planted in the earth, and the earth takes care of them; she is genial to them and does not kill them. It suggests a certain vitality and intelligence in the earth, which I had not realized. This mother is not merely inanimate and inorganic. Though the immediate mother turtle abandons her offspring, the earth and sun are kind to them. The old turtle on which the earth rests takes care of them while the other waddles off. Earth was not made poisonous and deadly to them. The earth has some virtue in it; when seeds are put into it, they germinate; when turtles' eggs, they hatch in due time.

Acknowledgments

We thank Dr. S.J. Shettleworth and S.F. Kingsmill for comments. Support came from the Natural Sciences and Engineering Research Council of Canada and NIH grant HDO3834.

Literature Cited

Ackerman, R. A., and H. D. Prange
1972. Oxygen diffusion across a sea turtle (*Chelonia mydas*) egg shell. *Comparative Biochemistry and Physiology* 43A:905–09.

Balasingam, E.
1965. The giant leathery turtle conservation programme——1964. *Malayan Nature Journal* 19:145–46.
1966. The ecology and conservation of marine turtles in Malaya with particular reference to *Dermochelys coriacea*. Abstracts, 11th Pacific Science Congress, Tokyo 1966, Proceedings, vol. 5, Science Council of Japan.

Bustard, H. R.
1972. *Sea Turtles*. Glasgow: Collins.

Bustard, H. R., and P. Greenham
1968. Physical and chemical factors affecting hatching in the green sea turtle, *Chelonia mydas* (L.). *Ecology* 49:269–76.

Carr, A., and H. Hirth
1961. Social facilitation in green turtle siblings. *Animal Behavior* 9:68–70.

Dimond, M. T.
1965. Hatching time of turtle eggs. *Nature* 208:401–02.

Fretey, J.
1978. Raising green turtles at Réunion Island. *IUCN/SSC Marine Turtle Newsletter* 8:3.

Garnett, M. C.
1978. The breeding biology of hawksbill turtles (*Eretmochelys imbricata*) on Cousin Island, Seychelles. Research Report, International Council for Bird Preservation (British Section).

Hendrickson, J. R.
1958. The green sea turtle, *Chelonia mydas* (Linn.) in Malaya and Sarawak. Proceedings of the Zoological Society of London 130:455–535.

Marquez, R.
1978. The Atlantic ridley in Mexico: 1978 season and conservation programme. *IUCN/SSC Marine Turtle Newsletter* 9:2.

Mrosovsky, N.
1968. Nocturnal emergence of hatchling sea turtles: control by thermal inhibition of activity. *Nature* 220:1338–39.
1978. Editorial. *IUCN/SSC Marine Turtle Newsletter* 7:1–2.
1980. Thermal biology of sea turtles. *American Zoologist* 20:531–47.

Owens, D. W., and J. R. Hendrickson
1978. Endocrine studies and sex ratios of the green sea turtle, *Chelonia mydas*. *Florida Marine Research Publications* 33:12–14.

Pieau, C.
1971. Sur la proportion sexuelle chez les embryons de deux Chéloniens (*Testudo graeca* L. et *Emys orbicularis* L.) issus d'oeufs incubés artificiellement. *C. R. Acad. Sci.* (Paris), 272:3071–74.
1972. Effets de la température sur le développement des glandes génitales chez les embryons de deux Chéloniens, *Emys orbicularis* L. et *Testudo graeca* L. *C. R. Acad. Sci.* (Paris), 274:719–22.
1973. Nouvelles données expérimentales concernant les effets de la température sur la différenciation sexuelle chez les embryons de Chéloniens. *C. R. Acad. Sci.* (Paris), 277:2789–92.
1975. Différenciation du sexe chez les embryons d'*Emys orbicularis* L. (Chélonien) issus d'oeufs incubés dans le sol. *Bull. Soc. Zool.* France, 100:648–49.

Pritchard, P. C. H.
1978. Comment on Tim Cahill's article "The Shame of Escobilla." *IUCN/SSC Marine Turtle Newsletter* 7:2–4.

Schulz, J. P.
1975. Sea turtles nesting in Suriname. *Zool. Verhandelingen* 143:1–143.

Servan, J. P.

1976. Ecologie de la tortue verte a l'île Europa (Canal de Mozambique). *La Terre et la Vie* 30:421–64.

Simkiss, K.

1962. The sources of calcium for the ossification of the embryos of the giant leathery turtle. *Comparative Biochemistry and Physiology* 7:71–79.

Thoreau, H. D.

1967. Thoreau's Turtle Nest: From the Journal Notes of Henry David Thoreau. Worcester, Massachusetts: Achille J. St. Onge.

Yntema, C. L.

1968. A series of stages in the embryonic development of *Chelydra serpentina. Journal of Morphology* 125:219–51.

1976. Effects of incubation temperatures on sexual differentiation in the turtle, *Chelydra serpentina. Journal of Morphology* 150:453–62.

1978. Incubation times for the eggs of the turtle *Chelydra serpentina* (Testudines: Chelydridae) at various temperatures. *Herpetologica* 34:274–77.

1979. Temperature levels and periods of sex determination during incubation of eggs of *Chelydra serpentina. Journal of Morphology,* 159:17–28.

Yntema, C. L., and N. Mrosovsky

1979. Incubation temperature and sex ratio in hatchling loggerhead turtles: a preliminary report. *IUCN/SSC Marine Turtle Newsletter* 11:9–10.

1980. Sexual differentiation in hatchling loggerhead sea turtles (*Caretta caretta*) incubated at controlled temperatures. *Herpetologica* 36:33–36.

Anne Meylan
Department of Zoology
University of Florida
Gainesville, Florida 32611

Behavioral Ecology of the West Caribbean Green Turtle (*Chelonia mydas*) in the Internesting Habitat

ABSTRACT

The present study investigates ecological aspects of behavior of green turtles (*Chelonia mydas*) in the internesting habitat at Tortuguero, Costa Rica. The longshore waters off the nesting beach are occupied for up to 3 months of each reproductive cycle; turtles nest during this period from 1 to 7 times, at approximately 12-day intervals. Twenty-six female turtles were tracked by visual techniques from shore-based triangulation stations after emergence on the nesting beach. Travel plots were used in conjunction with bathymetric and current charts to determine habitat relations and spatial and periodic characteristics of internesting travel. Most turtles traveled parallel to shore and within the 24-m contour line; maximum longshore travel distances of 10.1 km south and 10.0 km north of sites of emergence on the beach were recorded. Various lines of evidence suggest that this is a conservative estimate of the extent of internesting travel. The habitat was found to be relatively restricted in seaward extent; with one exception, all sightings were made within 4.8 km of shore. Turtles showed no initial preference in the direction of longshore travel. In most cases, they left the nesting area within 24 hours. The maximum travel speed recorded was 4.5 km/hr; mean continuous travel speeds approached 2 km/hr. Accurate and direct returns to specific locations in the internesting habitat were observed. Simultaneous tracking of turtles and current drogues showed the ability of turtles to hold courses independent of the current. The sites of recapture of 3 turtles away from the nesting beach suggest that homeward migration from Tortuguero may occur both along the coast and via the open sea. Two departure courses were against prevailing currents.

Introduction

With few exceptions, studies of marine turtles at the nesting grounds have been conducted on the beach itself. The ecology and behavior of turtles in the longshore waters of the rookery—the internesting habitat—have received little attention. Occupation of this habitat is an important stage in the life cycle. For up to several months, turtles take up residence in a habitat that may be ecologically very different from the home foraging grounds. Courtship and copulation take place here, and females carry out their season's nesting routine. The internesting period may be a time of increased vulnerability for the colony, because a large contingent of the reproductive members of a population is concentrated within a relatively small area. The potential impact of commercial harvesting, incidental catch, and ecological disasters such as oil spills to turtles in this habitat is considerable. Knowledge of the habitat's dimensions, of the activities that take place within it, and of the ecological requirements of turtles during this period is prerequisite to effective conservation and management programs.

Investigations conducted by Carr (1967, 1972) at Tortuguero, and Carr, Ross, and Carr (1974) at Ascension Island, were the first attempts to monitor the travel of internesting turtles. The track plots obtained provided the first information on spatial and temporal characteristics of travel, and showed the promise of tracking as a tool for future investigations of this kind. Booth and Peters (1972) made a substantial contribution to knowledge of courtship and copulation in their study of internesting green turtles at Fairfax Island, Australia. They introduced the concept of a female refuge area within the internesting habitat, where females may elude the attentions of courting males.

With this background, the present study of the internesting ecology of *Chelonia mydas* was initiated at Tortuguero, Costa Rica, in 1976. Results of that investigation will be presented in this paper. A similar study of internesting ecology of green turtles at Ascension Island has subsequently been conducted by J. Mortimer. Comparison of results at this midoceanic rookery with those from mainland Tortuguero should provide new insight into this stage of the life cycle. Studies involving other genera are ongoing: the Office of Endangered Species of the U.S. Fish and Wildlife

Table 1. Range of movements of green turtles in the internesting habitat, Tortuguero, Costa Rica

Tag no.	Maximum distance offshore (km)	Maximum longshore distance from site of emergence (km)		Observed range of longshore travel (km)
		North	South	
4346	1.9	—	1.2	4.2
6277	4.8	—	4.7	4.8
	3.0	1.3	4.8	—
7652	3.0	0.5	2.2	4.8
9171	—	—	—	3.4
9915	1.5	0.6	—	6.4
10156	3.6	—	5.2	5.2
10359	—	—	—	0.6
11914	—	—	—	4.8
12006	2.6	—	3.7	4.6
12012	1.5	—	1.3	5.8
12181	1.2	—	1.9	6.0
12224	2.2	0.6	—	0.8
12426	—	3.0	—	3.4
12531	1.7	—	10.1	10.7
12690	1.8	10.0	2.4	16.1
13017	—	7.0	—	7.0
13018	0.5	0.5	—	0.4
13019	14.5	1.6	—	1.2
13206	0.5	0.6	—	1.2
13209	3.4	1.1	1.6	2.6
13280	2.0	—	5.2	5.2
13285	1.6	—	—	—
13295	0.9	1.2	—	1.2
13361	—	2.0	—	2.0
13366	2.8	—	5.0	5.4
13370	1.9	—	4.0	4.2

— No data.

ANNE MEYLAN

Service (Albuquerque, New Mexico) initiated an investigation of the internesting ecology of Kemp's ridley (*Lepidochelys kempi*) at Rancho Nuevo, Mexico, in 1979; the South Carolina Wildlife and Marine Resources Department is studying the internesting ecology of *Caretta* at Georgetown, South Carolina; and the National Park Service is studying *Caretta* off Little Cumberland Island, Georgia. Knowledge of internesting ecology will undoubtedly be greatly advanced when full results of these projects become available.

This study was conducted at Tortuguero, Costa Rica, the major breeding site for *Chelonia mydas* in the western Caribbean. The open-sea beach extends 35 km and is bounded at each end by major rivers. The rookery has been previously described by Carr and Ogren (1960) and Carr, Carr, and Meylan (1978). Turtles arrive at the rookery as early as June, and may stay as late as October. They nest from 1 to 7 times during a season, at approximately 12-day intervals; the average number of nests is 2.8. The colony is drawn from feeding grounds throughout the western Caribbean.

Methods

Observations of the movements of 26 green turtles in the internesting habitat were made by visually tracking tear-shaped polyurethane floats (19 × 12 × 7 cm; 2 kg) towed by the turtles on 18-m lines. The lines were attached to the rear of the carapace with ungalvanized iron wire to insure eventual disengagement. Previous tracking experiments using towed floats (Carr 1967, 1972; Carr, Ross, and Carr 1974) indicated that locomotion, orientation, and homing are not impaired by the technique. Evidence from the present study, including returns to previous sites of emergence by turtles with and without their floats, reinforces this supposition (Meylan 1978).

To allow the monitoring of nighttime movements, 6 floats were illuminated with chemical light sticks (Cyalume) and 9 with 6-volt electric lights (with and without flasher circuits). The lights were attached to the tip of a 140-cm mast. The possibility was considered that lights would disrupt the guidance mechanism of the turtle, either by evoking a positive or negative phototactic response or by confusing stellar patterns. That a phototactic response was not elicited is clearly shown in the tracking data. Ehrenfeld and Koch (1967) have cast considerable doubt on the possibility that green turtles employ a stellar navigation mechanism.

Twenty-one turtles were tracked after nesting and 6 after unsuccessful nesting attempts. One turtle was tracked after 2 separate emergences. Floats were attached to nesting females as they laid their eggs. They were then left undisturbed to finish their nests and return to the water. Non-nesting turtles were turned on their backs to facilitate float attachment, and were

released immediately or the following morning. In all cases, turtles returned to the water at the site of their emergence on the beach.

Simultaneous compass bearings of the floats were recorded from 2 tracking towers placed 0.8 km apart on the beach. Tracking towers were located at the northern end of the beach, approximately 1.6 km from the mouth of the Tortuguero River. Bearings were taken with Enbeeco compass-bearing monoculars, optical howitzer sights, and alidades. Positions of the turtles were determined by triangulation. Additional tracking stations were used to monitor extended longshore movements. Patterned searches of the offshore waters were made by boat to relocate turtles that had moved outside the tracking range.

The direction and speed of currents in the internesting habitat were determined by optically tracking current drogues, placed 0.5 to 3 km offshore from the tracking towers. Sixteen measurements were made from 3 September to 25 September of 1976 and 1977. Current velocities were calculated from straight-line distances between initial and final positions of the drogues and total elapsed times. Depth transects were made in the longshore waters with a Sonar Self-Recording Fathometer (Model DC-250).

The V Test, or modified Rayleigh Test (Batschelet 1972; Schmidt-Koenig 1975), was used to test directional correlation between simultaneous movements of turtles and current drogues. The mean vector of the current was considered the preferred direction, and the directions of the subvectors of the track of the turtle were compared to it. Subdivision of the turtle track into its component subvectors, as in the Hodges and Ajne test (Batschelet 1972), supposes independence of the subvectors. This is not a wholly valid assumption, because future choices of direction in the travel of any animal are influenced by previous choices. Nevertheless, the test appears to be the best statistical tool that is presently available.

Results and Discussion

Periodic and Spatial Aspects of Internesting Travel

Table 1 presents data on the range and direction of travel of turtles observed in the internesting habitat. The most distant offshore position recorded for 23 of 24 turtles observed after release was 4.8 km. Most turtles traveled parallel to shore and within the 24-m contour line; only 1 sustained seaward course was recorded. Previous tracking experiments at Tortuguero also recorded predominantly longshore travel (Carr 1967, 1972). No minimum distance from shore was maintained in internesting travel. There was no evidence that female turtles used the shallows as a refuge from courting males, as recorded at 1 rookery in eastern Australia (Booth and Peters 1972).

With 1 exception, both nesting and non-nesting females left the waters adjacent to the emergence site within 24 hours. Maximum longshore travel distances of 10.0 km north and 10.1 km south of release sites were recorded (Table 1). Four turtles were observed to travel both north and south of release sites. The total range of longshore travel, indicated by both tracking and beach emergence data, is also shown in the table. Both measures of the extent of longshore travel are extremely conservative, and were influenced by the limits of observation.

A better indication of the extent of longshore travel is that turtles were rarely within range of observation from the tracking towers for more than 3 days after release. Most disappeared on southward courses. Travel paths followed the shoreline, but the lack of roads and vehicles made it difficult to establish southerly travel limits. Some turtles had presumably finished nesting for the season and departed from the rookery. Others, however, returned to the vicinity of the tracking towers after varying intervals. One turtle traveled at least 10.1 km south of her emergence site during the 2 days after her release and returned to the northern end of the rookery during the next 2 days. Two others traveled at least 4.7 and 5 km south of release sites before returning to nest. During a previous tracking experiment at Tortuguero, Carr (1967) observed a turtle 7.9

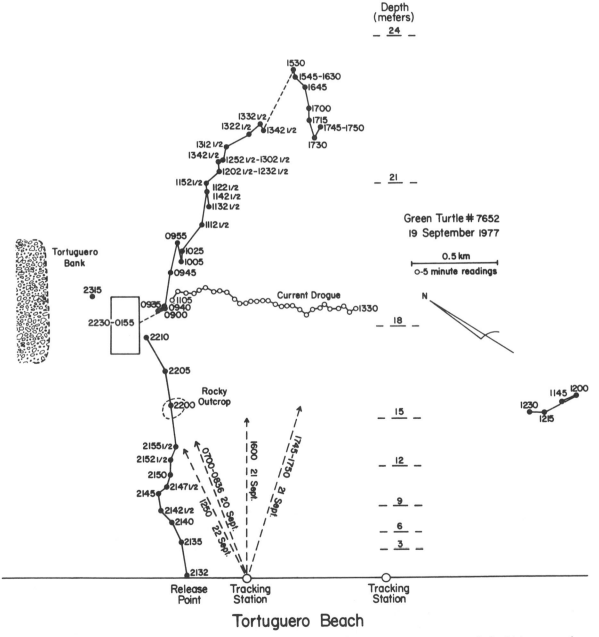

Figure 1. Track plot, turtle 7652. Turtle was released on September 19 after nesting. Dashed arrows represent single bearings. Positions recorded within rectangle were too close together to be individually shown.

ANNE MEYLAN

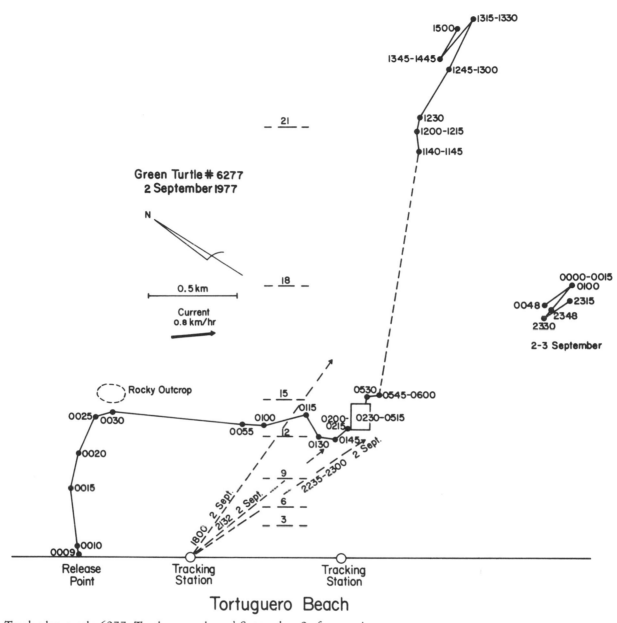

Figure 2. Track plot, turtle 6277. Turtle was released September 2 after nesting.

km south of her future nest site during the day preceding nesting.

The purpose of extended travel up and down the coast is not clear. No food source has been identified in the longshore waters. Evidence presented later clearly shows that southward travel is not an obligatory response to the current, which flows southward along the beach at approximately 1 km/hr. Currents may, nevertheless, be involved. Longshore travel in the direction of the current, periodically corrected by active up-current swimming, may represent a solution to the

problem of avoiding extensive displacement from the site of future nesting emergencies. At Tortuguero, the average distance between successive emergence sites is only 1.2 km (Carr and Carr 1972). The alternative solution, that of steadily holding a position off the nesting beach, may be precluded at Tortuguero by the lack of sufficient topographic features that could provide anchorage.

A possibly relevant discovery is that the internesting travel of Kemp's ridley (*Lepidochelys kempi*) is also characterized by extensive movements up and down the coast. Five ridleys tracked with radio transmitters at their mainland nesting beach at Rancho Nuevo, Mexico, traveled more than 14 km south of their nest sites immediately after (daytime) release, only to return the next morning (Diderot Gicca, personal communication). This pattern was repeated for 2 to 4 days, until they disappeared to the south for the rest of the internesting period. The current at Rancho Nuevo flows north, so the relationship between longshore travel direction and current is exactly the reverse of that at Tortuguero. That is, turtles at Rancho Nuevo initially traveled against the current, and later returned with it. The purpose of longshore travel in the case of these ridleys is even more difficult to explain than for green turtles at Tortuguero, because food and bottom shelter appear to be locally available.

There is some evidence that the Tortuguero River, with its conspicuous plume of effluent, serves as a boundary of the internesting habitat. That it marks the northernmost extreme of nesting has long been known (Carr and Ogren 1960). No turtle was monitored on a course which continued north of the river mouth, despite its proximity to the tracking towers. Only 2 turtles were sighted during routine searches of the waters north of the rookery. Two turtles captured by fishermen 24 km and 43 km north of the river at the end of the season are believed to have been intercepted during homeward migration.

Tracking data showed that turtles traveled largely within the 24-m contour line, which is 3.2 km offshore. This limit does not correspond to any major topographic change in the seafloor; the bottom is of uniform

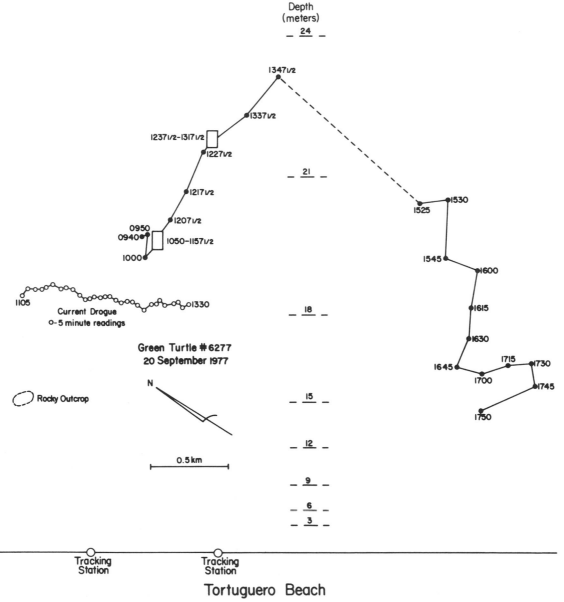

Figure 3. Track plot, turtle 6277. Turtle was released between the tracking towers on 15 September after a nesting attempt in which no eggs were laid. The turtle and current drogue were tracked simultaneously from 1105–1330.

ANNE MEYLAN

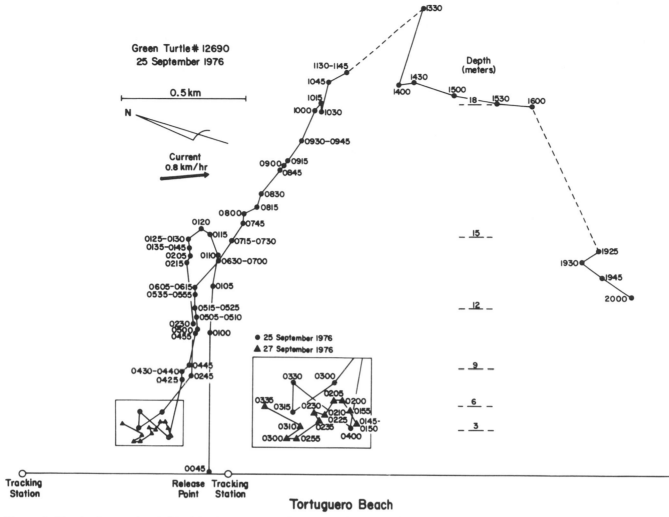

Figure 4. Track plot, turtle 12,690. Turtle was released September 25 after laying only 68 eggs. Enlargement shows travel courses recorded later the same night and on September 27, when the turtle returned to the waters off her previous nest site.

relief to at least 4.8 km off the coast. There was no evidence that turtles followed isobaths in their long-shore travel. The maximum depth at a position recorded for a float in the internesting habitat was 30.5 m.

Nine of 26 turtles traveled to Tortuguero Bank, a localized volcanic or old corraline outcrop located 1.5 km offshore and just south of Tortuguero River. The bank is approximately 1-km wide and varies irregularly in depth from 15 to 18 m. A shelf 20 m deep extends seaward from it for 0.8 km. With the exception of a few isolated, rocky patches, the bank constitutes the only topographic relief in the longshore seabottom. Reconnaissance of the waters off the nesting beach by boat showed that the bank was the only place where green turtles, both singly and in pairs, could predictably be sighted. It is possible that turtles go there to find bottom shelter. At Ascension Island, internesting green turtles have been observed to use rocky ledges and caves for resting sites (Carr, Ross, and Carr 1974).

A pronounced difference between daytime and nighttime activity levels of turtles was observed. Resting periods—times when no displacement was observed—were only once observed to exceed 1 hour during daylight hours. At night, periods of up to 210 min were recorded (Figures 1 to 3, 5; see also Meylan 1978). Because many of the nighttime tracking data were collected soon after turtles had nested, it is possible that some observed resting periods represented periods of recuperation from the nesting exercise rather than a regular feature of diel activity. Nighttime resting periods were, however, also recorded several days subsequent to release. On the feeding grounds, green turtles are reported to be diurnal (Carr 1954).

A common pattern observed in the track plots is a gradual movement of turtles on oblique seaward courses during the day, and a return to inshore waters as evening approached (Figures 1 to 4). An interesting detail of this pattern is the close correlation between the vector of offshore movement and the sun's bearing on the horizon at sunrise. It seems possible that the azimuth position of the sun is of use as a positional reference. In a tracking experiment conducted previously at Tortuguero, Carr (1972) reported that a turtle which had been quiescent for a long period suddenly resumed activity at the moment that the sun showed on the horizon. In the present study, no activities were observed to coincide exactly with sunrise, but resumption of activity in the early dawn hours was frequently observed (Figures 2, 5; see also Meylan 1978).

Most turtles returning to sea adhered to courses almost perfectly perpendicular to the shore for approximately 1 km. Once away from the beach, they either stopped and held their position for a while, or changed their course of travel (Figures 1, 4, 5; see also Meylan 1978). The orientation mechanism involved in maintaining such unerringly straight offshore headings is unknown. It seems unlikely that the shore could provide cues to turtles swimming at night, underwater, and in an offshore direction. Wave orientation, piloting by bottom topography, or some as yet unidentified

mechanism may be involved. Few initial departure courses showed any sign of current deflection, although the current flowed at right angles to the travel paths. No differences were observed between the departure courses of turtles released by day or night, or between nesters and non-nesters.

The repeated observation of individual turtles on small subsections of the beach, both within and between seasons, has long suggested that green turtles are well oriented during their internesting travel (Carr and Ogren, 1960; Carr and Carr, 1972; Carr, Carr and Meylan 1978). Tracking data, which provide detailed information on actual travel courses, confirm this. Both during the day and at night, turtles appear to be capable of controlling their spatial position within the habitat. That they are capable of accurate and direct returns to specific longshore locations is illustrated by Figure 6. This track plot shows the travel of a turtle (6277) on 2 separate nights. Exact positions were obtained for only the final hours of each period of tracking, but on both nights the turtle's course was monitored on foot

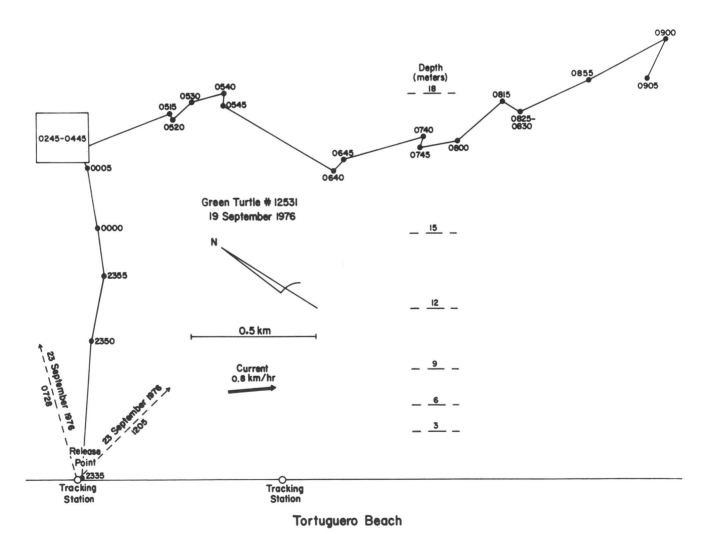

Figure 5. Track plot, turtle 12,531. Turtle was released September 19 after nesting.

ANNE MEYLAN

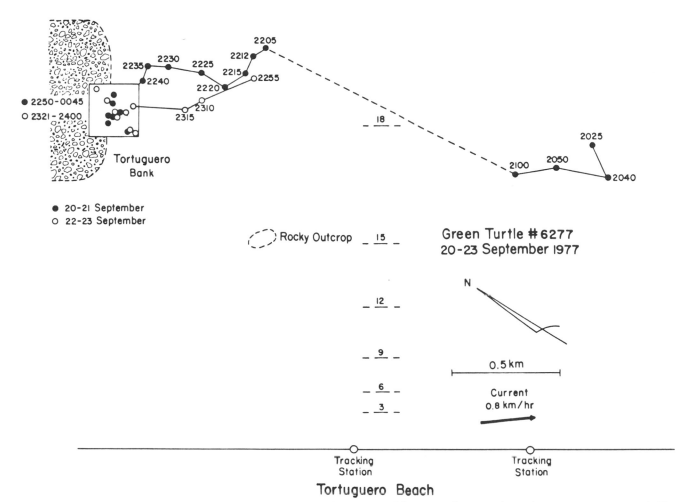

2205

2212

2230 2225 2215

2235 2220

2240 2255

2250-0045

2321-2400 2310

2315

Tortuguero
Bank

2025

_ 18 _

2100 2050 2040

● 20-21 September
○ 22-23 September

Rocky Outcrop   _ 15 _

Green Turtle #6277
20-23 September 1977

N

_ 12 _

_ 9 _

0.5 km

_ 6 _   Current
_ 3 _   0.8 km/hr

Tracking
Station

Tracking
Station

Tortuguero Beach

Figure 6. Travel courses of turtle 6277 on the nights of September 20 and 22. On both nights, the turtle swam more than 4 km on a direct and continuous course to reach Tortuguero Bank. Only the final periods of tracking are shown.

for several of the preceding hours. On both occasions the turtle swam for more than 4 km against the current on a continuous and direct course to her resting position on Tortuguero Bank. Her arrival times there were only 30 min apart, and her positions there were spatially overlapping.

Current Relations

One facet of the present investigation was to determine the influence of currents on the travel behavior of turtles. To do this, responses to currents were observed directly, by simultaneously tracking turtles and current drogues traveling in juxtaposition, and indirectly, by using average values of current parameters to evaluate the body of tracking data.

DIRECT OBSERVATIONS

Two turtles (6277 and 7652) and a current drogue were tracked simultaneously for 2.4 hr at a position 1.6 km offshore from the tracking towers (Figure 7). Table 2 compares resultant vector directions, net displacements, and mean speeds. The V Test (Batschelet 1972) indicates no significant directional correlation between the current and the travel path of either turtle ($u = 1.210$ for turtle 7652; $u = 0.901$ for turtle 6277). Moreover, travel of both turtles was characterized by frequent stops, whereas the movement of the drogue was continuous.

In a second experiment, 1 turtle (12690) was tracked simultaneously with 2 current drogues (Figure 8; Table 2). The V Test indicates that there is directional cor-

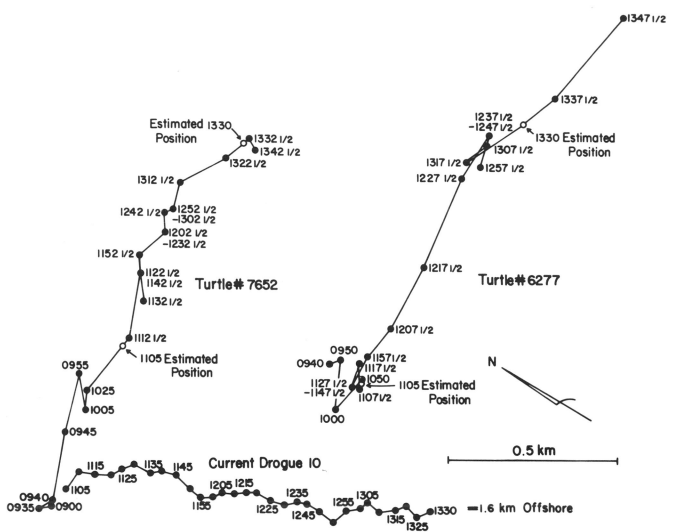

Figure 7. Travel paths of turtles 6277 and 7652 and of a current drogue tracked simultaneously 1.6 km offshore Tortuguero, Costa Rica.

relation between the turtle's path and the mean direction of both drogues (for drogue 5, u = 2.604; for drogue 11, u = 3.406). The turtle, however, traveled only 0.4 km during the tracking period, while the drogues traveled 1.1 and 1.7 km.

INDIRECT OBSERVATIONS

Sixteen current drogues placed offshore from the tracking towers traveled southeast in an average direction of 154° (range 127° to 186°, SD = 17.8), which is roughly parallel to the shoreline. Drogues moved at an average speed of 0.88 km/hr (range 0.5 to 1.9 km/hr, SD = 0.34), which is comparable to current speeds measured previously at Tortuguero (Frick 1976). Assuming these values to be representative, the tracking data reveal a wide range of current responses. Turtles

traveled down the current (Figure 5), against it (Figure 9), and at right angles to it (Figures 1, 3, 4). Additional track plots showing current responses can be found in Meylan (1978). Turtles were observed to hold positions for up to 210 min. Initial departure courses from the shore, made at right angles to the current, showed no sign of southward deflection. No initial preference in the direction of longshore travel was observed, nor was there found to be any correlation between current direction and placement of subsequent nests. Mean continuous travel speeds of up to 2.7 km/hr were recorded for travel with the current, compared to 1.3 km/hr for against-current travel. Against-current travel was sustained by some turtles for several hours.

The emergence records of turtles that fail to nest on one night and return to do so within the next few nights also provide information on responses to current. Fig-

ANNE MEYLAN

Table 2. Resultant travel vectors of turtles and current drogues tracked during simultaneous periods

Item	Direction of resultant vector (degrees)	Net displacement (km)	Time (hr)	Mean speed (km/hr)
Turtle 6277	91	0.9	2.42	0.4
Turtle 7652	90	0.7	2.42	0.3
Current drogue 10	153	1.1	2.42	0.5
Turtle 12690 (drogue 5)	93	0.4	1.75	0.2
Current drogue 5	146	1.1	1.75	0.6
Turtle 12690 (drogue 11)	95	0.5	2.25	0.2
Current drogue 11	127	1.7	2.25	0.8

ure 10 shows the distances between successive emergence sites for turtles recorded back on the beach within 6 days. The data indicate a strong tendency for turtles to return close to the site of their previous emergence. Emergence sites that do not correspond to previous sites are evenly distributed down-current and up-current, which suggests that turtles correct for current set in their approach to the beach.

Short-term recoveries of 2 turtles with floats at points 24 km and 43 km north of the Tortuguero River suggest that migration back to resident feeding grounds may, in some cases, entail movement against the current. One turtle that was last sighted 14 km straight out from the mouth of the Tortuguero River after a travel time of approximately 5 hr had traveled a course perpendicular to prevailing currents. Additional details of these and other short-term recoveries are presented by Meylan (1978).

Speed of Travel

Only a few direct observations of swimming speeds of adult green turtles in the wild have been made (Carr, 1972; Carr, Ross, and Carr, 1974). Travel speeds are usually calculated from point-to-point tag recovery data. They therefore suffer from the uncertainty as to when the turtle actually arrived at her destination, and from the possibility that both travel and rest periods may be represented.

Travel speeds recorded in the internesting habitat at Tortuguero are shown in Table 3. Travel was defined as movement with no known stopping periods. The maximum speed of travel, recorded during a turtle's initial departure from shore, was 4.5 km/hr. Carr (1972) observed maximum travel speeds of female green turtles at Tortuguero of 2.3 and 3.9 km/hr. The greatest speed yet recorded for *Chelonia* is 7.2 km/hr (Carr,

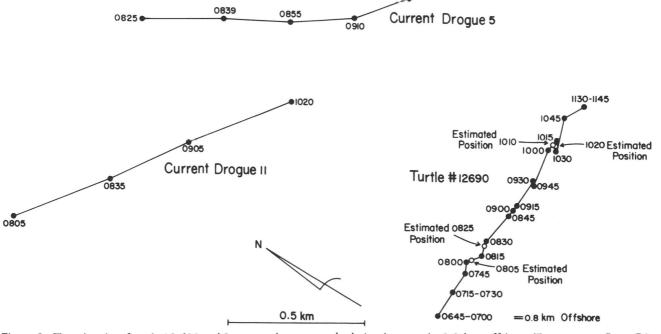

Figure 8. Travel paths of turtle 12,690 and 2 current drogues tracked simultaneously 0.8 km offshore Tortuguero, Costa Rica.

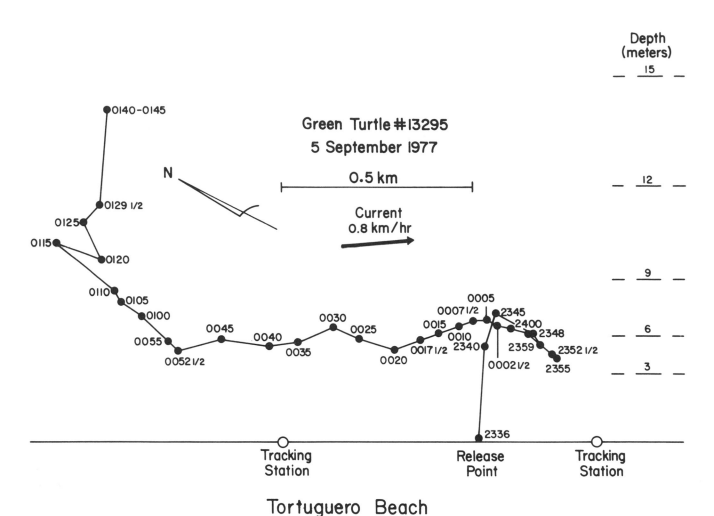

Tortuguero Beach

Figure 9. Travel plot, turtle 13,295. Turtle was released September 5 after nesting.

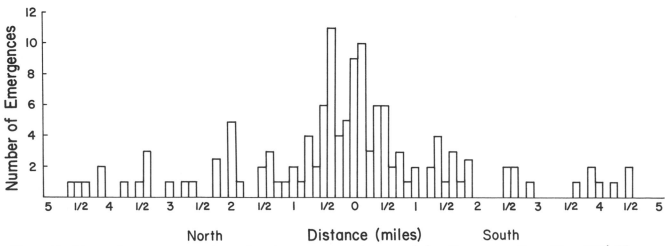

Figure 10. Distance between non-nesting and nesting emergence sites recorded at Tortuguero, Costa Rica, 1976–77. Only emergences made within 6 days are included

Table 3. Travel speeds of turtles tracked in the internesting habitat, Tortuguero, Costa Rica

			Mean speed, continuous travel		
Tag no.	Maximum speed (km/hr)	Current	Speed (km/hr)	Time interval (hrs)	Current
6277	2.7	across	0.9	1.6	with
			1.0	2.8	across
			1.1	0.8	against
			1.3	0.5	against
			2.2	0.8	with
7652	3.2	across	—	—	—
9915	2.4	across	—	—	—
10156	3.2	across	2.0	3.4	with
12006	1.6	with	2.7	1.1	with
12012	2.6	with	1.5	1.5	with
12181	2.5	with	1.4	1.8	with
12224	3.7	across	—	—	—
12531	3.1	across	0.7	3.8	with
12690	2.2	across	—	—	—
13206	4.5	across	—	—	—
13209	3.8	across	0.8	1.7	against
13295	3.0	against	1.2	1.3	against
13366	2.6	with	1.3	1.1	with
			2.3	1.6	with
13370	3.3	with	2.5	1.0	with

Note: Average prevailing current speed is 0.88 km/hr.
— No data.

Ross, and Carr, 1974). Mean continuous travel speeds ranged from 0.7 to 2.7 km/hr; all speeds over 2 km/hr were recorded for travel in the same direction as the current. Speeds of 1.1, 1.2, and 1.3 km/hr were observed against the current, which, if average current projections are accurate, would mean actual travel speeds of approximately 2 km/hr.

Implications for Conservation

Studies of internesting ecology can contribute greatly to the protection of turtles at the nesting grounds. Females emerging on the beach are already afforded good protection at many rookeries, but it is equally important to protect them, and the males, in the longshore waters off the nesting beach. The potential impact of commercial harvesting, incidental catch, and oil pollution on the survival outlook of a colony is greatly increased by the fact that large numbers of turtles may be affected, and that it is the reproductive contingent of the population that is jeopardized.

An important goal is the establishment of the dimensions—at least the core areas—of the internesting habitats of major rookeries. These dimensions undoubtedly vary among species and among habitats with different physical characteristics. There are even variations to be found within a species. For Hawaiian *Chelonia,* for example, the internesting habitat includes terrestrial basking sites (Balazs 1980). Behavioral characteristics of a species may alter the temporal patterns of habitat use, as in the case of arribada formation by *Lepidochelys.* Spatial and temporal differences in habitat use must be determined for each colony and taken into consideration in the formulation of conservation and management programs.

Acknowledgments

This work was jointly funded by the Caribbean Conservation Corporation and the National Science Foundation (OCE 77–09842, Princ. Invest. A.F. Carr). I am greatly indebted to Steven Carr, Peter Meylan, Durham Rankin, and a long list of volunteers from the Green Turtle Station at Tortuguero for help in carrying out the tracking work. I am especially grateful to Dr. Archie Carr for his invaluable assistance, advice, and encouragement throughout the project. Dr. D.W. Johnston and Dr. Walter Auffenberg read the manuscript in its original form as a master's thesis and made constructive suggestions. Dr. Tom Carr offered helpful advice for plotting the tracking data. I thank Alma Lugo for drafting the illustrations.

Literature Cited

Balazs, G.
1980. Synopsis of biological data on the green turtle in the Hawaiian Islands. National Oceanic and Atmospheric Administration, National Marine Fisheries Service, NOAA-TM-NMFS-SWFC–7, 141 pp.

Batschelet, E.
1972. Recent statistical methods for orientation data. NASA SP–262:61–91.

Booth, J., and J. Peters
1972. Behavioral studies on the green turtle (*Chelonia mydas*) in the sea. *Animal Behavior* 20:808–12.

Carr, A.
1954. The passing of the fleet. *A.I.B.S. Bulletin* 4:17–19.
1967. Adaptive aspects of the scheduled travel of *Chelonia*. In *Animal Orientation and Navigation*, pp. 33–55. Corvallis: Oregon State University Press.
1972. The case for long-range chemoreceptive piloting in *Chelonia*. NASA SP–262:469–83.

Carr, A., and M. H. Carr
1972. Site fixity in the Caribbean green turtle. *Ecology* 53:425–29.

Carr, A.; M. H. Carr; and A. B. Meylan
1978. The ecology and migrations of sea turtles, 7. The West Caribbean green turtle colony. *Bulletin of the American Museum of Natural History* 162:1–46.

Carr, A., and L. Ogren
1960. The ecology and migrations of sea turtles, 4. The green turtle in the Caribbean Sea. *Bulletin of the American Museum of Natural History* 121:1–48.

Carr, A.; P. Ross; and S. Carr
1974. Internesting behavior of the green turtle, *Chelonia mydas*, at a mid-ocean island breeding ground. *Copeia* 1974:703–06.

Ehrenfeld, D. W., and A. Koch
1967. Visual accommodation in the green turtle. *Science* 155:827–28.

Frick, J.
1976. Orientation and behavior of hatchling green turtles (*Chelonia mydas*) in the sea. *Animal Behavior* 24:849–57.

Meylan, A. B.
1978. The behavioral ecology of the West Caribbean green turtle (*Chelonia mydas*) in the internesting habitat. Master's thesis, University of Florida.

Schmidt-Koenig, K.
1975. *Zoophysiology and Ecology*, vol. 6, *Migration and Homing in Animals*. Berlin: Springer-Verlag.

ANNE MEYLAN

George R. Hughes
Natal Parks Board, Pietermaritzburg
South Africa

Nesting Cycles in Sea Turtles—Typical or Atypical?

ABSTRACT

Some tagging programs are in their second decade and at least one is in its third, but there is still uncertainty regarding the proportion of any population that nests in more than 1 season. There are wide variations in recoveries between populations of the same species and between species. Nesting cycles exist in all species, some animals displaying regular or unmodulated cycles and others irregular or modulated cycles. There appears to be no fixed pattern, and it is dubious to extrapolate data received from small numbers of animals to apply to whole populations, or indeed species. It is considered that the major cause of the uncertainty is the monel tag which is currently in widespread use. Although superior to plastic in many ways, the monel tag has serious shortcomings. The loss of these tags has resulted in uncertainty about an extremely important facet of the population dynamics of sea turtles. If tag loss is not as serious as thought, it must be concluded that either most sea turtles nest only once in their lifetime or that the mortality of sea turtle females following nesting is in many cases extremely high and remigrations to beaches in subsequent seasons are by the fortunate few survivors.

Introduction

Since the late Tom Harrisson started his tagging program in 1953 (Harrisson 1956) many thousands of sea turtles have been tagged and released. The initial returns stimulated considerable interest in nesting cycles and resulted in the appearance of simplistic interpretations of the nesting behavior of sea turtles (Hendrickson 1958, Hirth 1971, Pritchard 1967, Rebel 1974).

Following these publications, some authors (Hirth and Schaffer 1974, Pritchard 1971) have postulated that nesting cycles can be assumed to apply to whole populations, thus forming a valid basis for calculating the size of a breeding population. Other authors have

81

emphasized the need for caution in extrapolating results of recaptures of marked turtles (Hughes 1976, Bustard 1972, Limpus 1978). The fact is that in Natal, for instance, some loggerheads and some leatherbacks have exhibited nesting cycles, but this applies only to a minority of the animals marked. Even Archie Carr's Tortuguero program after 20-odd years of work shows a limited return of tagged turtles to the breeding beaches (see below). In the last decade more and more tagging programs have started, and the feature of limited returns seems to be common to all.

There is no question that nesting cycles exist, whether regular or irregular, modulated or unmodulated. What is in doubt is whether nesting cycles occur in the majority of a population, or merely in a small, rather exceptional portion of each population.

Materials and Methods

In virtually all programs the site of tagging is on the trailing or distal edge of the foreflipper (Figure 1). With the leatherback in Tongaland, after considerable tag loss (suspected by virtue of the fact that there were very few recoveries, as callosities or scars are not readily recognizable in leatherbacks [Hughes 1974]) tags are placed on the inside trailing edge of the hindflipper well under the carapace. Apart from the early attempts at marking using plates, for example, (Schmidt 1916, Moorhouse 1933, Carr and Giovannoli 1957), the vast majority of tagging programs have been using either monel metal clinch tags (#49) from the Kentucky Band and Tag Company, Newport, Kentucky, or plastic tags known as Jumbo Rototags (Dalton, Henley, England).

Various problems have resulted from monel tags. Sometimes they do not clinch properly as there is a fairly critical depth of flesh that can be punched without deflecting the blade of the tag. If the tag is punched too deeply into the flipper, it may not clinch properly, and the weak link, being buried in flesh, may not be

visible. The tag can then work its way out and be lost although some unclinched tags have been known to stay in place for several years.

Even if clinched totally, monel tags, especially on loggerheads, show the most remarkable corrosion even after relatively brief periods (Figure 2). This corrosion may rapidly remove all the wording on the tag or, by removing the narrow bridge, cause it to unclinch and eventually fall out.

Plastic tags have the problem of cracking around the hole of the female half and breaking off later, leaving only the male part of the tag. This normally falls out although some have been seen on loggerheads found in Tongaland.

An added problem with loggerheads is the apparent habit of other loggerheads (or something else!) biting the tag in place on the flipper. In Tongaland, loggerheads have been found the night following tagging with the tag flattened and twisted by some very substantial force. A plastic tag would have been destroyed.

In virtually every program, research workers have been aiming at the highest patrol efficiency possible in order not to "miss" emerging turtles. Most patrols are carried out on foot, but motorcycles and beach buggies have been used to great advantage as patrol areas have increased.

Current Data

Flatback Turtle, Chelonia depressa

Only limited data are available on this species and all have been derived from the work by Colin Limpus (for example Limpus 1971). From the 1973 tag-year summary of data from Mon Repos, Queensland, (Limpus, personal communication) it would appear that flatbacks can renest after 2-, 3-, or 4-year intervals. No doubt as more results are obtained it will be found that variability is characteristic of this species. Data available are insufficient to comment further.

Green turtle, Chelonia mydas

This species has long been the most intensively studied. For detailed results on remigrations see Carr and Ogren (1960), Carr and Carr (1970) and Carr, Carr, and Meylan (1978).

Table 1 summarizes the overall remigration data from Tortuguero. It can be seen that green turtles only rarely nest in consecutive years, a higher proportion may nest at 2-year intervals, highest at 3-year and some nest at 4-year or longer intervals. The Tortuguero data (Table 7 in Carr, Carr and Meylan 1978) indicate that those remigrants displaying regular cycles of nesting (such as 2–2, 3–3) are equaled by those displaying irregular or modulated remigrations, for example, 86 definite

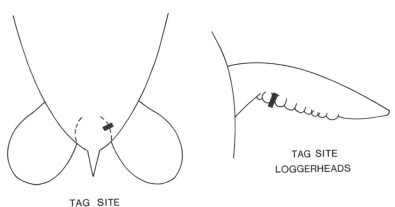

TAG SITE
LOGGERHEADS

TAG SITE
LEATHERBACKS

Figure 1. Tag sites used on loggerhead and leatherback turtles in Tongaland. The loggerhead site is widely used for all species.

GEORGE R. HUGHES

Table 1. Overall summary of green turtle remigrations to Tortuguero beach, Costa Rica

Absence (years)	Remigrations (numbers)	Percentage remigration
1	6	0.4
2	396	21.0
3	746	49.0
4	274	18.0
+	101	28.6
Total	1,523	100.0

Note: 1,412 individual turtles are responsible for the 1,523 remigrations recorded above. A return of 11.8 percent from ±12,000 turtles tagged since 1955.
Source: Carr, Carr, and Meylan, 1978.

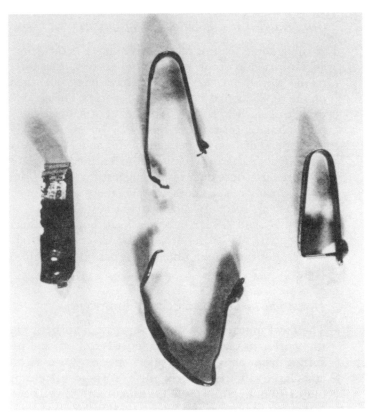

Figure 2. Monel tags showing a normal clinched tag compared to those showing corrosion and damage caused by an unknown but considerable force.

regular versus 85 irregular remigrations. Assumed modulated or unmodulated remigrations are discounted although even if they were not, the difference would remain insignificant.

Table 2 compares the Tortuguerò results with those from other green turtle colonies. The point to consider here is simply that the number of remigrations in total apparently fails to constitute 17 percent of the population. Even with the improved efficiency of beach patrols since 1967, a maximum of 20 percent has been achieved in recent years at Tortuguero.

The variability is remarkable and hard to explain. Carr (1975) suggests that low Ascension Island remigration rates may in part be due to heavy exploitation on the Brazilian feeding grounds, but this would not apply in Australia. In addition many Surinam green turtles also visit Brazilian waters, and that population overlaps the feeding distributions of the Ascension green turtles.

Our present state of knowledge indicates that green turtles can nest 2 or more times up to a maximum of 5, and this they may do in consecutive years (although only rarely) or at 2, 3, 4 or more year intervals. However, turtles that are tagged and depart the nesting beaches as a single-time nester, never to be seen again, far exceed those returning as remigrants.

Kemp's ridley turtle, Lepidochelys kempi

This endangered species of turtle nesting only in Mexico has been studied more intensively in recent years, but from limited published data (Pritchard and Marquez 1973, Zwinenberg 1977), it would appear that Kemp's ridley can nest in consecutive years after 2-year or 4-year intervals. Although only 17 returns (1.6 percent) are recorded out of 1,038 females tagged the intense pressure on this very limited population may well be the cause of such low returns. The general pattern, however, does not appear to differ much from other species in that there are only limited remigrations.

Table 2. Total remigration rates of green turtles recorded at 5 separate nesting colonies

Program	Number of years	Total number ♀♀ tagged	Number of individual remigrants	Percent remigration	Reference
Tortuguero	21	12,000	1,965	16.4	Carr, Carr, and Meylan 1978; Carr, unpub. data
Ascension Island	10	1,300	24	1.8	Carr 1975
Heron Island, Australia	4	859	9	1.0	Bustard 1972
Surinam	5	2,206	532	24.0	Schulz 1975
Sarawak	3	1,514	14	0.9	Harrisson, 1956; Hendrickson 1958

Olive ridley turtle, Lepidochelys olivacea

A remarkable feature of the olive ridley is that it occurs, or used to occur, in vast numbers in various parts of the world, but very few data have been published on the reproductive biology of the species. The notable exception is Schulz (1975) who reports remarkably high remigration rates of nearly 1,460 out of 2,733 (53.4 percent) tagged between 1966 and 1972. Early individual-year group remigrations reached as high as 69 percent. These data would suggest that in Surinam at least many olive ridleys are annual nesters although such a high recovery rate was not sustained and 2-, 3-, 4-, 5-, and 6-year intervals have been recorded. Zwinenberg (1976) reports annual, 2- and 3-year absences, but this is undoubtedly a repetition of Schulz's results.

Hawksbill turtle, Eretmochelys imbricata

The hawksbill is probably the least studied species in the world, its tagging being almost fortuitous and resulting from activities on other species. Carr, Hirth and Ogren (1966) reported only 2 (2.8 percent) returns out of 70 individuals tagged; 1 female after 3 years and 1 after 6 years. Carr and Stancyk (1975) updated their records from Tortuguero with 6 recoveries (4.6 percent) out of 130 hawksbills tagged. Remigrations occurred after 3 years (n = 3), 4 years (n = 1) and 6

years (n = 2). The lack of consecutive-year nesting appears confirmed by Diamond (1976) although the number of hawksbills being protected on Cousin Island appears too dramatically small (23 to 27 hawkbills) to be accorded too much weight. Nevertheless, as it is virtually the only colony of hawksbills receiving individual attention and it is thought that hawksbill colonies are remarkably faithful to their breeding beaches, (in Torres Straits it is claimed that certain islands have identifiable hawksbill colonies based on the colors in the shell), valuable data may yet emerge from this modest endeavor.

Loggerhead turtle, Caretta caretta

As early as 1967 Hughes and Mentis suggested that the reproductive remigrations of loggerheads were not a simple cyclical phenomenon. In 1974 Hughes expanded his argument and again in 1976. Data from other areas appeared to support his results (Limpus 1973, Richardson et al. 1978). The latter authors incidentally suggested that Hughes (1974) had reported 22.0 percent annual remigrations. This is misleading in that it suggests that 22.0 percent of an annual population remigrates the following year. In fact no more than 5.5 percent of any annual group of females has nested the previous year.

Table 3 summarizes all the recoveries of remigrating turtles in Tongaland over 16 years.

Table 3. A summary of loggerhead remigrations to Tongaland

Season	No. turtles	New	Per-cent	1 yr	Per-cent	2 yr	Per-cent	3 yr	Per-cent	4 yr	Per-cent	5 yr
1963–64	82	82	100.0	—	—	—	—	—	—	—	—	—
1964–65	223	220	98.6	3	1.4	—	—	—	—	—	—	—
1965–66	200	186	93.0	5	2.5	9	4.5	—	—	—	—	—
1966–67	221	199	90.1	3	1.4	16	7.2	3	1.4	—	—	—
1967–68	293	252	86.0	5	1.7	12	4.1	4	1.4	9	3.1	—
1968–69	184	156	84.8	—	—	3	1.6	3	1.6	5	2.7	1
1969–70	285	211	74.0	—	—	—	—	—	—	1	0.4	1
1970–71	241	160	66.4	8	3.3	—	—	—	—	—	—	1
1971–72	321	191	59.5	10	3.1	18	5.6	—	—	—	—	—
1972–73	262	139	53.0	6	2.3	41	15.7	9	3.4	—	—	—
1973–74	332	175	52.7	12	3.6	48	14.5	25	7.5	6	1.8	—
1974–75	310	156	50.3	17	5.5	35	11.3	27	8.7	13	4.2	3
1975–76	348	173	49.7	7	2.0	54	15.5	22	6.3	11	3.2	8
1976–77	319	176	55.0	14	4.4	37	11.6	19	6.0	1'	4.7	9
1977–78	339	216	63.7	12	3.5	38	11.2	11	3.2	17	5.0	7
1978–79	408	282	69.1	8	2.0	21	5.1	28	6.9	11	2.7	2

GEORGE R. HUGHES

Table 4. Recorded renestings in separate seasons of loggerhead and leatherback females encountered in Tongaland, Natal, from 1974–75 to 1978–79

Year	Loggerhead seasons							Leatherback seasons					
	1	2	3	4	5	6	n	1	2	3	4	5	n
1974–75	156	107	35	12	—	—	310	55	6	1	1	—	63
1975–76	171	111	38	18	3	—	341	46	18	1	—	—	65
1976–77	181	90	30	16	7	—	324	33	23	2	—	—	58
1977–78	214	71	30	14	4	3	336	34	29	5	1	1	70
1978–79	282	77	32	11	4	2	408	35	20	7	1	—	63

During the past 5 years, in particular in Tongaland, numerous turtles have shown that remigrations are of significance. Table 4 shows that loggerheads can nest up to 6 separate seasons spread over as long as 9 years, and the proportions of each remigrant group are remarkably similar although there now appears to be an increase in neophytes. This is perhaps better illustrated by Figure 3 showing the smooth climb and fall of remigration percentages since 1963/64. One notable feature of loggerhead remigrations to Tongaland is their highly irregular or modulated nature, (Hughes 1976) but a more notable feature is that, like the Tortuguero program and Heron Island studies on green turtles, large numbers of loggerheads are tagged and never return in subsequent seasons. Since 1969, 2,014 loggerhead females have been tagged and, although some

have certainly lost tags before subsequent returns, only 467 (23.2 percent) have returned with tags, some many times. After 10 seasons of using monel tags one would expect, if remigration were typical of the population as a whole, to have a much higher recovery rate even allowing for tag loss and mortality.

Leatherback turtle, Dermochelys coriacea

Not unlike the olive ridley, the leatherback is one of the most spectacular sea turtle species, if not in numbers then in size, the world's largest. It is so spectacular that it was hardly credible that some major breeding areas remained undiscovered until fairly recently. It is even less credible that so few data are available on the cyclic nesting behavior of the species. Pritchard (1971),

Percent	6 yr	Percent	7 yr	Percent	8 yr	Percent	9 yr	Percent	Calloussed	Percent	Remigrations (percent)
—	—	—	—	—	—	—	—	—	—	—	—
—	—	—	—	—	—	—	—	—	—	—	1.4
—	—	—	—	—	—	—	—	—	—	—	7.0
—	—	—	—	—	—	—	—	—	—	—	9.9
—	—	—	—	—	—	—	—	—	11	3.8	14.0
0.5	—	—	—	—	—	—	—	—	16	8.7	15.2
0.4	4	1.4	—	—	—	—	—	—	68	23.9	26.0
0.4	2	0.8	—	—	—	—	—	—	70	29.1	33.6
—	1	0.3	—	—	—	—	—	—	101	31.5	40.5
—	—	—	—	—	—	—	—	—	67	25.6	47.0
—	—	—	—	—	—	—	—	—	66	19.9	47.3
1.0	—	—	—	—	—	—	—	—	59	19.0	49.7
2.3	6	1.7	—	—	—	—	—	—	67	19.3	50.3
2.8	2	0.6	—	—	—	—	—	—	47	14.7	45.0
2.1	1	0.3	2	0.6	1	0.3	—	—	34	10.0	36.2
0.5	3	0.7	4	1.0	0	0	1	0.2	48	11.8	30.9

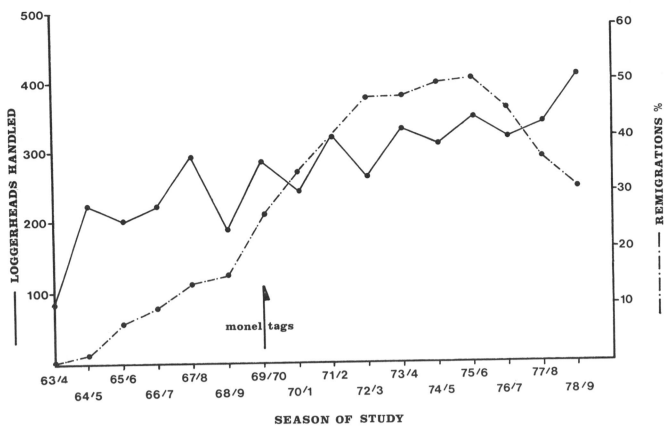

Figure 3. Loggerhead numbers handled in Tongaland from 1963 with total annual remigrant percentage recorded each season.

reviewing all data available at that time, makes no direct mention of year-to-year remigrations. Some very limited data were available, however, from Tongaland, and these were later reviewed by Hughes (1974). Schulz (1975) reported 9 (3.5 percent) returns out of 257 taggings in Surinam with internesting intervals of 1 (n = 1), 2 (n = 6) and 3 (n = 3) years with only 1 turtle recorded in 3 separate seasons. Pritchard (quoted in Schulz 1975) found a predominance of 2-year returns: 23 out of 26 records. Given such limited data, Table 5 and Figure 4 are worth including showing the total remigration records for Tongaland from 1963/64 to the present. Apart from showing an encouraging increase in turtles handled, from a low of 5 females in 1966/67 to 70 in 1977/78, there has been a recent improvement in tag recoveries following the 1974 change of tag site from the foreflipper to the inner side of the hindflipper (Figure 1).

Table 4 also summarizes the remigration records of individual leatherback females since 1974, and it is noteworthy that the odd females have returned in 4 or 5 separate seasons. One female was tagged first in 1964 and another 1965, giving reproductive lifetimes of 15 and 14 years respectively.

The renesting frequencies are irregular as in log-

gerheads and so far it can be said that annual nesting is rare; 2-year and 3-year intervals are most common. As with other species, however, in Tongaland many leatherbacks have been tagged and never seen again. Only 94 (29.3 percent) have ever returned out of 321 tagged.

Discussion

It must be fairly clear from the data presented above that virtually all sea turtle populations display some degree of remigration for breeding purposes. It has been well demonstrated that most populations have specimens that display both regular and irregular renesting behavior. The main question that remains unanswered is just how typical is the reproductive cycle?

From a conservation point of view it is imperative that we be certain of this and do not merely assume from the rather limited returns so far recorded. In Tongaland it appeared that our steady climb of returns would stop, level out as more and more turtles were tagged and thus provide a fairly accurate recruitment rate and the return data to ensure certainty regarding how many of each group was returning. This has not happened; in fact, the relative remigration rate is drop-

GEORGE R. HUGHES

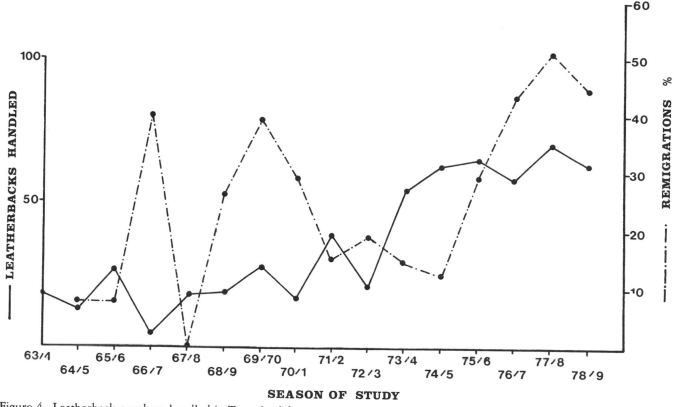

Figure 4. Leatherback numbers handled in Tongaland from 1963 with total annual remigrant percentage recorded each season.

ping sharply (Figure 2).

Tom Harrisson (1956) first expressed concern over knowing whether he had a large population of green turtles laying 1 clutch of 100 eggs a season each or a much smaller population laying numerous clutches.

Twenty-three years later we must express similar concern in that we are still very uncertain about a fundamental fact. Can we accept that green turtle populations—as an entire unit—nest every 2, 3, or 4 years? I firmly believe we cannot and the same applies to loggerheads and leatherbacks. In fact, most turtles (of any species) tagged are never seen again.

Three possibilities must be considered:
• Beach patrols are extremely inefficient; or
• Tag losses are astronomical; or
• The mortality rate of sea turtles is extremely high, what does return as a remigrant is in reality a survivor against substantial odds.

Point number one is unlikely although most programs could be improved upon, Tongaland's for one. Some programs like the Heron Island study, record every animal and can hardly be improved upon, yet they too have low returns. It must be assumed that all workers try to be as efficient as possible.

The possibility that tag losses are unacceptably high,

despite claims by Hughes (1974) and Richardson et al. (1978) that they record callosities, cannot be overlooked. The monel tag is obviously the best to date, but it is not the best that could be made. George Balazs is trying out small inconel tags, and we await his results with interest. Tag loss is something that could and must be reduced by using a better design and better metal. It is ridiculous to spend thousands of man-hours using an inferior tag that gives inferior results.

High mortality rates must be seriously considered and if, following the removal of the uncertainty of the inferior tag, recovery rates remained low, this would answer the very question that Tom Harrisson asked. Do we have in essence a new and discrete nesting population each year with a few notable survivors from previous years? Carr, Carr, and Meylan (1978) reported their tagging of 12,000 turtles at Tortuguero of which 1,110 (9.2 percent) have been recaptured elsewhere. These are only a part undoubtedly of the kills or captures for which man is responsible—how many more die from natural causes and natural predation?

Or is it simply that most sea turtle females do nest once only, and the remigrants are exceptions? This may not seem acceptable to many workers, but neither is

Table 5. A summary of leatherback remigrations to Tongaland, 1969–78

Season	Total numbers	New	Percent	1 yr	Percent	2 yr	Percent	3 yr	Percent	4 yr	Percent	5 yr	Percent	6 yr	Percent	Calloused[a]	Percent[a]	Percent remigrations
1963–64	18	—	—	—	—	—	—	—	—	—	—	—	—	—	—	—	—	7.7
1964–65	13	12	92.3	1	7.7	—	—	—	—	—	—	—	—	—	—	—	—	7.7
1965–66	27	25	92.6	—	—	2	7.7	—	—	—	—	—	—	—	—	—	—	7.7
1966–67	5	3	60.0	—	—	1	20.0	1	20.0	—	—	—	—	—	—	—	—	40.0
1967–68	18	18	100.0	—	—	—	—	—	—	—	—	—	—	—	—	—	—	—
1968–69	19	14	73.7	—	—	—	—	4	21.0	1	5.3	—	—	—	—	—	—	26.3
1969–70	28	17	60.7	—	—	3	10.7	—	—	2	7.1	1	3.6	—	—	5	17.9	39.3
1970–71	17	12	70.6	—	—	3	17.7	—	—	—	—	—	—	—	—	2	11.8	29.4
1971–72	39	33	84.6	—	—	3	7.7	—	—	—	—	—	—	—	—	3	7.7	15.4
1972–73	21	17	81.0	—	—	1	4.7	—	—	—	—	—	—	—	—	3	14.3	19.1
1973–74	54	46	85.2	—	—	4	7.4	2	3.7	—	—	1	1.9	—	—	1	1.9	14.8
1974–75	63	55	87.3	—	—	3	4.8	1	1.6	2	3.2	—	—	—	—	2	3.2	12.6
1975–76	65	46	70.8	—	—	13	20.0	2	3.0	—	—	1	1.5	—	—	3	4.6	29.2
1976–77	58	33	56.8	1	1.7	11	19.0	8	13.8	—	—	—	—	—	—	5	8.6	43.1
1977–78	70	34	48.6	—	—	12	17.1	23	32.9	—	—	—	—	—	—	1	1.4	51.4
1978–79	63	35	55.6	—	—	9	14.3	16	25.4	—	—	—	—	—	—	3	4.8	44.4

a. Very dubious data because of difficulty of recognition.

it acceptable to state that: "green turtles nest *every* 4 years in Australia (Bustard and Tognetti 1969); *every* 3 years in the South China Sea (Harrison 1956 [sic]; Hendrickson 1958) and *every* 2 or 3 years in the Caribbean and South Atlantic with the triennial cycle *predominant*, (Carr and Ogren 1960, Carr 1965), (Hirth 1971)" (*Emphasis* added, G.R.H.).

That there is uncertainty after the length of time that serious programs of research have been going on is most unfortunate. It is not enough to blame tag loss and extrapolate. No beast is more faithful to its nesting ground than the sea turtle (Carr and Carr 1972, Hughes 1974) and yet there are low returns. With a more trustworthy tag, the question posed may soon be answered: are sea turtles predominantly multiseason nesters or not?

Acknowledgments

Primarily thanks must go to the many workers in Tongaland and in other parts of the world whose endeavors have made this report possible. There is no lack of enthusiasm among workers on turtle beaches.

Secondly may I thank the Director of the Natal Parks, Game and Fish Preservation Board for permission to deliver the report. My thanks, too, to the Principal Scientific Officer, R. S. Crass of the Natal Parks Board, who criticized the report. Attendance of the Conference would not have been possible without the financial assistance of the Southern Africa Nature Foundation and the Natal Parks Board. Finally my thanks to Maxie Holder and Cindy Pringle for their help.

Literature Cited

Bustard, H. R.
1972. *Sea Turtles: Their Natural History and Conservation.* London: Collins.

Bustard, H. R., and K. P. Tognetti
1969. Green sea turtles: a discrete simulation of dependant population regulation. *Science* 163:939–41.

Carr, A.
1975. The Ascension Island green turtle colony. *Copeia* 1975:547–55.

Carr, A., and M. H. Carr
1970. Modulated reproductive periodicity in *Chelonia. Ecology* 51:335–37.
1972. Site fixity in the Caribbean green turtle. *Ecology* 53:425–29.

Carr, A.; M. H. Carr; and A. B. Meylan
1978. The ecology and migrations of sea turtles, 7. The West Caribbean green turtle colony. *Bulletin of the American Museum of Natural History* 162:1–46.

Carr, A., and L. Giovannoli
1957. The ecology and migrations of sea turtles, 2. Results from fieldwork in Costa Rica, 1955. *American Museum Novitates* 1835:1–32.

Carr, A.; H. Hirth; and L. Ogren
1966. The ecology and migrations of sea turtles, 6. The hawksbill turtle in the Caribbean Sea. *American Museum Novitates* 2248:1–29.

Carr, A., and L. Ogren
1960. The ecology and migrations of sea turtles, 4. The green turtle in the Caribbean Sea. *Bulletin of the American Museum of Natural History* 121:1–48.

Carr, A., and S. Stancyk
1975. Observations on the ecology and survival outlook of the hawksbill turtle. *Biological Conservation* 8:161–

GEORGE R. HUGHES

72.

Diamond, A. W.

1976. Breeding biology and conservation of hawksbill turtles, *Eretmochelys imbricata* L. on Cousin Island, Seychelles. *Biological Conservation* 9:199–215.

Harrisson, T.

1956. The edible turtle (*Chelonia mydas*) in Borneo. (5). Tagging turtles (and why). *Sarawak Museum Journal* 8:504–15.

Hendrickson, J. R.

1958. The green sea turtle (*Chelonia mydas* Linn.) in Malaya and Sarawak. *Proceedings of the Zoological Society of London* 130:455–535.

Hirth, H. F.

1971. Synopsis of biological data on the green turtle *Chelonia mydas* (Linnaeus) 1758. *FAO Fisheries Synopsis* 85:1–76.

Hirth, H. F., and W. M. Schaffer

1974. Survival rate of the green turtle, *Chelonia mydas*, necessary to maintain stable populations. *Copeia* 1974:544–46.

Hughes, G. R.

1974. The sea turtles of South East Africa, 2. *Investigational Report of the Oceanographic Research Institute Durban* 36:1–96.

1976. Irregular reproductive cycles in the Tongaland loggerhead sea turtle, *Caretta caretta* L. (Cryptodira: Chelonidae). *Zoologica Africana* 11:285–91.

Hughes, G. R., and M. T. Mentis

1967. Further studies on marine turtles in Tongaland, 11. *Lammergeyer* 7:55–72.

Limpus, C. J.

1971. The flatback turtle *Chelonia depressa* Garman in Southeast Queensland, Australia. *Herpetologica* 27:431–46.

1973. KGCTE: Turtle Research, 1973 Report. Kelvin Grove Teacher's College, Brisbane. Typescript report, 2 pages.

1978. The reef. In *Exploration North*, ed. H. J. Lavery, pp. 187–222. Melbourne: Richmond Hill.

Moorhouse, F. W.

1933. Notes on the green turtle (*Chelonia mydas*). *Report of the Great Barrier Reef Committee* 4:1–22.

Pritchard, P. C. H.

1967. *Living Turtles of the World*. Jersey City: T.F.H. Publications.

1971. The leatherback or leathery turtle, *Dermochelys coriacea*. *IUCN Monograph, Marine Turtle Series* 1:1–39.

Pritchard, P. C. H., and R. Marquez

1973. Kemp's ridley turtle or Atlantic ridley, *Lepidochelys kempi*. *IUCN Monograph: Marine Turtle Series* 2:1–30.

Rebel, T. P.

1974. *Sea Turtles and the Turtle Industry of the West Indies, Florida and the Gulf of Mexico*. Coral Gables, Florida: University of Miami Press.

Richardson, T. H., J. I. Richardson, C. Ruckdeschel, and M. W. Dix

1978. Remigration patterns of loggerhead sea turtles (*Caretta caretta*) nesting on Little Cumberland and Cumberland Islands, Georgia. *Florida Marine Research Publication* 33:39–44.

Schmidt, J.

1916. Marking experiments with turtles in the Danish West Indies. *Meddr. kommn. Havunders. Fiskeii* 5:1–26.

Schulz, J. P.

1975. Sea turtles nesting in Surinam. *Zoo. Verhand. Rijksmus. Nat. Hist. Leiden* 143:1–144.

Zwinenberg, A. J.

1976. The olive ridley, Lepidochelys olivacea (Eschscholtz, 1829): probably the most numerous marine turtle today. *Bulletin of the Maryland Herpetological Society* 12:75–95.

1977. Kemp's ridley *Lepidochelys kempii* (Garman, 1880) undoubtedly the most endangered marine turtle today, with notes on the current status of *Lepidochelys olivacea*. *Bulletin of the Maryland Herpetological Society* 13:170–92.

Anne Meylan
Department of Zoology
University of Florida
Gainesville, Florida 32611

Sea Turtle Migration—Evidence From Tag Returns

ABSTRACT

In this paper, summaries of available tag-return data bearing on the migration of adult marine turtles are presented. Low frequencies of tag-return for most colonies studied have hindered efforts to establish migratory patterns. For some species, tagging has provided good evidence of periodic travel between resident foraging grounds and the nesting beach. No turtle has yet been monitored during a migratory journey, and navigatory mechanisms and travel routes remain unknown. For some species, there is evidence of group migration, both from direct observation of migrating schools, and from the simultaneous long-distance recovery of turtles tagged together on a nesting beach. Tag-return data show that average travel speeds of 20 to 40 km/day are sustainable for long periods, and that migration may be made against prevailing currents. Feeding and developmental migrations remain largely unstudied, due to the difficulty of tagging and recovering turtles during these phases of the life cycle.

One of the principal goals of sea turtle research during the last few decades has been to determine the extent and character of migratory travel. Tens of thousands of turtles have been tagged worldwide, primarily on nesting beaches, with the expectation that subsequent recaptures of tagged turtles at sea would reveal the migratory pathways and destinations of each species. Unfortunately, most tagging projects have recorded singularly few recaptures away from the nesting beach, and only a handful of geographic patterns of migratory travel has so far been determined. Moreover, virtually nothing has been learned of the behavior of turtles during migration. Tagging has contributed far more to our knowledge of other aspects of the life history, notably the reproductive ecology, than to that of migratory behavior.

Despite low recapture rates, data obtained by tagging quickly revealed unexpected complexity in the migratory behavior of sea turtles. Turtles have been shown to journey thousands of kilometers to nest, even when suitable nesting beaches are locally available. In some localities, there is an exchange of turtles during the nesting season: local populations migrate to distant beaches to nest, while migrants from other areas arrive to nest on the beaches left behind. Limpus (1978) reported that the number of loggerheads (*Caretta caretta*) residing in the waters of Heron Island, Australia, is approximately equal to the number that nest on the island's beaches, but that, in each case, different turtles are involved. In the Galápagos Islands, tagging has suggested that both migratory and nonmigratory members may be included within a species, and perhaps even within a population (Green 1979). Shared occupancy of feeding grounds has also been revealed by tagging studies. Different populations of the same species have been shown to share a resident foraging habitat, but to migrate to widely separate nesting beaches to reproduce (Pritchard 1973, 1976; Carr 1975).

Many aspects of the migratory behavior of sea turtles have remained beyond our grasp. Although poorly documented by tagging evidence, the occurrence of developmental and feeding migrations for some species is unquestioned. The difficulty of monitoring these phases of the life history has prevented our obtaining any clear idea of the routes and schedules of these migrations.

Tag-return data have provided good evidence of periodic travel between nesting beaches and forage grounds for some species. In the case of green turtles (*Chelonia mydas*), these 2 habitats are nearly always spatially separate. Recapture sites are usually algal or seagrass pastures, which constitute prime foraging grounds for the species. For Kemp's ridley (*Lepidochelys kempi*) tag-returns cluster in coastal areas rich in crustaceans and molluscs. The feeding grounds for some species, such as the olive ridley (*Lepidochelys olivacea*) and the leatherback (*Dermochelys coriacea*), appear to be less well defined. Patchy, drifting, or locally fluctuating food resources may dictate more or less itinerant feeding strategies.

Evidence of 2-way migration between specific foraging and nesting grounds, although convincing, remains largely circumstantial because turtles tagged on the nesting beach and captured upon their return to the feeding grounds are usually killed. Recent studies have provided several cases of two-way travel between forage grounds and nesting beaches. Balazs (1976, 1980) has documented migration from the feeding grounds to the nesting beach by tagging turtles while they were basking on islands in the feeding area. A male green turtle, tagged while basking at Pearl and Hermes Reef in the Hawaiian Archipelago in 1964, was observed at the colonial nesting site at French Frigate Shoals in 1976, and then recorded back at Pearl and Hermes Reef 1 year later. Limpus (1978) has also tagged turtles

on their feeding grounds and recently reported a remarkable travel itinerary for an Australian loggerhead. The turtle was tagged while nesting at the Mon Repos rookery in 1972, and was captured 2 years later in a lagoon 160 km away; in 1975 the turtle nested again at Mon Repos, and in 1977 she was recaptured in the same lagoon. Continued monitoring of the feeding grounds has provided several additional records of regular commuting between nesting and foraging habitats (Limpus, personal communication).

Most tagging projects do not monitor the feeding grounds, and it is usually necessary to rely on other sources of data to interpret tag recoveries. Capture of tagged turtles in one limited area during all months of the year, and the existence of very long and very short periods of time between the tagging and recapture of individual turtles at a single location can be used to support an assumption that primary sites of grouped tag recovery represent final destinations for the migrants. In order for nesting beach–forage ground patterns to be reliably determined, however, a considerable number of tag recoveries must accumulate because migrants are frequently caught in transit. In some areas, fishermen who anticipate the migration of turtles along their coast purposely set nets perpendicular to it during the appropriate season to intercept the migrants (Carr 1954).

To date, no sea turtle has been directly monitored for a significant portion of a migratory journey, and, as a consequence, the navigatory mechanisms employed and the travel routes followed remain unknown. When carefully interpreted, in-transit recoveries may provide clues about travel routes. Recaptures made soon after turtles leave the nesting beach, especially in areas where a species is not generally common, suggest points along a migratory pathway. Unfortunately, recapture data reveal very little about open-sea routes, due to reduced fishing efforts away from shore.

Tag-return data have contributed very little to our knowledge of the behavior of turtles during migration. Basic information such as the daily schedule of activity is lacking. There is some evidence that green turtles may stop to feed during migration. Turtles bearing Tortuguero tags are frequently caught in quiet lagoons near Bocas del Toro, Panama, at times coinciding with the onset and cessation of nesting in Costa Rica. Fishermen there told the author that groups of turtles enter the sheltered waters of Chiriqui Lagoon and Almirante Bay to feed for 2 or 3 days before continuing their migration. Mortimer (1981) reported that green turtles traveling from Miskito Cays, Nicaragua, to Tortuguero, sometimes travel close to shore and feed on *Syringodium filiforme*, red algae, and terrestrial debris deposited at river mouths. Green turtles crossing stretches of open ocean probably forego feeding enairely, inasmuch as their usual forage—seagrass or algae—is unavailable. It is probable that other species feed during migration when food is available. Migrating leatherbacks, for example, are likely to have constant access to their pelagic coelenterate prey.

Both male and female turtles migrate to the nesting grounds. Copulation may occur during this time (Mortimer, personal communication), although most mating takes place off the nesting beach. It is not known whether the arrivals and departures of the two sexes at the nesting beach are synchronous. Many authors have reported a decline in mating activity as the nesting season progresses, but could not, of course, determine if the males had actually departed. Márquez, Villanueva, and Peñaflores (1976) reported that male olive ridleys are the last to leave the nesting grounds on Mexico's Pacific coast.

Cornelius (1975) suggested that immature green turtles may accompany adults to the nesting beach, basing

the possibility on his finding many subadult turtles among 73 that had died from unknown causes during the nesting season in Costa Rica. A fragment of evidence supporting this hypothesis is the report made to the author that subadult green turtles are frequently caught near Bocas del Toro, Panama, during the seasonal migration of adult Tortuguero-bound migrants. No examination of the catch made by harpooners off Tortuguero has ever been made to investigate this possibility, and it has never been corroborated elsewhere.

Carr, Carr, and Meylan (1978) summarized published data on the speeds of migratory travel for the various genera (Table 1). These speeds, calculated from tag-return data, represent minimum averages, and are based on the assumption that turtles follow direct travel courses and are captured immediately upon arrival at the capture site. Travel speeds of up to 90 km/day have been reported for turtles tagged at Tortuguero, although most are on the order of 20 to 40 km/day (Carr,

Table 1. Representative minimum speeds of travel previously recorded for *Chelonia* and other genera of sea turtles

Species	Interval (days)	Approximate distance (km)	Minimum average speed (km/day)	Reference
Caretta caretta	11	442	40.2	Bell and Richardson (1978)
	63	1,770	28.1	Bustard and Limpus, 1970
	91	2,655	29.2	Hughes, Bass, and Mentis, 1967
	66	2,640	40	Hughes, 1974
	76	2,640	34.7	Hughes, 1974
	82	2,400	29.3	Hughes, 1974
Chelonia mydas	31	713	23	Balazs, 1976
	73 ± 15	3,085	53–35	Carr, 1975
	81 ± 22	2,661	33–26	Carr, 1975
	83	2,201	27	Carr, 1975
	68	2,302	34	Carr, 1975
(2 individuals)	85	1,915	22.5	Hirth and Carr, 1970
	48	1,200	25	Hughes, 1974
	29	1,010	34.8	Pritchard, 1973
	41	1,250	30.5	Pritchard, 1973
	37	1,070	28.9	Pritchard, 1973
	32	2,100	66	Schulz, 1975
	—	—	53	Schulz, 1975
	—	—	53	Schulz, 1975
	43	2,000	46.5	Schulz, 1975
Lepidochelys kempi	32	945	29.5	Chavez, 1968
	32	769	24	Chavez, 1968
Lepidochelys olivacea	12	440	36.7	Pritchard, 1973
	32	910	28.4	Pritchard, 1973
	23	1,900	82.6	Schulz, 1975
	12	650	54.2	Schulz, 1975
	16	450	28.1	Schulz, 1975

— No data. *Source:* Carr, Carr, and Meylan (1978).

Carr, and Meylan 1978). Recently, de Silva (this volume) recorded a minimum average speed of 17.8 km/day for a hawksbill (*Eretmochelys imbricata*) that was tagged in Sabah, East Malaysia, and recaptured 713 km away in the Philippines. An impressive feature of the data shown in Table 1 is the indication that relatively rapid travel speeds are sustained for long periods. It is also of interest that in many cases migration appears to be made against prevailing currents (Bustard and Limpus 1970; Bustard 1974, 1976; Balazs 1976; Carr, Carr, and Meylan 1978; Meylan, this volume).

Several authors (Carr and Giovanolli 1957; Hirth and Carr 1970; Bustard 1974; Limpus 1979; Brongersma, this volume) have suggested the occurrence of group migration. The evidence most commonly cited is the simultaneous recovery on a distant feeding ground of two or more turtles which were originally tagged together while nesting. Surprisingly few direct observations of groups of migrating turtles have been published (Oliver 1946; Leary 1957) although unpublished records are numerous. The phenomenon may prove hard to substantiate by tag-return data due to the unlikelihood of simultaneous capture, but its occurrence seems likely for several genera.

Tag-return Data Summary

The following accounts briefly summarize available tag-return data bearing on the migratory travel of adult marine turtles. No data on captive-reared turtles are included.

The Green Turtle (Chelonia mydas)

By the end of the 1979 season, approximately 15,000 green turtles had been tagged at Tortuguero, Costa Rica. This colony has been under study for 25 years (Carr and Giovannoli 1957; Carr and Ogren 1960; Carr, Carr, and Meylan 1978). Foreign recoveries of 1,335 turtles show the colony to be derived from feeding grounds throughout the western Caribbean. The principal site of tag recovery is the continental shelf of Nicaragua where extensive pastures of *Thalassia testudinum* provide rich forage. Other recaptures have been made chiefly in Yucatan, Mexico; Panama; Colombia, and Venezuela. Tracking evidence suggests that at the end of the nesting season some departing migrants follow inshore routes, swimming against the current, to return to northern feeding grounds (Meylan, this volume).

Aves Island, 180 km west of Guadeloupe, is the site of the second major colonial nesting aggregation of the green turtle in the Caribbean. Of 800 to 1,000 turtles tagged from 1971 through 1976, 19 have been subsequently recaptured, largely in the Dominican Republic and the Windward and Leeward Islands (Rainey,

personal communication). Recaptures of 2 Aves turtles at Miskito Cays, Nicaragua, and 1 at Isla Mujeres, Mexico, sites which are major feeding grounds for Tortuguero turtles, suggest that some feeding areas are shared with the Tortuguero colony.

The green turtle colony nesting on Bigisanti and Galibi beaches of Surinam has been under study since 1963 (Pritchard 1969, 1973, 1976; Schulz 1975). All but one of the 91 recoveries of green turtles tagged there have been made along the coast of Brazil (Pritchard 1976). Recoveries cluster off the states of Ceará and Alagoas; nearly one-third of the total number of recaptures was made along a 45-km stretch in the Cearan county of Itapipoca. The distance between Surinam beaches and the feeding grounds in Ceará is over 2,000 km, and migrants cross the equator in their journey. Schulz (1975) proposed an average minimum speed of travel for Brazil-bound migrants of 35 to 80 km/day and suggested that most migrants are away from their feeding grounds for from 2 to 5 months. Tagging studies have revealed that the Brazilian foraging grounds are shared with green turtles that nest on Ascension Island in the South Atlantic Ocean (Pritchard 1973, 1976; Carr 1975). The 2 populations feed side by side in Brazilian waters, but migrate to separate locations to nest. No turtle tagged in Surinam has ever been observed nesting at Ascension, and vice versa. Moreover, since mating occurs en route to the nesting beach or at the nesting beach itself, there is little opportunity for genetic exchange between the 2 populations. Pritchard (1973) speculated that the Brazilian feeding grounds may also be shared with turtles nesting on Trindade, Marajó, and on the Brazilian mainland.

The Ascension Island green turtle colony has been monitored intermittently since 1960 (Carr and Hirth 1962; Carr 1975). Recoveries of 56 turtles tagged on the island document an exclusively Brazilian origin for this colony (Carr 1975, unpublished data). Recovery sites span 4,000 km of coast line, from Parnaiba (Piauf) to Vitoria (Espirito Santo). Migrants must cross 2,200 km of open ocean to reach this small island. The return journey to Brazil for both the adults and hatchlings may be facilitated by the west-trending Equatorial Current. Carr and Coleman (1974) hypothesized that the Brazil-Ascension pattern evolved when the distance to be traversed was slight, and that the present situation resulted from gradual seafloor spreading during the early Tertiary.

Green turtles have been tagged at several localities in the Indian Ocean. Hughes (1974, this volume) reported the recaptures of 5 green turtles tagged while nesting on Europa Island: 4 were taken in Madagascar, at distances of 320 to 1,400 km; and 1 was caught off mainland Africa, at Maputo, Mozambique. Tagging data suggest that Tromelin Island is also a nesting ground for turtles that feed in the waters of Madagascar. Two

ANNE MEYLAN

turtles tagged at Tromelin have been recovered there (Hughes, this volume). Tromelin turtles have also been recaptured at Reunion Island and Mauritius.

Approximately 50 green turtles have been tagged on Aldabra and Astove islands, but none has been recaptured away from the nesting beach (Frazier, in press). Hornell (1927) and Hirth and Carr (1970) postulated that Aldabran turtles migrate to feeding pastures in the Mozambique Channel.

Nine green turtles tagged on Musa and Sharma beaches in the People's Democratic Republic of Yemen have been recaptured away from the nesting beach (Hirth and Carr 1970; Hirth and Hollingsworth 1973; Hirth, personal communication). Seven of these were taken along the coast of Somalia, and 2 at feeding grounds within the country, at Khor Umaira and Ras al Ara. A minimum travel speed of 22.5 km/day is calculated for 2 of the turtles captured in Somalia which had presumably traveled together for 85 days.

Ross (this volume) reported the first international recapture of a green turtle tagged in Oman. The turtle nested at Ras al Had, and was recovered 3 months later at Assab, Ethiopia, in the Red Sea.

One of the earliest tagging studies of green turtles was conducted in the 1950s in Sarawak, East Malaysia. Disappointingly few recaptures away from the nesting beach were ever reported. One turtle tagged in 1953 was recaptured off North Borneo in 1959, a minimum travel distance of 800 km (Harrisson 1960).

Five green turtles tagged in the Turtle Islands National Park, Sabah, East Malaysia, have been recaptured away from the nesting beach (de Silva, this volume). Three were taken in the Philippines and 2 in Indonesia. Three of the recoveries indicate migrations of over 1,000 km.

Tagging studies of green turtles in Australia have been conducted at Heron Island and other islands in the Capricorn and Bunker Groups since 1964. Recaptures have been made primarily along the Queensland coast although 5 turtles have been taken in New Caledonia and 1 in southern Papua New Guinea (Bustard 1974, 1976; Limpus, 1980). One male green turtle tagged at Heron Island was recaptured at Hervey Bay, Queensland (Limpus, personal communication).

Forty-seven recaptures of green turtles tagged at Raine Island and nearby Pandora Cay in Australia's Great Barrier Reef suggest that this colony is drawn primarily from feeding grounds in Torres Strait (Limpus, 1980). Recoveries have also been made in the southern Cape York Peninsula, off the Northern Territory, at Aru Island in Indonesia, and in southern Papua New Guinea.

Migrations of green turtles tagged at Scilly Island in French Polynesia are among the longest ever reported for this species. Four turtles have been recovered in Fiji, 3 in the New Hebrides, 2 in New Caledonia, and 1 in the Kingdom of Tonga (Galenon 1979). The re-

coveries in New Caledonia represent migrations of over 4,000 km.

French Frigate Shoals is the only aggregate breeding site for *Chelonia* in the Hawaiian Archipelago (Balazs 1976, 1980). Migrants converge at this central locality from feeding grounds at both ends of the archipelago. Since tagging was begun there, 52 long-distance migrations of adult turtles have been recorded (Balazs 1980). There are records of 28 migrations from French Frigate Shoals to the main populated islands in the southeast, and 3 in the reverse direction. From the northwest islands, 15 turtles have migrated to French Frigate Shoals, and 6 in the opposite direction. Minimum travel distances for 16 of the migrations were more than 1,000 km. Recoveries indicate fidelity to specific feeding grounds. No recoveries of tagged adult Hawaiian green turtles have been made outside the archipelago, and it appears likely that this breeding colony is reproductively isolated from other Pacific *Chelonia*.

Nineteen international recaptures have been reported of green turtles tagged in the Galápagos Islands. These include recovery sites in Costa Rica, Panama, Colombia, Ecuador, and Peru (Green, personal communication). While most green turtles leave the archipelago after the nesting season, one contingent of the population appears to be resident (Green 1979).

The Leatherback Turtle (Dermochelys coriacea)

The pelagic habitat preference of this species, an extraordinarily high rate of tag loss, and a dearth of commercial fisheries with impact on leatherbacks have contributed toward a singularly low frequency of tag recovery. Nevertheless, there is good evidence that *Dermochelys* is a strong, though possibly vagrant, migrator. The frequent sighting of leatherbacks in extremely cold, northern waters, far from their mainly tropical nesting beaches, is in itself fairly good evidence that they undertake breeding migrations.

The few existing long-distance recaptures of tagged leatherbacks document some of the longest migrations ever recorded for any reptile. Five of 6 leatherbacks tagged in the Guianas and subsequently recaptured had traveled over 5,000 km (Pritchard 1976). Recapture sites include: Ghana, West Africa; Campeche, Mexico; the Gulf of Venezuela, and South Carolina, Texas, and New Jersey. There have been too few recoveries to determine whether the postnesting movements of this colony represent travel to specific home foraging-grounds, but the recovery of 2 tagged leatherbacks in the Gulf of Mexico and frequent sightings of untagged specimens there suggest this may be a preferred destination (Pritchard 1976).

During the 1977–79 seasons, 10,221 leatherbacks were tagged in French Guiana. One long-distance recapture had been recorded as of September 1979: a female tagged while nesting at a beach at Kawana was taken by fishermen in Venezuela (Fretey, personal communication).

To date, only 1 leatherback tagged at Tongaland, South Africa, has been retaken away from the nesting beach (Hughes 1974, personal communication). The turtle was captured at Beira, Mozambique, which indicates a northward migration of 1,000 km.

The tagging project at Trengganu, West Malaysia, has provided the only evidence of a migratory pattern for *Dermochelys*. Approximately 35 international recaptures have been made, most of them in the Philippines (Siow, personal communication). Other recovery sites are Japan, Taiwan, Hainan Island (China), and Kalimantan (Indonesia).

The Hawksbill Turtle (Eretmochelys imbricata)

It has been suggested that, for some species, nesting and feeding requirements do not impose the necessity of migration. Carr (1952) reported a general, though undocumented, belief that hawksbills do not migrate to any extent. This opinion was recently restated by Bustard (1979), who described the species as "virtually sessile," nesting on beaches adjacent to the reefs on which it lives. This view of the hawksbill as a parochial nester is founded largely on the knowledge of its ecological requirements. For this coral-reef dweller, suitable nesting beaches are nearly always locally available. Some caution is urged in making the inference that turtles always use the beaches closest at hand. Results of recent tagging programs have shown that the simplest model of sea turtle migration is not always applicable. Limpus (1978) demonstrated that the loggerheads living on the reefs near Heron Island, Australia, are not the same ones that use the island's nesting beaches, although the number nesting is approximately equal to the number foraging on the reefs. Although loggerheads and hawksbills are far apart ecologically, the possibility must be kept in mind that a similar pattern exists, at least for some populations.

Additional tagging data are needed. The diffuse nesting habits of this species have mitigated against any concerted tagging efforts. The meagre tagging evidence available is inconclusive. There are some data that suggest that long-range travel does occur. Twelve hawksbills tagged at Tortuguero, Costa Rica, have been recaptured away from the nesting beach (Carr and Stancyk 1975; Carr, unpublished data). Seven of these were taken in the Miskito Cays region of Nicaragua, the primary site of recapture of green turtles tagged at Tortuguero. Two others were captured north of Tortuguero, perhaps in transit to Miskito Cays, at Barra del Colorado, Costa Rica. Turtles have also been recaptured south of the nesting beach—at Matina and

Limon, both in Costa Rica, and at Colon, Panama. Only 20 or so hawksbills are tagged yearly at Tortuguero, and thus quite a long time will be required to substantiate the Tortuguero-Miskito Cays pattern.

Three of 60 hawksbills tagged on their foraging grounds in eastern Nicaragua were subsequently recorded on a nesting beach—2 in the vicinity of the tagging site, and one at Pedro Cays, Jamaica, a minimum travel distance of 628 km (Nietschmann, in press). A tagged male was recaptured 443 km away at Almirante Bay, Panama, which is near the hawksbill nesting beach at Rio Chiriqui.

De Silva (this volume) reported a single long-distance recovery of a hawksbill tagged at the Turtle Islands National Park, Sabah, East Malaysia and recaptured in the Philippines. The turtle traveled 713 km in 40 days, representing a minimum average travel speed of 17.8 km/day.

A hawksbill tagged while nesting on Santa Isabel Island, Solomon Islands, was recovered off Port Moresby, Papua New Guinea (Spring, personal communication). This represents a journey of approximately 1,600 km, possibly the longest recorded for the species.

Pritchard (1973) recorded a single recapture from 33 hawksbills tagged in Surinam and French Guiana; the turtle was taken 80 km from the tagging site.

Three of 55 hawksbills tagged at Cousin Island in the Seychelles have been recaptured at sea, all within 27 km of the island (Frazier, in press).

It is obvious from the tagging evidence reported to date that we do not yet have sufficient data on which to base a conclusion about the migratory behavior of the hawksbill. Considering the importance of this information to any conservation program, tagging projects focusing on this species warrant high priority.

The Olive Ridley (Lepidochelys olivacea)

As of March 1975, there had been 72 at-sea recaptures out of the 3,359 olive ridleys tagged on the beaches of the Guianas through the 1973 season. Recapture sites span the mainland coast of South America from eastern Venezuela to northern Brazil, and additionally include Trinidad and Barbados (Pritchard 1976). Recoveries of 28 turtles off the coast of Surinam and French Guiana, many of which were made during the non-nesting season, indicate that some turtles remain in the general region of the nesting beach. The proportionately large number of returns from Trinidad (8), Isla Margarita (4), and eastern Venezuela (13) are perhaps due to the presence of a rich food source at the Orinoco mouth (Pritchard 1973). One of the turtles tagged in Surinam was recaptured 1,900 km away after only 23 days, a journey requiring a minimum

ANNE MEYLAN

travel speed of 82 km/day. The migration was presumably made against the Guiana Stream (Schulz 1975).

Olive ridleys have been tagged extensively along the Pacific coast of Mexico. Recoveries have been primarily within Mexican waters, with the exception of 2 in El Salvador, 1 in Colombia, and 1 in Ecuador (Vargas 1973; Márquez, Villanueva, and Peñaflores 1976). Márquez, Villanueva, and Peñaflores (1976) suggested the possibility that a portion of the turtles that nest in Oaxaca remain in that region after nesting while others travel south to Central and South America. Turtles nesting in the states of Guerrero and Jalisco appear to disperse principally to feeding areas in the north, in the southern Gulf of California and along the western coast of the Baja peninsula.

Two olive ridleys tagged in Costa Rica have been recaptured away from the nesting beach. A female tagged at Playa Nancite was recovered in Ecuador (Cornelius, personal communication). A second was tagged at Ostionál, and caught in a tuna net 950 km due west of northernmost Costa Rica (11°24′N 94°37′W)(Robinson, personal communication).

There have been 3 international recoveries of olive ridleys tagged on Nicaragua's Pacific coast. All were tagged in 1972 while nesting in the Pochomil–La Boquita area near Masachapa. One was observed nesting 19 days later at Playa Nancite, Costa Rica. The other 2 were taken in 1977 at Manta, Ecuador, and Tumaco, Colombia (Nietschmann, personal communication).

Kemp's Ridley (Lepidochelys kempi)

Kemp's ridleys have been tagged at their sole breeding locality at Rancho Nuevo, Mexico, since 1966. Tag recoveries indicate that after the nesting season turtles return to feeding areas in the northern Gulf of Mexico, principally off Louisiana, and off Campeche, Mexico (Chavez 1968; Pritchard and Márquez 1973). All recoveries have been made close to shore, and it is speculated that migrations of this species do not normally involve open-sea crossing. One unusual recovery was that of a female ridley tagged at Rancho Nuevo and reported nesting 5 years later at Playa de Guachaca, Colombia (Chavez and Kaufmann 1974). Not only did this represent a spectacular violation of nesting site fidelity, but mature ridleys had not previously been recorded in the Caribbean (Pritchard and Marquez 1973).

The Loggerhead Turtle (Caretta caretta)

An intensive tagging program for *Caretta* has been conducted for over a decade at Tongaland, South Africa. Recaptures away from the nesting beach indicate postnesting dispersal to northern feeding areas along the mainland coast of East Africa. Recoveries have been made principally off Tanzania and Mozambique, with a small number of returns from Madagascar and South Africa (Hughes 1974, personal communication). The most distant recovery was made at Zanzibar, 2,880 km from the tagging site; several other recoveries indicate migrations of over 2,000 km. The rapid travel times evidenced by some of the recaptures support the hypothesis that postnesting movements of this colony represent purposeful, nonrandom travel (Hughes 1974).

Bustard and Limpus (1970) reported the first international recovery of a loggerhead tagged in Australia. A female that had nested at the Mon Repos rookery in Queensland was recovered 63 days later in the Trobriand Islands, Papua New Guinea, a straight-line distance of 1,770 km. Bustard and Limpus (1971) later recorded travel of another Mon Repos loggerhead that had apparently rounded the Cape York peninsula. A substantial number of additional recoveries of turtles tagged in the Capricorn and Bunker Groups (including Heron Island) and Mon Repos from 1966 to the present have been made along the Queensland coast, the eastern Gulf of Carpentaria, and Papua New Guinea (Bustard 1974; Limpus, this volume).

The loggerhead colony nesting in the southeastern United States is the subject of several tagging investigations. The first long-distance recovery involved a female tagged while nesting at Hutchinson Island, St. Lucie Co., Florida, and retaken 1,600 shoreline km away, at the mouth of the Mississippi River (Caldwell, Carr, and Ogren 1959).

Eighteen of the 647 loggerheads tagged in Georgia at Little Cumberland Island from 1964 to 1976 have been captured away from the area of the nesting beach (Bell and Richardson 1978). Recoveries have been made primarily along the eastern seaboard of the United States as far north as New Jersey. One recapture was made on Florida's west coast, at Tampa Bay. The distribution of returns suggests that, after nesting, turtles head north toward Cape Hatteras and the Chesapeake Bay. The fact that few are caught in these waters in winter months is thought to indicate that they eventually depart for warmer areas.

All 4 international recaptures of loggerheads tagged at the Cape Kennedy Space Center in Brevard Co., Florida, have been made in the Bahamas (Ehrhart 1976, 1979). Recapture sites include Grand Bahama, Abaco, and Eleuthera. One turtle that had been tagged at the Cape washed up dead at Sanibel Island on Florida's west coast.

Loggerheads tagged by Billy Turner at Melbourne Beach, Brevard Co., Florida, have been recovered along the eastern seaboard, in the Florida Keys, the Gulf of Mexico, the Bahamas, the Dominican Republic, and Cuba (Meylan, Bjorndal, and Turner, in prep.).

Caretta tagged at Sebastian Inlet, just south of Melbourne Beach, Florida, have been recaptured at Little Abaco, Bahamas; Chesapeake Bay, Virginia; Virginia Beach, Virginia, and Atlantic City, New Jersey (Le-Buff, personal communication).

Recoveries of loggerheads tagged at Jupiter Island, Florida, have been made principally along the eastern seaboard and in the Bahamas. A small number of returns has come from the U.S. Gulf coast, Cuba, Yucatan, and Belize (Lund, personal communication).

Loggerheads have also been tagged on Florida's west coast, largely in Lee and Collier counties. Three were recaptured at shoals northwest of Key West, and 1 was recovered at Cabo Catoche, Yucatan (LeBuff, personal communication). A loggerhead released at Indian Shores Beach, Pinellas Co., Florida, was retaken in the Chandeleur Islands, Louisiana (LeBuff, personal communication).

As might be expected, there is quite a bit of overlap in the tag-return data for the various loggerhead tagging projects in the southeastern United States. Notable is the lack of Bahamian recoveries from the project at Little Cumberland, Georgia. The data demonstrate the importance to this colony of feeding areas along the eastern seaboard.

The loggerhead colony nesting at Buritaca, Colombia, has been tagged since 1970. Kaufmann (1975) reported the recapture of a turtle 100 km east of the nesting beach 6 months after tagging. Several other recaptures have since been made along the north coast, especially in the Guajira peninsula, and in the vicinity of Santa Marta (INDERENA, undated).

The Flatback Turtle (Chelonia depressa)

Two long-distance recaptures have been recorded for this species, but details have not yet been published (Limpus, personal communication).

Summary and Conclusions

As the preceding review of tag return data plainly shows, our understanding of the migratory behavior of sea turtles is still in a very preliminary stage. For some species, it is not even certain that migration is a common feature of the life cycle. For others, there is only fragmentary evidence of the geographic patterns of migratory travel. It is apparent that tagging, as a tool for studying migratory behavior, has its limitations. Moreover, only intensive, long-term projects can be expected to yield useful results. Even then, many important aspects of migratory behavior cannot be addressed. It is perhaps time to begin to explore other possible sources of information that could advance our knowledge of this important aspect of the life history of marine turtles.

Acknowledgments

I would like to acknowledge all the people who have so generously contributed tag-return data for inclusion in this paper. I thank A.F. Carr and K. Bjorndal for their constructive comments and assistance in preparing the manuscript.

Literature Cited

Balazs, G. H.
1976. Green turtle migrations in the Hawaiian Archipelago. *Biological Conservation* 9:125–140.
1980. Synopsis of biological data on the green turtle in the Hawaiian Islands. National Oceanic and Atmospheric Administration, National Marine Fisheries Service, NOAA-TM-NMFS-SWFC–7, 141 pp.

Bell, R., and J. I. Richardson
1978. An analysis of tag recoveries from loggerhead sea turtles (*Caretta caretta*) nesting on Little Cumberland Island, Georgia. *Florida Marine Research Publication* 33:1–66.

Bustard, H. R.
1974. Barrier reef sea turtle populations. In *Proceedings of the Second International Coral Reef Symposium 1*, pp. 227–34. *Great Barrier Reef Committee*, Brisbane.
1976. Turtles of coral reefs and coral islands. In *Biology and Geology of Coral Reefs*, eds., O. A. Jones and R. Endean vol. 3, Biology, 2, pp. 343–68. New York: Academic Press.
1979. Population dynamics of sea turtles. In *Turtles: perspectives and research*. eds. M. Harless and H. Morlock. New York: Wiley and Sons.

Bustard, H. R., and C. J. Limpus
1970. First international recapture of an Australian tagged loggerhead turtle. *Herpetologica* 26:358–59.
1971. Loggerhead turtle movements. *British Journal of Herpetology* 4:228–30.

Caldwell, D. K.; A. Carr; and L. Ogren
1959. The Atlantic loggerhead sea turtle, *Caretta caretta caretta* (L.), in America. 1. Nesting and migration of the Atlantic loggerhead turtle. *Bulletin of the Florida State Museum* 4:293–348.

Carr, A.
1952. *Handbook of Turtles*. Ithaca, New York: Cornell University Press.
1954. The passing of the fleet. *AIBS Bulletin* 4:17–18.
1975. The Ascension Island green turtle colony. *Copeia* 1975:547–55.

Carr, A., M. H. Carr, and A. B. Meylan
1978. The ecology and migrations of sea turtles, 7. The West Caribbean green turtle colony. *Bulletin of the American Museum of Natural History* 162:1–46.

Carr, A., and P. J. Coleman
1974. Seafloor spreading theory and the odyssey of the green turtle. *Nature* 249:128–30.

Carr, A., and L. Giovannoli
1957. The ecology and migrations of sea turtles, 2. Results of field work in Costa Rica, 1955. *American Museum Novitates* 1835:1–32.

ANNE MEYLAN

Carr, A., and H. F. Hirth
1962.   The ecology and migrations of sea turtles, 5. Comparative features of isolated green turtle colonies. *American Museum Novitates* 2091:1–42.

Carr, A., and L. Ogren
1960.   The ecology and migrations of sea turtles, 4. The green turtle in the Caribbean Sea. *Bulletin of the American Museum of Natural History* 121:1–48.

Carr, A., and S. Stancyk
1975.   Observations on the ecology and survival outlook of the hawksbill turtle. *Biological Conservation* 8:161–72.

Chavez, H.
1968.   Marcado y recaptura de individuos de tortuga lora (*Lepidochelys kempi* Garman). *Instituto Nacional de Investigaciones Biológico Pesqueras* 19:1–28.

Chavez, H., and R. Kaufmann
1974.   Informacion sobre la tortuga marina *Lepidochelys kempi,* con referencia a un ejemplar marcado en Mexico y observado en Colombia. *Bulletin of Marine Science* 24:372–77.

Cornelius, S.
1975.   Marine turtle mortalities along the Pacific coast of Costa Rica. *Copeia* 1975:186–87.

Ehrhart, L. M.
1976.   Studies of marine turtles at Kennedy Space Center and an annotated list of amphibians and reptiles of Merritt Island. Final Report to NASA, March 2, 1976.
1979.   Threatened and endangered species of the Kennedy Space Center. Part 1. Marine turtle studies. Final Report to NASA, August 21, 1979.

Frazier, J.
In press.   Marine turtles in the Seychelles and adjacent territories. In *Biogeography and Ecology of the Seychelles Islands.* D. R. Stoddart, editor.

Galenon, M.
1979.   Tagging and rearing of the green turtle *Chelonia mydas* conducted in French Polynesia by the Department of Fisheries. Paper presented at the Joint SPS-NMFS Workshop on Marine Turtles in the Tropical Pacific Islands, Noumea, New Caledonia, 11–14 December 1979.

Green, D.
1979.   Double tagging of green turtles in the Galapagos Islands. *IUCN/SSC Marine Turtle Newsletter* 13:4–9.

Harrisson, T.
1960.   Notes on the edible green turtle (*Chelonia mydas*). 8. First tag returns outside Sarawak, 1959. *Sarawak Museum Journal* 1960:277–78.

Hirth, H. F., and A. Carr
1970.   The green turtle in the Gulf of Aden and Seychelles Islands. *Verhand. Konin. Nederl. Acad. Weten. Natur.* 58:1–41.

Hirth, H. F., and S. L. Hollingsworth
1973.   Report to the government of the People's Democratic Republic of Yemen on marine turtle management. FAO publication TA 3178.

Hornell, J.
1927.   *The Turtle Fisheries of the Seychelles Islands.* London:

H. M. Stationery Office.

Hughes, G. R.
1974.   The sea turtles of South-East Africa. 2. The biology of the Tongaland loggerhead turtle *Caretta caretta* L. with comments on the leatherback turtle *Dermochelys coriacea* L. and the green turtle *Chelonia mydas* L. in the study region. *Investigational Report of the Oceanographic Research Institute,* Durban, South Africa, 36:1–96.

Hughes, G. R.; A. J. Bass; and M. T. Mentis
1967.   Further studies on marine turtles in Tongaland, I. *Lammergeyer* 7:5–54.

INDERENA
Undated.   La tortuga gogo *Caretta caretta caretta* en la costa norte Colombiana, Operacion Tortuga Marina 1974–1975. Instituto Nacional de los Recursos Naturales Renovables y del Ambiente, Barranquilla, Colombia.

Kaufmann, R.
1975.   Studies on the loggerhead sea turtle *Caretta caretta caretta* (Linn.) in Colombia, South America. *Herpetologica* 31:323–26.

Leary, T. H.
1957.   A schooling of leatherback turtles *Dermochelys coriacea coriacea* on the Texas coast. *Copeia* 1957:232.

Limpus, C. J.
1978.   The reef. In *Exploration North: Australia's Wildlife from Desert to Reef,* ed. H. J. Lavery, pp. 187–222. Richmond, Australia: Richmond Hill Press.
1979.   Observations on the leatherback turtle, *Dermochelys coriacea* (L.), in Australia. *Australian Wildlife Research* 6:105–16.
1980.   The green turtle *Chelonia mydas* (L.) in eastern Australia. Management of turtle resources. Research monograph 1:5–22. Townsville, Australia: James Cook University of North Queensland.

Márquez, R.; A. Villanueva; and C. Peñaflores
1976.   Sinopsis de datos biológicos sobre la tortuga golfina (*Lepidochelys olivacea* Eschscholtz, 1829). INP Sinopsis sobre la Pesca numero 2. Instituto Nacional de Pesca, Mexico.

Mortimer, J.
1981.   The feeding ecology of the West Caribbean green turtle (*Chelonia mydas*) in Nicaragua. *Biotropica* 13:49–58.

Nietschmann, B.
In press.   Following the underwater trail of a vanishing species: the hawksbill. *National Geographic Society Research Reports, 1972.*

Oliver, J. A.
1946.   An aggregation of Pacific sea turtles. *Copeia* 1946:103.

Pritchard, P. C. H.
1969.   Sea turtles of the Guianas. *Bulletin of the Florida State Museum* 13:85–140.
1973.   International migrations of South American sea turtles (Cheloniidae and Dermochelidae). *Animal Behavior* 21:18–27.
1976.   Post-nesting movements of marine turtles (Cheloniidae and Dermochelidae) tagged in the Guianas. *Copeia* 1976:749–54.

Pritchard, P. C. H., and R. Márquez
1973.   Kemp's ridley turtle or Atlantic ridley, *Lepidochelys*

kempi. IUCN Monograph, Marine Turtle Series 2:1–
30.

Schulz, J. P.

1975.   Sea turtles nesting in Suriname. *Zoologische Verhandelingen, uitgegeven door het Rijksmuseum van Natuurlijke Historie te Leiden* 143:1–144.

Vargas, M. E.

1973.   Resultados preliminares del marcado de tortugas
marinas en aguas mexicanas (1966–1970). *Serie Informacion INP SI/i* 12:1–27.

ANNE MEYLAN

Nutrition, Growth, and Hibernation

Jeanne A. Mortimer
Department of Zoology
University of Florida
Gainesville, Florida 32611

Feeding Ecology of Sea Turtles

ABSTRACT

This paper reviews the feeding ecology of the world's sea turtles. Much of the present information is based on qualitative observations of stomach contents. Four of the 5 genera—the leatherback (*Dermochelys*), the ridley turtles (*Lepidochelys*), the loggerhead (*Caretta*) and the hawksbill (*Eretmochelys*)—are carnivorous, and their feeding habits are poorly documented. The diet of the herbivorous green turtle (*Chelonia mydas*) has been studied in more detail. Throughout most of its range, this species forages primarily on seagrass meadows or, where seagrasses are lacking, algae.

In an attempt to analyze the diet of green turtles more quantitatively, I examined the stomach contents of 243 turtles taken at their foraging grounds off the eastern coast of Nicaragua. Turtle grass, *Thalassia testudinum*, accounted for an average of 80 percent of the dry weight of the samples. The turtles graze at the bases of the *Thalassia* plant where they obtain the youngest growth. In decreasing order of abundance, the remaining food items consisted of other species of seagrass (10 percent), algae (8 percent), benthic substrate (2 percent), and animal matter (1 percent). In the northern part of the Nicaraguan foraging range, *Thalassia* accounted for nearly 90 percent of the diet while in the more southerly regions red algae predominated. When the turtles are in habitats where good forage is scarce, such as near their nesting beaches, they appear to consume material of little or no food value. No difference between the diets of the two sexes was recorded.

Feeding Habits of Sea Turtles

Although conspicuous advances in the study of sea turtle ecology have been made in the last two decades, the feeding ecology of these animals is still poorly known. It can be said with confidence that the mature green turtle is mainly herbivorous. The other 4 genera are chiefly carnivorous, but their feeding regime and habitats have not been clearly revealed. In no case is the diet of the hatchlings well known although it is assumed that during their first months they consume floating organisms.

The first section of this paper reviews the feeding habits of sea turtles. Most of the observations summarized are qualitative descriptions of stomach contents. The remainder of the paper describes a study in which I attempted to quantitatively analyze the diet of the green turtle on its foraging pastures off the east coast of Nicaragua.

Leatherback or Trunkback

The leatherback or trunkback, *Dermochelys coriacea,* is the most pelagic of the sea turtles. Although leatherbacks often weigh well over 500 kg, they draw their sustenance chiefly from a diet of jellyfish (Schyphomedusidae) and tunicates, together with crustacean parasites and symbiotic fish that are associated with the jellyfish (Brongersma 1969, Bleakney 1965). This diet is reflected by their sharp-edged jaws which completely lack the massive construction and crushing plates found in the jaws of the loggerhead or ridley turtles (Pritchard 1971a).

Ridley Turtles

In view of the vast numbers of ridley turtles that have been killed in slaughterhouses in recent years, it is astonishing that so little is known of their feeding habits. Review papers by Pritchard and Marquez (1973) and Zwinenberg (1977) indicate that the food of Kemp's ridley, *Lepidochelys kempii,* is primarily crabs. Hildebrand (this volume) reports that they primarily eat 2 genera of portunid crabs, *Ovalipes* and *Callinectes.* Available published information, as reviewed by Marquez, Villanueva, and Penaflores (1976) and Zwinenberg (1976) indicates that shrimp predominates in the diet of the olive ridley, *Lepidochelys olivacea.* However, Alfredo Martinez (personal communication) observed that the stomachs of ridleys killed off the nesting beaches at Escobilla, Mexico, contain crabs and jellyfish as well as shrimp. The guts of 10 olive ridleys captured 30 to 50 km off the coast of Ecuador contained 100-percent crabs (Derek Green, personal communication). Rice (undated report) examined the guts of 12 olive ridleys killed in shrimp trawls off Costa Rica. In addition to a relatively small amount of shrimp, he found crabs, sessile and pelagic tunicates, and numerous other small invertebrates. The abundance of both benthic fauna and substrate suggested to Rice that *L. olivacea* is primarily a bottom feeder. Reports reviewed by Hughes (1974a) of olive ridleys captured in prawn trawls at depths ranging from 80 to 110 m indicate that they are capable of foraging in very deep water.

Loggerhead

The loggerhead turtle, *Caretta caretta,* eats a variety of benthic invertebrates (including molluscs, crustaceans and sponges) which they crush before swallowing. Off the Natal coast, molluscs (especially *Bufonaria crumenoides* and *Ficus subintermedius*) predominate in the diet although occasionally more pelagic items including fish and even hatchling loggerheads have been recorded (Hughes 1974b). Similarly, at Heron Island at the south end of the Great Barrier Reef the turtles eat mostly horn shells (Cerithidae), ear shells (*Haliotus* spp.) and turban shells (*Turbo* spp.). They occasionally consume jellyfish (Limpus 1978). However, off the beach at Mon Repos, Queensland, female loggerheads feed predominantly on prawns and fish during the nesting season (Limpus 1973). The gut of a loggerhead captured off the New Jersey coast (Fowler 1914) was filled with the remains of hermit crabs (*Pagurus pollicaris*) and borers (*Natica duplicata*). Mary Mendonça (personal communication) found that 95 percent of the diet of subadult and adult loggerhead turtles living in Mosquito Lagoon, Brevard County, Florida, consists of the horseshoe crab, *Limulus polyphemus.* The remaining food items include blue crabs, *Callinectes,* and occasionally mullet. The loggerheads that Norine Rouse (personal communication) has observed while diving off Palm Beach, Florida, seem to prefer encrusting organisms such as sponges, although she frequently observes them eating basket stars (Ophiuroidea). Because *Caretta* has very powerful jaws, it is capable of crushing the shells of such seemingly invulnerable prey as the queen conch, *Strombus gigas,* in the Caribbean (Babcock 1937), and the giant clams, *Tridacna maxima* (Limpus 1973; 1978) and *T. fossor* (Bustard 1976), in Australia.

Hawksbill

The hawksbill, *Eretmochelys imbricata,* inhabits tropical reefs, where it feeds on encrusting organisms such as sponges, tunicates, bryozoans, molluscs and algae which it scrapes off the reef faces (Carr and Stancyk 1975, Limpus 1978). It is also reported to eat Portuguese man of war, *Physalia* (Babcock 1937). Occasionally, stomachs of individuals are found full of plant material, including algae and seagrass (True 1884; Deraniyagala 1939; Pritchard 1977; Colin Limpus, personal communication) or fruits of the red mangrove (Carr 1952).

Flatback Turtle

Virtually nothing has been published on the diet of the flatback turtle, *Chelonia depressa,* of Australia, and it is not even clear whether the species is primarily carnivorous or herbivorous (Williams, Grandison, and Carr 1967). Although he obtained no firsthand data,

Bernard Nietschmann (in litt.) reports that the Torres Strait islanders say the flatback feeds on seagrasses and some algae. However, because flatbacks are captured so often in shrimp trawls (Cogger 1969; Spring, this volume), it is possible that their diet may consist largely of shrimp. Colin Limpus (personal communication) believes that the jaw structure of *C. depressa* is more similar to that of the carnivorous olive ridley than to that of the herbivorous green turtle.

Green Turtle

Most of what we know of the feeding habits of the species of sea turtles just discussed is based on qualitative descriptions of stomach contents from relatively few individuals. The diet of the green turtle, *Chelonia mydas,* is better known. In certain parts of its range, stomach contents of large numbers of individuals have been examined. Methods of pumping the stomachs of freshly captured sea turtles under field conditions recently devised by Balazs (1979a) in Hawaii and Mary Mendonça (personal communication) in Florida promise to increase this data base dramatically. The green turtle is herbivorous, but not averse to eating animal matter, which it readily accepts in captivity. Throughout most parts of its range it forages primarily on seagrass pastures. Where seagrasses are lacking, algae are the mainstay of the diet.

WESTERN ATLANTIC

In the western Atlantic north of Brazil, *Chelonia* forages primarily on the seagrasses *Thalassia testudinum* and *Syringodium filiforme* (Hirth 1971, Mortimer 1976). Immature green turtles in Mosquito Lagoon, Brevard County, Florida, ranging in size from 7 to 50 kg, seem to be grazing exclusively on the seagrasses (*S. filiforme, Halodule wrightii* and *Halophila* sp.) and avoiding the abundant algal growth available (Mary Mendonça, personal communication). Ferreira (1968) analyzed contents of the stomachs of 94 green turtles captured off the east coast of Brazil, an area where seagrass is not abundant, and found that they subsist almost entirely on algae, especially Rhodophyceae.

EASTERN PACIFIC

The green turtles of the Galapagos Islands feed mainly on algae, especially green algae of the genera *Ulva* (Derek Green, in litt.) and *Caulerpa* (Pritchard 1971b). Green (in litt.) has also seen them foraging on the algae (*Bostrychia calliptera, B. radicans* and *Caloglossa leprieurii*) that grow on the stilt roots of the red mangrove, *Rhizophora mangle.* They sometimes consume the mangrove leaves (Pritchard 1971b; Green, in litt.).

Some *Chelonia* populations along the Pacific coast of Central America forage on algae, others on seagrasses. In the Bay of Fonseca, Honduras, Carr (1952) found kelp and sponge in the stomachs of some butchered green turtles, but found seagrass in others. A similar dichotomy in foraging habits exists near Tiburon Island along the northwest shore of Sonora, Mexico, in the Gulf of California. Green turtles of the Infiernillo region, between the island and the mainland feed on eelgrass, *Zostera marina,* while those foraging only a few kilometers away, off the west coast of Tiburon Island (Felger and Moser 1973) and islands nearby (Carr 1961) eat algae.

CENTRAL PACIFIC

The diet of the turtles of Tahiti (Hirth 1970a) and Hawaii (Hirth 1970a, Balazs 1979a, 1979b) consists mostly of algae. In Tonga, the food appears to be mainly *Syringodium isoetifolium* and *Halodule ovalis* (Hirth 1970b). In one part of the Suva region of Fiji, *S. isoetifolium, Halodule pinifolia* and occasionally *H. ovalis* are consumed while in a nearby locality the turtles eat red and green algae (Hirth 1970b). Pritchard (1977) found *Chelonia* to be foraging on seagrasses in northern Palau, and on seagrasses and algae in the Truk District.

WESTERN PACIFIC

In Torres Strait, Australia, green turtles primarily consume red and some green algae (Bernard Nietschmann, in litt.; Nietschmann, in press). However, there appear to be two feeding regimes, one in which the algae are dominant, and in the other, seagrasses. Green turtles foraging in the waters of nearby Papua New Guinea (Spring, this volume) forage on seagrass.

At the southern end of the Great Barrier Reef, the diet of the green turtles is determined by the habitat in which they are living. In the reef habitats the turtles feed mostly on the dominant algal species present (Colin Limpus, in litt.). On the reef fronts a range of algae is consumed, including soft ones such as the green algae, *Chlorodesmis* sp., coarse brown algae like *Turbinaria* sp., and the red alga *Amansia* sp. Occasionally, however, they will consume macrozooplankton, including jellyfish (Bustard 1976; Limpus 1978; C. Limpus, in litt.). Those turtles living over sandy bottom habitats feed extensively on the soft algae *Enteromorpha* sp., *Polysiphonia* sp., *Champia* sp., and *Dictyota* sp. Colin Limpus (in litt.) reports that in nonreef situations (bays, for example), seagrasses (including *Zostera* sp. and *Thalassia* sp.) appear to be the predominant food, along with small amounts of mangrove fruits (*Avicennia* sp., *Rhizophora* sp.).

At One Arm Point, north of Broom, Australia, Archie Carr (personal communication) found the guts of 2 butchered females packed with an unidentified species of algae. Deraniyagala (1939) reports that the green turtles of Ceylon feed on *Cymodocea* sp., *Thalassia* sp., *Zostera* sp., *Halophila* sp. and algae. On the Aldabra Atoll, Frazier (1971) found the turtles eating the seagrass, *Cymodocea* sp., and algae, *Gelidium* sp. and *Laurencia* sp., growing in association with it. He felt they may have been selectively consuming *Caulerpa* sp. According to Hornell (1927), the turtles of the Seychelles also forage primarily on *Cymodocea* sp. Green turtles off Mozambique (George Hughes, in litt.) feed on seagrasses such as *Cymodocea* sp. and *Halodule* sp. Because most types of seagrass do not grow in the colder African waters, the turtles in that region consume mostly algae, including *Gelidium* sp., *Codium duthieae* and *Caulerpa filiformes* (Hughes 1974b), and occasionally eat *Zostera* sp. (G. Hughes, in litt.).

In the Bay of Khor Umaira of the Gulf of Aden (Peoples Democratic Republic of Yemen), Hirth (Hirth and Carr 1970; Hirth, Klikoff, and Harper 1973) examined the stomach contents of 100 green turtles taken on their feeding grounds. Chiefly 2 kinds of seagrasses, *Posidonia oceanica* and *Halodule uninervis* were found along with small amounts of red and brown algae. The stomachs of turtles caught about 19 km west of the site contained other seagrasses, *Syringodium* sp. and *Cymodocea* sp. In the eastern part of the country, the stomachs of 100 adult females captured at their nesting grounds were examined. About half of these were empty while contents of the remainder were mostly brown and green algae, with very little seagrass.

Near the island of Masirah, Sultanate of Oman, the green turtles forage mostly on the bases of the seagrasses, *Halodule uninervis*, *Halophila ovalis* and *H. ovata*. The algae, *Sargassum illicifolium* and *Chaetomorpha aerea* is also consumed (Perran Ross, in litt.).

Diet of the Nicaraguan Green Turtle

In 1975, I initiated a study to analyze in a detailed, quantitative manner, the stomach contents of turtles captured on foraging pastures off the eastern coast of Nicaragua, which are the major feeding grounds of the green turtle in the western Caribbean. Most accounts in the literature had stated simply that in the Caribbean, *Chelonia* feeds predominantly on *Thalassia*. I wanted to determine whether the green turtle was as completely herbivorous as the literature suggested. I also wanted to ascertain what importance the animal and plant species living epiphytically on the blades of *Thalassia* might have in the nutrition of the green turtle.

Composition of the Stomach Samples

I examined the stomach contents of 243 adult and subadult turtles, which were in most cases of known sex and site of capture. No differences in the diets of the two sexes were found. The turtle grass, *Thalassia testudinum*, was the main item in the diet of the turtles, accounting for an average of 80 percent of the dry weight of the stomach samples. In decreasing order of importance, the remaining food items comprised: other species of seagrasses, *Syringodium* and *Halodule* (10 percent); algae of 40 different species (8 percent); benthic substrate (2 percent); and animal matter (1 percent) (Mortimer 1981).

Foraging Patterns of the Turtles on the Feeding Grounds

I found no evidence that the diet of the herbivorous green turtle is substantially augmented by the ingestion of invertebrate material living on the surface of the seagrass blades (Mortimer 1981). *Thalassia* grows from a basal meristem. Thus, the bases of the blades consist of fresh, green growth while the older, distal portions are often dead and brown. Seagrass epiphytes are mostly confined to the older parts of the blades.

I collected benthic quadrat samples on the *Thalassia* pastures at several sites on the feeding grounds off the coast of Nicaragua, and determined that on a dry-weight basis, the older, brown portions accounted for 56 percent of the total above-ground *Thalassia* biomass. This figure was consistent in both deep water (13 m) and shallow water (1 to 4 m) localities. I examined the *Thalassia* present in the stomach contents of the turtles captured near these sites, and determined that only 5 percent of the *Thalassia* material present in the stomachs consisted of the older, brown blades. It appeared that the turtles were selectively consuming the new growth at the base of the *Thalassia* plants. Bjorndal (1980) found evidence that her semiconfined turtles in the Bahamas actually keep certain patches of grass cropped short by continuously returning to feed in one place, thus insuring to themselves a steady supply of the young, more proteinaceous shoots. The turtles at the Nicaraguan feeding grounds may be foraging in a similar fashion.

Foraging Patterns During Migration and Nesting

The lush, submarine vegetation typical of the Miskito bank feeding grounds of *Chelonia* grows only in quiet, sheltered waters. The nesting beaches of the population, on the other hand, occur only in areas of high-energy surf, which are frequently devoid of food. The 2 habitats of the species are often located some distance from each other, and in their migrations, the turtles may have to travel hundreds of kilometers through

JEANNE A. MORTIMER

territory lacking good forage.

The local turtle people report that turtles migrating between their Nicaraguan feeding grounds and Costa Rican nesting grounds apparently take a longshore route. They believe that the turtles move inshore to use the river effluents as piloting cues, on their 325 km migration to the nesting beach in Costa Rica. Along one section of the Nicaraguan coast, migrating turtles can be captured within 1.5 km of the shore. In the stomachs of turtles taken in this manner I found quantities of highly lignified terrestrial debris that had been deposited offshore in the effluents of the river mouths. Similarly, at the Tortuguero nesting beach, the nesting females are known to eat water hyacinth debris and other flotsam deposited at the mouth of the Tortuguero River. Under these circumstances the turtles are consuming material of little or no apparent food value.

Relative Importance of Algae and Seagrass

In the northern part of the Nicaraguan foraging grounds near the Miskito Cays, where the turtle grass pastures occur in nearly monospecific stands, *Thalassia* constitutes almost 90 percent of the diet (Mortimer 1981). Here the sponge *Haliclona rubens* is also frequently consumed. In the southern part of the range, near Set Net Cays, red algae predominate in the diet. Conversations with local people suggested that turtles captured in this region often "taste rank." The same thing is said of the meat of algae-eating green turtles in other parts of the world (Pritchard 1971b). For example, in the Gulf of California near Tiburon Island (Felger and Moser 1973) the Seri Indians contend that turtles that eat seagrass taste "sweet," while those which live only a few kilometers away and eat algae taste "bitter."

The Torres Strait Islanders of Australia (Nietschmann, in press) draw a similar distinction between algae-eating turtles and those whose diet is predominantly seagrass. The fat of the algae-eaters is said to be thin and tar black, and they weigh about 20 percent less than their seagrass-eating counterparts. Nietschmann (in press) suggests that the differences between these 2 types of turtles may be attributable to age, competition over scarce resources (seagrasses), or perhaps disease or parasites.

The fact that the flavor of the meat of individual turtles within the same population can differ from place to place along a short length of coast has broader implications. It suggests the possibility that, even within a relatively limited geographic area, consistent dietary variation between individual turtles may exist. It also supports other evidence indicating that turtles may remain resident in one area on their feeding grounds for extended periods of time (Carr 1954, Balazs 1979a, Bjorndal 1980, M. Mendonça, personal communication, Nietschmann, in press).

Bjorndal (1979) has demonstrated that the guts of seagrass-eating Nicaraguan green turtles contain a microflora which enables them to digest the cellulose of *Thalassia* plants. It is possible that algae-eating turtles require a different microfloral assemblage to digest those species of algae whose cell walls are not composed of cellulose. If so, then the intestinal flora of a turtle whose diet consists primarily of seagrass may be different from that of an individual within the same population that consumes mostly algae. Similarly, for a turtle to change from being predominantly a consumer of seagrass to a consumer of algae, as has been suggested by Nietschmann (in press), a corresponding change in intestinal flora would also have to occur (Bjorndal 1980).

Conclusions

The foraging habits of individual turtles are not as static as the literature suggests. The diets of the Nicaraguan green turtles vary according to the habitat in which they are feeding, and individual animals undergo complex seasonal movements at their feeding grounds. These movements are influenced by environmental conditions such as changes in ocean currents or turbidity and the presence of fresh water effluents during the rainy season. More work needs to be done on the feeding ecology of all the sea turtles, but especially on that of the four carnivorous genera. For example, we know that different populations of loggerhead turtles consume rather different types of food. As we gather more data on this subject, the foraging patterns will certainly reveal themselves to be even more complex.

Acknowledgments

I am grateful to the following people for allowing me to include their unpublished data in this paper: George Balazs, Derek Green, George Hughes, Colin Limpus, Mary Mendonça, Bernard Nietschmann, Richard Rice, Perran Ross, and Norine Rouse. Helpful suggestions regarding the content of the manuscript were received from Archie Carr, Jack Ewel, Karen Bjorndal, Anne Meylan, and Steve Cornelius. The work in Nicaragua was funded primarily by the Caribbean Conservation Corporation (much of it as a special donation from the late Jane Frick). Additional funding was received from Sigma Xi and the National Science Foundation (Biological Oceanography Program, grant GA 36638 to Archie Carr).

Literature Cited

Babcock, H. L.
1937. The sea turtles of the Bermuda Islands, with a survey of the present state of the turtle fishing industry.

Proceedings of the Zoological Society of London 107:595–601.

Balazs, G. H.

1979a. Growth, food sources and migrations of immature Hawaiian *Chelonia. IUCN/SSC Marine Turtle Newsletter* 10:1–3.

1979b. Synopsis of biological data on marine turtles in the Hawaiian Islands. National Oceanic and Atmospheric Administration, National Marine Fisheries Service, contract 79–ABA–02422. 76 pages.

Bjorndal, K. A.

1979. Cellulose digestion and volatile fatty acid production in the green turtle, *Chelonia mydas. Comparative Biochemistry and Physiology* 63A:127–33.

1980. Nutrition and grazing behavior of the green turtle, *Chelonia mydas. Marine Biology* 56:147–54.

Bleakney, J. S.

1965. Reports of marine turtles from New England and eastern Canada. *Canadian Field Naturalist* 79:120–28.

Brongersma, L. D.

1969. Miscellaneous notes on turtles, IIA. *Proc. Kon. Ned. Akad. Wetensch. Amsterdam* C 72:76–102.

Bustard, H. R.

1976. Turtles on coral reefs and coral islands. In *Biology and Geology of Coral Reefs,* eds. O. A. Jones and R. Endean, vol. 3, Biology, 2, pp. 343–68. New York: Academic Press.

Carr, A.

1952. *Handbook of Turtles.* Ithaca, New York: Cornell University Press.

1954. The passing of the fleet. *A.I.B.S. Bulletin* 4:17–19.

1961. Pacific turtle problem. *Natural History* 70:64–71.

Carr, A., and S. Stancyk

1975. Observations on the ecology and survival outlook of the hawksbill turtle. *Biological Conservation* 8:161–72.

Cogger, H. G.

1969. Marine turtles in northern Australia. *Australian Zoologist* 15:156.

Deraniyagala, P. E. P.

1939. *The Tetrapod Reptiles of Ceylon,* vol. 1, *Testudinates and Crocodilians.* London: Dulav and Company.

Felger, R., and M. B. Moser

1973. Eelgrass (*Zostera marina*) in the Gulf of California: Discovery of its nutritional value by the Seri Indians. *Science* 181:355–56.

Ferreira, M. M.

1968. Sobre a alimentacao da aruana, *Chelonia mydas* Linnaeus, ao longo da costa do Estado do Ceara. *Arq. Est. Biol. Mar. Univ. Fed. Ceara* 8:83–86.

Fowler, H. W.

1914. The food of the loggerhead turtle (*Caretta caretta*). *Copeia* 1914:4.

Frazier, J.

1971. Observations on sea turtles at Aldabra Atoll. *Philosophical Transactions of the Royal Society of London,* B, 260:373–410.

Hirth, H.

1970a. Report of marine turtle survey in Tucson, Hawaii, Tahiti, Western Samoa, American Samoa and New Caledonia. FAO, 7 September-19 October 1970. Mimeo, 18 pp.

1970b. Report of marine turtle survey in Tonga Islands and Fiji Islands. FAO, 19 October-15 December 1970. Mimeo, 19 pp.

1971. Synopsis of biological data on the green turtle, *Chelonia mydas* (Linnaeus) 1758. *FAO Fisheries Synopsis* 85:1:1–8:19.

Hirth, H., and A. Carr

1970. The green turtle in the Gulf of Aden and the Seychelles Islands. *Verh. K. Ned. Akad. Wet. (AFO Nat. Tweede Sect.)* 58:1–44.

Hirth, H.; L. G. Klikoff; and K. T. Harper

1973. Sea grasses at Khor Umaira, People's Democratic Republic of Yemen with reference to their role in the diet of the green turtle, *Chelonia mydas. Fishery Bulletin* 71:1093–97.

Hornell, J.

1927. *The Turtle Fisheries of the Seychelles Islands.* London: H. M. Stationery Office.

Hughes, G. R.

1974a. The sea turtles of South-East Africa. 1. Status, morphology and distributions. *South African Association Marine Biology Resources Oceanographic Research Institute Investigations Report* 35:1–144.

1974b. The sea turtles of South-East Africa. II. The biology of the Tongaland loggerhead turtle, *Caretta caretta* L., with comments on the leatherback turtle, *Dermochelys coriacea* L., and the green turtle, *Chelonia mydas* L., in the study region. *South African Association Marine Biology Resources Oceanographic Research Institute Investigations Report* 36:1–96.

Limpus, C. J.

1973. Loggerhead turtles (*Caretta caretta*) in Australia; food sources while nesting. *Herpetologica* 29:42–45.

1978. The reef. In *Exploration North: Australia's Wildlife from Desert to Reef,* ed. H. J. Lavery, pp. 187–222. Richmond, Australia: Richmond Hill Press.

Marquez, R.; A. Villanueva; and C. Penaflores

1976. Sinopsis de datos biologicos sobre la tortuga golfina. *Instituto Nacional de Pesca, Sinopsis sobre la Pesca* 2:1–61.

Mortimer, J. A.

1976. Observations on the feeding ecology of the green turtle, *Chelonia mydas,* in the western Caribbean. Masters thesis, University of Florida.

1981. The feeding ecology of the West Caribbean green turtle, *Chelonia mydas,* in Nicaragua. *Biotropica* 13:49–58.

Nietschmann, B.

In press. Hunting and ecology of dugongs and green turtles, Torres Strait, Australia. *National Geographic Society Research Reports.*

Pritchard, P. C. H.

1971a. The leatherback or leathery turtle (*Dermochelys coriacea*). *IUCN Monograph, Marine Turtle Series* 1:1–39.

1971b. Galapagos sea turtles—preliminary findings. *Journal of Herpetology* 5:1–9.

1977. *Marine Turtles of Micronesia.* San Francisco: Chelonia Press.

Pritchard, P. C. H., and R. Marquez

1973. Kemp's ridley turtle or Atlantic ridley, *Lepidochelys kempi*. *IUCN Monograph, Marine Turtle Series* 2:1–30.

Rice, R. E.

Undated. A preliminary investigation of the Pacific ridley sea turtle (*Lepidochelys olivacea*): sex ratio, external dimensions, reproductive development, feeding habits and the effect of shrimp fishing on them in Costa Rican waters. Report submitted to the ACM Central American Field Studies Program Director. Mimeo, 58 pp.

True, F. W.

1884. The useful aquatic reptiles and batrachians. *Fisheries of Indiana*, United States Section 1, Part 2, pp. 137–62.

Williams, E. E.; A. G. C. Grandison; and A. F. Carr

1967. *Chelonia depressa* Garman re-investigated. *Breviora* 271:1–15.

Zwinenberg, A. J.

1976. The olive ridley, *Lepidochelys olivacea* (Eschscholtz, 1829): probably the most numerous marine turtle today. *Bulletin of the Maryland Herpetological Society* 12:75–95.

1977. Kemp's ridley, *Lepidochelys kempii* (Garman, 1880), undoubtedly the most endangered marine turtle today (with notes on the current status of *Lepidochelys olivacea*). *Bulletin of the Maryland Herpetological Society* 13:170–92.

Karen A. Bjorndal
Department of Zoology
University of Florida
Gainesville, Florida 32611

The Consequences of Herbivory for the Life History Pattern of the Caribbean Green Turtle, *Chelonia mydas*

ABSTRACT

The Caribbean green turtle, *Chelonia mydas*, feeds primarily on the seagrass *Thalassia testudinum*, which has a high fiber content and thus a low forage quality. The green turtle has two adaptations for its low quality diet: a hindgut fermentation and a selective grazing pattern. Despite these adaptations, the green turtle is nutrient-limited, which results in low growth rates, delayed sexual maturity, and a low annual reproductive effort. The energy required by a female Tortuguero green turtle is approximately 805,800 kJ per year. The Tortuguero green turtle, which feeds on *T. testudinum*, is able to allocate only 10 percent of its annual energy budget for reproduction, while the Surinam green turtle, which feeds on algae, allocates 24 percent of its annual energy budget for reproduction.

The carrying capacity of *T. testudinum* for the green turtle is estimated at 138 adult female green turtles per hectare. The instantaneous death rates for the cohorts of Tortuguero turtles increased from 1959 to 1972, which suggests that the Tortuguero green turtle population has not been maintaining itself. The nutrient limitations imposed on the green turtle by its low-quality diet make it vulnerable to, and slow to recover from, overexploitation of its adult population.

Introduction

The seagrass *Thalassia testudinum* comprises 87 percent (by dry weight) of the diet of the Caribbean green turtle, *Chelonia mydas* (Mortimer 1976). Throughout the year, *T. testudinum* is constant in production (Greenway 1974) and nutrient quality (Bjorndal 1980a), forming the basis of one of the most productive ecosystems (Westlake 1963). Despite this high, constant productivity, few species utilize *T. testudinum* as a food source (Kikuchi and Peres 1977). Less than 10 percent of the leaf production of *T. testudinum* in the Caribbean is consumed by herbivores (Ogden, personal communication). One of the factors that limits herbivory

111

on *T. testudinum* is its high cellulose content (45 percent of dry weight), making *T. testudinum* a low-quality forage (Bjorndal 1980a). Thus, the herbivorous green turtle has a food source that is constant in abundance, constant in nutrient quality and relatively free of competitors. The specialization of the green turtle on this dependable but low-quality diet has had consequences for the green turtle's life history pattern. The consequences are discussed in this paper.

Herbivory: Adaptations and Limitations

Green turtles have 2 adaptations for utilizing their low quality diet. First, they have a high digestive efficiency resulting from a microbial fermentation in their hindgut that digests approximately 90 percent of the cellulose in their diet (Bjorndal 1979a). This fermentation produces volatile fatty acids (acetate, butyrate, and propionate, in order of decreasing concentration), which are a significant energy source for the turtle. The volatile fatty acids produced in the cecum alone provide approximately 15 percent of the green turtle's energy budget (Bjorndal 1979a). Secondly, green turtles consistently recrop areas of *T. testudinum,* leaving adjacent stands untouched. The young blades they consume are more digestible because of lower lignin levels and are higher in protein by 6 to 11 percent than the ungrazed stands (Bjorndal 1980a).

Despite this abundant food source, high digestive efficiency and selective grazing, the green turtle is nutrient-limited because of the low quality of its diet. At least 2 factors cause this limitation. One is the green turtle's low food intake. Green turtles consume the equivalent of only 0.24 to 0.33 percent of their body weight each day (dry weight to wet weight) (Bjorndal 1980a). Similar values for terrestrial, mammalian, nonruminant herbivores are 1.7 to 8.3 percent (Lloyd, McDonald, and Crampton 1978, National Research Council 1978). The green turtle's low food intake is a result of the high fiber (cellulose) content of its diet. A high fiber content increases the passage time of food through the gut of an herbivore with a cellulolytic fermentation (Lloyd, McDonald, and Crampton 1978). A longer passage time necessarily decreases the rate of intake. Because intake cannot be increased without decreasing digestive efficiency, the green turtle cannot simply eat more in order to increase nutrient and energy assimilation.

The second limiting factor is that green turtles feeding on *T. testudinum* have an apparent digestibility of protein of only 50 percent (Bjorndal 1980a). This is a very low value for an herbivore. Domestic herbivores have an apparent digestibility of protein of 75 to 80 percent (Lloyd, McDonald, and Crampton 1978).

Nutrient and energy limitations result in low growth rates, delayed sexual maturity and a low energy allocation for reproduction. Low growth rates in juvenile wild green turtles have been reported (Balazs 1979, Limpus 1979), and these data are supported by mark and recapture records for a wide size range of green turtles from the southern Bahamas (Bjorndal, unpublished data). Growth rates in captive-reared green turtles fed a high protein, animal diet are much more rapid (Caldwell 1962; J. Wood, personal communication). Thus, the low growth rates in wild green turtles are a result of nutrient limitations. These low growth rates result in a long juvenile period and delayed sexual maturity. Wood and Wood (1977) reported that green turtles on a high protein diet reached sexual maturity after 9 years. Ages at sexual maturity based on extrapolations from wild green turtle growth rates are much higher (Balazs 1979; Bjorndal, unpublished data).

Annual Energy Budget

The allocation of energy to reproduction can be more easily discussed using the equation for an animal's basic energy budget:

$$(J)(I) = M + A + R + G$$

where J is the apparent digestibility of the food energy in *T. testudinum,* I is ingestion, M is maintenance, A is nonreproductive activity, R is reproduction, and G is growth. Energy digestibility, ingestion, maintenance and nonreproductive activity of a green turtle just prior to sexual maturity probably do not differ significantly from those of an adult green turtle. For example, a 48 kg subadult green turtle has the same digestive efficiency as a 126 kg adult (Bjorndal 1980a). When sexual maturity is reached, growth essentially stops (Carr and Goodman 1970, Bustard 1972), and that portion of the energy budget that had been channeled into growth in the subadult turtle can now be allocated for reproduction. However, since growth in subadult turtles is slow, the amount of energy thus given to reproduction would seem to be small.

In order to estimate the energy allocation for reproduction, the following estimations of the reproductive effort (the percentage of the annual energy budget used for reproduction) of adult female green turtles in the Tortuguero, Costa Rica, breeding population have been calculated (Bjorndal, in prep.). Because there is no food for green turtles at the nesting beach at Tortuguero, each female green turtle must deposit enough fat while on the feeding grounds to provide energy for the migration to and from the nesting beach, the production of eggs, the excavation of nests, as well as maintenance and daily minimum activity for the period of stay at the nesting beach.

The cost of migration can be estimated using the formula given by Prange (1976) for the Ascension Island green turtle colony. Substituting the values of the Tortuguero green turtle breeding population for mean carapace length (Carr and Hirth 1962), mean swimming speed (Carr, Carr, and Meylan 1978), and mean round-trip migration distance (Carr, Carr, and Meylan 1978) gives a value of 76,000 kJ for the cost of a round trip migration (Bjorndal, in prep.).

The mean clutch size for recruits (turtles in their first nesting season) is 111.4 (N = 1706) and for remigrants (turtles that have nested in previous seasons) is 116.8 (N = 334). These 2 means are significantly different (t-test; p<0.001). Carr, Carr, and Meylan (1978) gave a mean number of nests per season of 2.7 for recruits and 3.4 for remigrants. The mean weight and energy value of 10 Tortuguero eggs was 8.8026 g organic matter and 29.5 kJ/g organic matter, respectively (Bjorndal, in prep.). These values give a mean energy content per egg of 259.7 kJ. Recruits, with a mean of 300.8 eggs at 259.7 kJ/egg, expend 78,117.8 kJ in eggs every nesting season. Remigrants produce 397.1 eggs at a cost of 103,126.9 kJ per season.

Digging a nest involves a considerable expenditure of energy by the green turtle. Prange and Jackson (1976) reported a mean maximum oxygen consumption of 0.206 l O_2/kg·hr and presented observations that suggest a green turtle sustains this maximum oxygen consumption throughout the nesting process, which averages 2 hours duration (Rebel 1974). The mean weight of a Tortuguero female green turtle is 126 kg (N = 234). Since the energy source is fat, Prange and Jackson's oxygen consumption figure can be transformed to an estimate of the cost of nesting activity: 1020 kJ. Recruits, with a mean of 2.7 nests (Carr, Carr, and Meylan 1978), expend 2750 kJ in nesting, while remigrants, with a mean of 3.4 nests (Carr, Carr, and Meylan 1978), use 3470 kJ (Bjorndal, in prep.).

Finally, the energy used for maintenance and activity during the internesting interval can be estimated. In the southern Bahamas, green turtle time budgets averaged 6 hours per day of activity and 18 hours per day of inactivity (Bjorndal 1979b). The mean minimum time each female spends off Tortuguero Beach is 1 minus the mean number of nests times the mean internesting interval of 12.1 days (Carr, Carr, and Meylan 1978). Thus, the average recruit spends at least 370 inactive hours off Tortuguero Beach, and the average remigrant spends a minimum of 523 inactive hours. Using the minimum oxygen consumption value of 0.024 l O_2/kg·hr given by Prange and Jackson (1976), it is estimated that recruits expend 22,000 kJ and remigrants 31,100 kJ in internesting inactivity. The energy utilized in internesting activity can be estimated from the value calculated above for swimming. Recruits use 20,000 kJ and remigrants use 28,200 kJ. Thus, a recruit active for 6 hours and inactive for 18 hours each day expends a minimum of 42,000 kJ, and a remigrant with

the same activity schedule expends at least 59,300 kJ in internesting maintenance and activity in 1 breeding season (Bjorndal, in prep.).

The total reproductive effort per reproductive season of a green turtle recruit is 198,900 kJ, whereas remigrants expend 241,800 kJ. The above reproductive parameters and their associated energy expenditures are summarized in Table 1. The average energy investment in reproduction is lower for the recruit than for the remigrant. However, because recruits lay fewer eggs, the energy expended per egg is greater for the recruit (661 kJ/egg) than for the remigrant (609 kJ/egg).

Referring to the energy-budget equation given above, the annual energy budget of an adult female green turtle can now be estimated. Maintenance (M) and activity (A) during the nonreproductive period can be calculated as described above for the reproductive interval, and equal 379,900 kJ and 345,300 kJ, respectively (Bjorndal, in prep.). As already discussed, green turtles stop growing at any appreciable rate after reaching sexual maturity, so G is zero. To put the energy expended for reproduction on an annual basis, the total reproductive effort must be divided by the number of years in the remigration interval. Since the average remigration interval for the Tortuguero green turtle is 3 years (Carr, Carr, and Meylan 1978), that value will be used in these calculations, resulting in an annual R of 80,600 kJ.

If the values for maintenance, activity, reproduction, and growth are added, a value of 805,800 kJ is obtained for the amount of energy required to balance an adult female green turtle's annual energy budget. The amount of energy expended on reproduction is only 10 percent of the annual energy budget. Much of this energy is used for activities associated with reproduction. The cost of the eggs is only 4.3 percent of the annual energy budget.

Table 1. Reproductive efforts of Tortuguero green turtles

Item	Recruits		Remigrants	
	kJ	Percent[a]	kJ	Percent[a]
Eggs	78,100	39.3	103,000	42.6
Migration	76,000	38.2	76,000	31.4
Nesting	2,750	1.4	3,470	1.4
Internesting[b]	42,000	21.1	59,300	24.5
Total	198,900		241,800	
Reproductive effort per egg	661		609	

a. Percentage of total reproductive effort.
b. Includes maintenance and activity.

The hypothesis that feeding on *T. testudinum* limits the energy allocated to reproduction is supported by comparing the reproductive efforts of the Tortuguero green turtle breeding population and the Surinam green turtle breeding population that feeds primarily on algae (Schulz 1975, Ferreira 1968). Using data provided by Schulz (1975), the above calculations for the reproductive effort have been repeated for the Surinam green turtle (Bjorndal, in prep.). The Surinam green turtle is able to channel 24 percent of its annual energy budget into reproduction compared with 10 percent for the Tortuguero turtle. The average migration distance of the Surinam turtle is over 4 times (4,000 km) greater than the migration distance of the Tortuguero green turtle (828 km), and Surinam turtles lay more eggs per season (Schulz 1975; Carr, Carr, and Meylan 1978). However, the most common remigration interval is 2 years for the Surinam population and 3 years for the Tortuguero population (Schulz 1975; Carr, Carr, and Meylan 1978). That is, the Surinam green turtle, feeding on algae, is able to deposit sufficient fat stores for a much greater energy output for reproduction in a shorter period of time than is the Tortuguero green turtle, feeding on *T. testudinum*.

Carrying Capacity

The constant, high productivity of the extensive beds

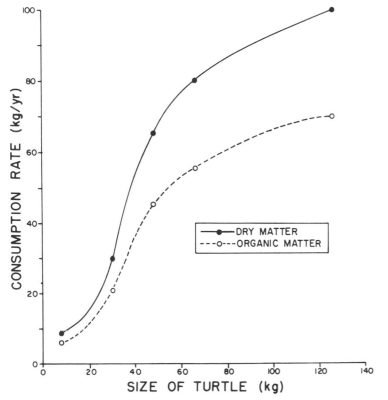

Figure 1. Consumption rates of green turtles feeding on *Thalassia testudinum* blades. Data for 126 kg turtles are from this paper; data for 8, 30, 48 and 66 kg turtles are from Bjorndal (1980a).

of *T. testudinum* in the Caribbean once supported large green turtle populations (Parsons 1962). The green turtle life-history pattern fits the pattern of a species that has evolved at or near the carrying capacity of its environment's resources: long life, delayed sexual maturity, low natural predation on adults, and extended reproductive lifespan. The carrying capacity of *T. testudinum* for the green turtle is an estimate of the maximum potential density of green turtle populations.

The carrying capacity can be estimated in the following manner. Given the value of 805,800 kJ as estimated above as the amount of energy required to balance an adult female green turtle's annual energy budget and a mean energy apparent digestibility coefficient (J) of 62 percent (Bjorndal 1980a), each turtle must ingest 1,300,000 kJ per year. *T. testudinum* has an energy content of 18.7 kJ per gram of organic matter (Bjorndal 1980a). Thus, an adult female green turtle consumes approximately 70 kg organic matter, or 100 kg dry weight, of *T. testudinum* each year. These ingestion rates are reasonable extensions of the rates measured in 8, 30, 48 and 66 kg green turtles (Figure 1). An average standing crop of 250 g dry weight of *T. testudinum* blades per square meter has been measured in both Jamaica (Greenway 1976) and Venezuela (Gessner 1971). Cropping stimulated blade growth, but continual recropping led to a decrease in growth rate, presumably as rhizome stores decrease (Greenway 1974). Working with *T. testudinum* in Jamaica, Greenway found a blade biomass turnover rate of 8.8 per year in ungrazed stands (Greenway 1976) and 5.5 turnovers per year in stands cropped to 2.5 cm above the leaf base (Greenway 1974), the average cropping height of green turtles (Bjorndal 1979b). Multiplying the 5.5 turnover rate by the standing crop of 250 g dry weight per square meter, an average 1.38 kg dry weight of *T. testudinum* blades are produced per square meter each year in a grazed area. The carrying capacity of 1 hectare of *T. testudinum* is thus 138 adult female green turtles, or 1 turtle per 72 square meters. This estimate of the carrying capacity would be improved if we had an estimate of the size-class distribution of a natural green turtle population and, using the feeding rates from Figure 1, calculated a carrying capacity based on all size classes. If the Seagrass Ecosystem Study Group of the National Science Foundation is successful in its attempt to chart and quantify the extent of seagrass beds in the Caribbean, the carrying capacity of the Caribbean for the green turtle, and thus the maximum potential population for the green turtle, can be estimated.

Survivorship Curves

The present populations of green turtles are greatly reduced from their past numbers (Parsons 1962). For

KAREN A. BJORNDAL

the green turtle to successfully maintain its population levels with its slow growth rates, delayed sexual maturity, widely spaced reproductive intervals and high predation on the young, there must be a long reproductive lifespan. In an effort to assess the stability of the Tortuguero green turtle breeding population, survivorship curves for 14 cohorts of female green turtles have been plotted, based on tag returns from turtles caught on their feeding grounds or seen again in later years as remigrants at Tortuguero (Bjorndal 1980b). Each cohort is composed of those turtles that were tagged for the first time in each year from 1959 to 1972. The slope of the survivorship curve represents the instantaneous death rate of that cohort. The death rates increase in the later cohorts due to the continued overexploitation of Tortuguero green turtles at their nesting beach and on their feeding grounds. If the death rates are plotted against their cohort year (Fig. 2), the resulting linear trend suggests that the Tortuguero green turtle population will be extinct within 40 years if the pattern of increasing death rates is not reversed. It may be that the protection granted the green turtle by the Costa Rican Government in 1975 and by the Nicaraguan Government in 1976 has come soon enough to allow the Tortuguero green turtle population to stabilize.

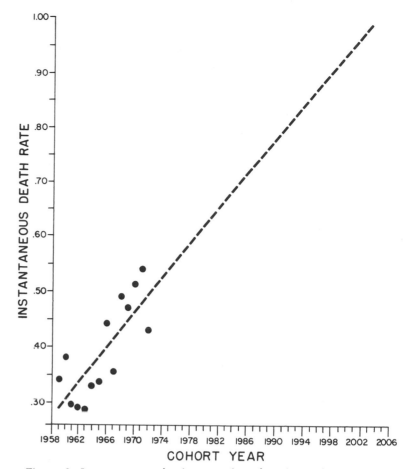

Figure 2. Instantaneous death rates plotted against cohort year for 14 cohorts of the Tortuguero green turtle breeding population. Data are from Bjorndal (1980b).

Conclusions

Thalassia testudinum is a dependable but low-quality food source. The consequences for the green turtle specializing on this seagrass have been slow growth rates, delayed sexual maturing and a low annual reproductive effort. Green turtles feeding on this low-quality diet and subject to high juvenile mortality depended upon a long reproductive lifespan to maintain the former high population levels. However, after the arrival of European man in the Caribbean and the subsequent overexploitation of green turtle populations, the reproductive lifespans of green turtles have been shortened, and recruitment into the breeding population has been decreased. The green turtle is locked into its life-history pattern by limited nutrient assimilation caused by its low food intake of a low-quality diet and low protein digestion. Therefore, the green turtle cannot increase the number of progeny produced per reproductive female in order to compensate for the overexploitation of its adult population and is slow to recover once the adult population has been severely reduced.

Acknowledgments

I want to thank Archie Carr for his continuing support and inspiration; Alan Bolten and Jeanne Mortimer for their helpful discussions and assistance; and John Moore for introducing me to the study of quantitative nutrition. Financial support for the studies summarized here was provided by the Inagua Project of the Caribbean Conservation Corporation.

Literature Cited

Balazs, G. H.
1979. Investigations of the growth, food sources and migrations of immature Hawaiian *Chelonia. IUCN/SSC Marine Turtle Newsletter* 10:1–3.

Bjorndal, K. A.
1979a. Cellulose digestion and volatile fatty acid production in the green turtle, *Chelonia mydas. Comparative Biochemistry and Physiology*, 63A:127–33.

1979b. Nutrition and grazing behavior of the green turtle, *Chelonia mydas*, a seagrass herbivore. Ph.D. dissertation, University of Florida.

1980a. Nutrition and grazing behavior of the green turtle, *Chelonia mydas. Marine Biology* 56:147–54.

1980b. Demography of the breeding population of the green turtle, *Chelonia mydas*, at Tortuguero, Costa Rica. *Copeia* 1980:525–30.

Bustard, H. R.
1972. *Australian Sea Turtles*. London: Collins.

Caldwell, D. K.

1962. Growth measurements of young captive Atlantic sea turtles in temperate waters. *Contributions in Science* 50:1–8.

Carr, A. F., M. H. Carr, and A. B. Meylan

1978. The ecology and migrations of sea turtles, 7. The West Caribbean grean turtle colony. *Bulletin of the American Museum of Natural History* 162:1–46.

Carr, A. F., and D. Goodman

1970. Ecologic implications of size and growth in *Chelonia*. *Copeia* 1970:783–86.

Carr, A. F., and H. F. Hirth

1962. The ecology and migrations of sea turtles, 5. Comparative features of isolated green turtle colonies. *American Museum Novitates* 2091:1–42.

Ferreira, M. M.

1968. Sobre a alimentacao da aruana, *Chelonia mydas*, ao longo da costa do estado de Ceara. *Arq. Est. Biol. Mar. Univ. Ceara* 8:83–86.

Gessner, F.

1971 · The water economy of the seagrass, *Thalassia testudinum. Marine Biology*, 10:258–60.

Greenway, M.

1974. The effects of cropping on the growth of *Thalassia testudinum* in Jamaica. *Aquaculture* 4:199–206.

1976. The grazing of *Thalassia testudinum* in Kingston Harbour, Jamaica. *Aquatic Botany* 2:117–26.

Kikuchi, T., and J. M. Peres

1977. Consumer ecology of seagrass beds. In *Seagrass Ecosystems: A Scientific Perspective*. eds. C. P. McRoy and C. Helfferich, pp. 147–93. New York: Marcel Dekker, Inc.

Limpus, C.

1979. Notes on growth rates of wild turtles. *IUCN/SSC Marine Turtle Newsletter* 10:3–5.

Lloyd, L. E., B. E. McDonald, and E. W. Crampton

1978. *Fundamentals of Nutrition*. Second Edition. San Francisco: W. H. Freeman and Company.

Mortimer, J. A.

1976. Observations on the feeding ecology of the green turtle, *Chelonia mydas*, in the western Caribbean. Masters thesis, University of Florida.

National Research Council

1978. Nutrient requirements of horses. Fourth edition. Washington, D.C.: National Academy of Sciences.

Parsons, J. J.

1962. *The Green Turtle and Man*. Gainesville: University of Florida Press.

Prange, H. D.

1976. Energetics of swimming of a sea turtle. *Journal of Experimental Biology* 64:1–12.

Prange, H. D., and D. C. Jackson

1976. Ventilation, gas exchange and metabolic scaling of a sea turtle. *Respiratory Physiology* 27:369–77.

Rebel, T. P., editor

1974. *Sea Turtles and the Turtle Industry of the West Indies, Florida and the Gulf of Mexico*. Miami, Florida: University of Miami Press.

Schulz, J. P.

1975. Sea turtles nesting in Surinam. *Zoologische Verhandelingen, uitgegeven door het Rijksmuseum van Na-*

tuurlijke Historie te Leiden 143:1–144.

Westlake, D. F.

1963. Comparisons of plant productivity. *Biological Review* 38:385–425.

Wood, J. R., and F. E. Wood

1977. Captive breeding of the green turtle, *Chelonia mydas*. Presented at the World Mariculture Society Meeting, San Jose, Costa Rica, January 1977.

KAREN A. BJORNDAL

George H. Balazs
Hawaii Institute of Marine Biology
P. O. Box 1346
Kaneohe, Hawaii 96744

Growth Rates of Immature Green Turtles in the Hawaiian Archipelago

Introduction

The formulation of sound management strategies for sea turtles is dependent in part upon an understanding of the rates of growth and age at sexual maturity of naturally occurring members of each population. However, these aspects have received comparatively little attention due to the difficulties of capturing and tagging immature turtles directly from the sea in their resident foraging areas. Most research has instead been directed at the colonial nesting beaches where usually only the adult females are available for tagging and observation. This has resulted in considerable insight into the reproductive ecology, migrations, and growth of the adult female, three critically important but nevertheless limited aspects of the animals' life history. Nesting beaches also offer easy access to large numbers of hatchlings, but the absence of a suitable tag for this size category has hampered research aimed at determining natural growth rates and maturation age. Growth studies conducted with captive turtles, or with turtles released into the wild after being raised for a period of time in captivity, cannot be considered representative of natural conditions. Determining the age of sea turtles from annuli present in bones offers some potential, but reliable analyses are not possible at the present time (Balazs 1979b).

Tag and recapture studies of naturally occurring immature sea turtles were first carried out by Schmidt (1916) on green turtles (*Chelonia mydas*) in the Virgin Islands. Subsequent investigations have been conducted on the west coast of Florida by Carr and Caldwell (1956) and at Bermuda by Burnett-Herkes (1974). Preliminary results of my own work in the Hawaiian Archipelago (Balazs 1979a) have been presented concurrently with those of Limpus (1979) working off Heron Island on the Great Barrier Reef. Other studies of immature sea turtles are also known to be in progress on the east coast of Florida at Mosquito Lagoon (Ehrhart and Yoder 1978), on the west coast of Mexico (Felger, Cliffton, and Cornejo 1978), in the Galapagos

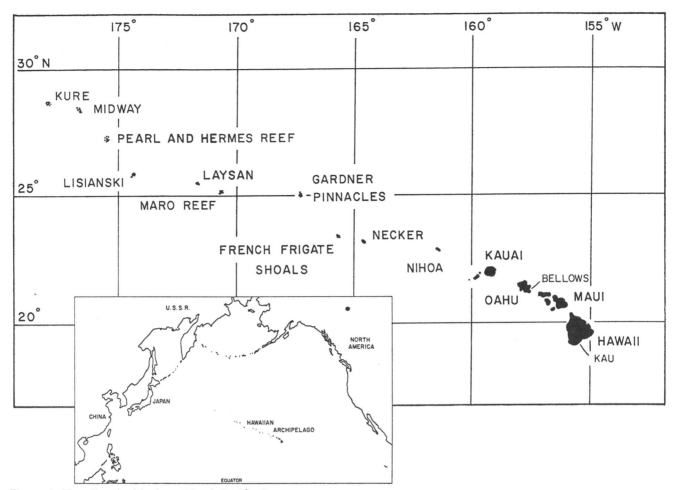

Figure 1. Hawaiian Archipelago, North Pacific Ocean.

Islands (D. Green, in litt.), at Bermuda (Frick 1977), and in the Bahamas (K. Bjorndal, in litt.).

Study Area

The Hawaiian Archipelago consists of 132 islands, islets and reefs under United States jurisdiction that extend for 2,450 km across an isolated region of the North Pacific Ocean from 18°54′N, 154°40′W to 28°15′N, 178°20′W (Figure 1). Eight main and inhabited islands (Hawaii, Maui, Kahoolawe, Lanai, Molokai, Oahu, Kauai, Niihau) located in the southeastern segment of the archipelago comprise over 99 percent (16,650 km²) of the total land area. The remainder consists of offshore islets and the small islands extending to the northwest of Kauai and Niihau known as the Leeward or Northwestern Hawaiian Islands. Except for Kure and Midway, the islands in this segment of the chain comprise the Hawaiian Islands National Wildlife Refuge.

A population of green turtles occurs throughout the Hawaiian Archipelago, with mixed aggregations of adults and immature individuals residing at foraging areas along the 1,210 km of coastal waters. Turtles measuring less than 35 cm in straight carapace length are not normally found in these resident foraging areas and are assumed to be living somewhere in the pelagic environment. In excess of 90 percent of all nesting takes place on 6 small sand islands in the middle of the archipelago at French Frigate Shoals. Adults periodically undertake long-distance migrations to this location for reproduction. Hawaiian *Chelonia* exhibit the rare behavioral trait among sea turtles of coming ashore to bask or rest, but only at certain undisturbed sites in the Northwestern Hawaiian Islands. This includes both adult males and females and, to a lesser extent, immature members of the population (Balazs 1976, 1979b).

Although immature green turtles have, at various times, been tagged at 16 different foraging sites throughout the Hawaiian Archipelago, 7 representative study areas have been selected for repetitive and long-term sampling. The locations of these areas are shown in Figure 1 and described as follows:

Kau District, Hawaii (19°08′N, 155°30′W). The Kau

District consists of a lava rock coastline that lacks protective reefs and has numerous areas where fresh water enters the ocean from underground springs. Although a few partially sheltered bays are present, most foraging occurs close to shore under turbulent conditions resulting from exposure to tradewind-generated waves.

Bellows Air Force Station, Oahu (21°10′N, 157°43′W). Bellows Air Force Station has a sand beach coastline within Waimanalo Bay that is protected from large surf by subtidal reefs. Foraging occurs close to shore, particularly near the exits of 2 fresh-water streams.

Necker Island (23°35′N, 164°42′W). Necker is an uninhabited lava rock island consisting of 17 ha with a maximum elevation of 85 m. Foraging occurs at the base of the island's cliffs and at a partially sheltered reef area adjacent to a rock ledge where basking takes place.

French Frigate Shoals (23°45′N, 166°10′W). French Frigate Shoals is a 35-km long crescent-shaped atoll that has shallow foraging areas located near the islands of East (4.0 ha), Whale-Skate (6.8 ha), and Tern. Tern was enlarged from 4.5 to 23 ha by dredging and landfill in 1942 to serve as a U. S. Naval Air Station. The station was later abandoned, but from 1952 to 1979 a small U.S. Coast Guard Loran Station was located at this site.

Lisianski Island (26°02′N, 174°00′W). Lisianski is an uninhabited 182-ha sand island with protective reefs and shallow foraging areas located close to shore.

Midway Islands (28°13′N, 177°21′W). Midway is an atoll 11 km in diameter with 2 sand islands (Sand and Eastern) and a well defined fringing reef. Foraging areas exist throughout the lagoon and adjacent to the islands. A U.S. Naval Station is located at this site.

Kure Atoll (28°25′N, 178°10′W). Kure is an atoll 9.5 km in diameter with 2 sand islands (Green and Sand) and a well-defined fringing reef. Foraging areas exist throughout the lagoon but are mostly adjacent to Green Island. Kure is the world's northernmost atoll and, since 1960, the site of a small U.S. Coast Guard Loran Station.

Sampling Methods

Four basic methods have been employed to capture immature green turtles in the Hawaiian Archipelago.

Scoop net. At Necker, French Frigate Shoals, Lisianski, and Kure, a scoop net 1 m in diameter with a 3.5-m long handle has been used to catch turtles, both from a small boat and from shore by walking out into shallow foraging areas up to 1.3 m deep.

Tangle nets. At Kau on Hawaii and Bellows on Oahu, turtles have been captured with tangle nets set vertically through the water column. These nets are made of 2-mm diameter cotton or nylon line and measure 3.5 by 20 m with a square mesh of 23 cm. Multiple sections of net can be easily tied together to form different lengths. The nets are usually set from shore extending out perpendicular to the coastline. Most captures are made at night, frequently from 1 to 3 hours before sunrise during periods of incoming tides. Periodic monitoring of the net must take place as a precautionary measure to prevent the turtles from drowning once they have become entangled. Tangle nets have not been used at foraging areas in the Northwestern Hawaiian Islands due to the possibility of catching resident Hawaiian monk seals (*Monachus schauinslandi*).

Diving. At Kau, Midway, and Kure, captures have been made by hand while diving, both with and without the aid of scuba. Turtles are usually found resting on the bottom in partially sheltered areas of rock and calcareous substrate.

Basking. At Necker, French Frigate Shoals and Lisianski turtles have at times been captured when they come ashore to bask.

Morphometric data recorded for immature Hawaiian *Chelonia* have included straight and curved carapace length along the midline, straight and curved carapace width at the widest point (usually at the sixth marginal), straight plastron-length along the midline, and body weight.

The identification tags that have been used since September of 1976 were specially manufactured by the National Band and Tag Company (Newport, Kentucky) from Inconel 625, an alloy consisting principally of nickel and cadmium (Balazs 1977). The change to Inconel was made following the determination that substantial corrosion was occurring in the Monel 400 (copper- and nickel-alloy) tags that had previously been in use. Both the Inconel and Monel tags were made in the manufacturer's series number 4–1005–681 self-piercing tag (8 by 29 mm, 3.5 g), which has a simplified locking mechanism and is suitable for use on both immature and adult turtles. Two tags have been applied to each turtle through the folds of flesh located proximal to the body on the trailing edges of the front flippers. Application is made so that some space remains between the end of the tag and the flipper to accommodate growth. No signs of corrosion have thus far been found in tags made of Inconel.

Sampling procedures have also included the recovery of food from the mouths of turtles captured while foraging and the retrieval of stomach contents using a flexible plastic tube inserted through the esophagus (Balazs in press). This has been carried out to identify food sources and to determine if differences that could influence growth exist between foraging areas.

A total of 629 immature *Chelonia* has thus far been

Table 1. Growth rates of immature green turtles sampled at 7 study areas in the Hawaiian Archipelago

Location and tag number	Straight carapace length, cm	Interval in months	Growth rate cm per month	Mean growth rate cm per month
Kau, Hawaii (19°08'N, 155°30'W)	—	—		.44
2520	47.6	17.5	.49	—
2934	50.8	8.5	.52	—
2934	55.2	7.5	.38	—
2887	54.0	7.5	.38	—
Bellows, Oahu (21°10'N, 157°43'W)	—	—		.20
2340	44.1	13.5	.21	—
2332	57.2	22	.19	—
Necker Island (23°35'N, 164°42'W)	—	—		.14
2391	48.3	20	.14	—
French Frigate Shoals (23°45'N, 166°10'W)	—	—		.08
2266	36.5	10.5	.06	—
2266	37.1	11.5	.11	—
1632	38.1	12.5	.08	—
2735	39.4	26	.09	—
1628	39.7	10	.06	—
2682	40.6	14	.09	—
2682	41.9	11.5	.08	—
1629	41.9	10	.03	—
1730	41.9	3	.11	—
3050	43.8	34	.08	—
3134	43.8	35.5	.04	—
2403	44.5	21	.01	—
1626	44.5	7.5	.13	—
1660	45.1	11.5	.03	—
1666	48.9	9	.11	—
1641	49.8	10	.10	—
1663	52.7	3	.11	—
1645	53.3	7	.09	—
1634	53.3	7	.05	—
Lisianski Island (26°02'N, 174°00'W)	—	—		.13
2854	35.9	2	.13	—
2850	40.0	2	.13	—
2858	44.8	2	.13	—
Midway Islands (28°13'N, 177°21'W)	—	—		.09
2454	40.0	7.5	.08	—
404	40.6	21	.03	—
873	41.9	37	.04	—
1538	45.7	11.5	.11	—
472	48.9	6	.21	—
1539	50.4	14	.12	—
1539	52.1	8	.04	—
407	55.1	28	.11	
Kure Atoll (28°25'N, 178°10'W)	—	—		.08
2991	46.0	24	.04	—
2469	59.4	13	.12	—

— Not applicable.

tagged throughout the archipelago, 524 of which were captured at the 7 study areas. These turtles range from 29.5 to 79.4 cm in straight carapace length. Prior to the initiation of my research program in 1973, 185 immature *Chelonia* had been tagged since 1967 by the Koral Kings Diving Club at Midway, using series 4–1005–49 self-piercing Monel tags (9 by 39 mm, 7 g) supplied by the U.S. Fish and Wildlife Service. Turtles tagged at Midway have often been released up to 6 km from the site of capture while at the other study areas the turtles have been released where they were captured.

Results

Thirty-five turtles, ranging from 35.9 to 59.4 cm in straight carapace length, have thus far been recaptured, in which growth could be detected after intervals of 2 to 37 months in the wild. Four of these turtles were recaptured on 2 occasions, thereby providing a total of 39 growth measurements. The rates of growth found at each of the 7 study areas were not dependent on the size of the turtles (Table 1). Thirty-four other turtles that were recaptured after intervals of 2 to 20 months showed no measurable growth. This included 1 turtle at Necker, 24 at French Frigate Shoals, 3 at Lisianski, and 6 at Midway. One of the turtles at French Frigate Shoals measured 68 cm and was recaptured after 20 months while the turtle at Necker measured 42 cm and was recaptured after 17 months. All of these turtles were vigorous and appeared to be in good health.

GEORGE H. BALAZS

None of the recaptured turtles showed evidence of tail enlargement indicative of males.

The mean rates of growth found at the 7 study areas ranged from .08 to .44 cm/month in straight carapace length. Growth rates at the 2 study areas in the main islands (Kau and Bellows) were considerably greater than at the 5 study areas in the Northwestern Hawaiian Islands (Table 2). The most rapid growth was recorded at Kau (range .38 to .52 cm/month), and the slowest growth at French Frigate Shoals (range .02 to .13 cm/month), and Kure (range .04 to .12 cm/month).

The use of straight carapace length has been found to be the most reliable index of growth for Hawaiian *Chelonia.* Curved carapace length is subject to greater error from variability in positioning the flexible measuring tape along the carapace in comparison to the use of calipers for straight line measurement. The use of body weight has also been found to be of reduced value. This is probably due to differences in the amount of food material present at various times in the gastrointestinal tract, a component that can comprise up to 18 percent of the weight of immature Hawaiian *Chelonia.*

All 69 of the turtles that were recaptured and remeasured were found in the same resident areas where they were originally tagged. Except at Midway, most of the turtles were either foraging or resting within 50 m of the original capture site. At French Frigate Shoals, no movements of tagged immature turtles were found between East, Whale-Skate, and Tern, even though the distance between any 2 of these islands is only 8 to 11 km.

Of the 629 immature turtles tagged in the Hawaiian Archipelago, only 2 long-distance movements have been reported, with 1 of these being of questionable validity. One recovery involved a 38-cm turtle tagged at Midway and found 6 months later at Wake Island (19°18′N, 166°36′E), a distance of 1,900 km. However, the weak and apparently pathological condition of this turtle, reported both at the time of original capture and at recovery, suggests that it may have passively drifted there with prevailing winds and currents. The other long-distance movement was a 40-cm turtle also tagged at Midway that was reported 7 months later by a fisherman as having been recaptured and released alive in Hilo Bay on the island of Hawaii. This involves a distance of 2,300 km against prevailing winds and currents present in the latitudes of the Hawaiian Archipelago. Although 2 Moncl tags were originally placed on the turtle, only 1 tag was found at the time of recovery.

Table 2. Summary of growth rates and projected number of years to maturity for immature green turtles sampled at 7 study areas in the Hawaiian Archipelago

Location, number tagged and size range	Growth rate cm per month			Interval in months	Years to maturity (35 to 81 cm)		Years to maturity (35 to 92 cm)	
	Mean	Range	N		Mean	Range	Mean	Range
Kau, Hawaii N = 72 37.7–79.4 cm	.44	.38–.52	4	7.5–17.5	8.7	7.4–10.1	10.8	9.1–12.5
Bellows, Oahu N = 21 38.1–61.6 cm	.20	.19–.21	2	13.5–22	19.2	18.3–20.2	23.8	22.6–25.0
Necker N = 7 39.4–48.3 cm	.14	—	1	20	27.4	—	33.9	—
French Frigate Shoals N = 130 36.4–67.9 cm	.08	.02–.13	19	3–35.5	47.9	29.5–191.7	59.4	36.5–237.5
Lisianski N = 23 35.9–53.3 cm	.13	—	3	2	29.5	—	36.5	—
Midway N = 250 36.5–59.4 cm	.09	.03–.21	8	6–37	42.6	18.3–127.8	52.8	22.6–158.3
Kure N = 21 29.5–61.6 cm	.08	.04–.12	2	13–24	47.9	31.9–95.8	59.4	39.6–118.8

The possibility therefore exists that the tag number may have been misread due to corrosion or other causes, and that this turtle was not the one tagged at Midway.

The major food sources of immature green turtles that have been identified at each of the 7 study areas consist of the following benthic algae:

Study area	Major food source
Kau, Hawaii:	*Pterocladia capillacea*
Bellows, Oahu:	*Codium edule, Codium arabicum, Codium phasmaticum, Ulva fasciata*
Necker:	*Caulerpa racemosa*
French Frigate Shoals:	*Codium arabicum, Codium phasmaticum, Codium edule, Caulerpa racemosa, Ulva fasciata, Turbinaria ornata*
Lisianski:	*Caulerpa racemosa, Turbinaria ornata*
Midway:	*Codium edule, Spyridia filamentosa*
Kure:	*Codium edule*

In the Northwestern Hawaiian Islands, particularly at Midway and Kure, immature turtles have also been recorded voraciously feeding on the invertebrates *Physalia physalia, Velella velella* and *Janthina exigua* that periodically drift into coastal areas.

Discussion and Conclusions

The growth rates of naturally occurring immature green turtles reported by workers in other areas have ranged from .05 to 5.26 cm per month (Table 3).

Using curved carapace length, Schmidt (1916) found a mean growth rate of .43 cm/month (range .10 to .69 cm/month) in the Virgin Islands, while Limpus (1979) reported growth rates ranging from .05 to .27 cm/month at Heron Island. These values are similar to the ones found at the 7 study areas in the Hawaiian Archipelago. Limpus (1979) also recaptured green turtles in which no growth could be detected. The apparent absence of growth over extended periods has also been found for some immature *Chelonia* in the Galapagos Islands (D. Green, *in litt.*). In contrast with the findings of other workers, Carr and Caldwell (1956) reported a growth rate estimated to be from .75 to 5.26 cm per month for a green turtle recaptured off the west coast of Florida (Table 3).

Estimates of the maturation age of green turtles that have appeared in the literature (summarized by Hirth 1971 and Rebel 1974) have ranged from 4 to 13 years. However, these values were based on growth rates obtained in captivity where conditions are substantially different from the natural environment. It is of interest to note that none of the estimates for age at maturity have been based on the natural growth rates resulting from Schmidt's pioneering work published in 1916.

If the growth rates determined at the study areas in the Hawaiian Archipelago remain constant until maturity, as available data suggest, then turtles measuring 35 cm that are new recruits would require from 8.7 years (at Kau) to 47.9 years (at Kure and French Frigate Shoals) to reach 81 cm, the minimum size at which nesting takes place in the population. From 10.8 to 59.4 years would be required to grow from 35 to 92 cm, the mean size of nesting Hawaiian *Chelonia*. Table 2 presents similar projections for 35-cm turtles that establish residency at each of the foraging areas investigated. All of these estimates are based on the assumption that residency is maintained at the same general foraging area, and that the turtles do not at some point prior to maturity move long distances to other foraging areas where different growth rates result. Except for the 2 long-distance recoveries previously described, all evidence accumulated to date indicates that residency continues for extended periods, and may be permanent except for reproductive migrations undertaken as adults. This concept is supported by the 69 stationary tag recoveries that have thus far been made after intervals ranging up to 37 months and by the fact that all sizes of turtles from 35 cm to mature adults are present at most of the foraging areas in the Hawaiian Archipelago.

Green turtles are believed to mature at different sizes (Carr and Goodman 1970), therefore age at maturity would be expected to differ, even among individuals at the same foraging area where similar growth rates are taking place. Carr and Carr (1970) have found that after reaching maturity the growth rate of green turtles nesting at Tortugero, Costa Rica is only approximately .02 cm/month. A similar slow growth rate of .04 cm per month (range .01 to .12 cm/month) has been found for 17 females nesting at French Frigate Shoals after intervals of 24 to 75 months. Consequently, some females (and presumably males) appear to mature at a small size and then reach a large size after many years of slow growth while other females do not mature until reaching a large size. Turtles that mature at 81 cm in the Hawaiian Archipelago would require an additional 23 years to reach the mean size of 92 cm. However, the size of adult green turtles is believed to be more heavily influenced by differences in maturation size than by growth (Carr and Goodman 1970), and most Hawaiian *Chelonia* that are 92 cm (or larger than 81 cm) probably grew to that size before achieving maturity (Table 2).

The natural growth rates of immature Hawaiian *Chelonia* less than 35 cm cannot be determined at the present time due to an absence of human contact with this size category following the departure of hatchlings

GEORGE H. BALAZS

Table 3. Summary of growth rates reported by other workers for naturally occurring immature green turtles

Location and reference	Number tagged	Type of measurement	Growth rate cm per month			Size range, cm	Interval in months
			Mean	Range	N		
Virgin Islands 18°20′N, 64°55′W Schmidt (1916)	65	curved carapace length	.43	.10–.69	8	29–57	3.5–11
West Florida 28°54′N, 82°35′W Carr and Caldwell (1956)	43	straight carapace length	——	.75–5.26[a]	1	44–58	3–3.5
Bermuda 32°20′N, 65°45′W Burnett–Herkes (1974)	19	straight carapace length	.04[b]	——	2	<50	12–17
Heron Island 23°27′S, 151°57′E Limpus (1979)	——	curved carapace length	——	.05–.27	45	40–90	<51

a. Range of possible growth rates — loss of tag prevented individual identification
b. Mean includes two naturally occurring turtles and one headstarted turtle

from the nesting beaches at French Frigate Shoals. During this unknown time period, the turtles are thought to be living in the open ocean where they feed on invertebrates occurring at or near the surface. In pelagic waters surrounding the Hawaiian Archipelago this could include *Physalia, Velella, Janthina,* the megalops stage of some portunid crabs, and immature individuals of certain oceanic squids that come to the surface at night in large numbers (*Symplectoteuthis oualiensis, Onychoteuthis banski,* and *Hyaloteuthis pelagica,* for example). Carnivorous foraging habits of this nature should produce growth rates that exceed those found at coastal areas where mostly algal food sources are utilized. In captivity, Hawaiian *Chelonia* have been found to require at least 19 months to grow from hatchlings to 35 cm.

The differences in growth rates found between the study areas in the Hawaiian Archipelago are thought to be a function of the sources and abundance of acceptable food. At Kau, where the most rapid growth has been recorded, dense pastures of the principal food source, *Pterocladia capillacea,* are present along the coastline. In the Northwestern Hawaiian Islands this is a rare species known to occur only in small quantities at Lisianski. Three other algal species, *Caulerpa racemosa, Turbinaria ornata,* and *Spyridia filamentosa,* identified as principal food sources in the Northwestern Hawaiian Islands, have never been found as dietary components in the main islands, even though they occur at a number of locations. This would suggest that green turtles in the Northwestern Hawaiian Islands feed on these 3 species out of necessity due to an absence or limited supply of other more desirable algae. The genus *Caulerpa* contains the toxic constituents caulerpicin and caulerpin. Caulerpicin can produce symptoms in humans similar to ciguatera fish poisoning, and caulerpin is toxic to mice (Doty and Aguilar-Santos 1966, 1970). Both of these compounds can be transferred along the food chain and concentrated in the process by certain herbivores. However, the effects on green turtles, if any, are presently unknown. *Caulerpa* has also been reported as a food source of green turtles in the Galapagos Islands (Pritchard 1971), South Africa (Hughes 1974), Aldabra Atoll (Frazier 1971), and Rose Atoll (Girard 1858).

Seawater temperature would be expected to exert some influence on growth at the study areas, but this is not evident based on the available data. At Kure and Midway, and probably extending to the southeast as far as Lisianski, mean monthly sea-surface temperatures range from a low of 20.5° C during February to a high of 26.2° C during August and September (Seckel 1962). This is 2 to 3° C cooler than at French Frigate Shoals (range of monthly means 23.8–28.3° C) where similar slow growth rates have been recorded. In contrast, sea-surface temperatures at French Frigate Shoals closely resemble those in the main islands where considerably faster growth has been found at Kau and Bellows.

Further evidence that food is the limiting factor for growth at some study areas, rather than seawater temperature or other environmental factors, has resulted from the recapture of a 46-cm green turtle raised from a hatchling in captivity and released at French Frigate Shoals. This turtle established a home range near Tern Island where it was regularly fed fresh fish scraps (*Caranx ignobilis, Caranx melampygus*) by personnel of the U.S. Coast Guard Loran Station. Over an 8-month period prior to the turtle's disappearance, a growth rate of .71 cm per month was recorded. This is far greater than the growth rates recorded for turtles feeding on natural food sources at French Frigate Shoals and, in fact, the most rapid growth thus far documented for a Hawaiian green turtle living in the wild. In assessing the growth rates and adaptability of captive-raised turtles returned to the wild (headstarting) the need exists to determine if food sources other than those used by naturally occurring turtles are being exploited.

The toxic properties of corroding Monel tags on turtles did not appear to be a factor affecting growth rates at the study areas where they were applied. Both Monel and Inconel tags have been used at Kau, French Frigate Shoals, and Midway. No relationship has been found at these locations between the type of tag and rate of growth. Only Inconel tags have been used at Bellows, Necker, Lisianski, and Kure. Nevertheless, the introduction of heavy metals into a turtle's system from Monel tags corroding at the unhealed piercing site could result in long-term adverse effects, in addition to the eventual loss of the tag.

The slower growth rates, and in many cases apparent cessation of growth, found at the study areas in the Northwestern Hawaiian Islands could have far reaching implications with respect to mortality rates of immature turtles and recruitment to the breeding colony. Tiger sharks (*Galeocerdo cuvier*) are essentially the only known natural predators of Hawaiian *Chelonia* at resident foraging areas throughout the archipelago (Balazs 1979b). At the locations where slower growth takes place, immature turtles would be exposed to this predation as small turtles for longer periods of time, thereby resulting in comparatively higher mortality rates. This is assuming, of course, that an increase in protection from tiger shark predation is afforded as a turtle grows larger. While such an inherent protective mechanism seems plausible and has been widely accepted as fact (Hirth 1971), it should be noted that the remains of full-size adults, in addition to immature individuals, have been periodically found in tiger sharks captured at French Frigate Shoals. If higher mortality rates of immature turtles do in fact occur in areas where slower growth takes place, then lower rates of recruitment of adults to the breeding colony would be expected. Many years may therefore be required for some green turtle populations to build up large breeding colonies due to these low adult recruitment rates and the protracted ages at maturity. Furthermore, these populations would be more susceptible to overexploitation and less able to undergo recovery once such declines have taken place.

Acknowledgments

This research was jointly funded by the State of Hawaii, Office of the Marine Affairs Coordinator, and the University of Hawaii Sea Grant College Program under Institution Grant Numbers 04-7-158-44129, 04-8-MOI-178 and NA79AA-D-0085 (National Oceanic and Atmospheric Administration, Office of Sea Grant, Department of Commerce). I would also like to gratefully acknowledge the support and assistance of the U.S. Fish and Wildlife Service, the National Marine Fisheries Service (Honolulu Laboratory), the Hawaii State Division of Fish and Game, the Fourteenth U.S. Coast Guard District, the Fifteenth Air Base Wing of the U.S. Air Force, the U.S. Naval Station Midway, the Koral Kings Diving Club at Midway, the New York Zoological Society, and A. K. H. Kam and A. L. Howard. This is contribution 574 of the Hawaii Institute of Marine Biology and conference paper UNIHI-SEA-GRANT-CP-80-03 of the University of Hawaii Sea Grant College Program.

Literature Cited

Balazs, G. H.

1976. Green turtle migrations in the Hawaiian Archipelago. *Biological Conservation* 9:125-40.

1977. Comments on Inconel tags. *IUCN/SSC Marine Turtle Newsletter* 2:7-8.

1979a. Growth, food sources and migrations of immature Hawaiian *Chelonia*. *IUCN/SSC Marine Turtle Newsletter* 10:1-3.

1979b. Synopsis of biological data on the green turtle in the Hawaiian Islands. National Oceanic and Atmospheric Administration, National Marine Fisheries Service, contract 79-ABA-02422.

In press. Field methods for sampling the dietary components of green turtles, *Chelonia mydas*. *Herpetological Review*.

Burnett-Herkes, J.

1974. Returns of green sea turtles (*Chelonia mydas* Linnaeus) tagged at Bermuda. *Biological Conservation* 6:307-8.

Carr, A., and D. K. Caldwell

1956. The ecology and migrations of sea turtles. 1. Results of field work in Florida, 1955. *American Museum Novitates* 1793:1-23.

Carr, A., and M. H. Carr

1970. Recruitment and remigration in a green turtle nesting colony. *Biological Conservation* 2:282-84.

Carr, A., and D. Goodman

1970. Ecologic implications of size and growth in *Chelonia*. *Copeia* 1970: 783-86.

Doty, M. S., and G. Aguilar-Santos

1966. Caulerpicin, a toxic constituent of *Caulerpa*. *Nature* 211:990.

1970. Transfer of toxic algal substances in marine food chains. *Pacific Science* 24:351-55.

Ehrhart, L. M., and R. G. Yoder

1978. Marine turtles of Merritt Island National Wildlife Refuge, Kennedy Space Center, Florida. In *Proceedings of the Florida and Interregional Conference on Sea*

GEORGE H. BALAZS

Turtles, ed. G. Henderson, pp. 25–30. St. Petersburg: Florida Department of Natural Resources.

Felger, R. S., K. Cliffton, and D. Cornejo

1978. Conservation of the sea turtles of the Pacific coast of Mexico. Research proposal supported by the World Wildlife Fund U.S.A.

Frazier, J.

1971. Observations on sea turtles at Aldabra Atoll. *Philosophical Transactions of the Royal Society of London,* B, 260:373–410.

Frick, J.

1977. Restocking green sea turtles. Newsletter, *Bermuda Biological Station for Research* 6:2.

Girard, C. F.

1858. *Herpetology—U.S. Exploring Expedition,* vol. 20. Philadelphia: Lea C. Blanchard.

Hirth, H. F.

1971. Synopsis of biological data on the green turtle, *Chelonia mydas* (Linnaeus) 1758. *FAO Fisheries Synopsis* 85:1:1–8:19.

Hughes, G. R.

1974. The sea turtles of South East Africa. 2. The biology of the Tongaland loggerhead turtle *Caretta caretta* L. with comments on the leatherback turtle *Dermochelys coriacea* L. and the green turtle *Chelonia mydas* L. in the study region. *Investigational Report of the Oceanographic Research Institute* 36:1–96.

Limpus, C.

1979. Notes on growth rates of wild turtles. *IUCN/SSC Marine Turtle Newsletter* 10:3–5.

Pritchard, P.C.H.

1971. Galapagos sea turtles—preliminary findings. *Journal of Herpetology* 5:1–9.

Rebel, T. P.

1974. *Sea Turtles and the Turtle Industry of the West Indies, Florida and the Gulf of Mexico.* Miami, Florida: Univ. of Miami Press.

Schmidt, J.

1916. Marking experiments with turtles in the Danish West Indies. *Meddelelser Fra Kommissionen For Havundersogelser* 5:1–26.

Seckel, G. R.

1962. Atlas of the oceanographic climate of the Hawaiian Islands region. *Fishery Bulletin* 61:371–427.

Larry Ogren
National Marine Fisheries Service, NOAA
Panama City Laboratory
3500 Delwood Beach Road
Panama City, Florida 32407

Charles McVea, Jr.
National Marine Fisheries Service, NOAA
Pascagoula Laboratory
P. O. Drawer 1207
Pascagoula, Mississippi 39567

Apparent Hibernation by Sea Turtles in North American Waters

ABSTRACT

Only 2 documented occurrences of aggregations of torpid sea turtles overwintering in North American waters have been reported. These areas are Baja California, Mexico, and Cape Canaveral, Florida, and include populations of the Pacific green turtle or black turtle (*Chelonia mydas aggassizi*) and the Atlantic loggerhead (*Caretta caretta*), respectively. Fishermen have also reported that hibernation by sea turtles apparently exists elsewhere in the Gulf of Mexico at approximately the same latitude, 29° N. These observations remain to be confirmed but should be given serious consideration, especially those on the overwintering populations of Kemp's ridley (*Lepidochelys kempi*) and Atlantic green turtles (*Chelonia m. mydas*) at Cedar Key, Florida.

It is evident that not all individuals within a population of sea turtles or residing in a particular geographic area hibernate in response to low water temperature. Dormancy occurs at temperatures below 15° C, both in the Baja California population of green turtles and the Cape Canaveral population of loggerhead sea turtles. The Cape Canaveral sea turtles do not appear to hibernate every year. Observations of cold-water stunning of sea turtles in the lagoon systems of Florida's central east coast suggests that the lower lethal temperature may occur below 8° C.

Additional information is needed to determine the degree and extent of this little-known aspect of sea turtle life history.

Introduction

Winter dormancy or aggregations of torpid sea turtles has been reported by Felger, Cliffton, and Regal (1976), and Carr, Ogren, and McVea (1980) for the Pacific green turtle or black turtle (*Chelonia mydas agassizi*) and the Atlantic loggerhead (*Caretta caretta*), respectively. Much earlier, Carr and Caldwell (1956) mentioned fishermen's reports of overwintering populations of Atlantic green turtles (*Chelonia m. mydas*) and Kemp's ridley turtle (*Lepidochelys kempi*). Scuba divers with the Georgia Office of Coastal Resources have observed lethargic sea turtles during winter months off the Georgia coast on reefs at depths of 30 to 36 meters (J. Richardson, personal communication). More recent reports by fishermen encountering lethargic, mud-covered sea turtles in their trawls exist for the northern Gulf of Mexico but remain to be documented. Thus, the only documented records of apparent hibernation by sea turtles are for the Baja California population of green sea turtles and the Cape Canaveral, Florida, population of loggerhead sea turtles (Felger, Cliffton, and Regal 1976; Carr, Ogren, and McVea 1980). In the light of these recent discoveries, serious consideration should be given to the reports by fishermen of Cedar Key, Florida, of overwintering green and ridley turtles in that area. More evidence will be required to document other reported occurrences of apparent hibernation elsewhere in the northern Gulf of Mexico and off the Georgia coast.

Indications are that all individuals within a population of sea turtles or residing in a particular geographic area do not hibernate in response to periods of low water temperatures (Carr, Ogren, and McVea 1980). During these periods, some individuals apparently migrate to warmer latitudes or depths. Dormancy occurs at temperatures below 15° C (Felger, Cliffton, and Regal 1976; Carr, Ogren, and McVea 1980). If the temperature drops much lower, however, and if suitable bottom type and depths for seeking refuge are lacking locally, or are too distant or blocked by physical barriers, cold-stunning or immobilization and death occur. This latter phenomenon was reported in detail by Ehrhart (1977 and 1978) and occurred in the Mosquito Lagoon–Indian River estuarine system on Florida's east coast during the winter of 1976–77 and to a lesser extent in 1977–78. The majority of sea turtles affected were immature greens; the others were loggerheads except for a single Kemp's ridley. The low temperature of 4° C was recorded for this shallow bay area, but it was believed that the lethal temperature was reached at a somewhat higher value but below 8° C (Ehrhart 1978). Schwartz (1978), in temperature tolerance observations performed under semicontrolled conditions, for example, concrete tanks and ambient seawater temperatures, determined that death occurred for 3 species of sea turtles exposed to 4° C to 5° C for 12 to 24 hours. Wilcox (1898) also reported cold-stunning and mortality of sea turtles in the Indian River area during the unusually cold winter of 1894–95. Another very cold winter in 1899, along the Texas coast, resulted in an almost total loss of the green sea turtle net fishery (H. Hildebrand, Texas A&I Biological Station, 19 September 1978, personal communication).

The occurrence of turtle cold-stunning and subsequent mortality within the principal nesting, foraging, or developmental areas for the loggerhead, green, and Kemp's ridley sea turtles does little to reinforce our belief in the apparent hibernation by these sea turtles——in fact, they suggest the opposite. However, it is important to note that these mortalities have occurred in shallow estuarine waters, such as those of Florida's central east coast. Cold fronts or "northers" pass through these areas during the winter months and lower the water temperature rapidly. Escape to deeper and warmer waters offshore is blocked by long barrier islands, and exposure to numbing and sometimes lethal temperatures results. Not all the winters are severe, however, and exposure to this limiting factor is infrequent. Apparently, some of the turtles that occupy this habitat have not evolved a strategy to avoid low temperatures.

Offshore, where the only 2 records of apparent hibernation have been reported, conditions are different from the coastal lagoons. Sea turtles occupying offshore habitats have a ready access to deeper and warmer waters at this latitude (29° N) during unusually cold winters. Water depths and minimum temperature observed for torpid green turtles in Baja California in January 1975, were 8 to 10 m and 14° C (Felger, Cliffton, and Regal 1976). Similarly, torpid loggerhead turtles were found in the Port Canaveral ship channel in February 1978, at a depth of 15 m and a mud substrate (bottom) temperature of 13.9° C (Carr, Ogren, and McVea 1980); the deep cloacal temperatures of the loggerheads were nearly identical (Figures 1 and 2).

In the nearshore waters of more northern latitudes off Georgia and the Carolinas the overwintering Atlantic loggerhead would be regularly exposed to seawater temperatures lower than 10° C. Offshore movements and southern migrations of these turtle populations during winter months to avoid these low temperatures are suggested by Scuba diver reports of turtles occupying deepwater reefs and fishermen's observations of active turtles at the surface along the western edge of the Florida Current (Gulf Stream). The absence of turtle sightings in coastal waters and bays and observations of torpid turtles in the deeper waters offshore support the belief that seaward movements of coastal-dwelling loggerheads occur at these latitudes during winter months. The extent of this behavioral response to cold water exhibited by sea turtles offshore the southeastern United States needs to be thoroughly explored.

Methods

Monthly trawl surveys of the Cape Canaveral ship channel and bight were begun in October 1978 by the National Marine Fisheries Service (NMFS) to determine if sea turtles would be found hibernating as they apparently had the preceding winter. These monthly

LARRY OGREN

surveys were continued through the summer of 1979. In February 1979, NMFS initiated additional trawl surveys of potential overwintering sites in selected bays, nearshore sloughs, channel entrances to bays, and channels that transect shallow lagoons in coastal waters from Florida to South Carolina.

Winter cruises of Florida waters included both the Gulf of Mexico and Atlantic coasts. Ninety-seven trawl stations were made from Cedar Key on the west coast south to Sanibel Island; Cape Sable to Key West; and Jupiter Island to Ponce de Leon Inlet on the east coast (including Mosquito Lagoon and the Cape Canaveral bight). The trawl tow times were limited to 10 to 30 minutes duration to prevent the accidental drowning of any captured turtles.

The survey of Georgia and South Carolina coastal waters was contracted out to Southeastern Wildlife Services, Inc. of Athens, Georgia. Despite the extremely foul weather conditions during the survey period and the wreckage and rock strewn trawling stations selected for investigation, 49 stations were completed. The main objective of this survey was to search for sea turtles that might be seeking refuge from low temperatures in the deeper channels leading to the bays——potential habitats that might be similar to the Cape Canaveral hibernaculum. Man-made channels were primary objectives of the survey. The area surveyed included 8 study sites: St. Mary's River entrance, Georgia-Florida; St. Andrew Sound, Georgia; St. Simon's Sound, Georgia; Savannah River, South Carolina-Georgia; Calibogue Sound, South Carolina; Port Royal Sound, South Carolina; Cooper River, South Carolina; and Winyah Bay, South Carolina.

Results from these trawl surveys, the monthly Cape Canaveral investigations, and a winter survey of coastal habitats from Florida to South Carolina are summarized below.

Results

In the survey of Florida's coastal waters 28 turtles were caught; all but 2 were captured in the Cape Canaveral area. These 2 exceptions were a loggerhead from Ponce de Leon Inlet and a juvenile green turtle from Florida Bay. None of the turtles caught appeared to be torpid or to have been in a hibernating state. Their deep cloacal temperatures ranged from 17.1° C to 18.7° C. The winter of 1978–79 was not as severe as the preceding ones. Difficulty in pulling the net at many of the selected stations may have resulted in the low capture rate outside the Canaveral area. Submerged vegetation, an abundance of sessile, attached benthic organisms, such as sponges and tunicates, and rock (limestone) outcrops frequently hampered the collecting effort on both coasts and the Keys.

The survey of Georgia and South Carolina coastal

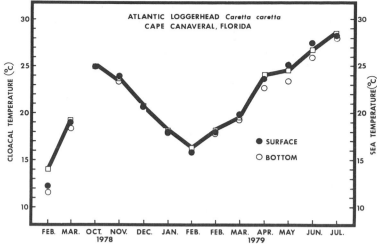

Figure 1. Mean monthly deep cloacal temperatures (open squares) of loggerhead sea turtles (*Caretta caretta*) captured by trawling in the Port Canaveral ship channel (N = 366). Circles are seawater temperatures.

habitats yielded no evidence that hibernating sea turtles occur in the nearshore environment of Georgia-South Carolina despite the fact that the observed winter seawater temperatures fall mostly within and below the range of temperatures reported by Felger, Cliffton, and Regal (1976) and Carr, Ogren, and McVea (1980). The 49 station temperatures recorded in this survey for these latitudes ranged from 7.8° C to 17.0° C. Only 3 stations were recorded above 14.5° C (Richardson and Hillestad 1979).

It is premature to draw any conclusions from these winter surveys because of vast coastal and estuarine areas that remain to be explored. Some of the likely or potential hibernacula located in deep channels, many dredged greater than normal depths of the surrounding waters, simply could not be sampled because of numerous wrecks, debris, and various other trawl obstructions. However, strong currents and extremely

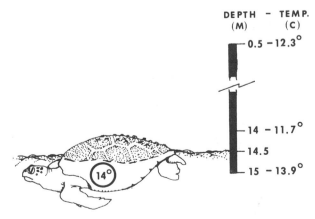

Figure 2. Sketch of a loggerhead sea turtle (*Caretta caretta*) buried in the Port Canaveral ship channel, February 1978. The pattern of staining on the turtles captured suggests this position.

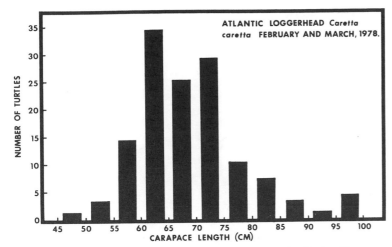

Figure 3. Maximum carapace lengths (straight line) of loggerhead sea turtles (*Caretta caretta*), captured in the Port Canaveral ship channel, February and March 1978 (N = 139).

low water temperatures encountered at the more northern stations in dredged channels may not afford a suitable refuge as is the apparent case at Cape Canaveral with its slight currents and warmer temperatures. In addition, none of the local fishermen and watermen interviewed from the areas sampled north of Florida could provide any information on the occurrence of sea turtles in nearshore waters during the winter months. However, many reported seeing active turtles at the surface along the western edge of the Florida Current (Gulf Stream) over 50 nautical miles offshore during the winter months.

As stated earlier, the monthly trawl surveys at Cape Canaveral were begun in October 1978, in the anticipation that overwintering turtles could be observed for the 1978–79 winter season. It was surprising to discover that loggerheads were not only abundant in the channel locations during February and March as they were the preceding winter, but they were also abundant October through January. In fact, many of them were stained black as described for the previous winter specimens (Carr, Ogren, and McVea 1980). Local fishermen reported that turtles were common in the channel during the summer, with many being captured in September 1978 during shrimping operations. The winter of 1978–79 was mild, however, and sea temperatures in the ship channel remained above 15° C. None of the 23 turtles captured in February, the coldest month, appeared as torpid as they had the previous winter. Their deep cloacal temperatures ranged from 15.7° C to 16.8° C (mean, 16.3° C); bottom seawater temperature was 16° C (Figure 1). There was no indication that they were hibernating in the bottom mud. Observations of turtles at the surface were frequently made at this time; the only difference in their behavior in February as compared to the other warmer months was that they were slower to respond to the approach

of the vessel.

Various physiological measurements were made on the trawl-captured specimens during this period to determine more precisely what the effects of low temperatures are on sea turtles. These studies were continued throughout the monthly surveys at Cape Canaveral. A preliminary report of these investigations (Lutz and Dunbar-Cooper 1979) suggests that the blood chemistry of the loggerhead is markedly sensitive to seasonal changes despite the fact that hibernating turtles were not found. Indication of an apparent preparatory state for hibernation was noted by a markedly reduced hematocrit, a fall in blood sodium and blood osmotic pressure, and a rise in blood magnesium. These studies are scheduled to continue.

The Cape Canaveral population of loggerhead sea turtles that was observed hibernating during winter of 1977–78 consisted primarily of subadults (85 percent) (Figure 3). Meristic data obtained from the loggerheads captured at Cape Canaveral during the monthly surveys that followed reinforced the above observation and, in addition, emphasized the bimodal length-frequency distribution that was suggested in the earlier data. The reason for the relatively small numbers of trawl captures of a particular size class (ca. 80 to 85 cm, carapace length) of loggerheads is not clearly understood at this time. The bimodality of length-frequencies from data collected on Atlantic loggerhead size distribution, other than from nesting females, was first observed in Georgia by Hillestad, Richardson, and Williamson (1977 and 1978). Their data were obtained from turtles caught by shrimp trawlers and dead specimens found stranded on the beach. A majority (88 percent) of these turtles were also classified as subadults.

Discussion

Many of the questions raised by Felger, Cliffton, and Regal (1976) and Carr, Ogren, and McVea (1980) concerning the degree and geographical extent of hibernation in sea turtles remain to be answered. Little is known about the various physiological responses and adaptations specifically required by sea turtles while submerged for long periods of time. The observed responses to decreased water temperatures appear to be inconsistent among individuals within a population or species and between age groups; some obviously migrate, some apparently hibernate. The occasional widespread mortality affecting sea turtle populations resulting from exposure to low temperatures is difficult to explain when part of the same population is apparently hibernating in the same general area. The observations of large aggregations of lethargic or moribund sea turtles passively floating miles offshore North Carolina and Florida during winter months are an enigma

(Schwartz 1978; Carr, Ogren, and McVea 1980).

The options available to sea turtles residing in the coastal waters of the United States in response to cold water are summarized in Carr, Ogren, and McVea (1980). The need still exists to investigate other areas, such as mud sloughs offshore, bottom disconformities, wrecks and reefs, deep channels in lagoons, and bay systems, to determine the extent and ecology of overwintering behavior. The locations of the few observations that have been made of hibernating sea turtles in North American waters may be significant and restricted to a narrow latitudinal zone (Figure 4).

Experience acquired during sea turtle surveys conducted over the past year necessitates some re-evaluation of our earlier statements concerning the supposedly unique appearance and behavior of turtles stated in Carr, Ogren, and McVea (1980). The degree of physical activity observed among individuals captured by trawls is 1 example. For this reason, reports of lethargic turtles being captured during winter months by trawlers are not necessarily related to the torpid condition of hibernating turtles. They may have been held underwater too long and become fatigued or comatose. The capture of mud-covered turtles and observations of mud plumes washing from the backs of turtles at the surface are additional examples, which, by themselves, indicate that the turtles are spending considerable time on the bottom. This information is important, however, when correlated with low water temperature, infrequent surface observations, and lethargic behavior. The degree of discoloration of the scutes and integument may be also a result of the time spent on the bottom. Black-stained individuals were collected from October 1978 to September 1979 at Cape Canaveral. The color appeared less intense for these turtles than those observed in February 1978, however. This may represent a combination of long-term exposure to bottom sediments during less-active periods in the winter months and repeated exposures during the rest of the year.

Difficulties in investigating potential and known sea turtle hibernacula are compounded not only by the unpredictability of the event but by sampling limitations. Use of passive gear, such as gill or tangle nets, would not be a successful method for capturing dormant turtles. Trawls, not less than 12 to 15 m headrope length, are very effective provided they can be used on unobstructed bottoms. When the trawling method is used, however, tows must be of less than 30 minutes to avoid stressing or drowning nonhibernating turtles. Use of Scuba may be excellent for observing offshore reefs and some inshore areas. In all cases, precautions must be taken to prevent mortality when capturing dormant turtles for study and tagging. Releasing these turtles immediately to sea might result in cold-stunning and death. Specimens captured under these conditions

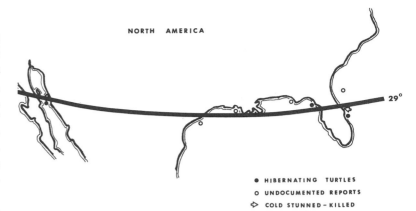

Figure 4. Latitudinal occurrences (including unconfirmed reports) of winter dormant or hibernating sea turtles in North American waters

may have to be held in tanks or pools at higher ambient temperatures for extended periods. This was done by Ehrhart (1977) after he had collected cold-stunned turtles from Mosquito Lagoon and Indian River.

The importance of acquiring additional information on this little known aspect of sea turtle life history cannot be over stressed. Such information would be extremely useful in decisions concerning the conservation of threatened and endangered populations, especially if dormant turtles are located in areas where man's activities could adversely affect them. For example, the Baja California population of green turtles has been severely depleted by divers who overfished dormant turtles during the winter (Felger, Cliffton, and Regal 1976). If knowledge of dormancy had been known, regulations might have prevented overfishing. Another example is the vulnerability of dormant turtles to trawling. This knowledge was applied by NMFS to protect overwintering loggerheads in the Port Canaveral navigation channel. Trawling was declared unlawful at this locality from November 1978 to March 1979.

Further protection of these endangered and threatened species of sea turtles will result from conservation measures based on new data relating to the degree and extent of hibernation in temperate waters.

Acknowledgments

We thank Frederick Berry and Wilber Seidel for their generous program support and field assistance. Alan Bunn and Douglas Harper collected the information on the 1979 winter trawl surveys of Florida coastal habitats. Thanks are also due to all the collaborators and students who helped us in many ways aboard the FRV LADY WEESA.

This paper is Contribution No. 80–21PC, Southeast Fisheries Center, National Marine Fisheries Service, NOAA, Panama City Laboratory, Panama City, Florida 32407.

Literature Cited

Carr, A., and D. Caldwell
1956. The ecology and migrations of sea turtles, 1. Results of field work in Florida, 1955. *American Museum Novitates* 1793:1–23.

Carr, A., L. Ogren, and C. McVea
1980. Apparent hibernation by the Atlantic loggerhead turtle off Cape Canaveral, Florida. *Biological Conservation* 19:7–14.

Ehrhart, L. M.
1977. A continuation of base-line studies for environmentally monitoring space transportation systems at John F. Kennedy Space Center. VI. Threatened and endangered species of the Kennedy Space Center. Annual report to NASA.

Felger, R. S., K. Cliffton, and R. Regal
1976. Winter dormancy in sea turtles: independent discovery and exploitation in the Gulf of California by two local cultures. *Science* 191:283–85.

Hillestad, H. O., J. Richardson, and G. Williamson
1977. Incidental capture of sea turtles by shrimp trawlermen in Georgia. Final report to NMFS. 106 pp. Contract 03–7–042–35129 Southeastern Wildlife Services, Inc., Athens, Georgia.

1978. Incidental capture of sea turtles by shrimp trawlermen in Georgia. 32nd Annual Conference of the Southeast Association of Fish and Wildlife Agencies. Hot Springs, Virginia, 5–8 November 1978, 24 pp.

Lutz, P. L., and A. Dunbar-Cooper
1979. Physiological studies on loggerhead sea turtles caught on shrimp trawl surveys in the Cape Canaveral ship channel, Florida, December 1978 to April 1979. Final Report to NMFS, 21 pp. Contract FSE 43–9–12–40. Division of Biology and Living Resources, Univ. of Miami, Rosenstiel School of Marine and Atmospheric Science, Miami, Florida.

Richardson, J., and H. O. Hillestad
1979. Survey for wintering marine turtles in South Carolina and Georgia. Final Report to NMFS, 50 pp. Contract 03–78–D08–0062. Southeastern Wildlife Services, Inc., Athens Georgia.

Schwartz, F. J.
1978. Behavioral and tolerance responses to cold water temperatures by three species of sea turtles (Reptilia, Cheloniidae) in North Carolina. *Florida Marine Research Publication* 33:16–18.

Population Dynamics

Anne Meylan
Department of Zoology
University of Florida
Gainesville, Florida 32611

Estimation of Population Size in Sea Turtles

ABSTRACT

Population censusing is a valuable conservation tool for monitoring the stability of sea turtle populations and assessing the efficacy of conservation and management practices. Seasonal and ontogenetic changes in habitat occupation make sea turtle populations particularly difficult to census. At present, the most feasible approach is to monitor the arrivals of females at nesting beaches. Aerial and ground surveillance techniques are here evaluated. Large annual fluctuation in nesting arrivals necessitates multiyear sampling of beaches to derive average values for population estimates. The apparent dichotomy within populations of cyclical remigration vs 1-season nesting must be taken into account in relating yearly nesting totals to the total female reproductive population. Ignorance of natural sex ratios and age structure prevents extrapolation of nesting beach censuses to total population.

Introduction

Seasonal and ontogenetic changes in habitat occupation make seaturtle populations particularly difficult to census. After leaving the nesting beach, hatchlings remain virtually out of human sight for a year or more, a phenomenon that has been called the lost year. The developmental habitats of juveniles are also poorly known and, for some species (for example, *Dermochelys coriacea, Lepidochelys olivacea*), have not even been identified. As adults, itinerant feeding habits may dictate wide and unpredictable distribution which makes censusing impossible. Species with more or less fixed food resources such as the green turtle (*Chelonia mydas*) can be censused on resident foraging grounds, but only with great logistical difficulty; in any event, results are of limited use in estimating total population size because most populations are distributed on a large number of feeding grounds, and many of these have not been identified.

The nesting beach is perhaps the only practical place

135

to conduct a census of sea turtles. An estimate of the number of yearly arrivals of females can be made for most species without excessive logistical difficulty, and this figure can be used as an index of population size, provided that proper caution is exercised in making the calculations. Essentially, three steps are involved: 1) determining the total number of female turtles nesting in a season; 2) relating the yearly number of nesting females to the total number of reproductive females in the population; and 3) relating the total number of reproductive females in the population to the total number of turtles of both sexes and all age classes.

Estimating Seasonal Nesting Totals

Aerial surveys of tracks, ground patrols or a combination of these methods can be employed to determine the number of turtles nesting on a beach during a season. Aerial surveying offers the advantage of covering large stretches of beach in a short time and makes it possible to survey beaches that are otherwise inaccessible. The main disadvantages are the relatively high cost of operation and the risk of inaccuracy inherent in incomplete sampling methods. The number of turtles arriving nightly on a beach fluctuates strongly, and unless aerial surveys are conducted on an intensive basis, there is a good chance that serious errors in estimating the number of seasonal arrivals will be committed. Stancyk, Talbert, and Miller (1979) recommended interflight intervals of the order of 5 days, based on their surveys of loggerheads (*Caretta caretta*) nesting in South Carolina.

A second potential source of inaccuracy in aerial surveys is the estimation of the number of tracks seen during any one survey. Very dense accumulations of tracks are difficult to decipher from a plane, and, if several species are nesting on the same beach, it is hard to differentiate among them. Smaller species leave only a light track, and estimates may tend to be conservative. LeBuff and Hagan (1978) found that tracks were sometimes obliterated by human recreation and motor vehicle operation. The visibility of tracks from an airplane varies with the time of day, making it necessary to standardize the timing of flights. Track longevity varies with both weather and tides, and it is often difficult to judge how many nights' emergences are represented in a single tally. In certain environments, tracks may persist for months, and thus the potential exists for gross overestimation of the number of turtles nesting on a beach. Stancyk, Talbert, and Miller (1979) found that experience was a definite factor influencing the ability of observers to recognize and correct for these variables. These authors emphasized the need for concurrent ground surveys to provide an estimate of the accuracy of aerial counts. With the "ground truth," aerial estimates can be appropriately corrected.

In order to use aerial track counts to estimate the total number of females nesting within a season, it is necessary to know both the proportion of tracks that represent nesting emergences and the average number of nests that each female makes. Non-nesting emergences, sometimes called half-moons or false crawls, may constitute a significant portion of the tracks observed. Servan (1975) reported that 47 percent of the green turtles nesting at Europa Island emerged at least twice before successfully completing a nest. At Ascension Island, Mortimer (personal communication) observed that approximately one-third of all emergences did not result in nesting. One approach in determining actual nesting totals is to tally only tracks associated with body pits. This is not a particularly accurate solution because some species, for example the hawksbill (*Eretmochelys imbricata*), prefer to nest under vegetation and their body pits are therefore not always visible from the air. Also, the presence of a body pit does not always indicate that eggs were deposited. A better alternative is to conduct ground sampling, which allows the opportunity for careful examination of individual tracks. Determination of the average number of nests a female makes during a season can best be accomplished by a tagging program. Data from studies conducted elsewhere can be used, but variation of these parameters among populations must be considered.

As an alternative to aerial surveys, daily ground monitoring represents a substantially larger investment of effort and funds. It provides, however, far more accurate data on the number of yearly arrivals at a nesting beach. The proportion of tracks resulting in nests can be accurately determined by ground inspection and, if a tagging program is conducted, the average number of nests per female can be determined. The biggest drawback of ground monitoring is the limitation that it imposes on the extent of beach that can be covered. Unless arrivals of all nesting females of a population can be monitored, additional extrapolations have to be made to calculate the seasonal total.

There is mounting evidence that, whatever the censusing method employed, estimation of seasonal nesting totals should be based on data from more than 1 season. Tremendous fluctuation in the number of yearly arrivals of nesting turtles has been observed at rookeries around the world. Limpus (this volume) reported that the number of green turtles arriving nightly at Raine Island, in eastern Australia, dropped from over 11,000 to about 100 in 2 consecutive years. Parallel fluctuations were noted throughout the Great Barrier Reef Provence. Similar phenomena have been recorded at Tortuguero, Costa Rica, where a 7-fold fluctuation in the number of arrivals has been observed on the study beach in consecutive years (Figure 1) (Carr, unpublished data). During the 1976–79 seasons, the 2 highest and 2 lowest seasonal nesting totals in

ANNE MEYLAN

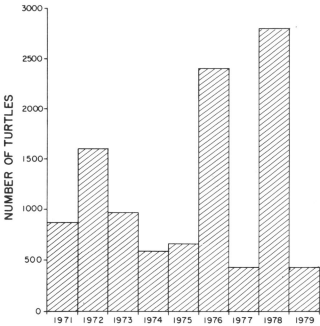

Figure 1. Yearly changes in the number of female green turtles observed on the 8 km study beach, Tortuguero, Costa Rica, 1971–1979.

of these females are destined to nest in future seasons. Hughes (this volume) showed that low remigration percentages are worldwide and discussed possible causes for the phenomenon. At Tortuguero, tag loss, incomplete beach surveillance, and mortality away from the nesting beach seem inadequate to account for the deficit. Because there is fairly conclusive evidence that these turtles are not nesting on other beaches, their continued status as reproductive members of the population seems doubtful. Although the assumption that they nest in only 1 season has puzzling biological implications, it is perhaps justified by the complete failure of long-term projects, such as Tortuguero, Costa Rica, and Tongaland, South Africa, to produce evidence indicating otherwise. A refinement of the population estimation method suggested by Carr, Carr and Meylan (1978) would be to incorporate this assumption by applying the remigration interval formula to only the remigrant fraction of the yearly nesting total. To the figure thus derived would then be added the number of turtles that are nesting for the first time that season. This would yield an instantaneous estimate of the number of reproductive females in the population.

the history of the 25-year project were recorded. Such fluctuation does not appear to be restricted to green turtle populations. Hughes (1974), Richardson and Richardson (1978) and Davis and Whiting (1977) noted large fluctuations in the number of nesting loggerheads (*Caretta caretta*) at rookeries in South Africa, and in Georgia and southern Florida in the United States. In none of these cases have the causes of the fluctuations been identified. Until the underlying biological basis is understood, nesting beaches will have to be monitored for a number of seasons and an average seasonal nesting total obtained if reliable population estimates are to be made.

Estimating the Female Reproductive Population

The next step in calculating population size is to convert the estimate of the number of females nesting within a season into the total number of reproductive females in the population. Females of most marine turtle species do not nest every year. Tagging studies have documented remigration intervals predominantly of the order of 2 to 4 years for most species. Carr, Carr, and Meylan (1978) provided an equation that weights the various remigration interval frequencies characteristic of a particular population, to take into account females that are of reproductive age but not present at the nesting beach. An admitted defect of the equation is that it assumes that all females of the population adhere to the established cyclical pattern of remigration, when actually fewer than 50 percent

Estimating Total Population

The final step in calculating population size is to relate the total number of reproductive females to the total number of turtles in the population. At present the tremendous gaps in knowledge of sea turtle biology make this impossible. One obstacle is ignorance of the natural sex ratios of marine turtle species. These badly needed data may soon be available as a result of newly developed endocrinological and histological sexing methods (Owens et al., 1978; Mrosovsky and Yntema, 1980). A more difficult problem is our lack of knowledge of the age structure of sea turtle populations. Without knowing the relationship between the size of the reproductive contingent of a population and that of its various immature categories, estimation of total population size based on nesting estimates is impossible.

The need for basic research in sea turtle population biology is critical for conservation. Population censuses are the only tool by which the stability of populations can be monitored and dangerous trends detected. Censuses can serve both to measure the negative effects of factors such as commercial exploitation and habitat alteration, and to evaluate the efficacy of conservation and management practices.

Literature Cited

Carr, A. F.; M. H. Carr; and A. B. Meylan
1978. The ecology and migrations of sea turtles, 7. The West Caribbean green turtle colony. *Bulletin of the*

American Museum of Natural History 162:1–46.

Davis, G. E., and M. C. Whiting
1977. Loggerhead sea turtle nesting in Everglades National Park, Florida, U.S.A. *Herpetologica* 33:18–28.

Hughes, G. R.
1974. The sea turtles of South-East Africa. 2. The biology of the Tongaland loggerhead turtle *Caretta caretta* L. with comments on the leatherback turtle *Dermochelys coriacea* L. and the green turtle *Chelonia mydas* L. in the study region. *Investigational Report of the Oceanographic Research Institute*, Durban, South Africa, 36:1–96.

LeBuff, C. R., and P. D. Hagan
1978. The role of aerial surveys in estimating nesting populations of the loggerhead turtle. *Florida Department of Natural Resources Marine Research Laboratory* 33:31–33.

Mrosovsky, N., and C. L. Yntema
1980. Temperature dependence of sexual differentiation in sea turtles: implications for conservation practices. *Biological Conservation* 18(4):271–280.

Owens, D. W., J. R. Hendrickson, V. Lance, and I. P. Callard
1978. A technique for determining sex of immature *Chelonia mydas* using a radioimmunoassay. *Herpetologica* 34:270–73.

Richardson, J. I., and T. H. Richardson
1978. Population estimates for nesting female loggerhead sea turtles (*Caretta caretta*) in the St. Andrew Sound area of southeastern Georgia, U.S.A. *Florida Department of Natural Resources Marine Research Laboratory* 33:34–38.

Servan, J.
1975. Ecologie de la tortue verte à l'île Europa (Canal de Mozambique). Ph. D. dissertation, University of Paris.

Stancyk, S. E., O. R. Talbert, and A. B. Miller
1979. Estimation of loggerhead turtle nesting activity in South Carolina by aerial surveys. *American Zoologist* 19:952 (abstract).

ANNE MEYLAN

Stephen E. Stancyk
Department of Biology
and
Belle W. Baruch Institute for Marine Biology and Coastal
Research
University of South Carolina
Columbia, South Carolina 29208

Non-Human Predators of Sea Turtles and Their Control

ABSTRACT

Predators of marine turtles eat eggs, hatchlings, juveniles, and adults. The most important egg predators are ghost crabs (*Ocypode* spp.) and small mammals (like feral dogs, pigs, raccoons). Predation by ghost crabs may be heavy on some beaches, but mammals appear to be the most effective egg predators. Land-based predators on hatchlings include mammals, crabs and birds. Loss of hatchlings to these predators is slight compared to the impact of nearshore aquatic predators, including numerous fish, sharks, and invertebrates. Predators on juveniles and adults are chiefly sharks, especially *Galeocerdo cuvieri*, the tiger shark. Nearshore predation of hatchlings is assumed to be quite heavy, and adult predation is thought to be relatively light, but supporting data are incomplete or lacking.

Control methods include predator removal, deterrence, and protection of eggs and hatchlings. Hunting or trapping is effective, especially for small mammals, but requires regular intensive activity. Poisons or aversive chemicals are too dangerous to use or are ineffective. Egg and hatchling protection includes hatcheries, nest transplantation and offshore release. Hatcheries are effective, but require equipment, maintenance and regular monitoring. Hatch success is often lower than in nature. Nest transplantation is simpler and more natural, requiring less maintenance and monitoring. However, this new method needs to be tested on different beaches with a variety of predators. Offshore release is relatively new and untried, and should be done with extreme care to insure proper behavioral development of hatchlings.

Non-human predation is insignificant at most turtle rookeries relative to other mortality pressures impinging on marine turtle populations. However, when predation is heavy due to human interference or combines with other factors to reduce offspring production, control methods are recommended. Such methods can be effective and should be designed for individual situations within economic and manpower constraints. Controls should be performed with care, and careful records should be kept.

Introduction

Predation has been a natural mortality source affecting turtle populations throughout their evolution. Many aspects of the behavior and life histories of sea turtles (such as elaborate nest-covering, nocturnal hatchling emergence, protective sleeping positions, production of large numbers of offspring) can be viewed as adaptations to predation. Unless environmental changes gave predators advantages over the adaptations of the turtles, predation alone has probably not threatened turtle populations with extinction. However, ecosystem modifications are often caused by humans, either by enhancing survival of predators or by placing additional mortality pressures on turtle populations. When individuals that avoid predators die by other causes, predation can become an important component in a suite of mortality factors threatening a population. Today, largely because of human influence, this circumstance is occurring in several turtle populations. Examination of predation in turtle populations worldwide provides perspective, and a survey of control methods shows which methods are most promising and where further research is necessary.

Although humans may be considered natural predators of sea turtles in some areas, they are not included here. Feral animals such as hogs or dogs will be discussed, since they often replace or exacerbate existing predation patterns. Finally, predation relative to the different species of sea turtles will not be discussed, except where noted, because predation generally appears to follow prey availability regardless of species.

Non-Human Predators and Their Effects

Predators of sea turtles can be classified according to the age or life stage of their prey. Tables 1 to 3 list predators of eggs, hatchlings, and juveniles and adults, respectively. The tables attempt to summarize our knowledge of non-human predators, and should not be taken as complete listings of every known predator or locality where predation has occurred. In each class, predator effects may differ between localities and species.

Predators of Eggs

Table 1 shows that the most notable predators of sea turtle eggs are crabs and small to mid-size mammals. Other predators are incidental, local, or of minor importance. Beach-dwelling crabs, particularly members of the genus *Ocypode,* are widely accused of egg predation, probably because they inhabit most of the

beaches where sea turtles nest. They burrow into nests and devour or destroy various numbers of eggs, depending upon the degree of infestation (Hughes 1974, Diamond 1976, Fowler 1978, Hopkins et al. 1979). Schulz (1975) noted that they had a greater effect on the more shallow nests of *Lepidochelys* in Surinam. However, ghost crabs may provide access to turtle nests for secondary predators such as vultures, crows, or insects. In Colombia, for instance, Tufts (1972) found that fly larvae and foxes both attacked nests containing *Ocypode* burrows, and there are many instances of complete destruction of nests by secondary predators after access was provided by ghost crabs (Fowler 1978).

Ocypode depredation generally appears to have a minor effect on hatchling production. *Ocypode* attacked about 2 percent of surveyed nests on several South Carolina beaches (Hopkins et al. 1979 and unpublished). Fowler (1978) found that crabs burrowed into only 13 of 450 observed nests at Tortuguero, Costa Rica, and destroyed a maximum of 21 eggs in any nest. Hughes (1974) found no evidence of direct crab predation on eggs in Tongaland, South Africa, and Talbert (personal communication) found no destruction of eggs in a hatchery on Kiawah Island, South Carolina, despite the fact that several *Ocypode* burrows reached nests. However, there are instances where crab predation was very heavy. *Ocypode* was the most important non-human predator of turtle eggs at Buritaca, Colombia (Tufts 1972). Ghost crabs attacked 60.9 percent of the nests on Cape Romain, South Carolina, in 1939 (Caldwell 1959), where a reduction in the population of crab predators (raccoons, *Procyon lotor*) could have allowed increased crab predation. In Surinam, Hill and Green (1971) found that crabs attacked approximately 60 percent of all nests and typically destroyed 11.8 percent of the eggs in a clutch. However, Green (personal communication) noted that artifacts of sampling could have caused some of the attacks. Such examples are uncommon, and many citations of crab predation are too poorly quantified to ascertain effects on the turtle populations.

Some predators are important only in specific geographical areas or rookeries. Depredation by monitor lizards (*Varanidae*) is restricted to Africa, Australia, Malaysia, the Philippines, and areas around the Indian Ocean (Table 1). Bustard (1972) stated that monitor predation could be the reason why most Australian rookeries occur on islands, and Limpus (in press) found that there was almost complete destruction of nests by monitors on Lacey Island, on the Great Barrier Reef. However, this occurred on only 1 of many small islands on the reef, and the effects of monitors in most places are probably minor, although poorly quantified. Insects may be locally important predators at some rookeries. Green (Green and Ortiz, this volume) found a scarabid beetle, *Trox suberosus,* in the Galapagos which

STEPHEN E. STANCYK

Table 1. Non-human predators of sea turtle eggs

Predator	Locality (source)
Ants (*Dorylus* sp.)	Tongaland, South Africa (McAllister et al. 1965; Hughes 1972)
Flies[a] (esp. larvae)	Surinam (Tufts 1972); Oman (Ross, personal communication); Seychelles (Frazier, personal communication)
Trox suberosus (Scarabidae)	Galapagos Is. (Green and Ortiz, this volume).
Crabs (*Ocypode* spp.)	Southeastern United States (Caldwell 1959; Hopkins et al., 1979; Talbert et al., 1980, etc.) Mexico (Pritchard and Marquez 1973, Caldwell 1966) Tortuguero, Costa Rica (Fowler 1978) Surinam (Schulz 1975, Hill and Green 1971) Colombia (Tufts 1972) So. Yemen (Hirth and Carr 1970) Oman (Ross, personal communication) Australia (Bustard 1972) Seychelles (Honegger 1967, Frazier 1971, Diamond 1976) Sarawak and Malaya (Hendrickson 1958)
Snakes (*Boa,* elapids)	Latin America (Carr, personal communication)
Varanids	Tongaland (McAllister, Bass, and van Schoor 1965) Angola (Hughes, Hustley, and Wearne 1973) Australia (Bustard 1972) Great Barrier Reef (Limpus, in press) Sarawak and Malaya (Hendrickson 1958) Philippine Is. (Hirth 1971) West Pakistan (Minton 1966)
Birds[a]	
Black Vulture, Turkey Vulture	Costa Rica (Fowler 1978; D. Robinson, personal communication) Southeastern U.S. (Talbert, personal communication) French Guiana (Pritchard 1971)
Ibis	Seychelles (Honegger 1967)
Crows (various)	Seychelles (Honegger 1967) Southeastern U.S. (Talbert, personal communication)
Mammals	
Rats (*Rattus* spp.)	Seychelles (Honegger 1967) Malaya and Sarawak (Hendrickson 1958)
Opossums	Tortuguero (Carr 1956, 1967)

Table 1. Non-human predators of sea turtle eggs (cont.)

Predator	Locality (source)
Coatis	Tortuguero (Fowler, 1978)
Raccoons	Southeastern U.S. (Caldwell 1959; Klukas 1967; Gallagher et al. 1972; Worth and Smith 1976; Richardson 1978; Hopkins et al. 1979; Talbert et al. 1980; Stancyk et al. 1980)
Mongooses	Angola (Hughes et al. 1973) Australia (Bustard 1972)
Genets	Tongaland (McAllister et al. 1965) Kenya (Frazier, personal communication)
Feral cats	Seychelles (Frazier 1971)
White-lipped Peccary	Tortuguero (Carr 1956 1967)
Pigs, Hogs	Southeast U.S. (Richardson 1978) Galapagos (Hirth 1971) Costa Rica, Pacific Coast (Zahl 1973) Australia (Bustard 1972) Tortuguero (Carr 1956, 1967)
Jackals	So. Yemen (Hirth and Carr 1970)
Dingoes	Australia (Bustard 1972)
Foxes	So. Yemen (Hirth and Carr 1970) Southeast U.S. (Hopkins, personal communication) Australia (Bustard 1972) Oman (Ross, personal communication)
Coyotes	Mexico (Pritchard and Marquez 1973; Caldwell 1966)
Dogs, feral	Tongaland (McAllister et al. 1965) So. Yemen (Hirth and Carr 1970) Malaya and Sarawak (Hendrickson 1958) Galapagos (Hirth 1971) Seychelles (Honegger 1967) Tortuguero (Fowler 1978) Surinam (Schulz 1975) Oman (Ross, personal communication)

a. Secondary predators; not seen excavating nests.

entered up to 90 percent of the turtle nests on Quinta Playa Beach. Infested nests had a greatly reduced hatching success. Hughes (1972) found ants (*Dorylus* sp.) attacking nests in a hatchery.

The most destructive egg predators by far are small to medium-sized mammals. The species may vary, but common mammalian predators include mongooses, ge-nets, canids (coyotes, dingoes, dogs, foxes), procyonids (raccoons, coatimundis) and hogs (Table 1). These organisms are able to find turtle nests with ease, probably by detecting olfactory and other cues released by the nesting female (i.e., cloacal fluid) or pre-emergent hatchlings (Stancyk, Talbert, and Dean 1980). Although many authors cite mammal predation as being

STEPHEN E. STANCYK

important, the most complete studies are those of Fowler (1978) and Hopkins et al. (1979). Fowler found that the natural mammalian predator at Tortuguero, the coati, had a relatively small effect compared to dogs. However, a direct assessment of coati predation was complicated by overlapping predation by dogs and vultures.

In the southeastern United States the raccoon, *Procyon lotor,* can take up to 96 percent of the nests of *Caretta caretta* on some beaches (Klukas 1967; Davis and Whiting 1977; Hopkins et al. 1979 and unpublished; Talbert et al. 1980; Stancyk, Talbert and Dean 1980). Talbert et al. (1980) found that there was a 2.5- to 3-week lag period during which predation estimates varied from 15.6 to 26.7 percent compared to rates of 50 to 97 percent later in the season. A brief lag period between first nesting and first predation has also been noted by Hopkins et al. (1979) and Gallagher et al. (1972). Generally, nests which are attacked by raccoons suffer 100 percent mortality on the same or subsequent nights (Stancyk, Talbert and Dean 1980; Hopkins, personal communication). Predation by raccoons usually occurs soon after nests are deposited in the beach. Gallagher et al. (1972) found that 34 percent of 398 nests taken by raccoons at Hutchinson Island, Florida, were discovered within 48 hours of laying, and Davis and Whiting (1977) reported 87 percent first-night predation at Cape Sable, Florida. Hopkins et al. (1979) found first-night predation as high as 51 percent, and noted that predation rate drops as olfactory and visual cues fade. However, Hopkins et al. (1979) and Klukas (1967) noted that raccoon predation increased as nests neared hatching, a phenomenon also seen in foxes (Bustard 1972) and dogs (Fowler 1978). The reasons for this could be that pre-emergent clutches release cues of their own before emerging or that predator searches become more thorough and discover previously overlooked nests in the latter part of the nesting season, when early nests are about to hatch and freshly deposited nests are less abundant.

In many parts of the world, domestic or feral animals take the place of natural mammalian predators, often with severe effects on hatchling production. Bustard (1972) noted that hogs are extremely effective egg predators in Northern Australia. In Georgia, they can destroy all nests on a given beach (C. Blanck, personal communication). Domestic pigs are released on *Lepidochelys olivacea* nesting beaches in Costa Rica (J. Frick, personal communication; Zahl 1973). Dogs are a serious menace to eggs in most places where turtle rookeries and human settlements are close together (Honegger 1967, Hughes and Mentis 1967, Hirth and Carr 1970, Tufts 1972, Schulz 1975, Fowler 1978), but rigorous quantification of their effects is often lacking. However, Fowler (1978) found that 39.7 percent of 450 marked nest sites were destroyed by predators,

chiefly dogs, at Tortuguero. The percentage of nests initially destroyed by dogs was difficult to determine due to subsequent depredations by black vultures, coatis, and other dogs. Feral cats have also been cited as egg predators (Frazier 1971).

Studies of heavily utilized rookeries with a high percentage of small mammal predation have provided insight into some of the theories concerning sea turtle nesting ecology. Bustard (1979) postulated that "High-density nesting must also reduce overall percentage destruction of incubating eggs by nest predators. . .predators become satiated after devouring only a percentage of the production, the percentage decreasing as production increases." In South Carolina Hopkins et al. (1979 and unpublished) found that predation usually paralleled the spatial distribution of nesting, and densely nested areas or islands had the heaviest predation. There was no evidence of predator satiation although it might have occurred at some of the massive arribadas of *Lepidochelys* in the past.

Bustard (1979) also postulated that nests in less-dense rookeries might suffer greater mortality. But in South Carolina, raccoons appeared to concentrate where the rewards were greatest, and nests deposited on beaches underutilized by turtles could have less chance of being depredated. Hopkins et al. (1979) found that predation was lowest on North Island, where nesting was also lowest. Talbert et al. (1980) found a decreasing percentage of wild nests depredated as a larger proportion of all nests was moved to a hatchery. Although these findings differ with Bustard (1979), it should be remembered that predator densities in South Carolina today may be higher (due to human interference) than those under which *Caretta* reproductive patterns were evolving, and could represent an unnatural situation.

Predators of Hatchlings

Hatchling predators (Table 2) can be divided into land-based and aquatic groups. Land-based predators include mammals, birds, and crabs. Mammals have less impact on hatchlings than on eggs (Hopkins et al. 1979, Fowler 1978, Hughes 1974), possibly because hatchlings are available to them for a shorter time than the eggs, and cues left by the nesting female are long absent. As noted earlier, however, hatchlings themselves may provide cues for mammalian predators (Fowler 1978, Hopkins et al. 1979).

The most commonly mentioned predators of hatchlings on land are diurnal birds (especially vultures, frigate birds, gulls, and crows), but their role is probably overstated because most turtle hatchlings emerge at night. Limpus (1973) estimated that bird predation accounted for less than 2 percent of *Caretta* hatchlings which emerged at dawn, and less than 0.1 percent of all loggerhead hatchlings produced at Mon Repos,

Table 2. Predators of hatchling sea turtles[1]

Predator	Locality (source)
Crabs	
Ocypode spp.	Southeast United States (Caldwell 1959)
	Sarawak and Malaya (Hendrickson 1958)
	Mexico (Caldwell 1966, Pritchard and Marquez 1973)
	South Africa (McAllister, Bass, and van Schoor 1965; Hughes 1974)
	South Yemen (Hirth and Carr 1970)
	Surinam (Hill and Green, 1971)
	Australia (Bustard 1972)
	Seychelles (Diamond 1976; Honegger 1967)
Hermit crabs (*Coenobita* spp.)	Europa and Tromelin Is. (Hughes 1974)
	Seychelles (Honegger 1967)
Coconut crabs (*Birgus* sp.)	Seychelles (Honegger 1967)
Sharks	Southeast United States (Caldwell 1959)
	Surinam (Schulz 1975)
	Malaya (Hendrickson 1958)
	South Yemen (Hirth and Carr 1970)
	South Africa (Hughes 1974)
	Australia (Bustard 1972)
Other fish[2]	
Centropristes striatus, Coryphaena hippurus, etc.	Southeast United States (Caldwell 1959, Witham 1974)
Arius sp., etc.	Surinam (Schulz 1975)
Lutianus argentimaculatus, Germo albacora, Moray eels, etc.	South Africa (Hughes 1974)
Lutjanus bohar, Caranx ignobilis, sharks, barracuda, etc.	Seychelles (Honegger 1967)
Rock cod	Ascension Is. (Hirth 1971)
Groupers	South Yemen (Hirth and Carr 1970)
	Galapagos (Hirth 1971)
Caranx hippos, Sciaenops ocellatus, etc.	Mexico (Pritchard and Marquez 1973)
Various spp.	Tortuguero, Costa Rica (Carr 1956)
	Malaya (Hendrickson 1958)
Snakes (*Boiga dendrophila, Python reticulatus*)	Malaya (Hendrickson 1958)
Varanids	Malaya (Hendrickson 1958)
Birds[2]	

STEPHEN E. STANCYK

Table 2. Predators of hatchling sea turtles (cont.)

Predator	Locality (source)
Larus novaehollandiae, Haliastur indus, H. sphenura, Falco cenchroides, Corvus orru, etc.	Australia- Mon Repos (Limpus 1973)
Fregata spp., *Corvus albus*	Europa, Tromelin Is. (Hughes 1974)
Fregata sp.	Caribbean (Carr and Meylan 1980)
Night herons; *Fregata* sp.	Galapagos (Hirth 1971)
Vultures (black, turkey)	Costa Rica (Fowler 1978; Robinson, personal communication)
Threskiornis aethiopia, Corvus albus, Fregata minor, F. ariel, etc.	Seychelles (Honegger 1967)
Milvus aegyptus	South Africa (Hughes 1974)
Gulls, crows, vultures	Southeast United States (Caldwell 1959; Talbert, personal communication)
Larus novaehollandiae, crows	Australia (Bustard 1972)
Gulls, vultures	Surinam (Schulz 1975)
Mammals	
Rattus spp.	Seychelles (Honegger 1967) South Africa[3] (Hughes 1974)
Mongooses	South Africa (Hughes 1974) Australia (Bustard 1972)
Genet[3] (*Genetta rubiginosa*)	South Africa (Hughes, Bass, and Mentis 1967)
Cats, feral	Galapagos (Hirth 1971)
Raccoons	Southeast United States (Caldwell 1959, Hopkins et al. 1979, Talbert et al. 1980)
Coatis[3]	Tortuguero (Fowler 1978)
Dogs, feral or domestic	Tortuguero (Fowler 1978) South Africa (Hughes 1974) Mexico (Hirth 1971) South Yemen (Hirth and Carr 1970)
Foxes	South Yemen (Hirth and Carr 1970)
Hogs	Costa Rica (Carr 1956; Robinson, personal communication) Southeast United States (Richardson 1978) Mexico (Hirth 1971)

1. Hirth (1971) contains extensive lists of predators of *Chelonia mydas* hatchlings.   3. Suspected predator.
2. In most localities, numerous other species are potential predators.

Australia. Diamond (1976) reached a similar conclusion in the Seychelles. Even so, bird predation in some localities may be quite heavy on nests which do emerge in the daytime. Black vultures (*Coragyps atratus*) prey heavily on the numerous *Lepidochelys olivacea* hatchlings which emerge or are excavated during the day at Ostional on the Pacific coast of Costa Rica (D. Robinson, personal communication). These birds also attack hatchlings which emerge on moonlit nights. Fowler (1978) found that black and turkey vultures were frequent and efficient predators of hatchlings at Tortuguero, Costa Rica. Hughes (1974) noted that frigate birds decimated all daylight emergences (for example, 133 hatchlings devoured by 40 birds in 15 minutes) on Europa and Tromelin Islands.

Crabs capture hatchlings as they emerge at night, but their effects on hatchling production are probably minor. Coconut crabs (*Birgus* spp.) and land hermit crabs (*Coenobita* spp.) have been noted (Frazier 1971, Hughes 1974), but *Ocypode* spp. are the most common crab predators. However, only larger crabs are capable of capturing and holding a struggling hatchling (Caldwell 1959), and Hughes (1974) noted that they probably captured only unfit individuals, such as stragglers or turtles with poor orientation faculties. Bustard (1979) stated that the larger size of *Chelonia depressa* hatchlings made them less susceptible to crab predation than *Chelonia mydas* hatchlings.

The greatest predation of hatchlings probably takes place after they have entered the water (Hendrickson 1958, Bustard 1979). Hatchlings are taken by numerous inshore predators, especially such fish as small sharks, barracuda, snook, jackfish, and snappers. Cuttlefish (Deraniyagala 1930) also capture hatchlings, and birds have been seen taking them from the water. For instance, Carr and Meylan (1980) reported the capture of a hatchling green turtle from sargassum weed by a frigate bird. The familiar diving response of turtles to overhead shadows may be an adaptation to such aerial predation.

Since relatively few of the many hatchlings which enter the water return to nest, mortality during maturation is assumed to be great (Bustard 1979; Hirth 1971; Richardson and Richardson, this volume). Loss of newly emerged hatchlings in the nearshore environment is thought to be especially severe, and there are many examples (Hendrickson 1958, Honeggar 1967, Hughes 1974) of inshore predators captured with their digestive tracts filled with hatchlings. Hatchlings leaving beaches with rocky approaches suffered higher predation losses than those leaving beaches with sandy approaches on Ascension Island (J. Mortimer, personal communication). Wingate, in his oral presentation at this Conference, reported higher predation on daytime-released hatchlings than on night releases in Bermuda. However, actual mortality rates in the nearshore environment are unknown, as are the mortality rates of hatchlings and juveniles between the time they reach open water and the time they mature. Determination of these parameters is an intractable problem whose solution would add considerably to our knowledge of the role of natural predation in marine turtle life history patterns.

Predators of Adults and Juveniles

Growth of marine turtles renders them immune to most predators. Sharks, however, remain a menace throughout their lives (Balazs 1979a). In particular, the tiger shark, *Galeocerdo cuvieri*, is the most commonly observed predator of adults and juveniles (although many other species of sharks eat turtles, Table 3). Other large marine predators, such as killer whales (Caldwell 1969) take adult turtles. Antipredator refuging behavior has been observed (Bustard 1972, 1979), and an account of an attack on *Dermochelys* by a white shark (Cropp 1979) suggests that there may be stereotyped escape behaviors among turtles. A few large terrestrial predators, such as jaguars (Schulz 1975), tigers and wild dogs (Hendrickson 1958) or feral and domestic dogs (Caldwell 1959, Hughes 1974) may attack nesting adults, but their impact on turtle populations is thought to be minimal.

Recent studies (Bjorndal 1980; Richardson and Richardson, this volume) indicate that adult mortality may be higher than was previously thought. Unfortunately, rates of predation on adult turtles are difficult to quantify, so the magnitude of this mortality factor on turtle populations is unknown. The number of nesting females which have evidence of shark damage varies from about 2 to 4 percent to 21 percent (Hendrickson 1958; Hughes 1974), but these may only be the individuals that escaped predation, telling us nothing about the unknown number of turtles which were victims of successful attacks (Hendrickson 1958). Counting turtle parts in the guts of captured sharks reveals nothing about feeding rates, since the duration of retention of such parts is unknown (Balazs 1979a). Only direct, long-term observations of turtle populations in the field such as those being carried out by Balazs (this volume) will provide information on natural adult mortality rates.

Control of Predators

A variety of methods have been used to control nonhuman predation, chiefly of eggs and hatchlings. Attempts vary in rigor, the amount of effort applied, and the degree of success. In general, methods are designed to meet a particular problem, and not all methods are equally effective in all situations. Many of the results of control efforts are unpublished or buried in local or

STEPHEN E. STANCYK

Table 3. Non-human predators of juveniles and adult sea turtles

Predator	Locality (source)
Fish	
Sharks (unspecified)	Southeast U.S. (Caldwell 1959[1])
	Seychelles (Frazier 1971)
Hammerhead Shark (*Sphryna* sp.)	Australia (Bustard 1972)
Tiger Shark (*Galeocerdo cuvieri*)	Hawaiian Islands, Pacific (Balazs 1979a)
	Malaya and Sarawak (Hendrickson 1958)
	South Yemen (Hirth and Carr 1970)
	Australia (Bustard 1972)
Lemon Shark (*Negaprion brevirostris*)	Bahamas (Bjorndal, personal communication)
White Shark (*Carcharodon carcharias*)	Australia (Cropp 1979)
Bull or Zambezi Shark (*Carcharhinus leucas*)	Tortuguero, Costa Rica (Gilbert and Kelso 1971)
	South Africa (Hughes 1974)
Oceanic White Tip (*Carcharhinus longimanus*)	South Africa (Hughes 1974)
Brindle Bass (*Promicrops lanceolatus*)	South Africa (Hughes 1974)
Grouper (*Epinephelus* spp.)	Hawaiian Islands (Balazs 1979a)
	South Yemen (Hirth and Carr 1970)
Mammals	
Killer Whales (*Orcinus orca*)	Lesser Antilles (Caldwell 1969)
Dogs, feral or domestic	Southeast U.S. (Caldwell 1959)
Dogs, wild (*Cuon javanicus*)[1]	Malaya and Sarawak (Hendrickson 1958)
Hyenas[1]	Angola (Hughes et al. 1973)
Jaguars	Surinam (Schulz 1975)
	Tortuguero, Costa Rica (Carr 1956)
Tigers[1]	Malaya and Sarawak (Hendrickson 1958)
Leopards[1]	Angola (Hughes et al. 1973)

1. Suspected predators

intra-agency reports. Success is measured in terms of increased hatchling production from the beach, usually in a single year, without regard to the ultimate fate of hatchlings or the long-term benefits of control. Such benefits are admittedly difficult to determine, given the problems inherent in tagging hatchlings and waiting for them to reach adulthood. The following survey covers some of the most commonly used or suggested control methods, but is by no means complete. Where possible, examples will be discussed, and recommendations made.

Chemical Controls

These include both poisons, which are designed to kill predators, and aversive chemicals (like lithium chloride, LiCl), which are meant to have sufficiently unpleasant sublethal effects to deter predators. Both methods are used primarily to reduce egg predation, particularly by mammals. Poisons are effective to a degree and have been shown to result in the deaths of feral dogs at Tortuguero, Costa Rica (Carr, personal communication). But local domestic dogs were also killed, which illustrates the general problem with the use of poison. While only 1 or 2 species may be responsible for initiating predation of turtle nests, poisons will affect secondary predators as well, including numerous birds, crabs, and domestic animals. The unspecific nature of such poisons makes their use inadvisable except in very special circumstances.

Aversive, nonlethal chemicals such as LiCl could avoid the problems associated with poisons. Hopkins (unpublished) has evaluated the effectiveness of this chemical as a deterrent to nest predation by raccoons. In a total of 306 laboratory trials with 15 raccoons, 1 g of LiCl (sufficient to produce illness in raccoons) was given in 2 or 4 turtle eggs. Eggs were left untouched in less than 3 percent of the trials, and no aversion was induced. In the field, 30 nests containing enough LiCl each to cause illness in at least 3 raccoons were buried at the apex of false crawls on a heavily depredated beach over a 5-week period. All treated nests were attacked the night after burial, and no subsequent reduction in nest destruction was noted (90.4 percent vs 86.3 percent the previous year). Whether longer-term exposure to treated nests would have eventually caused a reduction in predation is not known, but the lack of aversion in the laboratory trials makes the possibility unlikely. Klukas (1967) tried a commercial dog repellent, moth crystals, and seawater over nests, with similar negative results.

Chemical treatment of nests appears to be at best marginally effective, and is fraught with difficulties and potential side effects. Not only might relatively harmless scavenging species be killed but the effects of artificial chemicals on the developing hatchlings are also unknown. These real and potential problems should preclude the use of toxic chemicals as management practices to reduce egg predation.

Trapping or Shooting

These methods are used primarily to reduce predation by small mammals, especially raccoons, dogs, and hogs. In South Florida, Klukas (1967) found that daily trapping reduced first-night predation by raccoons from around 80 percent to 25 percent and 44 percent in 2 different years. Caldwell (1959) reported a low raccoon predation rate (5.6 percent) on Cape Romain in 1939, but it was considered high enough by the 1950s to justify employment of a professional trapper. In years when trapping was conducted, predation rates dropped to around 10 percent or less (L. West, personal communication). In 1979, the removal of 82 raccoons from Cape Island, mostly prior to the nesting season, helped reduce first-night predation by about half. Unless trapping is carried out intensively each year, however, immigrants may quickly replace the removed raccoons. At Cape Romain, predation increased to nearly 100 percent 3 to 6 years after trapping had ceased (L. West, personal communication).

Control of mammalian nest predators by trapping or shooting is effective in situations where immigration of new predators is limited (for example, on islands or beaches backed by barriers) or where manpower and equipment is available to carry out an intensive effort. In areas where the predator population is large or where new recruits are readily available, trapping and shooting may be very expensive, and there may be more efficient and economical ways to increase hatchling production. In some cases, although certainly not those where feral predators are involved, predator elimination could have negative effects on the rest of the ecosystem or conflict with previously established conservation guidelines (Davis and Whiting 1977; L. West, personal communication). Removal of natural predators could increase predation of eggs and hatchlings by organisms which were formerly controlled by the predators; for example, *Ocypode* increases after raccoon or fox removal (Caldwell 1959; Tufts 1972).

Hatcheries

Many hatcheries (where clutches are moved to a single protected site) have been designed specifically to reduce the effects of non-human natural or feral predators (Richardson 1978; Talbert et al. 1980; L. West, personal communication). The design and effectiveness of hatcheries is discussed elsewhere (Talbert et al. 1980), but will be summarized here. On occasion, weak or poorly designed hatchery enclosures can be broken into by hogs (Richardson, personal communication) or dug into by raccoons or crabs (Talbert, personal communication), but in-beach and in-box hatcheries are generally effective in reducing predation on eggs and hatchlings. On Cape Romain, for instance, erosion and predation combined to destroy nearly 100 percent of all nests in 1978, so the release of 10,500 hatchlings from an in-beach hatchery in 1979 (L. West, personal communication) is a considerable improvement. If production of hatchlings from a hatchery exceeds that of nests left on the beach, hatcheries may be justified.

There are many potential dangers to the use of hatcheries, however, and these must be carefully considered before the technique is adopted as a means of predator control. Although hatch rates may be quite high (Raj 1976), the percentage hatch of clutches in hatcheries is generally lower (55 to 85 percent, usually about 65 percent) than natural hatch rates (50 to 95 percent,

STEPHEN E. STANCYK

usually about 80 percent) (Talbert et al. 1980). In addition, there have been instances (Ragotzkie 1959; Talbert et al. 1980; L. West, personal communication) when flooding from heavy rains or passing storms has caused 100 percent mortality of unhatched clutches. Eggs placed in styrofoam boxes (Talbert et al. 1980) require continuous maintenance, and slight variations in temperature or other physical properties could have negative effects on the hatchlings (Yntema 1976; Mrosovsky 1978; Mrosovsky and Yntema, this volume). Styrofoam or similar materials could release substances which would modify normal development of the hatchlings in subtle, nearly undetectable ways (Carr, personal communication). Hatcheries require maintenance effort, and must be regularly attended to effect release of hatchlings. Although experiments with automatic release mechanisms are now being tested (L. West, personal communication), further work is necessary. Finally, hatchlings are often released from hatcheries during the day or in concentrations exceeding those occurring during natural emergences. Either of these treatments could increase loss to aquatic predators in the surf, but this has not been tested.

Nest Transplants

Based on the assumption that egg predators, particularly small mammals, find clutches through clues left by the nesting female, experiments have been conducted that remove whole or partially depredated clutches from their original nest pit to a similar cavity dug at a suitable site in the beach nearby (Stancyk, Talbert, and Dean 1980). Experiments were conducted over 4 years on 2 different islands in South Carolina, where nest destruction by raccoons is high. Predation of 464 natural nests ranged from 55.1 to 93.8 percent ($\bar{x}$ = 72.3 percent); predation of 123 transplanted nests was significantly lower (P = 0.01), ranging from 6.1 to 18.7 percent ($\bar{x}$ = 9.7 percent). Hatching success of transplants, determined by post-hatch shell counts, were not significantly different from natural or hatchery rates (60 to 81 percent). C. Blanck (unpublished) transplanted clutches on 3 beaches on Ossabaw Island, Georgia, where hogs and raccoons are common predators. Loss of transplants due to predation and erosion varied from 45 to 75 percent on different beaches vs 100 percent loss of natural controls. Percentage hatch per nest ranged from 64 to 78 percent, compared to a 72 percent hatch success in an in-box hatchery.

There are several attractive features of the transplant method as a conservation measure. It is inexpensive and simple, and can be performed by a single individual on regular beach patrols. When performed correctly, eggs are moved short distances soon after deposition (within 24 hours, 48 hours maximum), before egg membranes have attached to the shell (C. Blanck, personal communication). Nests threatened with erosion can be moved to higher sites, and transplanted nests can be spread out along the beaches to reduce the chance of massive mortality by flooding or concentrations of predators. Eggs develop naturally in the beach, and unless histories of individual nests are being monitored, nest sites need not be visited after the clutch is buried. Hatching can take place in a natural manner, without human interference.

However, the transplant method is new and requires further testing, especially with a variety of predators on different beaches. In particular, the success of transplants relative to the late-season predation observed by Fowler (1978) and Hopkins et al. (1979) has not been examined. Transplants are still susceptible to erosion, and some are depredated. Predators other than raccoons might not be fooled. Finally, predators might learn to find nests with fewer natural cues or with human-associated cues after an organized transplant program has operated for a few years.

Other Methods

Many other methods to control predation of eggs and hatchlings have been suggested or tried, usually without rigorous testing. Raccoon predation has been reduced by placement of fixed screens over nests on the beach at Cape Romain. However, such screened nests are still susceptible to erosion, may be dug into by crabs, and must be revisited at the time of hatching to remove the screen. Presence of humans on the beach may help reduce predation; Talbert et al. (1980) found that less-heavily patrolled sectors of Kiawah Island experienced significantly higher egg depredation by raccoons than regularly patrolled sectors. On the other hand, raccoons often sit nearby, apparently unafraid, while people perform various activities around turtles or nests (Talbert, personal communication; Hopkins, personal communication). Supplemental feeding of predators or track erasure (Hopkins, personal communication) may have potential, but require further testing. D. Robinson (personal communication) found that loud noise from a cannon net kept black vultures away for up to a week. None of the above methods appear to be more successful, economical or labor-saving than trapping, hunting, hatcheries, or transplants.

Little has been done to reduce aquatic predation, although some hatchlings have been released beyond the surf line (Frick, personal communication). Preliminary rearing ("head start") programs have also been conducted (Balazs, 1979b; Klima and McVey, this volume) in which hatchlings are released after a period of growth. Neither the effect of these procedures on normal hatchling development nor their potential for increasing hatchling survival is known.

Conclusions

In most marine turtle rookeries, destruction of eggs, hatchlings and adults by non-human predators is probably not the most important factor limiting offspring production. A possible exception to this generalization is predation of hatchlings by shallow-water aquatic organisms, which may be a universal threat off all nesting beaches. Such mortality has probably always been a pressure on turtle populations, and becomes a limiting stress only after other factors have reduced offspring production and survival. Quantification of aquatic mortality is lacking, and control is difficult. Raising hatchlings to a larger size and releasing them offshore may be one solution, but the method is fraught with potentially harmful consequences. For example, if young turtles gain their orientation sense by traveling down the beach (Frick, personal communication), they must be allowed to do so before they are released. The success of offshore-release programs will be difficult to monitor, and results will not be known for years. Non-human predation of adults is also difficult to quantify and even more difficult to control. At present, it appears to be a relatively unimportant mortality source compared to hunting and accidental death in fishing nets.

In areas where non-human predation has become a crucial limiting factor on hatchling production, such as the southeastern United States or Lacey Island on the Great Barrier Reef (Limpus, in press), control methods are advisable. The method should be designed to fit the situation within economic and labor restraints. If manpower is available, thorough programs of trapping and hunting appear to be very effective in reducing natural or feral mammal predation, and have the least effect on the normal development of eggs and hatchlings. However, effective control requires considerable effort and continuous long-term operation.

When natural predation is heavy, and hunting or trapping methods are impractical, methods of egg and hatchling protection are recommended. Such operations should be conducted with extreme care and advance planning. To avoid damaging eggs by agitation, they should not be moved or reoriented more than 48 hours after deposition (C. Blanck, personal communication). Hatcheries can be effective, but require equipment, careful maintenance, and regular monitoring. They may also induce undetectable sublethal modifications of normal development. Dividing eggs into 2 or 3 hatcheries placed at different locations on a beach might reduce chances of 100 percent loss to uncontrollable factors such as flooding.

Transplantation of clutches away from cues left by the nesting females is a less labor-intensive method of egg protection which may be effective against erosion, as well as such predators as crabs, varanid lizards, and small mammals. The transplant method has high conservation potential because it can be performed with little manpower during normal beach surveys and does not require regular monitoring. However, the method must be carried out with extreme care and requires additional testing before its general utility can be ascertained.

When predation severely reduces the production of offspring, control methods are justified, especially if other factors cause additional turtle mortality. Although many methods have been tried, those listed above appear to be the most effective relative to the effort expended. In any case, controls should be designed to fit the particular situation and should be carried out with care. Controls which are conducted in a haphazard manner may prove to be a bigger detriment to offspring production than the predation they are meant to prevent.

Acknowledgements

Thanks are due to the many individuals who provided unpublished information to help make this review more comprehensive, including J. Andre, G. Balazs, A. Carr, L. Fowler, D. Green, S. Hopkins, J. Mortimer, D. Robinson, J. P. Ross and L. West. Thanks also are due to the many people who helped with nest transplants, especially R. Aiken, O. R. Talbert, B. Talbert, and A. Miller. Partial financial support was provided by South Carolina Sea Grant, and logistic and secretarial help was provided by the Belle W. Baruch Institute for Marine Biology and Coastal Research, University of South Carolina. This is contribution number 317 of the Belle W. Baruch Institute for Marine Biology and Coastal Research.

Literature Cited

Balazs, G. H.
1979a. Synopsis of biological data on the green turtle in the Hawaiian Islands. National Oceanic and Atmospheric Administration, National Marine Fisheries Service, Contract 79–ABA–02422.
1979b. An additional strategy for possibly preventing the extinction of Kemp's ridley, *Lepidochelys kempi. IUCN/SSC Marine Turtle Newsletter* 12:3–4.

Bjorndal, K. A.
1980. Demography of the breeding population of the green turtle, *Chelonia mydas,* at Tortuguero, Costa Rica. *Copeia* 1980:525–30.

Bustard, H. R.
1972. *Australian Sea Turtles.* London: Collins.
1979. Population dynamics of sea turtles. In *Turtles: Perspectives and Research,* eds. M. Harless and H. Morlock, pp. 523–40. New York: Wiley and Sons.

Caldwell, D. K.
1959. The loggerhead turtles of Cape Romain, South Carolina. *Bulletin of the Florida State Museum* 4:319–48.

STEPHEN E. STANCYK

1966. A nesting report on the American ridley. *International Turtle and Tortoise Society Journal* 1:10–13, 30.

1969. Addition of the leatherback sea turtle to the known prey of the killer whale, *Orcinus orca. Journal of Mammalogy* 50:636.

Carr, A.

1956. *The Windward Road.* New York: Alfred Knopf Inc.

1967. *So Excellent a Fishe.* New York: Natural History Press.

Carr, A., and A. B. Meylan

1980. Evidence of passive migration of green turtle hatchlings in sargassum. *Copeia* 1980: 366–68.

Cropp, B.

1979. Where ocean giants meet. *Oceans* 12:43–47.

Davis, G. E., and M. C. Whiting

1977. Loggerhead sea turtle nesting in Everglades National Park, Florida, U.S.A. *Herpetologica* 33:18–28.

Deraniyagala, P. E. P.

1930 The testudinata of Ceylon. *Spolia Zelan* 16:43–88.

Diamond, A. W.

1976. Breeding biology and conservation of hawksbill turtles, *Eretmochelys imbricata* L. on Cousin Island, Seychelles. *Biological Conservation* 9:199–215.

Fowler, L. E.

1978. Hatching success and nest predation in the green sea turtle, *Chelonia mydas*, at Tortuguero, Costa Rica. Master's thesis, University of Florida.

Frazier, J.

1971. Observations on sea turtles at Aldabra Atoll. *Philosophical Transactions of the Royal Society of London, B,* 260:373–410.

Gallagher, R. M.; M. L. Hollinger; R. M. Ingle; and C. R. Futch

1972. Marine turtle nesting on Hutchinson Island, Florida in 1971. *Florida Department of Natural Resources, Special Scientific Report,* 37:1–11.

Gilbert, C. R., and D. P. Kelso

1971. Fishes of the Tortuguero area, Caribbean Costa Rica. *Bulletin of the Florida State Museum* 16:1–54.

Hendrickson, J. R.

1958. The green sea turtle, *Chelonia mydas* (Linn.) in Malaya and Sarawak. *Proceedings of the Zoological Society of London* 130:455–535.

Hill, R. L., and D. J. Green

1971. Investigation of the damage by the crab, *Ocypode quadrata*, to the eggs of the green turtle, *Chelonia mydas. Surinam Turtle Notes* 2:11–13.

Hirth, H. F.

1971. Synopsis of biological data on the green turtle *Chelonia mydas* (Linnaeus) 1758. *FAO Fisheries Synopsis* 85:1–76.

Hirth, H. F., and A. Carr

1970. The green turtle in the Gulf of Aden and the Seychelles Islands. *Verh. K. Ned. Akad. Wet.* 58:1–44.

Honegger, R.

1967. The green turtle (*Chelonia mydas japonica*) in the Seychelles Islands. *British Journal of Herpetology* 4:8–11.

Hopkins, S. R.; T. M. Murphy, Jr.; K. B. Stansell; and P. M. Wilkinson

1979. Biotic and abiotic factors affecting nest mortality in the Atlantic loggerhead turtle. *Proceedings of the 32nd Annual Conference of the South East Association of Fish and Wildlife Agencies* 32:213–23.

Hughes, G. R.

1972. The marine turtles of Tongaland, 6. *Lammergeyer* 15:15–26.

1974. The sea turtles of South East Africa. II. The biology of the Tongaland loggerhead turtle Caretta caretta L. with comments on the leatherback turtle Dermochelys coriacea L. and the green turtle Chelonia mydas L. in the study region. *South African Association of Marine Biology Research Investigational Report* 36:1–96.

Hughes, G. R., and M. T. Mentis

1967. Further studies on marine turtles in Tongaland, II. *Lammergeyer* 7:55–72.

Hughes, G. R.; A. J. Bass; and M. T. Mentis

1967. Further studies on marine turtles in Tongaland, I. *Lammergeyer* 7:5–54.

Hughes, G. R.; B. Hustley; and D. Wearne

1973. Sea turtles in Angola. *Biological Conservation* 5:58.

Klukas, R. W.

1967. Factors affecting nesting success of loggerhead turtles at Cape Sable, Everglades National Park. File N1415. National Park Service.

Limpus, C. J.

1973. Avian predators of sea turtles in south-east Queensland rookeries. *Sunbird* 4:45–51.

In press. Observations on the hawksbill turtle, *Eretmochelys imbricata*, nesting in north-eastern Australia. *Herpetologica.*

McAllister, H. J.; A. J. Bass; and H. J. van Schoor

1965. Marine turtles on the coast of Tongaland, Natal. *Lammergeyer* 3:12–40.

Minton, S. A.

1966. A contribution to the herpetology of West Pakistan. *Bulletin of the American Museum of Natural History* 134:27–184.

Mrosovsky, N.

1978. Editorial. *IUCN/SSC Marine Turtle Newsletter* 9:1.

Pritchard, P. C. H.

1971. The leatherback or leathery turtle, *Dermochelys coriacea. IUCN Monograph, Marine Turtle Series* 1:1–39.

Pritchard, P. C. H., and R. Marquez

1973. Kemp's ridley turtle or Atlantic ridley, *Lepidochelys kempi. IUCN Monograph, Marine Turtle Series* 2:1–30.

Ragotzkie, R. A.

1959. Mortality of loggerhead turtle eggs from excessive rainfall. *Ecology* 40:303–05.

Raj, U.

1976. Incubation and hatching success in artificially incubated eggs of the hawksbill turtle, *Eretmochelys imbricata* (L.). *Journal of Experimental Marine Biology and Ecology* 22:91–99.

Richardson, J. I.

1978. Results of a hatchery for incubating loggerhead sea turtle (*Caretta caretta*) eggs on Little Cumberland Island, Georgia. *Florida Department of Natural Resources Publications* 33:1–15.

Schulz, J. P.
1975.   Sea turtles nesting in Surinam. *Zool. Verh.* (Leiden) 143:1–144.

Stancyk, S. E.; O. R. Talbert, Jr.; and J. M. Dean
1980.   Nesting activity of the loggerhead turtle *Caretta caretta* in South Carolina, II. Protection of nests from raccoon predation by transplantation. *Biological Conservation* 18:289–98.

Talbert, O. R., Jr.; S. E. Stancyk; J. M. Dean; and J. M. Will
1980.   Nesting activity of the loggerhead turtle *Caretta caretta* in South Carolina, I. A rookery in transition. *Copeia,* 1980:709–19.

Tufts, C. E.
1972.   Report on the Buritaca marine turtle nesting reserve, with emphasis on biological data from "Operation Tortuga 1972" and recommendations for the future. Report to INDERENA, 73 pp.

Witham, R.
1974.   Neonate sea turtles from the stomach of a pelagic fish. *Copeia* 1974:548.

Worth, D. W., and J. B. Smith
1976.   Marine turtle nesting on Hutchinson Island, Florida in 1973. *Florida Department of Natural Resources and Marine Research Publication,* 18.

Yntema, C. L.
1976.   Effects of incubation temperatures on sexual differentiation in the turtle, *Chelydra serpentina. Journal of Morphology* 150:453–62.

Zahl, P. A.
1973.   One strange night on Turtle Beach. *National Geographic* 144:570–81.

STEPHEN E. STANCYK

René Márquez M.
Cuauhtémoc Peñaflores S.
Aristóteles Villanueva O.
Juan Díaz F.
Departamento de Pesca
Instituto Nacional de la Pesca
Mexico 7, D.F.

A Model for Diagnosis of Populations of Olive Ridleys and Green Turtles of West Pacific Tropical Coasts

ABSTRACT

Sea turtle populations along the west coast of Mexico have been studied for several years. The seasonal abundance of both *Lepidochelys olivacea* and *Chelonia mydas agassizii* increase during the reproductive period in front of the respective nesting beaches, during summer and autumn. There are clear differences between the nesting patterns of ridleys and greens. The olive ridley follows clear internesting cycles every 28 days, while the green turtle nests in semi-cycles of 14 days. Nevertheless, the total number of eggs laid per individual is similar for both species.

The mean clutch size by nest is: 115, 95, and 80 (average 95) for *Lepidochelys* and 80, 70, 60, and 50 (average 66) for *Chelonia*. The period between nesting is nearly annual in the ridley (1.3 years) and nearly biennial for the green (1.8 years). The ridley is estimated to take at least 8 or 9 years to reach sexual maturity.

The natural mortality rates for eggs, hatchlings, and adults were analyzed and determined from values of survival rates in each stage of the life cycle ($S_{0.17} = 0.592$, $S_1 = 0.581$, $S_7 = 0.497$, etc.). Using these data, a population model was constructed; changes in the model must be fitted year by year in accordance to the size of the nesting population in the involved season.

Introduction

From all the information published on sea turtles there are no thorough studies on population dynamics. Nevertheless, the actual knowledge nowadays of these species permits us to construct and apply a theoretical model for diagnosing the situation of populations of *Lepidochelys olivacea* and *Chelonia mydas agassizii*. Often it is said that they are over-exploited species, but to what extent have they been or are they being exploited? What is the present abundance? How many individuals can be captured for commercial purposes and when? To reach more accurate conclusions about

153

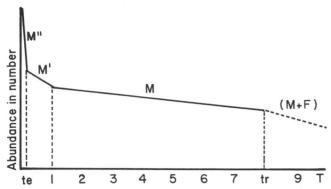

Figure 1. Abundance of population affecting the total number. Where: t_e = eclosion time and t_r = age of recruitment to the fishery.

these questions and improve the model, more information will have to be gathered on past and present capture, fishing effort, nesting sites, abundance of adults, fecundity, sex ratios, recruitment, breeding cycles, age, growth, and so on. Because of this it seems necessary to have a simple model useful for quick assessment of population size with the minimal data now available. This model should also help in the management of these vulnerable resources.

In the present work, as a first step we cover subjects of vital importance: age-specific rates of mortality, derived from tagging, nesting beach census of breeding females and total number of eggs laid in the season. Based on this a table for making predictions of changes

in the populations of *Lepidochelys olivacea* was constructed. A similar table for *Chelonia mydas agassizii* will be possible in the near future.

Model

In accordance with Gulland (1971) there are various models for representing the population dynamics of marine species. Considering their behavior with respect to capture, sea turtles could be included in either of the following groups described by Schaefer and Beverton (1963): one as simple units, subject to the law of population growth (Leslie 1957) where the fishing acts as an additional predator in the predator-prey system (Schaefer 1954); the other considers a group of individuals, where changes in biomass are determined by juvenile recruitment in a unit time to the population, by the growth and by the mortality. In this case fishing acts as another source of mortality (Beverton and Holt 1957; Ricker 1975).

Figure 1 represents the change in the abundance of various stages in the life cycle of a population. For the first stage, from egg laying until the hatchlings reach the sea, natural mortality (M'') is considered to be equal to total mortality (Z''). In the second stage, from the arrival of the hatchlings to the sea until they are 1-year old M' is considered equal to Z'. At this point in the life cycle it is assumed that habits change slowly until the individual reaches sexual maturity, and it is also assumed that mortality is constant from 1 year of age to shortly before sexual maturity. Natural mortality (M) is assumed to be equal to total mortality (Z). When

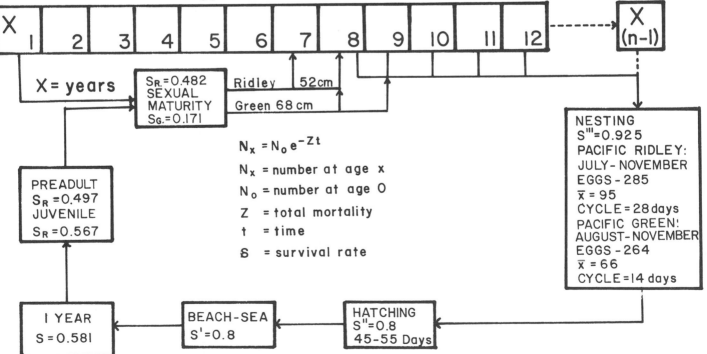

Figure 2. Life cycles of the olive ridleys and Pacific green turtles.

RENÉ MÁRQUEZ M.

the subadults mature they are subjected to additional pressure from fishery activities, then mortality is affected by fishing (F), and in this stage: $Z = M + F$.

The stochastic model is based on the life cycle, which is described as follows.

Life Cycle

In order to analyze and model population dynamics it is necessary to describe the life cycle. Figure 2 outlines the principal stages and a few important parameters without introducing too many variables which would confuse the process at this stage. In the future it will be necessary to include more variables, in more detail. In accordance with the observations, the olive ridley, is assumed to reach sexual maturity between 7 and 8 years of age with a minimum size, straight carapace length, of nesting females of 52 cm. *Chelonia*, which is larger, is assumed to be sexually mature at 8 or 9 years of age (Márquez and Doi, 1973) with a minimum breeding size of 68 cm. The reproductive cycles vary between the two species, but both seem to depend on lunar effect, most notably in the phases of first quarter and third quarter. Thus, the ridley shows a cycle of about 28 days and the green turtle about 14 days (unpublished data). Seasonal fecundity is also very similar. The ridley lays a greater number of eggs per nest, but less frequently, 2 to 4 times per season; the average clutch is 115, 95, 80 eggs, with an overall average of 95 and a seasonal total of 285 eggs. In green turtles there are fewer eggs per nest, but more nests per season; the average clutch is 80, 70, 60, 50 per nest, with an average of 65 and a seasonal total of 264 eggs (unpublished data).

Lepidochelys olivacea forms nesting aggregations or *arribazones* during which tens of thousands of females may nest in a day or so. The crowding affects egg survival principally at two levels: 1) eggs laid during an *arribazon* may be dug up by nesting females in the same arribazon; 2) eggs from one *arribazon* may be destroyed during a subsequent *arribazon*. These two sources of egg mortality were measured in the 1975 and 1976 seasons, and a mean mortality due to intraspecific nest disturbance was estimated to be $\hat{Z} = 0.0778$, and a mean survival of $\hat{S} = 0.925$.

Natural survival rate of undisturbed nests throughout the incubation period, which may last 45 to 55 days, was estimated from empirical data at $S = 0.8$. Also, the survival of the hatchlings, from leaving the nest to reaching the sea was estimated to be $S = 0.8$. The survival rate for this first stage in the life cycle, some 60 days or 0.17 years, was calculated to be $S = 0.592$ (Márquez et al., in prep.) and mortality is $Z_{0.17} = 0.521$.

Survivorship and mortality values for subsequent years in the life cycle are presented in Table 1. Observed annual survival, S_a, and the calculated survival, $\hat{S}_a$, are quoted for the population of *Lepidochelys olivacea* in the Pacific. Calculated annual survival, $\hat{S}_a$, was derived from the equation:

$$\ell n \, N = -0.0143t + 0.597, \text{ and } Z = \ell n \, \hat{S}_a.$$

The data were derived from the life cycle and fitted by the least squares method. Rate of survival was derived from the formula $S = e^{-Zt}$; the graphical representation is shown in Figure 3. This calculation was fitted with the first year value of total mortality, $\hat{Z}_a = 0.543$, and the theoretical results are quoted in the last column of Table 1.

Evaluation of the *Arribazones*

One of the principal problems in the study of marine turtles is in establishing population size. This can be estimated by several methods including catch effort,

Table 1. Theoretical estimations of annual survival rate and total mortality for *Lepidochelys olivacea* derived from the life cycle

	S_a	$\hat{S}_a$	$\hat{Z}_a$	$S = e^{-Zt}$*
S_v	1.000	1.000	0.000	1.000
$S_{.17}$	0.597	0.592	0.524	0.912
S_1	0.580	0.581	0.543	0.581
S_2	0.567	0.567	0.567	0.337
S_3	0.553	0.553	0.592	0.196
S_4	0.539	0.539	0.618	0.114
S_5	0.526	0.525	0.644	0.0662
S_6	0.512	0.511	0.671	0.0385
S_7	0.498	0.497	0.699	0.0223
S_8	0.482	0.483	0.728	0.0130
S_9	—	0.469	0.757	0.0075
S_{10}	—	0.456	0.785	0.0044
S_{11}	—	0.442	0.816	0.0025
S_{12}	—	0.428	0.848	0.0015

— No data.

* Was fitted with $\hat{Z}_a = 0.543$, for full calendar year.

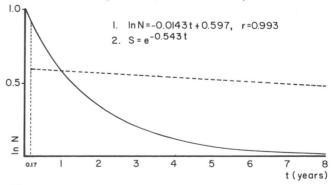

Figure 3. Annual survival rate (broken line) and instant rate of decrease of the population (solid line) for *Lepidochelys olivacea*, derived from the life cycle.

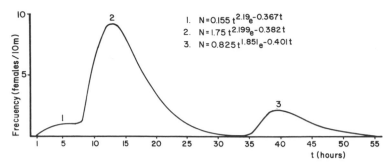

I. $N = 0.155\,t^{2.19}e^{-0.367t}$
2. $N = 1.75\,t^{2.199}e^{-0.382t}$
3. $N = 0.825\,t^{1.851}e^{-0.401t}$

Figure 4. Relation between time and mean number of nesting female *Lepidochelys olivacea* in a sample strip of 10 m on the beach of La Escobilla, Oaxaca, during 11, 12, 13 of August 1979.

Table 2. Yearly data on tagging and tag recoveries of *Chelonia mydas agassizii* in Mexico

Marked		Recaptured			
Year	Number	1975	1976	1977	1978
1975	394	6	3	0	0
1976	1245	—	4	25	2
1977	391	—	—	8	9
1978	583	—	—	—	2
	2613	r_0	r_1	r_2	
		20	37	2	

tagging, and direct censuses. The last method was used in the present study and is described in the following section.

Direct evaluation

This is carried out on the nesting beaches using a method of probability. Instantaneous random sampling consisted of quantifying the turtles that nested each hour in 10-m wide strips, perpendicular to the water line every 200 m along the beach. The action of the turtles was classed in one of three different situations: ascending the beach, descending the beach, and those that were in any stage of nesting (digging, ovipositing, and covering the nest). To estimate the quantity of turtles nesting, the number ascending was added to the number in any nesting phase from which the number descending the beach was subtracted. The data on turtle abundance from each sample group were recorded at hourly intervals during the *arribazon*. It is assumed

that the average turtle spends nearly one hour nesting on the beach. To determine the hourly average and variation in turtle number in each strip of beach, the hourly mean values were used to generate the exponential equations, which were used to estimate the number of turtles nesting in each sampled strip.

Figure 4 illustrates the number of nesting females in a 10-m sample strip during the *arribazon* in 1979 showing changes in number through the day and between days. To estimate the total nesting for the *arribazon* the above values were used for each 200 m of beach. This permits an estimate of the number of female turtles which came out to lay eggs in each *arribazon*. This was analyzed with average fecundity per female and annual survival rate to evaluate the size of the population in the sea (Márquez, ms.).

Total Mortality and Survival

Tagging individuals is used principally to study migra-

Table 3. Yearly data on tagging and tag recoveries of *Lepidochelys olivacea* in Mexico and the Central East Pacific coast

Year	Number	66	67	68	69	70	71	72	73	74	75	76	77	78
1966	153	2	11	2	1	1	0	0	0	0	0	0	0	0
1967	87	—	8	6	5	1	1	0	0	0	1	0	0	0
1968	30	—	—	2	0	0	1	1	0	0	0	1	0	0
1969	26	—	—	—	1	2	0	0	1	0	0	0	0	0
1970	874	—	—	—	—	4	4	2	11	2	5	1	1	1
1971	—	—	—	—	—	—	—	—	—	—	—	—	—	—
1972	16	—	—	—	—	—	—	—	—	—	—	1	—	—
1973	516	—	—	—	—	—	—	—	3	4	8	5	3	5
1974	513	—	—	—	—	—	—	—	—	2	4	1	2	6
1975	1,485	—	—	—	—	—	—	—	—	—	71	2	16	11
1976	2,543	—	—	—	—	—	—	—	—	—	—	8	17	41
1977	2,452	—	—	—	—	—	—	—	—	—	—	—	63	25
1978	3,564	—	—	—	—	—	—	—	—	—	—	—	—	183
		r_0	r_1	r_2	r_3	r_4	r_5	r_6	r_7	r_8				
	12,660	347	118	77	32	16	10	1	1	3				

RENÉ MÁRQUEZ M.

Table 4. Yearly changes in total mortality and survival rate, gathered from tag recoveries in Mexico and the Central East Pacific Coast

Species	1977		1978	
	Z	S	Z	S
Chelonia mydas agassizii	2.004	0.134	1.766	0.171
Lepidochelys olivacea	0.824	0.438	0.729	0.482

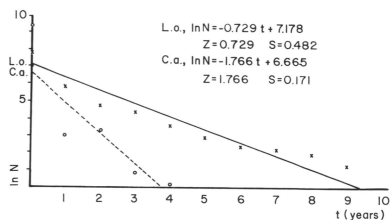

Figure 5. Total mortality (Z) and survival rate (S) for adults of *Lepidochelys olivacea* and *Chelonia mydas agassizii*, derived from tag recoveries.

tion and to evaluate a population. It can be used to determine total mortality (Z) and survival rate (S) (Lucas 1975). In this case both Z and S are presented.

The mark-recapture data for *Chelonia* comprise 4 years, 1975–78, and for *Lepidochelys* 12 years, 1966–78. Recaptures, away from the nesting beach, are shown for both species in Tables 2 and 3. This information, notably the values for r_0, r_1, r_2 etc., which represent the number of tagged turtles marked and recaptured the same year, the next year, and in following years, were analyzed by the logarithmic method of least squares to estimate total mortality (Z) and survival rate (S), with assumed 10 percent annual tag loss (Doi 1974). The predicted values are plotted in Figure 5. In Table 4 these parameters are compared with those obtained until 1977. The reliability of data on *Lepidochelys* seems to be better than those for *Chelonia*, probably because of the greater number of marked ridley turtles (4.8 times more) and time worked. The calculated mortality in *Chelonia* is considerably higher, Z = 2.004 and 1.766 compared with *Lepidochelys* (Z = 0.824 and Z

= 0.729) and those obtained by Márquez and Doi (1973) for *Chelonia* from the Gulf of California, Z = 0.223 and S = 0.807 (which was derived from ages and size composition).

Change in the Population of *Lepidochelys olivacea*

In treating this subject it is necessary to consider some characteristics of behavior because this is necessary in defining "stocks." Apparently the populations are clearly defined, despite extensive dispersion of adults during migration from the breeding area. In summer and autumn, they aggregate at nesting areas in Mexico and Central America while in winter and spring they are in known feeding areas from South America to Mexico. Consequently, we should define at least the reproduc-

Table 5. Theoretical Population of *Lepidochelys olivacea*

Year	Spawning size	0.912 $N_{0.17}$	0.0223 N_7	0.0130 N_8	0.00754 N_9	0.00438 N_{10}	0.00255 N_{11}	0.00148 N_{12}	0.00086 N_{13}	0.00050 N_{14}	0.00029 N_{15}	0.00017 N_{16}	0.00009 N_{17}
72	22243	20285	496.0	289.1	167.1	97.4	56.7	32.9	19.1	11.1	6.4	3.8	2.0
73	24575	22412	548.0	319.5	185.3	107.6	62.7	36.4	21.1	12.3	7.1	4.2	2.2
74	19867	18119	443.0	258.3	149.8	87.0	50.7	29.4	17.1	9.9	5.8	3.4	1.8
75	28999	26448	646.7	377.0	218.6	127.0	73.9	42.9	24.9	14.5	8.4	4.9	2.6
76	21706	19796	484.0	282.2	163.7	95.1	55.3	32.1	18.7	10.8	6.3	3.7	1.9
77	9587	8743	213.8	124.6	72.3	42.0	24.4	14.2	8.2	4.8	2.8	1.6	0.9
78	18059	16470	402.7	234.8	136.2	79.1	46.0	26.7	15.5	9.0	5.2	3.1	1.6
79[a]	12497	11808	288.7	168.3	97.62	56.7	33.0	19.2	11.1	6.5	3.7	2.2	1.2
80[b]	11118	10140	247.9	144.5	83.8	48.7	28.3	16.4	9.6	5.5	3.2	1.9	1.0
81[b]	9289	8471	207.1	120.7	70.0	40.7	23.7	13.7	8.0	4.6	2.7	1.6	0.8
82[b]	7461	6804	166.4	97.0	56.2	32.7	19.0	11.0	6.4	3.7	2.2	1.3	0.7
83[b]	5632	5136	125.6	73.2	42.5	24.7	14.4	8.3	4.8	2.8	1.6	0.9	0.5
84[b]	3804	3469	84.8	49.4	28.7	16.7	9.7	5.6	3.3	1.9	1.1	0.6	0.3
85[b]	1975	1801	44.0	25.7	14.9	8.6	5.0	2.9	1.7	0.9	0.6	0.3	0.2

Note: The total population size of any year can be followed after maturity age (8 years) in addition with the others next ages. Example: The class 1979 will be formed after 8 years (1987) with ages 8 through 15, equal to 480,500 adults of both sexes.

a. Data not yet reconfirmed.

b. Fitted with: R = 0.649   m = 1828.629   b = 25747

tive stock of Mexico and Central America. Between them there is probably a certain amount of exchange, not only in individuals, but also genetically. However, there are no known means for differentiating individuals from these two populations from morphological characteristics. However, there may be differences that are natural as well as those induced by man: fishing mortality may have tremendous effect, although in a different manner from nonhuman pressures especially in breeding and feeding areas, where fishing causes selective mortality in the population principally on breeding-size females.

Despite the fact that breeding and feeding areas may be separated by several thousand kilometers, it is necessary to determine whether or not these populations should be analyzed as the same or different stocks. There is evidence from tagging returns that Mexican and Costa Rican stocks are mixed in South America, and it is necessary to define what proportion of new recruits represent population interchange. These data must be made available for management of the species in an international cooperative manner, not in the unilateral, isolated procedure as is done today.

As regards hatchlings, juveniles, and subadults, there is little information. However, they probably interact ecologically, although they occupy separate levels as indicated by the absence of animals below adult size in areas where adults normally occur.

Estimates of annual numbers of eggs produced by *Lepidochelys olivacea* in Mexico were derived from estimates of the number of seasonal nesting females (Figure 4), and the fecundity at the time of nesting (in prep.).

Annual survival rates with the changes during the life cycle (Table 1, Figure 3) were adjusted with the analysis of tagging and recapture data (Tables 2 and 3). These estimates were used to predict population size of various classes, especially for future populations of adults (Table 5). Estimates for age classes after sexual maturity have been presented in detail, this table can be used as a tool in diagnosing the population under investigation. In summary, the prediction for the relevant age class gives the theoretical adult population size for a given year. For example, the composition of the population in 1987 would have 168,300 8-year olds, plus 136,200 9-year olds, plus 42,000 10-year olds, etc.

It can be observed that in 1980 the population would contain 289,100 8-year olds of both sexes and in 1981 there would be 319,500 8-year olds plus 167,100 9-year olds. These values depend on the number of eggs hatched out 8, 9, 10, etc. years before, illustrating that the protection given now greatly determines the future of the species.

Acknowledgments

This work would not have been possible without the help of innumerable persons, principally the cooperative fishermen. In the last 3 years, after forming the "Turtle Fund" for the investigation of marine turtles, we have been able to count on an increase in material and financial help, and it was possible to increase the work on the nesting beaches. The Turtle Fund was formed not only by the cooperatives but also by private industry. Fundamental work and support were carried out by the personnel of the Direccion General de Regulacion Pesquera and the personnel of the Mexican navy. Special acknowledgment to Dr. J. Frazier of the Smithsonian Institution for his critical reading of the manuscript.

Literature Cited

Beverton, R. J. H., and S. J. Holt
1957. On the dynamics of exploited fish populations. *Min. Agric. Fish. and Food Fish. Invest.* 19 (series 2):1–533.

Doi, T.
1974. Outline of mathematical analysis on fish populations for practical use in front. In *Fisheries Biology and Population Dynamics of Marine Resources.* Japan: JICA.

Gulland, J. A.
1971. Ecological aspects of fishery research. *Advances in Ecological Research* 7:115–76.

Leslie, P. H.
1957. An analysis of the data for some experiments carried out by Gause with populations of the protozoa *Paramecium aurelia* and *P. caudatum. Biometrica* 44:314–27.

Lucas, C.
1975. A method for estimating mortality rates from tag recoveries when fishing is not constant. *Australian Journal of Marine and Freshwater Research* 26:75–79.

Márquez M., R., and T. Doi
1973. A trial of theoretical analysis of populations of Pacific green sea turtle, *Chelonia mydas carrinegra* Caldwell, in the waters of the Gulf of California, Mexico. *Tokai Reg. Fish. Res. Lab.* 73:1–22.

Ricker, W. E.
1975. Computation and interpretation of biological statistics of fish populations. *Fish. Research Board of Canada* 191:1–382.

Schaefer, M. B.
1954. Some aspects of the dynamics of populations important to the management of the commercial marine fisheries. *Inter-American Tropical Tuna Commission* 1:26–56.

Schaefer, M. B., and R. J. H. Beverton
1963. Fishery dynamics—their analysis and interpretations. *The Sea* 2:464–83.

RENÉ MÁRQUEZ M.

René Márquez M.
Aristóteles Villanueva O.
Manuel Sanchez Perez
Departmento de Pesca
Instituto Nacional de la Pesca
Mexico 7, D.F.

The Population of the Kemp's Ridley Sea Turtle in the Gulf of Mexico— *Lepidochelys kempii*

ABSTRACT

In this work on Kemp's ridley sea turtle, its life cycle is briefly described and used as a basis for the model. In it are included parameters of survival rates for different stages of its development, principally for egg survival during nesting and incubation, and from hatching until adult. It shows the individual fecundity or the seasonal average total number of eggs (140.8), the frequency of nesting (1.304 nests per turtle per season), and the cycle of reproduction which is repeated with yearly, biennial, and triennial, etc. patterns. Recruitment is approached in a preliminary way, and an R = 0.0572 was obtained. This is considered still somewhat low, but it has been improving in recent seasons.

A simple model is being developed to evaluate this seasonal nesting population through the fecundity index and following cohorts affected by the instantaneous survival rate. A curve of decrease in the population is developed and used for the assessment of the adult population size in the sea.

Introduction

The principal and only beach that now exists for the nesting of the Kemp's turtle, *Lepidochelys kempii* is located near the upper section of the Tropic of Cancer, between the mouths of the San Rafael River and the sand-bar of Ostionales and 4 km east of the village of Rancho Nuevo, in the municipality of Aldama, Tamaulipas.

Before the sixties it was clear in Mexico City that the trade and consumption of turtle eggs originated from this region, but it was not until 1966 that the Instituto Nacional de Investigaciones Biológico Pesqueras of the Direccion General de Pesca, initiated a systematic protection for sea turtles with the establishment of turtle camps (Chavez, Contreras, and Hernandez 1967). Since then the work has continued without interruption.

Later, to assure the future of the species and to strengthen the investigations, an official decree was

159

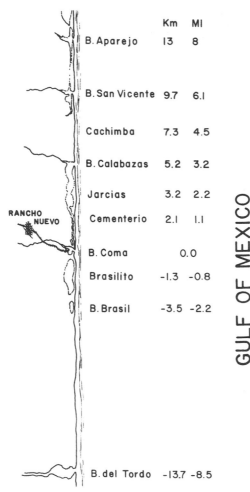

Figure 1. Kemp's ridley sea turtle nesting beach at Rancho Nuevo, Tamaulipas, Mexico.

written (6/1/77) designating the beach at Rancho Nuevo (Figure 1) as a zone of Natural Reserve, prohibiting any fishery under the range of 4 miles around it. This protected the ecology of the zone, which covers a strip of shore about 17.6 km of sandy beaches, and its corresponding ocean up to 50 fathoms. More details in respect to this were quoted by Márquez (1976).

During 1978 an agreement was reached between Mexico (Departamento de Pesca) and the United States (FWS/NMFS). Mexico would provide around 2,000 eggs and a given number of new hatchling turtles. The eggs and turtles are provided for the establishment of a new nesting population on the beach at Padre Island, Texas, and consequently an even greater chance for survival of the whole population.

The information on this species acquired over 14 years of work facilitates the development of the present model for the evaluation of the population of Kemp's ridley.

Life Cycle

Reproduction

The first results of unpublished data indicate that the Kemp's ridley has a reproductive cycle described as follows: turtles nesting every year, 58 percent = 1.000; turtles nesting every 2 years, 29 percent = 0.500; turtles nesting every 3 years, 13 percent = 0.224.

Fecundity

The total fecundity per turtle is obtained easily from the information on the number of times each turtle nests and how many eggs are in each clutch. Table 1 data are applied for the 1979 nesting season.

Nesting Population Size

The average number of eggs per clutch (h) is 105.48 (for the 1979 season) and, applying the following formula, gives a total number of turtles per season:

$$N = \frac{H}{I \times h}$$

where: H is the total number of eggs laid during the season, I is the index of fecundity, and h is the mean number of eggs per female per season.

Survival

In the life cycle (Figure 2) information was presented regarding the survival (Márquez et al., in prep.). This information was obtained by direct observation on the beach (for eggs and hatchlings) and from data during tagging and recapture of adults from the incidental catch and renesting observation (Tables 2 and 3). The survival rate (S_a) for intermediate stages were obtained by extrapolation and afterwards were adjusted in a straight line (Figure 4). The information obtained is shown in Table 5.

Total Mortality

Applying the logarithmic method (Doi 1974) to obtain the total mortality and using data of Tables 2 and 3, two similar results were obtained (Figure 3). The average values of annual survival (S) and mortality (Z) obtained for adults in 1979 are quoted in Table 4, and compared with those obtained in former years. It should be remarked that the species has improved slowly, year by year.

Table 1. Seasonal fecundity index for nesting females of Kemp's ridley sea turtle, *Lepidochelys kempii*

Situation	Frequency	Index	Average Eggs	σ
1	411	1.000	110.16	14.5
2	115	0.2798	101.90	17.1
3	16	0.0389	93.65	22.9
4	3	0.0073	85.39	
		1.3260	140.81	

RENÉ MÁRQUEZ M.

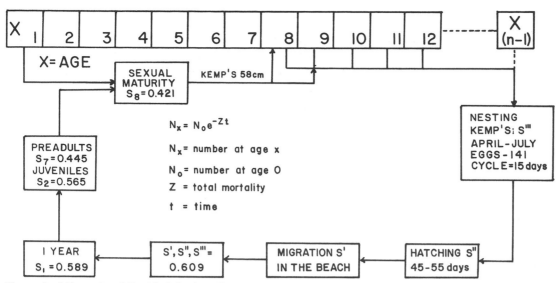

Figure 2. Life cycle of *Lepidochelys kempii*.

Table 2. Yearly information from tagging and tag recoveries of Kemp's ridley sea turtle, *Lepidochelys kempii*, in the Gulf of Mexico

Year	Number	66	67	68	69	70	71	72	73	74	75	76	77	78	79
1966	285	12	7	4	1	3	1	—	—	—	—	—	—	—	—
1967	271	—	9	5	1	1	—	—	1	—	—	—	—	—	—
1968	326	—	—	7	1	2	1	1	—	—	—	—	—	—	—
1969	86	—	—	—	1	3	—	—	—	—	—	—	—	—	—
1970	133	—	—	—	—	5	1	—	—	—	—	—	—	—	—
1971	—														
1972	41	—	—	—	—	—	—	—	1	—	—	—	—	—	—
1973	76	—	—	—	—	—	—	—	1	—	—	1	—	—	—
1974	77	—	—	—	—	—	—	—	—	1	—	1	—	1	—
1975	105	—	—	—	—	—	—	—	—	—	2	—	—	—	—
1976	127	—	—	—	—	—	—	—	—	—	—	1	1	—	—
1977	81	—	—	—	—	—	—	—	—	—	—	—	—	1	—
1978	251	—	—	—	—	—	—	—	—	—	—	—	—	4	—

1859

		r_0	r_1	r_2	r_3	r_4	r_5	r_6	—	—	—	—	—	—	—
Annual tag recoveries		43	20	8	4	5	1	1	—	—	—	—	—	—	—

Table 3. Yearly nesting and renesting information of Kemp's ridley sea turtle, *Lepidochelys kempii*, in Rancho Nuevo beach, Mexico[a]

Year	Number	t_0	t_1	t_2	t_3	t_4	t_5	t_6
1966	285	12	9	6	4	3	2	1
1967	271	9	5	2	1	—	—	1
1968	326	9	5	3	2	2	1	—
1973	76	1	8	—	5	—	—	—
1974	77	7	8	11	1	4	3	—
1975	105	8	16	1	2	2	—	—
1976	127	16	5	12	6	—	—	—
1977	81	10	9	13	—	—	—	—
1978	251	40	30	—	—	—	—	—
1979	371	97	—	—	—	—	—	—
	1970	209	95	48	21	11	6	2

a. Where, t_0 = same year, t_1 = next year, etc.

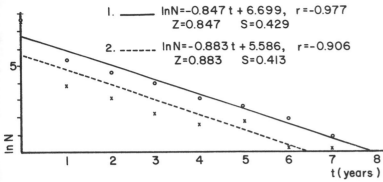

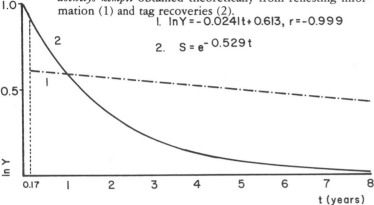

I. $\ln N = -0.847t + 6.699$, $r = -0.977$
$Z = 0.847$ $S = 0.429$

2. $\ln N = -0.883t + 5.586$, $r = -0.906$
$Z = 0.883$ $S = 0.413$

Figure 3. Total mortality (Z) and survival rate (S) for *Lepidochelys kempii* obtained theoretically from renesting information (1) and tag recoveries (2).

I. $\ln Y = -0.0241t + 0.613$, $r = -0.999$

2. $S = e^{-0.529t}$

Figure 4. Annual survival rate (broken line) and the instant rate of decrease of the population (solid line) for *Lepidochelys kempii*, derived from the life cycle, tagging and tag recoveries and virgin stock (eggs produced by season).

Table 4. Yearly changes in total mortality and survival rate in adults, gathered from tagging and tag recoveries

Parameter	1977	1978	1979
Z	0.897	0.883	0.847
S	0.408	0.413	0.428

With the theoretical information from the last column in Table 5, a curve was made in Figure 4 which represents the instantaneous change of population and utilizes the value of $Z = 0.529$ for total mortality (Table 5). Starting with the first calendar year, all details of the instant decreasing rate of the population curve were adjusted; later on this curve was used to form the theoretical table of the annual change of population.

Conclusions

Density of the Breeding Population

An estimation of the number of turtles in the sea may be obtained by several methods. The most adequate in this case, with the available information, was the general survival equation, already mentioned in the above section and used by authors like Ricker (1975) and Doi (1974). This equation being multiplied by the number of individuals in the virgin stock gives the population number at a given time as follows (Figure 2):

$$N_x = N_o e^{-Zt} = N_o S$$

where N_o is the total spawning size, Z is the total mortality, t is time in years and N_x is the number of individuals at age x.

The results of these calculations are shown in Table 6, which can be used to get the population diagnosis at any moment, within the limits that are quoted in it and following the cohorts. For example, in the year 1978, 97,900 eggs were laid on the beach. If 7 years is considered enough to reach the age of sexual maturity, by 1985 there will be around 2,410 turtles that are 7-years old, plus the previous cohort that will then be 8-years old plus those of 9 years, etc. Adding all these individuals gives a population total of 4,272 adults of both sexes in 1985.

It is necessary to clarify certain points on this table in relation to the first two columns of H_r and N_v. H_r is the total annual number of protected hatchlings liberated on the beach (thousands of individuals). N_v is a back calculation and standardized data, as the calculation is based on virgin stock, and indicates thousands of eggs theoretically laid on the beach, in accordance with actual liberated hatchlings. $N_v = H_r/Z$ where $Z = 0.496$ (mortality during nesting and incubation). Of course Z changes every year, but this figure is used in a general mode to develop Table 6.

Table 5. Theoretical estimations of annual survival rate and total mortality for *Lepidochelys kempii* derived from the life cycle and tag recoveries

	S_0	S_a	$\hat{S}_a$	$\hat{Z}_a$	$S = e^{-Zt*}$
S_v	1.000	1.000	1.000	0.000	1.000
$S_{0.17}$	0.608	0.608	0.609	0.496	0.914
S_1	—	0.590	0.589	0.529	0.589
S_2	—	0.565	0.565	0.571	0.347
S_3	—	0.541	0.541	0.614	0.205
S_4	—	0.518	0.517	0.660	0.121
S_5	—	0.492	0.493	0.707	0.0710
S_6	—	0.469	0.469	0.757	0.0418
S_7	—	0.444	0.445	0.810	0.0246
S_8	0.421	0.421	0.421	0.865	0.0145
S_9	—	—	0.397	0.924	0.00856
S_{10}	—	—	0.373	0.986	0.00504
S_{11}	—	—	0.348	1.056	0.00297
S_{12}	—	—	0.324	1.127	0.00175

— No data.

* Was fitted with $Z = 0.529$, for full calendar years.

RENÉ MÁRQUEZ M.

For this reason some data in column N_v and $N_{0.17}$ seem unreal.

Recruitment

Certain preliminary results can be indicated with regard to recruitment (R). This parameter was obtained through the change of population which is analyzed in Table 6, column N_7 and it was adjusted by a logarithmic linear regression. The results are shown in Figure 5. For this calculation, an estimated average of three year periods was used and the annual rate of recruitment was 0.0572, for the period between 1966–79.

Discussion

For a complete description of the life cycle it is necessary to gather more information on habits, behavior and biological phases of each species. With *L. kempii,* and in general for all the marine turtles, little is known of the survival rate from the time when the hatchlings first go to the sea until they return to coastal waters. Hence, in the present study these values were estimated. Data from tagging returns, and hatching success from the hatchery were used to estimate the instantaneous decrease of population, from the virgin stock, that is, total number of eggs laid per season, up to 15 years of age. Table 6 was prepared from these estimates. Total population size of adults is estimated by adding respective values of population sizes for each existing cohort. For example, the estimate for the 1978 population is the sum of the adults recruited in 1978, plus those surviving from previous cohorts, as indicated by the diagonal line.

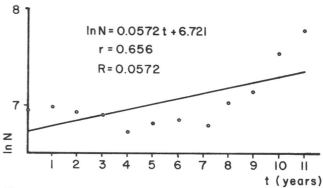

Figure 5. Theoretical annual rate of recruitment for *Lepidochelys kempii,* adjusted by logarithmic linear curve. Period 1967–1978.

On this occasion, valuable information was obtained on the fecundity, cycle of nesting, and the average number of eggs per clutch. This is basic for population size estimation, through the use of the evaluated number of females nesting every year. It is worth mentioning that in this work, the corresponding analysis of the information is not completed, but the method for doing this is described.

Regarding the recruitment, as may be observed, it is still low, but it should be clarified that in these last years a positive change has been seen from R = 0.0453 to R = 0.0572, which is a good indication of the recuperation of the species. It should also be noted that during the present year (1979) there was an extraordinary case, on May 20, a small arrival of 20 to 30 nesting females came ashore on Lauro Villar (Washington Beach) near the border between Mexico and Texas, about 400 km north of Rancho Nuevo. This event may be encouraging with regard to the situation of the species or it may be an accidental case due to

Table 6. Theoretical change of populations of *Lepidochelys kempii*

Year	H_r	N_v	S = 0.914 $N_{0.17}$	S = 0.0246 N_7	S = 0.0145 N_8	S = 0.0086 N_9	S = 0.0050 N_{10}	S = 0.0030 N_{11}	S = 0.0017 N_{12}	S = 0.0010 N_{13}	S = 0.0006 N_{14}	S = 0.0004 N_{15}
1966	22.1	44.56	40.73	1096	646	381	225	132	78	46	27	16
1967	20.7	41.73	38.14	1027	605	357	210	124	73	43	25	15
1968	20.8	41.94	38.33	1032	608	359	211	125	73	43	25	15
1969	23.5	47.38	43.31	1166	687	406	239	141	83	49	29	17
1970	18.2	36.69	33.53	903	532	314	185	109	64	38	22	13
1971	17.7	35.69	32.62	878	518	306	180	106	62	37	22	13
1972	14.4	29.03	26.53	714	421	248	146	86	51	30	18	10
1973	22.3	44.96	41.09	1106	652	385	227	134	79	46	27	16
1974	20.3	40.93	37.41	1007	593	350	206	122	72	42	25	15
1975	11.2	22.58	20.64	555	327	193	114	67	40	23	14	8
1976	36.9	74.40	68.00	1830	1079	637	375	221	130	77	45	27
1977	28.9	58.27	53.26	1433	845	499	294	173	102	60	36	21
1978	48.6	97.98	89.55	2410	1421	839	494	291	171	101	60	35
1979	65.4	131.85	120.51	3244	1912	1129	665	392	231	136	80	47

Note: The total population size of any year can be estimated after maturity (7 years) by adding succeeding year classes. Example: the class of 1978 will be formed after 7 years (1985) with ages 7 through 15, equal to 4272 adults of both sexes.

the environment. It is clear, however, that recruitment must be studied more fully as it is a definite basis for its protection.

Acknowledgments

We appreciate the valuable help and support of the Delegacion de Pesca y Centro de Investigacion Pesquera de Tampico, through its representatives Sr. Jose Garcia and Biol. Sergio Garcia Sandoval.

Also, special mention should be made of the people of Rancho Nuevo who every year have been most understanding with respect to the project, and especially in memory of Sr. Juan Gonzalez and the Gonzalez family, who, apart from their help, have been most friendly.

These last years, a group of 5 volunteer North American students coordinated by Dr. Peter Pritchard has given valuable help. These students have reinforced the work and given it a new impulse for the protection of the species.

Special mention should be made of Sra. Aliveth Suarez for her patience, understanding and typing of the work, and Biol. Mirna Cruz for her review of the manuscript.

Literature Cited

Chavez R., H.; M. Contreras G.; and E. Hernandez D.
1967. Aspectos biologicos y proteccion de la tortuga lora, *Lepidochelys kempi* (Garman), en la Costa de Tamaulipas, Mexico. *Inst. Nal. de Inv. Biol. Pesq.* 17:1–40.

Doi, T.
1974. Outline of mathematical analysis on fish population for practical use in front. In *Fisheries Biology and Population Dynamics of Marine Resources*. Japan:JICA.

Márquez M., R.
1976. Reservas naturales para la conservacion de las tortugas marinas de Mexico. *INP Serie Informacion* 83:1–22.

Ricker, W. E.
1975. Computation and interpretation of biological statistics of fish populations. *Fish. Research Board of Canada* 191:1–382.

RENÉ MÁRQUEZ M.

James I. Richardson
Thelma H. Richardson
Institute of Ecology
University of Georgia
Athens, Georgia 30602

An Experimental Population Model for the Loggerhead Sea Turtle (*Caretta caretta*)

ABSTRACT

Sixteen consecutive years of intensive tagging surveys on Little Cumberland Island, Georgia, provide a data base for establishing a population model for nesting female loggerhead sea turtles (*Caretta caretta*). The model incorporates frequency of remigration intervals, probabilities of remigration, and fecundity. The model predicts annual recruitment (39 percent of nesting females), mean longevity of nesting females (3 years as adults), and turnover of nesting females (6 years). The model represents a hypothetical population with stationary age distribution; its similarity to the Little Cumberland population is implied. A survivorship curve is constructed for a hypothetical cohort of female turtles, although age to maturity and survivorship of the juveniles is not known. The cohort replaces 50 percent of itself during the first 3 nesting seasons, 90 percent during the first 13 nesting seasons. Computer simulation through time provides predictions of population doubling times resulting from various changes in hatchling recruitment, juvenile survival, and age to maturity. Observed recruitment to the seasonal nesting population appears to be a potentially sensitive indicator of changes in hatchling production, if hatchlings return to nest on their natal beach. The model can be used to simulate the effects of periodic destruction of the eggs by predators or adverse beach conditions.

Introduction

Population models, particularly those which predict population numbers within statistically defined confidence limits, have been noticeably lacking from the marine turtle literature. Population models for other species of animals are generally constructed from life table data, but certain essential parameters, such as survivorship and age to maturity, continue to elude marine turtle investigators (Bustard 1979). Portions of sea turtle population models have been appearing in the literature for a number of years. Hughes (1974) used annual egg production, egg survival, and observed

165

recruitment to a population of adult nesting female loggerheads to estimate juvenile survival rates. Various investigators have been using remigration-interval frequencies and seasonal population counts to estimate total population size of nesting females. Carr, Carr, and Meylan (1978) provides a general equation for calculating this parameter estimate. Bustard and Tognetti (1969) developed a model for density-dependent population regulation through the mechanism of intraspecific nest destruction.

In this paper we collect existing information on various life history stages of a Georgia loggerhead population and incorporate these data into a population model (Hillestad, Richardson, and Williamson 1977; Richardson and Hillestad 1978). The data upon which this population model is based have been gathered primarily from Little Cumberland Island, Georgia. Our information is derived from beach tagging studies of nesting females and the survival of their eggs on the beach. We do not know the survivorship of the hatchlings to maturity or the average number of years required to reach maturity. Information on the larger (carapace length 45 to 90 cm) juvenile loggerheads has been derived entirely from records of turtles found dead on the beach and those caught incidentally by

shrimp trawlermen (Hillestad et al, this volume).

Because of our uncertainty of the juvenile life history stages, the Little Cumberland population model is still incomplete. It simulates survivorship and fecundity of nesting females after they enter the nesting population, but the model cannot predict population response to changes in hatching success without knowing survivorship and age to maturity of the juveniles. In this respect, the Little Cumberland population model, in its present form, cannot be used as a management device to determine acceptable predation levels or harvest quotas. Survivorship of Little Cumberland nesting females is undoubtedly determined by a complex blend of mortality factors: distant (sharks, food supplies, oceanic weather patterns), local (boat collision, drowning in shrimp trawl nets), and other mortality factors, (parasites, for example). The similarity of this survivorship to the survivorship of loggerheads from other geographic areas or to other species of sea turtles remains to be seen. The Little Cumberland population model represents a preliminary effort to predict population behavior from existing data. The model is meant to stimulate further investigation into marine turtle population models and not to provide definitive predictions of animal numbers, although the latter is the ultimate goal of any population modeling effort.

The Little Cumberland loggerhead project, now entering its eighteenth year of operation, is sponsored by the National Audubon Society and funded in its entirety by the owner-members of the Little Cumberland Island Association. This nesting population study owes its continuity and longevity to the continuing financial support from association members and to the dedicated work of the 24 research assistants who were involved in the all-night field work required by an intensive sea turtle nesting survey. We would also like to recognize the other Georgia nesting studies, particularly those on the adjacent islands of Cumberland and Jekyll. These federal, state, and privately supported studies place the Little Cumberland population data within the context of the entire Georgia coast (Figure 1) and provide critical information concerning nesting overlap between islands.

Results and Discussion

Population Parameters for Nesting Females

Seventeen consecutive years (1964–1980) of population tagging data have been collected from adult female loggerheads nesting on Little Cumberland Island (Richardson et al 1978). During the first 10 years, a 3-year cycle in the numbers of nesting females appearing each year was fairly consistent (Figure 2a). Significant damping of this cycle occurred for unknown reasons at some point proximate to the 1973 season. Nesting studies on other Georgia islands were not in

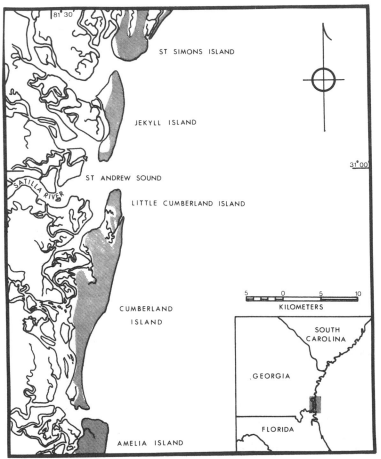

Figure 1. Map of coastal Georgia, showing locations of islands with populations of nesting loggerhead sea turtles.

JAMES I. RICHARDSON

existence for enough years at the time to corroborate this apparent crash in numbers of nesting females recorded for Little Cumberland. Recent evidence (1973–80) suggests that the Little Cumberland population of adult females may be rebuilding from the post-1973 low, as well as maintaining a 3-year cycle (Figure 2b).

It required 6 years to tag the majority of existing members of the Little Cumberland nesting female population, an exercise that must be completed before recruitment (the percentage of unmarked individuals appearing each season) can be identified (Figure 2b). The proportion of recruits to total turtles has remained approximately between 30 percent and 40 percent a season since the 1973 crash (Figure 2c). We are confident, because of the other Georgia tagging programs, that the observed recruitment in recent years on Little Cumberland Island is real and not the result of wandering nesting from adjacent populations. Furthermore, evidence from the Georgia tagging program supports our contention that loggerheads rarely, if ever, shift to another nesting beach when the preferred or "home" nesting beach is altered or destroyed (Bell and Richardson 1978).

The Little Cumberland population model is derived from the following data (Richardson et al. 1978):
• Types of remigration intervals occur with characteristic frequencies.

Interval	Frequency percentage
1-year	3
2-year	56
3-year	31
4-year	7
5-year or more	3

• A turtle appearing on the nesting beach for the first time (neophyte turtle) will remigrate (return) to nest during a subsequent nesting season with a 49 percent probability.
• A turtle returning for at least a second nesting season (remigrant turtle) will remigrate for at least a third nesting season with 70 percent probability.

Confidence in these properties of loggerhead population behavior is an essential part of the predictions that follow. Because of the intensity of the beach coverage, virtually all nesting females on Little Cumberland are observed (tagged) each season. Frequencies for Little Cumberland remigration intervals are based on 453 observations from 1964 to 1975. Remigration rate probabilities have been adjusted to compensate for error due to tag loss (Richardson et al. 1978). Nesting beach surveys on either side of Little Cumberland have been recording the presence of Little Cumberland tagged turtles since 1972 (Figure 1). An analysis of interisland nesting overlap has shown that less than 2

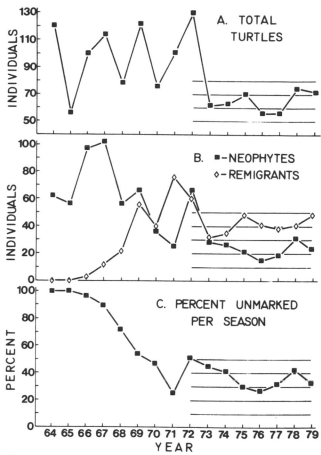

Figure 2a. Fluctuation in the annual number of nesting female loggerhead turtles at Little Cumberland Island, Georgia.

Figure 2b. Total number of neophyte and remigrant nesting loggerheads observed per season at Little Cumberland Island, Georgia.

Figure 2c. Percentage of the annual number of nesting loggerheads not previously observed on Little Cumberland Island, Georgia.

percent of Little Cumberland turtles nest on distant (≥10 km) beaches where they might be missed by tagging crews (Richardson et al. 1977).

The above behavioral characteristics of Little Cumberland nesting loggerheads are used to generate the model, a pattern of predicted nesting appearances for a theoretical population of 1,000 neophyte turtles followed for 25 years (Table 1, column A). Appendix I provides a detailed description of this and all other calculations involved in the development of Table 1. The decision to limit the lifespan of a sea turtle to 25 years is arbitrary, although at least 1 loggerhead has now nested on Little Cumberland Island for a period spanning 17 nesting seasons.

The Little Cumberland model, in its most simple form, simulates a population with a stationary age distribution. Under this hypothetical situation, the pro-

Table 1. A 25-year remigration and survivorship table for cohorts of 1,000 and 389 neophyte female loggerhead turtles entering the breeding population

Years of reproductive maturity	A Seasonal remigration (1,000 cohort)	B Seasonal remigration (389 cohort)	Survivorship		
			C Turtles alive at end of year (389 cohort)	D Average mortality per year (percentage)	E Number of eggs produced (389 cohort)
0	1,000	389	389	—	116,700
1	15	6	296	24	1,800
2	274	106	234	21	31,800
3	163	63	190	19	18,900
4	147	57	157	17	17,100
5	143	56	134	15	16,800
6	110	43	116	13	12,900
7	103	40	101	13	12,000
8	86	33	88	13	9,900
9	76	30	77	13	9,000
10	68	26	67	13	7,800
11	57	22	58	13	6,600
12	51	20	50	13	6,000
13	44	17	44	13	5,100
14	39	15	38	13	4,500
15	33	13	33	13	3,900
16	30	12	29	13	3,600
17	26	10	25	13	3,000
18	23	9	22	13	2,700
19	20	8	19	13	2,400
20	18	7	17	13	2,100
21	15	6	14	13	1,800
22	13	5	13	13	1,500
23	11	4	11	13	1,200
24	9	3	9	13	900
25	0	0	0	100	0
	2,574	1,000	2,231		300,000

portion of the population in each age class remains constant and the total number of animals in the population does not change. Natural populations rarely maintain stationary age distributions, and the reduction of total nesting females ca 1973 implies the same is true of the Little Cumberland population. However, much can be implied about the behavior of a natural population by comparing its behavior to a model with known properties.

If a stationary age distribution is assumed, the model predicts that neophytes will constitute 39 percent of the seasonal population (1,000/2,574 from Table 1, column A). This prediction agrees with the independ-

ent observation that neophytes have accounted for approximately 30 percent to 40 percent of the seasonal nesting population for the last five nesting seasons (Figure 2c) on Little Cumberland Island. A predicted remigration pattern for 389 neophytes (Table 1, column B) satisfies the more general case of 1,000 nesting turtles per season. Hughes (1974) predicted that annual recruitment occurred somewhere between zero and 56 percent for South Africa loggerheads.

Since annual remigrant turtles are seen every year, 2-year remigrants seen every other year (50 percent of these animals are seen each year), and 3-year remigrants seen every third year (33 percent of these

JAMES I. RICHARDSON

animals are seen every year), and so on, a predicted 44 percent of the total Little Cumberland population of nesting females will be on the nesting beach each season. Thus, a seasonal population of 1,000 nesting females would represent a total population of 2,273 adult females. Carr, Carr, and Meylan (1978) provides a general formula for predicting this value.

Based on the remigration pattern predicted for a cohort of 360 neophytes (Table 1, column B), a survivorship table can be generated (Table 1, column C) with associated mortality rates which will be sustained during each year of maturity (Table 1, column D). Mortality, in this sense, means disappearance from the nesting population and not death per se. Actual mortality is only implied. Mortality rates during the first 5 years of maturity have been adjusted to fit the observed remigration probabilities of neophytes (49 percent) and remigrants (70 percent) on Little Cumberland Island.

The Little Cumberland model considers only female turtles; males do not enter the nesting population and are, therefore, lumped with mortality. Adjusting the model to reflect a specific sex ratio (if known) would be a simple task and not significantly affect the model's structure or predictive abilities.

The model predicts the following population attributes:
• Fifty percent of a cohort of neophyte adult females will be gone from the nesting population by the third year of their initial appearance, and 90 percent will be gone after 14 years.
• Nesting turtles observed during a single season represent 44 percent of the existing population. In other words, 56 percent of the total adult female population does not nest in any given year. Little Cumberland nesting loggerheads do not appear in the vicinity of Little Cumberland during their off-nesting years (Bell and Richardson 1978).
• Neophytes are recruited to the existing population at the rate of 39 percent of the seasonal nesting population per year. If total population numbers are stationary, an equivalent number of adult females would be expected to leave the nesting population each year.
• Population turnover occurs every 6 years (2,231/389 = 5.7). In other words, a number of individuals equivalent to the total population size is replaced every 6 years.

Fecundity and Survival of Eggs

Little Cumberland loggerheads lay approximately 2.5 clutches of eggs per turtle per season, and clutches average 120 eggs per clutch (unpublished data). The model predicts that an average Little Cumberland turtle will produce 771 eggs during her lifetime (Table 1, column E). Individual performances vary widely;

some turtles appear on the beach but never nest, while others nest repeatedly during a season and over several seasons, accounting individually for several thousand eggs per turtle (Richardson et al. 1978). Hughes (1974) predicts that an average South African loggerhead will lay 4.5 times per season for 4 nesting seasons. This amounts to a lifetime production of 2,052 eggs, given 114 eggs per clutch.

If a population could maintain a stable age distribution, the survival rate of hatchlings reaching maturity would be equal to the number of adults entering the nesting population, divided by the number of hatchlings produced per season. Similarly, the survival rate of eggs becoming mature turtles is equal to the product of the survival rate of eggs on the beach and the survival rate of hatchlings reaching maturity. Hughes (1974) estimated in this manner that no more than 1 or 2 South African loggerhead hatchlings per 1,000 reach maturity after entering the sea, regardless if annual recruitment of nesting females to his population is 10 percent or 50 percent.

The Little Cumberland population model suggests that a seasonal population of 1,000 nesting females would be expected to lay 300,000 eggs during a season, from which 389 females must survive to maturity to satisfy the stationary age distribution of the model. Figure 3 illustrates the relationship between survival of eggs on the beach and the survival to maturity of hatchlings at sea, under this hypothetical situation. If 20 percent of the eggs fail to hatch (optimum hatchery conditions), then 2 hatchlings per 1,000 must survive

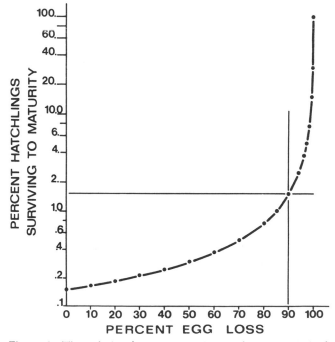

Figure 3. The relation between egg loss and the survival of hatchling to maturity, for a theoretical case where 389 female loggerheads reach sexual maturity per 300,000 eggs laid.

to enter the nesting population as mature females. If 90 percent of the eggs fail to hatch, then 13 per 1,000 of the remaining hatchlings must survive. Clearly, the model suggests that Little Cumberland loggerheads are capable of sustaining considerable variation in egg loss without having to compensate these losses with large variations in hatchling survival rates. However, the actual survival of Little Cumberland hatchlings is not known, so the model cannot yet be used to make statements about sustainable egg loss.

Age to maturity has not been measured for free-ranging Georgia loggerheads, but preliminary estimates from Australia exceed 20 years (Limpus 1979). If mortality of the juveniles is independent of population density and if hatchlings return to their natal beach, then the Little Cumberland population model predicts that the hatchery program (70 percent hatching success; an average 8,000 hatchlings released per season for the last 15 seasons) should ultimately increase recruitment by a factor of 2.3 if previous egg loss was 70 percent, 3.5 if egg loss was 80 percent, and 7.0 if egg loss was 90 percent. If hatchlings, upon reaching maturity, disperse to surrounding nesting

beaches, then evidence of recruitment will be more difficult to identify. There has been no evidence to date of increased recruitment to the Little Cumberland population (Figure 2c).

Survivorship Curve for Georgia Loggerheads

A survivorship curve provides a graphic interpretation of mortality sustained by a cohort of even-aged organisms through time. Figure 4 depicts the survivorship curve for the Little Cumberland population model. The total number of eggs (300,000) produced during a nesting season by 1,000 female lggerheads (2.5 clutches per turtle; 120 eggs per clutch) is taken as the initial cohort.

An 89 percent loss of eggs on the beach has been arbitrarily chosen since it approximates observed natural mortality (Hopkins et al. 1979). We believe that predation of the eggs on Little Cumberland Island was very high (± 90 percent) for many years prior to the initiation of the hatchery program in 1965 and that the effects of the hatchery on recruitment to the nesting population have yet to be observed because of the many years it apparently takes for wild sea turtles to reach maturity. An 89 percent loss of eggs serves, also, to illustrate a useful attribute of survivorship curves. The vertical axis of the plot (Figure 4) is projected on an exponential scale; absolute vertical distances on the figure are a measure of relative importance. The vertical distance between eggs and hatchlings in the figure is half the vertical distance between hatchlings and neophytes. In relative terms, this means that the loss of 89 percent of the eggs on the beach is only half the magnitude of the loss sustained by hatchlings during the juvenile years at sea. The survivorship curve illustrates the resilient nature of the modeled population to perturbations from egg predation. The similarity of the model to the natural Little Cumberland loggerhead population is suggested.

Little is known about actual survivorship of juvenile loggerheads. Several alternative, theoretical survivorship patterns for the juvenile years are presented in Figure 4. Alternative I illustrates attrition primarily to the very young juveniles, while alternative II illustrates attrition primarily to the older juveniles. The age class most commonly drowned in shrimp nets along the Georgia coast (over-the-curve carapace length 55 to 70 cm) has been arbitrarily located on Figure 4 for illustrative purposes. If alternative I represents the natural population, then shrimping losses would be causing a maximum impact on the nesting population; sustainable losses would be low. If alternative II best represents the real population, then the number of turtles drowned in shrimp nets would probably not have as much effect on population structure; sustainable losses could be higher. Actual survivorship is

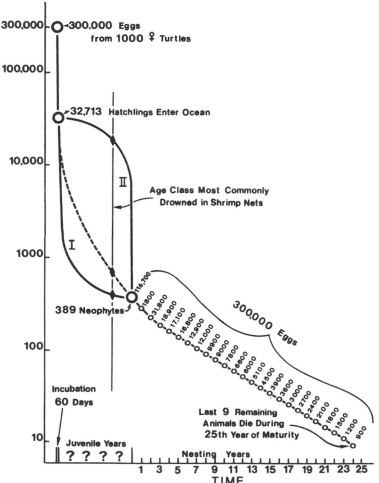

Figure 4. A survivorship curve for a cohort of 300,000 eggs deposited by 1,000 female loggerheads during a single breeding season in Georgia.

JAMES I. RICHARDSON

probably closer to a pattern suggested by the dotted line (Figure 4).

The Little Cumberland model predicts that 389 females from the original cohort of 300,000 eggs will reach sexual maturity and enter the nesting population as neophyte female turtles. These 389 animals must produce 300,000 eggs during their combined lifetimes in order to replace the original cohort. The survivorship curve predicts that 50 percent of the original cohort (300,000 eggs) will be replaced within 3 nesting seasons and that 75 percent will be replaced within 7 nesting seasons. The model clearly illustrates that long-lived adults (10 or more years in the reproductive population) are of reduced ecological importance relative to the combined reproductive potential of the younger members of the population. The last 10 remaining animals are arbitrarily terminated in the survivorship curve during their twenty-fifth year of maturity. At this age, their combined reproductive effort for each additional season survived is less than 0.3 percent of the cohort's total reproductive effort.

Model Simulation

The survivorship curve can be used to simulate changes in population structure for any number of generations, with the help of a computer. The population model will maintain a stationary age distribution for an indefinite number of generations, provided that egg mortality and juvenile mortality combine to produce a recruitment of 389 neophyte nesting females from each 300,000 eggs deposited on the beach. If recruitment is altered, then population numbers will change accordingly. The following examples illustrate the variety of questions (hypothetical conditions) that can be tested with the model. The answers (model predictions) are not meant to imply actual conditions but to provoke discussion of possible population behavioral characteristics.

Figure 5 illustrates the predicted change in seasonal numbers of nesting females if juvenile survival is held at 0.4 percent and egg survival is raised from 30 percent and maintained at 70 percent, which might be expected from the sudden application of a hatchery program. Age to maturity has arbitrarily been set at 15 years, and the introduction of neophytes into the adult nesting population is distributed over a 5-year period around this 15-year maturation date. The effect of increased egg survival is first noticeable on the fourteenth year, particularly the sudden rise in annual recruitment from 39 percent to greater than 50 percent. The population doubles in size by the twenty-second season and doubles again by the thirty-eighth season. Population size grows at a geometrically increasing rate, while recruitment ultimately stabilizes at 48 percent.

Table 2 lists doubling times for selected simulations

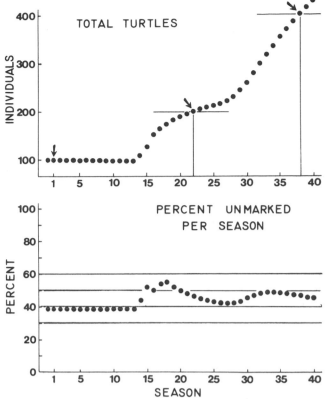

Figure 5. A simulation of loggerhead population behavior through 40 generations. The term "individuals" relates to the number of adult females observed on the nesting beach per season. Conditions are as follows: 70 percent egg survival; 0.4 percent juvenile survival; 15 years for age to maturity.

Table 2. Population doubling times or halving times (−) in years, for selected simulations of the Georgia loggerhead population model

Egg survival	Juvenile survival	Years to maturity	Doubling time in years			
			1	2	3	4
.30	.004	—	∞	∞	∞	∞
.11	.012	—	∞	∞	∞	∞
.70	.004	10	17	11	11	11
.70	.004	15	22	16	16	16
.70	.004	20	27	21	21	21
.70	.010	15	15	5	10	6
.20	.020	10	12	10	8	9
.00	—	10	12−	5−	4−	4−

of the population model. The results reinforce the conclusions drawn from Figure 3 that very small changes in juvenile survivorship have greatly magnified effects on population growth, while changes in egg survival have less effect. If pre-hatchery conditions on Little Cumberland Island were characterized by 6 percent survival of the eggs and 2 percent survival of the juveniles, then a hatchery program (70 percent survival

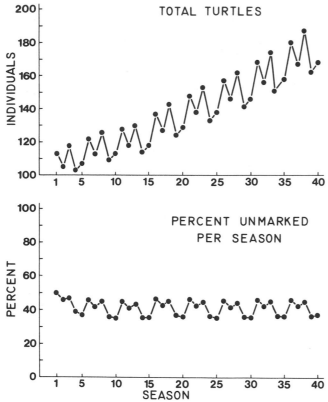

Figure 6. A simulation of loggerhead population behavior through 40 generations. The term "individuals" relates to the number of adult females observed on the nesting beach per season. Conditions are as follows: egg survival on a 5-year cycle (5 percent, 5 percent, 33 percent, 5 percent, 32 percent;) 1 percent juvenile survival; 10 years for age to maturity.

of the eggs) would be expected to exert a dramatic effect on nesting female numbers as soon as the first cohort of juveniles reached maturity. If, on the other hand, juvenile survival were 0.4 percent and age to maturity 15 years, a similar hatchery program would require 22 years for the population to double its original numbers and 16 years for each doubling time thereafter.

Population simulations are also useful for investigating intermittent recruitment into a population such as might occur from periodic predation on the eggs or other natural disasters that might befall eggs and young. In Figure 6, juvenile survival is 1 percent, age to maturity is 10 years, and survival of the eggs shifts on a 5-year cycle (5 percent, 5 percent, 33 percent, 5 percent, 32 percent). The resultant behavior of the model is meant to simulate the observed 2- and 3-year cyclical behavior of numbers of nesting females in natural populations. Note that the predictions fail to simulate observed behavior, corroborating a statement by Hughes (1974) that natural population fluctuations in numbers of nesting females are not logically caused by periodic variations in nesting success and juvenile survival. Specifically, the model is unable to simulate natural changes in numbers of adult females (Figure 2a) by changes in

hatchling survivorship, and recruitment to the simulated adult population varies more than is observed in natural populations. A feeding or migratory strategy of the adults is the logical alternative explanation.

A significant fluctuation in the numbers of nesting females can be simulated with intermittent recruitment pulses, if such pulses are large and separated by several years of reduced egg survival, as might occur on a coastal island with periodic dieoffs of natural predators (raccoons). In Figure 7, juvenile survival is 1 percent, age to maturity is 15 years, and egg survival varies over a 10-year cycle (5 percent, 5 percent, 5 percent, 5 percent, 5 percent, 5 percent, 5 percent, 5 percent, 70 percent, 50 percent). A change in the recruitment of adult females of this magnitude would probably be observable in a natural population. An additional observation from this simulation is that total numbers of nesting females are actually increasing through time, even though population numbers exhibit decreasing trends which last for 7 years at a time. A 4-year study of a population such as this could indicate either dramatic increases or catastrophic declines in population numbers, depending on which year the study is initiated. The need for long-term population studies is apparent.

Conclusions

Marine turtles may prove to be unique among wildlife species, in that a management decision by one man may not become apparent in the turtle population until an entire human generation has passed. Present-day changes in population numbers of nesting loggerheads at Cape Romaine National Wildlife Refuge, South Carolina, may be reflecting predator management policies of the 1950s (S. Hopkins personal communication). Little Cumberland Island has maintained an expensive, time consuming hatchery program since 1965, but no measurable effect has yet to appear. The question is what, if any, these efforts and activities have had on population numbers. Predictive simulation models are one way to suggest management and research approaches to populations with unusually long time lags.

Data from the Little Cumberland nesting survey have been used to derive a preliminary population model that simulates long-term population responses to changes in recruitment. The strength or usefulness of the model depends upon the reliability of the data upon which it is built. Some parameters, such as remigration rate, are approximations. Others, particularly age to maturity and juvenile survivorship, are entirely unknown and can only be surmised at the present time. Inspite of its uncertainties, the Little Cumberland model provides predictions that can be tested against real data and focuses attention on needed areas of research. How closely this model approximates the actual Little Cum-

JAMES I. RICHARDSON

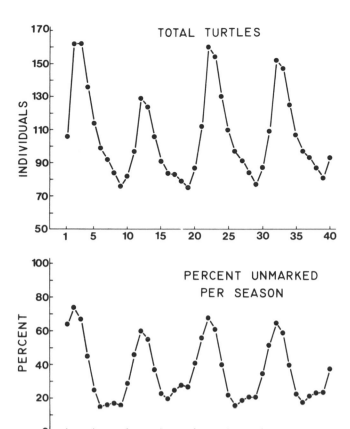

Figure 7. A simulation of loggerhead population behavior through 40 generations. The term "individuals" relates to the number of adult females observed on the nesting beach per season. Conditions are as follows: egg survival on a 10-year cycle (5 percent, 5 percent, 5 percent, 5 percent, 5 percent, 5 percent, 5 percent, 5 percent, 70 percent, 50 percent); 1 percent juvenile survival; 15 years for age to maturity.

berland population, or other sea turtle populations, remains to be seen.

Every effort should be made to derive life table information from existing, long-term population studies. The results could revolutionize the application of population models as tools for the management and protection of sea turtle populations throughout the world. Population parameters, particularly those which have been approximated, must be discussed and challenged by the various groups involved in long-term population studies. Only in this way will there be developed a sense of confidence concerning population models and their predictions.

Most population data have been collected from nesting female turtles and hatchlings. These data will become increasingly reliable with the continuing development of persistent flipper tags. In contrast, juvenile loggerheads, with carapace lengths of 50 to 90 cm, represent a critical life history stage about which almost nothing is known. These young animals represent the majority of carcasses found dead on U.S. beaches; most

of them are presumed drowned in trawl nets. Population studies of juveniles, though requiring expensive boat time, would be worth the investment in terms of increased management expertise and understanding. Growth rates, age to maturity, and a knowledge of juvenile population stocks are critical unknowns, blocking all present attempts to model population behavior over multiple generations with realism. Quantitative studies of juvenile populations should receive the highest priority, if realistic population models are to be realized in the near future.

Literature Cited

Bell, R., and J. I. Richardson
1978. An analysis of tag recoveries from loggerhead sea turtles (*Caretta caretta*) nesting on Little Cumberland Island, Georgia. *Florida Marine Research Publication* 33:20–24.

Bustard, H. R.
1979. Populations dynamics of sea turtles. In *Turtles: Perspectives and Research,* eds, M. Harless and H. Morlock, pp. 523–40, New York: John Wiley and Sons.

Bustard, H. R., and K. P. Tognetti
1969. Green sea turtles: a discrete simulation of density-dependent population regulation. *Science* 163:939–41.

Carr, A.; M. H. Carr; and A. B. Meylan
1978. The ecology and migrations of sea turtles, 7. The West Caribbean green turtle colony. *Bulletin of the American Museum of Natural History* 162:1–46.

Hillestad, H. O.; J. I. Richardson; and G. K. Williamson
1977. Incidental capture of sea turtles by shrimp trawlermen in Georgia. Report to the National Marine Fisheries Service, Contract Number 03–7–042–35129.

Hopkins, S.; T. Murphy, Jr.; K. Stansell; and P. Wilkinson
1979. Biotic and abiotic factors affecting nest mortality in the Atlantic loggerhead turtle. *Proceedings of the 32nd Annual Conference of the Southeastern Association of Fish and Wildlife Agencies,* 32:213–223.

Hughes, G. R.
1974. The sea turtles of South-east Africa. II. The biology of the Tongaland loggerhead turtle *Caretta caretta* L. with comments on the leatherback turtle *Dermochelys coriacea* L. and the green turtle *Chelonia mydas* L. in the study region. *Oceanographic Research Institute Investigational Report,* Durban, South Africa, 36:1–96.

Limpus, C.
1979. Notes on growth rates of wild turtles. *IUCN/SSC Marine Turtle Newsletter* 10:3–5.

Richardson, J. I., and H. O. Hillestad
1978. Ecology of a loggerhead sea turtle population in Georgia. *Proceedings of Rare and Endangered Wildlife Symposium.* Georgia Department of Natural Resources, Game and Fish Division, *Technical Bulletin* WL4.

Richardson, J. I.; C. Ruckdeschel; H. O. Hillestad; and M. May
1977. Within-season interisland nesting overlap by loggerhead sea turtles in the St. Andrew Sound area of

Georgia. *Proceedings of the 57th Annual Meeting of the American Society of Ichthyologists and Herpetologists,* abstract.

Richardson, T.; J. I. Richardson; C. Ruckdeschel; and M. Dix
1978. Remigration patterns of loggerhead sea turtles (*Caretta caretta*) nesting on Little Cumberland and Cumberland Island, Georgia. *Florida Marine Research Publication* 33:39–44.

Appendix I. Procedures for calculating remigration, survivorship, and fecundity values appearing in Table 1 and Figure 4

1. A year of reproductive maturity runs from nesting season to nesting season. A neophyte, by definition, nests at the beginning of her first year of maturity.

2. A cohort of neophytes (Table A1) will remigrate according to the following intervals and frequencies: 1-year (3 percent), 2-year (56 percent), 3-year (31 percent), 4-year (7 percent), 5-year (3 percent). Only 49 percent of neophytes will remigrate; predicted numbers of remigrating neophytes are reduced to 49 percent of maximum values. Seventy percent of remigrants will remigrate again; predicted numbers of remigrating remigrants are reduced to 70 percent of maximum values. Figures in Table A1 are summed by rows to generate column A, Table 1.

3. Column B, Table 1, is derived by multiplying values in column A by 38.9 percent, the predicted annual recruitment rate:

$$\text{Recruitment rate} = \frac{1,000}{2,574}$$
$$= 38.9 \text{ percent (from column A, Table 1).}$$

There will be 389 neophytes in a seasonal nesting population of 1,000 adult female loggerheads.

4. Survivorship (column C, Table 1) is calculated from Table A2. It is predicted from observed remigration intervals and frequencies that 44 percent of the total population of adult nesting females will nest each season.

Remigration interval	Remigration frequency		Proportion observed annually		Annual nesters
1-year	3%	×	1.00	=	3%
2-year	56%	×	.50	=	28%
3-year	31%	×	.33	=	10%
4-year	7%	×	.25	=	2%
5-year	3%	×	.20	=	1%

Total annual nesters = 44%

Total population values (column A', Table A2) are derived by dividing seasonal values (column A, Table A2) by 44 percent. The percent mortality (loss from reproductive population) experienced during each year (column A", Table A2) is derived by comparing changes in total numbers for all pairs of consecutive years (column A', Table A2). Sixteen years of predicted annual mortality were arbitrarily chosen to derive a mean annual survival rate of 13 percent. Due to the constraints of observed remigration intervals and frequencies placed on a theoretical cohort of neophytes, the model appears to require approximately 7 years to stabilize; it begins to lose stability after 23 years because of small numbers of individuals left in the original cohort.

5. Annual mortality, averaged and adjusted (column D, Table 1), is taken to be 13 percent, the mean for 16 years of column A" (Table A2). To reflect the observed lower remigration rate (49 percent) of neophytes, annual mortality is increased during the first 5 years, according to the following formula:

$$\text{Annual mortality} = 13 \text{ percent } (1.13^{6-\text{years of maturity}}).$$

The choice of 1.13 as a multiplier of annual mortality is arbitrary. The objective is to reduce in numbers a cohort of 389 neophytes during the first 5 years of reproductive maturity, such that total surviving individuals approximate numbers of individuals in column A (Table 1) on a year by year basis.

6. Survivorship within a cohort of 389 neophytes (column C, Table 1) is generated by applying appropriate annual mortality rates (column D, Table 1). To satisfy the model, seasonal nesting totals (column B, Table 1) should be approximately 44 percent of total population size (column C, Table 1).

7. A Little Cumberland Island turtle lays an average of 2.5 clutches per season and 120 eggs per clutch. Fecundity (column E, Table 1) is derived by multiplying seasonal nesting totals (column B, Table 1) by 300 eggs per turtle.

Table A1. Iterative procedure for predicting remigration of a cohort of 1,000 neophytes

Years	Iterations							Total individuals
0	1,000							1,000
1		15[a]						15
2		274[a]						274
3		152[a]	6	5				163
4		34[a]	3	107	3			147
5		15[a]	1	60	64	3		143
6			—	14	36	57	3	110
7	2			5	8	32	56	103
8	43	2			3	7	31	86
9	24	40	2			3	7	76
10	6	23	34	2			3	68
11	2	5	19	30	1			57
12		2	4	17	27	1		51
13		2	4	15	22	1		44
14	1			2	3	13	20	39
15	17	1			1	3	11	33
16	10	15	1			1	3	30
17	2	9	13	1			1	26
18	1	2	7	12	1			23
19		1	2	7	10	—		20
20			1	2	6	9	—	18
21	—			1	1	5	8	15
22	7	—			1	1	4	13
23	4	6	—			—	1	11
24	1	3	5	—			—	9
25	—	1	3	4	—			0

| | | | | | | Predicted population total | 2574 |

Note: Remigration figures reduced to 49 percent of maximum for neophyte remigrants (a) and 70 percent of maximum for all other remigrants.

Table A2. Procedure for calculating survivorship of a cohort of 389 individuals

Years of reproductive maturity	A Seasonal nesting population	A' Total population	A'' Annual mortality during year (percentage)	D Annual mortality averaged and adjusted (percentage)	C Numbers of survivors of 389 cohort
0	1000				389
1	15			24	296
2	274			21	234
3	163	370		19	190
4	147	334		17	157
5	143	325		15	134
6	110	250		13	116
7	103	234		13	101
8	86	195	17	13	88
9	76	173	11	13	77
10	68	155	10	13	67

Table A2. Procedure for calculating survivorship of a cohort of 389 individuals (cont.)

Years of reproductive maturity	A Seasonal nesting population	A' Total population	A" Annual mortality during year (percentage)	D Annual mortality averaged and adjusted	C Numbers of survivors of 389 cohort
11	57	130	16	13	58
12	51	116	11	13	50
13	44	100	14	13	44
14	39	89	11	13	38
15	33	75	16	13	33
16	30	68	9	13	29
17	26	59	13	13	25
18	23	52	12	13	22
19	20	45	13	13	19
20	18	41	9	13	17
21	15	34	17	13	14
22	13	30	12	13	13
23	11	25	17	13	11
24	9	20		13	9
25	0	0		100	0

Note: By definition, all remaining individuals die during the twenty-fifth year of maturity.

Status of Sea Turtle Populations

Historical Review

F. Wayne King
Florida State Museum
Gainesville, Florida 32611

Historical Review of the Decline of the Green Turtle and the Hawksbill

Few wild species of vertebrates have played a more persistently important role in the European exploration and settlement of the western hemisphere and parts of Asia than has the green turtle, *Chelonia mydas*. Found in all tropical and temperate seas (between 35° N and S latitudes), the green turtle was readily exploited to supply protein (meat, calipee, calipash, eggs) to early explorers and coastal settlers (Carr 1956). It was netted or harpooned on the grass and algae-covered shoals, submerged banks, and in coastal estuaries, and females were captured in large numbers when they came ashore in nesting aggregations. This behavioral phenomenon of gathering together in hundreds or thousands to nest made the green turtle a ready source of fresh meat before mechanical refrigeration was invented. Turned on its back on the beach or in the hold of a boat, the turtle could be kept alive for weeks on end, if kept in the shade away from the heat of the sun (Carr 1956). The nesting aggregations also gave the mistaken impression the marine turtle was a superabundant, inexhaustible source of meat.

Quite the contrary, when those same numbers of turtles are dispersed over the entire range of the population, spread over the migratory routes, the turtle grass feeding flats, and across the shallow estuaries where green turtles grow to maturity, it soon becomes apparent that this species is not so abundant (Ross 1978). Thousands of marine species are more plentiful. Under the pressure of systematic commercial exploitation, a number of populations have become extinct, and most of those remaining are depleted and in danger of disappearing.

The Green Turtle

To understand historically what happened to the green turtle in most parts of the world, it is necessary to examine in detail what happened to 1 or 2 populations.

Possibly the largest green turtle rookery that ever existed was located in the Cayman Islands, West Indies.

One of the first accounts of this rookery is found in the narrative of Christopher Columbus's fourth and final voyage to America. On 10 May 1503, while sailing from Panama to Hispaniola, Columbus's ships came, ". . . in sight of two very small and low islands, full of tortoises, as was all the sea about, insomuch that they look'd like little rocks, for which reason those islands were called Tortugas . . ." (Lewis 1940; Carr 1956). It is interesting to note the green turtles were on the islands during the daytime, a phenomenon that occurs today only in the western Hawaiian Islands.

Spanish, French, and English ships sporadically began to visit the renamed Cayman Islands to capture turtles, but it was not until the mid-1600s that the rookery was systematically plundered. In 1655, the British colony in Jamaica needed meat, so the admiral dispatched one of his ships to the Caymans to acquire turtles (Long 1774; Lewis 1940). From that date on, the rookery was under constant pressure as the British fleet was victualled with Cayman turtles, and Jamaica was supplied with fresh meat and eggs. The laying season, from May through September, and the migratory routes used by the multitude of turtles to reach the island were well known to sailors. It is reported the number of migrants was so great that when,

"The greater part of them emigrate from the gulph of Honduras . . . it is affirmed, that vessels, which have lost their latitude in hazy weather, have steered entirely by the noise which these creatures make in swimming, to attain the Cayman isles . . . In these annual peregrinations across the ocean they resemble . . . herring shoals . . . Thus the inhabitants of all these islands are, by the gracious dispensation of the Almighty, benefited in their turn; so that, when the fruits of the earth are deficient, an ample sustenance may still be drawn from this never failing resource of turtle, or their eggs, conducted annually as it were into their very hands . . ." (Long 1774).

By 1688, a total of 40 sloops from Jamaica were engaged full time, year-round in bringing turtles from the Caymans to Jamaica (Lewis 1940). The turtles were turned on the beach during the summer nesting season, and during the rest of the year they were captured among the shoals and cays along Cuba's south shore. One of these sloops could be filled with 30 to 50 nesting females in as little as a single night, but it usually took 6 weeks if the turtles had to be pursued to the feeding grounds off Cuba. This fleet of turtlers returned upwards of 13,000 turtles a year to Jamaica.

By 1711, turtles were becoming sufficiently scarce that a law was enacted prohibiting the collecting of turtle eggs on any island belonging to Jamaica. Although the Cayman Islands were under Jamaican jurisdiction, the law was never enforced there or for Cayman eggs returned to Jamaica (Long 1774). In 1730, green turtles were reported to supply the principal

meat eaten in Jamaica (Catesby 1731–43). Forty sloops were still engaged in supplying the Jamaica market with turtles, and permanent settlers on the Caymans made their living by turtling.

Green turtles were in danger of becoming extinct in the Cayman Islands by the late 1700s. Only a few nested there each year (Lewis 1940; Carr 1952), rather than the huge aggregations that formerly crawled ashore night and day during the summer. Most turtles were captured among Cuba's southern cays. The fishery supported only 8 or 9 boats that carried the catch to Jamaica.

By 1840, the Cayman Island turtles were so near extinction the islanders were forced to sail to the Miskito Cays off Nicaragua to make their catch (Lewis 1940). The Cayman nesting population was extinct by 1900. Only immature green turtles were found feeding in the shallow water around the islands. The Cayman turtle fishery depended entirely on greens brought from the Nicaraguan coast, from which the vessels seldom returned without a full catch. The Cayman fleet consisted of 12 to 17 schooners in the 1940s, each of which could hold approximately 200 turtles (Rebel 1974). The fleet produced an annual catch of 2,000 to 3,000 Nicaraguan green turtles, more of it exported to the United States than to Jamaica.

To supply newly constructed abattoirs and freezing plants in Bluefields and Corn Island, Nicaragua began large-scale commercial exploitation of green turtles in 1970. An average of 10,000 turtles were processed in these plants each year, and the Miskito Cays turtle population declined drastically, as documented in the survival of females returning to its Tortuguero, Costa Rica, rookery (Bjorndal 1980; see also Nietschmann, this volume).

It took just over a hundred years from the time the Cayman Island rookery first came under systematic, wholesale exploitation in the 1650s until it was destroyed and no longer significant as a commercial source of turtles in the late 1700s. It took another century to kill the few remaining females that matured and straggled ashore to lay eggs, but by 1900, green turtle nesting on the Caymans was a thing of the past. As the population disappeared, the fishery moved first to the coast of Cuba and then to Nicaragua, where it exploited and endangered another population which has its rookery in Costa Rica.

Similar depletion and extinction has occurred wherever commercial exploitation of green turtles replaced a subsistence take, or where turtles were exploited for an international market (IUCN 1975; Ross 1978).

In the early 1800s, the Halifax and Indian river estuaries on the east coast of Florida, United States, contained many green turtles in spring and summer. The population supported a commercial turtle fishery (Audubon 1926; Carr and Ingle 1959). In 1886, 1 fisherman landed 2,500 greens using 8 nets (Wilcox 1896). Only 738 turtles were landed in 1890 despite the 168 nets attended by 24 fishermen. In 1895, 519 turtles were landed, and by 1900 the Halifax-Indian river fishery was finished for lack of turtles (Rebel 1974).

In Bermuda, green turtles were caught in nets, harpooned, and captured when they came ashore to nest. By 1620 they had become sufficiently rare for the Bermuda assembly to pass an act prohibiting the killing of turtles less than 18 inches (45.7 cm) in width or length (Garman 1884; Carr 1952, 1967; Rebel 1974). The turtle fishing continued and the rookery aggregations became extinct, but two boats were still able to capture 40 turtles a day (Garman 1884). Fifty years later, no turtles at all nested in Bermuda and only 20 to 60 immature turtles were netted annually (Rebel 1974). In 1970, 25 immature turtles were captured.

Similarly, nesting populations in the Dry Tortugas, Florida, were extirpated within a hundred years of the initiation of commercial exploitation. Other populations have been severely depleted or have become endangered.

In Sarawak, Malaysia, the green rookeries on Talang Talang Besar, Talang Talang Kechil, and Satang islands have long been exploited for their eggs by the local Muslims, although the adult turtles have not been systematically killed (Harrisson 1962, 1967). Government records from egg collecting leases show a steady decline from the 1930s when more than 2,000,000 eggs were collected each year until the mid-1960s when only one-tenth that number was being laid (Figure 1). During the period of Japanese occupation, 1941 to 1945, adult turtles were eaten and Talang Talang Kechil was used as a bombing target, but the decline cannot be attributed solely to cause-and-effect during this 5-year period. Throughout the history of egg exploitation on the Sarawak turtle islands, virtually 100 percent of the eggs have been collected. During at least the 1960s and 1970s, a tiny fraction of the eggs (less than 1 percent) has been hatched and the hatchlings released in an attempt to restock the resource (Chin 1970, 1975). This token effort has been insufficient to stem the continuing decline, for during the 1970s the number of eggs laid has continued to decrease and has never exceeded 300,000 a year (de Silva, this volume). It is another example of commercial exploitation endangering a green turtle population, and in less than a hundred years.

Declines of populations can be even more dramatic under some circumstances, as demonstrated by the recent commercial exploitation of the green turtles in the northern end of the Gulf of California (Felger, Cliffton, and Regal 1976). Prior to 1970 that population had been exploited primarily by the Seri Indians, who harpooned turtles hibernating on the bottom. These adults were taken "as needed" and accounted for an

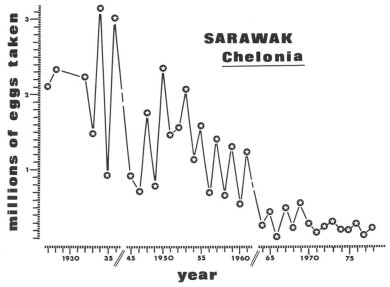

Figure 1. Decline in Sarawak *Chelonia mydas* population as reflected in the numbers of eggs collected for sale each year (Chin 1969, 1970, 1975; de Silva 1979; Harrisson 1962, 1967).

estimated 25 percent of the animal protein in the Seri diet. Then Mexican skindivers accidentally discovered the hibernating turtles in the winter of 1972–73 and immediately began scouring the bottom for all the turtles they could find. During the 1974–75 season, divers were averaging 5 turtles/hr of diving time. Five boats were capturing 80 turtles a week. By 1978, the winter dormant population was already so depleted as to be endangered (Felger and Cliffton, personal communication, 1978).

Most green turtle populations in the Atlantic Ocean, Mediterranean Sea, Indian Ocean, and Pacific Ocean are depleted or endangered as a result of direct exploitation or incidental drowning in trawl nets (IUCN 1975). The only populations not now declining, and which seem not to be threatened with extinction, are those that nest on Europa, Tromelin, and Glorious Isles in the Mozambique Channel; and on Raine Island, Australia. The Surinam population may also be safe, but recent increases in the numbers of shrimp trawlers operating in those waters will increase losses due to incidental drownings and may threaten the turtles, as might the present exploitation of the rookery for eggs.

While green turtles and their eggs are hunted primarily for human consumption, in the last 25 years the adults have come under increasing exploitation for leather (though not to the same extent as olive ridleys, *Lepidochelys olivacea*). The major international markets for meat and soupstock are the Federal Republic of Germany, the United Kingdom, and Japan (Mack, Duplaix, and Wells, this volume). The major consumers of leather are France, Italy, and Japan.

The Hawksbill Turtle

Hawksbill turtles, *Eretmochelys imbricata,* have been variously exploited for thousands of years as one culture after another coveted the horny scutes of their shells for making tortoiseshell jewelry and objects of art (Deraniyagala 1939; Holbrook 1842; Carr 1952). While this tortoiseshell trade has been the prime object of hawksbill exploitation, the species also is hunted for its eggs, leather, and for immature specimens which are stuffed, lacquered, and sold to tourists. In some regions the turtles are eaten, but in others their meat is poisonous (Deraniyagala 1939; Carr, this volume).

The hawksbill turtle is found in all tropical seas between about 30° N and S latitudes, where the water is less than 16-m deep and reefs, shoals, and estuaries are present. The hawksbill is less of a long-distance migrant than other species of marine turtles (Carr 1952; Carr, Hirth, and Ogren 1966; Pritchard 1979). Most hawksbill populations tend not to concentrate their nesting efforts into a few localized rookeries. Instead, nests are dispersed along many kilometers of undisturbed beach, including rookeries of other species of sea turtles. This dispersed nesting habit probably has saved many populations from extinction, since turtle fishermen largely have been unable to concentrate their efforts on rookery aggregations. Instead, they have been forced to intercept the few nesters they could find and to net or harpoon the rest of their catch on the feeding grounds. At the same time, the lack of rookery concentrations has hampered efforts to accurately document the conservation status of hawksbill populations. Population declines are not immediately reflected in fewer females returning to one or two beaches where their numbers can be counted. Frequently one of the first signs of overexploitation is a reduction in the numbers of adults, or even their elimination, leaving a population consisting of immature turtles (Carr 1952).

Government records of tortoiseshell imports and exports may reveal declines in local populations (Japanese Tortoise Shell Association 1978; Mack, Duplaix, and Wells, this volume), but often they do not do so. Turtle fishermen may be catching fewer adults, but supplementing their catch with more juvenile turtles. Government statistics rarely indicate the numbers of hawksbills taken, only the weight or value of scutes. Turtlers may also be traveling to new or more distant fishing areas to fill their catch, or staying away longer to capture the same number of turtles as in previous years (De Celis, this volume; Frazier, this volume). When these things happen, government records of the amount of tortoiseshell traded may remain relatively unchanged for years, or even increase temporarily, despite depletion of local turtle stocks. To complicate things further, government statistics often do not distinguish between raw scutes and worked shell products (Mack, Duplaix, and Wells, this volume).

During the 1930s and 1940s, hawksbills were given a temporary reprieve, especially in the Caribbean, as plastics replaced many of the utilitarian uses of tortoiseshell (Carr 1952; Carr, Hirth, and Ogren 1966; Rebel 1974). Eyeglass frames, pocket combs, and the backs of hand mirrors are seldom fabricated from genuine tortoiseshell even today, but the use of hawksbill shell for the manufacture of luxury items, jewelry, and art objects, enjoyed a resurgence in the 1950s that is continuing to the present. Of even greater concern is an enormous increase in volume of raw and worked tortoiseshell traded in the last 2 or 3 years (Mack, Duplaix, and Wells 1979).

Most populations of hawksbills are depleted or endangered (I.U.C.N. 1975). Small, but relatively dense rookeries (unusual for the species) exist on Cousin Island, Seychelles; Aziz Island, People's Republic of Yemen; Masirah Island, Oman; Shitvar and Lavan islands, Iran; in the Suakin Archipelago, Sudan; Nangka and Belitung islands, Indonesia; and on the islands of the Torres Straits, Australia. What may well have been the largest hawksbill rookery in the world, on Chiriqui beach, Panama, has been severely depleted as a result of commercial exploitation.

Tortoiseshell remains a much sought-after commodity, and hawksbill populations are under heavier exploitation pressure than ever before. Kilogram for kilogram, tortoiseshell is more valuable today than elephant ivory, and the last 5 years has seen the volume of the international trade in scutes increase dramatically (Mack, Duplaix, and Wells, this volume). For example, Indonesian exports, after increasing from less than 10,000 kg a year between 1971 and 1977, jumped to 219,585 kg in 1978 alone (Mack, Duplaix, and Wells, this volume). Exports from India, the Philippines, and Thailand also increased, as they did from a number of the Latin American countries. Imports into Taiwan soared from less than 3,000 kg in 1974 to over 128,000 kg in 1978. Japan's imports were generally less than 40,000 kg prior to 1969, then fluctuated between 40,000 and 92,000 kg a year from 1970 and 1973, and have since been about 42,000 to 47,000 kg a year (Mack, Duplaix, and Wells, this volume). Japan and Taiwan are the 2 largest consumer-importers, accounting for 75 percent to 80 percent of world production each year.

Simple economics suggests the tortoiseshell trade will yield reluctantly, if at all, to conservation efforts that would curtail or eliminate it. The rate of increase in the price of tortoiseshell makes it a good investment. However, cultural values may transcend economic considerations in some countries. For example, in Japan, tortoiseshell is used primarily in the artisanal production of hairpins, jewelry, and art objects that have great significance in Japanese cultural tradition (Japanese Tortoise Shell Association 1978). As a consequence,

Japan can be expected to continue importing tortoise-shell regardless of any consideration of the need to conserve the wild resource.

During the 1973 conference in Washington, D.C., that drafted the Convention on International Trade in Endangered Species of Wild Fauna and Flora (CITES), considerable debate surrounded the placement of various species on Appendix I. Such placement prohibits commercial trade, allowing only exchange of specimens for scientific or conservation purposes (King 1974). Not all delegations initially agreed on whether or not particular species were being threatened by commercial trade, and that was the subject of the debate. However, negotiations led to unanimous agreement on the inclusion or deletion of some species, and compromises on many. The only major dispute which reached the point of being recorded as a formal objection in the summary record of the conference was the placement of the hawksbill, *Eretmochelys imbricata,* in Appendix I.

During the discussion of this species, Japan argued against any curtailment of trade. After the assembled delegations had examined the evidence, they seemed to unanimously agree that commercial trade in tortoiseshell should be eliminated. Rather than stand alone against the majority, Japan did not speak out further. Instead, a representative of the Japanese delegation persuaded Panama (a major Latin American exporter of tortoiseshell to Japan) to lodge the objection (personal observation). Until approached by the Japanese delegation, Panama had supported protection for the hawksbill. Panama ratified the CITES in 1978 without reservations. During 1979, it was continuing to export hawksbill scutes. When Japan joins the CITES, it almost certainly will lodge reservations (i.e., announcing nonacceptance of certain restrictions) on the hawksbill and other sea turtles and will continue to export these species regardless of what happens to the wild populations.[1] Green turtles[2] and the other species of sea turtle are also protected by being on Appendix I of the CITES. To date, over 50 nations have joined the CITES. Two of those nations—France and Italy—have taken reservations on sea turtles and are importing all the sea turtle leather and tortoiseshell they can buy. It is clear they are more interested in protecting their commercial interests than in conserving wild species. This is particularly distressing since no definite evidence exists that any wild populations of green or hawksbill or other species of sea turtle have ever been returned to abundance after depletion through over-exploitation. The historical trend points toward depletion, endangerment, and extinction. If consuming nations are unwilling to cooperate with their CITES partners and the producing nations in avoiding threat to wild species resulting from uncontrolled exploitation, there seems little hope of conserving marine turtles for this and future generations of man.

Literature Cited

Audubon, J. J.
1926. The turtlers. In *Delineations of American Scenery and Character,* pp. 194–202. New York: G. A. Baker and Company.

Bjorndal, K. A.
1980. Demography of the breeding population of the green turtle, *Chelonia mydas,* at Tortuguero, Costa Rica. *Copeia* 1980:525–30.

Carr, A. F.
1952. *Handbook of Turtles: the Turtles of the United States, Canada, and Baja California.* Ithaca, New York: Comstock Publishers.
1956. *The Windward Road.* New York: Alfred Knopf.

Carr, A. F.; H. Hirth; and L. Ogren
1966. The ecology and migrations of sea turtles, 6. The hawksbill turtle in the Caribbean Sea. *American Museum Novitates* 2248:1–29.

Carr, A., and R. M. Ingle
1959. The green turtle (*Chelonia mydas*) in Florida. *Bulletin of Marine Science in the Gulf and Caribbean* 9:315–20.

Catesby, M.
1731–43. *The Natural History of Carolina, Florida and the Bahama Islands.* 2 vols. London.

Chin, L.
1969. Notes on turtles and orangutans (1969). *Sarawak Museum Journal* 17:403–04.
1970. Notes on orangutans and marine turtles. *Sarawak Museum Journal* 18:414–15.
1975. Notes on marine turtles (*Chelonia mydas*). *Sarawak Museum Journal* 23:259–65.

Deraniyagala, P. E. P.
1939. Tetrapod reptiles of Ceylon. *Ceylon Journal of Science* 1939:1–412.

Felger, R.; K. Cliffton; and P. Regal
1976. Winter dormancy in sea turtles: independent discovery and exploitation in the Gulf of California by two cultures. *Science* 191:283–85.

Garman, S.
1884. Contributions to the natural history of the Bermudas: Reptiles. *Bulletin of the United States National Museum* 25:1–353.

---

1. Japan ratified the CITES in May 1980 and took reservations on the hawksbill, green turtle, olive ridley turtle, estuarine crocodile, 3 species of monitor lizard, musk deer, and fin whale.

2. Because green turtles are still abundant in Australia, where exploitation for international commerce is not occurring, that population of green turtle was not on Appendix I of the CITES. It was on Appendix II, which requires export permits from Australia before importation into another CITES nation is allowed. In 1981, all populations of green turtles were placed on Appendix I at the request of Australia when it was discovered Mexico was exporting greens with false documentation they claimed they originated in Australia.

Harrisson, T.

1962. Notes on the green turtle, *Chelonia mydas*. 2. West Borneo numbers, the downward trend. *Sarawak Museum Journal* 10:614–23.

1967. Notes on marine turtles. 18. A report on the Sarawak turtle industry (1966) with recommendations for the future. *Sarawak Museum Journal* 15:424–36.

Hirth, H. F.

1971. Synopsis of biological data on the green turtle, *Chelonia mydas* (Linnaeus) 1758. *FAO Fisheries Synopsis*, 85:1–71.

Holbrook, J. E.

1842. *North American Herpetology*. Reprinted in 1976. Lawrence, Kansas: Society for the Study of Amphibians and Reptiles.

IUCN (International Union for the Conservation of Nature and Natural Resources).

1975. *Red Data Book: Volume 3, Amphibia and Reptilia*. Morges Switzerland: IUCN.

Japanese Tortoise Shell Association.

1978. Preliminary report on the hawksbill turtle (*Eretmochelys imbricata*) in Indonesia, Philippines, Malaysia, and Singapore. English translation by G. Balazs and N. Nozoe. Xeroxed.

King, W.

1974. International trade and endangered species. International Zoo Yearbook, 14:2–13.

Lewis, B.

1940. The Cayman Islands and marine turtle. *Bulletin of the Institute of Jamaica Science Series* 2:56–65.

Long, E.

1774. *The History of Jamaica, or General Survey of the Ancient and Modern State of that Island*. London: T. Loundes.

Pritchard, P.

1979. *Encyclopedia of Turtles*. Neptune, New Jersey: T. F. H. Publications.

Rebel, T. P.

1974. *Sea Turtles and the Turtle Industry of the West Indies, Florida and the Gulf of Mexico*. Coral Gables, Florida: University of Miami Press.

Ross, P.

1978. Present status of sea turtles: a summary of recent information and conservation priorities. Report to the International Union for the Conservation of Nature and Natural Resources. Manuscript, 45 pp.

Secretariat of CITES

1977. *Convention on the International Trade in Endangered Species of Wild Fauna and Flora: Proceedings of the First Meeting of the Conference of the Parties*. Morges, Switzerland: IUCN.

Wilcox, W. A.

1896. Commercial fisheries of Indian River, Florida. *Report of U.S. Commercial Fish and Fisheries*, 22:249–62.

F. WAYNE KING

James Perran Ross
Museum of Comparative Zoology
Harvard University
Cambridge, Mass. 02138

Historical Decline of Loggerhead, Ridley, and Leatherback Sea Turtles

ABSTRACT

The distribution of nesting populations of *Caretta caretta*, *Lepidochelys olivacea* and *Dermochelys coriacea* is reviewed. Populations that are known to have declined in numbers and populations that appear to be threatened are identified.

Each species has a large proportion of the individuals concentrated in a very few nesting populations. The total number of populations and the number of large populations is small.

A major identifiable cause of population decline is excessive commercial exploitation. Commercial exploitation must be reduced while management programs are being devised.

Introduction

This review is difficult to present because there is very little historical information on populations of these species with which present populations can be compared. A further problem is the taxonomic confusion between *Caretta* and *Lepidochelys* which, until recently, led to many misidentifications. (Nishimura 1967 for discussion) Also, even rough estimates of population size are lacking from the literature; many early reports only record presence and absence data. Another problem is the large amount of useful recent information that is hidden away in obscure unpublished reports. I would therefore urge all turtle researchers to publish their reports and include estimates, no matter how rough, of population sizes.

The Loggerhead (*Caretta caretta*)

This species is an omnivorous turtle which is often reported in temperate waters (Brongersma 1971). Juveniles and subadults disperse into major oceanic circulations although Hughes (1974b) suggests that the availability of benthic feeding areas may limit recruitment. Loggerheads often show a less rigid site fixity

189

Table 1. *Caretta caretta* nesting sites

Location	Approximate number of females/yr	Reference
Southeast U.S.A.[a] Cape Sable Merrit Is., Hutchinson Is., Jupiter, Cumberland Is., Cape Romain	6000–25,000[b]	Lund 1974 Carr & Carr 1978
Cuba	—	Bacon 1973
Honduras	—	Bacon 1973
Quintana Roo	500[b]	Marquez 1976
Santa Marta, Colombia	400	Kaufman 1973
Turkey	—	Carr 1952
Cyprus	—	
Cape Verde Islands	—	Schleich 1979
Almadies, Senegal	—	Cadenat 1949
Tongaland, South Africa	500[b]	Hughes 1974a, b
Paradise Island, Mozambique	300	Hughes 1974a, b
Fort Dauphin, Malagasy Republic	300	Hughes 1974a, b
Masirah Island, Oman	30,000	Ross 1979
Honshu, Japan	—	Nishimura 1967
Kyushu, Japan		Nishimura 1967
Mon Repos, Australia	200?[b]	Bustard and Limpus 1971
Wreck Island, Australia	1000	Limpus, personal communication
Capricorn and Bunker Islands, Australia	1000[b]	Bustard 1972

— No data.

a. Loggerheads nest on many small beaches along this coast, only larger sites listed.

b. Beaches with some protection.

than *Chelonia* and some nesting aggregations are strung out over kilometers of coastline (Carr and Carr 1972).

Table 1 shows nesting sites of *Caretta*. The southeastern United States location consists of numerous small nesting sites; only the more prominent are listed. A complete listing is found in Dodd (1978). Most nesting populations are of the order of 1,000 females per year. Exceptions are the unusually large population in Oman (30,000) and the populations in the Caribbean and Mediterranean, most of which are reduced to 100 females or less. Only nesting locations in the United States, Australia, Mexico, and South Africa are protected.

Loggerhead populations throughout the world are under severe pressure from local exploitation. Loggerhead meat is a favored item in the Mediterranean, Africa, and South America. Other pressures come from the taking of eggs and accidental capture in commercial shrimp trawls.

Populations which are known to have declined are Honduras (Parsons 1962; Bacon 1973), the Mexican populations of Quintana Roo (Márquez 1976), and the nesting population of Colombia (Kaufman 1973; Ramirez 1976).

Numerous other populations are subject to heavy predation but there are no comparative data to assess the effect. In the southeastern United States accidental catch in shrimp trawlers is estimated to remove 4,000 to 12,000 turtles a year from the population (Ross, unpublished). Most mortality is of subadult turtles. There has been considerable loss of nesting habitat to coastal development, but the remaining populations are protected by the Endangered Species Act. In Cuba, loggerheads are caught at sea for commercial use although nesting turtles are protected. Recent evaluations of the nesting populations suggest they are declining (Abascal 1971). In the Mediterranean the remaining populations are small and subject to disturbance and exploitation. Brongersma (1971) reported an annual catch of 1,000/yr at the Azores, mostly subadults that may be derived from the U.S. population. Similar levels of exploitation are reported from Cape Verde Islands (Schleich 1979), and the eggs of the small nesting population there are often taken.

Capture of loggerheads for food is extensive in West Africa (Carr 1952) and Mozambique (Hughes 1971, 1972). The situation in Japan is unclear, but Nowak (1974) reports extensive exploitation and disturbance are causing the population to decline. Nishimura (1967) reports extensive killing of females and taking of eggs.

Although *Caretta* remains abundant, it is clearly under pressure from local exploitation and accidental capture. Those populations which are undisturbed (for example, Masirah) reach large size. The small size of most populations may well be a result of continual pressure.

Populations afforded protection have tended to increase, for example, Tongaland (Hughes 1974a), Cumberland Island (Richardson, Richardson, and Dix 1978). This may indicate a more resilient life history strategy than other sea turtles, and some recent results confirm this. Hughes (1976) suggests many loggerheads nest only once, and annual recruitment is high. Richardson et al. (1978) have similar data from Cumberland Island, United States. More research is urgently needed to establish life history parameters of all sea turtles.

The anomalous absence of nesting grounds of *Caretta* in the central and western Pacific is unexplained. Hirth (1971a) reported loggerheads from Tonga and Fiji, but nesting is not reported.

JAMES PERRAN ROSS

The Olive Ridley, (*Lepidochelys olivacea*)

The olive ridley is the most extreme example of the strategy of aggregated nesting. This species nests in huge concentrations on a few days each year at a very restricted number of locations. For example, it is likely that the nesting grounds in Pacific Mexico and Central America support all the ridley turtles in the entire East Pacific. The biology of this species is reviewed by Márquez, Villanueva, and Peñaflores (1976).

Sixteen large nesting grounds of the olive ridley are listed in Table 2. Smaller nesting populations of a few hundred individuals are reported from Venezuela (Nowak 1974), Senegal (Miagret 1977), Angola (Nowak 1974), Oman (Ross 1979), Malaya (Moll, personal communication) and New Britain (Pritchard, personal communication). The population estimates shown indicate the nesting populations before recent extensive exploitation and in many cases are greatly reduced at present. Several populations are known to have virtually collapsed in recent years and are cause of great concern. Three populations in Mexico—Tlacoyunque, Mismaloya and Chacahua—have been greatly reduced in numbers since 1977 (Felger and Cliffton 1977; Anonymous 1979). A fourth, Escobilla, is reported to be currently collapsing (Anonymous, 1979). However, good estimates of population size are not available. The apparent cause of decline is exploitation of nesting female ridleys for an industrial commerce in turtle leather and turtle meat. Contributing factors to the disastrous effect of this exploitation appear to be an inadequate population model from which capture quotas are calculated and the disproportionate influence of the businessmen who control the turtle export trade.

Recent evidence from tag returns shows that these populations disperse southward to the coasts of Panama and Ecuador. Green and Ortiz Crespo (this volume) reports very heavy exploitation of these migrating ridleys in Ecuador. There is no rational program or control of this fishery. These populations are therefore being overexploited throughout their range and their future is in jeopardy.

A population of ridleys in Orissa province, India, estimated at 150,000 (Anonymous 1976) was reported not to be present in 1977 following heavy exploitation for food (Davis, Bedi, and Oza 1978). However, Kar and Bhaskar (this volume) reports that 150,000 females nested in 1979. On the west coast of India the large diffuse nesting population of ridleys is subject to heavy predation of eggs and many thousands of nesting females are killed (Whitaker 1977; Valliappan and Whitaker, 1975).

Ridleys are caught in substantial numbers in Madagascar (Hughes, this volume). Selm (1976) reports that the available nesting area on Sandspit and Hawkes Bay beaches in Pakistan is being reduced by building development. Schulz (1975) considers the accidental

Table 2. *Lepidochelys olivacea* nesting sites

Location	Approximate number of females/yr (thousands)	Reference
Mismaloya, Mexico	20–50[a]	Marquez 1976
Tlacoyunque, Mexico	20–50[a]	Marquez 1976
Chacahua, Mexico	20–50[a]	Marquez 1976
Escobilla, Mexico	50–100[a]	Marquez 1976
Nancite, Costa Rica	200	Hughes and Richard 1974
Ostional, Costa Rica	100	Hughes and Richard 1974
Eilanti, Surinam	1	Schulz 1975
West Africa	—	Carr 1952
Mozambique	1?	Hughes 1974a, b
Malagasy Republic	1?	Hughes 1974a, b
Hawkes Bay, Pakistan	1?	Selm 1976
Coramandel Coast, India	10?	Valliappan and Whitaker 1975
Orissa Coast, India	150	Anon. 1976
Burma	—	Smith 1931
North East Australia	1?	Bustard 1972
Trengganu, Malaysia	1	Moll, personal communication

— No data.

a. Estimated numbers up to 1974. These numbers are known to be greatly reduced in 1977–1979.

take of ridleys in shrimp trawlers a serious danger to the Suriname population, and recent coastal development at Ostional Beach in Costa Rica may pose problems for the ridleys nesting there.

The only large, unassailed ridley population is at Playa Nancite in Costa Rica. This beach is subject to occasional egg poaching, but the level is low. Recent overtures from the ridley exploiters in Mexico to begin operations on this beach are thought to have been rejected by Costa Rican authorities. Dialogue between critics and proponents of turtle exploitation in Central America has recently begun.

The Leatherback or Luth, (*Dermochelys coriacea*)

Dermochelys, the largest of the sea turtles, nests widely but often sparsely in tropical regions. Leatherbacks range to north and south temperate zones where they feed on a variety of soft-bodied marine organisms (Brongersma 1969). The species has been described as a temperate zone form with a tropical nesting range (Carr, personal communication).

Table 3. *Dermochelys coriacea* nesting sites

Location	Approximate number of females/yr	Reference
Matina, Costa Rica	500[a]	Carr, personal communication
Gulf of Uraba, Colombia	100	Mrosovsky, personal communication
St. Croix, Virgin Islands	30[a]	Dodd 1978
East Florida	25[a]	Carr, personal communication
Trinidad	400–500	Bacon 1971
Culebra, Puerto Rico	—[a]	Dodd 1978
Bigisanti, Surinam	300–400[a]	Schulz 1975
Silebache, French Guiana	6000	Fretey 1977
Dominican Republic	100?	Carr, personal communication
Tongaland, South Africa	70[a]	Hughes, this volume
Angola	—	Hughes, this volume
Ceylon	100?	Selm 1976
Phuket, Thailand	—	Polunin 1977
Indonesia	—	Suwelo 1971
Trengganu, Malaysia	1,000–2,000[a]	Siow 1974
Manus Island, Papua New Guinea	100?	Pritchard, personal communication
Solomon Islands	—	Pritchard, personal communication
Chacahua, Mexico	2000	Marquez 1978
Tierra Colorada, Mexico	3000	Marquez 1978
Macconi, Sicily	—	Bruno 1978

— No data.

a. Beaches with some protection.

Table 3 lists 19 known nesting beaches. Of these only 4 are larger than 1,000 females nesting each year. Many nesting populations are small (25 to 100 females/yr). Seven populations receive some protection. Ten of these populations, including the large populations in Mexico, have been first reported within the last 5 years.

Leatherbacks are caught and eaten in Peru (Hays and Brown, this volume) and the Caribbean (Carr, personal communication). Leatherback eggs are taken universally. In many areas this is a subsistence activity, but in Asia it is highly organized on a large scale. Leatherbacks are caught in Arabia and India and rendered for oil which is used to treat boat timbers.

Populations which are known to have declined occur on the west coast of India (Cameron 1923; Kar and Bhaskar, this volume), Ceylon (Deraniyagala 1939; Selm 1976) and Thailand (Polunin 1977). The main reason appears to be excessive removal of eggs by people. The large population in Trengganu, Malaysia is subject to intense egg harvest. Nearly 100 percent of eggs laid are taken, but up to 15 percent of these are acquired by authorities for hatching and release (Siow and Moll, this volume). It is not clear whether the present management program is effective (Chua and Furtado 1979).

Other areas where there are problems of unknown extent are the Dominican Republic where subsistence take of eggs is high, Mexico where there are occasional large scale removal of eggs, and Peru where a local industry catches nonbreeding leatherbacks for meat. Anonymous (1975) reports widespread slaughter of nesting turtles and egg taking in Trinidad. Leatherbacks are reported to be taken in large numbers as incidental catch in commercial fishing operations in the Mediterranean and occasionally elsewhere. Fretey and Lescure (1976) express concern that legislated protection of the large nesting ground in French Guiana is not preventing large-scale poaching of leatherback eggs. Populations in Suriname, Florida and South Africa which do have adequate protection have increased in recent years (Schulz 1975; Hughes 1974a).

Discussion

From the preceding accounts several general points on the status of these sea turtles can be derived. Table 4 summarizes the information from Tables 1–3. There is a surprisingly small number of large populations of any species. This may reflect the incomplete information available. Several of the large populations were discovered recently: *Lepidochelys* at Nancite, Ostional and Orissa, India, and *Caretta* at Masirah. It is possible that future discoveries will add more large populations. However, for the present we can only proceed with the available information.

I have made no attempt to extend estimates of the number of females nesting each year to total population size estimates. Annual numbers fluctuate by factors of 5 to 20 from season to season for the various species. Examples are given in Hughes (1974b), Carr, Carr, and Meylan (1978), and Chin (1970). Therefore these numbers are only general indications, and a priority for research is to refine them. Although the estimates of nesting females are crude or absent in many cases, it is clear that there are surprisingly small numbers of *Caretta* and *Dermochelys* known at present.

The most striking point from Table 4 is the degree

JAMES PERRAN ROSS

Table 4. Summary of the available data on populations of *Caretta caretta*, *Lepidochelys olivacea*, and *Dermochelys coriacea*. Population estimates are not based on firm data and the record of known populations is incomplete. The very concentrated and vulnerable nature of these species is demonstrated.

Species	Large popula- tions 1000 ♀/yr +	Total ♀/yr[a]	Percentage of individuals in 4 largest populations	Populations Protected	Reduced	Threatened
Caretta caretta	2	41,000	93	5	3	5
Lepidochelys olivacea	5[b]	518,000	97	4	3	6
Dermochelys coriacea	4	14,325	84	7	3	5

a. Populations for which estimates are not available were assumed to have 100 individuals (*Caretta* & *Dermochelys*) or 1000 individuals (*Lepidochelys*).

b. Three formerly large populations in Mexico are greatly reduced and are omitted. See Table 2 and text.

to which the numbers of turtles are concentrated in a very few large nesting populations. Between 84 percent and 97 percent of the females of each species are accounted for in the 4 largest nesting grounds currently known. This is an astounding degree of concentration and brings into focus the fact that much of our past research has concentrated on a very small proportion of each species.

The number of populations that receive effective protection is small, particularly when compared with the number of populations threatened with exploitation. In many cases those populations that are effectively protected are not the large populations.

Doubt is sometimes expressed that turtle populations have actually declined. Even with the scarce information available I have still been able to show that 3 populations of each species have suffered serious reduction in numbers. There are probably other populations that have disappeared or that persist in greatly reduced numbers about which we have no data.

The cause for the declines is unambiguous. In every documented case it is large-scale exploitation by man that has led to drastic reduction in numbers. For this reason, turtles now being intensively exploited are listed as threatened. The exact form of human activity leading to population reduction varies for each species. Loggerheads have been reduced by hunting for food for both local market and export consumption. The olive ridley has been decimated by commercial exploitation for its leather in the Americas. In India it is the large scale commerce in turtle meat to big cities which is critical. Leatherbacks seem most affected by the intense exploitation of their eggs. Accidental capture in commercial fishing operations is a problem for all species. In all cases, the extension of commerce in turtle products beyond the immediate subsistence needs of the local people has been followed by reduction of turtle numbers. It can be argued that I have only demonstrated loose correlations which may not be causal. They may only be correlations, but they are compelling

ones and certainly sufficient to cause concern. They merit intense further study and cannot be dismissed.

The question arises, what action can we take to ensure that these demonstrably vulnerable animals do not disappear? To effectively manage sea turtles we need to know as a minimum, population sizes, age structure, and reproductive output. When these parameters are known rational management can be planned. A model for such treatment is the paper by Márquez and Doi (1973). The authors recognize the need to make assumptions about some parameters because real data are not available. The assumptions can now be seen to be unrealistic, and the practical application of their results is limited. Nevertheless, the technique is exemplary.

It is quite clear from recent changes in the accepted knowledge of sea turtles that we do not have adequate data for any species. The green turtle *Chelonia mydas* is the most extensively studied sea turtle. In recent papers estimates of the number of nests laid by a green turtle in one season were 1 to 3 (Carr, Carr, and Meylan 1978; Schulz 1975) rather than early figures of 3 to 7 (Hirth 1971b; Hendrickson 1958.) The age at maturity is now thought to be 15 to 50 years (Balazs 1979, Limpus 1979) not 5 to 10 years as previously (summarized in Hirth 1971b). Hughes (1976) has challenged a number of previous ideas of the total reproductive output of loggerheads. Uncertainty abounds, which is the very reason for this conference. There are indications that these factors are different and largely unknown for other species. We do not have enough correct information to manage sea turtle populations today.

There is one indisputable fact: when sea turtle populations have been exploited, they decline; when they have been protected, they increase. The political realities in developing nations and pressure from the commercial vested interests make complete protection an unpalatable and perhaps an impossible goal. We must ensure that the necessary compromise is based on caution and what little good data we have. I believe that the correct strategy for preventing the extinction of sea turtles is to limit exploitation to the lowest take that is compatible with the real social needs of indigenous people who eat turtle products. This may give us the time to do the research that is necessary to achieve sustainable management policies worldwide.

Literature Cited

Abascal, J.
1971. Extinction of the loggerhead turtle imminent. *Mar y Pesca*, Havana, 1971:21–27.

Anonymous
1975. Turtle slaughter in Trinidad. *Oryx* 13:6–7.
1976. Huge turtle slaughter in India. *Oryx* 13:325.
1979. Mexico: the turtles are gathering for their nesting season massacre. *IUCN Bulletin*, June 1979:42–43.

Bacon, P. R.
1971. Sea turtles in Trinidad and Tobago. *IUCN Publications, New Series, Supplemental Paper*, 31:79–84
1973. Review on research, exploitation and management of the stocks of sea turtles. *FAO Fish circ.* FIRS/ C 334:1–19.

Balazs, G. H.
1979. Growth, food sources and migrations of immature Hawaiian *Chelonia*. *IUCN/SSC Marine Turtle Newsletter* 10:1–3.

Brongersma, L. D.
1969. Miscellaneous notes on sea turtles IIA. *Kon. Ned. Akad. van Wet. Proc. Ser. C*, 72:76–102.
1971. Ocean records of turtles (North Atlantic Ocean). *IUCN Publications, New Series, Supplemental Paper*, 31:103–8.

Bruno, S.
1978. La taratughe nei mari italiani e nel Mediterraneo. *Natura et Montgna*, 25:5–17.

Bustard, H. R.
1972. *Australian Sea Turtles*. London: Collins.

Bustard, H. R., and C. Limpus
1971. Loggerhead turtle movements. *British Journal of Herpetology* 4:228–30.

Cadenat, J.
1949. Notes sur les tortues marines des côtes du Senegal. *Bull. IFAN* 11:16–35.

Cameron, T. H.
1923. Notes on Turtles. *Journal of Bombay Natural History Society* 29:299–300.

Carr, A.
1952. *Handbook of Turtles*. Ithaca, New York: Cornell University Press.

Carr, A., and M. H. Carr
1972. Site fixity in the Caribbean green turtle. *Ecology* 53:425–29.

Carr, A.; M. H. Carr; and A. B. Meylan
1978. The ecology and migrations of sea turtles, 7. The West Caribbean green turtle colony. *Bull. Am. Mus. Nat. Hist.* 162:1–46.

Carr, D., and P. Carr
1978. Report on loggerhead turtles of Southeastern USA. NMFS. Manuscript, 15 pp.

Chin, L.
1970. Notes on orangutan and marine turtles. *Sarawak Museum Journal* 18:403–4.

Chua, T. H., and J. I. Furtardo
1979. Conservation of the leathery turtle in peninsula Malaya. *Malaysian Applied Biology* 8:97–101.

Davis, T. A.; R. Bedi; and G. M. Oza
1978. Sea turtle faces extinction in India. *Environmental Conservation* 5:211–12.

Deraniyagala, P. E. P.
1939. *The Tetrapod Reptiles of Ceylon*. Colombo: Colombo Museum.

Dodd, C. K.
1978. Terrestrial critical habitat and marine turtles. *Bulletin of the Maryland Herpetological Society* 14:233–40.

Felger, R. S., and K. Cliffton
1977. Marine turtles in the Gulf of California. IUCN/WWF Project 1471 report. 12 pp.

Fretey, J.
1977. Causes de mortalite des tortues luths adultes sur le littoral guyanais. *Le Courrier de la Nature* 52:257–66.

Fretey, J., and J. Lescure
1976. Guyane française: les infortunes de la tortue marine. *La Recherche* 7:778–81.

Hendrickson, J. R.
1958. The green sea turtle *Chelonia mydas* (Linn) in Malaya and Sarawak. *Proceedings of the Zoological Society of London*, 130:455–535.

Hirth, H.
1971a. South Pacific islands—marine turtle resources. *IUCN Publications, New Series, Supplemental Paper*, 31:53–56.
1971b. Synopsis of biological data on the green turtle, *Chelonia mydas*. *FAO Fisheries Synopsis*, 85:1:1–8:19.

Hughes, D. A., and J. D. Richard
1974. Nesting of the Pacific ridley *Lepidochelys olivacea* on Playa Nancite, Costa Rica. *Marine Biology* 24:97–107.

Hughes, G. R.
1971. Sea turtles—a case study for marine conservation in Southeast Africa. In *Proceedings of the Symposium on Nature Conservation*, pp. 115–23. SARCCUS, Mozambique.
1972. The olive ridley sea turtle in South East Africa. *Biological Conservation* 4:128–34.
1974a. The sea turtles of South East Africa I. *Investigational Reports of the Oceanographic Research Institute*, Durban, South Africa 35:1–144.
1974b. The sea turtles of South East Africa II. *Investigational Reports of the Oceanographic Research Institute*, Durban, South Africa, 36:1–96.
1976. Irregular reproductive cycles in the Tongaland loggerhead sea turtle *Caretta caretta*. *Zool. Africana*, 11:285–91.

Kaufman, R.
1973. Studies on the loggerhead sea turtle *Caretta caretta* in Colombia, South America. *Herpetologica* 31:323–26.

Limpus, C.
1979. Notes on growth rates of wild sea turtles. *IUCN/SSC Marine Turtle Newsletter* 10:3–5.

Lund, F.
1974. Marine turtles nesting in the United States. Report to U. S. Fish and Wildlife Service, 39 pp.

Maigret, J.
1977. Les tortues de mer au Sénégal. *Bulletin AASNS* 59:7–14.

Márquez, R.
1976. Reservas naturales para la conservation de las tortugas marinas en Mexico. INS/SI i83, 23 pp.
1978. *Dermochelys* in Pacific Mexico. Manuscript, 4 pp.

Márquez, R., and T. Doi
1973. A trial of theoretical analysis on population of Pacific green turtle in Mexico. *Bull. Tokai Reg. Fish Res. Lab* 73:1–23.

Márquez, R; A. Villanueva; and C. Peñaflores
1976. Sinopsis de datos biologicos sobre la tortuga golfina. INP/S2:1–62.

Nishimura, S.
1967. The loggerhead turtles in Japan and neighbouring waters. *Publ. Seto Mar. Biol. Lab.*, 15:19–35.

Nowak, R. M.
1974. Status of the green sea turtle, loggerhead and olive ridley turtle. Report to the U.S. Office of Endangered Species, 56 pp.

Parsons, J. J.
1962. *The Green Turtle and Man.* Gainesville; University of Florida Press.

Polunin, N. V. C.
1977. Cons. News, Assoc. Cons. *Wildlife Thailand*, 1977:6–10.

Ramirez, E.
1976. Report to INDERENA Colombia. Manuscript, 7 pp.

Richardson, J. I.; T. H. Richardson; and M. W. Dix
1978. Population estimates for nesting female loggerhead sea turtles *Caretta caretta* in the St. Andrews Sound area of Georgia, southeastern USA. *Florida Marine Research Publication*, 33:34–38.

Ross, J. P.
1979. Sea turtles in the Sultanate of Oman. World Wildlife Fund, Project 1320 report: 53 pp.

Schleich, H. H.
1979. Sea turtle protection needed at the Cape Verde Islands. *IUCN/SSC Marine Turtle Newsletter* 12:12.

Selm, R. V.
1976. Marine turtle management in Seychelles and Pakistan. *Environmental Conservation* 3:267–68.

Schulz, J. P.
1975. Sea turtles nesting in Surinam. *Zool. Ver. uit. door het Rijkmuseum voor Nat. Hist. Leiden* 143:1–144.

Siow, K. T.
1974. A report on the conservation of the leathery turtle. Mimeo, 1 p.

Smith, M. A.
1931. *The fauna of British India including Ceylon and Burma.* Vol., pp. 1–185.

Suwelo, I. S.
1971. Sea turtles in Indonesia. *IUCN Publications, New Series, Supplemental Paper*, 31:85–89.

Valliapan, S., and R. Whitaker
1975. Ridley sea turtle on the Coromandel coast of India. *Herpetological Review* 6:42.

Whitaker, R.
1977. A note on the sea turtles of Madras. *Indian Forester* 103:733–34.

Eastern Pacific Ocean

Kim Cliffton
Arizona-Sonora Desert Museum
Route 9 Box 900
Tucson, AZ 85743
and
Florida Audubon Society

Dennis Oscar Cornejo
Richard S. Felger
Arizona-Sonora Desert Museum

Sea Turtles of the Pacific Coast of Mexico

The west coast of Mexico once provided food, protection, and nesting beaches for one of the largest and most diverse assemblages of sea turtles in the world. Their history spans earth ages while their decimation spans but a few decades. In a rapidly developing country, social, political, and economic concerns seemingly outweigh biological realities. Of millions of sea turtles that occurred on the Pacific coast of Mexico, only a few hundred thousand remain. The sea turtles of the Pacific coast of Mexico are on a rendezvous with extinction.

The Pacific coast of Mexico and offshore islands still provide important breeding grounds for the green, ridley, and leatherback turtles. While many green and olive ridley turtles remain in Mexican seas all year, others migrate thousands of kilometers to feeding grounds in Central and South America (René Márquez, personal communication; Green and Ortiz, this volume). Five species of sea turtles are native to the Pacific coast of the Americas; some range as far north as the United States (Stebbins 1966).

The following is a species-by-species account of the present status of the sea turtles of the Pacific Coast of Mexico. Our approach is heuristic; we compare the more recent quantitative information with the older qualitative knowledge that predates heavy exploitation. Our quantitative information was compiled from existing literature, the Mexican Fisheries Department (PESCA), and 5 years of our fieldwork. Qualitative information was collected from our ethnobiological fieldwork along the Pacific Coast of Mexico and existing literature.

The Leatherback (*Dermochelys coriacea*)

The leatherback is called *Laud, De Altura, or Del Canal* in southern Mexico, and *Siete Filos* in the northwest. Its nesting beaches are known from the states of Michoacán, Guerrero, and Oaxaca. The recently discovered rookery at Tierra Colorado, Guerrero, may be the largest for this species in the world, with an esti-

199

mated 500 females nesting per night during the season (René Márquez and Peter Pritchard, personal communication). This breeding population was discovered in 1976, when a shipment of many thousands of eggs was confiscated en route to Mexico City. The Tierra Colorado nesting beach does not have sufficient protection.

The greatest nesting densities at Tierra Colorado occur in December and January, following the peak breeding season of both the ridley (July through October) and the green turtle (October through November). Leatherback nesting outside the Tierra Colorado area seems to coincide with green turtle nesting activities (most leatherback nesting in Michoacán occurs in October and November). These nestings may be females migrating towards the Tierra Colorado beach. Pritchard (1973) reports similar behavior for leatherbacks in Surinam.

In 1978 the eggs sold for 10 to 12 pesos (US $0.52) each in Mexico. Generally egg poachers wait for the turtle to lay her eggs. However, on beaches being protected by Mexican marines, egg poachers are known to kill turtles in order to hurriedly remove the eggs. Nesting animals are also killed for their oil, which is used in skin lotion and for medicinal purposes (for example, as a remedy for respiratory disorders). The meat is of negligible value.

These pelagic animals are uncommon visitors elsewhere along the Mexican coast. They are occasionally seen in the coastal areas of Sinaloa, the Gulf of California, and the Pacific coast of Baja California and the United States. Dead, stranded leatherbacks are occasionally seen along the shores of the Gulf of California (e.g., near El Golfo, Sonora, in March 1977, Cliffton, Felger, and Nabhan; Puertecitos, Baja California Norte, in November 1980, Brian Brown).

No male leatherback has been positively identified in the northern part of the Gulf of California. The Seri Indians, who revere the animal's spirit power, and Mexican fishermen at Kino Bay, Sonora, recall having encountered less than 2 dozen of them during the past 4 to 5 decades.

The Pacific Hawksbill (*Eretmochelys imbricata squamata*)

The hawksbill is called *Carey* in Mexico. It was once common along the Pacific coast of Mexico. Nesting may have occurred in scattered localities south of the desert coast of Sonora. Today it is rare, and there are no known nesting beaches for it on the Pacific coast.

Tortoiseshell has been traded along the coast of Mexico since ancient times. During the Spanish colonial era the shell was traded between coastal Indians, such as the Seri and those from Baja California, and Spaniards (Del Barco 1980; Hardy 1829). In the late nineteenth and early twentieth centuries, the hawksbill was abundant and heavily exploited, particularly in the southern part of the Gulf of California (Townsend 1916). Twenty years ago it was still common (Caldwell 1962).

Older fishermen from the Gulf of California region also tell of an abundance of hawksbills only 20 to 30 years ago. They say that the increase in price of tortoiseshell has doomed this creature. Hunting activities were concentrated in La Paz and Concepcion Bay on the east coast of Baja California, and the Infiernillo Channel on the Sonora coast north of Kino Bay. Seri Indians say that large hawksbills were frequent visitors into the Infiernillo Channel but are now seldom seen.

Mexican fishermen from the east coast of Baja California told us that 30 years ago the crew of a single fishing canoe (1 harpooner and 2 or 3 men paddling) could capture 5 to 7 hawksbills in 1 night (3 to 4 hours' work) in calm seas. They sold the tortoiseshell to local prisons, where the inmates were famous for working the material into combs, pendants, rings, and other ornaments. One fisherman told us of a hawksbill from Concepcion Bay weighing 100 kg which yielded 5 kg of tortoiseshell. A large population was reported from the Tres Marias Islands which may have been a major breeding ground (Parsons 1962).

Today, even though the hawksbill is rare, the fishermen consider it fair game regardless of size. When one is taken it is carefully cleaned and the meat eaten. It is then stuffed and sold as a curio to North American or Mexican tourists, or kept in the fisherman's home as a trophy. During the past 6 years we have seen about a dozen hawksbills captured by Seri and Mexican fishermen in the Kino Bay region, and 3 in a single season from the Michoacán coast. Only one showed adult characteristics.

The Pacific Loggerhead (*Caretta caretta gigas*)

The Pacific loggerhead is called *Perica* in Sonora and *Javelina* in Baja California. No nesting beaches are known for this species on the Pacific coast of Mexico. Though now relatively rare, it is occasionally taken incidentally in shrimp trawls in the Gulf of California and on the Pacific coast of Baja California. Shrimpers say that loggerheads are sighted floating on the surface of waters 20 to 60 m deep. They also enter the shallow bays, channels, and estuaries of northwestern Mexico.

The Seri Indians recognized 2 ethnospecies of loggerheads in their region, indicating greater diversity and abundance in the past (Felger and Edward Moser, unpublished notes). Mexican fishermen often refer to the loggerhead as *caguama mestiza* (sea-turtle half-breed), a designation given to any turtle that does not fit readily into their folk classification.

The loggerhead is generally not worth as much money as the green turtle. Although the fresh meat is consid-

KIM CLIFTON

ered "rank" by Sonoran epicures, it is palatable when prepared as *machaca* (meat which is salted, dried, shredded and fried). It has less meat in relation to body weight than does the green turtle. Juvenile loggerheads are stuffed and sold as curios by the fishermen.

No loggerhead taken in the Kino Bay region has been found with eggs in any stage of development. In recent decades there has also been no confirmed capture of a sexually mature loggerhead. The largest one we have seen, an immature male, weighed 65 kg.

According to the Seri and Mexican fishermen, large loggerheads once occurred in the Gulf of California. During the summer months lobster divers in the midriff island region of the Gulf of California sometimes encounter these turtles resting underwater among rocks along shallow lava reefs called *tepetates*. One diver saw 9 loggerheads during 1 summer. He estimated the largest animal to weigh 90 kg. During October and November 1976, Seri Indians harpooned 6 loggerheads in the Infiernillo Channel. It is considered unusual nowadays to see so many in such a short time.

Cliffton found 2 torpid (probably overwintering) loggerheads at 15-m depths, 1 in the Infiernillo Channel in March 1978, and the other off Margarita Island at Magdalena Bay, Baja California Sur, in February 1979. Unlike most overwintering green turtles, which have thick epizoic algae on their carapaces, the 2 loggerheads were heavily covered with bryozoans and associated invertebrates.

Mexican fishermen say that the loggerhead occurs at the Tres Marias Islands, the Revillagigedo Islands, and the Pacific coast of Baja California. They say that it does not occur on the Pacific coast of southern Mexico, and we have not seen it in that region. However, there are reports of loggerhead rookeries in Panama (Cornelius, this volume), and they also occur in northern Chile (Hays and Brown, this volume).

The Olive Ridley (*Lepidochelys olivacea*)

In Mexico, the smallest and most prolific Pacific Ocean sea turtle is called *Golfina*. The Nahuatl Indians call it *chiwanini* (the little one). The ridley is sought after for its quality leather, eggs, and meat. The most economically valuable sea turtle in Mexico, the ridley is the prime target and an easy catch for Mexico's industrialized turtle fisheries because it gathers in huge numbers during the breeding and nesting season.

The most dramatic feature of its life history is the simultaneous nesting of as many as hundred of thousands of turtles, known as the *arribazón* (great arrival), *arribada* (arrival), or the *moriña* (homecoming). The arribadas used to occur once during each lunar cycle of the nesting season, June through November, with peak nesting lasting 5 to 6 days in August, September, and October. The arribadas often follow strong and persistent south winds (in 1979 2 small arribadas occurred at Maruata Bay and Colola, Michoacán, when strong south winds signaled the daylight emergence of 30 to 100 females). After the nesting season, the turtles return to feeding grounds in Ecuador, Central America, and Mexico (René Márquez, personal communication).

The major olive ridley rookeries in Mexico were El Playón de Mismaloya, Jalisco; Piedra del Tlacoyunque, Guerrero; and Bahía Chacahua, La Escobilla, and El Moro Ayuta, Oaxaca. Nesting also occurred in the states of Sinaloa, Colima, Michoacán and Baja California Sur (Márquez, Villanueva and Peñaflores 1976).

According to a report by the National Marine Fisheries Service (1976), an average of 137,000 ridleys nested in the largest arribadas of 3 major beaches in the years 1968, 1969, and 1970, with considerable year to year variation. However, reliable informants in these areas say there was not as much variation in numbers of nesting turtles as indicated by the Fisheries report. Indeed there were tens of thousands of ridleys nesting at La Escobilla, Oaxaca, in 1969 (Antonio Suarez, personal communication), whereas the publication reports only a few hundred. It is apparent that the estimates presented in the report do not include all the arribadas from those years. Therefore, we use these estimates to calculate the lower limits of the adult female ridley populations of western Mexico prior to 1969. We make this calculation as follows:

$$N = (x)(m)/(p)(t)$$

Where: N *equals* the estimated total number of adult female turtles in the Mexican populations in 1969; x *equals* the average number of female turtles in the largest arribadas at the 3 nesting beaches (La Escobilla, Piedra de Tlacoyunque, and El Playón de Mismaloya) in 1968, 1969, and 1970 (x = 137,000); m *equals* the average remigration period (m = 1.3 years; Márquez et al., this volume); p *equals* the proportion of females nesting in each arribada (p = 0.6) [This proportion was used because in the past there were more arribadas than the average number of clutches laid by a turtle per season—each female did not participate in each arribada (Márquez et al., this volume)]; and t *equals* the proportion of all olive ridley nesting in Mexico that occurs at these three beaches (t = 0.5)(Márquez, Villanueva, and Peñaflores 1976). Therefore:

$$N = \frac{(137,000)(1.3)}{(0.6)(0.5)} = 593,667 \text{ female turtles.}$$

In the early 1970s, these 3 beaches were the site of approximately one-half of the olive ridley nestings in Mexico (Márquez, Villanueva, and Peñaflores 1976).

Thus, assuming a 50:50 sex ratio, at the lowest limits, there were approximately 1,185,000 adult (including males) olive ridleys in the breeding populations of Mexico in 1969.

During the peak of ridley turtle exploitation, the late 1960s, the harvest far exceeded any damage natural mortality could do (or that natural recruitment could make up for). By adding the number of turtles harvested over a short period of time to our 1969 population calculation we can estimate a population level that predates heavy exploitation.

The reported take of olive ridleys in the 5-year period to 1969 was approximately 27,800 metric tons (Mexican fisheries records as reported by Márquez, Villanueva, and Peñaflores 1976). This represents almost 700,000 individuals, most of which were females (based on Márquez's unpublished data from San Augustinillo, Oaxaca, 1977: $\bar{x}$ = 40.9 kg/turtle). This is an underestimation.

For catch data given by Márquez, Villanueva, and Peñaflores (1976) for 1968 alone, our calculation ($\bar{x}$ = 40.9 kg) yields 316,127 adult turtles. However, the fisheries records of this period have underestimated total catch. Carr (1972) states that more than 1,000,000 olive ridleys were landed in Mexico during 1968. Using this proportionality (the actual take being 3 times larger than the reported take), at least 2,000,000 olive ridley turtles would have been landed during the 5 years prior to 1969. Antonio Suarez, an undisputed expert on ridley harvest data in Mexico, corroborates this figure (Antonio Suarez, personal communication). The estimate of approximately 1,185,000 adult turtles in the breeding populations in 1969 coupled with 2,000,000 turtles killed in the 5 preceding years yields a conservative estimate of 3,185,000 olive ridleys in the seas of western Mexico in the mid-1960s.

We estimate the pre-1950 population size of olive ridleys in this area by another method. If the 3 major olive ridley nesting beaches in Mexico had the density of turtles that our informants claim (approximately one nesting turtle per square meter) there were on the order of 10,000,000 adult turtles prior to the recent intensive predation and destruction of habitat. This calculation is probably conservative considering that actual reported nesting densities of large ridley populations in western Costa Rica are 14 nests/m² (Cornelius, this volume) and that all nesting beaches are not included in the calculation.

For the past 15 years 3 human forces have affected the olive ridley in Mexico: the Mexican government's sea turtle program, industrialized sea turtle fisheries, and poachers. In the 1960s, Mexico's sea turtle program was heralded as a conservation model for the world. However, it has shown signs of weakness as seen in the following translation from Tecnica Pesquera (1970:7–8):

The obstacles that the Mexican turtle program is encountering (funding and organization) and the increased clandestine exploitation of eggs, meat, and turtle skin during the closed season are causing Mexican and foreign biologists to worry over the fate of the Pacific ridley turtle, Lepidochelys olivacea, which is intensively exploited.

Until a short time ago, it was believed that the survival of the species was guaranteed by the program of breeding and protection established by the Mexican government, and by the precautions taken to insure rational exploitation. The situation now seems less promising, however, and could be said to be very disturbing.

Mexico's answer to the pleas for rational exploitation was the industrialized sea turtle fishery, developed by Antonio Suarez Gutierrez. He began by selling Oaxaca ridley skins to the European markets. By 1969 Mexican law allowed exploitation of sea turtles only by those equipped to utilize the entire animal. Suarez's company had one of the only legitimate processing plants in Mexico. Prior to Suarez's entry into the business, it was common practice to utilize only a portion of the hide for the leather trade. The rest of the animal was discarded.

Overshadowing both industrialized fishing and government protection programs are the poachers. They take millions of eggs and countless thousands of animals each year. Suarez said that despite his well organized protection of the Oaxaca beaches, the poachers took about 1 million eggs from La Escobilla in 1979. Sr. Suarez claimed that he championed the sea turtles' cause in Mexico by combating the poachers in Oaxaca. Of the 3 major olive ridley rookeries in Mexico, only the Oaxaca beaches still have arribadas. Suarez claims this as a testimonial to his practices. However, his theories are based on the effects of exploitation. In 1976, the Mexican Department of Fisheries estimated a total population of 485,000 olive ridleys breeding off the Pacific coast (Márquez, Villanueva, and Peñaflores 1976). Suarez disagreed openly with this figure in June 1979, when he stated, "The ridley cannot be in danger of extinction if we (Mexico and Ecuador) have taken 400,000 turtles in 2 years (1977–78)" [This probably includes turtles that nest in Costa Rica and feed in Ecuador (see Cornelius, this volume)].

In fact there is no biologically sound formula for the continued exploitation of Mexico's ridleys. The huge breeding populations at El Tlacoyunque, Guerrero; El Playón de Mismaloya, Jalisco; Sinaloa; Michoacán; and Baja California Sur have crashed. The Oaxaca population is severely depleted and is showing signs of collapse.

In 1977 the size (58,000 female ridleys in the largest arribada) and the frequency of arribadas (only 2 per season) was one of the latest indications of the downward trend. In 1977, Suarez's company, PIOSA (Pes-

KIM CLIFTON

quera Industria de Oaxaca), took 70,000 ridleys (90 percent of them gravid) from this population, and 58,000 in 1978. At a meeting with Suarez in Mexico City in 1979, Peter Pritchard said, "Then in fact, the largest part of the arribada is taken each year."

Since then, with Suarez's approval, the government quota for capture of olive ridleys was cut in half. Observers at La Escobilla estimated that 36,000 female ridleys participated in the largest arribada in the 1979 season. PIOSA took 24,500 of them. There is little doubt that Suarez's protection of the Oaxaca nesting beaches has postponed the total collapse of this population. However, coupled with the uncontrolled exploitation in Ecuador, an estimated 100,000 ridleys harvested per season (Green and Ortiz, this volume; Antonio Suarez, personal communication), these great populations seem doomed.

Much of the industrialized fisheries protection program is based on the incubation of ovaductal eggs. Calcified eggs are taken from the ovaries of slaughtered gravid females, and less developed eggs are discarded. The calcified eggs are then transplanted to artificial nests along the beach or placed in styrofoam boxes with sand. Some of the problems with this method are: low hatch rates due in part to unskilled handling techniques (7 to 22 percent in 1977, Oaxaca; Antonio Suarez and others, personal communication), unnatural temperature regimes during incubation, and the absence of natural environmental stimuli that might be important if imprinting is required.

The East Pacific Green Turtle (*Chelonia mydas agassizi*)

The East Pacific green turtle is called *Caguama negra* or *Caguama prieta* in northwestern Mexico, and *Sacasillo* in Oaxaca. This species is vanishing from all its major feeding grounds in the Gulf of California and the Pacific Coast of Baja California. The last major nesting sites for the species in North America are 2 nearly adjacent beaches at Maruata Bay and Colola, Michoacán, nearly 3,000 km from the northern feeding grounds (Figure 1). Our tag returns indicate that the turtles traverse the Mexican coastline in their migrations, as well as travel to distant feeding grounds in Central and South America (Instituto Nacional de Pesca, unpublished data).

This species once provided coastal peoples with a major source of protein. Today, the meat is of very high commercial value in northern Mexico (100 pesos or US $4.44/kg, liveweight). The eggs are illegally collected and sold in Mexican cities. There is a popular belief that consumption of sea turtle eggs enhances virility.

The green turtle is the only Pacific turtle with both major feeding and breeding areas in Mexico. The green turtle section is divided into 2 subsections: the north-

ern feeding grounds and the southern breeding grounds.

The Green Turtle's Northern Feeding Grounds

Earlier in the century the green turtle was prevalent throughout the Gulf of California. It was abundant along the coasts and in the large bays, great saltwater lagoons, and deltas of Baja California, Sinaloa, and Sonora. It even entered the Colorado River, traveling upstream some 80 km to the Cocopa Indian village of El Mayor (Felger and Rea, unpublished notes). In these places the green turtle grazed on eelgrass (*Zostera marina*), ditchgrass (*Ruppia maritima*), marine algae (Fucales, Gelidiales, and Ulvales), white mangrove leaves (*Laguncularia racemosa*), and marine invertebrates (Felger and Moser 1973; Felger, Moser, and Moser 1980). Isolated groups of green turtles occurred along the rocky shores and bays of Nayarit, Jalisco, Oaxaca, and Chiapas, and it was abundant around the Revillagigedo and Tres Marias Islands. These were the major historic feeding grounds of the green turtle in western Mexico (Parsons 1962).

An indication of their former abundance on the Pacific coast is found in the report of the visit of the Albatross *to San Bartolome, or Turtle Bay in April 1889, when a remarkable catch of 162 green turtles, many of large size, was made in a single haul of a 600-foot-long seine. Half as many again were believed to have escaped from the seine before it was beached (Parsons 1962:73).*

On the Pacific coast green turtles were first heavily fished at the turn of the century in Baja California (O'Donnell 1974). An estimated 1,000 live turtles/month were sent to San Diego, California, from Scammon's Lagoon. By the 1930s, the market for fresh sea turtle meat in the United States dwindled but found added vigor in the border towns of Tijuana, Mexicali, Nogales, and the major cities in Baja California and Sonora (O'Donnell 1974). Demand for green turtle meat within Mexico grew steadily as the meat became equated with increased physical vitality and stamina, or virility.

In 1947, Jose León, Enrique Alverez, and Rafael Valenzuela were returning to Kino Bay, Sonora, and anchored at an abandoned Seri Indian camp on the shore of the Infiernillo Channel. Using a gasoline lamp and harpoon, they drifted into the canal that night in search of green turtles. In 4 hours the fishermen captured 9 green turtles whose total weight was 1.2 metric tons, an average of 133 kg per animal. The large turtles were preferred because it was easier to salt and dry the meat of a few big turtles than many smaller ones. The meat was sold for 50 centavos/kg, but salting and drying is a laborious process, and the turtles were readily sold alive for 10 to 30 pesos each.

Since ancient times the Seri Indians of the Sonora

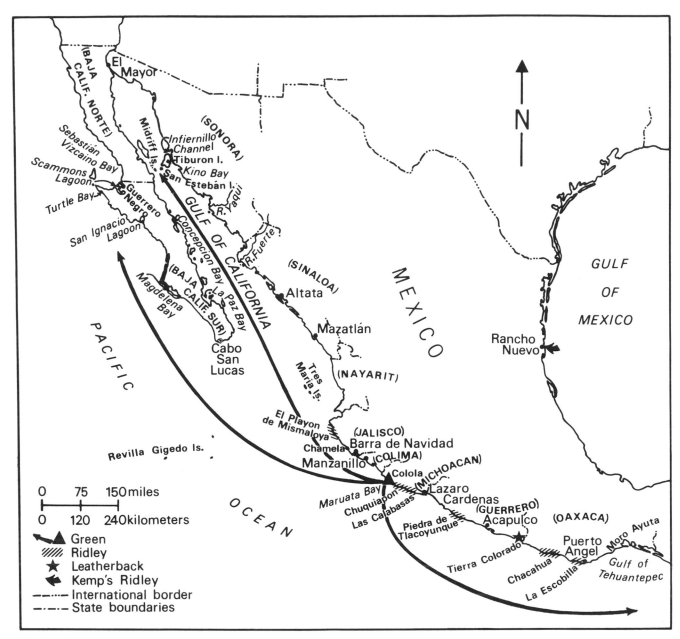

Figure 1. Map of Mexico showing the major sea turtle nesting beaches of the Pacific coast. Arrows leading away from Maruata-Colola indicate general migratory path of green turtles returning to feeding grounds. The arrow to Baja California is hypothetical since there are no tag returns from the Pacific coast of Baja California.

coast have hunted sea turtles in the northern part of the Gulf of California (McGee 1898; Smith 1974; Felger, Cliffton, and Regal 1976; Felger, Moser and Moser 1980). Certain groups of Seri Indians depended upon the abundance of large green turtles the year round. Their knowledge of marine turtles is rich and reveals detailed biological observations. The Seri name 7 ethnotaxa of green turtles while the Mexican fishermen from Kino Bay name 5 (Felger and Edward Moser, unpublished notes; Cliffton and Felger, in prep.). Today both the Seri and the Kino Bay fishermen see only 2 of these ethnotaxa.

In the 1950s, Chui Montaño, a Seri Indian, was enticed by a Mexican fish buyer to hunt green turtles commercially. In those days Chui could drift lazily with the current through the Infiernillo Channel and singlehandedly harpoon and boat 12 to 18 large (70 to 100 kg each) green turtles a day. He sold them for 10 pesos apiece.

Once a fixed price per turtle was set, the fishermen stopped hunting large turtles because more smaller ones, each at the same price as a large turtle, could fit in their boats. The catches were made entirely with harpoons until cloth tangle nets were introduced in the early 1960s. The nets reduced the investment of human energy. Despite the relatively small size of these

KIM CLIFTON

nets (100-m long), an average of 22 green turtles/day was caught with little effort. In 1965 the price was set at 1 peso/kg of live turtle. Once again the pressure on large turtles was increased.

In the 1960s the Seri readily adapted their *pangas* (6- to 7-m plank boats) to accommodate powerful outboard motors (25 to 50 h.p.). They combined ancestral knowledge and skill (the use of harpoons 7- to 10-m long) with modern techniques (monofilament turtle nets 160- to 200-m long) to virtually extirpate the remaining green turtles from their region by the late 1970s. In 1965, Guadalupe Lopez, the best Seri turtle hunter, could harpoon 5 metric tons of green turtle a week within 50 km of his home at El Desemboque. As larger turtles became scarce the Seri simply caught more juveniles (18 to 25 kg) at rates of 25 to 30 turtles/*panga*/day. There were 10 to 12 Seri *pangas* in operation in those days. Seri turtle hunting efficiency was demonstrated soon after we tagged and released 13 adult and subadult green turtles in the Infiernillo Channel in March, 1977. The Seri proudly returned 7 of the tags by May 1977.

In 1975 compressor diving was introduced as a new technique for hunting sea turtles in the Gulf of California. Mexican divers from Kino Bay accidentally discovered sluggish, overwintering green turtles off the south shore of Tiburon Island while hunting lobsters (Felger, Cliffton, and Regal 1976). These turtles were present in surprisingly large numbers, lying motionless at depths of 10 to 30 m. Exploitation of the overwintering populations quickly became a cottage industry. By the winter season of 1975 5 turtle boats or *pangas* were landing 4 to 5 metric tons of turtles/week at Kino Bay from late November through early March ($\bar{x}$ = 29.0 kg/turtle, S.D. = 8.31, n = 161). In 1975 it took 5 boats of fishermen with diving equipment to catch as much turtle as 1 boat of Seri Indians with harpoons in the 1960s.

One after another of these overwintering sites were located and depleted. The fishermen then began to exploit more distant areas in the northern part of the Gulf of California. By winter 1979–80 Kino Bay divers traveled 3 times the distance and invested many times the hunting effort but obtained smaller catches. The rapidly rising cash value of these increasingly rare animals provided ample incentive to the hunters.

Southern Breeding Grounds of the Green Turtle

The major nesting beaches were located on the sparsely inhabited mainland coast of Michoacán. Today the only remaining major nesting site for the green turtle on the Pacific coast of North America is Colola-Maruata Bay, between the Río Tikla and Río Balsas, Michoacán. Other less important nesting beaches may occur on the Pacific coast of Baja California Sur near Todos Santos

and western Chiapas (Pritchard 1979). Green turtles also nested at the Revillagigedo and Tres Marias Islands (Parsons 1962).

Maruata Bay was used by British privateers while raiding Spanish shipping lanes near Acapulco and Manzanillo. They replenished stores of fresh meat by capturing green turtles which abounded in the area, and obtained fresh water at the perennial river which flowed into a deep estuary and out to sea. William Funnel, mate to Captain Dampier, wrote on 22 November 1704, "Here we watered our ship and found in a small river a great many large green turtles, the best I ever tasted" (Dampier 1906, *in* Peters 1956:21).

In August 1950 Peters (1956:22) journeyed into the "isolation and comparative primitiveness" of Maruata Bay and made the following observations 2 months prior to the peak nesting:

The evidences of their activity were everywhere . . . In a half mile stretch between two stringers I counted 472 such tracks (nesting turtles), which would indicate 250 individual turtles . . . Many turtles returned to the ocean via the track they made so I'm sure my estimate is not high.

By extrapolating these nesting densities over the entire 3-km length of beach, we estimate that there were on the order of 900 turtles nesting at Maruata Bay within several days of Peters' observations. (The tracks may remain for 2 to 5 days during this time of year which is the rainy season). In contrast, we estimate that only 600 turtles nested at Maruata Bay during the entire 1979 season. Furthermore, nesting green turtles are no longer common in August.

Native (Nahuatl Indian) informants claim that 500 to 1,000 green turtles a night nested at Colola only 10 years ago during the heaviest nights of nesting. At first these claims seemed to be impossible; they were saying that there were 10 to 20 times as many turtles in 1970 as there were in 1977. That would be approximately 25,000 females nesting in Michoacán yearly, possibly 150,000 adult turtles (males and females) in the population, assuming a 3-year remigration period and a 50:50 sex ratio. Informants who have hunted for more than 25 years in the northern feeding grounds of the Gulf of California and the Pacific coast of Baja California report a similar decrease in the abundance of green turtles in their area.

According to the Instituto Nacional de Pesca, in a 5-year period from 1966 through 1970, 4,618 metric tons of green turtle were landed on the Pacific coast of Mexico (Márquez, Villanueva, and Peñaflores 1976). We calculate that this represents approximately 125,000 turtles, adults and subadults [36.5 kg average weight (Caldwell 1962)]. While 125,000 is an impressive figure insomuch as there are not that many green turtles left in Mexico's seas, it is an underestimation of the total number of turtles landed during that period. Until the mid- to late-1970s there were no roads to the major nesting beaches in Michoacán. Meat and the skins were taken out by boat or plane, and many fishermen in these underdeveloped areas never reported their catches. Carr (1972) indicates that the fishery records for this period represent about a third the number of turtles landed. In other words, about 375,000 green turtles were taken in that 5-year period, of which at least half were adults.

The number of green turtles nesting each year in Michoacán fluctuates dramatically, as it does elsewhere in the world (Carr, Carr, and Meylan 1978). For example, 7 to 30 turtles nested per night at Colola during October and November 1977, as compared to 50 to 150 per night on the same beach in October and November 1979.

We estimate that there are now between 5,114 and 8,523 adult female green turtles in the Michoacán breeding population. We make this calculation as follows:

$$y = (u)(r)/(n)(s)$$

Where: y *equals* the number of adult female green turtles in the Michoacán population, n *equals* the number of clutches they lay per season (n = 4, Márquez *et al.*, this volume), s is the average number of eggs per clutch (s = 66, Márquez et al., this volume; and our data), u *equals* the average number of eggs laid per season at Colola and Maruata Bay (u = 750,000, estimated average from 1977–79), and r is the average remigration period (r = 1.8 years, Márquez, et al., this volume, for the lower limit; and r = 3.0 years for the upper limit because it is so common for the green turtle elsewhere in the world).

For the past 15 years the Nahuatl people have increased their population and settlements on the coastal plain encompassing the green turtle breeding grounds. Their slash and burn agriculture in the nearby hills greatly increased the rate of erosion. This, coupled with a hurricane in 1970 resulted in the near destruction of the estuary at Maruata Bay, changed the course of the river (now a dry arroyo for most of the year), and reshaped the western beach.

As international markets for sea turtle leather expanded in the 1960s and 1970s, so did human predation on the green turtle rookeries in Michoacán. Fishermen from Colima, Jalisco, and Guerrero began capturing breeding turtles with nets and by pouncing on mating pairs (a hunting technique called *al brinco*).

At first the Nahuatl Indians gazed upon this carnage in disgust. They even killed an occasional outside fisherman because they equated the wholesale slaughter with a threat to their own commercial interests. Beginning in the early 1970s, a market for sea turtle eggs was introduced to the Nahuatl. They were offered 20 pesos per 100 eggs. They claimed that each night 15,000 to 20,000 eggs were collected at Maruata Bay, 70,000 eggs at Colola. The people became "rich" overnight, and settlements around the nesting beaches increased rapidly.

In 1974 President Luis Echeverria's administration expanded the development of fishing cooperatives along the Pacific coast. The Nahuatl were given, on loan, new boats, motors, and nets to increase the production of coastal fisheries. However, the young men who organized the Pomaro fishing cooperative of Maruata Bay turned their new found technology on the green turtle. On Cliffton's first visit to Maruata Bay in August 1976, he witnessed the illegal capture of 40 to 80 green turtles a day. Most of the animals were drowned in shark nets set directly in front of the nesting beaches. Though of low quality, the hides were stripped from the animals to be sold for 50 pesos each (US $2.20). The carcasses were thrown into the sea.

Government control in this remote area was non-existent until Mexican marines were sent to protect the Colola beach in 1977. The marines collected nearly 50,000 sea turtle eggs, and protected them in a hatchery. However, they were relieved of their duty prior to the hatch-out, and within weeks the entire hatchery was destroyed by pigs and dogs.

In 1977 Suarez's company, IPOSA (Industria Pesquera Occidental), a subsidiary of PIOSA, began exploiting the breeding green turtles in Michoacán. In 1978, IPOSA began full-scale processing of male and female green turtles legally captured by the Pomaro fishing cooperative in Maruata Bay. The majority of the animals were drowned in shark nets set by the fishermen. The incubation of oviduct eggs, which has had poor results in Oaxaca for the ridley, had no success when eggs were taken from turtles dead for some hours, then transported in trucks for 6 hr to the IPOSA plant in Barre Navidad, Jalisco (Figure 2). Regional fisheries inspectors described the operation as "scandalous" and prohibited the use of shark nets and the capture of any more female turtles. The final tally for the season was 500 female and 10,000 male green turtles taken in Michoacán and Jalisco by IPOSA (Antonio Suarez, personal communication). These figures do not include losses to poachers for 1978 that probably figure in the hundreds of female green turtles. After analyzing the IPOSA catch data for 1978 which indicated a 20:1 male-to-female sex ratio for the green turtle, Suarez decided that there was an abundance of still exploitable male green turtles.

For good reasons, the male sex ratios appear higher in the breeding grounds. Nahuatl fishermen can capture male green turtles in the breeding grounds all year. Tagging data indicate that the male turtles wander hundreds of kilometers to and from the breeding beaches (male green turtles tagged at Colola were recaptured

KIM CLIFTON

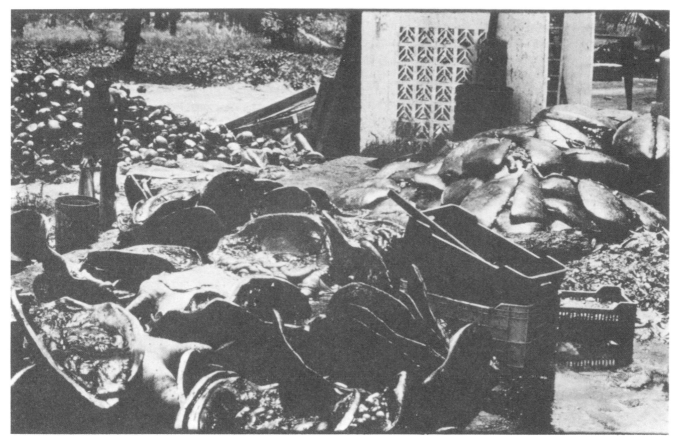

Figure 2. Carapaces, bones, and entrails of female green turtles from Colola-Maruata at the IPOSA plant awaiting conversion into chicken feed, July 1978.

in Chamela, Jalisco within 2 weeks of release). The males are extremely active during the breeding season in comparison with the females. Offshore from the nesting beaches in rocky areas (20 to 60 m deep) they feed on crabs, jelly fish, sponges, hydroids, algae, and debris washed down from the rivers. Fishermen in the feeding and breeding grounds describe male green turtles as "skinny." A few tag returns indicate that some of the males return to mate annually. In rut they are almost completely beyond any concern for self-preservation and are extremely vulnerable to fishermen hunting them *al brinco*.

The females spend their internesting periods in semimoble aggregations within a few hundred meters of the nesting beach in shallow water (1- to 6-m deep). Consequently, they are infrequently captured by the Nahuatl Indians who traditionally fish for the male greens in the offshore areas previously described.

The new coastal highway reached Maruata in 1978, passing within 200 m of both major nesting beaches. Virtually overnight a viable commerce in green turtle contraband sprang up, supplying meat to Sonora and Sinaloa markets, already depleted of their once abundant local reserves. Members of the Pomaro fishing cooperative soon discovered it was far more profitable to sell to the smugglers than to IPOSA.

Since the 1978 season, the Mexican Government has tried two methods to gain control of the green turtle fisheries. In 1979 a closed season was declared for the nesting beaches. However, the fishing cooperative continued to catch turtles and sell them to smugglers. We estimate that 2,000 female and 1,000 male green turtles were illegally taken at Colola and Maruata Bay in 1979.

In 1980 the Government changed strategy and established a quota of 250 male turtles a month throughout the season (September through December) for the Pomaro cooperative. The setting of a quota has greatly changed the attitudes of the co-op members. No longer do they sell to poachers or take nesting turtles. Indeed, for the first time they have begun to accept the green turtle recovery program.

The East Pacific green turtle recovery program, designed and implemented in 1978 by the Arizona-Sonora Desert Museum and the Instituto Nacional de Pesca, began with the release of over 70,000 hatchlings. It would be extremely difficult to control poaching on the Michoacán coast which now has open access from land, sea, and air. The 6 marines stationed at Colola, without a vehicle, could not be expected to control the contraband. Consequently, to combat poachers coming from towns on the Colima border, we saturated the remote sections of beach with egg collectors loyal to our cause.

In 1979 Suarez was convinced that the East Pacific green turtle was endangered and began actively to support their protection. With the addition of his financial support in 1979, 326,000 eggs were recovered and transplanted to protected hatchery-corrals at Colola and Maruata Bay by the Nahuatl men, women, and children. Most of the eggs were *Chelonia*, but there were also several thousand *Lepidochelys* and *Dermochelys*. In the decade prior to 1978 virtually the entire reproductive efforts of the green, ridley, and leatherback turtles at Maruata Bay and Colola were destroyed by man and his domestic animals.

The hatcheries at Colola and Maruata Bay have been set up to follow as closely as possible the natural pathways of the turtle's life history tactics by transplanting the eggs into a corral situated on the native beach. When the hatchlings emerged they were collected around a small kerosene lamp. Within a few hours they were released on diverse sections of the beach to avoid the build up of predatory fish (e.g., *Nemastistius*, rooster fish; *Coryphaena*, dolphin; and *Caranx*, jack). The hatchlings were allowed to crawl to the sea so as not to interfere with possible imprinting.

Conclusions

In the late 1970s a spirit of cooperation was reached among the Mexican government, commercial interests, local people, and the international sea turtle conservation community. It was agreed that conservation efforts are best served by open communication, honesty, and mutual respect. Yet these successes seem negligible when compared to losses to poachers. The mania for green turtle meat in Sonora and Sinaloa, and the demand for all sea turtle eggs in Mexico and leather for the luxury trade continue to drive the prices up, cultivating corruption at local levels.

In recent years a philosophy of "rational exploitation" has emerged in Mexico, and relies on modern processing to make maximum use of natural resources. Suarez and others involved in legal exploitation are strong adherents to this point of view. They see the sea turtle situation as influenced by a triangle of factors: socioeconomic, political, and biological. Consistant with the concept of rational exploitation, Suarez said, "The surest way to drive a species to extinction [in Latin America] is to give it total protection." He felt that a total moratorium on sea turtle hunting in the past benefited those operating outside of the law and crippled those wishing to work in compliance with it. This is why he insisted that, "You must exploit something, even a token quota, to maintain control of the fishery." This philosophy has been at odds with most international conservation activities.

Márquez (1976) proposed that the major breeding beaches of Mexico should be made into National Reserves. Organization of a private citizen's committee to fund research and conservation programs for sea turtles in Mexico seems a logical step to the protection of these magnificent animals. Excessive take, poaching and loss of habitat have brought the sea turtles of Mexico to the brink of extinction. The social and economic pressures that have led to this ecological disaster need to be assuaged by setting new priorities in the very near future. We believe that Mexico's sea turtles can probably still be saved.

Acknowledgments

Many people and institutions have generously helped with our sea turtle research. In this regard we especially thank Holt Bodinson, Dr. Jorge Carranza Fraser, Dr. Sylvia Earle, Fernando Fuerte, Dr. John Hendrickson, Roberto Herrera T., George Huey, Dr. Wayne King, Mervin Larson, Linda S. Leigh, Jose León, Dr. Paul Licht, Dr. René Márquez, Clay J. May, C. Allan Morgan, the late Edward Moser, Mary Beck Moser, Carlos Nagel, Rosa Martin de Oliver, Robert Perrill, Dr. Peter Pritchard, William Rainey, Lynn Ratener, Dr. Phillip J. Regal, the late Alexander Russell, David Russell, Paul B. Schneider, Vivian Silverstein, Charles A. Stigers, Antonio Suarez Gutierrez, Barbara Tapper, Alfredo Topete, Robert Truland, Aristotle Villanueva, Donald Way, and R. Curtis Wilkinson. We are grateful for support from World Wildlife Fund-U.S., the Roy Chapman Andrews Research Fund of the Arizona-Sonora Desert Museum, the Chelonia Institute, New York Zoological Society, Antonio Suarez Gutierrez, and National Science Foundation (BNS 77–08–582).

Literature Cited

Caldwell, D. K.
1962. Sea turtles in Baja Californian waters (with special reference to those of the Gulf of California) and the description of a new subspecies of Northeastern Pacific green turtle. *Contributions in Science* 61:1–31.

Carr, A.
1972. Great reptiles, great enigmas. *Audubon* 74:24–34.

Carr, A.; M. H. Carr; and A. B. Meylan
1978. The ecology and migrations of sea turtles, 7. The West Caribbean green turtle colony. *Bull. Am. Mus. Nat. Hist.*, 162:1–46.

Del Barco, M.
1980. *The Natural History of Baja California.* Translated by F. Tiscareno. Los Angeles: Dawson's Book Shop.

Felger, R. S.; K. Clifton; and P. J. Regal
1976. Winter dormancy in sea turtles: independent discovery and exploitation by two local cultures. *Science* 191:283–85.

Felger, R. S., and M. B. Moser
1973. Eelgrass (*Zostera marina* L.) in the Gulf of California:

discovery of its nutritional value by the Seri Indians. *Science* 181:355–56.

Felger, R. S.; M. B. Moser; and E. W. Moser

1980. Seagrasses in Seri Indian culture. In *Handbook of Seagrass Biology: An Ecosystem Perspective*, eds., R. C. Phillips and C. P. McRoy, pp. 260–76. New York: Garland STPM Press.

Hardy, R. W. H.

1829. *Travels in the Interior of Mexico*. London: Henry Colburn and Richard Bently.

Márquez, R.

1976. Reservas naturales para la conservación de las tortugas marinas de Mexico. *Instituto Nacional de Pesca, Subsecretaria de Pesca, Serie Información*, 83:1–22.

Márquez, R.; A. Villanueva O.; and C. Peñaflores S.

1976. Sinopsis de datos biologicas sobre la tortuga golfina *Lepidochelys olivacea* (Eschscholtz, 1829). *Instituto Nacional de Pesca Sinopsis Sobre la Pesca*, No. 2.

McGee, W. J.

1898. The Seri Indians. Seventeenth Annual Report of the Bureau of American Indian Ethnology, pp. 1–344. Washington, D.C.

NMFS (National Marine Fisheries Service)

1976. Proposed listing of the green sea turtle (*Chelonia mydas*), loggerhead sea turtle (*Caretta caretta*), and Pacific ridley sea turtle (*Lepidochelys olivacea*) as threatened species under the Endangered Species Act of 1973.

O'Donnell, J.

1974. Green turtle fishery in Baja California waters: history and prospect. Masters Thesis, California State University, Northridge.

Parsons, J.

1962. *The Green Turtle and Man*. Gainesville: University of Florida Press.

Peters, J.

1956. The eggs (turtle) and I. *The Biologist* 39.21–24.

Pritchard, P.

1973. International migrations of South American sea turtles (Chelonidae and Dermochelidae). *Animal Behavior* 21:18–27.

1979. *Encyclopedia of Turtles*. Neptune, New Jersey: T.F.H. Publications.

Smith, W. N.

1974. The Seri Indians and the sea turtle. *Journal of Arizona History* 15:139–58.

Stebbins, R. C.

1966. *A Field Guide to Western Reptiles and Amphibians*. Boston: Houghton Mifflin Company.

Tecnica Pesquera

1970. In Marine turtles. Proceedings of the 2nd Working Meeting of Marine Turtle Specialists 8–10 March 1971 at Morges, Switzerland. *IUCN Publications New Series Supplementary Paper*, No. 31:7–8.

Townsend, C. H.

1916. Voyage of the *Albatross* to the Gulf of California in 1911. *Bull. Am. Mus. Nat. His.*, 35:399–476.

Stephen E. Cornelius
Department of Biology
Texas A&I University
Kingsville, Texas 78363

Status of Sea Turtles Along the Pacific Coast of Middle America

ABSTRACT

The current status and distribution of sea turtle populations, historical trends, manner of exploitation and conservation and research programs are described for each of the 6 countries of Middle America. Five species, listed in order of their estimated relative abundance, are: olive ridley (*Lepidochelys olivacea*), green turtle (*Chelonia mydas*), loggerhead (*Caretta caretta*), leatherback (*Dermochelys coriacea*), and hawksbill (*Eretmochelys imbricata*). The olive ridley is the most abundant turtle in El Salvador, Honduras, Nicaragua, and Costa Rica. Large synchronous nesting emergences (*arribadas*) of this species occur in Costa Rica and formally occurred in Nicaragua and Panama. The green turtle is the most abundant species in Guatemala. Unusually large numbers of loggerheads apparently comprise the majority of Panama's turtle population. All countries either have had in the past, now have or are preparing protective legislation for sea turtles and/or their eggs. Costa Rica's populations are the best known, but basic inventories of nesting beaches are needed in all countries to fill gaps in knowledge of population status, distribution, reproductive potential, and level of exploitation. The commercial overexploitation of eggs, inadvertent mortalities from shrimping operations, egg predation by domestic animals, and general inadequacy of regulatory mechanisms are all, to some degree, factors contributing to the decline of sea turtle populations in Middle America.

Introduction

This report discusses the current status of sea turtle populations along the Pacific Coast of Middle America. Each of the 6 countries—Guatemala, El Salvador, Honduras, Nicaragua, Costa Rica and Panama—is treated separately in the following general manner: physical description of the coastline, status and distribution of the populations, threats to survival, protective measures, and conservation/research programs.

Except for Mexico and Costa Rica, very little infor-

Figure 1. Map of Middle America showing the location of marine turtle nesting beaches, research stations, reserves, and major shrimping grounds.

mation has entered the turtle literature for the East Pacific region. It was necessary, therefore, to rely extensively on information solicited from those acquainted with the current situation in each country. This was especially important regarding the effectiveness of existing legislation.

Guatemala

Guatemala's Pacific coastline of about 250 km is straight and open with no natural harbors and relatively shallow offshore waters. The long stretches of black sand beaches are broken by 7 estuaries between its river borders with Mexico and El Salvador.

Probably less is known about the sea turtles in Guatemala than anywhere else in Middle America. Published information is typically incidental to other studies. Coe and Flannery (1967) listed both the meat and eggs of the *parlama (Chelonia mydas)* as a food item in the southwestern coastal villages near Ocós. The barrier beaches fronting the Chiquimulilla Canal (La Rosario, Las Lisas, El Hawaii) harbor a sizable population of green turtles (Figure 1). At least 20 individuals a night have been observed along a 15-km stretch during the nesting season of September and October (Herman

Kihn, personal communication). Olive ridleys are present in fewer numbers.

The past 20 years have witnessed a steady decline in Guatemala's turtle populations. This can be attributed to two factors: commercial harvests of eggs and the incidental catch of subadult and adult turtles in shrimp trawls. There is no demand for turtle skins or meat.

Although Carr (1961) found olive ridley eggs common in the markets in 1959, human exploitation probably did not adversely affect turtle populations until access to the remote southeast coastal region improved. With increased contact with the more populated highland interior, poaching and marketing have become vigorous businesses in areas where turtles nest. Eggs can be purchased at the nesting beaches for 50 cents a dozen by coastal families or 2 for 25 cents by tourists. In the large cities the price rises to $1.50 a dozen in the open markets and 2 for 50 cents in bars and restaurants (anonymous communication at request of informant).

The shrimp fishery, which began in the late 1950s, has expanded considerably causing an increase in the incidental catch of turtles. Approximately 150 to 200 carcasses wash ashore in southeast Guatemala annually

STEPHEN E. CORNELIUS

(anonymous communication). Their undecomposed and nonmutilated condition upon stranding implicated the shrimp fleet that patrols the entire coast in water less than 70 m. The center of the fishery is from San José to the Salvadorean border. Until recently, most turtles were returned to the sea. Serious consideration is now being given to developing a market for these incidental catches (Kihn, personal communication).

At present, the capture and commercial use of all sea turtles is prohibited by law (Bacon 1973; Kihn, personal communication), although the law is widely ignored in the case of eggs. In 1976, the government attempted to limit the harvest of eggs (James Richardson, personal communication). It is likely that commercial use has now returned to total exploitation. A hatchery was established on the southeast coast five years ago but is now unattended (Kihn, personal communication).

El Salvador

El Salvador has a 300-km coastline, about half of which fronts a narrow coastal plain. The south central plain widens to 30 km around the 2 major estuaries, Estero Jaltepeque and Bahia Jiquilisco.

Four species of sea turtles are known to utilize Salvadorean beaches: golfina (*Lepidochelys olivacea*), tortuga (*Chelonia mydas*), baul (*Dermochelys coriacea*) and carey (*Eretmochelys imbricata*) (Victor Rosales and Manuel Benítez, personal communication). Preliminary results of a government survey indicate the olive ridley and hawksbill are the most abundant and rarest, respectively. Dispersed nesting occurs from July through December on all sandy beaches.

Coastal fishermen report nesting has decreased abruptly in recent years (Benítez, personal communication). Few people kill nesting turtles, and there exists only a small craft industry in hawksbill shell. El Salvador appears to be the only country in Middle America where beach destruction and alteration are serious threats to nesting habitat. Construction of tourist facilities near the high tide line causing beach erosion and general pollution of the shoreline are blamed (Benítez, personal communication). Most exploitation centers, however, around a well-organized domestic egg trade. In three markets, 18,956 eggs were sold during September to December 1978 (Zelaya 1979). The high price of $2.50 to $3.00 a dozen was indicative of their scarcity.

Many turtles, mostly olive ridleys, are entangled by nets of the shrimp fleet that operates in nearshore waters from Estero Jaltepeque to the mouth of the Gulf of Fonseca (Benítez, personal communication). Observations of government personnel and coastal fishermen suggest a direct relationship between degree of shrimping activity and numbers of dead turtles stranding.

Although El Salvador offered nesting turtles some protection in the early 1970s, no legal instrument currently regulates the use and conservation of fisheries and wildlife (Serrano 1978; Benítez, personal communication). The civil code, in fact, considers all wildlife as *propriedad feudal*. Neither has the country ratified any international conventions on natural resource conservation. Responsibility for marine resources are jointly held by the Division of National Parks and Wildlife and the Fisheries Resources Service. Considerable effort has been made to rectify the situation, beginning with a study of coastal areas appropriate for national parks and marine biological reserves (Serrano 1978).

During 1978, Legislative Decree 427 (December 1977) prohibited the hunting, selling, buying, exporting and consumption of all marine turtles and their eggs. The law had a rather short lifespan, expiring after 1 year. During the time the law was in effect, fewer eggs were sold openly in the markets. Poaching at the beaches was uninterrupted, however, and most bars and restaurants continued selling eggs with few restrictions.

The Fisheries Resources Service established hatcheries at four locations in 1975. Beginning in 1978, the program was expanded and a cooperative effort between the Ministry of Agriculture (with assistance from the U.S. Peace Corps), Ministry of Education, University of El Salvador and the commercial fishing cooperative at Barra de Santiago (ACUARIO) resulted in adoption of a more scientific approach to the collection of eggs, identification of nesting turtles and care of transplanted nests (Rosales, personal communication).

El Salvador has initiated an ambitious study of their sea turtle populations. The absence of protective legislation, lack of a strong conservation ethic in the general populace, and insufficient funding and direction will make their effort difficult.

Honduras

Honduras has only a 65 km-long contact with the Pacific Ocean at the head of the Gulf of Fonseca, which it shares with El Salvador and Nicaragua. It also has jurisdiction over several nearshore, largely uninhabited, islands. The perimeter of the shallow gulf has only a few narrow and uncompacted sand beaches.

In spite of the apparent unsuitability of the gulf's shores for marine turtle nesting, most early information of Middle America's west coast populations has come from Honduras (Carr 1952). Carr (1948) reported green turtles, hawksbills, and olive ridleys in the market places of the capital, Tegucigalpa, and observed ridleys nesting on several islands in the gulf.

Recent estimates of the ridley population in Pacific Honduras are placed at 3,000 nesting individuals (Jonathan Espinoza, personal communication). They are common throughout the gulf from July through December with a peak in September and October. Most nesting activity is concentrated in the areas of Punta Ratón and Playa Cedeño (Paul Purdy, personal communication).

A noticeable decline in numbers of nesting turtles has been observed (Espinoza, personal communication). At the time of Carr's visits, egg collecting, conducted primarily by Salvadoreans, occurred on the islands from August through November. They were transported to La Union, El Salvador for sale. Ridleys probably suffered the most from such collecting, but nests of hawksbills were likely poached also. The sale of turtle eggs in local and countrywide markets and coastal restaurants is still a lucrative business for people residing near nesting beaches. It is estimated that at least 90 percent of all eggs are removed for the commercial trade (Purdy, personal communication). There is no exploitation of turtles for meat, oil or leather.

The National Fishing Law (1959) prohibits the capture of turtles and the taking of eggs, although there is little effort to enforce the regulation (Purdy, personal communication). Resolutions to prohibit the sale of turtle eggs are submitted annually to the congress but have never been approved (Espinoza, personal communication).

There is no developed shrimp industry in Pacific Honduras as few shrimp of commercial size are produced in the Gulf of Fonseca. The estuaries are excellent nursury grounds for many species of finfish, however, and a subsistence cast net and set net fishery probably takes turtles occasionally.

Sea turtle conservation and management is the responsibility of the Department of Wildlife and Environmental Resources, created in 1974. No biological reserves exist, although several are planned (Aguilar 1978). Some effort has been made to protect certain nesting beaches and operate artificial hatcheries but these have not been very successful (Espinoza and Purdy, personal communication).

Nicaragua

Nicaragua's 300-km shoreline on the Pacific is composed of 3 physiographic regions. Sand beaches suitable for sea turtle nesting are found in all regions.

Five species reportedly frequent the beaches of Pacific Nicaragua (Nietschmann 1975; Reynaldo Arostegui, personal communication): *paslama (Lepidochelys olivacea), tortuga tora (Dermochelys coriacea), tortuga caguama* or *falso carey (Caretta caretta), carey (Eretmochelys imbricata)*, and *tortuga verde (Chelonia mydas)*. The olive ridley is the most common species but relative abundances of the others are unknown. Known nesting areas by political departments are as follows: Chinandega—Potosí, Padre Ramos, Jiquilillo and Corinto; Leon—Peneloya, Salinas Grandes, El Tránsito and Puerto Sandino; Managua—Masachapa and Pochomil; Carazo—La Boquita and Casares; Rivas—El Astillero, Ostional and Brito (Figure 1). The most important beaches are in the departments of Managua, Carrazo and Rivas. The 20-km stretch between Masachapa and Pochomil and El Astillero Beach are well known sites for mass nesting olive ridleys and are the focus of year-round nesting by other species (Nietschmann and Arostegui, personal communication). Ridleys nest from September through November but are observed in nearshore waters during the remainder of the year.

Local people claim the ridley used to nest in much larger numbers, and that the size and frequency of mass arrivals are diminishing. Nesting turtles are rarely killed except by an occasional coastal family for fresh meat, and few are accidentally caught and killed in shrimping operations (Nietschmann 1975). Rather, the observed decline is traced to overexploitation by egg collectors. Total removal in a traditional, well-organized and legal manner occurs during August and September. Egg collecting is banned during October and November. Weak surveillance, however, permits continual exploitation, especially from Masachapa to La Boquita (Arostegui, personal communication). A small number are eaten by coastal dwellers while the vast majority are transported to the major cities and sold in the open markets, supermarkets, bars, and restaurants. Large shipments are also exported to other Middle American countries. Nietschmann (1975) reported 500,000 sent to Guatemala. El Salvador imported 568,000 eggs in 1975 and 648,000 in 1976 from Nicaragua (Arostegui, personal communication). Assuming an average clutch of 100 eggs and 2 nests a season, this represented the reproductive effort of 2,840 females in 1975 and 3,240 in 1976.

The apparently low rate of capture in shrimp nets may be due to the inability of boats to drag close to shore. Shrimpers operate along the entire coast but because the nearshore shelf is poorly known and rocky, most trawling takes place at depths greater than 20 m (Gross 1971). This probably reduces contact between trawlers and turtles near nesting beaches.

The protection and scientific study of sea turtles falls under the jurisdiction of the Department of Wildlife, created in 1956. It was not until Legislative Decree 625 was published in May 1977, however, that capture and hunting of all wild animals, including sea turtles, for commercial gain was prohibited (Salas 1978). This legislation also prohibited for 10 years the exportation of olive ridley eggs. It is poorly enforced and exporters are able to circumvent its intent (Arostegui, personal communication). The new revolutionary government

STEPHEN E. CORNELIUS

that came to power in 1979 has "expressed a concern for the protection and appropriate use of the country's renewable natural resources" (Arostegui, personal communication).

From 1959 until 1976, a hatchery was operated at Pochomil during the nesting season of the ridley. In 1975, 38,250 eggs, representing 425 nests, were transplanted. No information is available on percent hatch, but even if a 100 percent success rate were achieved, this attempt could only restore a very small portion of that lost through the egg trade. The only scientific study of Nicaragua's turtle populations was conducted by Bernard Nietschmann in 1972 at the beaches of Masachapa and La Boquita (Nietschmann, personal communication).

Costa Rica

The 575-km coastline of Costa Rica is broken by a series of peninsulas and gulfs. from its northern border with Nicaragua to the tip of the Peninsula of Nicoya, a multitude of cobblestone and compacted beaches have formed. Around the perimeter of the Gulf of Nicoya to the Peninsula of Osa, the coastal plain broadens and long beaches are broken intermittently by rock headlands and large estuaries. The precipitous nature of the northwest coastline returns at the Peninsula of Osa.

At least 4 species of turtles nest along the Pacific coast. They are: *lora* or *carpintera (Lepidochelys olivacea), tora* or *verde (Chelonia mydas). baula (Dermochelys coriacea)* and *carey (Eretmochelys imbricata)*. The loggerhead *(caretta caretta)* may nest around the Peninsula of Osa since it is reportedly common in Panama.

Ostional and Nancite beaches are well known for their large mass nesting populations of olive ridleys. Other beaches in Guanacaste Province used extensively by various species but not approaching the numbers at Ostional and Nancite are: Coloradas, Cabuyal, Matapalo-Zapotal, Langosta-Tamarindo, and San Miguel-Coyote (Richard and Hughes 1972). Except for Savegre and Matapalo beaches south of Puerto Quepos, only scattered nesting is recorded between the Nicoya and Osa peninsulas. Heavy nesting occurs on the long exposed beaches of the Osa, particularly Piro-Madreigal and Llorona-Sirena (Richard and Hughes 1972; Christopher Vaughan, personal communication) (Figure 1).

Leatherbacks nest sporadically along the length of the coast. Total numbers are small (Cornelius 1976). Local people claim the leatherback is more abundant during the dry season months of December to May. This nesting chronology coincides with reports from El Salvador, where leatherbacks nest in November and December after other species have finished (Benítez, personal communication).

The green turtle is locally common although far less abundant than ridleys. The nesting season is prolonged, possibly extending the year around (Cornelius 1976). The appearance of numerous moribund subadults on the northwest Guanacaste coast during 1972 (Cornelius 1975) suggests that Costa Rica's green turtles may be resident, or that juveniles participate in seasonal movements.

Hawksbills are more often seen in shrimp trawl catches than nesting on the beaches. Coastal people recognize the name *carey* and can reasonably describe the species.

Egg poaching is a time-honored tradition. Ostional has been the primary egg supplier to the markets of Guanacaste and the Central Highlands. At least 60,000 eggs left Ostional during a 3-day period in 1970 (Doug Cuillard, personal communication). Because of a surprisingly low hatching rate of eggs at arribada beaches (Hughes and Richard 1974) a controlled harvest of ridley eggs was proposed in 1977. This project was subsequently denied following a lively newspaper debate.

Wherever villages are situated on or near nesting beaches natural predation is augmented or replaced by domestic animals. Dogs destroy almost every nest along the Peninsula of Osa during certain times of the year (William Rainey, personal communication). At Ostional, pigs consume large quantities of eggs with the ironic result of instilling an undesirable flavor to the flesh. Guards were placed at Ostional in 1979 for the first time and have been successful in reducing the loss to domestic animals as well as stopping human poaching (Douglas Robinson, personal communication).

Large numbers of turtles are caught by shrimp trawlers most often along the coast of Guanacaste Province (Figure 1). Islas Negritos, Punta Guiones, Cabo Velas and the Gulf of Papagayo are cited by shrimpers as areas with high incidental catch rates. It is likely that turtles are also taken in the other major shrimping grounds in the open Colorados Bay and the Dulce Gulf. The Gulf of Nicoya is a declared nursury ground and shrimp trawling is prohibited in the upper reaches (Gross 1971).

Estimates of catch rates differ among shrimp captains and range from 600 to 2,000 annually by the fleet of 61 trawlers (Eduardo Lopez, personal communication). Most turtles captured are olive ridleys and many of these are females with shelled eggs. Catches of up to 45 in 1 haul, with averages of 4 to 5 per haul are also reported. Other estimates reach upwards of 200 juveniles and adults taken daily in late spring and early summer (B. Couper in communication with J. D. Richard).

Those that arrive on deck alive and some of the dead are returned to the sea. This mortality is probably the cause of the relatively large numbers of carcasses observed on Costa Rican beaches. Seventy-three green turtles and 2 ridleys in varying stages of decomposition were counted along 6 km of beach over a 75-day period in 1972 (Cornelius 1975). Ten ridleys, 3 greens and 1 hawksbill were observed on the same beach in August 1977 (Rainey, personal communication). The latter count was thought to represent 10 day's deposit.

During the late 1960s and early 1970s, the Costa Rican press reported foreign fishing vessels of unknown nationality operating off the Guanacaste coastline. Turtles, mostly ridleys, were shot on the beaches and transported to a "mother ship" which remained outside the country's territorial waters. This activity ceased shortly after news articles appeared.

The protection and management of sea turtles and their habitat in Costa Rica are the dual responsibility of the Executive Direction of Fisheries and Wildlife Resources and the National Park Service. Probably nowhere in Middle America has a framework for the conservation of renewable natural resources been established so concretely as in Costa Rica. This is fortunate for sea turtles, as nowhere on the Pacific side of the isthmus are there larger populations.

The legal basis for the protection of sea turtles is contained in the broad Wildlife Conservation Law (No. 4551), prohibiting the commercial exploitation of animals and their products for both the domestic and export markets. For a species to be covered it must first be declared either endangered or threatened (López 1978). The taking of sea turtles and their eggs is prohibited on the beach and in the territorial waters. It remains legal to land turtles killed on the high seas and to import them from foreign countries (Anonymous 1978). This applies to the domestic green turtle fishery on the Caribbean. No permits have been granted, or perhaps requested, for capture of sea turtles on the Pacific coast since 1977 (Fernando Víquiz, personal communication).

The government of Costa Rica, though maintaining the excellent attitude toward natural resource conservation initiated by two previous administrations, has been less generous with funding as a result of a governmentwide budget-tightening process (José Rodriquez, personal communication). No additional protection can be expected in the immediate future but neither will there be any relaxation of present laws. A law to permit Caribbean turtle fisherman to operate closer to nesting beaches was recently vetoed by the President.

Sea turtles have long been the subject of interest and scientific research in Costa Rica. Most field studies have been directed at the West Caribbean populations nesting in northeast Limon Province (Carr and Giovannoli 1957; Carr and Ogren 1959; Carr and Hirth 1962; Carr, Hirth, and Ogren 1966).

Except for a few notes on solitary nesting olive ridleys in Guanacaste and Puntarenas provinces (Carr 1961; Caldwell and Casebeer 1964), little was known until recently of the Pacific coast sea turtles. In 1970,

Richard and Hughes (1972) conducted a low level aerial reconnaissance of Costa Rica's shores. In the Pacific, major nesting beaches of the olive ridley were discovered, each having over 100,000 turtles aggregated offshore. Both Nancite beach, within Santa Rosa National Park, and Ostional beach, located 90 km to the south, host periodic mass nesting emergences (variously called *arribada, salida de flota,* or *cardumen*) from mid-summer through December (Hughes and Richard 1974). Smaller arribadas sometimes begin as early as April at Ostional, where the University of Costa Rica has maintained a field station for nine years (Robinson, McDuffie, and Cornelius 1973).

Two- to three-hundred-thousand ridleys participate in the arribadas at Nancite and Ostional. Each mass emergence lasts 3 to 10 days, with daylight nesting sometimes observed. These 2 beaches are the most important sites of ridley reproduction in the western hemisphere, if not in the world. The adults suffer little intentional human exploitation since consumption of turtle meat is, to some extent, culturally unacceptable on the Pacific. Until recently, these populations were considered fairly secure because of the natural isolation of the major nesting beaches, legislated protection within Costa Rican waters, and the likelihood that they were distinct from ridleys nesting in Mexico.

The over-harvest and eventual demise of the Mexican fleets will force that industry to locate a new supply. Contact between Costa Rican officials and the Mexican turtle industry has been made. This event together with the development within the past 3 years of a massive commercial harvest of ridleys in Ecuador has left Costa Rica's population highly vulnerable.

Panama

Very little of Panama's sinuous and island-studded coastline of 1,100 km is unsuited for sea turtle nesting (Figure 1).

Five species reportedly occur in the coastal waters and beaches of Pacific Panama. In order of their relative abundances, they are: *tortuga cahuama (Caretta caretta), tortuga mulato (Lepidochelys olivacea), tortuga verde (Chelonia mydas), tortuga canal (Dermochelys coriacea)* and *carey (Eretmochelys imbricata)* (Argelis Ruiz, personal communication). The green turtle nests very little although it is frequently observed in coastal waters. Although no systematic inventory has been conducted of these populations, the loggerhead is considered to be by far the most abundant. Nowhere else along the Middle American coast is the species common, and in most countries it is listed only because of Carr's (1952) broad range description. It may be common in Colombia (Green and Ortiz, this volume).

Panamanian officials have identified 12 important

STEPHEN E. CORNELIUS

nesting beaches on the Pacific coast. The reproductive season for all species is from May to December (Nicholas Real, personal communication). Turtle populations have decreased drastically in the last 10 years. At least 30 beaches were formerly known to host large nesting aggregations. These were called *arribadas* but whether the term was used in the sense of mass emergences such as now occurs in Costa Rica and Mexico or whether it was a qualitative expression of large numbers of solitary nesters is unknown. In either event, the sizes of the populations at each of the 12 remaining beaches are not as large nor is the nesting process as prolonged.

There is no industry established in Pacific Panama for the exploitation of sea turtles. This is in contrast to the Caribbean coast, where several companies are involved in the importation, exportation, and transshipment of stuffed juveniles and hawksbill shell. From 1964 to 1976, over 96,000 kg of shell were officially exported from Panama (Vallester 1978). This represented approximately 55,670 hawksbills, most originating from Bocas del Toro and San Blas provinces.

Coastal villagers of the Pacific are permitted a small number of turtles for their own use, and egg consumption exists on a small scale but is illegal. The Panamanian shrimp fleet, largest in Middle America, reportedly releases the majority of the turtles caught incidental to their activities in the gulfs of Chiriquí and Panama (Real, personal communication).

The National Direction of Renewable Natural Resources (RENARE) has the principal responsibility of managing and protecting the country's wildlife, including sea turtles. Partial protection is now afforded the green turtle, loggerhead, and olive ridley by Executive Decree 23 (January 1967) and a closed season from 1 May to 30 September was declared by Executive Decree 104 (September 1974). A proposal to bring the leatherback and hawksbill some protection will soon become law (Vallester 1978; Real, personal communication). As elsewhere in Middle America, insufficient funds make strict vigilence of the beaches difficult, and the laws are often ignored.

There are now no sea turtle studies underway, but a program to identify nesting areas and mark populations is planned (Real and Ruiz, personal communications). In 1975, a biological reserve was established at Isla de Cañas along the coast of Los Santos Province where thousands of loggerheads, olive ridleys, and leatherbacks reportedly nest on 70 km of beach. At present, 2 additional turtle reserves have been established at Playas de la Barquete, Chiriquí Province and on the Caribbean at Playas Largas-Isla Bastimentos, Bocas del Toro Province. An artificial hatchery program has been in operation since 1975 at Islas de Cañas. The program was expanded in 1979, and egg transplants are expected to reach 65,000 (Real, personal communication). Guards patrol the beaches and are making an effort to eliminate the hordes of feral dogs that destroy up to 90 percent of the nests.

Conclusions

Undoubtedly, the total number of sea turtles in the East Pacific was once many times what it is today. The large concentrations seen off the Middle American coast as recently as the mid-twentieth century (Oliver 1946), were probably still largely unaffected by human exploitation.

Destruction and modification of nesting and feeding habitats have not been a significant factor adversely affecting sea turtles in Middle America, with the possible exception of recent events in El Salvador. Neither have these populations suffered from large-scale exploitation for meat, leather, oil, or tortoiseshell. That some populations of ridleys, which nest in Middle America, are endangered by such activities on their feeding grounds in Ecuador is now evident, however. Nesting turtles are rarely killed by coastal residents of the Pacific, as most do not consider the meat very good. In general, the Pacific coast populace do not have a strong marine component of their culture, except for the consumption of turtle eggs. This contrasts sharply with the Caribbean side of the isthmus, where sea turtles have traditionally been an important dietary item.

The regionwide population decline is likely the result of 2 factors, working in concert or alone depending on the specific locale. Poaching of eggs is a lucrative commerce in all 6 countries, regardless of protecting legislation. Nicaragua's diminished populations can be traced easily to the tremendous drain of eggs from the nesting beaches. Predation by domestic animals is normally associated with human poaching. Certain beaches of Costa Rica and Panamá are especially susceptible to this form of destruction.

The second major factor adversely affecting the region's sea turtles is the inadvertent mortality caused by the relatively new shrimping industry. This fishery began in 1941 in northern Mexico, but did not spread to the tropical East Pacific until the early 1950s. General development proceeded slowly, with Panamá and Costa Rica the first to exploit the extensive white and pink shrimp grounds (Gross 1971). In 1958, Costa Rica expanded considerably, and El Salvador began harvesting what has become the most productive grounds along the isthmus. Development of commercial fisheries in Nicaragua and Guatamala followed. The fishery operates throughout the year, with white shrimp and pink shrimp taken in the wet and dry seasons, respectively.

The incidence of moribund turtle strandings seems to be highest in Guatemala, El Salvador and Costa Rica where shrimping operations are fairly close to shore and overlap major turtle nesting areas. Nicaragua's small fleet may trawl far enough offshore to alleviate the problem. Panamá's large fleet probably kills more turtles than suspected as its grounds in the bays of Chiriquí and Panamá coincide with important nesting beaches. Studies of the incidental catch problem in the southeast United States, report subadults rather than gravid females as the primary group affected (Anonymous 1978). Both juveniles and adults, including females with eggs, are reported in the trawls in Middle America.

All 6 countries have or are preparing some legislated protection for sea turtles and eggs. The lack of means of enforcement, however, is the plight of well-intentioned but financially restricted governments. Protection of all beaches from egg poachers is impossible, even given multiple increases in funding for vigilence. A policy of permitting coastal communities to harvest reasonable numbers of eggs for their own use, while directing most law enforcement effort at the few beaches which supply the bulk of the commercial trade seems appropriate for each country.

Basic inventories of nesting beaches have yet to be conducted in most countries, resulting in obvious gaps in knowledge of distribution, relative abundances, nesting seasons, etc. As some populations appear to be resident, studies of nonreproductive activities are possible. With international funding and consultation from experienced turtle researchers, several countries, notably El Salvador, Nicaragua, and Panamá, appear prepared to undertake such investigations.

Literature Cited

Aguilar, W.
1978. El manejo de la vida silvestre en Honduras. Paper presented at the First Regional Wildlife Meeting of Central America. Matagalpa, Nicaragua. 25–29 July 1978.

Anonymous
1978. Final environmental impact statement: listing and protecting the green sea turtle (*Chelonia mydas*), loggerhead (*Caretta caretta*) and Pacific ridley (*Lepidochelys olivacea*) under the Edangered Species Act of 1973. Washington, D.C. July 1973.

Bacon, P. R.
1973. Appraisal of the stocks and management of sea turtles in the Caribbean and adjacent regions. Report to working group on fisheries. Results of the VIth International Coordinated Group Management, Cooperative Investigations Caribbean and Adjacent Regions. Cartagena, Colombia.

Caldwell, D. K., and R. S. Casebeer
1964. A note on the nesting of the East Pacific ridley sea turtle, *Lepidochelys olivacea*. *Herpetologica* 20:213.

Carr, A.
1948. Sea turtles on a tropical island. *Fauna* 10:50–55.

1952. *Handbook of Turtles*. Ithaca, New York: Cornell University Press.

1961. The ridley mystery today. *Animal Kingdom* 64:7–12.

Carr, A., and L. Giovannoli
1957. The ecology and migrations of sea turtles, 2. Results of field work in Costa Rica 1955. *American Museum Novitates* 1835:1–32.

Carr, A., and H. Hirth
1962. The ecology and migrations of sea turtles, 5. Comparative features of isolated green turtle colonies. *American Museum Novitates* 2901:1–42.

Carr, A.; H. Hirth; and L. Ogren
1966. The ecology and migrations of sea turtles, 6. The hawksbill turtle in the Caribbean Sea. *American Museum Novitates* 2248:1–29.

Carr, A., and L. Ogren
1959. The ecology and migrations of sea turtles, 3. *Dermochelys* in Costa Rica. *American Museum Novitates* 1958:1–29.

Coe, M. D., and K. V. Flannery
1967. Early cultures and human ecology in south coastal Guatemala. *Smith. Contributions to Anthropology*, vol. 3.

Cornelius, S. E.
1975. Marine turtle mortalities along the Pacific coast of Costa Rica. *Copeia* 1975:186–87.

1976. Marine turtle nesting activity at Playa Naranjo, Costa Rica. *Brenesia* 8:1–27.

Gross, G. B.
1971. Shrimp industry of Central America, Caribbean Sea and Northern South America. MFR reprint 971 (original—Foreign Fisheries Leaflet Number 73–1). National Marine Fisheries Service. International Activities. Washington, D.C.

Hughes, D. A., and J. D. Richard
1974. The nesting of the Pacific ridley turtle (*Lepidochelys olivacea*) on Playa Nancite, Costa Rica. *Marine Biology*, 24:97–107.

Lopez, E.
1978. Informe sobre les actividades de la Direccion General de Recursos Pesqueros y Vida Silvestre de Costa Rica. Paper presented at the First Regional Wildlife Meeting of Central America. Matagalpa, Nicaragua. 25–29 July 1978.

Nietschmann, B.
1975. Of turtles, arribadas, and people. *Chelonia* 2:6–9.

Oliver, J. A.
1946. An aggregation of Pacific sea turtles. *Copeia* 1946:103.

Richard, J. D., and D. A. Hughes
1972. Some observations of sea turtle nesting activity in Costa Rica. *Marine Biology* 16:297–309.

Robinson, D. C.; J. McDuffie; and S. E. Cornelius
1973. Observations of reproductive behavior of the Pacific ridley turtle (*Lepidochelys olivacea*). Abstract. 53rd Annual Meeting of the Society of Ichthyologists and Herpetologists. San Jose, Costa Rica.

Salas, J. B.
1978. Informe sobre las actividades que desarrolla el Departamento de Vida Silvestre de Nicaragua. Paper presented at the First Regional Wildlife Meeting of Central America. Matagalpa, Nicaragua. 25–29 July

STEPHEN E. CORNELIUS

1978.

Serrano, F.
1978.   Informe de actividades de la Unidad de Parques Nacionales y Vida Silvestre de El Salvador. Paper presented at the First Regional Wildlife Meeting of Central America. Matagalpa, Nicaragua. 25–29 July 1978.

Vallester, E.
1978.   Informe de Panama sobre la situacion de la fauna
silvestre. Paper presented at the First Regional Wildlife Meeting of Central America. Matagalpa, Nicaragua. 25–29 July 1978.

Zelaya, V. M.
1979.   Uso comercial de la fauna silvestre en El Salvador.
Unidad de Parques Nacionales y Vida Silvestre. Ministerio de Agricultura y Ganaderia. San Salvador.

Derek Green
Charles Darwin Research Station
Santa Cruz, Galápagos, Ecuador

Fernando Ortiz-Crespo
Departamento de Biología
Universidad Católica
Quito, Ecuador

Status of Sea Turtle Populations in the Central Eastern Pacific

ABSTRACT

Data relating to research, exploitation and conservation of sea turtles in Colombia, Ecuador, and the Galápagos Islands are presented. Data for Colombia are restricted to personal communications. In Galápagos, extensive data have been gathered continuously for 4 years, especially on the east Pacific green turtle, *Chelonia mydas agassizi*. There is no current research in mainland Ecuador, but results of earlier investigations by the authors are given. The east Pacific green turtle, *Ch. m. agassizi,* the Indo-Pacific hawksbill, *Eretmochelys imbricata* and the Pacific leatherback, *Dermochelys coriacea,* have been reported from all 3 regions. The Pacific loggerhead, *Caretta caretta,* has been reported only from Colombia and the Pacific ridley, *Lepidochelys olivacea,* only from Colombia and mainland Ecuador. Exploitation in Colombia is for domestic consumption and in the Galápagos Islands is at a subsistence level. In mainland Ecuador there is heavy commercial exploitation of *L. olivacea*; possibly as many as 100,000 will be killed by the end of 1979, and their meat and skins exported, mainly to Japan. Efforts to curtail this exploitation are described in detail, and suggestions for future research and conservation are made.

Introduction

For the purposes of this paper, the Central Eastern Pacific is defined as Colombia and Ecuador (including the Galápagos Islands). The aim is to update our knowledge of the status of sea turtles in these regions, especially in regard to exploitation, conservation and research. The data for Colombia are from personal communications only; neither author has first-hand knowledge of the status of the sea turtles in this region. Data for the Galápagos Islands are gleaned from the senior author's current research program at the Charles Darwin Research Station on Galápagos sea turtles. The data for mainland Ecuador are derived from 2 independent investigations; one by the senior author for the National Fisheries Institute (INP), Guayaquil, and

221

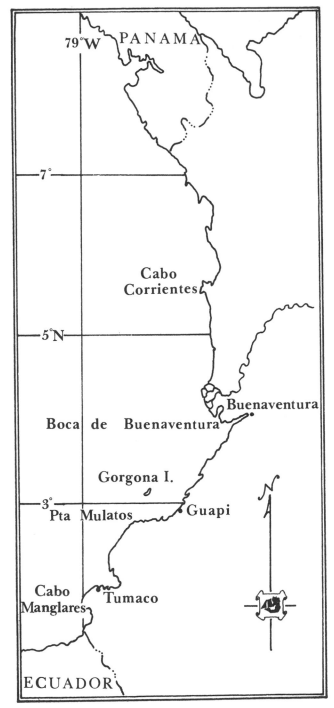

Figure 1. Map of the Pacific coast of Colombia showing the location of some of the areas mentioned in the text.

the other by the junior author for the Ministry of Agriculture and Livestock (MAG), Quito.

Colombia

Most efforts to contact the various Colombian bodies concerned with turtles proved fruitless. This, together with the paucity of turtle literature for this region, has left us very much in the dark as to the real state of affairs. The following account is based on information received by the senior author in letters from Henry von Prahl and Felipe Guhl from the Universidad de los Andes, Bogotá, and a single recapture of a Galápagos-tagged green turtle. Figure 1 shows the location of some of the areas mentioned below.

Species Present

The following species have been reported:

	English Name	Local Name(s)
Chelonia mydas agassizi	East Pacific green	Tortuga de mar, Caguama
Eretmochelys imbricata	Indo-Pacific hawksbill	Tortuga fina, Carey
Dermochelys coriacea	Pacific leatherback	Tortuga bufeadora
Caretta caretta	Pacific loggerhead	Tortugaña de mar, Tortuga
Lepidochelys olivacea	Pacific (olive) ridley	—

The only record we have of *Ch. m. agassizi* in Colombia is of a female tagged in Galápagos and recaptured in the Boca de Buenaventura (Green, unpublished data). The local name given for this specimen is *caguama,* which is also the common name used in Colombia for the Atlantic loggerhead, *Caretta caretta* (Kaufmann 1971). Von Prahl reports *E. imbricata* from the Guapi, Mulatos, and Sanquianga estuaries and on the *Pocillopora* reef off Gorgona Island, and *D. coriacea* from near Gorgona Island. According to Guhl, *C. caretta* is the most common species present around Gorgona Island. Von Prahl reports it also from southern Colombia near Tumaco and the Guapi, Sanquianga and Satinga estuaries. It thus appears to be the most common species present. The Pacific ridley, *L. olivacea,* has been observed in Colombia, but its abundance is unknown to us.

Exploitation

According to Von Prahl and Guhl, there is no international market. *C. caretta* and *Ch. m. agassizi* are caught for local consumption, and hawksbill shell is often used in craftwork. All 5 species are occasionally taken accidentally in shrimp and fishing nets.

Nesting

The nesting situation is unknown to the authors.

DEREK GREEN

The Galápagos Islands

The Galápagos Archipelago straddles the equator and lies approximately 1,000 km off the coast of mainland Ecuador. It consists of 17 major islands, volcanic in origin, with a total land area of approximately 8,000 km² (Wiggins and Porter 1971). The 2 seasons, a hot, wet season between December and May, and a cool, dry one between June and November, are influenced by 3 major ocean currents, themselves influenced by the tradewinds (cited by Wellington, 1975). These climatic factors possibly affect the distribution of the sea turtles in Galápagos.

Research

THE PAST

Prior to 1970, very little work of a scientific nature had been done on Galápagos sea turtles, although they are mentioned in the journals of several earlier visitors to the islands. Slevin (1905–06 California Academy of Sciences Expedition) was one of the first to record them on a scientific basis. He skinned numerous green turtles, took measurements of their flippers and noted the stomach contents. His collection is preserved in the California Academy of Sciences, San Francisco.

During the breeding seasons of 1970 through 1975, Dr. Peter Pritchard, Miguel Cifuentes, and Judy Webb tagged 867 nesting females and 1 male (Pritchard 1971a, 1971b, 1972, and 1975; Cifuentes 1975). In addition, Pritchard made extensive surveys of the shoreline of most of the islands in order to determine the extent of nesting within the Archipelago (Pritchard 1972 and 1975). In September 1975, the senior author initiated a year-round research program which has just entered its fifth consecutive year.

THE PRESENT

The three main objectives are to determine the status of sea turtle populations in Galápagos; to collect enough scientific data on which to base recommendations for future management policies both in Galápagos and mainland Ecuador; and to incorporate Ecuadorian university students into the program with the aim of extending, with their help, the research to mainland Ecuador.

Apart from monitoring the nesting beaches and thus gathering data only on nesting females and only during the nesting season, feeding grounds are also visited at intervals throughout the year. Here both males and immatures as well as females are caught with nets or by hand.

In addition, a study of the effects of a scarabeid beetle, *Trox suberosus,* on the nesting success of *Chelonia mydas agassizi* was made this year by Karin Allgöwer (1979).

The present research program has produced data on breeding cycles, multiple nesting, reproductive potential, hatching success, clutch size, incubation periods, migration, interisland movements, feeding potential, etc. In addition, extremely important data were gathered on growth rates of immatures in the wild. Some of these results are incorporated in the text below.

Species Present

The following species have been reported:

	English Name	Local Name
Chelonia mydas agassizi	East Pacific green	Tortuga negra
Eretmochelys imbricata	Indo-Pacific hawksbill	Carey
Dermochelys coriacea	Pacific leatherback	None

Chelonia mydas agassizi is the most common species and exists in large numbers. It is known locally as *tortuga negra* or black turtle (referring to the dark color of the carapace) in order to distinguish it from a rare yellow form, the yellow turtle or *tortuga amarilla*, of unknown scientific status, but considered by Pritchard (1971a and 1972) to be a sterile mutant of the black form. A more detailed description of this rare yellow turtle will appear in a future paper (Green, in prep.).

Eretmochelys imbricata is encountered occasionally but *D. coriacea* has been sighted only 3 times and is by far the rarest of all 3 species present. Figure 2 shows sightings of *E. imbricata* and *D. coriacea* by local inhabitants, visiting yachtsmen and scientists.

Nesting

Chelonia mydas agassizi is the only species that nests in the Galápagos Islands. These islands are probably the most important nesting area in the Central Eastern Pacific and perhaps even the whole Pacific South America for this species.

Turtles nest on most of the islands within the Archipelago. Six of the most important beaches, which correspond to the study sites, are shown in Figure 2. A total of 3,784 mature females and 10 mature males have been tagged from these nesting beaches since 1970.

Although eggs have been laid in every month of the year, a distinct season occurs between December and May, with a peak in February. This corresponds to the hot or wet season.

Quinta Playa, a 2-km-long beach on southern Isabela, is probably the most important single nesting beach in Galápagos. The maximum number of nests laid in any one night on this beach were 23, 20, 45, and 18 for the 1976, 1977, 1978, and 1979 seasons, respectively. For these same seasons, the total numbers of different nesting females recorded on this beach were 315, 308, 610, and 300, respectively.

Table 1 gives the percentage hatch (referring to the total number of hatchlings) and the percentage emergence (the number of hatchlings actually reaching the surface) for Las Salinas, Baltra, and Quinta Playa, Isabela, for the years 1976–79; and Bahía Barahona on Isabela, Las Bachas on Santa Cruz, Espumilla on James, and Bartolomé for 1979.

The low hatching and emergence rates for Quinta Playa are mainly due to nest destruction by feral pigs and egg predation by a scarabeid beetle, *Trox suberosus*. This beetle has also been found preying on turtle nests on nearby Bahía Barahona and on Bartolomé, and nests of the Galápagos land inguana, *Conolophus subcristatus*, on South Plaza Island (Howard and Heidi Snell, personal communication). As far as we are aware, there

is no record in the literature of this beetle's attacking turtle nests in other parts of the world.

The extremely low hatching and emergence rates of 1.88 percent for Espumilla are almost entirely due to feral pigs, which in 1979 totally destroyed most of the nests laid. Of 122 nests marked, only 8 resulted in hatchlings, and 5 of these had also been attacked by pigs. These results do not include 7 other marked nests which were partially destroyed by pigs and several other marked nests swamped by high tides, for which the hatching and emergence rates are unknown.

Feeding Grounds

The most important feeding grounds for the green turtles in Galápagos appear to be the algae beds in the western islands of Isabela and Fernandina. Stomach analyses and underwater observations have shown that these green turtles eat at least 15 different species of algae.

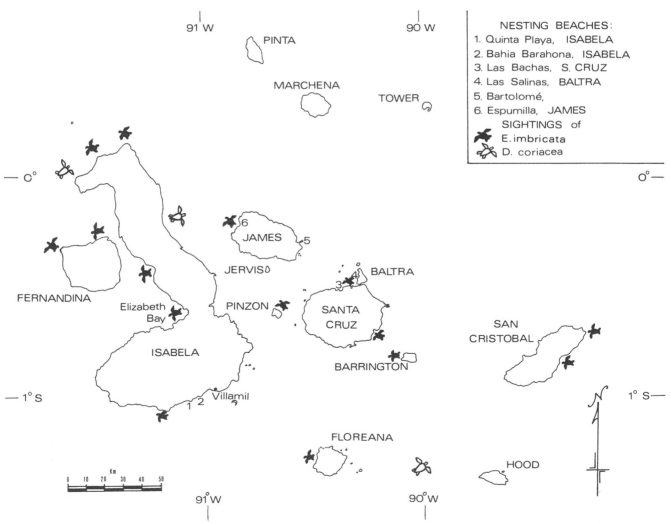

Figure 2. Map of the Galápagos Islands showing the principal nesting beaches of the east Pacific green turtle *Chelonia mydas agassizi*, and sightings of the Indo-Pacific hawksbill, *Eretmochelys imbricata*, and the Pacific leatherback, *Dermochelys coriacea*.

DEREK GREEN

Table 1. Comparison of percentage hatch and percentage emergence of the east Pacific green turtle, *Chelonia mydas agassizi*

Year	Nesting beach	Number of nests marked	Number of eggs	Percentage hatch	Percentage emergence
1976	Las Salinas, Baltra	94	8198	72.12	69.66
	Quinta Playa, Isabela	120	9651[a]	38.63	37.82
1977	Las Salinas, Baltra	21	1842	52.23	49.67
	Quinta Playa, Isabela	101	8018[a]	43.73[b]	40.47
1978	Las Salinas, Baltra	38	3334[a]	70.97	69.77
	Quinta Playa, Isabela	40	3385[a]	43.46	41.18
1979	Las Salinas, Baltra	22	1688	71.09	69.91
	Quinta Playa, Isabela	67	5363[a]	48.69	47.70
	Bahía Barahona, Isabela	69	5643[a]	74.23	72.92
	Las Bachas, Santa Cruz	22	1687	80.44	78.42
	Espumilla, James	122	9709[c]	1.88	1.88
	Bartolome	15	1142	50.00	47.20

Note: This comparison is based on observation of 731 natural nests on several different beaches in Galápagos between 1976 and 1979.
a. Samples include nests where the original clutch size was impossible to determine due to total destruction of the nest and where the percentage hatch and the percent emergence were therefore both zero. In these cases the mean clutch size for the respective beach is added for each nest destroyed in order to determine the total number of eggs in the sample.
b. A pig totally destroyed one nest while the hatchlings were in transit to the surface. The emergence rate was zero. The mean number of hatchlings per nest for Quinta Playa 1977 was added to the total in order to determine the total number of hatchlings in the sample.
c. In all 122 nests it was impossible to determine the original clutch size. Therefore the mean of 188 clutches laid on the other 5 beaches during 1979 ($\bar{x}$ = 79.58 eggs per clutch) is used to determine the original number of eggs in the sample.

Population: Resident vs Migratory

Recaptures indicate that some of the green turtles tagged on the local feeding grounds, especially immatures and virgin females, are present there year round. In addition, several females tagged on the nesting beaches were recaptured on these feeding grounds well after the season had ended. On the other hand, from these same beaches there have been 17 international recoveries (Peru, 7; mainland Ecuador, 4; Colombia, 1; Panama, 3; and Costa Rica, 2) of Galápagos-tagged female green turtles. It thus appears that part of the Galápagos population is resident and part is migratory. What percentage of the resident part consists of late departures from the previous nesting season or early arrivals for the next is unknown. Without more recapture data it is not possible to say what proportion is truly resident and what migratory, or even to give a worthwhile estimate of population size. The subject of migration of Galápagos green turtles will appear in a future paper (Green and Pritchard, in preparation).

Growth Rates

Recaptures of immatures on the feeding grounds have furnished some important data on growth rates of immatures in the wild. This topic will be the subject of a future paper (Green, in preparation) but results can be summarized by saying that growth rates are extremely slow. For example, 11 immatures with straight carapace lengths ranging from 46 to 59 cm showed a mean growth rate of only 0.53 cm/yr over periods of 4 to 33 months. These slow growth rates will certainly have to be taken into consideration when planning future management policies.

Exploitation and Conservation

There is very little exploitation in Galápagos at present. By law, only local inhabitants are allowed to fish for turtles and only on a subsistence basis. This privilege is rarely exercised in Galápagos because turtle meat is not highly esteemed, and because there are so many cheap foods available. Since the Galápagos Islands are a National Park, all nesting beaches are completely protected, even from locals, although the military on Baltra take up to six nesting females per season from the beaches on that island. However, because of the migratory habit of green turtles (see above), although protected in Galápagos, they are open to exploitation elsewhere.

The Future

Four research needs, for which adequate financing is crucial, must be met in the future.

First, the present research program needs not only to be continued but also to be expanded in order to include more nesting beaches and more feeding grounds. We are still ignorant, for example, of the total population size of the Galápagos green turtle. Second, the work on growth rates of turtles in the wild should continue. Third, a detailed study of the feral pigs is urgently required, with the overall objective of eradication. Fourth, the work on *Trox suberosus* should be continued in more detail and methods designed and tested for the prevention of egg predation.

FINANCIAL FEASIBILITY

Funds promised by the Ecuadorian National Fisheries Institute (INP) have not materialized; nor, it seems, are they likely to materialize. Unless funds become available immediately, the program is in grave danger of terminating. Present resources will allow research to continue only on a skeleton basis, and only for a few months more.

Mainland Ecuador

Research

Prior to 1978, the status of sea turtles of mainland Ecuador was unknown. From May to July 1978 the senior author, under the auspices of the National Fisheries Institute (INP), Guayaquil, made a survey of the mainland shoreline from the Peruvian border in the south to Rocafuerte in the province of Esmeraldas in the north, approximately 80 km south of the border with Colombia. This investigation also included several visits to turtle-processing factories and an analysis of their trade figures. In June of 1978, Drs. Fernando Ortiz-Crespo of the Catholic University and Galo Cantos of the Central University, both in Quito, were commissioned by the Ministry of Agriculture and Livestock (MAG) to investigate the activities of the turtle factories. Both investigations were carried out independently, and their respective reports reside with their respective ministries. Data from both reports have been used in this paper.

Due to lack of time and, later, lack of funds, the senior author's investigation terminated in Rocafuerte. Turtles are said to occur north of this point, for example, in La Tola and probably as far north as the border with Colombia.

Species Present

The following species have been reported:

	English Name	Local Name(s)
Chelonia mydas agassizi	East Pacific green	Numerous
Eretmochelys imbricata	Indo-Pacific hawksbill	Carey, peinilla
Dermochelys coriacea	Pacific leatherback	Numerous
Lepidochelys olivacea	Pacific ridley	Tortuga verde

Chelonia mydas agassizi, E. imbricata and *D. coriacea* all occur in small numbers from the Gulf of Guayaquil in the south to at least Rocafuerte and probably to the Colombian border in the north. In addition, *Ch. m. agassizi* and *E. imbricata* are found as far south as the border with Peru. Of these three species, *Ch. m. agassizi* is the most frequently encountered, and *D. coriacea* the least.

The Pacific ridley, *L. olivacea,* is the most common species and occurs in fairly large numbers from Anconcito, in the province of Guayas, to Esmeraldas and probably even farther north. This species is migratory in Ecuadorian waters and tends to remain farther offshore than the other species. Isla de la Plata (a small island 30 km from the mainland) and an area 30 to 50 km out to sea from Esmeraldas are the two areas where it is the most abundant.

There is no evidence at present for the occurrence of the Pacific loggerhead, *Caretta caretta,* in Ecuador.

Nesting

Chelonia mydas agassizi, E. imbricata and *D. coriacea* all nest in small numbers along most of the Ecuadorian coast but are more commonly found nesting between Manta and Cojimíes in the province of Manabí. Figure 3 shows the northern and southern nesting limits of these 3 species as they are known at present. For *Ch. m. agassizi,* the limits are Costa Rica Island near the Peruvian border to the south, and a small beach just north of Atacames in the province of Esmeraldas to the north. This same small beach is also the northern limit for *E. imbricata*; its southern limit is Ayampe in the province of Manabí. For *D. coriacea,* the southern limit is Río Chico in the province of Manabí, and the northern limit is a beach 14 km south of Esmeraldas, the northernmost record in Ecuador for any species. These limits do not take into account nesting which may take place in the extreme north of Ecuador between Rocafuerte and the Colombian border.

There is at present no authentic nesting record for *L. olivacea* in Ecuador.

The season starts in December and ends in April or May, with a peak in February. Though shorter, this basically coincides with the Galápagos nesting season.

DEREK GREEN

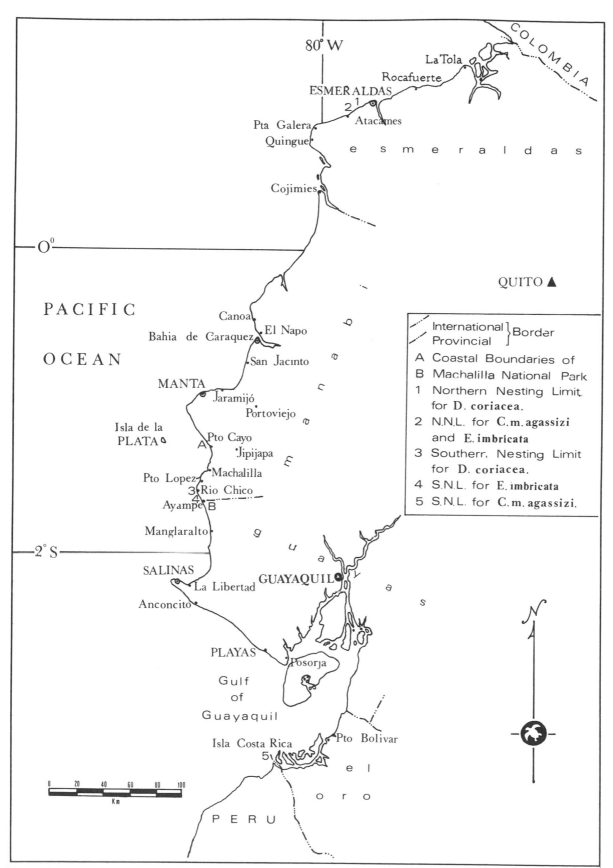

Figure 3. Map of the coastal region of mainland Ecuador showing the known nesting ranges of the east Pacific green turtle *Chelonia mydas agassizi*, the Indo-Pacific hawksbill, *Eretmochelys imbricata* and the Pacific leatherback, *Dermochelys coriacea*.

Since the senior author's survey was started after the nesting season had virtually finished, it is difficult to give an accurate estimate of the abundance of nesting females, although based on reports by local inhabitants, the numbers are certainly low. For example, the continuous 14-km stretch of beach between El Napo and Canoa in the province of Manabí, certainly one of the higher density nesting beaches, could only boast 10 turtles of all species per night during the peak of the season.

Exploitation

Chelonia mydas agassizi. This species is taken in very small numbers for local consumption and accidentally in shrimp or fishing nets all along its range, especially around Santa Rosa and La Libertad in the province of Guayas. Varnished carapaces are often sold in certain tourist spots such as Playas in the Guayas province. As with the other two species which nest in Ecuador, the numbers of nesting females are so low that local inhabitants do not bother to patrol the beach, although they will take both nesting females and eggs if encountered.

Eretmochelys imbricata. Being rarer than *Ch. m. agassizi,* it is taken less often and usually only incidentally in shrimp or fishing nets. The meat is not eaten, and the eggs only on the rare occasions they are encountered. The only commercial exploitation of this species is in certain tourist spots such as Playas, where varnished carapaces and plastrons, or even stuffed juveniles, can be found for sale.

Dermochelys coriacea. Being the rarest of all 4 species present, it is the one least taken. Nesting females or eggs are taken only on extremely rare occasions. It is not exploited commercially.

Lepidochelys olivacea. Although some, for example near Santa Rosa and La Libertad in the province of Guayas, are taken for local consumption, the vast majority are caught for export. This makes *L. olivacea* the only species of marine turtle found in Ecuadorian waters that is commercially exploited on an international level.

Since 1970, at least 6 companies (Neptuno, Expromar, Exporklore, Inexpac, Shayne, and Songa) have been involved in the exportation of turtle products, mainly in the form of frozen meat for human consumption and salted skin for the leather trade. Table 2 gives the annual exportation of turtle (*L. olivacea*) meat and skin since 1970. The maximum and minimum estimates of the corresponding numbers of turtles involved as presented in this Table and in Tables 3 to 5 are based on the following: 3 companies said they obtained on average 4.54 kg (10 lbs), 4.54 kg, and 5.45-6.81 kg (12 to 15 lbs) of meat, respectively, from an

Table 2. Six companies'[a] combined annual exports of skins and meat of the Pacific ridley, *Lepidochelys olivacea*, from Ecuador, 1970 to June 1979

	Skins			Meat			Total number of turtles	
		Estimated number of turtles			Estimated number of turtles			
Year	Weight (kg)	Min.	Max.	Weight (kg)	Min.	Max.	Min.	Max.
1970	0	0	0	2,398	352	528	352	528
1971	0	0	0	30,165	4,430	6,644	4,430	6,644
1972	0	0	0	45,124	6,626	9,939	6,626	9,939
1973	8,425	4,212	4,681	28,759	4,223	6,335	4,212	b
1974	32,703	16,351	18,168	114,053	16,748	25,121	16,351	b
1975	32,602	16,301	18,112	59,837	8,787	13,180	16,301	b
1976	49,556	24,778	27,531	137,223	20,150	30,225	24,778	b
1977	110,150	55,080	61,194	122,198	17,944	26,916	59,476[c]	67,788[c]
1978	161,070	80,535	89,483	62,967[d]	9,246	13,869	80,535	89,483
1979[e]	139,900	69,950	77,722	3,230[d]	474	711	69,950	77,722

a. Neptuno, Expromar, Exporklore, Songa, Shayne, and Inexpac.
b. Maximum estimates in these cases impossible to determine because details of which companies exported both skins and meat and which the meat only, were not available.
c. See Table 3.
d. This meat is from turtles already butchered for their skins and so does not alter the total number of turtles involved which is estimated from the weight of the skins.
e. Figures are for the first 6 months only.
Source: Figures by courtesy of the Instituto Nacional de Pesca, Guayaquil and Ing. Tuly Loor, Subsecretaría de Recursos Pesqueros.

DEREK GREEN

Table 3. Six companies'ᵃ combined monthly exportation of *Lepidochelys olivacea* products from Ecuador, 1977

| | Skins | | | | Meat | | | | | | Total number of turtles | |
| | | Estimated number of turtles | | | Weight from companies exporting meat only (kg) | Estimated number of turtles (meat only) | | Carapaces | Heads | Claws | (skin only + meat only) | |
Month	Weight (kg)	Min.	Max.	Total weight (kg)		Min.	Max.	Number of turtles	Number of turtles	Number of turtles	Min.	Max.
January	3,083	1,542	1,713	15,282	4,037	593	889	1,110	108	250	2,135	2,602
February	4,100	2,050	2,278	29,043	4,059	596	894	0	0	0	2,646	3,172
March	8,903	4,452	4,946	15,801	2,268	333	500	0	0	0	4,785	5,446
April	10,119	5,060	5,622	24,521	9,752	1,432	2,148	0	0	0	6,492	7,770
May	23,289	11,645	12,938	22,087	6,804	999	1,499	0	0	0	12,644	14,437
June	9,517	4,759	5,287	8,459	3,016	443	664	0	0	0	5,202	5,951
July	16,213	8,107	9,007	0	0	0	0	0	0	0	8,107	9,007
August	5,293	2,647	2,941	506	0	0	0	0	0	0	2,647	2,941
September	1,089	545	605	499	0	0	0	0	0	0	545	605
October	7,265	3,633	4,036	0	0	0	0	0	0	0	3,633	4,036
November	17,466	8,733	9,703	0	0	0	0	0	0	0	8,733	9,703
December	3,813	1,907	2,118	6,000	0	0	0	0	0	0	1,907	2,118
Totals	110,150	55,080	61,194	122,198	29,936	4,396	6,594	1,110	108	250	59,476	67,788

a. Neptuno, Expromar, Exporklore, Songa, Shayne, and Inexpac.
Source: Figures supplied by courtesy of the Instituto Nacional de Pesca and the Direccion General de Pesca, Guayaquil.

average sized turtle. Hence, the minimum number of turtles is estimated by using 6.81 kg per turtle, and the maximum number by using 4.54 kg per turtle. Two of these companies said they obtained 1.8 kg and about 2 kg of skin per turtle and so again both minimum and maximum estimates are given. These weights are slightly lower than those determined by Márquez, Villanueva, and Peñaflores (1976), who obtained average weights of 7.27 kg and 2.5 kg for the meat and skin, respectively, from 14 adult *L. olivacea* each weighing approximately 40 kg. Since there is less variation in the weights of skin per turtle than the weights of meat per turtle, estimates of the corresponding number of turtles based on the former are more accurate. Some companies export both the skin and the meat from the same turtle and others only the meat. Therefore estimates from companies exporting the meat only are added to the estimates from the skins to give the total number of turtles involved. In cases where companies export both, if an estimate from the skins exceeds the estimate for meat, it indicates that the company has a surplus of meat or stock in deep freeze. It will be seen later that this concept of stock is a stumbling block for any attempts at legislative control.

It can be seen from Table 2 that there has been a marked increase in the exportation of both skins and meat since 1970. Although exportation of skins started in only 1973, the demand from the leather trade has been so great that this has proved the most profitable

and hence most heavily exploited of all the sea turtle products. Table 3 presents the combined monthly exportation figures of turtle products of the 6 turtle companies for 1977. Table 4 presents the combined monthly exportation figures of the 3 surviving companies (Neptuno, Expromar, and Exporklore) for the first 6 months of 1978. Although only these 3 companies are still involved with sea turtle products, the amount of skins, some 107,714 kg, exported in the first 6 months of 1978, was almost as high as that for the whole of 1977, which in itself was double that for 1976. By the end of 1978 this 107,714 kg had increased to 161,070 kg— representing a minimum of 80,535 turtles.

The figures for 1979 are even more alarming. During the first 6 months of this year, 139,900 kg of skins (some 70,000 turtles) were exported. Judging by the 1977 figures (Table 3) and the 1978 figures (Tables 2 and 4), exportation could well reach the 200,000 kg mark by the end of the year. In other words, around 100,000 adult *L. olivacea* may be butchered in Ecuador during 1979 for the luxury goods trade. According to René Márquez (personal communication), only 150,000 to 200,000 female ridleys nested in Pacific Mexico this year (1979). How many will nest there next year?

Table 5 shows that most of the turtle products exported during 1977 were destined for Japan and Italy, both nonsignatories of the Convention on International Trade in Endangered Species of Wild Fauna and Flora (see below).

Table 4. Three companies'ᵃ combined exports of skin and meat of the Pacific ridley, *Lepidochelys olivacea*, January to June 1978

Month	Skins Weight (kg)	Skins Estimated number of turtles Min.	Skins Estimated number of turtles Max.	Meat Weight (kg)	Meat Estimated number of turtles (meat only) Min.	Meat Estimated number of turtles (meat only) Max.	Total number of turtles (skin only + meat only) Min.	Total number of turtles (skin only + meat only) Max.
January	6,189	3,094	3,438	0	0	0	3,094	3,438
February	9,302	4,651	5,168	0	0	0	4,651	5,168
March	9,647	4,823	5,359	0	0	0	4,823	5,359
April	9,904	4,952	5,502	0	0	0	4,952	5,502
May	35,632	17,816	19,796	4,000ᵇ	587	881	17,816	19,796
June	37,040	18,520	20,578	58,967ᵇ	8,659	12,988	18,520	20,578
Totals	107,714	53,856	59,841	62,967ᵇ	9,246	13,869	53,856	59,841

a. Neptuno, Expromar and Exporklore.

b. This meat is from turtles already butchered for their skins and so does not alter the total no. turtles involved which is estimated from the weights of the skins.

Source: Figures by courtesy of the Direccion General de Pesca and the Instituto Nacional de Pesca, Guayaquil.

Ridleys are most abundant, and hence most exploited, in Ecuadorian waters between April and June (see Tables 3 and 4 but note that there will always be a slight delay between catch and exportation). During two visits by the senior author to the Neptuno factory in Jaramijó (province of Manabí) on 31 May and 2 June 1978, totals of 390 and 1,003 *L. olivacea* respectively, were brought in for processing. In early June the ridleys start to move north. Several caught in Ecuador bore tags from Mexico and Costa Rican nesting beaches, and so it is assumed that the ridleys return there for the nesting season which starts in Mexico in June (Márquez 1976)

Conservation

The Past

When Ecuador became the eighth signatory of the Convention on International Trade in Endangered Species of Wild Fauna and Flora in 1976, the Department of National Parks and Wildlife, under the auspices of the Ministry of Agriculture and Livestock, was designated the Management Authority responsible for all legislation regarding marine turtles in Ecuadorian waters. Unknown to the Ministry of Agriculture and Livestock, both the General Fisheries Management (DGP) and the National Fisheries Institute (INP), subsidiaries of the Ministry of Natural Resources, were continuing to allow various companies to export turtle products. The Ministry of Natural Resources was not cognizant of the signing of the Convention by Ecuador.

In 1977, the senior author and others, including Craig MacFarland, who was then the Director of the Charles Darwin Research Station, Galápagos, wrote several reports and letters to the Ministries of Agriculture, Natural Resources, and other responsible bodies, informing them of the exploitation. More letters, reports, memoranda and several meetings in 1978 led to the independent investigations of the authors mentioned above.

Their reports and recommendations were discussed in July in Quito at a meeting of all interested parties, including representatives of the government bodies concerned, conservationists, and the factory owners. Basically there were two factions. One faction wanted to follow the advice of the authors and close the factories immediately, at least until further research, such as a periodic census of *L. olivacea* populations, had been conducted and quotas had been fixed. Ortiz and Cantos (1978) suggested that the turtle companies, in their own interest, should sponsor this research. In addition this faction maintained that Ecuador, being a signatory to an international convention, had an obligation to fulfill.

The other faction opposed closure of the factories and maintained that, because *L. olivacea* is migratory, if Ecuador did not utilize this resource someone else would and referred to the continued harvesting of these turtles by Mexico. In addition, because there is less demand internationally for the meat than for the skin, several factories had a meat surplus or stock. It was proposed that these factories be allowed to export this stock since the turtles were already dead. This could lead to some problems because the stock could be augmented illegally. Realizing this, Ortiz and Cantos (1978) had previously recommended that there be regular inspection of the factories, especially to establish that the amount of stock quoted on invoices agreed with the amount still in the refrigerators.

DEREK GREEN

Any possible friction between the two government bodies had already been somewhat forestalled by the very diplomatic proposal of Ortiz and Cantos (1978) that a joint, interministerial commission be formed between the Ministry of Agriculture and Livestock, and the Ministry of Natural Resources; this commission would deal with all aspects of the turtle problem, including centralization of data, legislation, and control.

The results of the meeting were disappointing, for no definite line of action was taken one way or the other.

The Present

The present situation with regard to the turtle factories is a stalemate. No definite action has been taken thus far and so this species is still being exploited on an international level and on a greater scale than ever. However, there has been a recent change in government and it seems that personnel of the Subsecretaría de Pesca of the new government are more receptive to conservation measures. Visits by the junior author in September 1979 to Mauricio Dávalos, the new Minister of Natural Resources and to Ing. Tuly Loor, the new Subsecretaría of Fisheries Resources resulted in the 1979 figures being put at our disposal and sincere promises to fully investigate and act upon the turtle problem.

There is no intention and indeed no need to curtail the present level of exploitation of *Ch. m. agassizi* and *L. olivacea* in areas such as La Libertad in the province of Guayas, nor the negligible taking of nesting females and eggs of *Ch. m. agassizi*, *E. imbricata* and *D. coriacea* in other areas.

Due to financial and administrative difficulties, all efforts by the senior author to establish a tagging program on the mainland beaches to determine the status of the 3 species known to nest in Ecuador have so far proved fruitless.

On the brighter side, in September 1979 a new National Park was declared. The 35,000 ha Machalilla National Park in the province of Manabí is the first to include any coastline of mainland Ecuador. It stretches from Ayampe in the south to Puerto Cayo in the north (Figure 3), with approximately 43 km of coastline. Although turtle nesting and hence exploitation on the beaches is negligible in this area, the new Park will offer complete protection for any turtles that do nest there. Even more significantly, the Park also includes Isla de la Plata, an important feeding area for the migratory *L. olivacea*. In addition, Park boundaries extend 2 nautical miles out to sea. Even if turtles are caught outside these limits, they cannot be disembarked inside the Park. This means that places such as Puerto Lopez, an important unloading area in the past, can no longer function as such.

The Future

Providing the law can be enforced, Machalilla National Park will provide some protection for the persecuted *L. olivacea*, and this in itself will have an immediate

Table 5. Destinations of six companies'[a] skin and meat exports of the Pacific ridley, *Lepidochelys olivacea*, from Ecuador, 1977

Country of destination	Skins				Meat				Total number of turtles (skin only + meat only)	
	Total weight (kg)	Price in US $ per kg	Estimated number of turtles Min.	Max.	Total weight (kg)	Price in US $ per kg	Estimated number of turtles Min.	Max.	Min.	Max.
Japan	67,000	4.18	33,500	37,222	1,000[b]	1.00	147	220	33,500	37,222
Italy	25,000	8.00	12,500	13,889	0	—	0	0	12,500	13,889
Panama	7,000	4.29	3,500	3,889	0	—	0	0	3,500	3,889
Hong Kong	2,000	8.50	1,000	1,111	0	—	0	0	1,000	1,111
United States	0	—	0	0	81,000[b]	3.02	11,894	17,841	[b]	[b]
Companies exporting meat only	0	—	0	0	30,000	[c]	4,405	6,608	4,405	6,608
Totals	101,000	x̄=5.22	50,500	56,111	112,000	x̄=2.99	16,446	24,669	54,905	62,719

— No data.

a. Neptuno, Expromar, Exporklore, Songa, Shayne, and Inexpac.

b. Companies exporting this meat to Japan and the United States had already butchered the turtles for their skin and so these turtles are excluded from the total number of turtles.

c. Prices not known.

Source: Figures by courtesy of the Instituto Nacional de Pesca, Guayaquil.

bearing on the exportation of ridley products. Even so, efforts must be redoubled in an attempt to reach a favorable conclusion regarding the activity of the turtle factories.

Population censuses of ridleys are urgently required. It is also of vital importance to establish a tagging program on the beaches similar to the one in Galápagos in order to determine the size of the nesting populations of *Ch. m. agassizi, E. imbricata* and *D. coriacea,* as well as to gather nesting data and details of migration. This research will provide the necessary scientific background on which to base recommendations for future management policies of sea turtles in mainland Ecuador.

Here there is another ray of hope—or rather 10; 10 Ecuadorian students from the University of Guayaquil were trained on the Galápagos nesting beaches in 1979. It is hoped, should funds become available, that some of these students will pioneer the much needed research on their own mainland shores.

Summary

Species Present

Chelonia mydas agassizi, E. imbricata and *D. coriacea* are present in all three regions. *Caretta caretta* occurs in Colombia but has not been recorded from either the Galápagos Islands or mainland Ecuador. *Lepidochelys olivacea* occurs in mainland Ecuador and Colombia, but not the Galápagos.

Nesting

The situation in Colombia is unknown to us. *Chelonia mydas agassizi* nests in both the Galápagos Islands and mainland Ecuador whereas *E. imbricata* and *D. coriacea* nest only on the mainland.

Exploitation

In Colombia, as far as we are aware, *Ch. m. agassizi* and *C. caretta* are taken only for local consumption. The carapace of *E. imbricata* is often used in craftwork. In mainland Ecuador, although there is exploitation of *Ch. m. agassizi* for local consumption and occasional use of *E. imbricata* in the tourist trade, by far the main exploitation is of the migratory *L. olivacea* for the international market. All species are also caught accidentally in shrimp and fishing nets in both Colombia and mainland Ecuador. Exploitation in Galápagos is virtually nonexistent.

Research and Conservation

The situation in Colombia is unknown to us. Intensive research has been conducted in the Galápagos Islands for 4 consecutive years. Because they are a National Park there is total protection of sea turtles. Research in mainland Ecuador was limited to the 2 independent investigations by the authors. The new Machalilla National Park will offer total protection for sea turtles, but it includes only a very small proportion of the coastline of mainland Ecuador.

Providing that the local consumption and accidental capture of *Ch. m. agassizi, E. imbricata, D. coriacea* and *C. caretta* are maintained at present levels, populations of these species should not decline. Hence, conservation and legislative measures are not of great urgency except to protect nesting females. In the case of the Pacific ridley, *L. olivacea,* however, if the exploitation in mainland Ecuador continues on its present scale, it will not be a question of whether this species will become seriously threatened with extinction, but when. Protection measures are needed now.

Acknowledgments

The research in Galápagos was supported by National Geographic Society grants 1432, 1666, and 1814 to Craig MacFarland with Derek Green as Principal Investigator. The National Fisheries Institute (INP) provided the funds for Green's survey of the sea turtles of mainland Ecuador. The Ministry of Agriculture and Livestock (MAG) supplied the funds for the Ortiz-Cantos investigation of the turtle factories. We are deeply grateful to these organizations for their financial support. In addition, we thank both the National Fisheries Institute and the Dirección General de Pesca (DGP), Guayaquil, and Ing. Tuly Loor the new Subsecretaría de Recursos Pesqueros for allowing us access to the trade figures. We are indebted to Srs. Henry von Prahl and Felipe Guhl from the University of Los Andes in Bogotá for their personal communications on Colombian sea turtles and to the various people, especially Coppelia Hays in Peru, who sent in tag returns. In Galápagos, our sincere thanks go to Darwin Station for its logistical support, the numerous field assistants, especially Mario Hurtado, for their long hours in the field, and the various people involved in the preparation and improvement of the manuscript, especially Warwick J. Reed for his penmanship with the maps.

Literature Cited

Allgöwer, K.
1979. Effect of the scarab beetle, *Trox suberosus,* on the hatching success of the East Pacific green turtle, *Chelonia mydas agassizi,* in the Galápagos Islands. Master's thesis, University of Heidelberg.

Cifuentes, M.
1975. La reproducción y varios aspectos de la ecología de

la tortuga negra, *Chelonia mydas agassizi,* de las Islas Galápagos. Thesis for Licenciado, Universidad Catolica, Quito.

Green, D.

1978. Investigación sobre las tortugas marinas en las costas del Ecuador continental. Report to National Fisheries Institute (INP), Guayaquil, 66 pp.

Kaufmann, R.

1971. Report on status of sea turtles in Colombia. *IUCN Publications, New Series, Supplemental Paper* 31:75–78.

Márquez, R.

1976. Reservas naturales para la conservación de las tortugas marinas de Mexico. *Inst. Nac. Pesca/SI* 83:1–22.

Márquez, R.; A. Villanueva; and C. Peñaflores

1976. Sinopsis de datos biológicas sobre la tortuga golfina, *Lepidochelys olivacea* (Eschscholtz, 1829). *Inst. Nac. Pesca/S2:*1–61.

Ortiz-Crespo, F., and G. Cantos

1978. El problema de la captura y comercialización de tortugas marinas. Report to the Ministry of Agriculture and Livestock (MAG), Quito, 20 pp.

Pritchard, P. C. H.

1971a. Galápagos sea turtles—preliminary findings. *Journal of Herpetology* 5:1–9.

1971b. Sea turtles in the Galápagos Islands. *IUCN Publications, New Series Supplemental Paper,* 31:34–37.

1972. Galápagos sea turtles—research and conservation. Progress report on WWF project number 606, 19 pp. On file at the Charles Darwin Research Station, Santa Cruz, Galápagos.

1975. Galápagos sea turtles. Progress report on WWF project number 790. On file at the Charles Darwin Research Station, Santa Cruz, Galápagos.

Slevin, J.

1905–1906. Unpublished field notes of the California Academy of Sciences Expedition, 1905–1906. Deposited in the California Academy of Sciences, San Francisco.

Wellington, G.

1975. The Galápagos coastal marine environments. Report to the Department of National Parks and Wildlife, Quito, 357 pp.

Wiggins, I., and D. Porter

1971. *Flora of the Galápagos Islands.* Stanford, California: Stanford University Press.

Coppelia Hays Brown
William M. Brown
Route 11, Box 29B
Bowling Green, KY 42101 USA

Status of Sea Turtles in the Southeastern Pacific: Emphasis on Perú

Introduction

Two species of marine turtles, *Chelonia mydas* and *Dermochelys coriacea,* were reported to occur along the coast of Perú. A preliminary survey of marine turtle populations was conducted from January to June 1979. The study area covered approximately 1,750 km along the Peruvian coast. The information on Chile was obtained from published literature and, consequently, is minimal.

Species Present

The following 5 species of sea turtles were found to occur along the southeast Pacific: *Chelonia mydas, Dermochelys coriacea, Lepidochelys olivacea, Eretmochelys imbricata* and *Caretta caretta.* The East Pacific green turtle (*Chelonia mydas agassizi*) is the most abundant species in Perú. They are locally known as *tortuga* or *tortuga blanca.* Its distribution is along the entire coastline and into Chile to Isla Desolación (52° S). This is the most southern report of a marine turtle.

Along the Peruvian coast, leatherbacks (*Dermochelys coriacea*) are reported 3 to 4 hours out to sea, being sighted closer to the coast in the central Departments of Lima and Ica. They are generally referred to as *tinglada, tortuga galápagos,* or *dorso de cuero.* According to the late Dr. Donoso-Barros (1966), they are frequently encountered in Chile; the southern limit of their distribution is the Island of Chiloe.

The olive ridley (*Lepidochelys olivacea*), referred to as *pico de loro* is uncommon in central Perú. It may be common in northern Perú where a nest of eggs was found. Donoso-Barros (1966) mentions them as relatively frequent in northern Chile and, consequently, they may also appear in southern Perú. The olive ridley prefers protected waters, especially the warm waters to the north, and can be found as far south as Talcahuano, Chile.

The hawksbill (*Eretmochelys imbricata*) is rarely found. Only 5 carapaces were recorded from northern Perú;

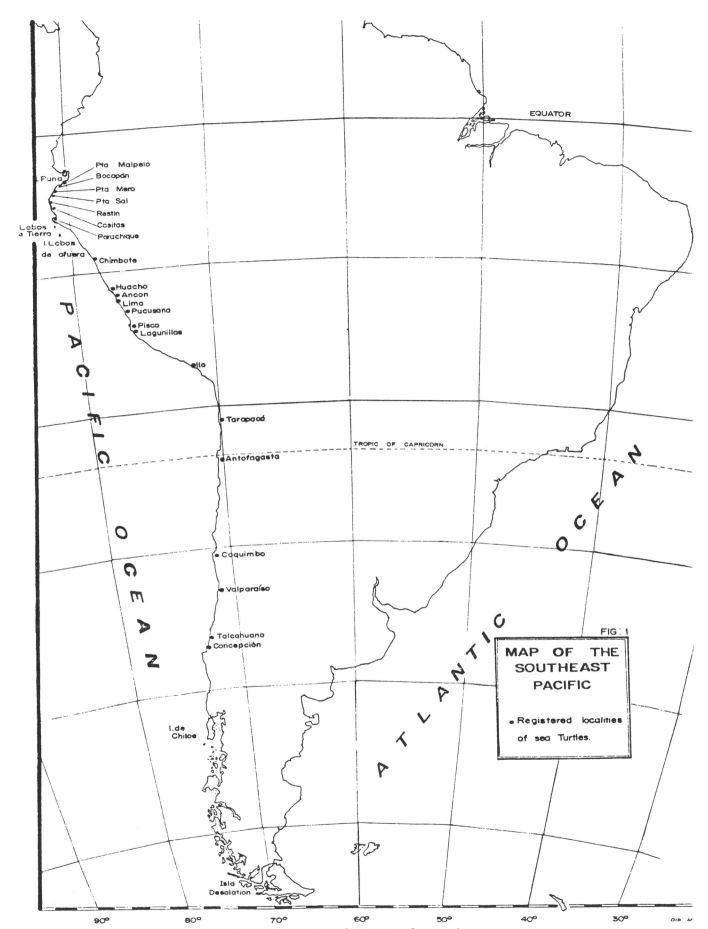

Figure 1. Map of the southeast Pacific showing registered localities of sea turtles.

COPPELIA HAYS BROWN

the last was encountered farther south at 4°30'S. This is the first documented evidence of their presence. As of yet, no specimens are known for Chile.

The loggerhead (*Caretta caretta*) has been cited by Donoso-Barros (1966) as relatively abundant along the coast of Tarapaca, northern Chile, where its meat is consumed. It is known as *tortuga boba*. Mention has been made for Perú, but documentation is lacking. An evaluation of the southern coastline would most likely reveal the occurrence of loggerhead (*Caretta caretta*) since it is abundant in northern Chile.

Feeding

Green turtles were frequently sighted feeding in offshore areas in significant numbers (at least 6 turtles). These feeding areas were common along the northern Peruvian coast. The locations where *Chelonia* was sighted most frequently were characterized by rocky outcroppings with an abundance of algal growth associated with rocky intertidal zones. Green turtles were also found in shallow bays and offshore island beaches (Isla Lobo de Tierra). Locations along the northern Peruvian coast where green turtles were feeding regularly were Bocapan, Punta Mero, Punta Sal, Restin, Casitas, Parachique, Islands of Lobos de Tierra and Lobos de Afuera, Pisco (Jaguay), and Lagunillas (Figure 1).

Stomach contents from 39 green turtles caught in Pisco were analyzed. Some of these turtles had been on their backs for several days, and only residues of their diet were encountered. Data for analyzed stomach contents in Table 1 indicate the occurrence of food content but does not reveal the quantity. Likewise, Paredes (1969) analyzed the contents of 20 Peruvian *Chelonia* stomachs. Algae were found to occur in 100 percent of the samples, jellyfish in 60 percent, fish in 60 percent, and various molluscs in 50 percent.

During interviews in the north, octopi were mentioned as part of the varied diet of green turtles. Only 3 stomach samples from *Chelonia* caught in northern Perú were analyzed, 1 of which contained an octopus.

Fishermen report that leatherbacks are present from December to March feeding on the jellyfish which occur off the coast in large quantities during the summer season. They are believed to follow the jellyfish, indicating that these leatherbacks are on a feeding migration.

Distribution

The greatest number of marine turtles are captured in the port of Pisco, next to the Paracas Reserve. The catch is mainly green turtles and, to a much lesser degree, leatherbacks. The olive ridley is taken on occasion. During the summer months of December to April, there is an increase in the green turtle popula-

tion. Leatherbacks are said to be most plentiful from December to March.

Measurements of 416 green turtles were taken. The straight carapace length (SCL) was recorded (Table 2). The captured population was found to be 89 percent immatures (80 cm and below) and 11 percent mature individuals (above 80 cm). Southern Perú may well be considered a developmental habitat since immatures comprise the majority of the population. No hatchlings were encountered.

Of the 416 green turtles, only 27 males were encountered with secondary sexual characteristics (elongated tail). The SCL of males varied from 65 to 92 cm, with the average being 84 cm.

Measurements were taken of 4 hawksbill carapaces. The SCL ranged from 30.5 cm to 41 cm. Six olive ridley carapaces ranged from 47.5 cm to 72 cm. At Pucusana 115 leatherbacks were measured. Mean carapace length was 135 cm with the carapace length varying from 112 cm to 168 cm. At Playa Naranjo, on the Pacific coast of Costa Rica, 18 nesting leatherback measurements ranged from 128 cm to 151 cm (Cornelius 1976). These data indicate that the East Pacific leatherback is mature at a smaller size than the Atlantic leatherback, which has a reported mean carapace length of 158 cm (Pritchard 1971). Considering 130 cm to be the size at which Pacific leatherbacks reach sexual

Table 1. Stomach content analysis from 39 *Chelonia mydas agassizi*, ranging from 51.5–89 cm straight carapace length

Stomach content	Number of samples	Percentage occurrence
Molluscs: mainly *Nassarius*, *Mytilus* and *Semele*[a]	25	64
Algae: macrocystis, *Rhodymenia*, and *Gigartina* predominate[a]	20	51
Annelids: polychaetes only[a]	19	49
Jellyfish and amphipods[b]	12	31
Fish (eggs included): sardine and anchovy	9	23
Distichlis (salt grass)[c]	7	18
Crustaceans[a]	5	13
Plastic bags[d]	9	23

Note: In addition to the 39 turtles mentioned in Table 1, 3 more were analyzed which contained empty stomachs and were not included in the data.

a. Majority derived from mesolittoral zone.

b. The amphipod *Hypenia medusarum* is commensal to jellyfish and was encountered in various stomach contents.

c. These turtles are feeding in the vicinity of the mouth of the Rio Chico so it is most probable that the *Distichlis* is being carried out.

d. Floating plastic bags may resemble jellyfish; reflects degree of pollution.

Table 2. Straight-line carapace lengths of 416 *Chelonia mydas*

SCL (cm)	Number of *Chelonia mydas*	Average SCL (cm)
≤ 50	26	46.7
51–60	122	55.7
61–70	148	65.6
71–80	75	76.1
> 80	45	84.4
Total	416	65.7

Table 3. Tag returns from Perú

Tag number	Sex	Year tagged in Galápagos	Year recovered at Pisco
Z 1582	F	1977	1979
Z 1948	F	1978	1979
Z 2047	F	1978	1979
Z 2298	F	1978	1979
Z 2482	F	1978	1979
Z 2739	F	1978	1979
376	M	—	1979
3578	F	1970	1978

— No data.

maturity, 71 percent of the measured Peruvian population were mature while 29 percent were immature individuals.

Nesting

Many northern beaches appear suitable for nesting. Fishermen report sea turtles nesting sporadically along the coast, more frequently in past years than in the present.

On one occasion evidence of nesting was recorded. At Punta Malpelo (3°30'S), on the south of the Rio Tumbes, a nest of 80 eggs was found which had been transplanted from their original site by local fishermen. One egg contained an embryo measuring 2.3 cm and identified as *Lepidochelys olivacea*. The remaining eggs showed no signs of development. The mean of the measured eggs was 4.3 cm in diameter.

The nesting season is apparently from December to February. Fishermen commented that 3 to 4 years ago more turtles nested in the area. During this past season other nests were reported, but due to extremely high tides all evidence had been erased. Nesting along this beach appears to be infrequent though not uncommon.

The coast of Punta Malpelo is a long stretch of white sand composed mainly of quartz and feldspar. The slope of the beach is gradual. Approximately 50 m above the high tide mark driftwood mounds up and, beyond that, there are mangroves. This beach is being developed. Breeding tanks for shrimp hatcheries are going to be built, as are beachside hotels.

South of Punta Malpelo we occasionally found turtle tracks which indicated nesting activity in this area. The local fishermen stated that turtle nests were seldom disturbed and that there had been more nesting in the past. Nesting seems to have declined due to recent development along this section of the coast.

There is no evidence that sea turtles nest in southern Perú. Turtle dealers report that the few olive ridleys captured usually bear unshelled eggs. Older fishermen of the area refer to turtles nesting, or rather coming ashore, at Caucato, above the Pisco River. Today this beach is subject to intense human disturbance.

At the University of Luis Gonzaga in Pisco, there are 5 preserved marine turtle eggs. These were found in a butchered female 6 years ago, species unknown. Other than this, no evidence of nesting exists.

Commensalism

Many green turtles bear barnacles of the species *Chelonibia testudinaria* on their carapaces. Occasionally, barnacles of the genus *Lepas* are also found on the tough leathery skin between the plastron and carapace.

The crab *Planes cyaneus* is found on the basal section of the tail of *Chelonia mydas agassizi*. These crabs are recovered when the turtle is captured at sea. Schweigger (1964) refers to an old fisherman's tale of a relationship between a crab and a turtle; the crab rides the turtle when it is ready to submerge to the depths.

Currents and Migrations

The coast of Perú is bordered by a continental shelf. The widest part, approximately 105 km, is located in the north at Pimentel (9° S); the narrowest part is at Punta San Juan in the south (15° S). The continental shelf inclines to depths of more than 3,000 m.

Currents flowing along the coast are responsible for the upwelling known to supply Peruvian waters with rich nutrients. This upwelling creates good feeding habitats for marine turtles.

By July, *Chelonia mydas* has left its nesting grounds in the Galápagos Islands for its feeding grounds (Green, this volume). The distribution of part of this *Chelonia* population is reflected by tag recoveries from Perú. The 7 tag returns from Perú reveal that *Chelonia mydas* nesting in the Galápagos have migrated as far south as Pisco, 13°5'S, (Table 3), a distance of 2,300 km. The recoveries are too few to indicate whether a large portion of this population regularly migrates to Perú.

COPPELIA HAYS BROWN

Part of the feeding population in Perú may have migrated from the nesting population on Puna Island, southern Ecuador (3° S).

During years of increased water temperatures there seems to be a direct relationship with an increase in the sea turtle population. Catch statistics, which are not very reliable for exact numbers but do reflect the relative proportions, indicate that during the years 1972, 1975, and 1976 there was an increase in the catch (Figure 2). These were years of the phenomenon known as *El Niño*, which occurs when the equatorial current pushes warm waters farther south than normal.

In Figure 2 the high catch in 1978 may reflect the fact that the Ministry of Fisheries is now more concerned about marine turtles than in past years. Stricter control in declaring the poundage caught has been enforced since the enactment of the Ministerial Resolution in 1977 protecting sea turtles. To date, catch statistics are still not specific as to species.

Legislation

In January 1977, the Ministerial Resolution, No. 01065, was enacted. Through it, protection is provided for the green turtle and the leatherback. It prohibits hunting of all leatherbacks, and only *Chelonia mydas* 80 cm or above are allowed to be taken. Only at one port was this resolution found to be enforced. It is now being enforced at Pisco where the greatest number of turtles is caught. The responsibility of enforcement is on the captain of each port.

In October 1978, 167 leatherback carapaces were found in a canyon near the port of Pucusana. They were captured during the summer months (January to March). A conservative estimate of the catch per season is 200 leatherbacks. As stated by Pritchard (1972) and Frazier (1979), central Perú has the largest known leatherback fishery in the world. Most appear to be adult and subadult and thus represent a considerable number of reproductive and near reproductive individuals.

Green turtles along the northern coast are incidently caught in fishing nets and shrimp trawlers. If the turtle is not consumed by the fisherman, he sells it to a turtle buyer who generally passes by every few days. This buyer will sell the meat at a central market. Main markets in the north are La Cruz, Piura, and Chiclayo. Meat is occasionally transported frozen to Lima.

In southern Perú there is a more sophisticated turtle traffic where fishermen have nets specially designed to catch turtles. They are made with 59 cm² mesh and function as tangle nets. In the area of Pisco there are approximately 7 to 10 boats dedicated to the capture of turtles, going out on 2 to 3 day ventures. The turtles are sold to the *tortuguera*—a woman whose business is buying and selling turtle products. The green turtles

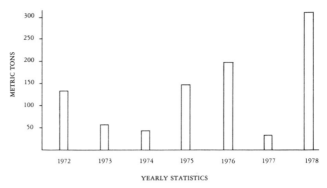

Figure 2. Catch statistics of marine turtles according to mass (metric tons).

are surreptitiously transported ashore during the night and early hours of the morning and slaughtered as the sun rises.

During the height of the season (December to April) there may be up to 70 green turtles lying on their backs, in the shade, waiting to be butchered. During this season, the average catch was 10 to 30 turtles per day. The family begins work around 4:30 A.M. and by 7:30 A.M. the meat is ready to be sold at the market. At Pisco, turtle meat sells for $2 per kg, which is half the market value of beef.

The entire turtle is utilized. The meat, liver, kidneys, heart, esophagus, and head are consumed. The fat is boiled to extract the oil which is believed to be a remedy for bronchial problems such as asthma. The hides of the front and hind flippers are salted and sold to a leather company in Lima. The blood is believed to fortify the body and is drunk soon after decapitating the turtle. The carapaces are used for artistic purposes, for feeding dishes for pigs and ducks, and for bowls to salt fish in. The calipee, which is so highly esteemed in other parts of the world, is of no use to Peruvian consumers. At seaside restaurants in Pisco, turtle meat is openly advertised. The meat is also transported to several first class restaurants in Lima. All turtle products are consumed within the country and are of no consequence for exportation.

Conclusion

The following 5 species of marine turtles are present in the southeastern Pacific: *Chelonia mydas, Dermochelys coriacea, Lepidochelys olivacea, Eretmochelys imbricata,* and *Caretta caretta.* Perú's highly productive coastal waters are an important feeding ground for immature green turtles and also may be an important feeding area for the East Pacific leatherback turtle.

An extensive tagging and recovery program is necessary to further establish what breeding populations supply these migrants in Perú. If adequate protection is given at the breeding grounds, a moderate scale exploitation program may also be conducted with the immature green turtles. The Reserve of Paracas should be proclaimed a "turtle sanctuary," thus providing partial protection of important feeding grounds.

Leatherbacks should be given total protection until further investigations are conducted with these poorly studied turtles. In addition to adults, we also find juvenile leatherbacks which are rarely seen elsewhere.

Further analysis of marine turtle stomach contents is recommended. This may lead to a greater understanding of the ecology of marine turtle populations in Perú.

Acknowledgments

There have been various institutions and persons to whom appreciation for their support and cooperation is extended. The New York Zoological Society has provided support for this project. Dr. Antonio Brack has supervised it and offered much valuable advice. The Institute del Mar Peruano has provided cooperation in all aspects throughout the study. The Ministry of Fisheries has shown much concern and has offered further assistance. The Museum of Natural History "Javier Prado" provided the use of their facilities. We are grateful to Dr. Alamo for his unremitting dedication throughout all phases of the study, and to Dr. Enrique del Solar for assisting with identifications. Our gratitude to Javier de los Rios, a volunteer who worked in Pisco. We are indebted to Mr. Bernard Peyton, conservation fellow of the New York Zoological Society, who supported both the research project and the preparation of this report.

Literature Cited

Cornelius, S.
1976.   Marine turtle nesting activity at Playa Naranjo, Costa Rica. *Brenesia* 8:1–27.

Donoso-Barros, R.
1966.   *Reptiles de Chile*. Santiago de Chile: Ediciones de la Universidad de Chile.

Frazier, J.
1979.   Sea turtles of Peru. Manuscript, 236 pp.

Paredes, R. P.
1969.   Introduccion al estudio biologico de *Chelonia mydas agassizi* en el Perfil de Pisco. Masters Thesis, Universidad Nacional Federico Villareal, Lima, Peru.

Pritchard, P. C. H.
1971.   The leatherback or leathery turtle, *Dermochelys coriacea*. *IUCN Monograph Marine Turtle Series*, 1:1–39.

1972.   *World Wildlife Fund Yearbook 1971–1972*. Morges: World Wildlife Fund, pp. 149–51

Schweigger, E.
1964.   *El Litoral Peruano*. Lima, Peru: Grafica Morsom S. A.

COPPELIA HAYS BROWN

Hawaii, Oceania, and Australia

George H. Balazs
Hawaii Institute of Marine Biology
P.O. Box 1346
Kaneohe, Hawaii 96744

Status of Sea Turtles in the Central Pacific Ocean

Except for the Hawaiian Archipelago, sea turtle populations in the Central Pacific Ocean and other areas of Polynesia have not been systematically surveyed and only limited information exists on their occurrence and present survival status. This report will summarize and review what is known for a number of locations within the region, specifically the Hawaiian Archipelago, Line Islands, Phoenix Islands, Cook Islands, American Samoa, Western Samoa, Tokelau, Tuvalu, Wake, Johnston, Howland, and Baker (Figure 1). While the information for most of these areas is clearly inadequate, there is nevertheless evidence to indicate that the numbers of turtles have declined within historical times. At those islands with indigenous human populations, the traditional conservation systems that served to protect turtles and other marine resources from overexploitation have deteriorated considerably, and in some cases vanished altogether. Three interrelated factors contributing to this breakdown have been the introduction of money economies, the decline of traditional authority, and the imposition of new laws and practices by colonial powers (Johannes 1978). In Polynesian societies, sea turtles are known to have played an important role in certain religious ceremonies, in mythology and art, in the production of implements and medication, and as high protein food sources generally reserved for chiefs and priests (Buck 1932; Emory 1933, 1947; Emory, Bonk, and Sinoto 1968; Kalakaua 1888; Pukui and Elbert 1971).

Of the islands covered in this report, only the ones under United States jurisdiction currently have governmental regulations for sea turtles. Under the U.S. Endangered Species Act, all sea turtles at these U.S. areas are fully protected.

Status

Hawaiian Archipelago (United States)

Three species of sea turtles occur in Hawaiian waters, the green turtle, *Chelonia mydas,* the hawksbill, *Eret-*

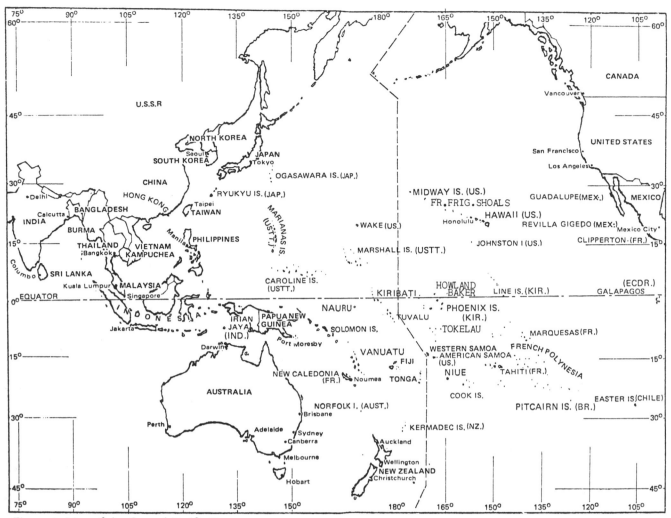

Figure 1. Map of the Central Pacific Ocean.

mochelys imbricata, and the leatherback, *Dermochelys coriacea.* The olive ridley, *Lepidochelys olivacea,* and the loggerhead, *Caretta caretta,* have been recorded, but only as rare visitors.

The Hawaiian hawksbill population is small and only known to occur in coastal waters of the 8 main and inhabited islands at the southeastern end of the 2,450-km-long archipelago. Several nestings have been documented on the island of Hawaii where black volcanic sand beaches are utilized. A single nesting has also been recorded on the island of Molokai (Ernst and Barbour 1972).

Leatherbacks are regularly sighted in offshore waters at the southeastern end of the archipelago, but nesting does not take place. During August 1979, at least 10 leatherbacks ranging from 60 to 120 cm in carapace length were sighted in pelagic waters to the northwest of the Hawaiian Archipelago between 40° to 43°N and 175° to 179°W (G. Naftel, in litt.).

Green turtles are by far the most abundant of Hawaiian sea turtles, with mixed aggregations of adults and immature individuals larger than 35 cm residing in coastal waters throughout the archipelago where they feed on several kinds of benthic algae. In excess of 90 percent of all nesting occurs on 6 small sand islands at French Frigate Shoals (23°45'N, 166°10'W), a 35 km long atoll situated in the middle of the archipelago. Tagging has demonstrated that long-distance migrations to this site are periodically undertaken by adults from numerous resident foraging areas, all of which are within the Hawaiian chain (Figure 2; Balazs 1976a, 1979). Hawaiian green turtles therefore appear to be genetically isolated from other populations in the Pacific. Systematic monitoring of the breeding colony at French Frigate Shoals was initiated in 1973 and has continued during each subsequent year. The number of females nesting annually has been found to fluctuate considerably, with the range extending from 94 in 1976 to 248 in 1978 (Figure 3). No population trends are apparent for the 7-year study period. The production of hatchlings since 1973 has ranged from approximately 12,500 in 1976 to 32,900 in 1978. Predation

244

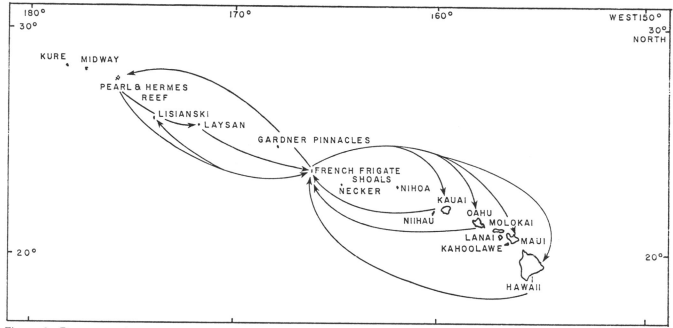

Figure 2. Documented migrations of adult green turtles in the Hawaiian Archipelago.

on eggs does not take place at French Frigate Shoals, and predation on hatchlings appears to be minimal. However, predation by tiger sharks (*Galeocerdo cuvier*) on both adults and immature turtles throughout the chain may be substantial (Balazs 1979). Prior to 1973, the annual breeding colony at French Frigate Shoals was incorrectly estimated by previous workers to contain as many as 2,600 to 5,200 turtles (Hendrickson 1969; Amerson 1971).

Hawaiian green turtles exhibit the rare behavioral trait among sea turtles of coming ashore to bask or rest, but only at certain undisturbed sand beaches or rock ledges in the uninhabited Northwestern Hawaiian Islands (Wetmore 1925; Balazs 1976b; Eliot 1978; Lipman 1978). This behavior provides access for the tagging of males as well as females, both at the breeding grounds and at a number of resident foraging areas. However, caution is being exercised in these research activities so that normal behavioral patterns will not be adversely affected.

Large numbers of green turtles were commercially exploited in the Hawaiian Archipelago until 1974 when the State of Hawaii adopted a protective regulation banning this activity. In 1909 all of the Northwestern Hawaiian Islands except Midway were designated as a Bird Reservation, which in 1940 became known as the Hawaiian Islands National Wildlife Refuge. However, the exploitation of turtles in these areas, particularly at French Frigate Shoals, periodically continued until at least 1969 (Amerson 1971; Balazs 1975a). Of concern at the present time is the well-documented, drastic decline of the foraging and basking aggregations in the Northwestern Hawaiian Islands at Laysan Island, Lis-

ianski Island and, to a lesser extent, at Pearl and Hermes Reef. Furthermore, the forthcoming development of various commercial fisheries in this segment of the chain represents a potential threat to the remaining aggregations. Terrestrial areas in the Northwestern Hawaiian Islands are under review by the Fish and Wildlife Service for designation as Critical Habitat under the U.S. Endangered Species Act (Dodd 1978; see also Balazs 1978).

Johnston Atoll (United States)

Johnston Atoll is located at 16°45′N, 169°31′W and

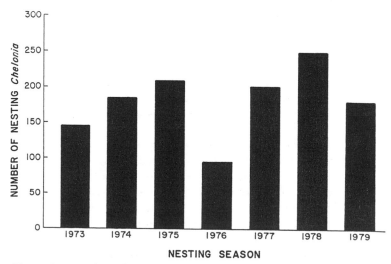

Figure 3. Number of turtles nesting annually at French Frigate Shoals.

contains 4 islands, 2 of them completely man-made. From the late 1950s until 1962, nuclear weapons' testing was conducted over the atoll. The area is administered by the Nuclear Defense Agency, but is now used prinicpally as a storage site for chemical munitions (Inder 1978). Johnston is concurrently managed as a National Wildlife Refuge.

Both immature and adult green turtles are regularly seen foraging in shallow waters, but nesting is not known to take place. Courtship behavior and possibly sustained copulation have, however, been periodically reported by resident personnel. Numerous species of algae occur within the atoll (Buggeln and Tsuda 1966), including *Caulerpa racemosa*, *Codium arabicum* and *Gelidium pusillum* which are known food sources of green turtles (Balazs 1979). Large sharks that are probably tiger sharks have been observed attacking and feeding on turtles (C. Cecrle in litt.).

Wake (United States)

Wake is an inhabited atoll located at 19°18′N, 166°35′E that is administered by the U. S. Air Force. Both immature and adult green turtles are regularly observed foraging in the lagoon and along the outside perimeter of the atoll. Nesting has never been recorded.

Howland and Baker (United States)

Howland (0°48′N, 176°38′W) and Baker (0°13′N, 176°28′W), 2 low coral islands, were designated National Wildlife Refuges in 1974 and are now uninhabited. Turtles were reported to be "abundant" in the waters around Howland by residents present in May and June of 1935 (Bryan 1974). No information on turtles exists for Baker. Feral cats, which are known in some areas to be predators of hatchlings and eggs, are present on both of these islands.

Line Islands

All of the Line Islands are low coral islands and atolls. The Northern Line Islands consist of Kingman Reef, Palmyra and Jarvis, under the jurisdiction of the United States; and Washington, Fanning and Christmas under the jurisdiction of the newly independent nation of Kiribati (formerly the Gilbert Islands). The Southern Line Islands are also under the jurisdiction of Kiribati and consist of Malden, Starbuck, Vostok, Caroline, and Flint, all of which are now uninhabited. Information on sea turtles exists only for the following locations.

- *Palmyra (5°53′N, 162°05′W).* From 1958 to 1965, green turtles were periodically seen in shallow waters at the eastern side of the atoll. On one of these occasions a group of 11 adults was observed foraging together (P. Helfrich and J. Naughton, personal com-

munication). Similar observations have also been made during recent years (M. Vitousek, personal communication).There are no reports of nesting. Algal collections at Palmyra have included *Pterocladia*, a known food source of green turtles (Dawson 1959; Balazs 1979). The atoll is now used as a copra plantation and has a small resident human population. Along with Midway and Wake, the U.S. government is considering Palmyra as an international storage site for nuclear wastes.

- *Jarvis (0°23′S, 160°01′W).* A low level of nesting, apparently involving green turtles, was recorded along the western coast of Jarvis by residents present in August of 1935 (Bryan 1974). The island was designated a National Wildlife Refuge in 1974 and is now uninhabited. Feral cats are present on the island.

- *Fanning (3°52′N, 159°20′W).* Turtles were reported to "abound" at Fanning in the 1850s (Burnett 1910, quoted by Wiens 1962). The atoll has been continuously inhabited since 1852 and used principally as a copra plantation. A small number of turtles are regularly sighted in the lagoon, and a low level of nesting still takes place. The residents capture turtles whenever possible.

- *Christmas (1°59′N, 157°30′W).* When Captain James Cook discovered uninhabited Christmas atoll in late December of 1777, between 200 and 300 green turtles were captured during the 8-day visit (Beaglehole 1967). Turtles were taken both in the shallow lagoon and on the beaches, with weights ranging from 20 to 90 kg. Publicity resulting from Captain Cook's visit caused numerous whaling vessels to stop at the atoll for provisions (Bryan 1942). Green turtles were still abundant in 1838 (Tresilian 1838, quoted by Wiens 1962). Christmas has been inhabited and used as a copra plantation since 1902. Nuclear weapons' testing was conducted over the atoll by the British from 1956 to 1958, and by the United States in 1962 (Inder 1978). In 1975 a visitor noted that some nesting was still taking place, but no details were available (D. Crear, personal communication).

- *Vostok (10°06′S, 152°23′W).* In June of 1965, Clapp and Sibley (1971a) saw several turtles in the waters surrounding Vostok, but no signs of nesting were found. M. Vitousek (personal communication) was informed that numerous turtle tracks were seen on the beaches during a visit made in recent years, but no details are available. Vostok is only 1.2 km².

- *Malden (4°1′S, 154°58′W).* No evidence of turtles has been found during several recent visits to Malden (E. Vitousek, personal communication).

- *Caroline (9°58′S, 150°14′W).* Dixon (1884) reported that turtles were seen at Caroline, but not in great numbers. No turtles were seen by Clapp and Sibley (1971b) during a 2-day visit in June 1965.

GEORGE H. BALAZS

Phoenix Islands

The Phoenix group is under the jurisdiction of Kiribati and consists of 8 low coral islands and atolls. Only Canton is now inhabited.

- *Canton (2°50'S, 171°43'W).* Green turtles nest along the northern, eastern and western shores of Canton throughout the year, but greater numbers are present during October and November (Balazs 1975b). The total annual number of nesting females using the atoll may involve as many as 200 turtles. Large populations of ghost crabs (*Ocypode* spp.) and hermit crabs (*Coenobita perlitus*) are present and probably prey heavily on hatchlings. It is not uncommon for nesting females and hatchlings to become disoriented and travel inland where they die of hyperthermia. Adult males and females in groups of up to 40 individuals have been observed foraging close to shore in water less than 50-cm deep (J. Hass, in litt.). Algal collections from Canton have included *Caulerpa racemosa, Codium arabicum, Gelidium pusillum,* and *Pterocladia* (Dawson 1959).

- *Enderbury (3°07'S, 171°03'W).* Green turtles nest along Enderbury's western and eastern shores (Balazs, 1975b; J. Keys, in litt.). King (1973) listed this island as one of the most important nesting sites for green turtles in the Central Pacific. Two Korean fishing vessels were wrecked on the island during recent years.

- *Phoenix (3°43'S, 170°43'W).* Turtle bones were found on Phoenix during a visit in 1924 and nesting was presumed to take place (Bryan, 1942). Feral rabbits are present on the island.

- *Birnie (3°35'S, 171°31'W).* During a low-altitude overflight of the Phoenix group in January 1978, the author found the beaches of Birnie to be covered with turtle tracks. Birnie is the smallest island in the Phoenix group (0.5 by 1.25 km) and the only one that has never been inhabited or mined for guano.

- *Hull (4°30'S, 172°10'W).* Green turtles nest along the northeastern and southeastern shores of Hull (Balazs 1975b; J. Keys, in litt.). When the U.S. Exploring Expedition visited this atoll in August 1840, a Frenchman and 11 Tahitians were found to have been stationed there to catch turtles (Wilkes 1845). In May 1974 large numbers of dead fish and an adult male green turtle were found washed ashore from inside the lagoon. The cause of this mortality could not be determined.

- *Sydney (4°27'S, 171°16'W).* Turtle tracks have been observed on Sydney's northwestern shore (Balazs 1975b). Evidence of trespassing by crewmembers of foreign fishing vessels has been found on this atoll during recent years.

- *Gardner (4°40'S, 174°32'W).* Turtle tracks have been observed on Gardner's southwestern shore (Balazs 1975b). Evidence of trespassing has also been found on this atoll.

- *McKean (3°36'S, 174°08'W).* No information on turtles exists for McKean. Several foreign fishing vessels have been wrecked on the island.

American Samoa (United States)

American Samoa consists of the mountainous volcanic islands of Tutuila and the Manua group (Ofu, Olosega, Tau), and Swains and Rose Atoll which are of coral origin. Approximately 94 percent of the Polynesian inhabitants reside on Tutuila (port city—Pago Pago). Rose is the only uninhabited island in the group. Certain terrestrial areas, including Swains and Rose, are under review by the Fish and Wildlife Service for designation as Critical Habitat under the U.S. Endangered Species Act (Dodd 1978).

- *Tutuila (14°16'S, 170°40'W)* and *Manua group (14°10'S, 169°35'W).* Green turtles and hawksbills occur in the waters surrounding these islands, but apparently only in small numbers. There is some indication that the hawksbill is the more abundant species. Occasional nesting on isolated beaches is thought to take place (Coffman 1977; S. Swerdloff, W. Pedro and R. Wass, personal communication).

- *Swains (11°03'S, 171°05'W).* Green turtles and hawksbills are known to nest at Swains (Swerdloff, personal communication). Turtle eggs were observed being gathered by the native inhabitants during July and August 1963 (Pedro, personal communication). The atoll is only 2 km in diameter.

- *Rose Atoll (14°33'S, 168°09'W).* Green turtles, and probably some hawksbills, nest on the islets (Rose and Sand) at Rose Atoll. An account in the 1800s stated that large numbers of turtles nest during August and September, and that numerous sharks prey on the hatchlings (Graeffe 1873, quoted by Hirth 1971a). On 7 October 1970, Hirth (1971a) counted 35 and 301 nesting pits of varying age on Sand and Rose Islets, respectively. Fishermen in Pago Pago confirmed that the peak nesting season is August and September.

On a low-altitude overflight in October 1974, 75 adult turtles were counted within the lagoon (P. Sekora, personal communication). During a 5-day visit in May 1976, only 3 adults and 1 immature green turtle were observed, and no nesting took place (Coffman 1977). During a daytime visit on 29 March 1978, Coleman (1978) recorded 1 recently excavated pit on Rose and 4 that he estimated to be 1-month old. Other older pits were noted, as well as a single adult green turtle in the lagoon and the rib bones of a turtle on Sand Islet. Numerous black-tipped sharks (*Carcharhinus melanopterus*) 20- to 40-cm long were present.

Direct observations of predation on hatchlings by rats have been made during recent years (Swerdloff, personal communication), but the extent and significance are unknown. Mayor (1921) was the first author to record rats at Rose. Hirth (1971a) stated that Rose Islet "swarms with rats (possibly *Rattus exulans*)." Coleman (1978) found that rats were "extremely abundant" and thought that black rats (*Rattus rattus*) might be present.

Following the recommendations of Hirth (1971a), Rose Atoll was designated a National Wildlife Refuge in 1974 (see also Rockefeller and Rockefeller 1974).

Western Samoa

Western Samoa is an independent nation consisting of 2 large islands of volcanic origin (Savaii and Upolu) and several offshore islets. The islands are located between 13° to 15°S and 168° to 173°W. Approximately 72 percent of the 152,000 Polynesian inhabitants reside on Upolu Island.

Green turtles and hawksbills occur in the surrounding waters of both Savaii and Upolu. The green turtle has been reported by fishermen to be the more abundant (Hirth 1971a). It is uncertain whether this species nests in the area. Hawksbills are known to nest, but now only on the offshore islets of Namua, Nuutele, and Nuulua located at the western end of Upolu. The nesting season extends from October to June, with most activity occurring in January and February. Nesting tracks counted by Witzell (1972a) suggest that not more than 45 females use these beaches each season. The number of hawksbills is believed to have declined considerably, due mostly to human exploitation of eggs and nesting females (Witzell 1972a, 1972b, 1974). The coasts of both Upolu and Savaii were reported to have abounded with turtles in the early 1800s (Williams 1837).

In 1971 a hatchery project was initiated by the Fisheries Division and 2 U.S. Peace Corps volunteers in an attempt to replenish the hawksbill population. This effort has continued until the present time. During each nesting season as many freshly laid eggs as possible are transferred from the 3 islets and reburied at a protected facility on the adjacent mainland shore of the Aleipata district. Hatchlings are held for up to 3 months in concrete tanks before being released into offshore waters. Marginal scutes have been notched for identification purposes. Hatchery data for the years 1973 through 1975 are as follows (from Anonymous 1974, 1975; A. Phillip, O. Gulbrandsen and T. Poutoa, personal communication):

Year	Eggs collected	Eggs hatched
1973	4,656	3,257 (70 *percent*)
1974	6,231	4,951 (79 *percent*)
1975	5,159	2,460 (48 *percent*)

This restocking effort has been considered at least a partial success by fisheries personnel because several marked immature turtles have been found for sale on Upolu (Anonymous 1975). Educational programs have also been periodically conducted to inform the populace of the need to conserve sea turtles (Witzell 1972b, 1974).

Based on the advice of an FAO sea turtle consultant, plans were prepared for a ranching industry in which hawksbill hatchlings would be grown to a size suitable for stuffing and export to Japan (Banner 1971). However, raising the turtles for more than a few months was not found to be feasible. Difficulties encountered included the need to frequently change the sea water in the rearing tanks, the presence of disease which caused serious tissue necrosis, the turtles' constant biting of one another, and the absence of a suitable, inexpensive food (Witzell 1972a; Anonymous 1974).

Tokelau

Tokelau is a New Zealand dependency consisting of three atolls (Atafu, Fakaofo, Nukunonu) located between 8° to 10°S and 171° to 173°W. The total Polynesian population is 1,600.

Hirth (1971a) reported that green turtles and, to a much lesser extent, hawksbills nest in Tokelau during September and October, but that their numbers were rapidly declining. In 1977, 1 of the remaining nesting sites was along the southern portion of Taulagapapa Islet at Nukunonu Atoll (N. Walton, personal communication).

Tuvalu

Tuvalu, formerly known as the Ellice Islands, is a newly independent nation comprised of 9 coral islands and atolls located between 5° to 10°S and 176° to 180°E. From north to south, the group consists of Nanumea, Niutao, Nanumanga, Nui, Vaitupu, Nukufetau, Funafuti, Nukulaelae, and Niulakita. The total Polynesian population is estimated to be 9,000. The capital of the group, Funafuti Atoll (8°30′S, 179°10′E), is 18 by 25 km and contains 30 islets (Inder 1978).

Hedley (1896), quoted by Wiens (1962), stated that "the green turtle was the only one found at Funafuti," but no additional information was provided. Carr (1965) listed the atoll as a minor green turtle nesting area, while Hirth (1971b) included it among important nesting sites in the western hemisphere. No other information on turtles is known to exist for Funafuti or other members of the group. However, in 1972 an intense tropical cyclone struck Funafuti and deposited an 18-km-long rampart of coral rubble along the atoll's southeastern outer reef (Maragos, Baines, and Beveridge 1973). The impact of this new formation on

GEORGE H. BALAZS

available nesting habitat could be substantial.

The coinage of Tuvalu includes a $1-piece displaying the green turtle. Furthermore, the commemorative stamps issue in 1976 features a leatherback and the uninhabited atoll of Niulakita. This would suggest that leatherbacks either nest or are sighted in the area.

Cook Islands

The Cook group, a self-governing state associated with New Zealand, consists of 15 volcanic islands and atolls located between 9° to 23°S and 156° to 167°W. The islands of volcanic origin include Rarotonga, Aitutaki, Atiu, Mitiaro, Mauke, and Mangaia, while the coral atolls, most of which are in the northern portion of the group, include Palmerston, Suwarrow, Pukapuka (not to be confused with Pukapuka in the Tuamotu Archipelago), Nassau, Manihiki, Rakahanga, Penrhyn (Tongareva), Manuae, and Takutea. There are approximately 18,000 Polynesian inhabitants, 54 percent of which live on Rarotonga. Information on the occurrence of turtles exists for the following locations.

• *Palmerston (18°04'S, 163°10'W)*. Powell (1957), quoted by Weins (1962), indicated that green turtle eggs were "fairly plentiful" at Palmerston and that both turtles and eggs were frequently used for food by the 85 inhabitants. Carr (1965) considered Palmerston to be a major Pacific nesting site for green turtles. Although Hirth (1971b) included the atoll in a list of important nesting sites in the western hemisphere, it was stressed that the number of turtles involved was unknown and that the situation warranted immediate research attention.

In 1977 each family on the atoll had a tradition of raising 15 hatchlings in floating cages for 1 to 3 months before releasing them as a restocking effort (S. Kavakana and D. Brandon, personal communication). This practice apparently started in the 1950s following recommendations offered by Powell (1957). However, other reports in 1977 indicated that, instead of being released, many of the turtles were gutted, injected with formalin, and sent as curios to relatives in New Zealand. From 1972 to 1977 a decline in the number of nesting turtles was observed by the inhabitants, thereby prompting the local Island Council to prohibit the use of spearguns (T. Wichman, personal communication). Approximately 4 to 5 turtles are sent from Palmerston each year to the market in Rarotonga where the meat is not readily accepted by the residents and sells for only US$0.45 to 0.90 per kg. Large shells, however, bring US$50.00 or more in the growing tourist trade. Hatchlings were reported to be present at Palmerston in January (Brandon 1977), but the range of months in which nesting takes place is unknown.

Prior to 1862 Palmerston was uninhabited. In that year an Englishman (William Marsters) and 3 women from Penrhyn settled on the atoll and founded the colony that now exists (Bryan 1942).

• *Pukapuka (10°53'S, 165°49'W)*. Green turtles and some hawksbills nest on one of the uninhabited islets at Pukapuka. Turtles and eggs that are taken from this location must be shared among the native inhabitants of the atoll (D. Clark, personal communication).

• *Manihiki (10°25'S, 161°01'W)*. Green turtles, and possibly some hawksbills, nest at Manihiki. The natives take both the turtles and eggs for food. Hatchlings are also raised for a few months and preserved with formalin for shipment to relatives in New Zealand (T. Wichman, personal communication).

• *Rakahanga (10°02'S, 161°05'W)*. Both Carr (1965) and Hirth (1971b) list Rakahanga as a nesting site for green turtles. Gill (1876), quoted by Wiens (1962), stated that "Several species of turtle—loggerhead, hawksbill, green turtle, etc.,—are very plentiful on Rakahanga in the breeding season." Although loggerheads (as well as leatherbacks) have occasionally been sighted in the Cook Islands (Brandon, 1977), this is the only known report of nesting. The northwestern point of Rakahanga is named Te Mata i Pahonu and relates to sea turtles.

• *Penrhyn (9°0'S, 157°59'W)*. Green turtles and some hawksbills are known to nest at Penrhyn and forage in the adjacent waters. During 1976 between 40 and 50 turtles of unknown sizes were taken principally for their shell. A few were sent to Rarotonga, but most were used for trading with Japanese, Korean, and Taiwanese fishing vessels that illegally visit the atoll (Brandon 1977 and personal communication).

• *Suwarrow (13°15'S, 163°06'W)*. Brandon (1977) lists Suwarrow as a nesting site for turtles, but no details are provided. One of the islets is named Turtle Island. Only one person lives on the atoll.

• *Manuae (19°16'S, 158°58'W)*. In May of 1975 the Cook Islands' Government donated Manuae for use as the first World Marine Park (Allen 1975; see also Shadbolt 1967). However, questions of ownership of the atoll have prevented this action from being carried out. Manuae is listed by Brandon (1977) as a nesting site for turtles, but details are not provided.

• *Takutea (19°49'S, 158°18'W)*. Takutea is also listed as a nesting site for turtles by Brandon (1977), but again no details are provided.

From 1974 to 1977, studies on the rearing of green turtles as a village industry were conducted at Rarotonga with financial support from the South Pacific Commission (Baird 1975; see also Powell 1957; Anonymous 1972). As part of this project, laboratory experiments were also carried out at the University of the South Pacific in Fiji (Raj 1975). The findings of this work were similar to those made in Western Samoa, in that problems of disease and a suitable, inexpensive food supply could not be resolved (Brandon 1977, Balazs 1977, Anonymous 1978).

Recommendations

The survival status of sea turtle populations occurring in the areas covered by this report can be enhanced through implementation of the following recommendations.

1. Where appropriate, island governments should attempt to reinforce the traditional conservation systems that formerly served as a buffer for sea turtles. The absence of governmental regulations for most of the areas covered makes it imperative that some protective action be undertaken, but in close consultation with native inhabitants.

2. Known rookeries should be intensively monitored during the peak period of at least 1, but preferably several, nesting seasons to determine the number of females present. As a minimum, this should include Rose Atoll, Enderbury, Birnie, Palmerston, Penrhyn, and the offshore islets of Namua, Nuutele, and Nuulua in Western Samoa. Tagging with durable tags (see Balazs, this volume) should be carried out as an integral part of this work to gain some insight into the occurrence of international migrations, hence shared usage of resources.

3. Based on the results of Recommendation 2, and in harmony with Recommendation 1, governments should be encouraged to designate certain islands and their surrounding waters as sea turtle sanctuaries. This would be a relatively uncomplicated undertaking in the Phoenix group and Southern Line Islands where most of the islands are currently uninhabited and unused by man.

4. The illegal landings on uninhabited islands by foreign fishing vessels should be investigated by an international task force. Such violations constitute a worldwide problem and usually involve the theft and destruction of natural resources that include sea turtles.

5. A comprehensive appraisal should be undertaken of the 9-year-old hawksbill hatchery in Western Samoa. The results of this little-known conservation experiment may be of considerable value to worldwide efforts aimed at saving endangered hawksbill populations.

6. The predation on hatchlings by rats at Rose Atoll should be quantified and control methods implemented if the conditions warrant.

7. Military agencies of the U.S. government administering islands covered in this report should undertake a thorough investigation of the aggregations of sea turtles occurring at such sites. This would include Johnston Atoll, Kingman Reef, Wake, and Midway in the Northwestern Hawaiian Islands. Any plans for the storage of nuclear wastes and other highly toxic substances on Pacific islands should include a careful evaluation of the potential impact on both nesting and foraging aggregations of turtles.

8. The Hawaiian Islands National Wildlife Refuge should encompass a substantial amount of surrounding marine habitat to serve as a buffer against the forthcoming development of commercial fisheries.

Acknowledgments

The research involved in preparing this report was jointly funded by the State of Hawaii, Office of the Marine Affairs Coordinator, and the University of Hawaii Sea Grant College Program under Institution Grant Numbers 04–7–158–44129, 04–8–MOI–178 and NA79AA–D–0085 (NOAA, Office of Sea Grant, Department of Commerce). I would also like to gratefully acknowledge the support and assistance of the U.S. Fish and Wildlife Service, the National Marine Fisheries Service (Honolulu Laboratory), the Hawaii State Division of Fish and Game, the South Pacific Commission, the Fourteenth U.S. Coast Guard District, the Fifteenth Air Base Wing of the U.S. Air Force, and the New York Zoological Society. Appreciation is also extended to Fisheries Departments of the Cook Islands and Western Samoa for their hospitality and assistance during my visits in 1977. This is contribution number 578 of the Hawaii Institute of Marine Biology and conference paper number UNIHI-SEAGRANT-CP-80–05 of the University of Hawaii Sea Grant College Program.

From 11 to 14 December 1979 a workshop on sea turtles in the Pacific islands was to be jointly held by the South Pacific Commission and the U. S. National Marine Fisheries Service in Noumea, New Caledonia.

Literature Cited

Allen, R.
1975. Marine Parks: the Cinderella of conservation. *New Scientist* 14 August:366–69.

Amerson, A. B.
1971. The natural history of French Frigate Shoals, Northwestern Hawaiian Islands. *Atoll Research Bulletin* 150:1–303.

Anonymous
1972. Cook Islands Marine and Atoll Development Agency (CIMADA)—total atoll production system. *Aquaculture* 1:231.

1974. Fisheries Division annual report 3. Turtle Project. Government of Western Samoa, Apia.

1975. Fisheries Division annual report—Turtle Project.

GEORGE H. BALAZS

Government of Western Samoa, Apia.

1978. Report on the South Pacific Commission turtle project in the Cook Islands. Noumea, New Caledonia: South Pacific Commission.

Baird, R., editor

1975. SPC inshore special project: turtles. *The SPC Fisheries Newsletter* 13:13–14.

Balazs, G. H.

1975a. Green turtle's uncertain future. *Defenders* 50:521–23.

1975b. Marine turtles in the Phoenix Islands. *Atoll Research Bulletin* 184:1–7.

1976a. Green turtle migrations in the Hawaiian Archipelago. *Biological Conservation* 9:125–40.

1976b. *Hawaii's Seabirds, Turtles and Seals.* Honolulu: World Wide Distributors.

1977. South Pacific Commission Turtle Project—a constructive review and evaluation with recommendations for future action. Report prepared for the South Pacific Commission, Noumea, New Caledonia. University of Hawaii, Hawaii Institute of Marine Biology, Kaneohe.

1978. Terrestrial critical habitat for sea turtles under United States jurisdiction in the Pacific region. *'Elepaio* 39:37–41.

1979. Synopsis of biological data on the green turtle in the Hawaiian Islands. National Oceanic and Atmospheric Administration, National Marine Fisheries Service, contract number 79–ABA–02422.

Banner, A. C.

1971. Proposal and plan for a turtle ranch at Aleipata to determine the feasibility of developing an export industry for hawksbill turtles. Aleipata, Western Samoa.

Beaglehole, J. C., editor

1967. *The Journals of Captain James Cook: The Voyage of the Resolution and Discovery, 1776–1780.* Cambridge: Hakluyt Society

Brandon, D. J.

1977. Turtle farming: progress report on the South Pacific Commission turtle farming project in the Cook Islands. *Proceedings of the SPC Ninth Technical Meeting on Fisheries,* Working Paper 21:1–12.

Bryan, E. H., Jr.

1942. *American Polynesia and the Hawaiian Chain.* Honolulu: Tongg Publishing Company.

1974. *Panala'au Memoirs.* Honolulu: Bernice P. Bishop Museum.

Buck, P. H.

1932. Ethnology of Tongareva. *Bernice P. Bishop Museum Bulletin* 92:1–225.

Buggeln, R. G., and R. T. Tsuda

1966. A preliminary marine algal flora from selected habitats on Johnston Atoll. *University of Hawaii, Hawaii Institute of Marine Biology, Technical Report* 9:1–29.

Burnett, F.

1910. *Through Tropic Seas.* London: F. Griffiths.

Carr, A.

1965. The navigation of the green turtle. *Scientific American* 212:79–86.

Clapp, R. B., and F. C. Sibley

1971a. The vascular flora and terrestrial vertebrates of Vostok Island, South-Central Pacific. *Atoll Research Bulletin* 144:1–10.

1971b. Notes on the vascular flora and terrestrial vertebrates of Caroline Atoll, Southern Line Islands. *Atoll Research Bulletin* 145:1–18.

Coffman, D. M.

1977. An inventory of the wildlife and wildlife habitat of the islands of American Samoa. Report to the United States Fish and Wildlife Service, Honolulu. Environmental Consultants, Inc., vol. 1.

Coleman, R. A.

1978. Trip Report—Rose Atoll National Wildlife Refuge, 28–30 March 1978. United States Fish and Wildlife Service, Honolulu. Manuscript, 3 pp.

Dawson, E. Y.

1959. Some marine algae from Canton Atoll. *Atoll Research Bulletin* 65:1–5.

Dixon, W. S.

1884. Notes on the zoology of Caroline Island. In Report of the operations of the American expedition to observe the total eclipse 1883, May 6 at Caroline Island, South Pacific Ocean. *National Academy of Sciences,* Memoirs Number 2:90–92.

Dodd, C. K., Jr.

1978. Terrestrial critical habitat and marine turtles. *Bulletin Maryland Herpetological Society* 14:233–40.

Eliot, J. L.

1978. Hawaii's far-flung wildlife paradise. National Geographic 153:670–91.

Emory, K. P.

1933. Stone remains in the Society Islands. *Bernice P. Bishop Museum Bulletin* 116:1–179.

1947. Tuamotuan religious structures and ceremonies. *Bernice P. Bishop Museum Bulletin* 191:1–101.

Emory, K. P.; W. J. Bonk; and Y. H. Sinoto

1968. Fishhooks. *Bernice P. Bishop Museum Special Publication* 47:1–62.

Ernst, C. H., and R. W. Barbour

1972. *Turtles of the United States.* The University Press of Kentucky.

Gill, W.

1876. Life in the Southern Isles. London: Religious Tract Society.

Graeffe, E.

1873. Samoa oder die schifferinseln. I. Topographie von Samoa. *Jour. Mus. Godeffroy* 1:1–32.

Hedley, C.

1896. General account of the atoll Funafuti. *Australia Museum Memoirs,* no. 3.

Hendrickson, J. R.

1969. Report on Hawaiian marine turtle populations. *IUCN Publications New* Series 20:89–95.

Hirth, H. F.

1971a. South Pacific islands—marine turtle resources. A report prepared for the Fisheries Development Agency Project. FAO, Rome.

1971b. Synopsis of biological data on the green turtle *Chelonia mydas* (Linnaeus) 1758. *FAO Fisheries Synopsis* 85:1:1–8:19.

Inder, S., editor

1978. *Pacific Islands Year Book*. Sydney: Pacific Publications.

Johannes, R. E.
1978. Traditional marine conservation methods in Oceania and their demise. *Annual Review of Ecology and Systematics* 9:349–64.

Kalakaua, D., King
1888. *The Legends and Myths of Hawaii*. New York: C. L. Webster and Company.

King, W. B.
1973. Conservation status of birds of Central Pacific islands. *The Wilson Bulletin* 85:89–103.

Lipman, V.
1978. Hawaii's endangered wildlife: what chances for survival? *Honolulu Magazine* 13:47–62.

Maragos, J. E.; G. B. K. Baines; and P. J. Beveridge
1973. Tropical cyclone Bebe creates a new land formation on Funafuti Atoll. *Science* 181:1161–64.

Mayor, A. G.
1921. Rose Atoll, Samoa. *Science* 54:390.

Powell, R.
1957. Breeding turtles for profit. *South Pacific Commission Quarterly Bulletin* 7:41–42.

Pukui, M. K., and S. H. Elbert
1971. *Hawaiian Dictionary*. Honolulu: The University Press of Hawaii.

Raj, V.
1975. U.S.P.-based turtle research: progress report 1974–1975 breeding season. Proceedings of the SPC Eighth Technical Meeting on Fisheries, Noumea, New Caledonia, Working Paper 23:1–8.

Rockefeller, M., and L. S. Rockefeller
1974. Problems in paradise. *National Geographic* 146:782–93.

Shadbolt, M.
1967. New Zealand's Cook Islands: paradise in search of a future. *National Geographic* 132:203–31.

Tresilian, F.
1838. Remarks on Christmas Island. *Hawaiian Spectator* 1:245–47.

Wetmore, A.
1925. Bird life among lava rock and coral sand. *National Geographic* 48:77–108.

Wiens, H. J.
1962. Atoll Environment and Ecology. New Haven: Yale University Press.

Wilkes, C.
1845. Narrative of the United States Exploring Expedition, 1838–1842, vol. 5. Philadelphia: Lea C. Blanchard.

Williams, J.
1837. A Narrative of Missionary Enterprises in the South Sea Islands, with Remarks upon the Natural History of the Islands, Origin, Languages, Traditions and Usages of the Inhabitants. New York: D. Appleton and Company.

Witzell, W. N.
1972a. The hawksbill turtle (*Eretmochelys imbricata squamata*) in western Samoa. Proceedings of the SPC Fifth Technical Meeting on Fisheries, Noumea, New Caledonia, Working Paper 3:1–22.
1972b. To live or not to live. *International Turtle and Tortoise Society Journal* 6:32–35.
1974. The conservation of the hawksbill turtle in Western Samoa. *South Pacific Bulletin, First Quarter*:33–36.

GEORGE H. BALAZS

Peter C. H. Pritchard
Florida Audubon Society
1101 Audubon Way
Maitland, Florida 32751

Marine Turtles of the South Pacific

The turtles of the Pacific Ocean have received extensive coverage in many fine presentations at this symposium. This paper will discuss the turtles of those islands and territories not specifically covered by other contributors. The area thus included comprises part of Melanesia (New Caledonia, the Solomons, New Hebrides, and Fiji), and southern Polynesia (New Zealand, the Kermadec Islands, Tonga, French Polynesia, Pitcairn and Henderson, and Easter Island).

New Caledonia

It has long been known that green turtles (*Chelonia mydas*) abounded in the waters and islands of the d'Entrecasteaux Reefs, north of New Caledonia. For example, William Billings, master of a sailing vessel that ran aground on the reefs in September 1856, recorded the presence of many turtles, one of which reportedly weighed 600 pounds (273 kg). The nesting season was supposed to take place from July to December. Eight turtles caught were reported to have an *average* weight of 5 cwt (254 kg). The turtles apparently basked both on land and in water, as do Hawaiian green turtles; Billings reported that "the Master turned twenty-seven one morning without wetting his feet, and he counted eighteen more asleep in six inches [15 cm] of water. . ." (Billings, in Chimmo, 1856).

Although the islands in the d'Entrecasteaux Reef system are visited at least annually by weather station personnel, no literature seems to have been published recently on the turtles there. However, in December 1979 I was part of an aerial survey group that overflew the reef at low altitude, and we confirmed that the 4 islands large enough to have scrub or tree cover in their interior (Isles Surprise, Le Leixour, Fabre, and Huon) were still attracting large numbers of turtles; the beaches and dunes of these islands gave the appearance of being nested to capacity. No turtles were seen on the land at the time of our survey, but many were seen in the adjacent shallow waters, and they appeared to be green turtles.

On 10–11 February 1980 the French Navy ship *La Dunkerquoise* visited Huon Island, which was then estimated to be 3 km in length (including the elongate sandbanks), a maximum of 200 m wide, and with only 12 of the 200 ha surface covered with vegetation (a single species of creeper). Turtle nesting density was impressive; 20 ha were used for nesting, of which a single 50 by 50-m sample contained approximately 140 nest pits, about 25 percent of which were 48 hours old or less. Evidence of much hatchling emergence was visible, and about 15 dead and decaying adult turtles were seen on land. The density of tracks of adult turtles along the shoreline averaged about 25 distinct departures every 100 m (only counting tracks estimated to be 24 to 48 hours old). Clutch sizes, reported as 60 ± 20, were surprisingly small.

Calculations based on these figures suggest $25 \times 60 = 1,500$ emergences over the entire 6-km perimeter of the island over a 2-night period. The numbers of nests in this same period calculated from the area mentioned above would be $35 \times 4 \times 20 = 2,800$. Actual observations during a 4-hour period 1 night revealed only about 50 nesting emergences. It therefore appears likely that the nests and tracks had accumulated over more than 24 to 48 hours; but still the island, together with its 3 neighbors in the d'Entrecasteaux Reef system, constitute the most important turtle nesting area among all the oceanic islands of the Pacific.

The reported average size of the turtles was very large—carapace length approximately 1.4 m, width approximately 0.9 m. These dimensions are entirely compatible with Billings' report of an average weight of 5 cwt, but even so this size is similar to that of the record size known for a green turtle, and is greatly in excess of the average size of any known population. It will be interesting to obtain confirmation and to salvage the dead specimens on the island.

The hawksbill is well-known in New Caledonia, and is probably ubiquitous on the reefs that virtually surround the island. On the Isle des Pins, at the southern end of New Caledonia, we were informed by a local fisherman that the loggerhead, locally known as *grosse tête* (big head), was in fact the commonest species; the presence of this species was confirmed by a photograph of an adult nesting and a young specimen, about 30 cm in length, in the Noumea aquarium.

A fourth species, that would appear to be the olive ridley, was familiar to some informants; it was reported to be rare. No one seemed to be familiar with the leatherback; this species avoids coral reefs, and would not be expected anywhere in New Caledonia except as an accident.

In the Loyalty Islands, a dependency of New Caledonia, our aerial survey in December 1979 revealed significant turtle nesting activity on the smaller islands in the northern and northwestern part of the Uvea atoll complex. By far the best nesting was on the island of Beautemps-Beaupré, where the more northern beach was literally covered with turtle tracks: this island is the only emergent point of a separate atoll system northwest of Uvea. It was not possible to determine the species responsible for the nests.

Turtle nesting in New Caledonia is concentrated in the summer months (November to March).[1] The sea turtles in New Caledonia receive considerable protection under the provisions of Ordinance No. 220 of 3 August 1977, adopted by the New Caledonia Territorial Assembly. This ordinance prohibits any collection of turtle eggs, or any marketing of turtle meat, mounted turtles, or shells. All species are totally protected during the nesting season (November to March, inclusive). Application may be made for special permission to take turtles during these times for scientific purposes or for Melanesian traditional feasts and ceremonies. Between 3 August 1977 and 22 August 1979, permits were issued for the taking of 77 turtles for traditional purposes during the closed season.

On Lifou Island, we were informed that only the chief can eat turtle meat, although he can authorize certain clans to eat the meat approximately once a year. The turtles were reported to be very abundant and virtually unhunted. Both the hawksbill and green turtle were present in the lagoon of Uvea. Melanesian traditional law is still respected, and turtles are essentially private property of certain clans, which provides for considerable protection.

Solomon Islands

A recent report by McKeown (1977) gives a great deal of valuable data on turtle nesting in the Solomon Islands.

The green turtle is thought to be the commonest species in the Solomons, although aggregated nesting does not seem to take place there. They are most commonly seen in the Roviana-Marovo Lagoon System; at Ontong Java Atoll; and in the Reef Islands, notably Nupani Atoll. Sporadic breeding has been reported from October to February at Kerehikapa; Oroa Island, Makira; Lilika Bay, Santa Ysabel; and the Hele Bar Islands, New Georgia. As with green turtles in the Bismarck Islands, the green turtles in the Solomons are apparently never colonized by barnacles.

Carapace lengths of 4 adult females ranged from 78 to 89 cm; weights from 84 to 95 kgs (means 85 cm and 89.78 kg). Mean clutch size for 5 clutches was 84.6 but the range was great—45 to 156.

Green turtles in the Solomons are subject to rather intensive pressure by local people, who catch them principally for special feasts. McKeown reports 22 caught

---

1. But Billings (op.cit.) refers to a somewhat earlier nesting season.

PETER C. H. PRITCHARD

in Baola in November 1976 for a feast, and 39 in Samasodu in July 1977; formerly, according to the Samasodu people, 100 was not uncommon for a day's catch. The turtles are caught by tangle nets, spearing, diving and holding while the turtles are sleeping, shooting with bows and arrows, with nets baited with pawpaw, on the nesting beaches, and by hook and line, using pawpaw as bait.

The hawksbill is the commonest nesting species in the Solomons, but although still ubiquitous, populations have declined considerably; even religious groups (such as the Seventh Day Adventists) who do not eat turtles will kill hawksbills for their shells, and considerable volumes are exported from the Solomons. Numerical data on the volume of this trade are scarce, but exploitation is clearly excessive; despite rising prices and greater hunting effort, traders found that the quantities of shell offered them decreased steadily. In July 1977, the price of turtle shell on Wagina Island, a center for the industry, was $6.50/lb (about $14.30/kg).

Hawksbills are known to nest on many beaches in the Solomons, but the most important is Kerehikapa, a horseshoe-shaped island about 4-km long in the Arnavon Group. Straight carapace lengths of nesting hawksbills there varied from 68 to 93 cm, although 1 female caught while copulating measured only 61 cm. Weights ranged from 42 to 77 kg; average values were 80.5 cm and 66 kg (N = 40). Nesting is almost always nocturnal; only 1 diurnal nesting was witnessed, although many local people claimed that this habit was common in former times, as it still is on many of the Torres Strait Islands. Some cases of renesting have been reported, though surprisingly few in view of the extent of beach coverage; of 91 individuals tagged at Kerehikapa, 4 were seen to come back to lay another clutch, at intervals of 13, 15, 17, and 28 days. Sixty-six hawksbill nests were unobserved during this period, so there may in fact have been more renesting than this.

The mean incubation period for 164 nests was 64 days. Clutch size varies from 75 to 250 eggs, the latter being apparently the largest clutch on record for any turtle—or indeed reptile—species: mean was 137.53. Nesting is year-round at Kerehikapa, though there may be an indistinct peak in June–July.

The leatherback turtle nests on numerous isolated beaches in the Solomons, preferring those that are located near river mouths, having a reefless approach, and being composed of black sand. Nesting takes place in the November–January season. The taking of this species is illegal (Regulation 17 of the 1972 Fisheries Regulations), but nevertheless, nesting animals are frequently butchered, and their nests raided.

The loggerhead is very rare in the Solomons, but it is recognized by villagers in many areas, and the iden-tification has now been confirmed by McKeown, who had the opportunity to examine a skull and a live specimen. Nesting in the Solomons by this species is unknown.

The olive ridley is known only from 3 individuals from the Solomons, 2 of which were a mating pair caught off northern Guadalcanal near Honiara in February 1976. The third individual was a juvenile found at Makira in July 1977.

A hawksbill turtle tagged while nesting on Santa Ysabel in December 1976 was caught on the reef at Fishermen's Island, Papua New Guinea, in March 1979 (P. Vaughan, personal communication). This journey of at least 1,800 km represents the longest reported movement of a hawksbill turtle.

Additional information on sea turtles in the Solomon Islands, including proposed population models for both the green turtle and the hawksbill, is given by McElroy and Alexander (1979).

New Hebrides

The turtles of the New Hebrides are in need of study. However, there is evidence that about 200 turtles were caught at Mota Lava, near the northern end of the New Hebrides system, in the first 6 months of 1979, and Reef Island nearby is an important green turtle nesting ground. The following information on sea turtles in the New Hebrides is taken from the report by McElroy and Alexander (1979).

The species named as nesting in the group are the green and hawksbill turtles. Information on the leathery turtle indicated that it occurs in some parts of the group but no nesting beaches were known. The green and hawksbill turtles are common in the extensive reefs and shallow areas of the group.

The most important nesting area in the group is at south Malekula Island. Important mainland nesting of greens occurs at South West Bay, and particularly Lambobe beach. Small numbers of hawksbills also nest here. A rough estimate of the numbers nesting each year is from 40 to 120 turtles. The Maskelynes form a group of offshore islands off the southern coast of Malekula where turtles are particularly plentiful. Regular nesting of both species also occurs within the group, particularly at Seior and Laifond islands. Sakau and 2 small islands close to Akam are used occasionally. Purposeful fishing at sea accounts for most turtles captured. The people of the Maskelynes group are reported to be the best fishermen in the New Hebrides. Estimates indicate a yearly catch here of 60 to 120 turtles, about evenly spread between both species. Other notable areas for nesting turtles were southeast of Epi, and in the north amongst the Torres group. Whenever found, nesting females and eggs are usually taken.

The nesting season for both species extends from

September to early January. The general picture gained indicated that nesting on main (populated) islands was still common. Hunting and fishing pressure was extremely localized and never intense. The lack of strong fishing pressure coupled with a small population and minimal developments along the coasts of the group, indicates that the normal nesting pattern of these 2 species in the group is virtually unchanged.

Though there is negligible trade in tortoise shell from the New Hebrides, in recent years turtles and turtle eggs have been appearing more frequently in the markets of Santo and Vila. Legislation has been passed (Joint Rules No. 7 of 1974) to prohibit the taking or selling of nesting turtles or their eggs. Thus islanders who catch turtles at sea can still use these for food.

These regulations were found to be poorly understood and little known despite a recent article in the New Hebrides News. However, the situation still seems to be good. The major long-term threat to turtles in the group is increased human predation brought about by the continuing population explosion.

Fiji

Turtles of Fiji are currently being studied by M. Guinea of the University of the South Pacific at Suva; but until these results are published, the principal source of information is Bustard (1970). The 2 most plentiful species in Fiji, as elsewhere in the South Pacific, are the hawksbill and the green turtle. The leatherback is also reported to nest in Fiji, though very rarely (Bustard 1970). A very low level of nesting of this species takes place on the southeastern coast of Vanua Levu (M. Guinea, personal communication). This is the most easterly record of nesting of *Dermochelys* in the Pacific Islands. Both Bustard and Guinea report that either the loggerhead or the olive ridley, or both, are occasionally found in Fiji, and Hirth (1971) saw a shell of the loggerhead at Malake Village, Fiji.

Some of the salient points regarding sea turtles in Fiji reported by Bustard (1970) are as follows:

1. Turtles feed in the passage between Taveuni and Vanua Levu but rarely nest in the area.

2. No significant nesting occurs on Qamea.

3. Hawksbill nesting on Matagi is now reduced to about 1 nesting per week.

4. Some nesting takes place on Nanuku Levu, mostly by greens, but numbers are now very depleted.

5. Reasonable levels of hawksbill nesting were observed on Nanuku Lailai.

In December 1979 I found carapaces of half-grown hawksbills·offered for sale in some numbers in the market in Sigatoka, on the south coast of Viti Levu.

In theory, sea turtles in Fiji are well protected by the Fisheries Act. Turtle eggs enjoy total protection; turtles less than 18 inches (46 cm) in length are also protected; and there is a completely closed season during the nesting months of November to February. Fijian law also requires that "no person shall harpoon any turtle unless the harpoon is armed with at least one barb of which the point projects not less than 3/8 inch [0.95 cm] from the surface of the shaft, measured at right angles to the long axis of the shaft." Prosecutions are sometimes made under the turtle laws, but basically violations are frequent and not even clandestine. Moreover, feral mongooses eat the eggs, or the baby turtles before they reach the sea.

Some of the salient points in the report of Hirth (1971) on marine turtles in Fiji are as follows:

1. Species known: Green (Vonu Damu, Vonu Loa, Mako Loa, Ika Damu); Hawksbill (Taku). Loggerheads and leatherbacks are present but extremely rare.

2. Hawksbills are fairly common in north loop of Great Astrolabe Reef in Kadavu, and nest there, the season peaking in January. Some green and hawksbill turtles nest in the southern Lau group. The largest nesting sites are likely to be the Yasawa Islands, the Mananutha Islands, and the Lau Islands.

3. Some kraals of green and hawksbill turtles are maintained for the benefit of tourists. In one such kraal on Vinivandra Island, hawksbills grew to 15 to 17 cm in 2 years.

4. Green turtles are caught in feeding pastures off north-central coast of Viti Levu; 80 were caught in October. The turtles subsist largely on red and green algae.

5. Many hawksbill curios are sold in Suva shops and in the market. Demand increases as tourists increase. Green turtle meat is erratically available in the market and in first class hotels.

6. Export of raw tortoise shell was proscribed as of September 10, 1969 to encourage local manufacture of curios and other products. In 1969, 137 kg of turtle shell were exported (88 kg of hawksbill shell); in 1968, 270 kg (78 kg hawksbill).

New Zealand

The leatherback turtle is fairly regularly encountered in New Zealand waters. McCann (1966a) reported this species to be seen most often in northern waters, but that it had been recorded as far south as Otago and Foveaux Strait. Subsequently, McCann (1966b) listed the following localities for leatherbacks in New Zealand: Bay of Islands, passing Cape Brett (1892); between Bay of Islands and Mangonui (1894); New

PETER C. H. PRITCHARD

Plymouth (1895); Palmer Head (ca 1930); Doubtless Bay (1924); Mayor Island (1939); off Cape Brett (1939); Pukerua Bay (1948); Waitarere Beach (1954).

Other records cited by Eggleston (1971) include a sight record for Foveaux Strait, off Saddle Point, on 20 February 1970; this is the southernmost known record for the species. Also, a specimen was sighted off Cape Palliser on 12 December 1969; and a specimen washed ashore at Gillespie's Beach, South Westland, in 1969.

Some more recent records that came to my attention on a visit to the National Museum (formerly the Dominion Museum) in December 1979 were as follows: Wairarapa Coast (26 May 1976; specimen mounted for National Museum); Poison Bay (15 February 1972); Three Stones Bay, 24 km north of Kennedy Bay (5 February 1970); Mangonui, north of Kaitata 46 m depth, 2.4 km offshore (2 April 1973) drowned by trawler; near mouth of Conway River (1 April 1970); near Raglan (23 January 1973); Ariel Reef, 11 km off Gisborne Coast (24 February 1975); Coromandel Coast (April 1971).

Fordyce and Clark (1977) gave the following records: Jacksons Bay, Westland, 4 October 1975; Great Barrier Island, Northland, 4 April 1976; Kaikoura, 3 specimens fouled in crayfish pot lines, March-April 1977; and Adderley Head, Banks Peninsula, 27 March 1977.

The green turtle is very rare in New Zealand waters. Specimens reported by McCann (1966b) were as follows: Kawau Islands, Hauraki Gulf (carapace 872 by 685 mm); Hauraki Gulf (1911); Waikato River, Taupiri (1936); Great Exhibition Bay, south of Parengarena Harbour (1895) "nearly three feet long"; and Ninety-mile Beach (January 1949, carapace 775 by 525 mm).

In addition to these records, the *Nelson Evening News* of 23 March 1971 illustrates an immature green turtle caught in 59 m in Tasman Bay by a trawler; and the *Christchurch Press* of 3 January 1948 reported a 36 kg green caught on a set line at Whangaparaoa, near Auckland.

The hawksbill is even rarer than the green turtle in New Zealand waters. McCann (1966b) reports on 3 specimens: an adult (90 cm in carapace length) taken at Muriwai Beach on 21 June 1949; a juvenile taken at the same location on 29 August 1956; and a specimen of unrecorded size taken at Ninety-mile Beach, about 18 km north of Waipapakauri, on August 15, 1956. The 1956 Muriwai Beach specimen contained fragments of barnacles, cephalopods, *Vellella, Salpa,* and pteropods.

Robb (1973) reported that a live hawksbill was found on Piha Beach, near Auckland, in August 1970.

A live turtle found at Uretiti Beach, 25 cm in length, was identified as an *Eretmochelys* in an article in the *Northern Advocate* of 23 July 1973, but the identification was corrected to *Caretta* in the 9 August 1973 issue of the same newspaper.

The available literature confuses the loggerhead and the olive ridley in New Zealand waters, but the photographs in McCann (1966b) make it clear that both species occur, although *Caretta* is probably much more common. The specimen 610 by 560 mm in size, 25 kg in weight, with enlarged ova present, and with 7 costal scutes on one side and 8 on the other, was clearly a *Lepidochelys olivacea.* On the other hand, the illustrated specimen caught at Whenupai in 1956, which exhibited 5 pairs of costal scutes and a carapace measuring 500 by 446 mm, was clearly a *Caretta caretta.*

Other *Caretta* in New Zealand include the following: a 33 cm immature washed up at Flat Point, 64 km from Masterton, in August 1966, reported in the *Wairarapa Times-Age* (29 August 66); a 72 kg specimen caught 3 km off the Wairarapa Coast in March 1973, and released the next day at Castlepoint Beach (*Wairarapa Times-Age,* 2 March 1973); and a juvenile netted by a Greymouth fishing boat in January 1975 (*Greymouth Evening Star,* 10 January 1975).

The possibility exists that *Caretta* may occasionally nest on the extreme northern beaches of New Zealand. Very small specimens, 8 to 10 cm long, are found on occasion, for example, 3 specimens taken at Ninety-mile Beach, 1 in 1949 and 2 in 1952, reported by McCann (1966b); these specimens ranged from 86 to 97 mm in length. Another specimen in this size range was caught on East Beach in July 1973 (*Northland Age,* 7 September 1973).

McCann (1966b) observed correctly that, although these specimens were indeed very small, they were probably around 6-months old rather than hatchlings; and the fact that they were caught in late winter would correlate with their having hatched about 6-months before. It might be concluded that the specimens had hatched on beaches in Queensland, Australia, and had reached New Zealand by passive drifting during the ensuing months. Nevertheless, an article in the *Northland Age* (27 July 1973) concludes that "marine turtles must occasionally breed on the coast of the Far North (of New Zealand)," based on the evidence not only of the very small loggerheads occasionally found in the region, but also on reports of "strange tracks like those made by turtles" in Spirits Bay and Ninety-mile Beach; and a report by Mrs. Peter Nilsson that turtle tracks had definitely been seen on several occasions in late summer leading from the sea to the sandhills of Tom Bowling Bay, adjacent to Spirits Bay at the extreme northern tip of New Zealand.

Kermadec Islands

Oliver (1910) reports that green turtles may be found feeding in substantial numbers between January and

March near Sunday Island, in the Kermadec Islands, north of New Zealand at a latitude of about 30°S, although no breeding occurs there. A 1908 photograph in the Archives of the National Museum in Wellington, New Zealand, shows an immature green turtle from the Kermadec Islands. Parsons (1962) postulated that these turtles might be derived from the colony that nested on Vatoa Island, the southernmost of the Fiji Group where Captain Cook found numerous nesting turtles. However, I am informed that there is no longer a nesting colony of green turtles on Vatoa, and in any case, it is dangerous to assume that a feeding population of green turtles is necessarily derived from the geographically closest nesting population.

Tonga

The only available information on the sea turtles of Tonga is unpublished; it appears in the reports of H. F. Hirth (1971) and Wilkinson (1979). Since this information is not generally available, some quotes are presented below.

According to Hirth, 3 species of marine turtles are found in the Tonga Islands: the green turtle (local names: Fonu, Fonu Tu'a'uli, Fonu Tu'akula, Fony Tu'apolata, Tuai Fonu); hawksbill (bonu Kolaa), and loggerhead (Tu Fonu). The vernacular names vary from village to village. Some fishermen have additional names for the color phases and various sizes and sex of each species. The green turtle definitely nests on islands in the Ha'apai group, chiefly on the uninhabited ones. The consultant was unable to determine whether the other 2 species nest in the Kingdom.

Interviews all lead to the conclusion that most marine turtles are found in the Ha'apai group and that the period of nesting is in the summer (November to February) with December being the peak month for oviposition.

Turtles, especially green turtles, can be caught throughout the year in the Ha'apai group as well as in the Tongatapu group.

Tongans eat turtle meat and eggs. Eggs are eaten by the inhabitants of the nesting beaches and no eggs are found in markets in Tongatapu. Turtle meat itself is rarely sold in the market in Nuku'alofa because turtles are usually butchered and sold on the beaches as soon as they are landed.

Green turtles (some loggerheads and rarely hawksbills) are caught on the feeding pastures off Tongatapu Island. One of the best pastures is north of Nuku'alofa and east of the islands of Paloa, Alakipeau, Tufata, and Atata. The most common method of capture is by spear gun. Occasionally, nets are set and only very seldom are special turtle fences bulit. The latter method, however, was common a couple of decades ago (see current turtle regulations). Fences are seldom used now be-cause turtles are rarely seen on the feeding pastures close to the shore. According to the best turtle divers, females are more common than males on the grass flats. Turtles can be caught all year on the pastures, and some turtle fishermen claim that the best months are from November to March when the "grass" is especially lush. Furthermore, again according to the local fishermen, the grass goes through cycles when it is very lush and then sparse. Divers also state that green turtles smaller than "plate-size" (about 25 cm in carapace length) are not seen on the feeding pastures.

Three men, working from a small boat and diving all day for turtles on the Nuku'alofa pastures will call it a "very good day" if they manage to get 2 or 3 green turtles. The most common method of catching turtles in the Ha'apai group is with nets. These usually have a 12-inch mesh. A few fishermen chase turtles with motorboats and after the reptile becomes tired or has been chased into shallow water, a man jumps overboard and catches the turtle by hand.

Turtle grass pastures (*Syringodium isoetifolium* and *Halodule uninervis*) encompassing at least a square mile, are found off the coast of Nukunukumotu. These spermatophyte pastures, along with those in Fiji, are the best the consultant has seen on the survey, although it should be emphasized that he did not make extensive surveys for pastures throughout the South Pacific.

According to some fishermen, a few turtles (the species are not known) still occur in the lagoon but they are not so abundant as in former days. The consultant could not determine the extent of turtle fishing in the lagoon at the time of writing these comments, but he is of the opinion that turtling here would be very unprofitable. In some areas the water is extremely polluted.

Fishermen on Pangaimotu Island used to keep adult turtles in a kraal feeding them with turtle grass (*Syringodium* and *Halodule*) which washed up on the beach. They declared that the turtles ate this grass readily, and the consultant saw large amounts of turtle grass of both species washed up on shore in October.

The consultant examined the stomachs of 2 green turtles caught during the week of 19 October on the feeding pastures in the main channel off Nuku'alofa and both were full of *Syringodium*. One stomach also contained a few pieces of *Halophila ovalis*. The Tongan name for turtle grass, as well as for algae, is *limu*.

Since green turtles are caught regularly on the feeding grounds off Tongatapu (but do not nest on Tongatapu) and since green turtles are known to nest in the Ha'apai group—it may be that there is a nesting-feeding migration between the 2 groups. The consultant dissected 2 adult green turtles (caught on the feeding pasture off Nuku'alofa during the week of 19 October) and both contained hundreds of small developing eggs which appeared to be about 1 month away from

PETER C. H. PRITCHARD

Table 1. Tongan names of marine turtles and their English and scientific equivalents

Tongan name	Equivalents
Fonu Koloa	Hawksbill (*Eretmochelys imbricata*)
Tuangange	Probably olive ridley (*Lepidochelys olivacea*), but may possibly be the Indo-Pacific loggerhead (*Caretta caretta gigas*).
Ika ta'one Hulemui	Males of the green turtle; different size, color. (*Chelonia mydas*)
Tu'a polata Tu'a 'uli Tongotongo Aleifua Tufonu	Females of the green turtle (*Chelonia mydas*). Different names due to size variation, age variation as well as color variation. Possibly the black turtle (*Chelonia agassizi*) is included here.

being shelled. Both turtles also possessed large amounts of fat which might indicate a premigratory condition.

The information from Wilkinson (1979) is as follows.

The hawksbill nests on the following islands: Kelefesia, Tonumea, Telekitonga, Lalona, Telekivava'u, Lalona, Telekivava'u, Fetokopunga, Nukufaiva, Meama (near Fonoifua), Fonuaika, Tokulu, Nukulai, Luanamu, Kito, Fetoa, Putuputua, Limu, Uonukuhahake, Tofanga, Uonukuhihifo, Luangahu, Hakauata, Tatafa, Luaheka, Nukutula, Nukupule, Meama (near Nukupule), Niniva (uninhabited), and Nukufaiau. One additional uninhabited island which is a probable hawksbill nesting site is Lekeleka.

The turtles are known to nest on 3 inhabited islands: Mango; 'Uiha, Liku side, and Ha'ano, Muitoa.

The Tongan names of the different marine turtles in Ha'apai and the English and scientific names which fit descriptions of the Tongan names are presented in Table 1.

Of the 10 Tongan names for marine turtles, 8 of these describe *Chelonia mydas* (the green turtle) with possibly 1 of the names describing *Chelonia agassizi* (black turtle). This is reasonable, however, since *Chelonia mydas* is really a combined species name for a number of yet unnamed races of the green turtle. There are small differences in color, head, flipper size, and overall size between these unnamed races.

Fishermen from the Ha'apai islands were asked whether they had seen (and not just heard it from others) any of the other turtles nesting besides the fonu koloa (hawksbill). The *tu'a 'uli*, one of the names describing the female green turtle (*Chelonia mydas*) has been seen nesting by fishermen from Tungua, 'O'ua and Ha'afeva islands (also the *aleifua*, the green turtle). Islands where the green turtle has been seen nesting include 2 uninhabited and 1 inhabited island: Nukufaiva, Fetoa, and Mango Island.

The most common and least common turtle names were asked of the fishermen. Almost consistently they mentioned the tu'a 'uli as being the most commonly caught, which is what we have observed as well. Many said the *fonu koloa* (hawksbill) was not the least common turtle in Ha' apai, so apparently the population is not down to its most critical level as yet. *Tuangange* (ridley) and *tu'a kula* (*Chelonia* sp.) were often mentioned as being the least common and maybe the black turtle. This is what was expected since no ridleys have been seen, nor their shells at the market or on the boats.

According to fishermen, nesting begins very sparsely in October, increases in November and reaches a peak in December-January, slackens off quickly, and nesting probably ceases sometime after the middle of January.

In the Vava'u group the composition of the turtle population is the same as identified in the Ha'apai group: tungange, most likely the Pacific ridley *Lepidochelys olivacea*; aleifua, probably the green turtle *Chelonia mydas*; fonu koloa, the hawksbill *Eretmochelys imbricata*; tofunu, possibly *Chelonia agassizi*. In discussions with fishermen, it was found that the main egg-laying season is from November to January, although the gathering of turtles on the sea, off the nesting sites, begins as early as October.

The main nesting islands are in the southwestern area. They are Fonua'one'one, Fangasito, Folifuka, Foeata, and Maninita. These islands are relatively accessible and all should be declared seasonal breeding sanctuaries. Fishermen admit the islands are visited by people and that eggs are taken. In the long term, this practice could be disastrous for the turtle population in Tonga. Enforcement of sanctuary regulations should not be difficult if a fisheries station and regulatory staff were stationed at Vava'u.

Turtle nesting is reputed to occur on Malinoa Island off Tongatapu. Nowhere else, on or around Tongatapu, do turtle nesting areas presently exist, mainly due to human predation and interference.

French Polynesia

French Polynesia includes several far-flung archipelagoes, including the Society Islands (Isles de la Société), the Marquesas (Iles Marquises), the Tuamotu Islands (Archipel des Tuamotu-Gambier), the Austral Islands (Iles Austral), and Rapa. The most commonly encountered turtle in French Polynesia is the green; the hawksbill is reported to be almost extinct, and other species are unreported.

Society Islands

Green turtles are reasonably plentiful in the Society Islands. The principal nesting islands, as in Micronesia, are uninhabited or only seasonally inhabited by man. One such island is Mopelia, locally renowned for turtle nesting. According to Eggleston (1953) and Legand (1950), the peak of the nesting season is around November. Nevertheless, by continental standards the nesting colony is small; it is considered a good week in which a dozen turtles nest. Parsons comments on some locally initiated conservation measures that have taken place, including a program of protection of the young turtles and their eggs. Regulations allow the local people to "head-start" the turtles in corrals in the lagoon, and to send a turtle to the Papeete Market for each one they release.

Hirth (1971) gives the following information for turtles in the Society Islands: The most common sea turtle in the area is the green turtle (French: Tortue; Tahitian: Honu). One of the principal nesting grounds is Scilly. Other important nesting sites are Mopelia, Bellinghausen, Tupai, and some of the Tuamotu atolls. The peak nesting season in Scilly, Mopelia, Bellinghausen, and Tupai is October through December. Reports indicate that some turtles can be found throughout the year off Scilly. The hawksbill turtle is sometimes taken by fishermen. There is 1 authentic record of a leatherback caught in a seine. On 24 September, the consultant counted 20 green turtles in the water around Mopelia (but there were no tracks on the beach), and 42 around Scilly, including 12 in a village kraal. He also noticed fresh tracks and nests on Motu Honu (Islet of Scilly).

Many males and some gravid females are speared as they mate off the nesting beaches on the atolls. Tahitian fishermen report a sex ratio in favor of males. The turtles sold in Papeete market in September were mostly males. Green turtle meat is considered a delicacy and sells for about $3 per kg in the Papeete market but there is no market for eggs. A few cured shells are sold in tourist shops at $25 each but the demand is

insignificant.

There are no regulations in French Polynesia concerning marine turtles.

Stomachs examined by the consultant were chiefly empty but a few contained a little green algae and one harbored a long piece of plastic. During his limited survey, the consultant did not find any extensive algae beds or grass flats.

Fisheries Department records indicate that between 1953 and 1967 from 24 (1954) to 262 (1962) turtles caught at Scilly were sold annually in the Papeete market.

Further information on green turtles in the Society Islands was provided by Anon (1979). This paper reports that the principal nesting island, Scilly, was declared a "protected area" on 28 July 1971, with a family appointed to watch over the nesting turtles. Nesting takes place primarily from September to December, but with significant year-round nesting. There is significant predation of hatchlings by frigate birds (by day) and hermit crabs (by night).

The Scilly green turtle breeding colony was studied and tagged intensively in 1972 and 1973 and, after several years' hiatus, operations were resumed in 1979. 364 female turtles were tagged and measured in 1972, and 42 more in 1979. The population has dwindled considerably in recent decades; only 20 to 30 years ago, it was reported that 100 to 150 turtles could be turned in a single night. The fact that such numbers not only could be, but were, turned resulted in a decline to the point where today about 20 nest on a typical night on the islet of Motu Rahi, 5 to 6 on Motu Honu, and 8 or 9 throughout the rest of the atoll. It was also reported that, with the decline in numbers, average size of the turtles had declined, carapace lengths now typically lying between 93 and 97 cm, with the maximum 106 cm. Maximum weight was now 175 kg although a few years earlier, turtles weighing over 200 kg were supposed to have existed.

These weights, although not so mentioned in the report, are unusually high for Pacific green turtles, especially when remembering that the green turtles nesting on the mainland Pacific coast of Mexico at Colola, Michoacan, average only 77.32 cm and 57.36 kg (females) and 72.68 cm and 43.19 kg (males) (Cliffton, unpublished data). The rather large size may well correlate with the extensive transoceanic migrations of this population. In the Atlantic, the transoceanic migratory green turtles of the Brazil-Ascension Colony are among the largest known anywhere in the world. Similarly, some extensive migrations to points hundreds or thousands of kilometers to the west have recently been reported for the Scilly green turtles. These recaptures of tagged turtles (Anonymous 1979), are summarized in Table 2. It should be noted that the turtles were kept in captivity for up to 4 months before re-

PETER C. H. PRITCHARD

Table 2. Recaptures of green turtles (*Chelonia mydas*) tagged on Scilly Island, Society Islands, French Polynesia

Number	Sex	Carapace length	Tagging date	Recapture date	Location of recapture
18	F	101 cm	30 Apr. 1972	9 Aug. 1972	Vavau Is., Tonga
26	F	102 cm	30 Apr. 1972	26 Jul. 1972	Rabi, Fiji
39	F	93 cm	30 Apr. 1972	14 Sept. 1973	Maskeline Is. (New Hebrides)
103	F	99 cm	5 Dec. 1972	15 Jan. 1975	New Caledonia
138	F	88 cm	5 Dec. 1972	Jul. 1974	Malekula, New Hebrides
151	F	86 cm	5 Dec. 1972	15 May 1975	Baie de Gomen, New Caledonia
173	F	98 cm	5 Dec. 1972	Oct. 1973	Anatom, New Hebrides
180	M	103 cm	5 Dec. 1972	3 Oct. 1974	Kandavu Is., Fiji
181	F	102 cm	5 Dec. 1972	15 Oct. 1974	Kandavu Is., Fiji
1330	M	102 cm	5 Dec. 1972	1 Aug. 1974	Druadrua Is., Fiji

lease. They were fed on green plants during captivity and released in the lagoon.

The recapture of 2 male turtles at great distances is very interesting. These and D. Green's recaptures of male Galapagos green turtles in mainland South America are the only recorded instances of long-distance migrations by male turtles.

Similarly, the reports of a male and female, released on the same day and in the same place, and recaptured almost 2 years later within a few days of each other at Kandavu Island, Fiji, is of great interest, although no definitive interpretation can be given at this time.

Tuamotu Archipelago

The scattered literature suggests that green turtles occur throughout the Tuamotu Archipelago. Beaglehole and Beaglehole (1938) reported on green turtles at Pukapuka Island (not to be confused with Pukapuka Island in the Cook Islands). The turtles there are commonly taken on the beaches or are seized in the lagoon by swimmers, who tie a rope around a foreflipper and pull the turtle ashore. At Pukapuka, a turtle is considered the property of the entire community, as is common in many unspoiled Pacific Island cultures, and a public feast is held when a turtle is brought ashore. One native offered the Beagleholes the observation that "it is only in recent times, since people have taken eggs of turtles from the nest, that turtles have been dying out," although in many other areas of the Pacific the eggs are sought even more assiduously than the turtle itself, and apparently always have been.

More recent information from Manihi atoll, also in the Tuamotus, by Hirth (1971), suggests that a fairly sophisticated turtle-ranching program has been developed by local people. Turtle eggs are collected and hatched, and the young turtles are raised in village kraals for later consumption. The turtles, fed on coconut meat and fish, reach a length of 50 to 71 cm in

3 to 3.5 years—a much more rapid rate of growth than seems to operate in the wild.

Very few data are available on turtles in other parts of French Polynesia. Turtles are apparently rare in the Marquesas, where capture of a turtle is now so infrequent that it is considered a special occasion. The Marquesas, Austral, and Gambier Islands all have rocky coasts with very few beaches, and turtle stocks appear to be very limited, although a hawksbill was reported from the southern Marquesas in 1978, and another in 1979.

Pitcairn and Henderson Islands

There is no information available to me on turtles in these islands; certainly there is no evidence that occurrence of turtles around these islands is anything more than sporadic.

Easter Island

The only available information on the turtles of Easter Island of which I am aware is the following paragraph in Harrisson (1971):

The hitherto neglected marine turtles around Easter Island may turn out to be of special importance. Some remarkably detailed petroglyphs, carapaces retained as heirlooms, and discussion with informants suggest that at least three species, possibly four, visit the beaches and sheltered bays, for food and/or nesting. Wonderfully well-made stone towers were erected along sections of the coastline and are still called turtle towers—though they have not been used in living memory. Again there are strong indications that in the past turtles were not indiscriminately slaughtered but respected, but those sanctions have not operated since the island went Catholic in the last century. Turtle visitors are now much more scarce and irregular.

Tom Harrisson did not mention the species he identified in his paper, but in a letter dated 27 March 1971, he informed me that "at least the ridley, hawksbill and green come up there," and in a subsequent letter (12 May 1971), he wrote as follows:

I am almost certain that four species have been reported and are well known to the islanders, who are extremely good naturalists, especially so as there is no resident fauna and they are extremely interested in the birds and reptiles that come in from the ocean, as they did themselves centuries ago.

Keep in close touch and remember this Easter Island thing could be very interesting indeed: this is the only piece of land a turtle can come up on or breed in 2,000 miles of tropical ocean. Please do not pass this word around to everybody else; I would like to have that little bit of territory for my old age.

This is all the information I could find on Easter Island turtles. Tom Harrisson's old age, alas, did not come to pass, so the area seems to be wide open for investigation. If the nesting by the ridley is confirmed, this would be of great interest; ridleys do not normally frequent remote or isolated midoceanic islands.

Acknowledgments

I am most grateful to the South Pacific Commission (Noumea) for generously funding my visit to New Caledonia, New Zealand and Fiji, and to Dr. Rene Grandperrin and George Balazs for the catalytic role they played in making possible my attendance at the SPC sea turtle conference in Noumea in November 1979. Karen Bjorndal's vigorous editing of the initial version of this paper resulted in a substantially improved manuscript.

Literature Cited

Anonymous.
1979. Tagging and rearing of the green turtle *Chelonia mydas* conducted in French Polynesia by the Department of Fisheries. Paper presented at Joint SPC-NMFS Workshop on Marine Turtles in the Tropical Pacific Islands, Noumea, New Caledonia, 11–14 December 1979. Manuscript, 21 pp.

1980. Observations of marine turtles on Huon Island (New Caledonia) by the navy ship *La Dunkerquoise* (February 1980). Manuscript, 2 pp.

Beaglehole, E. and P. Beaglehole
1938. *Ethnology of Pukapuka.* Honolulu: Bernice P. Bishop Museum.

Bustard, H. R.
1970. Turtles and an iguana in Fiji. *Oryx* 10:317–22.

Carr, A. F., and P. J. Coleman
1974. Seafloor spreading theory and the odyssey of the green turtle. *Nature* 249:128–30.

Chimmo, W.
1856. Narrative of the loss of the Chinese junk *Ningpo* on d'Entrecasteaux Reefs, near New Caledonia, with an account of the reefs. *Nautical Magazine and Naval Chronicle*, March 1856:113–21.

Eggleston, D.
1971. Leathery turtle (Reptilia: Chelonia) in Foveaux Strait (Note). *New Zealand Journal of Marine and Freshwater Research*, 5:522–23.

Eggleston, G. T.
1953. *Tahiti: Voyage Through Paradise.* New York:

Fordyce, R. E., and W. C. Clark
1977. A leatherback turtle (*Dermochelys*) from Kaikowa, New Zealand. *Mauri Ora* (Christchurch) 5:89–91.

Harrisson, T.
1971. Easter Island: a last outpost. *Oryx* 11:111–15.

Hirth, H. F.
1971. South Pacific islands—marine turtle resources. Report to Fisheries Development Agency Project, FAO. Manuscript, 33 pp.

Legand, M.
1950. Contribution à l'étude des méthodes de pêche dans les térritoires français du Pacifique Sud. *Journal Soc. Océanistes* 6:141–84.

McCann, C.
1966a. Key to the marine turtles and snakes occurring in New Zealand. *Tuatara* 14:73–81.

1966b. The marine turtles and snakes occurring in New Zealand. Records of the Dominion Museum, 5:201–15.

McElroy, J. K., and D. Alexander
1979. Marine turtle resources of the Solomon Islands region. Working Paper presented at the Joint South Pacific Commission and National Marine Fisheries Service Workshop on marine turtles. Noumea, New Caledonia, 11–14 December 1979.

McKeown, A.
1977. Marine turtles of the Solomon Islands. Min. Nat. Res., Honiara. 47 pp.

Oliver, R. B.
1910. Notes on reptiles and mammals in the Kermadec Islands. *Trans. New Zealand Institute* 43:535–39.

Parsons, J. J.
1962. *The Green Turtle and Man.* Gainesville: University of Florida Press.

Robb, J.
1973. Reptiles and amphibia. In *The Natural History of New Zealand*, ed. G. R. Williams, pp. 285–303. Wellington: A. H. and A. W. Reed.

Wilkinson, W. A.
1979. The marine turtle situation in the Kingdom of Tonga. Paper presented at Joint SPC-NMFS Workshop on Marine Turtles in the Tropical Pacific Islands, Noumea, New Caledonia, 11–14 December 1979. Manuscript, 8 pp.

PETER C. H. PRITCHARD

Peter C. H. Pritchard
Florida Audubon Society
1101 Audubon Way
Maitland, Florida 32751

Marine Turtles of Micronesia

The islands of Micronesia comprise 1 of the 3 great groups of Pacific Oceanic Islands. They are almost all located north of the Equator, being situated east of the Philippines and southwest of the Hawaiian Islands. The boundaries of Micronesia are almost identical to those of the U.S. Trust Territory, with the exception that Guam, an unincorporated territory of the United States, is not part of the Trust Territory, while the Gilbert Islands (part of the independent Kiribati), and the independent Nauru are considered part of Micronesia. Nukuoro and Kapingamarangi Atoll, though included in the Trust Territory, are culturally considered to be part of Polynesia. Moreover, the northern Marianas Islands have recently achieved Commonwealth status with the United States. The islands are all small and distances between them are large. Micronesia occupies an area equal to that of the United States, yet the land area is only half that of Rhode Island. Bryan (1971) calculates the total number of islands in Micronesia as 2,203. The 1973 population was 114,973 (excluding Guam), with an annual growth rate of 3.6 percent. The total land area is only 1,851 km².

Geologically the islands are all of volcanic origin, but differing age and subsequent weathering, subsidence, and coral formation have given them a very varied physiognomy. As a first-order approximation, the eastern islands are typically low atolls, often composed of many dozens of small, narrow islands surrounding a large central lagoon. The westernmost islands contain much weathered limestone and reach much higher altitudes. The highest islands, such as Ponape, attract an exceedingly high rainfall, with consequently lush vegetation. Shoreline vegetation throughout the Territory shows certain dominant species, such as coconut palms (*Cocos*), *Pandanus, Messerschmidtia, Portulaca, Sida,* and *Scaevola.*

Species Present

The hawksbill (*Eretmochelys imbricata*) and the green turtle (*Chelonia mydas*) are present throughout Micronesia and are widely recognized animals among those familiar with marine life in all districts. Nearly everywhere the green turtle is the more plentiful species, although in the Palau Lagoon area the hawksbill appears to be more common.

Two other species have been recorded on rare occasions. The olive ridley (*Lepidochelys olivacea*) was first recorded in Micronesia by Falanruw, McCoy, and Namlug (1975), who observed a mating pair in M'il Channel, northwest of Yap, on 30 November 1973. These authors also recorded a small (29 cm) *L. olivacea* from Lamotrek, in the eastern Yap District. Cushing (1974) reported 5 *L. olivacea* that were caught accidentally by long lines and plankton nets between 13 and 20 September 1974, in the southern Palau District (0° to 4°N, 131° to 137°E). In addition, I saw an immature stuffed *L. olivacea* for sale in a souvenir shop on Saipan in April 1976 that was said to have been locally caught and preserved.

The leatherback (*Dermochelys coriacea*) is reported occasionally in Micronesia, although it appears to be encountered only in deep water and has never been reported nesting in Micronesia. McCoy (1974) mentioned a very young leatherback, 69.4 cm in carapace length, that was captured near Satawal, in the eastern Yap District, on 2 September 1972. The turtle was tagged and released. McCoy also mentioned a leatherback caught at Woleai in 1971 that was captured and consumed by local people. I also have an unidentified newspaper cutting describing a large leatherback (444 kg in weight, 2.167 m in total length) caught by 2 Kapingamarangi fishermen off Parem Reef, Ponape Island.

Conservation Laws and Jurisdictional Background

Three completely different legal systems prevail concurrently in the Trust Territory: traditional law, vested in the hereditary chiefs; Micronesian law, as elaborated by elected delegates to the Micronesian Legislature; and U. S. federal law. As far as turtles are concerned, traditional law reflects patterns of hereditary ownership of the turtle resource, and the need for permission to be sought from traditional owners before turtles can be exploited. Micronesian law, as reflected in the Trust Territory Code (Title 45, Section 2) prohibits the capture of hawksbills less than 27 inches (69 cm) long, or green turtles less than 34 inches (86 cm) long (although only recently has the code differentiated between the 2 species). In addition, turtles are totally protected by Trust Territory law during the months of 1 June to 31 August and 1 December to 31 January, inclusive. They may also not be captured on the nesting beaches.

Federal law at present offers total protection to the hawksbill turtle, which is listed as an endangered species. The green turtle is listed as a threatened species, with certain populations, namely those of Florida and Pacific Mexico, being listed as endangered. The Department of the Interior Regulations recognize and permit the continuation of certain patterns of traditional subsistence use of turtles in the Trust Territory.

Traditional ownership patterns are still respected to a large extent in Micronesia, and flagrant violations of these rights may lead to protest or sanctions of one kind or another. The Trust Territory Code, however, is not widely respected; hawksbills, for example, tend to be chased and caught whenever seen, whatever their size or whatever the season of the year, and the nests too are frequently raided. The green turtle has traditionally been collected on nesting beaches in many parts of Micronesia, especially in the Yap District, and no attempts have been made to enforce that section of the Trust Territory Code that prohibits such activities.

Little attempt is made to enforce the Endangered Species Act in the Trust Territory, and the law is ignored throughout Micronesia. Indeed, some question exists as to whether provisions of the Endangered Species Act even apply in the Trust Territory, but most legal opinions now hold that it does; for purposes of import and export of listed wildlife, the Act specifically refers to the Trust Territory as having the status of a State of the Union. Reluctance to enforce federal endangered species law in the Trust Territory probably stems from several considerations:

1. The Trust Territory has for years had but a single American conservation officer, based in Palau, to whom local people have made clear that his life may be in danger if he insists on rigorous enforcement of turtle protection laws.

2. The United States has been sensitive to charges of colonialism in thrusting conservation laws passed in Washington, D.C. on peoples leading traditional subsistence life-styles in remote islands on the far side of the world.

3. The Trust Territory is not a permanent political entity, and in the years to come the various districts will be electing whether or not they wish to remain associated with the United States. The United States has not deemed it politic or appropriate to thrust unwelcome conservation obligations upon people who would be likely to reject them totally on reaching political independence.

A loophole that has resulted from the wording of the Endangered Species Act, which considers the Trust Territory to have the status of a state, is that products of the hawksbill turtle hand-carried by tourists entering Honolulu from the Trust Territory can no longer be confiscated. Such transportation of products is legally

PETER C. H. PRITCHARD

simply a case of carrying personal effects across state lines, unless it can be proven that the material is post-Act in origin.

Palau District

The hawksbill is more abundant than the green turtle in the Palau District, or is at least more conspicuous in the more accessible areas such as the Palau Lagoon. Douglas Faulkner, the underwater photographer, reports that hawksbills may be seen virtually every day in the Palau Lagoon by a competent scuba diver, and immature hawksbills are also reported to be numerous in the Kayangel Lagoon at the northern end of the Palau system. However, Robert Owen, conservation officer for Micronesia from 1949 to 1978, reports a gradual but steady decline in abundance. Natural predators are relatively few, and no natural egg predators have been reported, but the turtles are eaten by crocodiles (*Crocodylus porosus*), and the human pressure on eggs is intense—estimated at 80 percent by Jim McVey, who conducted a head-starting program for hawksbills in Palau in the early 1970s. Adult turtles too are highly persecuted. Hawksbill meat is eaten locally, but the economic pressure on the species is definitely from the shell trade. Tourism in the islands increased about 300 percent with the advent of regular air service in the early 1970s; a large proportion of tourists in Palau are from Japan, which of course offers no legal impediments to the free importation of hawksbill products.

The hawksbill turtle nests on small beaches on limestone islands in the Palau Lagoon. The principal nesting months are July and August, but some nesting takes place in June and September, and a few may nest in any month of the year. Their nesting site fidelity is reported to be strong, and they nest at approximately 15-day intervals, 2 or 3 times in a season. Favored islands include Eomogan, where Jeff June of the Peace Corps saw 3 turtles nesting in 1 night in late August 1975, and Ngerugelbtang Island, where the turtles often walk the length of a long spit before reaching a nesting area safe from tidal inundation. Other islands sometimes used for nesting include Aulong, Ngeangas, Ngobadangel, Unkaseri, and Abappaomogan.

Green turtles are not often seen in the Palau Lagoon, but achieve substantial populations in the northern and southern extremes of the Palau District. Richard Howell, district fisheries officer on Truk, reported that about 10 years ago he found fully mature green turtles to be plentiful in the Ngaruangl Lagoon, at the northern tip of the Palau complex. Villagers from Kayangel could catch 5 in 30 to 60 minutes. The turtles were resident there year-round, feeding on the large strands of turtle grass present especially on the western edge of the reef. Howell reported seeing only 1 male turtle in the area. Nesting (probably by greens) takes place on the

barely exposed Ngaruangl Island, since natives of Kayangel returned from the lagoon with fresh eggs. Raids on the turtles were sporadic, and could be made only during calm weather. The turtles were only used by Kayangel people for special occasions, although they were also used for trade with villagers on northern Babelthaup.

Green turtles also nest in small numbers on Honeymoon Beach, Pelelieu Island, and, on 1 occasion, a female was seen inside the reef on Morei Island. However, the best green turtle beaches by far in the Palau System are on the southern islands of Merir and Helen Reef, located many kilometers to the south; coordinates are 4°19′N, 132°19′E for Merir, and 2°48′ to 3°01′N, 131°44′ to 131°51′E for Helen Reef. Merir now unfortunately has a small permanent settlement, numbering 7 people in 1976. Even such a small group of people can cause havoc to the turtle population on such a tiny island. Helen Reef, whose single emergent point of land, Helen Island, is too small for permanent settlement, still has heavy pressure on its marine resources, especially by pirates, the majority of whom come from Taiwan. When caught, they may be jailed in Palau for variable periods of time. Another serious problem for turtles in the outlying islands is that the crew of the government field trip vessel, far from being a positive force for law enforcement, take advantage of their subsidized trip to Helen Island, Merir, and other turtle islands to gather as many turtles as they can for themselves, which can be taken back to markets for personal profit.

There is an extensive folklore and legend regarding turtles in Palau. For example, the disovery of the approximately 2-week nesting cycle for both the green and the hawksbill turtles is attributed to a chance discovery described as follows:

"A young couple arranged to spend the night on a remote beach on Pelelieu Island. They used the girl's grass skirt as a pillow, and after making love, went to sleep. When they woke the next morning, there were turtle tracks on the beach and a nest right beside them but, to their great embarrassment, the grass skirt had disappeared. Nevertheless, they decided to repeat the rendezvous two weeks later, and, just before they fell asleep, noticed a large turtle crawling ashore with the remains of the grass skirt still attached to a front flipper."

This story is a favorite subject of Palauan story boards.

Yap District

Chief informant on sea turtles in the Yap District is Mike McCoy (this volume), formerly of the Peace Corps and now chief fisheries officer for Ponape and associate of the Yap Institute of Natural Sciences. McCoy's 1974 paper "Man and Turtles in the Central Carolines" is

	Region												Total sightings	Number of months
	1	2	3	4	5	6	7	8	9	10	11	12		
FY 1979	4	1	1	1		1	6	2	43	31	18	77	185	12
FY 1978	6	3	1	9		6	14	3	10	1	15	15	83	12
FY 1977	0	3	1	1		4	1	5	10	0	8	8	41	2
FY 1976	7	5	6	6		35	8	14	44	10	12	42	189	9
FY 1975	14	5	18	3		23	11	9	37	16	6	143	285	6
Total	31	17	27	20		69	40	33	144	58	59	285	783	41
x̄/Region	6	4	6	4		15	8	8	31	12	13	59	—	—

Source: Molina, unpublished report.

one of the most valuable sources available on human attitudes to turtles in Micronesia. To avoid duplication, reference is made to McCoy's paper herein for information on turtles in the Yap District.

Marianas District and Guam

Hendrickson (in manuscript) quoted the following information, received from Isaac I. Ikehara, chief of the Guam Division of Fish and Wildlife, regarding the available information on sea turtles in Guam in 1968:

Green turtles and hawksbills are reported to occur in Guam waters. They apparently nest on the island beaches, but only sporadically; eggs were harvested more commonly during the time before the second World War, in many areas of the island, especially on the northern and southern ends (Tarague, Ritidian, Uruno, Orote, Cocos Island, Asiga Beach, and other localities).

It appears from local residents that sea turtles are a rarity on the local market and the consultant found none on three of his visits. Skin divers occasionally bring them back but they are not considered a normal commercial item although red turtle meat is reputed to sell at $0.75 (US) per pound. There is no export of turtle products from Guam. In 1968 there were reportedly two divers specializing in turtles each catching three or four turtles on a good day.

There is apparently no legislation protecting sea turtles or regulating the catch in any way, but there are some good catch statistics. All sizes from 15 lb. to 400 lb. are taken, but the informant estimates that the average size is around 60 lb. (the type most likely to be taken by divers). No special feeding grounds have been identified.

Harry Kami, enforcement officer for the Guam Fish and Wildlife Division, made a number of flights over the Guam coast during the last couple of years, and saw sea turtles—sometimes in concentrations of 40 or 50 individuals—off the northern coast of Guam, between Ritidian Point and Pati Point. Kami also sometimes saw 3 or 4 turtles off the coast near Inarajan Bay, on the southeast coast, and said that turtles formerly nested on Cocos Island, off the southwest coast, although the island was now too intensively visited for nesting to take place.

The north coast of Guam, near which the turtles were seen, was under Air Force control, and was rather little visited. However, despite the presence of a good beach, little nesting took place here. Factors that lessen the suitability of this beach for nesting may include the shallow reef (only 1 m submergence by high tide), and the presence of dense vegetation above the high tide line on the beach. Most of the turtles seen off Guam were of adult size, and indeed appeared to be very

Table 2. Summary of turtle sightings by month, Guam, Fiscal Years 1975 through 1979

	Month												Total sightings	Number of flights
	J	A	S	O	N	D	J	F	M	A	M	J		
FY 1979	12	3	6	6	7	12	18	52	24	14	20	11	185	24
FY 1978	7	6	10	4	16	17	7	5	0	3	4	4	83	24
FY 1977	23	—	—	18	—	—	—	—	—	—	—	—	41	4
FY 1976	—	20	28	24	20	42	16	10	7	22	—	—	189	18
	—	—	—	—	—	—	45	44	32	46	54	64	285	12
Total	42	29	44	52	43	71	86	111	63	85	78	79	783	82
x̄/Month	14	12	15	13	14	24	22	28	16	21	26	26	—	—

Source: Molina, unpublished report.

PETER C. H. PRITCHARD

large from an aircraft at 65 to 80-m altitude; but mating pairs had not been seen.

Kami found 1 green turtle nest on the east coast of Guam between Ylig Bay and Togcha in 1974. Because of the extensive human use of this beach, the eggs were moved and, while reburying them, several incomplete nests were found.

Dr. Lucius Eldredge informed me by letter (dated 12 July 1976) that Dick Randall of the University of Guam Marine Laboratory reported 6 recent turtle nests on June 26, 1976, at the north edge of Sella Bay on the southwest coast of Guam.

A recent unpublished report by Molina includes the results of 5 years of aerial surveys of turtles around Guam. The following section is extracted from this report:

The island of Guam was divided into 12 survey regions (Figure 1). Marine turtles have been sighted within every survey region (Table 1) and during all months of the year (Table 2). Region 5 has not been censused due to military restriction. Two flights were made each month in all cases. A total of 783 marine turtles have been sighted around Guam on 41 aerial surveys made during the past 5 years. Far more turtles were sighted within region 12 (Pati Pt.-Ritidian Pt.) than in any other (Table 1). Approximately 74 percent of the observed turtles were seen within regions 8 to 12. The most probable explanations for this distribution are the low levels of development and fishing pressure in these areas.

Marine turtle abundance appears to peak twice during the year (Table 2). In general, these peaks coincide with the winter (December to February) and late spring (May to June) months. This also loosely correlates with Guam's "dry," tradewind season which usually lasts from December to June. It is unclear at the present time whether or not the turtles are mating during the entire period, yet it seems likely. The time of nesting is also unclear. However, reports from local fishermen indicate that nesting occurs around June.

Reports have been made of larger than usual numbers of turtles visiting Guam about every 3 years. The last of these visits happened in 1976, and is reflected in our aerial survey data (Table 2). Another visit was expected this year. Again, our data show the winter increase in numbers.

Since it is difficult to make positive species identifications on turtles from a moving airplane, we have no reliable estimate of the species composition of Guam's marine turtle community. However, it is generally regarded that Chelonia mydas is by far the major component.

Human interference with nesting turtles is a serious problem at Tarague Beach. The majority of the problem lies with the friends and relatives of the Tarague landowners who use the beach for "4-wheeling" and who actively hunt for turtle eggs. Since Tarague Beach is privately owned and enjoys military isolation, there may be a good chance

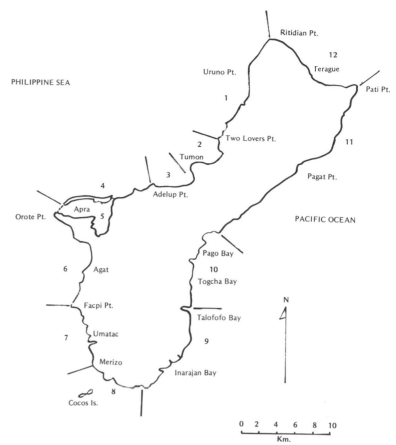

Figure 1. The island of Guam with its 12 aerial survey regions.

of controlling this problem, especially if the area could be designated as a marine turtle sanctuary. If it is not already too late, Tarague Beach may be Guam's only hope for such a valuable natural resource. Mr. Castro appears to be pro turtle conservation and has offered to do what he can in cooperation with our office to help protect these animals.

Turtle meat is occasionally sold in Guam, but is very expensive—although it can on occasion be purchased with U.S. government food stamps at Perez Market. There are no laws protecting turtles in Guam at the present time, and some opposition to establishing local laws because turtles protected in Guam may well be caught in the Trust Territory. However, because Guam is an unincorporated territory of the United States, federal law unquestionably applies, and hawksbills should already have legal protection. The green turtle too should soon receive nominal protection.

Very few Guamanians are expert at spearing sea turtles, with the exception of a few old-timers, and nets are never used nowadays for catching turtles.

North of Guam, the Northern Marianas Islands stretch in a slightly curved elongated chain. Few turtles appear to nest anywhere in the Marianas; to a large extent this may reflect shortage of nesting beach, most of the uninhabited islands having no beach whatsoever. Saipan has several kilometers of beach on the west coast, but the area is rather extensively developed with hotels and other beach facilities, and few if any turtles nest there. However, dense patches of turtle grass within a few meters of the beach suggest good feeding habitat for green turtles. Rota has several beaches, and Tinian 2 small, marginal ones, but I have no evidence that these are used by nesting turtles.

Stuffed turtles are for sale at several locations on Saipan. In a handcraft shop on Beach Road, 13 stuffed green turtles (half-grown to maturity) were for sale; also 3 hawksbills and 1 olive ridley. The turtles were reportedly all locally caught.

Turtles are being caught in increasing—and now rather large—numbers in the northern Marianas. The turtles were captured by divers for sale to hotels and gift shops, and 1 diver could easily catch 4 or 5 turtles in a day (Ben Sablan, personal communication).

Truk District

The islands of the Truk District lie to the east of the Yap District. Truk itself is composed of a large lagoon, roughly circular in shape, about 40 miles in diameter. The lagoon is fringed by a reef, broken in several places and reaching above sea level to form small, low islets, principally in the northern and southeastern sections. Most of the human inhabitants, who numbered 20,105 in 1970, however, do not live on the reef islands, but rather on several mountainous, large islands near the middle of the reef. The other islands of the Truk District—the Lower and Upper Mortlocks, to the southeast; the Hall Islands to the north; and the so-called Western Islands of Namonuito, Palap, Puluwat and Pulusuk—are low atolls.

Although only 2 days were spent on Truk during the survey reported here, I was able to learn a good deal regarding sea turtles in the District, through the kindness and efficiency of Mr. Richard Howell and Mr. Tawn Paul of the Fisheries Office. Informants for turtles in the outer islands were Mr. Casian Orik (Western Islands), Mr. Marion Henry (Mortlocks), and Mr. Appo Pius (Truk Lagoon).

Three species of sea turtles are recognized in the Truk District: the leatherback (locally called "mirang"); the green turtle ("winimon") and the hawksbill ("winichen"). The leatherback is seen only occasionally and always in deep water; there are no nesting records in the area. The other 2 species are both widespread, but the green turtle is generally more plentiful than the hawksbill.

The hawksbill is found principally in the Truk Lagoon and in the Mortlocks. On the northern fringe of the Truk Lagoon, hawksbills nest in small numbers on the islands of Holap, Tora, Ruac, Lap, Ushi, Onao, Tonelik, Pis, Alanenkobwe, Lemoil and Falalu. The largest of the islands, Pis, has human inhabitants, and

turtles nesting there are likely to get killed. Mr. Pius informed me that the casual nesting in this area (perhaps 1 or 2 turtles per night on each beach during May to October) had not diminished perceptibly during the last 50 years. In the Lower Mortlocks, Marion Henry reported casual hawksbill nesting in all 3 atolls (Etal, Lukunor, and Satawan), but not commonly on the inhabited islands (Kutu, Mor Satawan and Ta in Satawan atoll; Etal Islet; and Lukunor Islet).

The tiny island of East Fayu, about 100 km north-northwest of the Northeast Pass leading out of the Truk Lagoon, is an important one for green turtle nesting. About 6 or 7 turtles are reported to nest here each night during the season, which begins remarkably early (February), and lasts until about June. The island is elongate, less than 2-km long, and has a sandy beach with a deep water approach all around. The rights to the turtle resource are vested in the people of Nomwin Atoll, a few kilometers to the east. A few green turtles (1 to 3 per night) are also reported to nest on Fanang Islet, at the eastern end of Nomwin Atoll, and on a few tiny islets in adjacent Murilo atoll. A few also nest on northern Murilo Island.

Turtle nesting has not been reported in the Western Islands, all of which are inhabited. However, the people of Pulusuk, and also of Puluwat and Pulap (Tamatam Islet) take advantage of the March-April tradewinds to travel to Pikelot, in eastern Yap District, to collect turtles. This journey may be made 3 or 4 times during the 2-month period, and a typical catch is about 20 turtles, which are collected on the beach during a stay of 1 or a few nights. The eggs are also collected. It was estimated that about 30 turtles nest each night at Pikelot; however, from data obtained in the Yap District, I believe this to be a distinct exaggeration. Turtles were reported to be diminishing in Pikelot, but holding their own in the Hall Island–East Fayu region.

In the Truk Lagoon both species of turtle are found (mostly adults) and are about equally common; however, only the hawksbill is known to nest. Rather few turtle fishermen are operating, and the turtles are obtained by spearing. A turtle can be obtained on demand within 24 hours by certain fishermen. Reportedly, the hawksbills are killed for use of their shell, which is sold in souvenir shops; however, I did not see any for sale at the time of my visit.

In the Truk District, it was reported to me that the green turtle often weighs 300 to 350 lbs. (136 to 159 kg), and occasionally 400 lbs. (181 kg), and usually laid 80 to 120 eggs. The hawksbill weighed 100 to 150 lbs. (45 to 68 kg) and laid 110 or more eggs (maximum observed: 152). There is no reason to question the accuracy of these figures. I was also informed that *both* species eat seagrass and algae. When I questioned an informant (Appo Pius, a fisherman of 50 years' standing) on this, he appeared absolutely certain that stomachs of the hawksbill as well as the green contained such plant material, even when I pointed out that in most parts of the world the hawksbill is carnivorous.

Ponape District

The Ponape District is situated to the east of the Truk District. Ponape, the District Headquarters, is a large (129.04 square-mile), centrally located island, which is highly elevated, reaching an altitude of over 2,500 feet. Rainfall is heavy, and vegetation lush. The island is roughly circular in shape and is surrounded by a barrier reef penetrated by about twenty entrances. There are some sizeable offshore islands, including Sokehs, Langer, Parem, Mwahnd Peidi, Mwahnd Peidak, Takaiu, Dehpehk, and Temwen. The population of the island was estimated to be 14,520 in 1970.

Kusaie is the second largest single island of the Ponape District; it has an area of 42 square miles, an altitude of 2,064 feet, and a 1970 population of 3,743. It is situated approximately 300 miles east-southeast of Ponape.

The other islands of the District are all atolls. Mokil (population 1970, 411) and Pingelap (population 849) lie between Ponape and Kusaie. The atolls of Ant and Pakin lie close to the west coast of Ponape; Pakin had a population of 36 in 1973; Ant had 10. Oroluk atoll, which had a population of 42 in 1935, none in 1948 or 1970, but since mid-1973 inhabited by about 18 people, lies west-northwest of Ponape. Southwest of Ponape are the atolls of Ngatik (population 442) and Nukuoro (population 420). Far to the southwest, nearly 500 miles from Ponape, is the atoll of Kapingamarangi, inhabited by 432 people in 1970, but with a permanent overflow population now living on Ponape, and a few others on Oroluk.

I spent 5 days on Ponape, where my chief informant was Ben Sablan of the Fisheries Department. Valuable information was also received from Alan Millikan, the District Fisheries Specialist, and David Fullaway, the chief Forestry Officer.

Populations of sea turtles around Ponape itself appear to be relatively insignificant and very little nesting, if any, takes place. Indeed, Ponape has very few sandy beaches. Turtles used to provide an important source of food to the people of Kapingamarangi, but they are now rarely seen in that area (Niering 1963). Nesting does not take place on Pingelap and Mokil, but Mokil has a shallow lagoon in which small green turtles (less than about 50 cm long) are easily seen and caught. Ben Sablan observed 5 such turtles on an underwater survey of the 15.5 km² Mokil Lagoon in 1974.

Around Kusaie, Sablan found 31 green turtles and 6 hawksbills during a 3-day underwater survey in August 1973; nesting, however, appears to be sparse at best.

No details are available for nesting on Ngatik, but Sablan reports some nesting on the eastern islets of Peina, Bigen Karakar, Jirup, Bigen Kelang, Piken Mategan, Dekehnman, and Wat. Two green turtles were seen in the water during an underwater survey in September 1973.

Green turtles have been seen around Ant Atoll and it is rumored that daytime nesting occurs; but this needs confirmation.

Green turtles appear to be rather plentiful around Nukuoro, where Sablan counted 52 (but no hawksbills) during an October 1973 underwater survey. Nearly all were of adult size and were relatively inactive. They probably nest on the island, but there is no evidence of high-density nesting. Sablan also saw 3 green turtles underwater in Pakin.

Apparently the only nesting ground of importance in the Ponape District is the atoll of Oroluk, which once boasted as many as 19 islets, but apparently all but Oroluk Islet itself, at the extreme northwest of the atoll, have now disappeared. The District Administrator of Truk reported to me that he had seen green turtles nesting by daylight on Oroluk, on the lagoon side of the island, during a helicopter visit in November 1964. The island was uninhabited at that time, and the turtles reportedly showed no fear of the observers. Turtles in Oroluk are considered to have a split nesting season (December to January and June to July), and this may have been the original rationale for the split closed season throughout the Trust Territory. It is estimated that between 9 and 15 turtles nest on Oroluk on the average night, with up to 20 on a very good night. The local people, about 18 in number and resident on the island since mid-1973, catch a substantial proportion of the nesting turtles.

In a memorandum dated February 3, 1976, Sablan described the findings of a July 1975 visit to Oroluk. The islanders reported that since they first settled on the island 2 years ago, the number of turtles nesting had dropped considerably. This may well have been due to excessive predation, although Sablan also recorded the following human disturbances to the nesting beach during the night he was on the beach: 1) very active human activities until the early morning; 2) several campfires maintained until midnight; 3) copra operation with outboard motor until 9:00 a.m.; 4) ship generator and lights on until morning. However, at West Fayu in the Yap District, the tagging and hatchery crew in 1972 found that the turtles continued to nest even though a wrecked cargo ship containing 300 Toyotas was being salvaged on the island with extensive lights, noise and other disturbance by the salvage crew every night. The ship had spilled 600 tons of oil and was not completely defueled until more than 6 months after the wrecking.

Marshall Islands District

The Marshall Islands comprise a widespread District at the eastern end of Micronesia. With the exception of a few small isolated reef islands, such as Jemo, the Marshalls are comprised exclusively of atolls, most of which are made up of a few to many dozens of islets. The atolls are roughly aligned along 2 parallel axes, the northeastern being the Ratak Chain and the southwestern the Ralik Chain. None of the islands reaches a height of more than a few meters above sea level, and the total land area of the District is only 180.82 km². The human population, numbering 20,206 in 1970, is widely distributed, but only the atolls of Majuro, Kwajalein, and Ailinglapalap have more than a thousand people.

Bryan (1971) lists Taongi, Bikar, Taka, Jemo, and Erikub as the only atolls or islands that have never had human populations, while the people of Bikini and Enewetak were displaced after the second World War when these islands were used for atomic weapons testing. Rongerik is listed by Bryan as having 6 people in 1935 and 1948, but as being uninhabited in 1970; this island was used temporarily by the displaced people of Bikini, but proved unsatisfactory. The Marshall Islands are well described by Anonymous (1965), while excellent maps and directories to names of islands are provided by Bryan (1971).

Only Kwajalein and Majuro were visited during the present survey. However, much useful information on turtles elsewhere in the Marshalls was provided by Ben Sablan on Ponape, who was formerly resident in the Marshalls; by Major Ron Barnett and Rev. Elden Buck on Kwajalein; Jim Hiyane, the agricultural officer on Ponape; George Balazs in Hawaii; and Jobel Emos, a janitor at the Kwajalein Missile Range.

Bikar Atoll

The atoll of Bikar, one of the northernmost of the Marshalls, is generally thought to have the highest concentration of breeding green turtles in the District. The atoll is composed of several islets, the named ones being Jabwelo and Almani on the east, Bikar on the south, and the sandbank of Jaboero between Bikar and Almani. Bikar is the largest with an area of 0.063 miles².

Bikar has been thus described (Anonymous 1956): "Sea birds of many kinds are abundant, but the outstanding feature is the great number of turtles that come ashore to lay eggs on Bikar Islet." Fosberg (1969) recounted his experiences with the turtles of Bikar as follows:

On the night of August 6, a few baby black turtles were seen hurrying toward the sea. They were being attacked by large red hermit crabs (Coenobita perlata) and by rats (Rattus exulans). The hermit crabs bit through the carapace, the rats through the plastron. Almost all of the female turtles that visited Bikar Atoll, well over 300 in the seven nights, August 5–12, came ashore on Bikar Islet. One set of tracks and a pit were noted on Jaboero Islet, a few on the south part of Almeni Islet, but none on Jaliklik Islet, which is rocky and has no loose sand.

Judging by the numbers given in an earlier part of this paper, it is possible that the "over 300" turtles is a misprint for "over 30."

From the large numbers of tracks seen, the relatively light nesting observed and the observations on hatchlings, it appears that the season on Bikar reaches its peak probably around June and July.

In 1958 Bikar Atoll and Pokak (Taongi) Atoll, which lies to the north of it, were set aside as preserved natural areas by administrative decree by the then District Administrator, Maynard Neas. It is hoped that this protection may be strengthened, as clearly Bikar is the principal turtle nesting area in the Marshalls and should be kept as a stocking area for the rest of the archipelago.

Hendrickson (in manuscript) was able to visit Bikar on 2–3 July 1971 and made the following observations:

The consultant visited Bikar Atoll and all 3 of its islets judged suitable for green turtle nesting (Bikar, Arumeni and Jaboerukku). These are the only vegetated islets in the atoll, the remainder being barren bars and banks which are presumably swept by high wave action. The timing of the visit was particularly favorable, being at the end of a 7-day period of diminishing tides during calm weather. This left a series of high tide marks on the clear areas of beach where rocks had not confused the wave wash pattern and, for the most part, it was possible to identify the night on which recent beach ascents had been made by nesting turtles, by noting the particular high tide mark where the track ceased to be evident. It was possible to say with some confidence that 39 turtles had ascended the beaches during the preceding 6 days (78 tracks, half ascending, half descending). Thirty-five of the 39 turtles had used the beach on Bikar Islet, 1 had ascended the Arumeni and 2 had ascended Jaboerukku. One of the 35 tracks on Bikar was a hawksbill track (not ridley); all others were presumed made by green turtles (loggerheads have not been reported from the area).

Hendrickson made some calculations of the possible size of the nesting population on Bikar, concluding that the order of magnitude of the population was 711 sexually active adult female turtles in the Bikar breeding population. From these figures, he reasoned that "even the most favorable interpretation of the data available (granting the assumptions made) allows consideration of a population of only small size, not constituting an exploitable wild resource of any significant magnitude."

PETER C. H. PRITCHARD

Jemo Island

Jemo is an isolated, tiny island situated at 10°8′N, 169°32′E, located between the atolls of Ailuk and Likiep. The land area of Jemo is only 1.55 km². The turtles on Jemo were described as follows (Anonymous 1956): "Many turtles visited Jemo to lay their eggs. Jemo was formerly tabu for most of the year, being regarded as a bird and turtle reservation. Only during one month in the year were these animals hunted and their eggs taken."

Fosberg (1969) visited Jemo from 18 to 22 December 1951 and observed tracks corresponding to the nesting of 22 turtles during the past several days.

The Rev. Elden Buck of Kwajalein informed me that a boat from Likiep sometimes brings 10 to 15 turtles for sale on Ebeye. These turtles were presumably caught on Jemo, which is the closest turtle island to Likiep. Likiep itself has few turtles, according to Ben Sablan on Ponape. Further confirmation of the presence of nesting turtles on Jemo was provided by several informants during my survey.

Arno Atoll

Green turtles nest occasionally on the sandy beaches of Arno Atoll, but they are scarce and of no commercial importance (Hiatt 1951). Ben Sablan reported that nesting on Arno takes place on the islet of Ine, in the south and southwest.

Erikub Atoll

Erikub is an uninhabited atoll composed of 16 islets lying just south of the inhabited atoll of Wotje. Jim Hiyane, the agricultural officer on Ponape, informed me that he had seen turtles nesting on Erikub, and estimated that 6 or 8 turtles nested nightly. He mentioned that people from Wotje go to Erikub for copra, coconut, crabs, etc., and often picked up turtles when there, but did not go specifically for turtles.

Jobel Emos on Kwajalein confirmed that turtles nested on Erikub and pinpointed the northwestern islets of Enogo and Loj as being the most favored for nesting. Emos claimed that nesting on Erikub was year-round, but that the turtles were usually exploited during summer months because of the prevailing calm water at that season. He said that the Wotje people, when they caught a female turtle on Erikub, would tether it in shallow water so that it would attract males, which were captured as they mounted her. Emos' estimate was that 3 or 4 turtles nest nightly on Erikub.

On Kwajalein, the Rev. Buck showed me a photograph of a boatload of over twenty turtles that had been brought in from Erikub and Bikar for sale on Ebeye, the islet where the Marshallese workers on the Kwajalein Missile Range reside.

Taka Atoll

Taka is an uninhabited atoll lying very close to, and southwest of, the inhabited atoll of Utirik. It has five islets, the largest of which is Taka itself (2.5795 km²). According to the Rev. Buck, people from Utirik collect turtles and turtle eggs on Taka, but further details are not available.

Ebon Atoll

Ebon is the southernmost of the Marshall Islands. It is a roughly circular atoll composed of 22 islets, by far the largest of which is Ebon itself, an elongate island that makes up the southern side of the atoll; it is about 10-km long and has an area of 2.804 km². Bryan (1971) lists the 1970 population of Ebon as 480—substantially reduced from the 1935 and 1948 censuses. Ebon has a reputation for abundance of food of all kinds, and although no definite information on turtle nesting is available, it is considered to be the best area for catching turtles in the water. The turtles are nearly all of adult size and are caught with nets. Each night 2 to 4 can be caught. Rev. Buck said that if a turtle on Ebon is captured in a certain place, the next night it is often found that another turtle has moved to the same spot.

Kwajalein Atoll

Kwajalein is the largest atoll in the Marshalls, and reputedly the largest in the world. Ninety-three islets are listed by Bryan (1971). The islets of Kwajalein (at the southern tip) and Roi and Namur (now connected by a runway and called Roi-Namur) are devoted exclusively to U.S. military uses. The Marshallese residents live on Ebeye, a small and highly overcrowded islet a short distance north of Kwajalein, on the eastern edge of the atoll. Most of the other islets are very small, and in some parts the bounding reef is without islets for distances of 15 to 25 km.

Major Ron Barnett on Kwajalein gave me considerable information on turtle observations on Kwajalein. Turtles are often seen around Kwajalein Islet, and between Kwajalein and Ebeye. A few turtles appear to be extraordinarily static in range; a certain green turtle is reported to have resided at a certain coral head (known as K5) off the lagoon shore of Kwajalein for 2 to 3 years, and is very familiar to skin divers. Green turtles are also seen on the ocean side of Kwajalein at the end of the runway, where they scavenge for the kitchen scraps that are thrown in each day. They are usually of less than mature size. One turtle that I saw feeding on the kitchen scraps of Kwajalein, however, appeared to be of adult size.

No records are available for turtle nesting on Kwajalein, and indeed there is a shortage of good beaches. However, much of the atoll is poorly studied and a

Marshallese informant on Kwajalein informed me that turtles do nest sometimes on the islands at the northwestern end of the atoll.

Major Barnett, in a letter dated 16 July 1976, reported that on July 10 a green turtle had been found nesting on the ocean side of Bigej Island, about 19 km north of Kwajalein Islet.

Ujelang Atoll

Ujelang or Ujilang, is an elongate atoll about 20-km long located at the western extreme of the Marshalls, being closer to Ponape than to the population centers of the Marshalls. It had a small native population of about 40 people (plus 12 non-natives) in 1935. It was uninhabited in 1948 according to Bryan, but this is presumably in error, since Helfich (in manuscript) reports that the Enewetakese people displaced by atomic tests were settled on Ujelang in 1947. The 1970 population, according to Bryan (1971), was 281.

Ujelang is listed by Carr (1965) as a "minor nesting beach" for the green turtle. The source of this information was not quoted, but Carr informs me that he based this record on an observation made by the crew of a U.S. Naval vessel anchored off Ujelang one night in 1962. Baby green turtles were attracted to the lights of the ship in very large numbers—although at this point it is not possible to ascertain whether the numbers represented only 1 or 2 successful nests, or whether there were numerous nests erupting simultaneously. Two of these hatchlings were transmitted alive to Carr. Phil Helfich, in a brief manuscript, reports on an interview with Chief Johannes, chief of the exiled Enewetakese people on Ujelang: "Chief Johannes indicated that turtles nested all around the island Ujilang. Ujilang is the island which has been occupied by the Enewetakese since 1947, and it is difficult to visualize that they did not decimate the nesting turtle populations, because Ujilang is such a small island."

None of the informants on my survey had any information about turtles on Ujelang. The island is extremely remote and is not often visited. This would appear to be a priority for future studies.

Enewetak Atoll

Enewetak is a rather large, almost circular atoll in the western Marshalls. According to Bryan (1971), it is composed of 44 islets, has a land area of 2.26 miles² and had 128 people in 1948, but none in 1970. However, according to one writer (Anonymous 1972), 100 people, mostly civilians, live on Enewetak. Another report (Anonymous 1975) gives 1947 as the year in which the 136 Enewetakese residents were transferred to Ujelang; the island was used for nuclear tests between 1948 and 1958. Since 1954, the University of

Hawaii has operated the Mid-Pacific Marine Laboratory on Medren Island, Enewetak, which is financed almost completely by the U.S. Energy Research and Development Administration.

Helfich (in manuscript) quotes Chief Johannes of Enewetak, who lived on the atoll until 1946, as reporting turtle nesting (up to 1946) taking place from May through August on the islets of Alice, Bell, Runit (Yvonne), Glen through Keith, Leroy, Wilma, and Vero. The last 2 islands had the best nesting areas. Another islet by the name of "Vikai" was reported by Johannes to have abundant nesting turtles, but no island of this name is shown on available maps of Enewetak.

At the present time there appears to be little turtle nesting on Enewtak. However, George Balazs has prepared reporting sheets for observations of turtles by scientists at the Mid-Pacific Marine Laboratory and others, and valuable information may eventually be forthcoming from this program.

Majuro Atoll

Majuro, the District Headquarters, is an elongate atoll approximately 30-km long. The southern rim of Majuro was originally composed of a single extremely attenuated island, Majuro, and a series of much smaller islands to the west. However, these islands have now been connected in order to provide vehicular access between the principal town (known as D-U-D, from its constituent and now coalesced islets of Carrit, Uliga, and Dalap) and the airport; and the blockage of the former passages between islets, with no provision for bridges or culverts, has led to substantial pollution problems in the Majuro lagoon.

Turtle nesting has not been reported on Majuro, although turtles are spotted in the waters of Majuro relatively frequently. Ben Sablan informed me that large turtles are seen resting near the Windward Islands of Majuro, and on an afternoon dive one summer he had seen more than 15 turtles, all females.

Jaluit

Jaluit is a large, irregularly shaped atoll, about 30-miles long from north to south. It is composed of 91 islets. Bryan (1971) gives the 1970 population as 881, substantially reduced from former years. Ben Sablan informed me that turtles nest in small numbers on Lijeron Islet, near the northern end on the west side of the atoll.

Aur, Maloelap and Likiep Atolls

Ben Sablan reports that turtles may be found on each of these atolls, but that in no case were they plentiful.

Bikini and Taongi Atolls

Although my informants did not mention these atolls, both were recorded by Hendrickson (in manuscript) as being second in importance only to Bikar among the Marshall Island turtle nesting atolls. Hendrickson obtained his information about Bikini from Mr. Robert Ward, a heavy equipment maintenance supervisor for the Bikini Atoll Rehabilitation Project. Additionally, the popular movie *Mondo Cane* made several years ago showed rather large numbers of dead green turtles on Bikini, though the interpretation made that these had been disoriented by radiation damage and had wandered into the interior of the island to die is somewhat questionable. I have seen dozens of dead green turtles inland from the nesting beach on Baltra Island, Galapagos. This island appears to lack the normal sea-finding (or land-fleeing) cues that enable a turtle to identify the proper heading for the ocean.

Acknowledgments

I am most grateful to the World Wildlife Fund (United States National Appeal) for funding my field work in Micronesia; to R.M. (Chris) Christensen and *Chelonia* magazine for initial publication of the findings; and to Dr. Hal Coolidge for providing the initial impetus for my field work in Micronesia, as well as for vigorous support with all phases of the work.

Literature Cited

Agassiz, L.
1857. *Contributions to the Natural History of the United States, vol. 1.* Boston: Little, Brown and Company.

Anonymous
1956. Military geography of the Northern Marshalls. U.S. Army, Office of the Engineer, Army Forces Far East and Eighth United States Army.
1957. Notes on the present regulations and practices of harvesting sea turtles and sea turtle eggs in the Trust Territory of the Pacific Islands. Anthropological Working Papers, Trust Territory of the Pacific Islands, 18 pp.
1972. Tragedy of Eniwetok: One good bombing deserves another. *Pacific Islands Monthly* 43:28–29.
1975. Mid-Pacific Marine Laboratory. Enewetak, Marshall Islands, 25 pp.

Bates, M., and D. P. Abott
1959. *Ifaluk, Portrait of a Coral Island.* London: Museum Press Limited.

Brewer, Y.
1975. Micronesia: an historical summary. In *Trust Territory of the Pacific Islands:* The Annual Preservation Plan, vol. 3, 2nd. ed.

Bryan, E. H.
1971. Guide to place names in the Trust Territory of the Pacific Islands (the Marshall, Caroline, and Mariana islands). Pacific Science Information Center, Bernice P. Bishop Museum, Honolulu.

Carr, A. F.
1952. *Handbook of Turtles.* Ithaca, New York: Cornell University Press.
1965. The navigation of the green turtle. *Scientific American* 212:79–86.

Carr, A. F., and A. R. Main
1973. Turtle farming project in Northern Australia. Report of an enquiry into ecological implication.

Cushing, F. A.
1974. Observations on longline fishing methods conducted aboard the *R/V Bosei Maru. University of Guam Marine Laboratory Miscellaneous Reports,* no. 17.

Deraniyagala, P. E. P.
1939. The Tetrapod Reptiles of Ceylon, vol. 1.

Falanruw, M. V. C.; M. McCoy; and Namlug
1975. Occurrence of ridley sea turtles in the western Caroline Islands. *Micronesica,* 11:151–52.

Fosberg, F. R.
1969. Observations on the green turtle in the Marshall Islands. *Atoll Research Bulletin:* no. 135.

Girard, C.
1858. United States exploring expedition during the years 1838–1842, under the command of Charles Wilkes, U. S. N. *Herpetology* 20:1–496.

Helfich, P.
Undated. Report on the abundance of turtles and turtle nesting areas in the northern Marshall Islands. Manuscript, 2 pp.

Hendrickson, J. R.
Undated. South Pacific islands—marine turtle resources: a report prepared for the South Pacific Islands Fisheries Development Agency. Manuscript, 15 pp.

Hiatt, R. M.
1951. Marine zoology study of Arno Atoll, Marshall Islands. Scientific investigations in Micronesia: native uses of marine products. *Atoll Research Bulletin* 3–4:1–13.

Hirth, H. F.
1971. Synopsis of biological data on the green turtle *Chelonia mydas* (Linnaeus) 1758. FAO Fisheries Synopsis, no. 85.

Hirth, H. F., and A. F. Carr
1970. The green sea turtles in the Gulf of Aden and the Seychelles Islands. *Verhand. Konink Nederl. Akad. Wetensch., Afd. Natuurkunde.* 58:1–44.

Johnson, S. P.
1972. Palau and a seventy islands national park. *National Parks and Conservation Magazine* 46(4):12–17; 46(7):4–8; 46(8):9–13.

Marshall, M.
1975. The natural history of Namoluk Atoll, eastern Caroline Islands. *Atoll Research Bulletin* 189:1–53.

McCoy, M.
1974. Man and turtle in the Central Carolines. *Micronesica* 10:207–21.

Niering, W. A.
1963. Terrestrial ecology of Kapingamaringi Atoll, Caroline Islands. *Ecological Monographs* 33:131–60.

Pritchard, P. C. H.

1969. Studies of the systematics and reproductive cycles of the genus *Lepidochelys*. Ph. D. dissertation, University of Florida.

1971. The leatherback or leathery turtle. *IUCN Monographs,* 1:1–39.

Rüppell, E. W. P. E. S.

1835. Neue Wirbelthiere zu der Fauna von Abyssinien gehörig, vol. 3, Amphibien. *Frankfurt a. M.,* 1–18.

Smith, H. M., and E. H. Taylor

1950. An annotated checklist and key to the reptiles of Mexico, exclusive of the snakes. Bulletin of the United States Natural Museum 199:1–253.

Sowerby, J. de C., and E. Lear

1872. Tortoises, terrapins, and turtles. London, pls. i–1x, pages 1–16.

Viti, C.

1975. Voyage from the last century. *Glimpses of Guam* 15:17–32.

Wermuth, H. and R. Mertens.

1961. *Schildkroten, Krokodile, Bruckenechsen.* Jena: Gustav Fischer.

Wilson, P. T.

1969. Trust territory of the Pacific islands. In *Encyclopedia of Marine Resources,* ed. F. E. Firth, pp. 687–95. New York: Van Nostrand Reinhold Company.

1976. Conservation problems in Micronesia. *Oceans* 9:34–40.

PETER C. H. PRITCHARD

Mike A. McCoy
Micronesian Maritime Authority
Ponape, East Caroline Islands

Subsistence Hunting of Turtles in the Western Pacific: The Caroline Islands

ABSTRACT

The Caroline Islands comprise numerous low coral atolls and islands as well as high volcanic islands. Some of the remote coral islands are used by sea turtles, mostly the green turtle, *Chelonia mydas*, as nesting areas during the season which generally lasts from March to September. Green turtles, and to a lesser extent hawksbills, *Eretmochelys imbricata*, are found year-round in the lagoons of the high islands. The inhabitants of the coral atolls, the "outer islanders" have, for the most part, developed methods of capture and utilization exceeding those of the islanders residing in the administrative centers of Truk, Ponape, Yap, and Kosrae. However, population pressures, the emergence of a "money economy," and other factors have increased the pressures on turtles throughout the region. The decline in importance of traditional tabus and the preference for modern boats and motors over traditional canoes have led to the disappearance of the protective buffer these customs once provided. Turtles face increased harvesting, and there is a need for a conservation system to replace the original tabus. Any such system must be designed with the people in mind and worked out in partnership with them.

Stretching from 131° East to 163° East Longitude, the Caroline Islands comprise a series of both high volcanic and low coral islands and atolls placed in a rough line totaling 3,200 km in length. On the easternmost limit is the high volcanic island of Kosrae; the west is bounded by the small coral island of Tobi. The total land area does not exceed 1,193 km² of dry land. The lagoon area, on the other hand, encompasses 8,546 km² and a total of about 950 islands and islets. The ocean area inside the present political boundaries of the Federated States of Micronesia (Yap, Truk, Ponape, and Kosrae) and Palau exceeds 3.4 million km².

Within this vast expanse of ocean, sprinkled among

the island chain are a few uninhabited atolls and single islands, all coralline in structure, which serve as nesting beaches for green and, to a lesser extent, hawksbill turtles. These islands are sometimes visited by islanders in canoes from nearby inhabited islands, or by the crew from a passing fishing vessel. Little is known of the nesting turtle populations or of their capacity to withstand exploitation.

The inhabitants of the Carolines arrived many centuries before the first Europeans. They had already developed cultures allowing exploration of remote parts of the Pacific. Whether voyaging for discovery or due to social pressures, they inhabited the islands and were well established by the arrival of the first Spaniards in the sixteenth century. Serious colonization by Europeans did not take place until the Germans discovered how valuable the islands were for copra and other commodities, including turtle shell. There are, however, no reports in the literature of the early explorers or commercial entrepeneurs that would suggest that the resource ever existed on the scale reported for the Caribbean. While turtles were most certainly seen and occasionally eaten by early visitors and inhabitants alike, we can only speculate on their numbers.

The Carolines had a series of various colonial masters interested in different goals. First the Spanish arrived with soul-saving religion, and guns to back them up. They met with less than resounding success. Two notable events were the slaying of the Spanish Governor and troops in Ponape during the late 1800s, and the supposed eating by natives of a priest and his followers left on Ulithi atoll about the same period. The Germans administered the area from their western Pacific headquarters in Rabaul, but only for a relatively brief time. Their rule was cut short by the first world war and the almost immediate occupation of the major islands by the Japanese in 1914. The area was given to Japan as a mandated area under the League of Nations and remained so until 1945, when the United States occupied the main islands. The United Nations then gave the United States control over the area known as Micronesia as a Strategic Trusteeship, and today it remains the last trusteeship under U.N. control. Political talks are progressing with the United States on the one hand, and the separate delegations from the Marshalls, Palau, and the Federated States of Micronesia on the other.

Turtles are occasionally seen around the main islands of Yap, Koror, Truk, Ponape, and Kosrae, usually in the water, and never for very long. The turtles, mostly *Chelonia mydas*, are to be found in the uninhabited islands far from the dusty streets, bars, and tin shacks of the administrative and population centers. The turtles concentrate their nesting activities on small islands such as Oroluk, Pikelot, West Fayu, Gaferut, and Helens Reef.

Oroluk atoll is located midway between Truk and Ponape. Until the late 1960s it was uninhabited, but today a small band of Polynesians from Kapingamaringi occupy the island with government consent and cause disruption to what once was regarded as one of the largest turtle nesting areas in the Eastern Carolines.

The island of Gaferut, containing 0.111 km², has been used as a resource island for many years by the people of Faraulep, and to a lesser extent by those from Woleai and Ifaluk. Gaferut is now seldom visited, due to the decline in the use of voyaging canoes by those islanders. Another factor was the tragic loss of most of the able-bodied men of Faraulep in their canoes during a typhoon in the 1950s. The island is sometimes visited by the government field-trip ship from Yap, and occasionally passengers from Faraulep, Ifaluk, and Woleai take turtles to be carried to their home islands. The ship's visit is short, and the evening's take is usually never more than 8 or 10 green turtles. The island is visited in this manner perhaps 2 or 3 times a year. A unique feature of Gaferut is a reef extension on the northwest side of the island which contains a large, deep hole big enough to accommodate many large turtles. Turtles often stay in this natural hole during the day or days before nesting. The standard method of capture is to move silently to this depression and capture the turtles resting there. The island itself is heavily wooded and has a large population of sea birds, only 1 coconut tree, and no fresh water. This makes a rather inhospitable place for humans.

One recent visit to Gaferut was made by islanders in a canoe from Satawal who were returning home on a long sea voyage from Saipan in June 1979. The navigator of the canoe reported that the island was covered with tracks and nests. The canoe was heavily laden, and able to carry only 2 or 3 turtles on the continuation of its voyage. This points out the limitations placed on the taking of turtles by the traditional mode of conveyance: the voyaging sailing canoe. These canoes are capable of extended voyages of many hundreds of miles with capable navigators from the islands of Satawal, Puluwat, Pulap, Tamatam, and Elato. But the number of turtles taken is limited by the size of canoe (usually not exceeding 8.2 m in length) and by the winds encountered. This is in stark contrast to the government ship or stray Japanese or Taiwanese fishing vessel, which happen upon an island, with a capacity far exceeding that of the canoes.

The other important nesting areas in the central and western Carolines are the islands of Ulithi atoll. Traditional customs are still strong within the atoll, and the turtles are considered the property of the chiefs of Mogmog, the highest caste island in the atoll. Information about turtles from Ulithi is not readily available, but the relatively large numbers of nesting turtles reported on 2 small islands just outside the atoll war-

MIKE A. McCOY

rant a close study. A program sponsored jointly by the Peace Corps and the Micronesian Mariculture Center in Palau failed dismally in Ulithi during 1973 due to a number of factors, not the least of which was the personality of the Peace Corps volunteer assigned to the project. The resultant bad feelings have probably lessened the chances for serious investigators to do any work at Ulithi for some time.

Thus, except for the relatively few outer islanders who still possess the skills of their ancestors in bulding and sailing the large voyaging canoes, the great majority of the Micronesian population in the Carolines does not have access to turtle nesting sites. From a conservationist's standpoint this may be desirable, because the number of people exploiting turtles is reduced to manageable numbers (3,000 vs 90,000). Nevertheless, the remoteness of the islands makes the work more difficult than one would first imagine.

The inhabitants of the Caroline Islands are essentially of 1 Micronesian stock. Like other inhabitants of the Pacific, they can be grouped generally into 2 distinct groups: those that inhabit the low coral islands and atolls, and those who farm the higher volcanic islands of the chain. The languages are different from island to island, with the western Yapese and Palauan languages being distinct from the Carolinian dialects spoken from Ulithi to Kosrae. Ethnically and linguistically the outer islands of Palau are linked to Ulithi and the central Carolines, although administratively they are under the control of the administrative center in Koror, Palau. The rest of the islands' inhabitants live within 1 of the 4 states of the Federated States of Micronesia: Yap, Truk, Ponape or Kosrae. In Ponape, 2 outer islands, Kapingamarangi and Nukuoro, are inhabited by Polynesians rather than Micronesians, and make up the only distinctly different ethnic group in the Carolines.

It is important to note the basic differences between the coral atoll inhabitants and those living on volcanic islands. The former are mostly fishermen and "people of the sea," while the latter tend to concentrate their activities in farming and gathering crops on the more fertile high islands. As would be expected, the level of knowledge of the sea and its fauna is greatest in the coral islands.

The people of the coral islands, for the most part referred to as "outer islanders" by expatriates, have the greatest knowledge of turtle behavior, except for the inhabitants of Palau. Palau, in the western Carolines, is unique for many reasons; this uniqueness extends to knowledge of turtles and fishing activities in general. The Palauans have developed relatively exacting bodies of knowledge for much of the reef fauna, turtles included. Scientists studying there have remarked on the level of general knowledge and abilities of Palauans around the reefs of their home islands. But on the other high islands of Yap, Truk, Ponape, and Kosrae the knowledge of the sea is not on a par with the "outer islanders" of those states.

In the outer islands, knowledge of turtle behavior is wrapped up in traditional beliefs, altered somewhat by the advent of western schooling, the outboard motor, and other introductions. Population pressures in these islands have forced migrations to the administrative centers and, in some instances, colonization of atoll islands not usually used for habitation. In general, however, population pressures are not as extreme today as in the Gilbert Islands during the 1930s when the British Administration forced migrations and resettlement from traditionally inhabited islands to previously uninhabited ones.

These pressures will increase, particularly in the outer islands; and it appears that neither government nor local institutions are aware of or concerned with the problem. For example, the island of Satawal in the central Carolines had approximately 275 inhabitants at the end of the second world war. In 1968 the population of this 350-acre island was about 390. By 1978 it had risen to over 550, or a density of over 1,000 people per square mile. Marriage on other islands, employment in the administrative center and other factors tend to conceal the real growth, but an average of 22 births and only 3 deaths a year is quickly moving the island towards dangerous overpopulation.

These increasing pressures, in the case of Satawal, put increased pressure on traditional sources of protein, including turtles. Voyages in the traditional canoes will be made more often in search of turtles and fish on the nearby nesting islands, and perhaps the not-too-distant future will see the introduction of larger motorboats for this purpose. Indeed, on Lamotrek Island, 64 km west of Satawal, an 8.2 m diesel-powered skiff was recently purchased with funds granted by the District Legislature. The vessel reportedly travels to the atoll of Olimarao in search of turtles, and occasionally to Satawal for trading and social visits.

Turtles have suffered and will continue to suffer under such pressures. The atoll of Oroluk, located in western Ponape and already briefly described, was uninhabited until the late 1960s. When the Kapingamarangi people petitioned the government to allow colonization, a stable population resulted on Oroluk. While the numbers are not great, from 10 to 20 persons at any one time, the effects have been startling. The island itself is the only one in a large atoll enclosing 419 km^2 of lagoon area. The island has been known as a nesting ground for years during the season from March to September. The inhabitants have built a stone holding-pen, and turtles are placed within the pen to await the government field trip ship which calls about 6 times per year. Until recently turtles were loaded aboard the field trip vessel for return to Ponape, where they were either sold or eaten in the Polynesian village there. The enforcement of the Endangered Species Act has put a stop to commercialization, but subsistence use is still allowed under Federal law.

While there are no figures available on the numbers of nesting turtles at Oroluk, the inhabitants complain of a decline in numbers, and estimates of nesting females per year range from 40 to 100 individuals. This is not a great number, considering that, at least by reputation, Oroluk is considered one of the better "turtle islands" in the Carolines.

Because the physical environment of the outer islanders consists of coral atolls, and since the turtles prefer the beaches of the atolls and low islands to those found on the higher volcanic islands, they have the most contact with nesting turtles. They also develop the skills necessary for catching turtles in the lagoons and from boats and canoes. In the past 30 to 40 years, outer islanders in the administrative centers on the high volcanic islands, principally Truk, Ponape and Yap, often have shared these skills with friends and relatives there. Thus, techniques for catching turtles have been developed in the high islands which were absent in the past. As travel between islands is made easier by government-subsidized shipping, the chance for such minor technology-transfer increases.

In addition to sharing their own techniques and knowledge of turtles, the Micronesians learned much from the Okinawans who were brought to the islands prior to the second world war. The Okinawans came as laborers in the sea-oriented enterprises run by the Japanese, and many were excellent divers. They showed the Micronesians how to dive for turtles resting under the coral ledges, and how to gaff them with hooks embedded in long lengths of bamboo. The hooks were released from the gaff but remained tied to a long length of fishing line which was in turn tied to a floating log or other float. Some turtles were undoubtedly lost as they struggled to drag the float, and lines became entangled in the coral. But for the most part the method became an effective and successful way to catch green turtles, particularly in Yap.

The knowledge of turtle behavior possessed by outer islanders is limited, however. For example, in 1972 I inquired of the elders of an island their determination of periods between nesting. Some swore that females nest only once a season, others that she nests up to 10 times. Because they captured every nesting female they saw, there was no way for them to be sure. In another instance, a turtle was spotted nesting on a remote atoll away from the inhabited island. I asked the men how long they thought it would take for the eggs to hatch, assuming we left them in the ground. Some ventured 10 days, others 25. Nobody really knew, however, for they always dug up every nest they encountered and had no means of determining the time required. It was not until work at the West Fayu turtle hatchery showed the local crew that 58 to 60 days were required, that they began to understand some of the basics of turtle behavior.

Pressures on the turtles of the Carolines have been rising at an accelerated pace during the past 10 years. The main reasons for this seem to be the furtherance of the "money economy" in Micronesia, and the relative ease with which fishermen are now able to procure outboard motors, boats, and the gear required to hunt turtles. Amazingly, tangle nets such as are used in some places in the Caribbean are not used in Micronesia, and for that reason I have always hesitated to show the classic movie on the Miskito Indians produced a few years ago. Clarity of the water might be one of the reasons why people have never used nets, but the unavailability of materials might be another. Since most materials are now available, I felt it better not to show the movie and introduce the concept, rather than trust to *Chelonia's* eyesight.

Other factors have combined to increase the pressures on turtles. In the case of West Fayu, it was the island's flora. Until after the second world war, there were no coconuts on the island, which limited the amount of time voyagers could remain to await the turtles. Shortly after the war, a major infestation of an unidentified insect killed many of the bushy trees on the island which had prevented coconuts from receiving enough sunlight to survive. People from Satawal then transported copra nuts to the island and planted much of the island in coconuts. The coconuts have been the single most important change on the island (not counting a wrecked 9,000 ton freighter full of Toyotas in 1971). Man can now increase the length of a stay to hunt turtles.

Another factor that has increased the number of voyages is the improvement in materials used in the manufacture of the traditional canoes. Until the middle 1950s pandanus sails were used exclusively on all canoes on Satawal Island in the central Carolines. The introduction of cotton canvas sails greatly increased the speed and performance of the canoes. Recently, the introduction of dacron sailcloth has lessened the voyage time even more. Other improvements and introductions, such as the magnetic compass, have meant a greater confidence in voyaging and a strong probability that many more voyages are undertaken now than in the past.

Many of these improvements, including introduction of motor boats on other islands, have occurred during the years since 1965. They have greatly increased the pressure on the turtle populations in all of the areas visited by inhabitants of the central Carolines, with the possible exception of Gaferut.

The motorboats, usually under 6.7 m in length and powered by 25 to 40 horsepower outboard motors,

MIKE A. McCOY

are used mostly in the administrative centers (the high islands). In Truk, for example, motorboats are used to chase the turtles on moonlit nights across the shallows, and harpoons are used to spear and retrieve them. Turtles captured in the higher islands, where varieties of seagrasses are found, tend to be smaller and more variable in size than those in the nesting areas of the outer islands.

In the outer islands, the turtles are most often captured while mating or on the nesting beaches. The people of Satawal Island, and to a lesser extent those of the westernmost islands of Truk, Puluwat, Pulap, Tamatam, and Pulusuk, go to West Fayu to capture turtles. During the day a close watch is kept for mating turtles within the lagoon. If mating turtles are spotted, a canoe races to the position. The men affix large hooks to strong lines and then place the hook in a notch in the end of a piece of bamboo or stick approximately 2-m long. The ends of the lines are then tied to a large boom carried on the canoes, or to the canoe itself. Two men are given the responsibility of swimming up silently behind the mating turtles with the hooks. They then swim under the mating turtles, and each man places a hook into the skin on the turtle's neck. A sharp watch must be kept for sharks which occasionally cruise around mating turtles and take nips out of their flippers. For the most part, the mating turtles are oblivious to what is taking place around them. The swimmers are usually successful in their attempts. Once the turtle is hooked it immediately sounds and a tug-of-war ensues, with the turtle usually losing. Often the equipment for this type of capture is not available when islanders on fishing voyages sight mating turtles. In this case, the men swim up to the unsuspecting turtles and grab them in a "full nelson" hold from the underside. The man's hands are then placed under the chin of the turtle and force its head back, minimizing the chances of being bitten. Other men then jump off the canoes with whatever ropes and lines are available, and attempt to tie the front flippers in a manner that will allow them to drag the turtle aboard. This is obviously a much more dangerous and less successful operation than the hook and bamboo pole method.

During moonlit nights on West Fayu, it is also possible to tether a previously captured female to a tree, and allow her to swim in the shallows around the island. The nesting beach is not more than 15 m wide, and the tether is fairly short. Men then climb into the trees near the water's edge and wait for her to attract males. Once the males are attracted, the men chase them down.

Pikelot Island is perhaps one of the best known and most visited turtle islands in the central Carolines. Canoes from Puluwat, Tamatam, Pulap and Pulusuk in the western part of Truk visit the island during the summer months to capture turtles to take to their home islands for consumption.

Canoes from Satawal also visit Pikelot, as it is traditionally owned by them, and administratively it is the easternmost island in Yap State. In 1978 canoes journeying to Pikelot returned to Satawal carrying 18, 10, and 11 green turtles on 9 March, 8 May, and 16 June, respectively. Another trip by 4 canoes to West Fayu returned on 31 May the same year and brought with them 11 captured turtles from that island. This number of turtles is considered average to good. The yearly fluctuations in the total number harvested varies considerably. These excursions usually last from 1 to 3 weeks, depending upon the winds, weather, and food supply. An important consideration for the voyaging canoes is the weather, for the turtle nesting season coincides with the typhoon season in the central Carolines, and is also the time of the most variable and fickle winds.

Reports from Pikelot in 1977 showed canoes from Puluwat averaging 4 turtles a night (all nesting females) during a week's stay in May of that year. The report, published in Guam, further noted that because of the number of tracks on the beach, the 28 individuals taken could have represented only a portion of the population. What was not understood, however, was that during the good weather experienced on the island, the tracks and nests could have been made over a 1- to 2-month period by a relatively small number of turtles.

The total number of nesting females on the beaches at West Fayu and Pikelot is probably not very large, but their presence provides the incentive for inhabitants of the nearby islands to continue making the large sailing canoes primarily for the purpose of transporting live turtles back to the inhabited islands. There is a good chance that without adequate stocks of turtles, the canoe voyaging tradition would suffer, and with it an important component of local society. Thus, while actual numbers of turtles harvested may not be large, perhaps averaging 30 to 70 a year per island for the 6 major islands involved (Satawal, Puluwat, Pulap, Tamatam, Pulusuk, and Lamotrek-Elato), the turtles contribute much to their overall cultural stability, reinforcing their independence from the outside. The estimated maximum contribution to the protein, perhaps 18 kg a person a year, is not nearly as important as this cultural role.

An important buffer provided for the turtles were the past tabus and ceremonies surrounding the taking and consumption of turtles. Canoe travel provided an additional buffer, and has continued in the face of the lifting of traditional religious tabus. While it cannot be shown quantitatively that many of the tabus formed such a buffer, it is my opinion that a substitute is required to restore the balance and to enable the relationships to continue to exist.

The taking of turtle eggs was not traditionally covered by tabus. The exploitation of this resource has continued unchecked on almost all islands where there is nesting in the Carolines. The comparison of 25 g of protein to a possible 150 kg needs no elaboration here. However, the local inhabitants' belief must be remembered that the sea has been, and always will be, an adequate provider for all things. In my discussions with various inhabitants, none expressed great concern over the taking of eggs or, when concern was expressed, it was always countered by a bird-in-the-hand philosophy.

In the population centers such as Truk and Yap, the taking of marine turtles is an occurrence best equated with deer hunting in the United States. Often the hunting of turtles is undertaken with a form of sport in mind and, although the turtle is eventually consumed, it does not figure as prominently in the lives of the inhabitants as it does in the outer islands.

It has been my continued belief that efforts at conservation should be made with the people of the area firmly in mind and that assistance should be given to enable them to better understand the resource. This is not an easy task for scientists and others who themselves know so little of the behavior of turtles in the Carolines. Yet, the challenge presented must be met in the very near future if turtles are to remain a viable part of the island ecosystem.

MIKE A. McCOY

C. Sylvia Spring
Wildlife Division
Department of Lands, Surveys and Environment
P.O. Box 2585
Konedobu, Papua New Guinea

Status of Marine Turtle Populations in Papua New Guinea

ABSTRACT

There are 6 species of marine turtles found in Papua New Guinea (PNG). These are the green turtle *Chelonia mydas*, the hawksbill turtle *Eretmochelys imbricata*, the leatherback turtle *Dermochelys coriacea*, the loggerhead turtle *Caretta caretta*, the Pacific ridley turtle *Lepidochelys olivacea*, and the flatback *Chelonia depressa*. There is also 1 record of the subspecies *Chelonia mydas agassizi* from PNG (Pritchard 1979).

Village surveys indicate that populations of marine turtles are slowly declining in most areas. This decline is attributed to many things but mainly to the breakdown in traditional practices, introduction of modern fishing methods, population increase, and the introduction of a cash incentive in modern villages. It is hoped that education and extension work will result in the setting up of Wildlife Management Areas to allow populations of marine turtles to recover. Internationally, Papua New Guinea participates in the Convention on International Trade in Endangered Species of Fauna and Flora (CITES) which prohibits overseas exploitation of PNG marine turtle resources.

Introduction

Papua New Guinea is a recently independent country consisting of the eastern half of the island of New Guinea, including the Bismark Archipelago. The country is divided into 19 separate provinces, throughout which there are 717 spoken languages. The 3 main languages are English, Motu (spoken in Papua) and Pidgin (spoken in New Guinea).

Papua New Guinea has a rather unique biological resource. Of the 7 species of marine turtles alive in the world today, 6 are found in Papua New Guinea. Committed to the conservation of its natural resources, especially those of nutritional or cultural value to the people, the PNG government set up the Marine Turtle Project in 1977 to:

• ensure the survival of the marine turtle resource and its subsistence and cultural values for the coastal people of Papua New Guinea;

- map the distribution and abundance of marine turtle species in Papua New Guinea;
- collect traditional information concerning the subsistence and cultural values of marine turtles;
- educate the village people on the biology of their marine turtle resource and the need for conservation.

Methods

Village surveys and the closely related education and extension work are the main tools used in Papua New Guinea in marine turtle conservation.

Surveys

Village surveys provide the bulk of the information presented in this report. Most provinces have been visited, except the Morobe, Northern, and Gulf Provinces. However, parts of New Ireland, Bougainville, New Britain, and Milne Bay need to be more thoroughly investigated. It is hoped to eventually cover the entire coastline of Papua New Guinea and the islands for the sake of completeness. However this is a major and expensive operation involving travel to thousands of villages, many of them remote, and money, manpower, and time are limiting factors.

Aerial surveys have been carried out rather sporadically since 1977, and have been a useful means of reaching the more remote villages and of obtaining a general indication of nesting and feeding areas.

Market surveys are currently being conducted in Daru and Port Moresby.

Education and Extension Work

Education plays an important role in conservation of marine turtles, and extension work is tied in with village surveys. In Papua New Guinea, the onus for conservation work lies with the landowners. They will be responsible for introducing and enforcing conservation measures in the villages. The Wildlife Division provides the legal means for villagers to do this through the system of Wildlife Management Areas. Educational aids aimed at village, government, and general public include posters, badges and stickers produced by Rare Animals Relief Effort (RARE) Incorporated and T-shirts which are used as a reward for tag recoveries.

This paper summarizes the results of surveys into the distribution and abundance of marine turtles in Papua New Guinea.

Results

The Green Turtle (Chelonia mydas)

The green turtle (*Chelonia mydas*) is the most abundant and widespread turtle found in Papua New Guinea and is the most heavily utilized by villagers for food. It is identified with a vernacular name in nearly all parts of PNG.

Green turtle meat is considered by far the most tasty turtle meat. Adults and subadults are hunted in preference to juveniles, as are females to males, because of their higher fat content. In most places turtles are used for village consumption, however in the Central Province most turtles find their way into Koki Market where a large adult fetches between US $90 and US $115. The main turtle hunters in the Central Province are the Hula people. In March this year, 133 adult green turtles passed through Koki Market at a cash value of US $9500. Of these, 113 were caught and sold by Hula village people.

In comparison, in the Daru area, where turtles are much more abundant, an adult green turtle fetches between US $15 and US $30 in the market. In March this year, 47 turtles were sold at a cash value of US $1,080. The greater hunting pressure in the Port Moresby area is a result of the greater need for money and the higher prices of turtle meat.

The green turtle is found around the entire coastline of Papua New Guinea (Figure 1). The highest concentrations occur on the uninhabited island groups; in areas inhabited by Seventh Day Adventists and others who do not favor turtle meat as food; and in protected areas.

UNINHABITED ISLAND GROUPS

There is an abundance of good nesting beaches and uninhabited islands around the coast of Papua New Guinea. In most village interviews, the people could name a nearly uninhabited island or deserted stretch of beach where turtles nest. Turtles are no longer found nesting on beaches or feeding on reefs in front of villages, due to overhunting. People now have to go farther afield to catch turtles. There are several uninhabited island groups in the Manus Province which are important breeding areas for green turtles. These are the Sabben Islands (6 islands which are traditionally owned by a clan at Bipi Island), the Los Reyes Islands (3 islands traditionally owned by several nearby islands), the Purdy Islands, and the Johnson Islands. Nesting coincides with the southeasterly winds, from May to August. Eggs and females are collected from these islands, but is usually within limits imposed by distance and weather and by the clans who own the turtles. There are also the Kaniet Islands and the Anchorites, 2 uninhabited and isolated island groups in the Western Islands and rarely visited by people.

SEVENTH DAY ADVENTIST AREAS

Seventh Day Adventists do not eat meat and that inclubes turtle meat. Most Seventh Day villagers follow

C. SYLVIA SPRING

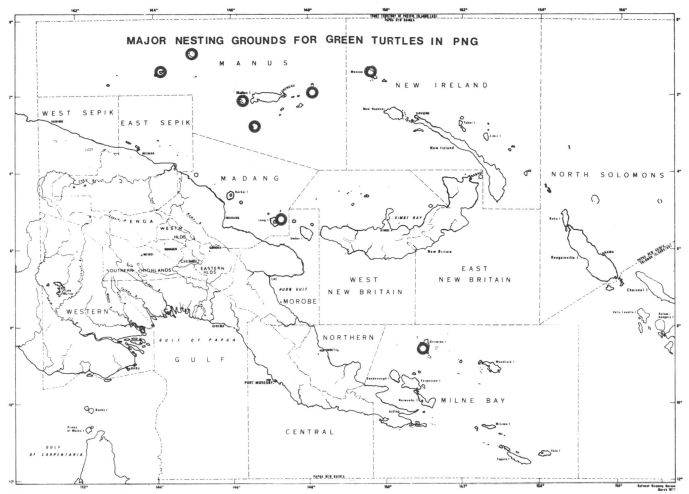

Figure 1. Major green turtle nesting grounds in Papua New Guinea.

this rule. In areas where this religion is established, villagers report a noticeable increase in the turtle populations over a period of 30 to 50 years at the most.

The Hermit Islands are an extremely isolated ring of coral islands, with a very small population (53 persons) who accepted the Seventh Day Adventist religion in the 1950s. Prior to that time they hunted and ate green turtles. There are 4 main turtle islands and turtles nest on all the islands from June to September: Pami, Makan, Planau, and Kocheran Islands.

The Ninigo Islands lie approximately 64 km to the west of the closest neighbor, the Hermits, and have a population of 567, spread over 6 islands, half of which are Seventh Day Adventists. Green turtles are not nearly so abundant here as in the Hermits, but are plentiful and nest on the uninhabited islands from May to September. Other important breeding areas for turtles occur on Lou Island in the Manus Province, and Mussau

Island of the St. Matthias Group in the New Ireland Province. The people in both areas converted to the Seventh Day Adventist religion in the 1930s.

PEOPLE NOT FAVORING TURTLES AS A FOOD

In the Trobriand Islands, the village people are cultivators of yams, a very important foodstuff to them. Turtles are not eaten, as it is believed that eating turtle meat will ruin the magic of the garden, and the yams will die.

Only 1 or 2 villages hunt turtles. Turtles nest on a few beaches on Kiriwina Island but mainly on the outlying uninhabited islands, for example, Tuma, Munuwata, and Simlindon islands from March to April. Many turtles are also reported on the Luscany islands and reefs and Simsim Island.

A very large feeding ground of approximately 52

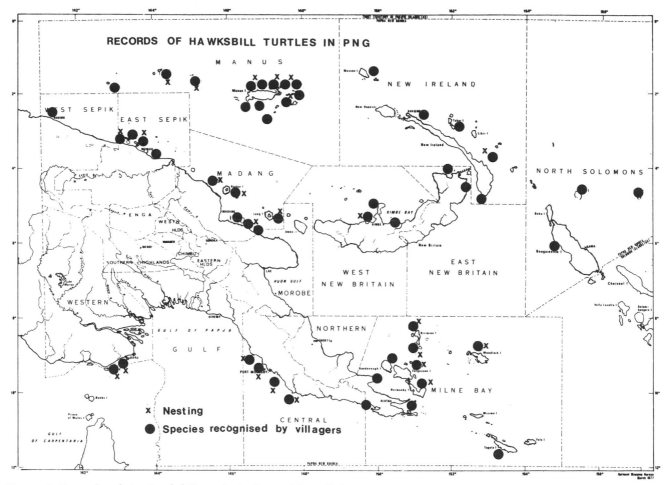

RECORDS OF HAWKSBILL TURTLES IN PNG

x Nesting

● Species recognised by villagers

Figure 2. Records of the hawksbill turtle in Papua New Guinea.

km² lies between Losuia and Vakuta Island. It consists of sandy flats full of seagrass and occasional coral heads, with a mean water level of 5.8 m low tide. Many turtles are seen feeding here.

PROTECTED AREAS

The other main breeding area for green turtles occurs at Long Island. The major nesting beaches occur along the northwest of the island from Malala village to Sororo. Nesting is reported to occur year round, except during the rough northwest season from November to February. This area has long been exploited by passing fishing boats and government trawlers. Reports expressing concern over the harvesting rate were received by the Wildlife Division as early as 1974. Long Island has since been declared a Wildlife Management Area with restraints on taking nesting females during the months of May, June, and July. Also, outsiders are no longer permitted to go to the area and take turtles.

It is known that green turtle stocks on the Papuan side are shared with Australia, especially in the Torres Strait area. Turtles tagged in the Torres Strait are regularly recovered around the Daru and sometimes in the Gulf and Central Provinces. There is no evidence of shared stocks on the New Guinea side. All green turtles seen in PNG are of the subspecies *Chelonia mydas mydas*. There is a single record of *Chelonia mydas agassizi* in PNG from the Manus Province (Pritchard 1979).

The Hawksbill Turtle (Eretmochelys imbricata)

Hawksbills, almost as abundant and widespread as green turtles, are found wherever there are coral reefs. In some places they are even more abundant than the green turtle for example, Lou Island in the Manus Province and Kairuri Island in the East Sepik Province. Hawksbills are eaten when found, apparently with little ill-effect. Poisoning is uncommon (Carr, this volume),

C. SYLVIA SPRING

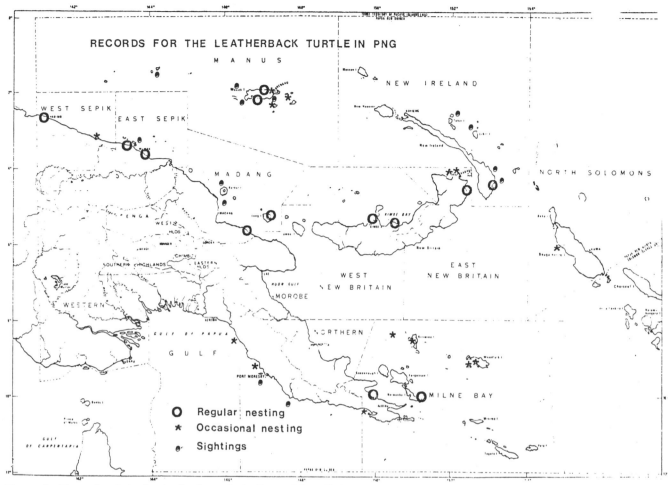

Figure 3. Records of the leatherback turtle in Papua New Guinea.

however in a recent incident on the Talasea Peninsula, 35 people were poisoned by eating hawksbill meat, and two children died.

Hawksbill shell is traditionally used to make combs, limesticks and jewelry and in some places, brideprice items. Shells are either kept as a decoration for the house or sold in the markets on the streets and in artifact shops to tourists. Apart from village use, there is one jeweler who specializes in tortoiseshell jewelry. Shells are bought in small quantities from the Marshall Lagoon area. There have been several incidents of illegal smuggling of hawksbill shells by Japanese tuna boats.

Nesting of hawksbills appears to coincide largely with that of green turtles (Figure 2). In the East Sepik Province, villagers report hawksbills nesting widely on the mainland and on small offshore islands from May to September. Raboin Island off the tip of Cape Wom is an uninhabited island where hawksbills nest in numbers. They are also reported to nest on the islands near

Vanimo in the West Sepik Province (Pritchard 1979).

In Manus, hawksbills nest on the mainland and small islands especially Onnita Island, Sabben, Purdy, and Johnson Islands from June to August.

Hawksbills also nest in the Hermit and Ninigo Islands from July to August. In Long Island, Madang Province, hawksbills are the third most common species after greens and leatherbacks and nest occasionally on the beaches from May to July. They are also reported to nest on the mainland of the Madang Province.

The hawksbill is the second most common turtle in the Trobriand Islands after the green and nests on parts of Kiriwina Island and outlying islands in March and April. Tortoiseshell earrings and limesticks are a common sight in the Trobriands. Hawksbills are reported to be as common as greens in the Woodlark Islands.

Hawksbills nest on several islands in the Central Province, Idia and Hoidana, Fisherman's Islands in low numbers, from December to January. Nesting of

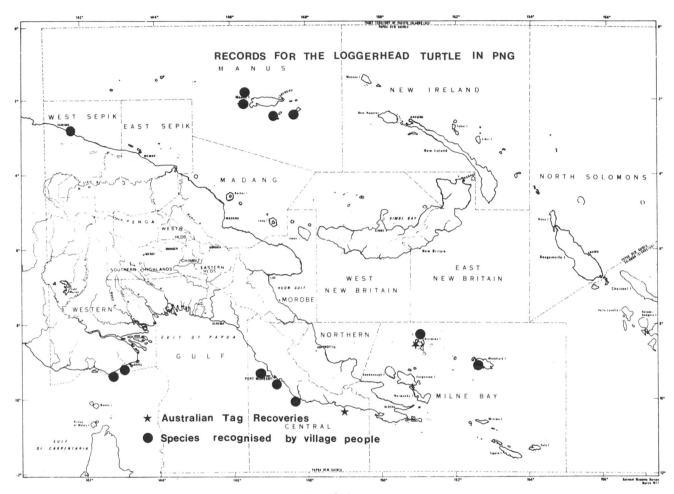

RECORDS FOR THE LOGGERHEAD TURTLE IN PNG

★ Australian Tag Recoveries
● Species recognised by village people

Figure 4. Records of the loggerhead turtle in Papua New Guinea.

hawksbills is also reported from the coastline near Daru, in the Western Province and at Garu Village in New Britain and on the southwest coast of New Ireland. Pritchard (1979) also saw a specimen that had been captured while nesting on the Gazelle Peninsula during the nesting season from September to January.

The Leatherback Turtle (Dermochelys coriacea)

The leatherback is easily identified from color photographs. Village people have no difficulty recalling this turtle.

Regular nesting is reported to occur widely along the north coast of New Guinea and on some of the larger islands but always in low densities (Figure 3). Reported regular nesting sites include Tulu village and Timonai village on Manus Island, Garu village, Kimbe Bay and Ganoi village in New Britain, along the southeast coast of New Ireland, on Long Island and parts of the mainland of the Madang Province, on Normanby

Island in the Milne Bay Province and along the coast from Boiken to Turubu in the East Sepik Province and around Aitape in the West Sepik Province.

Many occasional nestings have also been reported, for example, on Kiriwina and Simsim Islands in the Trobriands, in the Woodlark Islands, on Lou Islands and Tingos village in the Manus Province, at Pilapila Beach near Rabaul. Apart from nesting, many sightings of leatherbacks floating at sea have been recorded.

This turtle and its eggs are usually eaten by village people. In Tulu village in the Manus Province the people eat this turtle whenever it comes up to nest. There are several disadvantages in eating this turtle, for instance, oily meat and a fishy smell that stays with consumers for many days. The shell is also boiled down to collect oil for wick lanterns.

The Loggerhead Turtle (Caretta caretta)

Village people tend to confuse the green turtle, the

C. SYLVIA SPRING

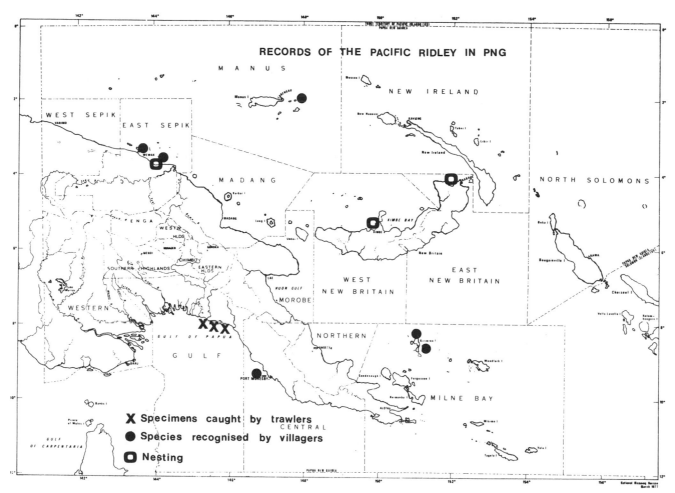

Figure 5. Records of the olive (Pacific) ridley in Papua New Guinea.

olive ridley, and the loggerhead. The only positive identification of this species in Papua New Guinea is from tag recoveries (Figure 4). There have been several tag recoveries of Australian-tagged loggerheads from the Trobriand Islands, where the village people report the mating and nesting of loggerheads on some of the outer islands. The loggerhead is also rather widely recognized and identified by a vernacular name along the coast of the Western Province; from Hula and Porebada villages and Fisherman's Island in the Central Province; in the Woodlarks in the Milne Bay Province; and from several locations in the Manus Province. They are always reported as uncommon at these locations with no known nesting.

The Olive Ridley (Lepidochelys olivacea)

Reported nesting sites for the olive or Pacific ridley in Papua New Guinea include: Turubu village in the East Sepik Province; Garu village in the West New

Britain Province (Kisokau 1972); and in Ataliklikun Bay in the East New Britain Province (Figure 5 and Pritchard 1979).

The species is also recognized at several locations around Papua New Guinea and identified by the village people with a vernacular name. These locations include: Nuguria Island, Bougainville Province; Mamuan, Kabilomo, Garu, Marem, and Nakanai villages on the island of New Britain; Porebada village and Fisherman's Island in the Central Province; Ahus and Pak islands in the Manus Province (Rhodin, Spring and Pritchard, in press), and Kitava Island and Kaibola village in the Trobriand Islands.

The Flatback Turtle (Chelonia depressa)

The only records of this species in Papua New Guinea waters comes from the incidental catch of flatbacks by prawn trawlers operating in the Gulf of Papua (Figure 6). To date there have been only 11 such records.

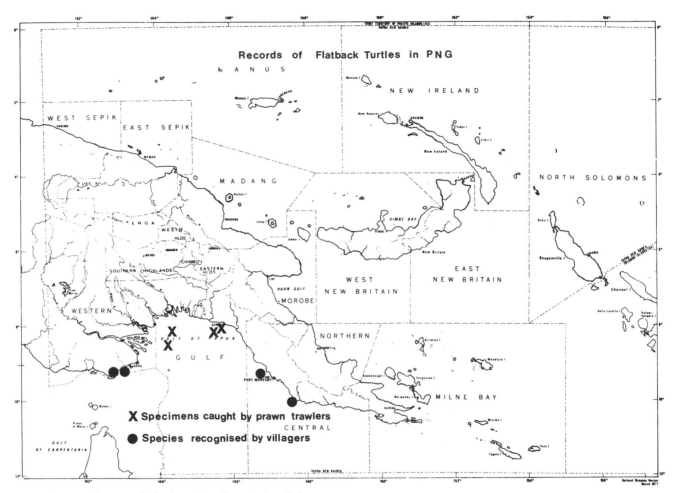

Figure 6. Records of the flatback turtle in Papua New Guinea.

There is no reported nesting of flatbacks along the coastline. This species is identified with a vernacular name from Fisherman's Island in the Central Province and from Tureture village in the Western Province (Rhodin, Spring, and Pritchard, in press).

Legal Status

Papua New Guinea signed the Convention on International Trade in Endangered Species of Fauna and Flora (CITES).

Under the *Fauna (Protection and Control) Act* of 1974, the export of all wildlife whether dead or alive, parts or products, requires an export permit issued by the Conservator of Fauna. Also under this Act, vertebrate wildlife may be exported only to approved institutions for legitimate scientific or zoological purposes (Parker 1978).

On a national level, the Wildlife Management Areas legislation allows landowners legally to protect their wildlife resources once the area and rules are gazetted in the PNG National Government Gazette.

Conclusions

The green turtle and the hawksbill are the most widespread and abundant turtles in PNG waters. The green turtle is the most heavily utilized for food. It is becoming obvious in areas where turtles are hunted for food that populations are declining. Many of the older people have observed the increased hunting of turtles and their subsequent decrease in numbers. There is a generation gap between the traditionally reared elders and the young people who have been exposed to a western way of life. Traditional respect and authority is eroding as a result of this generation gap, and traditional rules and regulations are being disregarded more and more by the younger people.

However, it is totally unrealistic to expect villagers to stop eating turtle meat. As an alternative, villagers

C. SYLVIA SPRING

should set aside beaches or islands as Wildlife Management Areas where the turtle populations could recover. Education through extension work will be important in stimulating village councils to consider the need for and to discuss ways to conserve marine turtles in their areas.

Acknowledgments

This paper could not have been prepared without the cooperation of the village people of Papua New Guinea who contributed information about their marine turtle resource. Nor could surveys have been conducted without the cooperation of the Provincial Wildlife staff. Also Fisheries biologists working in the Gulf have made available information on the incidental catches of ridleys and flatbacks in the Gulf of Papua. I would also like to thank Colin Limpus for making available tag-recovery data.

Literature Cited

Bustard, H. R., and C. Limpus
1970. First international recapture of an Australian tagged loggerhead. *Herpetologica* 26:358–59.

Kisokau, K.
1972. Turtle survey of Garu Village, West New Britain. Manuscript.

Parker, F.
1978. Collecting, export, research and filming involving wildlife in Papua New Guinea. *Wildlife Publication* 77/7.

Pritchard, P. C. H.
1979. Marine turtles of Papua New Guinea. Report on a consultancy. Unpublished report to the government of Papua New Guinea.

Rhodin, A. R.; C. S. Spring; and P. C. H. Pritchard
In Press. Glossary of turtle vernacular names used in the New Guinea region. *J. Poly. Soc.*

C. Sylvia Spring
Wildlife Division
Department of Lands, Surveys and Environment
P.O. Box 2585
Konedobu, Papua New Guinea

Subsistence Hunting of Marine Turtles in Papua New Guinea

ABSTRACT

In Papua New Guinea marine turtles are heavily utilized by coastal and island villagers as a source of food, for traditional feasts and exchanges, and for sale in local markets.

In general there are many traditional rules and regulations concerning the hunting and use of turtles, but these are dependent on respect for traditional authority. In most areas this authority is eroding as a result of the younger generation's exposure to the western economies and way of life. Young people are taking advantage of modern equipment to catch turtles for everyday use and for sale in the markets. Village elders, beginning to notice the subsequent decline in turtle numbers, attribute it to the disregard of old traditions.

This paper summarizes the findings of surveys on the subsistence and cultural significance of marine turtles in Papua New Guinea (PNG). These surveys include a postal questionnaire, and village and market surveys.

Methods

In 1977 the postal questionnaire was prepared in the 3 main languages of Papua New Guinea: English, Motu, and Pidgin and was distributed to various schools, colleges, missions and government organizations around the PNG coast. The information received in these questionnaires was carefully assessed and used as background information for village surveys. Village surveys are a valuable source of data on traditions. Interviews are conducted informally, with village elders, councillors and turtle hunters participating. Information is collected on traditions such as hunting methods and their associated rituals, the use of turtles in the village, and legends. Village leaders are advised of a visit by a *tok save*, a message sent over the local radio so that villagers will be present when the interviewers arrive.

Daily market surveys are currently being conducted in Daru and Port Moresby.

Results

Around the coast and islands of Papua New Guinea people rely on the sea as a major source of protein. Fish, turtles, and shellfish provide the main wealth of the village. Often gardens are very poor, and so the people traditionally exchange their fish and turtles for garden produce such as sac-sac (sago), taro, and greens from the mainland or island villages.

The major source of protein is fish which is eaten daily. Second in importance are turtles. The most heavily utilized turtle is the green turtle (*Chelonia mydas*). However hawksbills (*Eretmochelys imbricata*) are also widely eaten, usually with little ill effect (Spring, this volume).

In most parts of PNG, turtles are highly valued for traditional use and for their cash value. However, in the Trobriand Islands, they have no special significance for the people who cultivate and prize yams far beyond their nutritional worth. Turtles are eaten when found but are not a sought-after food and certainly not used on important occasions such as feasts.

In all other coastal areas, however, turtles are or were traditionally eaten in feasts, for example brideprice repayments, funerals, the building of a new canoe, the opening of a new *haus boi*, the birth of a first child. Today, feasts are also held for nontraditional special occasions relating to business, political and religious activities such as Christmas, Independence Day celebrations, the opening of a new church or business group. When turtles are required for a feast, the chief or leader organizes the hunters and canoes to go out and get turtles, up to 60 for a big feast. Turtles are either kept on their backs in the shade in the village or in *banis* in the sea. While they are in *banis* they are fed seagrasses, chopped clams and fish to ensure that they do not get too skinny in the meantime. When all preparations have been made for the feast, the turtles are killed. If guests from other villages are invited, they bring exchange presents such as sac-sac (sago), and other items of wealth (for example, *tambu*, dogs teeth), according to the number of turtles which are provided by the host village. Turtles are usually given a quick roast and then cut up and boiled in a pot with a few greens. All of the turtle is eaten including parts of the shell, bones, blood, and internal organs. When a hawksbill with a particularly beautiful shell is caught and eaten, the shell is saved for making into combs or preserved as a decoration for the house or sold to tourists. In the past, the hawksbill shell was used to make a number of everyday items such as spoons and knives, but these are now supplied by trade stores. Hawksbill shell was also used to make some items of traditional *bilas* such as belts, bracelets, earrings, limesticks and brideprice items but these are rarely seen today.

In areas close to town centers, turtles are being hunted with little restraint, for daily consumption and for sale in the town markets. A large green turtle will fetch between US $90 and US $115 in the Port Moresby market, for example. Shells are also sold to tourists at between US $15 and US $30 for a good size shell.

However, there are other areas where turtles are no longer hunted at all; these are the Seventh Day Adventist villages, where the people do not eat meat. Of the areas I have surveyed, there are only a few locations where turtles are still abundant; two of these are Seventh Day Adventist areas, Massau Island in the New Ireland Province and the Hermit Islands in the Manus Province.

Turtles also contribute to the oral history of the village. There are many legends and stories to explain the origin of turtles, why they entered the sea, how they got their shells, and so on. Some clans believe they are descended from turtles, and stories describe this relationship. Some magic men claim to possess powers over turtles. There are 4 such men, to my knowledge, in the widely separated provinces of East Sepik, Western, Manus, and Milne Bay. Each of these men is highly respected within his village and only uses his magic for very important occasions. For example, at Ponam Island in the Manus Province, the traditional net is only used in association with magic. It was last used in 1975 when there were several important occasions occurring together—Independence, the ordination of a local priest, and the opening of a church.

Methods of Hunting

In Papua New Guinea turtle hunting methods have been traditionally passed down from generation to generation, with a few modifications along the way. Hunting techniques and their associated rituals differ from area to area, but they can be roughly grouped as follows.

Netting

The traditional net is rarely used today, but it was rather widely used in the *taim bilong tambuna*, the olden days. The net is made from bush fibers. The art of making the *kapet*, the traditional net, belongs to certain families and is passed down from generation to generation. Most nets have disintegrated today. However, in the Manus Province several are left and are used for very special occasions. The one on Ponam is considered a sacred object and is stored in its own house and looked after by an elder who possesses magic powers and who is highly respected in the community.

When turtles are needed, the people concerned see the 2 leaders of the turtle net and discuss their requirements. The leaders then confer and set a date for the hunt. Twenty-four men are needed to cast the net, 12 on each side (each leader is responsible for his own side). The leaders pass the message among the 12 men that a hunt is on and to prepare according to the rules. On the day of the hunt the canoes gather together and leave at dawn. The net, weighted with stones, is carried across 2 large canoes. There are 10 canoes altogether, 4 small ones on each side. The canoes halt at a passage and wait until a turtle is seen. Then the net is cast and some hunters jump into the sea with it. When the turtle is caught in the net the men call to the large canoers who converge. The small canoes duck in and pick up the turtles. Up to 7 or 8 large turtles can be caught in 1 channel. The whole process is carried out according to strict ritual, and so the turtle hunt becomes quite an occasion in the village.

Harpooning

This, the most widely practiced technique, is traditional in some areas and introduced in others.

FIXED-SPEAR TIP

This consists of a wood or bamboo harpoon with a fixed iron tip. This is used in the East Sepik Province, Madang Province and the Trobriand Islands. Two or three men in a small canoe hunt turtles, usually at night, using a lantern. When the turtle is speared, 1 or 2 men jump in the water and pull the turtle on to the canoe. Only a few turtles are caught on these hunting expeditions as there is limited space on the canoes.

DETACHABLE SPEAR TIP

In the Manus Province this widely practiced method was taught to the villagers by Japanese fishermen prior to the second world war. It consists of a wooden harpoon with a detachable spear tip made from a 3-cornered file connected to a *perei*, a wooden float, by a nylon cord. When the turtle is speared either from the canoe or by a swimmer in the water, the harpoon detaches, and the turtle is allowed to swim until it is exhausted. Then it is picked up by the canoe. This technique is also used in the Western Province, where villagers have magnificent sailing outriggers. A spotter on the mast directs the harpoonist at the prow of the boat.

PLATFORM

This was the traditional way of spearing turtles and dugongs in the Western Province. It is no longer practiced, but was 40 or 50 years ago. Turtle hunters would build a platform made of bush materials over the reef and wait for turtles and dugongs to swim past. When one did, it was promptly speared. The turtle was allowed to run and was pulled in when tired.

By Hand

In the St. Matthias Group in the New Ireland Province, turtles were traditionally caught by hand. Today the people are Seventh Day Adventists and do not eat turtle meat. The village elders believed that drinking turtle blood would increase their swimming and diving powers so turtles had to be caught without spilling a drop of blood. Canoes would chase a turtle until it tired. A hunter would then leap into the water, wedge a wooden pole in the soft skin of the neck under the shell, and flip the turtle onto its back. The turtle was then lifted onto the canoe alive and unhurt.

In Bipi Island, turtles were also traditionally caught by hand for feasts. The village chief would call all the hunters and tell them to prepare their canoes to go and catch turtles. Each hunter would prepare his canoe and take along his supplies (some food, tobacco, and betel nut). On reaching the turtle islands the hunters would prepare all the food in one pot and offer it to the spirits of the reefs and beaches. Next morning all the canoes would go to sea in a line and look for turtles. When a turtle was spotted a competition would ensue to see who could catch the first turtle. Each canoe would average 4 or 5 turtles, depending on the skill of the hunters. Turtles are also traditionally caught by hand in the Western Islands, the Trobriands, and Woodlark Islands. In the Woodlarks turtles are hunted on a dark night with calm water full of phosphorescence. Canoes follow the turtle's phosphorescent trail, then hunters leap on the animal.

Nesting Females

Taking nesting females is a rather widespread practice today. In the Manus Province it is a tradition with associated rules. In other areas it is nontraditional with little or no regulation. In Manus, there is a practice of calculating when nesting females will return to lay a second clutch of eggs. When an individual needs a turtle for a household occasion, he asks the village elder if he can catch a nesting female using this method. If fresh tracks are seen on the beach, the nest is dug up and the number of eggs inside counted. According to a formula which varies from one location to another, a number of small sticks or *yakets* are planted in the ground, each stick representing 1 day. When 2 or 3 sticks are left, the hunter returns to the site of the original nest and awaits the female turtle. This technique, though still current, is practiced less often, as nesting females are more scarce than in the past.

The Tulu village in the Manus Province maintains a strong traditional tie between 2 clans and the leathery (leatherback) turtle. The people believe that the leatherback turtle belongs to these 2 clans and that the turtle will not return to nest if this ownership is not recognized. Only members of the 2 clans can use divining methods to predict the return of the nesting female. Every female coming ashore to lay its eggs is eaten, if found. When the turtle is killed, it is cut up and divided according to tradition. The front end and the head go to one clan and the back to the other. The pieces in between are divided among the rest of the village. All the turtle is eaten and oil is collected from the shell and used for wick lanterns. In 1978 1 leatherback was eaten out of 5 that were nesting. Three of these nests were dug up (Pritchard 1979). In 1979 2 nesting females came ashore. Both were eaten and their eggs dug up. When I visited Tulu recently, the people were worried about the decreasing numbers of nesting females. There are usually between 12 and 14 nesting females in a good year.

Other Methods

Turtles are also incidentally caught in fishing nets and by hook. A few are also shot by speargun, but in general this practice is frowned upon by the village elders. At Kitava village in the Trobriand Islands, mating pairs are caught with ropes during the breeding season.

Turtle-Hunting Ritual

Ritual still surrounds turtles where they are caught for feasts, and strict rules are associated with their capture and consumption. In areas where traditional authority and respect is breaking down, especially around city centers beset by a need for money, traditional restraints on taking turtles (and other wildlife) are becoming less effective.

Missionary activity has also resulted in a breakdown of traditional rituals, but not always to the detriment of turtle populations. In the Western Province, turtles, once hunted only for feasts, are now eaten as a daily food. On the other hand, where the Seventh Day Adventist Church is influential, the people no longer eat turtle meat, and the turtle populations are increasing. In the more remote provinces, traditional ways are still respected and practiced.

Traditional Ownership of Reefs and Beaches

In most places the right to fish certain reefs and beaches is controlled by individuals or by clans. This enables some measure of control over exploitation of turtles in these areas. However, this system relies heavily on traditional authority and respect within the village. Also, in the old days, traditional laws were defended effectively by force.

Today this is no longer possible. The Wildlife Management Area system of the Wildlife Division enables traditional owners to legally take any offenders to court, thereby enforcing traditional rules, and placing the onus for enforcement on the villagers themselves.

Social Restrictions

These restrictions while not primarily of a conservative nature often have a side benefit of conservation.

Hunting rituals are usually designed to discipline the hunting party into a well-organized and efficient hunt. To prepare for the hunt, hunters usually cannot sleep with their wives during the preparatory period. They must organize their personal effects and dress neatly and not indulge in any gossip or bad thoughts or pry into other peoples belongings. Silence is usually observed during the hunt, only the leader giving orders. If a man's wife is pregnant, he cannot participate in the hunt, or go near the hunting party.

Village restrictions are usually based on the superstition that unless these rules are observed the hunt will be poor or the hunters may have an accident. There are many restrictions on the hunters' wives. For example, they cannot sweep or work until the men return; they must sit down in their houses and not walk about. Children cannot play or make a noise until the hunt is over.

People or clans who believe themselves related to turtles cannot eat turtle meat (East Sepik, Trobriand Islands). All villagers are also prohibited from eating turtle meat during the yam planting season in the East Sepik. In the Trobriands, if a person has eaten turtle meat, he or she cannot go near the yam gardens for 3 days, or else the garden magic will be affected. Magic men who have powers over turtles do not eat turtle meat for fear of losing their magic powers (Manus, East Sepik, Western and Milne Bay Provinces).

Conclusions

Marine turtles play a significant role in the lives of coastal village people as an important source of food. The rules and rituals associated with turtle hunting and the legends explaining their origin also contribute to the cultural heritage of the people.

The greatest threat to turtle populations today is the breakdown of traditional restraints on catching turtles, the incentive to catch more turtles than was previously required, for sale in markets, and the use of modern fishing gear. An old man from Bipi Island said: "Before, in the old days, there were plenty of turtles; we used to hunt them only when our elders said so. Today the young people are following new ways, shooting turtles with spears from canoes with outboards and spearfishing with diving masks. In my opinion, if we

C. SYLVIA SPRING

still follow the old traditions, turtles will still be plentiful, but the new generation are killing them indiscriminately and turtles are getting scarce."

Acknowledgments

This paper could not have been prepared without the cooperation of the village people of Papua New Guinea who made traditional information freely available to the Marine Turtle Project of the Wildlife Division.

Literature Cited

Pritchard, P. C. H.
1979. Marine turtles of Papua New Guinea. Report on a consultancy. Unpublished report to the government of Papua New Guinea.

Colin J. Limpus
National Parks and Wildlife Service of Queensland
Northern Regional Centre, Pallarenda
Townsville 4810, Australia

The Status of Australian Sea Turtle Populations

ABSTRACT

Six species of sea turtles occur in Australia. *Chelonia depressa* is endemic to the Australian continental shelf and is widespread and abundant in Queensland and Northern Territory, while *Chelonia mydas* and *Caretta caretta* are widespread and abundant in Queensland and Western Australia. *Eretmochelys imbricata* nests commonly in Torres Strait and the northern Great Barrier Reef of Queensland. *Lepidochelys olivacea* is poorly known with only low density nesting recorded from isolated areas of Northern Territory and Queensland. *Dermochelys coriacea* migrates along the central east Australian coast in appreciable numbers, but very few nestings occur. The sea turtle resources of Western Australia and Northern Territory have yet to be completely surveyed.

Except for one rookery, no major change in nesting population levels in the past two centuries has been identified. However, marked species-specific and synchronous fluctuations have been recorded for the east Great Barrier Reef *Chelonia mydas* nesting populations. All Australian sea turtles are protected through most of the important feeding grounds and rookeries. No major conservation problems for the species are currently identified except for a possible overharvest of green turtles in Torres Strait and the adjacent areas of southern Papua New Guinea.

Introduction

Large populations of 6 species of sea turtles inhabit Australian waters and nest in the northern states (Queensland, Northern Territory, and Western Australia). In Queensland there has been extensive regular monitoring and general research of sea turtles for many years. As a result Queensland's sea turtle resources are well documented, and reliable statements can be made concerning their status during the past decade. Surveys to define the extent of the sea turtle resources of the Northern Territory and Western Australia have yet to

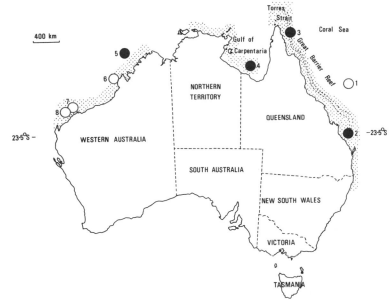

Figure 1. *Chelonia mydas* in Australia. ● = a known major rookery area for the species; ○ = a known rookery area which when surveyed may be of major importance for the species in Australia; ⁙⁙⁙ = known feeding grounds of immature and adult turtles. 1) Diamond Islets, Coral Sea; 2) Capricorn-Bunker Group Islands, e.g., North West Island, Wreck Island, Hoskyn Island; 3) Raine Island-Pandora Cay; 4) Wellesley Group, e.g., Bountiful Island, Pisonia Island; 5) Browse Island; 6) Lacepede Islands; 7) Dampier Archipelago; 8) Monte Bellow Islands and Barrow Island.

be made, and only a partial knowledge of sea turtle distribution and abundance exists for these states. The resulting bias towards Queensland in the following summary reflects this patchiness of knowledge.

Distribution

The flatback turtle, *Chelonia depressa*, which is endemic to the Australian continental shelf, is widespread and abundant in northern Australia. The few records for the species from outside Australia have all come from trawling captures in the Gulf of Papua, southern Papua New Guinea. The remaining species of sea turtles occurring in Australia are the 5 species with a pantropical distribution: *Chelonia mydas*, *Caretta caretta*, *Eretmochelys imbricata*, *Lepidochelys olivacea* and *Dermochelys coriacea*.

In southern Australian waters, sea turtles are not abundant. *Caretta caretta* and *Dermochelys coriacea* are the most frequently sighted species. However very occasional sightings have also been made of *Chelonia mydas*, *Eretmochelys imbricata* and *Lepidochelys olivacea*.

The Green Turtle (Chelonia mydas)

QUEENSLAND

The green turtle is a widespread and abundant species in the state of Queensland (Figure 1).

Rookeries. Three major rookery areas, each consisting of several small adjacent beaches or islands are known in Queensland: 1) Raine Island-Pandora Cay of the northern Great Barrier Reef; 2) Capricorn-Bunker Group Islands of the southern Great Barrier Reef (including North West Island, Wreck Island, and Hoskyn Island); and 3) Wellesley Group of Islands in the southern Gulf of Carpentaria (including Bountiful and Pisonia Islands).

The annual nesting population of each of these major green turtle rookery areas is usually thousands of females. Several other small Great Barrier Reef rookeries (Bramble Cay, No. 7/No. 8 Sandbanks, Bushy Island and Bell Cay) each support annual nesting populations of hundreds of green turtles. Lower density nesting occurs widely throughout the state.

There have been marked fluctuations in the annual nesting numbers of green turtles throughout the Great Barrier Reef Province in recent years. The cause of these fluctuations is unknown, and they have occurred only for green turtles. The populations of green turtles in the feeding grounds which support turtles from these rookeries have shown no equivalent fluctuations. The changes in the annual nesting population at Heron Island (Capricorn Group, southern Great Barrier Reef) is shown in Figure 2. The populations of all the other Great Barrier Reef rookeries have shown approximately synchronous fluctuations. Thus at Raine Island, the densest of these rookeries, in the peak nesting season of 1974–75 over 11,000 nesting turtles were ashore simultaneously on 1 night on the 1.7-km long beach. The following year, when all the rookeries were at their lowest nesting density, only about 100 turtles nested at Raine Island nightly. Several thousand nesting turtles can be expected ashore nightly in an average nesting season at Raine Island.

Feeding grounds. Tag recoveries have shown that green turtles nesting in the southern Great Barrier Reef use different feeding grounds than the green turtles nesting in the northern Great Barrier Reef. Turtles tagged while nesting in the southern Great Barrier Reef have been recaptured around the perimeter of the Coral Sea, including southern Papua New Guinea and New Caledonia, while those from the northern Great Barrier Reef have been recaptured throughout Torres Strait, southern Papua New Guinea, southern Cape York Peninsula, Northern Territory, and Aru Island in Indonesia. The distribution of feeding grounds supporting the Wellesley Group nesting turtles is unknown.

Throughout Queensland there are large populations of green turtles ranging in carapace length from about 40 cm to adult male and female in most shallow bays and reefs. In particular the waters of the Great Barrier Reef and Torres Strait are rich in resident green turtles. There is currently no knowledge of the dispersal pat-

COLIN J. LIMPUS

terns from Queensland rookeries of the presumably oceanic posthatchling stages.

OTHER STATES

No major green turtle rookery is known for the Northern Territory. In Western Australia numerous green turtle nesting areas are known. The relative importance of all these areas has still to be gauged, but Browse Island, Monte Bello Islands, Lacepede Islands, Dampier Archipelago and Barrow Island appear to be major rookeries. Large populations of green turtles occur in feeding grounds along the coastal areas of the northern part of Western Australia and throughout the Northern Territory.

AUSTRALIAN TERRITORIES

Green turtles nest on most of the islands of the western Coral Sea. The most important of these rookeries appears to be the Diamond Islets. In the Indian Ocean, the species nests in very low density on Christmas Island, and the small turtle rookeries on each of North Keeling Island and one of the Ashmore Reef Islands are assumed to be green turtle rookeries.

The Flatback Turtle (Chelonia depressa)

QUEENSLAND

This state supports several important rookeries for the flatback turtle and a major feeding ground (Figure 3).

Rookeries. Queensland's most important rookery for the flatback is Crab Island which supports many thousands of nesting turtles annually (approximately year-round nesting is a feature of this rookery). Another major rookery area is centred on Wild Duck and Avoid Islands. Low-density nesting for the species occurs throughout the coastal areas of the Gulf of Carpentaria, Western Torres Strait, and central to southern Queensland.

Feeding grounds. The major feeding grounds for the flatback turtle are the Gulf of Carpentaria and the shallow coastal waters sheltered by the Great Barrier Reef. Dispersal patterns for post-hatchlings are unknown.

OTHER STATES

In the northern Territory, Greenhill Island is a major rookery while low-density nesting is known throughout the state. Only isolated nesting records are known from Western Australia.

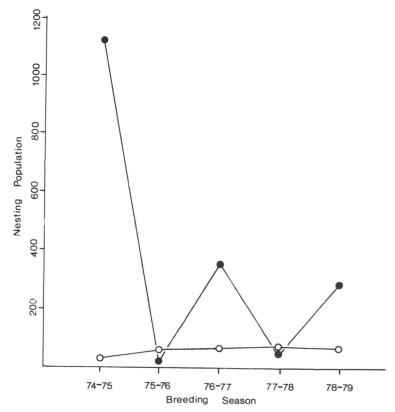

Figure 2. Annual fluctuations in the total nesting populations of sea turtles at Heron Island (Capricorn Group, Southern Great Barrier Reef). ● = *Chelonia mydas*, ○ = *Caretta caretta*.

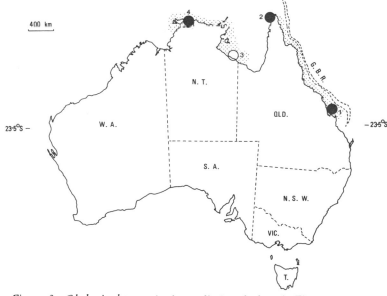

Figure 3. *Chelonia depressa* in Australia (symbols as in Figure 1). 1) Wild Duck Island-Avoid Island and adjacent areas; 2) Crab Island; 3) Sir Edward Pellew Islands; 4) Greenfield Island.

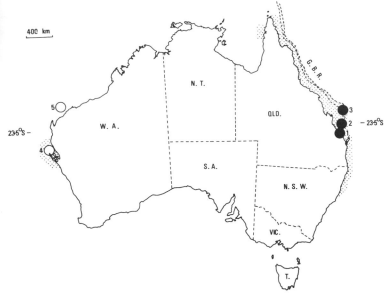

Figure 4. *Caretta caretta* in Australia (symbols as in Figure 1). 1) Bundaberg to Round Hill Head coast, e.g., Mon Repos, Wreck Rock beaches; 2) Crab Island; 3)Swain Reefs Islands; 4) Shark Bay area, e.g., Dorre Island, Bernier Island; 5) Barrow Island.

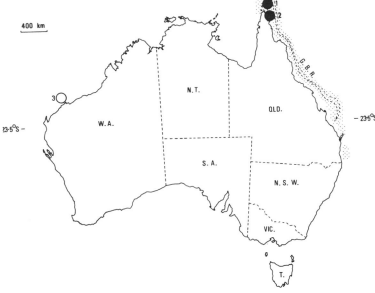

Figure 5. *Eretmochely imbricata* in Australia (symbols as in Figure 1). 1) Long Island and the islands of Torres Strait; 2) Milman Island and the inner shelf islands of the northern Great Barrier; 3) Dampier Archipelago.

The Loggerhead Turtle (Caretta caretta)

QUEENSLAND

In Queensland, the loggerhead is a widespread and abundant turtle (Figure 4).

Rookeries. The state has 3 major rookery areas, each consisting of several adjacent beaches or islands occurring in the subtropical to tropical areas in south to central Queensland: 1) Capricorn-Bunker Group Islands (including Wreck Island and Tryon Island); 2) Bundaberg to Round Hill Head coastline (including Mon Repos and Wreck Rock beaches); and 3) Swain Reefs of the southern Great Barrier Reef.

It is estimated that over 3,000 loggerheads nest annually in Queensland with approximately 1,000 of these nesting annually on Wreck Island alone. In addition low-density nesting occurs widely throughout the state south from Lizard Island (14°41'S).

Feeding Grounds. Tag recoveries indicate that loggerhead turtles nesting at the Capricorn-Bunker Groups and Bundaberg rookeries come from feeding grounds that extend widely along the entire Queensland east coast, eastern Gulf of Carpentaria, and Papua New Guinea including the Trobriand Islands. In particular, large populations of loggerhead turtles inhabit the Great Barrier Reef and the large shallow bays and estuaries.

OTHER STATES

Substantial numbers of loggerheads nest in the Shark Bay area (including Dorre Island) of Western Australia northward to at least Barrow Island. In eastern Australia, sporadic nesting occurs as far south as Newcastle (33°S) in New South Wales.

The Hawksbill Turtle (Eretmochelys imbricata)

QUEENSLAND

The hawksbill turtle is widespread but not abundant (Figure 5).

Nesting. Long Island of central Torres Strait is the only major rookery, and several hundred turtles are thought to nest there annually. Numerous small rookeries occur in 2 areas: 1) the islands of central to eastern Torres Strait; and 2) the inner shelf islands of the northern Great Barrier Reef.

Sporadic nesting extends throughout the Gulf of Carpentaria.

Feeding grounds. Although the species utilizes feeding grounds in the northern Great Barrier Reef and the reefs of Torres Strait, there are no tag recovery data to indicate the relationship between these feeding grounds and the rookeries.

OTHER STATES

Low-density nesting has been reported in several areas of the Northern Territory. There may be an important rookery area for the species in the Dampier Archipelago in Western Australia.

COLIN J. LIMPUS

The Olive Ridley Turtle (Lepidochelys olivacea)

The olive ridley is a poorly known turtle in Australia (Figure 6).

Rookeries. No major rookeries have been recorded although low-density nesting is expected to occur widely across northern Australia from Coburg Peninsula in the Northern Territory to Crab Island in Queensland.

Feeding grounds. Large feeding ground populations occur in the Gulf of Carpentaria and along the Arnhemland coast. In eastern Australia the species is known from immature turtles collected at Cairns (North Queensland) and Geelong (Victoria). There have been no records of the species from Western Australia.

The Leatherback Turtle (Dermochelys coriacea)

The leatherback turtle is not a common turtle for most of Australian waters (Figure 7).

Rookeries. Nesting in eastern Australia is restricted to approximately 160 km of coast northward from Bundaberg where 1, or possibly 2, turtles nest annually. A recent report (Lindner, personal communication) of an isolated nesting in northern Arnhemland is the only record of breeding by this species outside Queensland.

Feeding grounds. Numerous adult and near-adult leatherback turtles pass through coastal waters from south Queensland to central New South Wales each summer. The rookeries from which these turtles originate are assumed to be those of New Britain and the Solomon Islands to the north of the Coral Sea. The species is also regularly sighted in the shallow waters of Shark Bay, Western Australia. For the remainder of Australian waters the species occurs only sporadically.

Conservation

Legislation

All sea turtles are totally protected in 4 of the Australian states—Victoria, New South Wales, Queensland and Western Australia with the exception that indigenous people in Queensland and Western Australia may take turtles for their own use. There is no commercial harvest of turtles within these states. In the Northern Territory only green turtles receive limited protection that restricts the method of capture and the locations from where turtles may be taken for commercial use.

Within Australian territorial waters, all sea turtles are totally protected as are those turtles nesting in areas administered by Australia such as Christmas Island, Cocos-Keeling Islands, most islands of the western Coral Sea.

Figure 6. *Lepidochelys olivacea* in Australia. ● = known low density nesting; ░░░ = known feeding grounds. 1) Crab Island; 2) Coburg Peninsula.

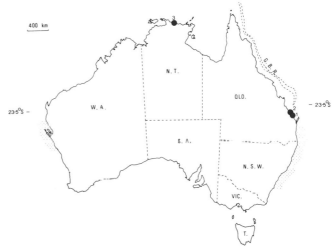

Figure 7. *Dermochelys coriacea* in Australia (symbols as in Figure 6). 1) Mon Repos, Bundaberg; 2) Wreck Rock; 3) Maningrida.

Negotiations are in progress between the Australian Government and the Papua New Guinea and Indonesian Governments concerning the fishing rights (which will include turtles) for fishermen from these countries to fish within traditional grounds in Australian waters in Torres Strait, and several Indian Ocean island areas, respectively.

Australia is signatory to the Convention for International Trade in Endangered Species (CITES) which has resulted in the abolition of trade in sea turtle products into Australia in recent years.

Habitat Protection

In Queensland none of the major sea turtle rookeries are included in land administered for habitat protection by the state's National Parks and Wildlife Service. However several of the less important rookeries (Lady Musgrave, Fairfax, Hoskyn, Heron, and Bushy Islands—green turtle and loggerhead nesting) are National Parks. Low-density sea turtle nesting occurs on many of the other island and coastal national parks throughout the state.

In Western Australia the many coastal and island nature reserves provide habitat protection for sea turtle rookeries which include a number of those currently thought to be the more important for this state. These areas include: Cape Range National Park (green turtles) and nature reserves at Bernier and Dorre Islands (green turtles and loggerheads), Barrow Island (green turtles and loggerheads), and Lacepede Islands (green turtles).

Utilization and Exploitation

The annual catch of turtles by indigenous peoples for their own use throughout northern Australia has been poorly documented. Brief surveys of the harvest in Torres Strait area (including the adjacent Papua New Guinea coast) indicate that an annual harvest rate of the order of 10,000 green turtles occurs in the area, the majority of these adults. Over 40 percent of this harvest appears to be large females, which are being selectively caught. Approximately half of this catch is taken by Papua New Guineans and then mostly from within Australian waters for sale through their local coastal markets. There are insufficient records available to determine if this harvest rate is increasing or otherwise. The author believes that this harvest needs to be monitored.

In addition turtles are regularly being taken in coastal Northern Territory and several Indian Ocean islands and reefs including Keeling Island and the Ashmore Reefs. In these latter areas the catch is principally by Indonesian fishermen.

There is no indication of turtles being illegally harvested at any significant level elsewhere in Australia.

The experimental turtle farms, established by the Australian government sponsored company, Applied Ecology Pty. Ltd., on several Torres Strait islands, are in the process of being closed. Captive rearing of green turtles has been found uneconomical by cottage farm methods in this area, and plans call for all the farms to cease operation by early 1980.

Incidental Catch

Appreciable numbers of turtles are caught in prawn trawls and large mesh set nets in many areas including Gulf of Carpentaria, Shark Bay, and Moreton Bay. In no area is the mortality rate of these incidentally caught turtles considered to be significant since so few of the turtles are killed. This potential problem needs monitoring as changes are made in existing fishing methods.

Predation

Predation by terrestrial fauna on most Australian turtle rookeries is minor. However there are some rookeries where localized high levels of predation of eggs and hatchlings occur, principally by introduced predators such as foxes, dingoes, and pigs. Varanids are significant predators on a few rookeries. In areas adjacent to communities of indigenous peoples, some rookeries are subjected to an almost total harvest of the eggs. This applies particularly to some of the small hawksbill rookeries of eastern Torres Strait. Each rookery with an unnaturally high egg predation rate will probably need to be considered separately when planning management measures.

Population Trends

With only scant historical data available on most Australian turtle rookeries, it is difficult to compare present day population levels with those recorded by the early navigators and visitors to these rookeries (1770–1930 in eastern Australia). However, in eastern Australia there are no rookeries (with the possible exception of Bramble Cay) which have changed significantly in magnitude of rookery size or species composition when present day data are compared with those from the earliest records for the area. There is some indication that in the past few decades the green turtle nesting density at Bramble Cay in the Northern Great Barrier Reef has decreased markedly to its current low level. Whether this is due to the intensive regular harvest to which these nesting turtles have been subjected in the past or whether it is due to the natural gradual movement and reduction in size of the island that is occurring remains to be seen. In the other states there are no turtle rookeries with historical records suggesting that a major decline in populations has occurred.

Acknowledgments

Much of the data summarized above was recorded within the Queensland Turtle Research Project of the National Parks and Wildlife Service of Queensland and was funded in part by the Australian National Parks and Wildlife Service. Many students and teacher voluntary assistants have participated in the field work. Dr. C. J. Parmenter, Applied Ecology Pty. Ltd., Torres

COLIN J. LIMPUS

Strait has provided opportunities for me to accompany him to many islands in the Torres Strait area. Dr. H. Cogger of the Australian Museum; Dr. A. Burbidge of the Western Australia Department of Fisheries and Wildlife; D. Lindner of the Territory Parks and Wildlife Commission, Darwin; and R. Jenkins of the Australian National Parks and Wildlife Service, Canberra, have all provided information on sea turtles. This assistance is gratefully acknowledged.

John Kowarsky
Department of Biology
Capricornia Institute of Advanced Education
Rockhampton, Queensland, Australia

Subsistence Hunting of Sea Turtles in Australia

ABSTRACT

Information sources included the results of a questionnaire survey, the biological and anthropological literature, and personal observations made in Torres Strait. The main findings were:

1. Hunting of turtles in Australia occurred north of 21°S on the east and west coasts by Aboriginal and Torres Strait Islander people for subsistence purposes.

2. The green turtle (*Chelonia mydas*) was the most preferred and most commonly taken species.

3. There were some regional differences in the sex and size composition of the catches, and in the seasonality of turtle hunting.

4. There was also regional variation in the apparent importance of turtles to the hunters, both from nutritional and cultural aspects.

5. The number of turtles caught seemed to depend on the extent of the desire of people to be involved in this activity.

6. An estimate of the total annual catch by Aboriginal and Torres Strait Islander people in Australian waters came to between 7,500 and 10,500 turtles.

7. Relative to turtle hunting, turtle egg collecting has insignificant impact on wild turtle stocks.

8. The hunting pressure on turtles before the time of white settlement was probably as great, if not greater, than at present.

9. Future levels of hunting pressure will depend more on changing socio-economic goals of Aboriginal people than on present trends of population increases.

Introduction

Australia's coastline north of the Tropic of Capricorn stretches for over 7,500 km through 3 major political

305

subdivisions: Queensland, Northern Territory, and Western Australia (Figure 1). Australian territorial waters include a zone of 3 international nautical miles from low water mark; in addition there is an Australian Fishing Zone of 200 international nautical miles from low water mark, defined by the Fisheries Amendment Act 1978. Several changes are pending with respect to delineation of sea borders (Appendix A).

The extant indigenous population of Australia can broadly be divided into 2 groups, the Aborigines and the Torres Strait Islanders, the latter people being defined in anthropological literature as having lived within the area bounded by latitudes 9°20′ and 10°45′S and longitudes 142° and 144°E (Beckett 1972). In 1971 there were approximately equal numbers of Aboriginals in Queensland, the Northern Territory, and Western Australia; however, about 95 percent of the total number of Torres Strait Islanders in these 3 political subdivisions lived in Queensland (1971 census of Population and Housing).

Legislation pertaining to the taking of sea turtles in Australian waters varies between political subdivisions (Appendix B). In summary, in both Queensland and Western Australia, noncommercial taking of turtles is permitted for Aborigines and Torres Strait Islanders, with the additional qualification of living on a reserve applying to those people in Queensland. In the Northern Territory, the taking of green turtles is permitted only within specified areas adjacent to some settlements. At the federal level, the taking of turtles in proclaimed waters (that is, outside the 3-mile territorial limit) is prohibited; there is apparently no provision made at present for noncommercial taking of turtles by Aboriginals or Torres Strait Islanders, but in practice this is permitted.

General accounts of aspects of subsistence hunting of turtles in Australian waters can be found in the biological and anthropological literature (for example, Beckett 1972; Cogger and Lindner 1969; Duncan 1974; Kowarsky 1978; McCarthy 1955; Moore 1972; Nietschmann, in press; Turner 1974); most of this information deals with the Torres Strait region. I have been privileged to have access to an unpublished manuscript on subsistence hunting of turtles in the north of Western Australia (Capelle 1979).

This investigation had 3 main aims: 1) collect and summarize information on subsistence hunting of turtles throughout Australia; 2) to assess the importance of turtles to those people who have traditionally hunted them; and 3) to provide information which, with data on the status of sea turtle populations in Australian and adjacent waters (documented elsewhere in these proceedings), might enable an assessment of the present, and future, impact of turtle hunting on sea turtles in the region.

Methods

A questionnaire (Appendix C) was distributed to persons and communities in coastal northern Australia. At the time of writing, 31 completed questionnaires had been returned. The areas of coastline of which respondents indicated knowledge are shown in Figure 1 (note: some respondents indicated familiarity with more than 1 region). Replies were from the following sources:

Source	Number of respondents
Aboriginal and/or islander community resident or administrator	9
Department of Aboriginal Affairs (federal)	3
Department of Aboriginal and Islanders Advancement (Queensland)	4
National Parks and Wildlife Service (Queensland)	4
Western Australian Department of Fisheries and Wildlife	4
Other	7
Total	31

Results

Turtle Hunters

All sources of information indicate that among Australians, turtle hunting is virtually the exclusive domain of Aboriginal and Islander people (with the proximity of the border to Papua New Guinea, residents of that country certainly would hunt turtles in Australian waters, but I have not documented these activities here). I have attempted to estimate the number of people living in reasonable proximity to the tropical coast who could legally, or who would traditionally, use turtles for subsistence (Appendix D).

Figures obtained from the above are certainly overestimates of numbers actually using turtles at present. In the Northern Territory, for example, most turtle hunting would be carried out by the 13 major coastal settlements of Aboriginal people; their numbers would be far lower than the figure estimated for that division. In Western Australia, the total number of persons who might hunt turtles between Pt. Hedland and Wyndham would not exceed 600 (Capelle 1979), yet over 10 times that figure is obtained by summing numbers of Aboriginals and Torres Strait Islanders living in coastal Local Government Areas along the same coastline.

Two factors could be advanced to explain these discrepancies. Firstly, the basic census unit, the Local Government Area, was too large and would have included many people living away from the coast. This would apply particularly to Western Australia and the

JOHN KOWARSKY

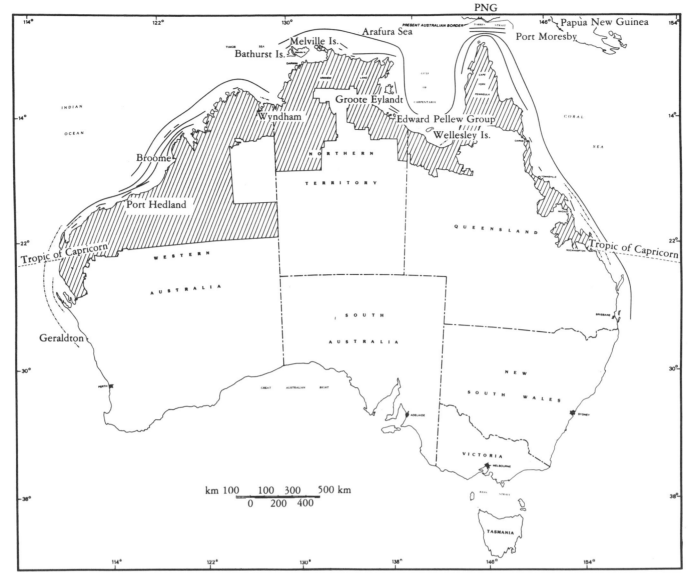

Figure 1. Map of Australia showing: 1) major political subdivisions; 2) regions of coastline about which respondents to the questionnaire indicated knowledge: continuous lines = turtle hunting reported, broken lines = no turtle hunting known; 3) local government areas north of the Tropic of Capricorn with a coastal border = shaded areas.

Northern Territory. Secondly, as will be discussed later, many Aboriginals and Torres Strait Islanders would now no longer hunt turtles.

Geographical Extent of Turtle Hunting

The survey indicated that turtle hunting was limited to north of about latitude 21° S on the east and west coasts of Australia (Figure 1). This distribution is very similar to that described by McCarthy (1955).

Hunting Methods

According to the survey, the most common method of capturing turtles is by use of a spear or harpoon from a boat. Less commonly, turtles are taken by hand, either in pools on the reef, while basking in very shallow water, or after chasing by boat. Turtles are also sometimes taken while nesting. Descriptions of hunting techniques can be found in McCarthy (1955), Turner (1974) and Capelle (1979). Today there is widespread use of aluminium dinghys, powered by outboard motors, as fishing platforms (Duncan 1974, Capelle 1979) as opposed to the traditional dugout canoes powered by sail or paddles (McCarthy 1955, Beckett 1972, Moore 1972).

Composition of the Catch

SPECIES

The survey indicated that green (*Chelonia mydas*), hawksbill (*Eretmochelys imbricata*), flatback (*Chelonia depressa*),loggerhead (*Caretta caretta*), and ridley (*Lepidochelys olivacea*) turtles were hunted, in order of decreasing frequency. There are apparently regional differences in the composition of catch, which may, at

Hunting in Australia

least in part, reflect differences in the relative abundances of species.

In Torres Strait predominantly green turtles are hunted; I found a widespread belief there that hawksbill turtles were sometimes poisonous and would be eaten only if they had been butchered by someone skilled in removing the poison parts. The reputed toxicity of the hawksbill is documented for Torres Strait (Beckett 1972, Bustard 1972) and for northwestern Australia (Capelle 1979).

In the Northern Territory *Chelonia mydas* was a staple item in the diet of coastal Aboriginals, being much preferred to *Chelonia depressa*, while the flesh of neither the loggerhead nor the ridley was locally esteemed, and these turtles were rarely hunted (Cogger and Lindner 1969). McCarthy (1955) reported the hunting of green, hawksbill, and leatherback (*Dermochelys coriacea*) turtles by Aboriginals.

SEX AND SIZE

Most respondents to the questionnaire reported that any turtle encountered would be taken; a reasonable proportion indicated that large turtles were preferred, but there was little to indicate a marked preference for a particular sex.

From other sources such a preference was apparent. In Western Torres Strait (Nietschmann, in press) and on Yorke Island in central Torres Strait (Kowarsky 1978) predominantly female turtles were taken. By contrast Capelle (1979) witnessed the release of captured female and juvenile turtles and the retention of a large male during a turtle hunting expedition by Aboriginals in the northwest of Western Australia.

A marked regional variation in size composition of the catch occurs in Torres Strait. On Yorke Island turtles taken were below reproductive size (Kowarsky 1978), while in western Torres Strait mature turtles, as adjudged by the presence of eggs in females, and by their carapace measurements, were commonly taken (Nietschmann, in press).

In western Torres Strait, Islanders distinguish between 2 general types of turtles, the more esteemed fat turtles and the poorer condition turtles; they claim that the former feed on seagrasses while the latter eat more algae (Nietschmann, in press). A similar distinction is reported by Pritchard (1976).

SEASONAL VARIATION IN CATCHES

In Torres Strait there appear to be regional differences in the seasonality of catches. In the eastern islands, turtles were taken only during their mating and egg-laying season (Duncan 1974). My records indicate that peak catches of turtles on Yorke Island between Oc-

tober 1975 and July 1976 occurred during the mid-summer months. However, in western Torres Strait, hunting activity took place throughout the year, and the recorded catches did not exhibit any clear seasonal trends (Nietschmann, in press). In northwestern Australia the main period of turtle hunting ran from mid-November through to March (Capelle 1979). Most respondents to the questionnaire reported either that turtles were taken all year or that the main period of capture was during the mating and nesting season.

REASONS FOR HUNTING AND USE OF TURTLE PRODUCTS

A large majority of respondents to the survey marked "subsistence" as the reason for hunting, while a few also marked "cash income" (this category was never exclusively given). In 1973 the price of turtles in Torres Strait varied between $10 and $15, but the extent to which sales took place was not documented (Duncan 1974). One respondent to the questionnaire reported a price of $160 for a full-grown adult turtle. Overall it would appear that the number of turtles hunted for cash income would be insignificant compared to the number hunted for subsistence. Systems of distributing meat among kin and neighbors have been described for Torres Strait (Duncan 1974; Nietschmann, in press).

Replies to the questionnaire indicated that the main turtle product used was the meat, with some respondents indicating that turtles were used for their shells and as curios. I had noticed this tourist trade occurring to some extent on Thursday Island in Torres Strait. Capelle (1979) reported some trade in turtle shells around the Broome area in Western Australia, stimulated by the increasing number of tourists visiting the region.

Most respondents did not regard turtle meat as a main source of protein in their diet, and most did not regard it as a very important food for traditional occasions. There were however, exceptions to these generalizations, with emphasis mainly on the cultural importance of turtles. In Torres Strait Nietschmann (in press) considered hunting (turtles and dugongs) to be important both culturally and nutritionally.

A different situation may apply to some turtle hunters living on mainland Australia or on some of the larger offshore islands. For example, along the Western Australian coast from Pt. Hedland to Wyndham there are permanent reserves of freshwater, easily obtainable vegetable foods, abundant game, available shellfish, clams, mussels, oysters and fish, and food from community stores which make turtles only one of a number of food resources (Capelle 1979). An account of subsistence hunting on Bickerton and surrounding islands in the Northern Territory (Turner 1974) presents a similar picture to that apparent in Western Australia,

JOHN KOWARSKY

with turtles being but one component in a diet of fish, birds, lizards, small marsupials, dugong and various vegetable foods like yams and berries.

FACTORS LIMITING THE CATCH

Almost all respondents to the questionnaire marked "lack of inclination by local people" as the major factor limiting the number of turtles caught; some added the comment that turtles were taken only in accordance with community needs. In Torres Strait, lack of access to boats and working outboard motors was not considered a limiting factor in the development of subsistence fishing (Duncan 1974). Nietschmann (in press) lists several factors influencing the frequency of hunting including money available to buy fuel, desire for fresh meat, occurrence of feast days, and environmental variation.

ESTIMATING THE ANNUAL CATCH

Most respondents indicated that fewer than 20 turtles a week are taken; some gave figures as low as 1 per fortnight or per month. The highest estimate was one for the Cape York Peninsula and Torres Strait of 50 to 100 turtles a week, 2,500 to 5,000 a year.

Some regional estimates of annual turtle catches are available. Nietschmann (in press) made a "very rough estimate" of the average annual catch of green turtles in Torres Strait as being just over 2,000 turtles. By extrapolating from catch records from one community in Torres Strait, an estimate of over 2,500 turtles annually for Torres Strait was obtained (Kowarsky 1978). Recently, Dr. John Parmenter, of Applied Ecology Pty. Ltd., made a similar extrapolation and arrived at a figure of about 4,000 turtles slaughtered annually by Torres Strait Islanders. He and I are aware of the number of untested assumptions implicit in such calculations, but nevertheless I consider such estimates useful in indicating the order of magnitude of the depletion of turtle populations by people of the region.

I have made estimates of the turtle catch per capita per year for 4 communities in Torres Strait using population figures from Duncan (1974) and turtle catch records from Kowarsky (1978) and Nietschmann (in press):

Community	Number turtles per capita per year
Yorke	0.55
Mabuiag	1.28
Kubin	1.99
Badu	0.80

In Queensland, excluding Torres Strait, the number of people living on reserves and former reserves near the coast within the area of turtle hunting as indicated by data in Figure 1 was taken to be about 6,500. Using the lowest estimate above of turtles taken per capita per year, a projected figure of about 3,500 turtles per year is obtained. Thus for the whole of Queensland, including Torres Strait, a broad estimate of the total annual catch could range between about 5,000 and 8,000 turtles.

To obtain an estimate of the extent of turtle hunting in the Northern Territory it has been necessary again to extrapolate from data of one community. At South Goulburn Island, the Aboriginal population of about 200 take about 3 turtles a week (D.L. Grey, personal communication 1979). In the Northern Territory there are about 13 major coastal settlements of such people. A projected total catch for the region of about 2,000 turtles a year can thus be obtained.

Capelle (1979) has made the following estimate of annual turtle catch on the coast of Western Australia between Pt. Hedland and Wyndham:

Locality	Annual catch
One Arm Point	48
Kalumburu	40
Other isolated communities (3)	10
Towns (4)	6
Total	104

Outside the area defined above, little turtle hunting would be expected on the basis of replies from the questionnaire survey (Figure 1).

The above estimates taken together would suggest that between about 7,500 and 10,500 turtles are taken annually by Aboriginal and Torres Strait Islander people in Australian waters. Assuming an average weight per turtle of 100 kg, between 750 and 1,050 tonnes of turtle would be taken.

EGG COLLECTING

Most respondents to the questionnaire reported that few, if any, eggs were taken during the nesting season. Weekly egg collections were commonly estimated at 20 or less, but 2 respondents gave estimates of about 1,000 per week. It would appear that green, hawksbill, loggerhead, and flatback eggs were eaten. Compared to exploitation of turtles by hunting, the collection of eggs in Australia would appear to have insignificant impact on the wild turtle populations.

Discussion

For many Aboriginals and Torres Strait Islanders today, sea turtles have little or no significance. Among those people still engaged in turtle hunting activities there appears to be a broad spectrum of attitudes. For some a captured turtle is a bonus food item; for others turtles are an integral part of their socioeconomic and cultural organization.

It would be useful to compare past hunting pressures with those of the present as one means of assessing the potential impact of such activities on the wild turtle populations. The absence of quantitative records of past hunting make such a direct comparison impossible; as an alternative, one can consider technological, population and socioeconomic changes which may have influenced the extent of hunting which takes place.

Modern technology, plus a cash income from employment opportunities and widespread social security payments, have provided the present day turtle hunter with a fast and efficient means of traveling over long distances at sea. This apparent increase in hunting power may be offset to some extent by dependence on factors such as fuel supplies and motor maintenance. Other developments such as increasing use of domestic refrigerators and increasing availability of fresh, frozen, and canned foods at community stores could act to remove some pressure on turtles as a food resource.

Australia's Aboriginal population at the time of arrival of white settlers is thought to have been about 300,000; since that time it drastically declined until in 1954 the total number of "full-blood" Aboriginals was about 40,000 (Anonymous 1965). The Torres Strait Islander population in Torres Strait apparently did not follow the same decline, with numbers in that region in the nineteenth century being estimated between 3,000 and 4,000 (Beckett 1972).

White settlement in Australia, as well as resulting in a decline in population, also brought profound socioeconomic changes to Aboriginal society, with a breakdown in tribal organization and religion, and often a forced movement from traditional homelands (Anonymous 1965, Brokensha and McGuigan 1977). In fairly recent times there has been a voluntary movement of Torres Strait Islanders from their home islands to the Australian mainland, and a decline in numbers in Torres Strait of about 5 percent in 5 years (Caldwell 1975).

On the basis of the above, and particularly since one of the main regions of Aboriginal settlement before the arrival of whites was the northern tropical coast (Brokensha and McGuigan 1977), it would be reasonable to presume that hunting pressures on turtles in the past were at least as great, if not greater, than those existing today.

Since the 1950s the Aboriginal population of Australia has been increasing. The total number of Torres Strait Islanders is also increasing, but this is due to increasing numbers on mainland Australia. In Torres Strait the Islander population will probably decline to under 3,000 by the end of the century (Caldwell 1975). A study of mainland Torres Strait Islanders found that their lifestyle changed markedly from that on reserves in Torres Strait and that they seemed to have a general lack of attachment for that region (Fisk, Duncan, and Kehl 1974).

If the projections of numbers of Torres Strait Islanders in Torres Strait are accurate, no significant increase would be predicted in hunting pressure on turtles in that region in the future. The extent to which increasing numbers of Aboriginals place additional hunting pressure on turtles would depend upon the proportion of those people maintaining a traditional lifestyle in future years.

This leads to a factor which is perhaps more likely to result in a rapid change in the pressure of hunting on turtles, and that is a change in the socioeconomic aims of Aboriginals. It is clear that in the general regions where turtles are hunted, there are far more Aboriginals than those now involved in hunting activities. Fairly recently, there has been a trend of movement away from settlements (established by government or missions) toward traditional clan territories which has been termed the "outstation" or "homelands" movement (Coombs 1974, Brokensha and McGuigan 1977, Anonymous 1979a). An example of this was the formations of 10 homeland centers by people formerly living at a mission (established in 1938) in northeastern Arnhem Land in the Northern Territory (Brokensha and McGuigan 1977). Such a movement, if widespread, could be expected to result in an increase of hunter-gatherer acitivities, including the subsistence hunting of sea turtles.

Appendix A. Changes in Present Australian Sea Borders

Recently the Commonwealth (federal), State, and Northern Territory governments agreed to changes in the division of responsibility in offshore areas, but the relevant legislation has not yet been passed. The present situation with regard to fisheries is that within the 3-mile territorial area, fisheries are subject to State or Northern Territory legislation, while outside this area they are subject to Commonwealth legislation.

Borders with Indonesia and Papua New Guinea present special problems and are yet to be settled (Prescott 1978). That with the latter country is of particular relevance to turtle biologists as the region in between, Torres Strait, includes significant nesting beaches and extensive reef systems and shallow waters which are apparently important feeding areas for turtles (Bustard 1972; Kowarsky 1978; Nietschmann, in press).The Torres Strait Treaty (summarized by Stanford, 1978) between Australia and Papua New Guinea was signed in December 1978 but is yet to be ratified. Among

JOHN KOWARSKY

other provisions, it delineates a Seabed Jurisdiction Line, a Fisheries Jurisdiction Line and a Protected Zone within the Torres Strait region; the combined area enclosed by these 3 lines does not differ markedly from the position of the present border (shown in Figure 1).

Appendix B. Legislation Related to the Taking of Sea Turtles in Australian Waters

Queensland

All species of the families of Chelonidae and Dermocholydae [sic] are "protected species" under the Fisheries Act 1976. This act does not, however, apply to "... the taking, otherwise than by the use of any noxious substance or explosive and for purposes other than commercial purposes, ... by any Aborigine or Torres Strait Islander who at the material time is resident on a reserve ..."

The terms Aborigine and Torres Strait Islander are defined in the Aborigine Act 1971 and the Torres Strait Islander Act 1971.

Northern Territory

The Fisheries Ordinance 1965–66 declares all waters of the Territory closed against the taking of green turtle except specified areas which correspond to the local people's more important traditional turtle hunting areas. Although commercial exploitation of turtles is not prohibited, this practice is discouraged by the fisheries authorities in the Northern Territory.

Western Australia

Under the Fauna Conservation Act 1950–70 turtles are wholly protected throughout the entire State at all times; however a person who is a native according to the definition in the Native Welfare Act 1963 may take turtles " ... sufficient only for food for himself and his family, but not for sale" on land other than that of a sanctuary.

Commonwealth (Federal) Legislation

The Fisheries Act 1952–74, through Fisheries Notice No. 48, prohibits the taking of turtles from proclaimed waters. This Act makes an exception for "traditional fishing," defined as "fishing by indigenous inhabitants of an external Territory" if the fish are taken in a traditional manner and landed in that Territory. It would thus appear that traditional fishing by Australian residents (Aboriginals or Torres Strait Islanders) is not excepted from the provisions of this Act. However, discussions I have had with Federal authorities indicate that a wider definition of traditional fishing than that

in the Act has been applied in practice, so that, effectively, the noncommercial taking of turtles by Aboriginals and Torres Strait Islanders would be permitted in proclaimed waters.

Appendix C. Turtle Hunting Questionnaire

1. Name
2. Organisation
3. Coastal region (including islands) of which you have knowledge (indicate, as far as possible, the extent of the region on the attached map).
4. Does turtle hunting take place in the above region? If yes, go straight on. If no, go to question 14.
5. Which people go turtle hunting? Aboriginal, Islander, Other (give details).
6. Reason for hunting: Subsistence, Cash Income
7. What species of sea turtles are caught? Green, flatback, hawksbill, loggerhead, ridley
8. What species are used for: meat, shells and curios, leather products, soup?
9. How are turtles caught? On beach while nesting, from boat with spear or harpoon, by rope after chasing by boat, other (give details)
10. What number of turtles are captured per week in the region? (If possible, make a guesstimate). Less than 20, 20–50, 50–100, more than 100
11. How important are turtles as: a main protein source in diet, a main source of cash income, a traditional food for important occasions?
12. Is there a preference for particular types of turtle? small, large, male, female, any turtle
13. What are reasons given for the preference above?
14. What is the major factor limiting the number of turtles caught? scarcity of turtles, lack of boats and equipment, lack of inclination by local people
15. Where in your region are the most significant aggregations of turtles, and what species are they?
16. What species' eggs are eaten by people within your region? (If possible make a guesstimate of number of eggs eaten per week). Green, hawksbill, loggerhead, other (give details)
17. Are there seasonal trends in the quantities of turtles/eggs collected by people in your region?

Note: Spaces for answers have been deleted to conserve space.

Appendix D. Population Statistics for Aboriginals and Torres Strait Islanders

The most recent comprehensive data in this regard is the 1976 Census of Population and Housing by the Australian Bureau of Statistics. Explanation of the question in that census relating to racial origin is given in Anonymous (1979b). The basic unit of area used in the census was the Local Government Area (LGA).

To gain an idea of the number of Aboriginals and Torres Strait Islanders who could be potential users of turtles, I extracted their numbers in each LGA north of the Tropic of Capricorn with a maritime border (shaded area, Figure 1) from the 1976 Census Data: Queensland, 19,300; Northern Territory, 12,931; Western Australia, 8,736; total, 40,427.

In Queensland, because of present legislation, more apposite in the present context would be the number of people resident on reserves near the coast. Excluding the Torres Strait region, for which recent data were not separately available, the number of persons resident on such reserves (either government or church) under the auspices of the Queensland Government Department of Aboriginal and Islanders Advancement was 7,552 in March 1978 (Anonymous 1978).

A special situation is found in Torres Strait itself. Here are 14 reserve islands with a total resident population of about 2,500 and some 2,000 Islanders living on Thursday Island, the region's Administrative center (Nietschmann, in press). From a stay of 4 months on Thursday Island, I gained a strong impression that turtle meat was available locally irrespective of the reserve-residency status of the Islander. It would probably be more realistic, therefore, to use the total number of Aborigines and Islanders in the region as the number of potential users of turtle products than to restrict the number to those actually resident on reserves in Torres Strait.

In Queensland, the total number of people who potentially could use turtle products would thus be 7,552 + 4,500 = 12,052, approximately 62 percent of Aboriginal and Islander people living in the LGA's already defined.

Acknowledgments

I am grateful to the respondents of the questionnaire survey, the individuals and organizations mentioned in the text, and many other people who helped me with my enquiries. I am especially indebted to Mikael Capelle for access to his most useful unpublished manuscript. John Marshall, CIAE cartographer, kindly prepared Figure 1. Tamara Kowarsky and Dr. Bryan Rothwell critically read the manuscript in preparation. Salary and expenses in preparation of this manuscript were borne by the Capricornia Institute of Advanced Education.

Literature Cited

Anonymous
1965.   Aborigines: the white contact. In *The Australian Encyclopedia*, ed. A. H. Chisholm, vol. 1, pp. 87–95. The Grolier Society of Australia Pty. Ltd.
1978.   *Annual Report: Department of Aboriginal and Islanders Advancement (Queensland)*. Brisbane: Government Printer.
1979a. The Outstation Movement. Background Notes, no. 6. Department of Aboriginal Affairs, Canberra, Australia.
1979b. 1976 Population Census Information Paper, no. 19, Racial Origin, cata. no. 2124.0. Australian Bureau of Statistics, Canberra.

Beckett, J. R.
1972.   The Torres Strait Islanders. In *Bridge and Barrier: the Natural and Cultural History of Torres Strait*, ed. D. Walker, pp. 307–26. Canberra: Australian National University.

Brokensha, P., and C. McGuigan
1977.   Listen to the dreaming: the Aboriginal homelands movement. *Australian Natural History* 19:118–123.

Bustard, H. R.
1972.   *Australian Sea Turtles*. Sydney: Collins.

Caldwell, J. C.
1975.   *The Torres Strait Islanders, vol. 4, The Demographic Report*. Canberra: Australian National University.

Capelle, M.
1979.   Subsistence turtle hunting on the Kimberley coast. Unpublished manuscript.

Cogger, H. G., and D. A. Lindner
1969.   Marine turtles in northern Australia. *Australian Zoologist* 15:150–59.

Coombs, H. C.
1974.   Decentralization trends among Aboriginal communities. *Search* 5:135–43.

Duncan, H.
1974.   *The Torres Strait Islanders, vol. 1, Socio-economic Conditions in the Torres Strait*. Canberra: Australian National University.

Fisk, E. K.; H. Duncan; and A. Kehl
1974.   *The Torres Strait Islanders, vol. 3, The Islander Population in the Cairns and Townsville Area*. Canberra: Australian National University.

Kowarsky, J.
1978.   Observations on green turtles (*Chelonia mydas*) in northeastern Australia during the 1975-76 nesting season. *Biological Conservation* 13:51–62.

McCarthy, F. D.
1955.   Aboriginal turtle hunters. *Australian Natural History* 11:283–88.

Moore, D. R.
1972.   Cape York Aborigines and islanders of western Torres Strait. In *Bridge and Barrier: the Natural and Cultural History of Torres Strait*, ed. D. Walker, pp. 327–43. Canberra: Australian National University.

Nietschmann, B.
In press.   Hunting and ecology of dugongs and green turtles, Torres Strait, Australia. National Geographic Society Research Reprints, 1976.

Prescott, J. R. V.
1978.   Drawing Australia's marine boundaries. In *Australia's Offshore Resources: Implications of the 200-mile Zone*, ed. G. W. P. George, pp. 22–45. Canberra: Australian Academy of Science.

Pritchard, P. C. H.
1976.   Post-nesting movements of marine turtles (Chelon-

iidae and Dermochelyidae) tagged in the Guianas. *Copeia* 1976:749–54.

Stanford, P. S. B.

1978. The Torres Strait Treaty. *Australian Foreign Affairs Records* December 1978:572–77.

Turner, D. H.

1974. *Tradition and Transformation: A Study of the Groote Eylandt Area Aborigines of Northern Australia.* Canberra: Australian Institute of Aboriginal Studies.

East and Southeast Asia

Itaru Uchida
Himeji City Aquarium
Hyogo, Japan

Masaharu Nishiwaki
University of the Ryukyus,
Okinawa, Japan

Sea Turtles in the Waters Adjacent to Japan

Five species of sea turtles are found around the southern districts of the Japan Islands: the green turtle (*Chelonia mydas*), the hawksbill (*Eretmochelys imbricata*), the loggerhead (*Caretta caretta*), the ridley (*Lepidochelys olivacea*), and the leatherback (*Dermochelys coriacea*). There are few sighting records of, and no report of laying eggs by leatherback and ridley turtles. However, the loggerhead, green turtle, and hawksbill have nested in Japan.

The Loggerhead Turtle (*Caretta caretta*)

Aka-umigame is the Japanese standard name for the loggerhead turtle, one of the very commonly distributed sea turtles along the coast of Japan (Figure 1). It breeds along the coasts farther north than any other species in the western Pacific. Along the Pacific coast of Japan's mainland, it nests occasionally as far north as Fukushima Prefecture, 37°N. While on the coast of the Sea of Japan, it nests as far north as Ishikawa Prefecture also about 37°N. These nesting areas can be considered margins of the loggerhead population in the western Pacific. In the Japan Islands, the number of nesting places is not large. The places where loggerheads nest in abundance are in Shizuoka Prefecture, Kii Peninsula, Shikoku, and the east coast of Kyushu. Nesting places in the Ryukyus and the Ogasawara (Bonin) Islands are rarer than those in the above places. Generally, it is considered that loggerheads or sea turtles are more abundant in the south. This report does not support that general consideration, although the loggerheads coming to these nesting places are considered to be recruits from subtropical waters. Between late May and August, their breeding season, loggerheads congregate in the offshore waters of the breeding places. This gregarious phenomenon is seen at the beginning of the season every year. In these areas, courting behavior is seen among rocks offshore

317

Caretta caretta

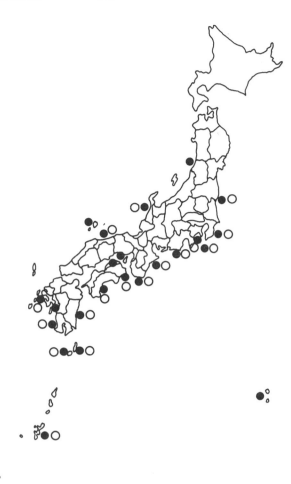

Figure 1. The distribution and nesting areas of the logger-head turtle (*Caretta caretta*) in Japan. ● Sighted positions; ○ Laying eggs and hatched.

in 20 to 30 m of water. The nesting season starts when 20°C isothermal waters approach the coast of Japan in the spring. Adults, subadults and juvenile loggerheads, over 1,000 in number, were tagged with "Roto-tag" (plastic cattle ear tag) in our study at Gamouda Beach, Tokushima Prefecture, during 1969–1979. Most of the females that came to the beach and were examined were between 72.0 cm and 107.5 cm in straight car-apace length, (89.0 cm in average; n = 118) and be-tween 53.0 kg and 125.0 kg (96.8 kg in average; n = 15).

The Green Turtle (*Chelonia mydas*)

Ao-umigame is the Japanese standard name for *Chelonia mydas*. As shown in Figure 2, the green turtle migrates to the southern coasts of the Japan Islands. Its nesting area is exclusively in the southern islands of Japan at about 30°N. Yakushima, Kagoshima Prefecture, is the

northernmost nesting site recorded. Thus, the green turtle shares most of the beaches with the loggerhead; both of them prefer warmer sandy beaches. The Oga-sawara (Bonin) Island at about 27°N is also an impor-tant nesting place.

The Hawksbill Turtle (*Eretmochelys imbricata*)

Figure 3 shows the distribution of the hawksbill turtle, whose Japanese standard name is *tamai*. There are only 2 records of nesting females of this species in Ishigak-ijima and Kuroshima in the Yaeyama Archipelago in the Ryukyus of Japan, at about 25°N. The area must be the margin of the hawksbill in Japan. Records of stranded turtles in other places are of subadults and juveniles. The hawksbill is the third least common spe-cies of sea turtle in Japan, and there is little knowledge about its nesting sites.

Chelonia mydas

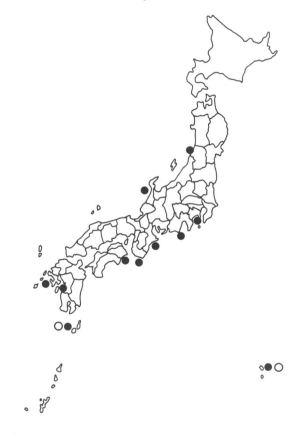

Figure 2. The distribution and nesting areas of the green turtle (*Chelonia mydas*) in Japan. ● Sighted positions; ○ Lay-ing eggs and hatched.

ITARU UCHIDA

Human Incursion on Sea Turtles

Recently, there has been worldwide concern about the nesting places of sea turtles. There are problems from human activities along the coast of Japan. Some measures should be taken to conserve the animals; otherwise we may soon see a decrease in the population. Only a few Japanese people, shore villagers, eat turtle meat and eggs, which do not appear at markets. At Yakushima, Kagoshima Prefecture, a popular nesting site, villagers take loggerhead and green turtle eggs to eat, and fishermen in Kagoshima, Wakayama and Kouchi Prefectures sometimes catch loggerheads or green turtles by hand harpoon for their own food during the nesting season. According to our investigation, about 50 to 100 adults are killed annually. These turtle catches seem to have begun after the second world war.

In Japan, the only turtle shell used for handicraft work is from the hawksbill, and all hawksbill shells are imported from Southeast Asia and tropical Atlantic and African countries. The major use of turtle shell in Japan is for making ornaments and accessories with long traditional workmanship. At present, about 1,200 people are engaged in this work. Their work is quite delicate and detailed, based on long training and tradition. If this work were to be interrupted once, a traditional and long-maintained skill should be lost forever. The hawksbill-shell workers organized an association to preserve their precious workmanship.

Japan and Sea Turtle Conservation

Japan's ratification of the Convention on International Trade in Endangered Species of Wild Fauna and Flora (CITES) has been postponed for a variety of reasons, but it is to be ratified by the Congress, maybe in 1980. As a result of this ratification, trade in turtle shell will be seriously limited. Because of this, some ill-advised people have tried to hoard the shell with the help of dealers. However, following criticism by the public these activities have become difficult to continue. Recently, taxidermy dealers have entered the turtle shell-purchasing business, which had hitherto been done by shell dealers connected to handicraft workers. Stuffed hawksbills and green turtles are now sold in resort areas. Stuffed turtles are loved by people as a symbol of good luck and longevity in Japan as well as in some Asian countries. Large, stuffed turtles are so profitable to dealers that the weight of traded turtles has increased rapidly and prices have shot up. Now people tend to prefer smaller turtles because big ones cost so much. Catching smaller turtles may have a very bad effect on the population. As mentioned before, there is no hawksbill or green turtle catch for profit along the coast of Japan, so almost all of these stuffed turtles

Eretmochelys imbricata

Figure 3. The distribution and nesting areas of the hawksbill turtle (*Eretmochelys imbricata*) in Japan. ● Sighted positions; ○ Laying eggs and hatched.

are imported. Importation of stuffed turtles has been done exclusively by dealers, and no handicraft workers on turtle shell are concerned with this business. The government of Japan announced an import ban on stuffed turtles on November 24, 1979.

At the same time, the Japanese government is trying to encourage the cultivation of hawksbills and other turtles to stabilize the supply of shells to protect the handicraft workers and their long-traditional skill, and to find a way to give aid to cultivation projects in the Southeast Asian countries for mutual understanding and interest. Cooperative conservation and cultivation projects have made progress, drawing attention to the Japanese and other shell-producing countries. To undertake this sort of cooperation, a fundamental biological knowledge, instead of purely a profit-slanted consideration, is indispensable to both sides. For conservation to succeed scientific reason must prevail over emotional or frantic action.

Huang Chu-Chien
Institute of Zoology
Academia Sinica

Distribution of Sea Turtles in China Seas

Introduction

To learn about the habits and geographical distribution of our sea turtles, we have made investigations from Haiyang Island of the Bohai Sea in the north through the Yellow Sea to the Zhoushang Islands of the East China Sea, then south along the coastal sea of Fujian (Fukien), Guangxi (Kwangsi), and Hainan Island of Guangdong (Kwangtung) to the Xisha Islands. The sea turtles increased in species and number as we sailed southward. Only a few scattered green turtles (*Chelonia mydas*), loggerhead turtles (*Caretta caretta*), and leatherback turtles (*Dermochelys coriacea*) were found in the Bohai Sea and Yellow Sea. Hawksbill turtles (*Eretmochelys imbricata*) appeared in the East China Sea. Sea turtles increased in number around Hainan Island, while only the islands in the South China Sea were the center of reproduction and habitat of the sea turtles in China.

Over the years, the fishermen of Hainan Island have often fished the Xisha and Nansha Islands. While sea turtles have been one of their quarries, their catches, made by hand from small sailing boats, have been small.

Individual Ecology and Distribution of the Sea Turtles

Sea turtles can be seen around Xisha Islands and Nansha Islands all year. One migrant population returns annually to reproduce beginning about April with the southwest warm currents. In addition, a local population resides there.

Four species of sea turtle are distributed in our country: the leatherback turtle, the loggerhead turtle, the green turtle, and the hawksbill turtle.

The Leatherback Turtle (Dermochelys coriacea)

The leatherback turtle's main egg-laying season is from May to June; 90 to 150 eggs are laid in each clutch. The young turtles hatch after 65 to 70 days and swim quickly out into the sea, but often a dozen eggs in each

321

clutch fail to develop. Leatherback nests are about 65 cm deeper than those of other sea turtles. This species continues to lay eggs even when disturbed while laying. Because leatherbacks cover their eggs with sand, their nests are hard to find. Leatherback turtles feed on algae, shrimps, crabs, molluscs, and fish.

In China, leatherbacks occur along the coastal seas of Guangdong (Kwangtung), Guangxi (Kwangsi), Fujian (Fukien), Zhejiang (Chekiang), Jiangxi (Kiangsi), Shandong (Shantung) and Liaoning.

The Loggerhead Turtle (Caretta caretta)

The loggerhead turtle nests from April to August, digging nests 33 to 65-cm deep and laying from 60 to 150 eggs in a clutch. The eggs generally hatch in about 2 months, depending on the season and the nest site. Loggerheads eat fish, shrimps, crabs and molluscs.

In China loggerheads are found along the coastal seas of Taiwan, Gungdong (Kwangtung), Guangxi (Kwangsi), Fujian (Fukien), Zhejiang (Chekiang), Jiangsu (Kiangsu), Shandong (Shantung), and Hebei (Hopei).

The Green Turtle (Chelonia mydas)

In the Xisha Islands, the green turtle's main breeding season is from May to July. Eggs can be found from April to December.

Large green turtles weigh about 450 kg (Figure 1). They often hold their head above water for 15 to 20 minutes when swimming. Mating is often observed from January to April. Males and females chase each other in a round-about way before mating, and they mate at the base of the reef. During the breeding season the female crawls up the beach beyond the high tide line,

Figure 1. Green turtles captured for market.

Table 1. Catch of sea turtles in Xisha region, 1959–70

Year	Catches (kg × 10³)
1959	130.63
1960	40.60
1961	54.02
1962	38.74
1963	103.01
1964	61.10
1965	181.35
1966	86.86
1967	149.91
1968	42.96
1969	104.78
1970	122.56

digs a big hollow pit with its fore limbs and then a nest pit 33 to 65-cm deep with its hind limbs. It will go back into the sea if disturbed immediately after landing but generally will not stop laying once it begins to lay. The female lays 3 times a year at intervals of about 2 weeks, 300 to 500 eggs annually. The spherical soft-shelled white eggs, 40 to 44 mm in diameter, are covered up with sand in the nest. Young turtles, 4-cm long, hatch in 40 to 50 days and quickly crawl into the sea. They eat algae, shrimps, crabs, molluscs, and fish.

In China green turtles are found along the coastal seas of Taiwan, Fujian (Fukien), Guangdong (Kwangtung), Guangxi (Kwangsi), Zhejiang (Chekiang), Jiangsu (Kiangsu) and Shandong (Shantung).

The Hawksbill Turtle (Eretmochelys imbricata)

The hawksbill, fierce and scarce, inhabits coral reefs, laying 3 clutches of eggs a year in the daytime in March and April. Each clutch consists of 130 to 200 eggs which hatch after 2 months or so. The hawksbill feeds mainly on fish, shrimps, crabs, molluscs, and algae.

In China, hawksbills are found along the coastal seas of Taiwan, Guangdong (Kwangtung), Guangxi (Kwangsi), Fujian (Fukien), Zhejiang (Cheking), Jiangsu (Kiangsu) and Shandong (Shantung).

Turtle Catch Records

The statistics of our Yong Xing purchasing station are shown in Table 1. Around the Xisha Islands, only the green turtle and the loggerhead turtle were objects of fishing, while the green turtle was the main object. Because the purchasing station has no classified statistics, the values shown in Table 1 represent both species. The amounts purchased in 1959–70 represent basically the catch of the area around the Xisha Islands.

Nelma Carrascal de Celis
Science Research Associate I
Outdoor Recreation and Wildlife Research Division
Forest Research Institute College
Laguna, The Philippines 3720

The Status of Research, Exploitation, and Conservation of Marine Turtles in The Philippines

ABSTRACT

Four marine turtles are found in Philippine territorial waters, namely: *Chelonia mydas, Eretmochelys imbricata, Lepidochelys olivacea* and *Dermochelys coriacea.*

Turtle research has been sporadic. Domantay did some work in the 1950s, and Alcala and de Celis in the 1970s. Preliminary studies show that turtle populations are fast diminishing due to poaching. The turtle population is low in the northern part of the country and dense in the southern part, especially around the Turtle Islands.

Government regulations are administered by the Ministry of Natural Resources through the Bureau of Forest Development. On 26 June 1979, President Ferdinand E. Marcos created the Task Force Pawikan through Executive Order 542 to conserve the marine turtles and develop turtle industries.

Introduction

Several decades ago in the Philippines, marine turtles abounded in most regions of the archipelago. Today, for various reasons, marine turtles are limited in distribution. In only a few remaining areas within the territorial waters of the country can the species breed. Turtles are harvested on a commercial scale, despite the existing rules and regulations because enforcement is weak. In the Philippines, there are no special marine areas where the species are truly protected.

In recognition of the precarious status of this marine resource and of the worldwide concern for its conservation, the Philippines created an interagency task force in order to focus activities calculated to mitigate the degenerating situation.

We share common problems and common goals of saving the marine turtles; we differ only in the extent of work. Many countries are already far advanced in their research and conservation measures. Sea turtle management in our country is just now receiving the attention it deserves. Since very few research studies are being conducted, we hope to profit from the experience of countries that have long been engaged in turtle conservation.

For an appreciation of the strategic position of the Philippines relevant to marine turtle conservation, its geographic setting is described here.

The Philippine archipelago is composed of approximately 7,107 islands and islets. It extends from north to south for 1,850 km, and from east to west for 1,060 km. Located in a most strategic position in the Pacific Ocean, it straddles major sea lanes of commerce. Surrounded by a vast water domain, only about 470 islands are each more than 1 km², while the vast majority are merely rocks, atolls, and reefs. The coast line is no less than 17,462-km long, 1,000 km of it sandy beaches offering turtles a potentially ideal nesting habitat. The remaining coastline is either mangrove or too steep or already appropriated for coastal towns, villages, fishponds, and coconut plantations.

Background

Four species of marine turtles are found in the Philippine territorial waters. Listed according to relative abundance and local importance, they are: *Chelonia mydas, Eretmochelys imbricata, Lepidochelys olivacea* and *Dermochelys coriacea*. All are commonly known as *pawikan* throughout the archipelago. Information on distribution is still limited to collection locations and to occurrence on nesting beaches.

The northern part of the country harbors few turtles; the populations increase towards the southernmost islands. The Pacific side of the country has the fewest records of capture and nesting sites; the south and southwest have the most.

Because this country is still in the initial stage of a systematic appraisal which will include an inventory and census of turtle populations and population trends, present findings are indicative rather than definitive. Unfortunately, there has been a marked decline in the turtle population, yet there has been a dearth of research work on them. Gomez (1979) cited the need to make known to the international community the existence of some sea turtle literature in the Philippines because reference to them has been lacking in foreign reviews.

Turtles were first recorded in the Philippines during the Spanish period (1521–1898). According to Taylor (1921), Elera (1895) listed 4 species of sea turtles in his systematic catalog of Philippine fauna, namely: *Dermochelys coriacea* Linnaeus, *Chelonia mydas* Linnaeus, *C. imbricata* Strauch, and *Thalassochelys caretta* Linnaeus.

During the American regime (1898–1946), Seale (1917, cited by Gomez 1979) mentioned 3 species of sea turtles as being of considerable commercial importance: the hawksbill (*Chelone imbricata*), the loggerhead (*Thalassochelys caretta*), and the green turtle

(*Chelone mydas*).

Domantay (1953) published a paper that was not taxonomic in nature, but he mentioned 3 species of marine turtles that breed around the Turtle Islands: *Chelonia mydas* (*C. japonica*), *Eretmochelys imbricata*, and *Caretta olivacea*. His paper presented a preliminary report on the turtle fisheries of the Turtle Islands, emphasizing the ecology of the small island groups and the feeding and breeding habits of the turtles. He also mentioned the value of turtle eggs produced annually, and described the marketing and preserving of turtle eggs. His paper also gave recommendations for conservation of the marine turtles of the Turtle Islands. The decline in turtle egg production compared to prewar production was already noted in his report.

After Domantay's work, turtle research in the Philippines made no progress for more than 20 years. In 1976 the Outdoor Recreation and Wildlife Research Division of the Forest Research Institute initiated a nationwide survey of the distribution of sea turtles. In 1978, Dr. Angel C. Alcala of Silliman University, Dumaguete City commenced a biological study of marine turtles in the Central Visayan waters (Alcala 1978).

Legislation

The green turtle has been listed since 1970 by the then Parks and Wildlife Office as one of the 20 wildlife species in the Philippines on the verge of extinction. Likewise, it has been reported lately that the hawksbill turtle (*Eretmochelys imbricata*) is also on the threatened list.

Marine turtle conservation was, up to the last decade, taken for granted. The responsibility was tossed from one agency to another, *i.e.* the Philippine Fisheries Commission (Fisheries Administrative Order No. 88—Regulations for the Conservation of Turtle, Turtle Eggs and Turtle Shells in the Philippines), then to the now defunct Parks and Wildlife Office (General Administrative Order No. 2 series of 1972—Transferring the Administration and Control of Turtles to the Parks and Wildlife Office and for Other Purposes), then to the Bureau of Forest Development with which the said defunct office was merged (Administrative Order No. 1).

Up to this writing, the only control on harvesting sea turtles is administered by the Ministry of Natural Resources through the Administrative Order No. 1 Series of 1974, implemented by the Bureau of Forest Development. It governs the gathering and disposing of marine turtles, turtle eggs and by-products. This provides a special permit to any individual or company (60 percent Filipino and 40 percent foreign capital) to collect marine turtles and provides a gratuitous permit for research institutions (Fontanilla 1979). The existing regulatory mechanism will be a good regulatory effort as soon as it is based on measurements of probable annual yield or useful statistics. Enforcement of this law has not been monitored rigorously due to lack of adequate manpower and necessary logistics. Another problem is the lack of cooperation of the permittees and poachers. In Mindanao alone, there are 5 registered traders, all based in Tawi-Tawi, and there are 132 permittees for live turtle collection. Several more applications are being held in abeyance because of the newly approved Executive Order No. 542. Added to these, are the 50 permit holders in other regions of the country. These permittees, however, do not report their catch against their quota of 25 heads a year per permittee. Only apprehended poachers are usually the ones reporting to the BFD Regional Office, after which they think that they have complied with government regulations and are so bestowed with rights and privileges to continue plying their trade.

There are indications of smuggling of live turtles and by-products; the latter are often misdeclared as shells, reptile leather, art crafts and ornaments.

Utilization and Exploitation

The green turtle is one of the most exploited species of wildlife in the Philippines. Since time immemorial, people have gathered this reptile for various economic purposes. The green turtle is a rich source of protein for thousands of Filipinos. Taylor (1921) reported turtle poisoning in 1917 in the central Philippines. He reported that *Chelonia virgata* (= *Chelonia japonica*) caused the 14 fatalities. Ronquillo and Caces-Borja (1968) reported another case of turtle poisoning in 1954 in the southern Philippines. In this case, ingestion of the boiled meat of a hawksbill caused the incident.

In January 1977, a kilo of turtle meat sold for US$.50 and the bones US$.25 in Zamboanga City. The bones (cartilage) make good turtle soup, a delicacy in some parts of the country. The eggs, both delicious and nutritious, also command a good price in the market. In Palawan, 3 eggs sell for US$.03. In the same place and year, the shell of a hawksbill (tortoiseshell) was sold at US$25 a kilo (Fontanilla and de Celis 1978). Almost all parts of the turtle are important. Even the blood and liver are used by country folk, as a cure for asthma. The high demand for these different products from sea turtles cause the coastal people to hunt them indiscriminately.

In addition to local consumption, hawksbill shells and shell products have been exported to various countries. Taylor (1921) reported the economic value of commerce in 1909 to be US$4,368 representing 2,040 kg of tortoiseshell. Seale (1917) reported that 2,296 kg of shell valued at US$4,369 were exported from the Department of Mindanao and Sulu in the southern Philippines. He estimated that 8,000 kg valued at US$12,500 were gathered in the country annually. In the 1960s, the Philippines exported less than 5,000 kg of raw tortoiseshell (Japan Tortoise Shell Association 1973), while from 1974 through 1978, we exported approximately 22,000 kg per year. Worked tortoiseshell exports have also increased from 425 pieces in 1974 to over 24,000 in 1976. Since then, exports have dropped, but in 1978 we exported 7,800 pieces. Exports went mainly to Japan, but in 1977, 3,000 pieces went to the United States and large numbers to Belgium and Italy. In 1978, over 1,000 pieces went to West Germany (Mack, Duplaix, and Wells, this volume).

Places of collection are nationwide as indicated by issued permits, although some of the major sources are the Philippine Turtle Islands, Lubang Island in Mindoro Occidental, Surigao del Norte, Negros, Antique, Sitangkai, Bungao, Davao, Basilan, Cotabato, Lanao del Norte, Quezon and Sorsogon (from BFD list of Permittees).

There is a full scale cottage industry for stuffing turtles in Cebu City, Mindoro and Zamboanga City. The processing center in Zamboanga City (Sinunoc) was checked by the Forest Research Institute, Philippines (FORI) research team in 1977. Thousands of stuffed green turtles and hawksbill turtles were noted. An average of 2,000 hawksbills and almost the same number of green turtles are processed each year. Negeri and Tow (1977) as cited by Mack, Duplaix, and Wells (this volume) reported that an average of 400 hawksbills and 100 green turtles are processed each year in Cebu City alone.

Historically, the green turtle has been the most abundant species in most parts of the country, while the others have occurred in relatively high numbers. However, a recent survey of the distribution of sea turtles, conducted by FORI, in the Philippines, showed a marked decline in the populations of green and hawksbill turtles as indicated by the decrease in their nest distribution (Fontanilla and de Celis 1978).

The status of the resource and the industry related to marine turtle exploitation has been evaluated recently by Mr. Ceferino P. Datuin (Team Leader of Wildlife Commodity of the Philippine Council for Agriculture and Resources Research) in the southern seas, west and southwest of Zamboanga City towards the southern borders (Datuin 1979). His choice of this survey area was based on the fact that this is one region where there has been heavy harvesting.

Datuin's survey, which included observations and interviews with collectors in the Tawi-Tawi group and the Philippine Turtle Islands, indicated that during the last 5 years, it has been extremely difficult to capture 1 turtle of marketable size (61-cm carapace length), each day for each collector, compared with as many as 4 per collector per day in previous years. This situation makes the turtle-dependent tribes explore other distant seas beyond their previous abundant hunting grounds, just within and off Cagayan de Sulu and Tawi-Tawi. He further reported that turtle egg gathering is becoming less lucrative, and the discovery of egg-laying spots is now "by luck." He estimated that only 300 eggs per collector is possible, in contrast with the 800 to 1,000 eggs per week per hunter in the early 1970s. In the past, 20 to 30 small islands in the southern seas, with a total of at least 70 to 100 linear km of vegetated sandy beaches were considered favorite nesting grounds for sea turtles. Datuin reported that, nowadays, only 20 km of these are good hunting grounds for eggs in the Tawi-Tawi group.

Local Plans and Strategies to Conserve Sea Turtles

In June 1979, President E. Marcos of the Philippines signed Executive Order 542 creating the Task Force Pawikan. The President sees the urgent need for the conservation of these economically important marine resources.

The Task Force Pawikan was created under the Office of the President to: 1) enforce existing rules and regulations pertaining to marine turtles; 2) conduct field investigations relevant to formulating and updating policies for the economic utilization of the species; 3) conduct information and extension work aimed at making people responsive to the marine turtle conservation movement; 4) conduct socio-economic surveys to be used as a basis in the formulation of substitute avocations for people currently utilizing marine turtles; and 5) coordinate with international agencies concerned with the conservation of marine turtles.

Conclusion

The Philippines' share of the international marine treasure, the sea turtles, shows a definite trend of population depletion as indicated by reports on recent observations, cursory surveys, and preliminary studies. The trend is attributed to the obvious imbalance between local exploitation and natural replenishment rates from both island-hatched and migratory sources. While the island-hatched group is adversely affected by unregulated collection of eggs and hatchlings; the migratory group is jeopardized by habitat disturbances that deplete sources of preferred foods. Poor food resources may discourage migrants which may sojourn here either for feeding or breeding.

The drastic decline of merchantable yields of even-sized or even-aged turtles indicates a resource crisis, an obvious case where the local conservation effort is outraced by harvest. This situation if not halted, will come to a point where the restoration of the population to productive levels may become untenable.

The pressure on the marine turtle resource is attributed to worldwide commercial demand; economic needs of local dependents; and the sociological, cultural and political maturity of the people in the archipelago.

Acknowledgments

I would like to thank our director, Dr. Filiberto S. Pollisco for encouraging me to attend this conference. I am also grateful to Mr. Ceferino P. Datuin and Conrado D. Fontanilla for helping me prepare this paper and to Dr. Edgardo Gomez for the improvement of the manuscript. Special thanks to Ms. Gladly D. Castillo for typing the manuscript.

Literature Cited

Alcala, A. C.
1978. Progress report on the biological studies on marine turtles in the Central Visayas. Manuscript, 12 pp.

Datuin, C. P.
1979. Executive brief for the Task Force Pawikan Council. Manuscript, 12 pp.

Domantay, J. S.
1953. The turtle fisheries of the Turtle Islands. *Bulletin of the Fisheries Society of the Philippines* 3 and 4:3–27.

Fontanilla, C.
1979. The marine turtle will not be a dodo. *Philippine Panorama* 1979:36–37.

Fontanilla, C., and N. C. de Celis
1978. The sea turtle that lays the golden egg is vanishing. *Canopy* 4:8–9,12.

Gomez, E. D.
1979. A review of sea turtle publications in the Philippines. Paper contributed to the workshop on Marine Turtles, Noumea, New Caledonia. Manuscript, 5 pp.

Japan Tortoise Shell Association
1973. Preliminary report on the hawksbill turtle (*Eretmochelys imbricata*) in Indonesia, Philippines, Malaysia and Singapore. G. Balazs and M. Nozoe, translators of English version, 1978.

Ronquillo, I. A., and P. Caces-Borja
1968. Notes on a rare case of turtle poisoning. *Philippine Journal of Fisheries* 8:119–24.

Seale, A.
1917. Sea products of Mindanao and Sulu, III. Sponges, tortoiseshell, corals, and trepang. *Philippine Journal of Science* 12:191–211.

Taylor, E. H.
1921. Amphibians and turtles of the Philippine Islands. *Bur. Sci. Manila* 15:1–193.

G. S. de Silva
State of Sabah, East Malaysia

The Status of Sea Turtle Populations in East Malaysia and the South China Sea

ABSTRACT

This paper confines its comments to sea turtles in the East Malaysian States of Sabah and Sarawak; principally the former. No information is available from Brunei, an independent Sultanate outside Malaysian jurisdiction. Comments are based on the writer's personal observations and experiences, reports from fishermen, islanders and seafarers, and antiquated records. The overexploitation of sea turtles, especially *Eretmochelys*, existed even 50 years ago. The slaughter has since accelerated catastrophically. This paper also describes the acquisition of 3 privately owned island rookeries by government for conservation purposes and their final evolution into The Turtle Islands National Park.

The economic importance of turtle eggs, meat, shell, and other products is discussed, and some export figures are provided. For the first time Sabah data on tagging, long distance recoveries, remigrations, and species are presented. Information regarding the Sarawak egg decline has been provided. Apparently, a similar variation occurs in Sabah, but the smaller Sabah islands yield larger harvests. There are brief notes on the Philippine situation. Finally, it is suggested that an International Marine Sanctuary be established by conservation agencies of both countries to save turtles from extinction. Their future is in our hands.

Introduction

The large land mass known as Borneo straddles the equator. Together the East Malaysian States of Sabah, Sarawak, and the independent Sultanate of Brunei comprise about a third of the island of Borneo. The remainder of the island is Indonesian territory.

The purposes of this paper are to comment on the status of *Chelonia* and *Eretmochelys* populations principally in the East Malaysian State of Sabah. To some extent, comments are made on the Philippine and Sarawak resources. All the Sabah rookeries mentioned in this paper were visited by the writer at various times. Out of the 7 species of marine turtles found in tropical

327

waters of the globe, only the green turtle (*Chelonia mydas*) and the hawksbill turtle (*Eretmochelys imbricata*) are known to nest in Sabah. In 1964, due to misidentification (Harrisson, personal communication 1969) the nocturnal nesting of loggerhead turtles (*Caretta caretta*) was reported on Pulau Gulisaan by Harrisson and quoted by de Silva (1969a). Another species reported to be found in Sabah waters will be commented on later. Green turtles, hawksbill turtles, and a few ridleys (*Lepidochelys olivacea*) are known to nest in Sarawak. In Philippine territory green turtles, hawksbills, and leatherbacks (*Dermochelys coriacea*) are found. Information regarding the 3 species in the Philippines has been gathered from barter traders and fishermen visiting various islands close to Sabah, but remotely situated from Manila. In this paper, unless otherwise indicated, the writer has drawn heavily from information passed on to him, as he has for various reasons been prevented from crossing the international boundary. On one occasion, pirates fired on the boat in which the writer was traveling. Another time, the writer witnessed the gunning down of a small fishing boat near Pulau Bakkungan Kecil, an island within the National Park. On 9 July 1977 the same island was the scene of a bloody battle between "pirates" and the Sabah police. During the past 2 years, various newspapers have reported several acts of piracy. Although naval and marine police patrols have greatly increased, the seas have to be traveled with trepidation.

Official Conservation Policy

In 1964, an excellent piece of legislation—the Fauna Conservation Ordinance, 1963 came into force. The control of turtle farms and all matters connected with turtle conservation passed firmly into the hands of the Chief Game Warden, and a conservation policy was formulated by the writer who was then Assistant Chief Game Warden. The issue of turtle hunting licenses ceased forthwith, and the closed season in March for turtle egg collection was to be vigorously enforced on the Turtle Farms to counteract the deleterious effects of the unwise use of a valuable natural resource. The former was carried out, but owing to the paucity of staff, inadequate transport, and the threat of piracy, the latter could not be enforced throughout the state. However, when the situation improved in 1966, the closed season was enforced on the Turtle Islands and a hatchery established on Pulau Selingaan. It is unknown why March was decreed under the legislation as the closed season. Only a few turtles nest during March on the turtle farms, and difficulties of enforcement arose due to monsoon conditions.

Turtle Farms

Under the Fauna Conservation Ordinance (Turtle Farms)

Regulations of 1964, 8 islands were constituted turtle farms. It was a legacy of the past and only procured revenue. Turtle rookeries or populations were insecure and changes had to be made gradually. On the turtle farms, the exclusive rights to collect turtle eggs could be granted to tenderers. As 5 islands were remotely situated, control was difficult. Control was possible only on the 3 islands near Sandakan.

Turtle Eggs—Native Reserves

Native rights were safeguarded under the Fauna Conservation Ordinance of 1963 to enable natives to collect turtle eggs, without a license from Turtle Eggs Native Reserves, off-shore along the mainland from Tanjong Nosong in Kimanis Bay to Kota Kinabalu Wharf, from Pulau Tiga and Pulau Gaya, from Pulau Sipadan in Tawau, and in the Kudat District. In 1968, native rights were extended to the whole of the Kota Belud District, including the islands of Mantanani Kechil, Mantanani Besar, Pulau Ukusan, Pulau Silar, and Pulau Pandan Pandan. Native rights, zealously guarded, are enshrined in the constitution.

Need for Conservation

In Sabah, millions of eggs have been harvested for several decades. Apart from being considered a delicacy by several ethnic groups, the eggs provide a very important source of food to an impoverished, protein-starved population in remote and underdeveloped areas. Harvests from rookeries close to markets provide a source of income. Along the coast where there is a Malay population professing Islam, turtles are not killed for food as custom forbids it. However, the eggs are eagerly collected. Apart from the eggs, turtle flesh is considered a delicacy among the Chinese, and the pagan Rungus Dusuns of Kudat who openly treat the game laws with utter contempt. The concept of exploiting fully an easily procurable resource is ingrained. At the time of the writer's visit, Sikquati Beach was littered with the skeletal remains of 13 turtles. Skeletal turtle remains were also found on other islands near Kudat. Immigrant Cocos islanders of Lahad Datu were, prior to 1964, issued with licenses to kill turtles as they considered its flesh very essential for their existence. After settling on the mainland, they did not need turtle flesh to satisfy their capacious appetites. According to reliable reports received from Saburi (personal communication 1966) and other islanders, large numbers of turtles were slaughtered by the Japanese militia during the second world war on the 3 islands in the Sulu Sea when food was scarce, and who can blame them for this? A similar situation existed in Sarawak during the same regime and period (Harrisson 1967; Chin 1976a). Indications are that turtles are becoming scarce on some well-known beaches.

The decrease in Sabah waters can be attributed to: 1) mass egg collection for over 50 years on every rookery, leaving insufficient nests left to hatch; 2) the constant frightening away of breeding turtles approaching the nesting beaches by brightly illuminated fishing vessels; 3) illegal hunting by mechanized fishing vessels in Sabah waters to supply the ever increasing demand on the mainland; 4) increased small boat activity off nesting beaches; and 5) the large scale slaughter of turtles outside the territorial waters of Sabah, in the Sulu Sea, the South China Sea, and the Celebes Sea by Filipino fishing vessels. In 1964, Harrisson indicated that the former policy of collecting only eggs was not being observed in the Philippines and that nesting turtles were being captured for meat. The Japanese, well aware of the demand, are raping the sea outside but near East Malaysian territory. It is, of course, impossible to prevent or control this pernicious slaughter. As nothing could be done about it, maximum care had to be taken to conserve what was presently available in Sabah. Therefore, as far back as 1965 it was decided to set up a turtle hatchery on Pulau Selingaan, but due to lack of funds and personnel it was postponed until 1966. Eggs were purchased from the licensees with great difficulty and there was a time when the government tendered the eggs to collectors at M = /02.5 cents each and purchased them for M = /09 cents each for conservation purposes. This situation went on for several years with protests from licensees to persons of authority. Moreover, the presence of rangers on the islands was resented. The collection of sand and coral for mainland construction projects was encouraged by the islanders for material gain. Sand was taken at all times of the year, and the few wild nests that had escaped the collectors were inadvertently dug up. However, conservation policy was implemented without fear or favor and rigorously enforced. Gradually staff were posted permanently on the 3 islands and hatcheries established. At times, due to security reasons or inclement weather, especially during the months of October to March, staff had to be withdrawn from the islands and hatchery operations discontinued. This took place several times during the period 1969–70.

Other Legislation Since 1964

Since 1966, trawler fishing close to the islands had considerably increased. Hatchlings and adult turtles were sometimes caught in fishing nets. As uncontrolled operations posed a threat to turtles, the Fisheries Department cooperated by banning trawling operations within 1 mile of the island. As in Sarawak (Hendrickson 1958) and elsewhere, in Sabah, turtles have a period of high-density nesting when egg collectors on the turtle farms harvested practically every egg. They re-fused to concede that the survival of turtles was dependent on mass egg production and concentration. Apart from mass egg collections, their habitat was threatened. Therefore, in 1971 the State compulsorily acquired the islands for M$89,000/= from private ownership. In 1972, the 3 islands were constituted Game and Bird Sanctuaries by the Yang di Pertuan Negeri (Governor).

Marine Turtle National Park

In late 1977, the government converted the 3 game sanctuaries into a 1,700-ha National Park to embrace not only the islands but also the surrounding coral reefs and sea between the islands. The move was essential to protect the coral reefs from commercial exploitation and the surrounding sea from illegal fish dynamiting. Apart from this, small fishing fleets anchor off the islands with bright lights to clean and pack fish for marketing. Their activities had several effects: 1) brightly illuminated fishing vessels frightened away turtles approaching the nesting beaches; 2) discarded fish, offal and edible refuse dumped into the sea, attracted large numbers of sharks and predatory fish to the vicinity of the islands, only to attack hatchlings entering the water after release; 3) survivors of the initial attack became disoriented and swam towards the brightly illuminated fishing vessels and were preyed upon by predators in the vicinity of trawlers; and 4) the foul discharge of bilges, toilets, empty cans, bottles, and plastic containers contaminated to some extent the coral reefs and island beaches.

Turtle Eggs—Economic Importance

The demand for turtle eggs in Sandakan is insatiable. Prior to 1973, the supply was obtained from the 3 turtle farms north of Sandakan, and augmented, seasonally, with harvests from the islands toward Kudat. At times the price of eggs fluctuated. During optimum laying months, the price fell, and the suppliers quickly ceased flooding the market. When the 3 islands became game sanctuaries, the egg supply ceased on 31 December 1972, but Sandakan continued to be supplied by Filipino barter traders with harvests from Bakkungan Besar and Taganak. At this writing, an egg costs M = / 25 cents in Sandakan. In 1967 and 1969, the egg harvest was 677,275 and 650,930, respectively. During those 2 years an uninterrupted physical check *in situ* could be maintained on egg harvests on the 3 islands. Since 1970, there has been a decline in the egg harvests.

Exploitation

Fortunately, up to now no export avenues for turtle

eggs or turtle products have been investigated by the commercial sector. However, demand for turtle eggs is heavy in Hong Kong's red-light district, although their efficacy as an aphrodisiac has not been scientifically shown (de Silva 1971). Their source of supply is undetermined, but Chinese traveling to Hong Kong from Sandakan take turtle eggs in small quantities as gifts which are highly appreciated in the Colony.

Kota Belud and Kudat get their supply from nearby beaches during the season. The sale of eggs contributes to the welfare of the local people in those places as they have very little or nothing to sell from their remote, barren and unproductive lands. In recent times, the egg supply in the aforementioned areas has dwindled, and fewer turtles come ashore to nest. In both areas, turtles are slaughtered by the local inhabitants for sale. The flesh, shell, and other parts are sold to Filipino barter traders. In Mengatal, turtle meat is obtained from an undetermined source and sold only to known customers. Collection records for the Sandakan islands during the period 1947–64 are very scanty. No records are available prior to 1947. However, it appears that even in 1933, exclusive licenses to collect *Chelonia* eggs for 3-year periods were issued by the Resident, Sandakan.

It is concluded, partly from conjecture, that even 45 years ago, the pressure on *Eretmochelys* was evident and appreciated. Laws were drafted in accordance with the Islamic custom of collecting only the eggs. Table 1 indicates the egg harvests for the period 1965–72 from the 3 privately owned islands. Although the islands ceased to be turtle farms in November 1972, licensees were permitted to harvest turtle eggs until the licenses expired in December. Since 1973, all eggs harvested are destined for hatchery. Table 2 indicates the egg harvests for the period 1973–78. Thieves from Pulau Liberan have, however, stolen eggs.

Revenue and Licenses

During the period 1950–64, the exclusive rights to collect turtle eggs were given by competitive tender,

and the price steadily increased until 1964 (Table 3). The very high tender rate indicates that rich harvests were obtained from the islands or harvests from the Philippines were cheaply procured and added to the licensees' harvest. The effects of the licensing system during 1950–64 were: 1) to increase the price of eggs to the consumer; 2) to cause tremendous dissatisfaction to the owners of the islands; and 3) to introduce a number of Chinese middlemen into the business.

From 1965 to 1971 licenses were given at reduced rates to the owners of the islands with a view to ameliorate their economic status. The rights were offered to the land owners of the 3 islands at the rate of M = /02.5 cents an egg multiplied by the island's estimated average yearly output.

Estimated Nesting Populations

Nesting occurs throughout the year on the National Park rookeries, but nesting densities vary monthly. Although attempts were made to record the activities of every nesting turtle, it has been impossible to do so. From available data, the average has been taken and the estimated nesting population for 1967, 1969, and 1973 through 1978 are given in Table 4.

Hatcheries

Since 1 August 1966, hatcheries have been in existence in the area now known as the Turtle Islands National Park. Three hatcheries are now in operation—1 on each island. Hatchery statistics for the period 1966–78 are presented in Table 5. During this period 2,705,903 eggs were transplanted and 1,792,350 hatchlings released.

Barter Traders and Turtle Products

In April 1970, 136 *Eretmochelys* and 126 *Chelonia* carapaces were brought to Sandakan by Filipino barter traders for shipment to Nansi Corporation, Osaka, Japan. In September, another consignment, comprising

Table 1. Egg harvest from privately owned islands

Year	Selingaan	Bakkungan Kecil	Gulisaan	Yearly total (3 islands)
1965	284,940	126,930	63,580	475,450
1966	236,191	73,617	55,622	365,430
1967	437,258	128,894	111,123	677,275
1968	175,097	60,052	63,648	298,797
1969	405,345	144,757	100,228	650,330
1970	359,848	97,140	82,605	539,593
1971	262,823	110,814	86,063	459,700
1972	218,847	120,803	66,409	406,059
Total	2,380,349	863,007	629,278	3,872,634

G. S. DE SILVA

Table 2. Egg harvest from the Turtle Islands National Park

Year	Selingaan	Bakkungan Kecil	Gulisaan	Yearly total (3 islands)
1973	271,380	161,416	77,476	510,272
1974	188,684	109,498	70,248	368,430
1975	196,730	105,991	77,573	380,294
1976[a]	(115,171)	(94,390)	(44,318)	(253,879)
1977	138,977	108,030	64,934	311,941
1978	118,407	137,472	66,223	322,102
Total	1,029,349	716,797	400,772	2,146,918

a. Figures for this year are inaccurate.

200 carapaces and 300 plastrons and flippers, was brought to Sandakan for shipment to Osaka, together with a certificate from the Fisheries Officer, Siasi, Sulu, indicating that the turtles were killed in Philippine waters (de Silva 1971). Their activities would certainly have encouraged natives or other interested persons to slaughter turtles when the value of shell was publicized. Furthermore, as investigations revealed that turtles were being slaughtered (within Sabah territorial waters) in the Celebes Sea and sent via barter traders to the Philippines for shipment to Sandakan, the import of *Chelonia* and *Eretmochelys* shell, skin, calipee, and oil was prohibited from all countries including West Malaysia by 2 Federal Gazette Notifications in 1971. The ban was effectively enforced without exception. Present indications are that the trade has gone elsewhere. According to Uchida (1977) over 1,000 kg of hawksbill turtle shells were imported to Japan in 1975 from the Philippines.

When Pulau Belian, which is outside the National Park, was inspected, large quantities of dry fish were found. As no nets were observed on the island or on the fishing boats, which were searched, the writer concluded that all the fish were dynamited. Although the fishermen emphatically denied killing turtles deliberately, they admitted that a few were killed "accidentally" by explosives. The carcasses were sold to Fili-

pinos. Dynamiting fish in this area has adversely affected breeding turtles found there. Explosives are purchased from Filipino fishermen, or from *kumpits* (boats) bringing immigrants to Sabah from the Philippines. Dynamiting has also been observed near Mantanani Besar and Mantanani Kechil and also takes place around other turtle rookeries situated in remote areas. This is impossible to control.

In the past, fish dynamiting frequently occurred in the area now within the jurisdiction of the National Park. Although curbed, this nefarious activity abounds between Pulau Bakkungan Kechil and Pulau Bakkungan Besar in Philippine territorial waters. Occasionally, small Filipino boats cross the border and dynamite fish near Pulau Bakkungan Kechil. These operations kill turtles. When action is taken to apprehend them, they take sanctuary in Philippine territory.

International Tagging Recoveries and Migration

The discussion of the status of marine turtles would be incomplete without some comment on international tag recoveries and migrations. The data from Sabah are made available for the first time but are too limited to determine whether *Chelonia* and *Eretmochelys* are pe-

Table 3. Tender prices, 1950–64

Year	Price (M$)	Year	Price (M$)
1950	500.00	1958	7,600.00
1951	500.00	1959	10,600.00
1952	500.00	1960	13,400.00
1953	600.00	1961	15,863.00
1954	1,000.00	1962	15,863.00
1955	2,300.00	1963	15,200.00
1956	2,860.00	1964	20,050.00
1957	4,550.00		

Table 4. Estimated nesting population on Turtle Islands during 1967, 1969 and 1973–78

Year	Number of eggs harvested	Number of nesting turtles
1967	677,275	806
1969	650,330	774
1973	510,272	607
1974	368,430	438
1975	380,294	452
1976[a]	—	—
1977	311,941	371
1978	322,102	383

— No data.

a. Accurate figures for 1976 are not available.

Table 5. Hatchery statistics, Pulau Selingaan, Pulau Bakkungaan Kechil, and Pulau Gulissan, 1966–78

Period	Number of eggs transplanted	Hatchlings released	Percentage hatch	Remarks and explanatory notes
1 Aug–30 Sept 1966	21,092	15,005	71.14	a. 1975 total number of eggs harvested = 380,294
27 July–30 Sept 1967	37,493	33,966	90.59	Less Selingaan figures 196,730
4 Mar–4 Nov 1968	137,500	96,951	70.50	Less eggs stolen 34,741
1 Feb–31 Dec 1969	50,053	31,729	63.39	Used for hatchery 148,823
1 Jan–30 Oct 1970	75,362	49,181	65.25	
1 Jan–31 Dec 1971	110,115	59,971	54.46	b. 1976 total number of eggs harvested = 253,879
1 Jan–31 Dec 1972	403,159	232,906	57.77	Less Gulisaan eggs not
1 Jan–31 Dec 1973	510,272	317,410	62.20	accounted for in hatchery
1 Jan–31 Dec 1974	368,430	304,889	82.75	(records missing) 44,318
				Used for hatchery 209,561
1 Jan–31 Dec 1975	148,823[a]	94,438	63.45	
1 Jan–31 Dec 1976	209,561[b]	114,665	54.71	c. 1978 – Stolen eggs not included
1 Jan–31 Dec 1977	311,941	205,591	65.90	
1 Jan–31 Dec 1978	322,102[c]	235,648	73.15	
Total	2,705,903	1,792,350		

riodic long distance migrants or whether they commute between the National Park and distant feeding grounds in the Philippines and Indonesia. So far 8,980 turtles have been tagged, but only 6 international recoveries have been reported: 4 from the Philippines and 2 from Indonesia (Table 6). The most distant recovery was from Kai Kechil (Indonesian territory). The green turtle had traveled 1,556 km from Selingaan prior to capture. The shortest distance 713 km, was traveled by a hawksbill slaughtered at Culasi. Three of the recoveries were made during the northeast monsoon, and it is unknown to what extent high velocity winds, currents, and heavy seas influenced their speed and direction. Apart from the insufficiency of recovery data to justify conclusions regarding turtle navigation, the period between tagging and recovery in 5 cases was too lengthy to indicate speed of travel. A hawksbill (No. 6634) tagged on Bakkungan Kechil was recovered in Culasi, Philippines after it had traveled 713 km in 40 days or 17 km per day, presuming that it left immediately after its last lay and was captured promptly on arrival. Of the many hawksbills tagged within the National Park since 1970, only 1 has so far been recovered. Carr and Stancyk (1975) record that out of 130 hawksbills tagged at Tortuguero only 7 have been recovered and adds that a female tagged at Miskito Cay, Nicaragua was observed nesting 496 km away at Pedro Keys, near Jamaica.

Available data indicate that 2 turtles tagged north of the equator traveled south of the equator into Indonesian territory. With one exception, all tag recoveries terminated with the capture of the turtles. Four of the international recoveries have been reported from the Philippines. Harrisson (1959) tagged turtles in Sarawak, and a *Chelonia* tagged at Talang Besar, Sarawak in 1953 was recovered 800 km away off Kimanis, British North Borneo (now Sabah), in 1959. Three Indonesians who had gone to Sematan, Sarawak, in August 1959 had reported to Harrisson (1959) that turtles bearing metal tags on their flippers had been observed nesting on the coast around the Sarawak border at Tanjong Datu during July and early August 1959. No other tagging data are available from Sarawak. As *Chelonia* from Sabah and *Dermochelys* from Trengganu (Polunin 1975) have been recovered mostly in Philippine waters, assistance of the Philippine authorities is vital to formulate conservation policy favorable to this region. Indonesia too will have to assist, especially as turtles tagged in Sarawak are said to be observed in Indonesian territory.

Tagging and Remigrations

During the period 1970–78, 8,980 turtles were tagged on the 3 islands. Accurate species records are unfortunately unavailable as some of the files are lost. From 1970–74, 5,000 turtles were tagged. Of these, 19 *Chelonia* returned in 2 years and 83 in 3 years. From 1975 to 1978 tagging was done intermittently. From available information, it is postulated that *Chelonia* remigrates to the National Park every 2 and 3 years with the triennial cycle dominating. Only a few 4-year returns have been recorded. Harrisson (1956) and Hendrickson (1958) record the existence of a triennial cycle in Sarawak. Schulz (1975) has recorded 1-, 2-, 3-, and 4-year returns in Surinam. In Sabah 4-year returns have been recorded only 5 times, and there is the possibility that these may be 2-year cycle returns with the inter-

G. S. DE SILVA

Table 6. Long-range recoveries of turtles tagged at the National Park

Tag number	Tagging date	Species	Tagging place	Recovery place	Recovery date	Distance traveled (kilometers)	Time taken (days)	Average speed per day (kilometers/day)
2840	22/8/72	Green	P. Selingaan 6°11'N, 118°04'E	Barrio Alegria, *Caluya*, Antique Philippines. 11°56'N, 121°33'E	18/5/73	740	296	2.7
3429	8/4/73	Green	P. Selingaan	Barrio Libas, *San Julian*, Eastern Samar, Philippines. 11°45'N, 125°27'E	2/11/73	1,056	209	5
3641	13/7/73	Green	P. Selingaan	Barrio Buli, San Augustine, *Rombolon*, Philippines. 12°35'N, 122°15'E	7/3/74	870	237	3
6634	12/2/77	Hawksbill	P. Bakkungan Kecil 6°10'N, 118°06'E	Tubungaan, *Culsasi*, Antique Philippines. 11°26'N, 122°03'E	23/3/77	713	40	17.8
4292	5/9/74	Green	P. Bakkungan Kecil	*Kai Kechil*, Malulucas, Indonesia. 5°45'S, 132°40'E	28/3/77	1,556	934	1.6
6766	10/5/77	Green	P. Selingaan	East Coast of *Cempedek* Island, S.E. Sulawesi, Indonesia. 2°28'S, 110°08'E	10/2/78	1,305	276	4.7

mediate remigration missed by the staff. One *Eretmochelys* remigrated after 2 years and 3 after an absence of 3 years. Again, lack of information precludes comment. Carr and Stancyk (1975) speculate that the 3-year remigration cycle predominates in Tortuguero.

Other Conservation Proposals

Pending is a proposal to convert an area of approximately 8,800 ha in the Darvel Bay area of the Celebes Sea including the islands of Bodegaya, Bokeydulang, Tetagan, Mantuban, Pulau Maiga and Pulau Sibuan including the coral reefs surrounding the islands into a National Park. If this park comes into existence more information on sea turtles in that area could be obtained and hatcheries set up.

Other Species

Although, at the moment, 2 species of turtles are definitely known to occur, there appears to be yet another species inhabiting Sabah waters. In 1969, when the writer investigated the occurrence of turtles on Banggi Island, inhabitants of Kampong Karakit reported that 3 species of turtles visited the beaches of Pulau Patanunan in October and November. One is known locally as *tohongan* (*Eretmochelys*), the other *penyu* (*Chelonia*), and the third *penyu bulu* (species unidentified) which is said to be a "reddish" (?) medium-sized turtle. According to K. K. Magimpat bin Kuyanga (personal communication 1969) of Kg. Minyak, Kudat District, 3 species visit Koromkunjaan beach towards the latter half of December and up to February in small numbers: *sisik pangal* (*Eretmochelys*), *timbau* (*Chelonia*), *morong* (species unidentified). He opined that the *morong* eggs were larger than *timbau* eggs. While in Kampong Sikqati, Kudat District, natives again reported the occurrence of 3 species of turtles. The unidentified turtle was referred to by the Rungus name *raya kaya*. Investigations are difficult, as the unidentified turtle only nests during the northeast monsoon (de Silva 1969b).

The leathery turtle or luth has not yet been recorded

in East Malaysian waters. However, on 13 July 1977, D. V. Jenkins, formerly Director, National Parks, Sabah and the crew of the MV *Sri Taman Negara* observed 2 "massive" turtles swimming about 100 m away from the launch. The writer observed only a large black flipper. From the description furnished by Jenkins and the crew, the animals were definitely *Dermochelys*. The writer has surveyed nesting beaches in the Sulawesi (Celebes) Sea but found no trace of *Dermochelys* even though fishermen and islanders were questioned (de Silva 1978).

Sarawak

The 3 Turtle Islands of Sarawak are situated near the island of Borneo. The 100-ha Satang Besar Island is located at approximately 110°9′E, 1°47′N. Talang Talang Besar (37 ha) and Talang Talang Kechil (12 ha) are located roughly at 109°46′E, 1°55′N. They are larger than any or all of the Sabah Turtle Islands.

Three species of marine turtle—*Chelonia*, *Eretmochelys*, and *Lepidochelys* (Hendrickson 1958; Harrisson 1969; Polunin 1975) are known to nest on the islands.

Harrisson (1969) indicated that the 3 island beaches are frequented almost exclusively by *Chelonia*, although a few *Eretmochelys* and *Lepidochelys* nest mainly during the early months of the year. Although Satang Besar is the largest of the 3 islands, it has the smallest nesting population and Talang Talang Besar the largest nesting population of *Chelonia* (Hendrickson 1958). Chin (1976b) reports a good *Chelonia* nesting beach between Sematan and Sungai Semunsan. Turtles also nest on beaches at Tanjong Similajan north of Bintulu.

Turtle eggs were harvested on the Sarawak Islands prior to the Brooke era. Harrisson (1962) who made an exhaustive study of turtles in Sarawak speculated that a century ago turtles had been numerous on the islands. Hirth (1971) quotes Banks's report of a 2,119,912 egg harvest in 1927. Of these 70 percent were consumed or sold locally. In 1935, 9 percent of 924,000 eggs collected were consumed locally. According to Harrisson (1962) 929,123 eggs were harvested in 1946 and 708,035 in 1947. This was during the Japanese occupation, and the figures are considered inaccurate. Yields in excess of 3 million eggs were produced in 1934 and 1936; 1950 and 1953 produced yields in excess of 2 million eggs. The records for 1954, 1955, 1957, 1959, and 1961 indicate that over 1 million eggs have been collected yearly. However during 1956 and 1958 the harvest dropped to 600,000 eggs per year. In 1960, only 519,677 were harvested. From 1927 to 1961 the variation from year to year is considerable. Harrisson provides the following information regarding average egg yields during the period 1927 to 1961:

Period	Number of years with data	Average green turtle eggs a year
1927–36[a]	7	2,184,095
1937–46[b]	—	—
1946–47[c]	—	—
1948–54	7	1,581,132
1955–61	7	1,038,129

a. Excludes 1929–31.
b. No data or unreliable data.
c. Inaccurate data (Japanese occupation).

Harrisson states that "a downward trend seems, on this data, evident. Inside 3 decades, the lay has halved. It appears fairly certain that the main decline began after the late thirties." From data on 1964–70 made available by Chin (personal communication 1979) the writer was able to add information for 1964–77, for the following average per annum green turtle egg yields from 1964 to 1978:

Period	Years with data	Average number per annum
1964–70	7	324,669
1971–77	7	271,895

These data are reflected in Table 7. The figures indicate that the downward trend continued. The 1964–70 figures indicate that the lay is less then half of the 1955–61 period and the 1971–77 lay is slightly more than half of the 1964–70 period. Harrisson (1967) submitted a lengthy statistical analysis of the egg trends to his Ministry and subsequently summarized it as follows:

1900/1940—slow decline; 1941/1945—serious interference by Japanese including eating turtles and using the exposed rocks off Talang Talang Kecil as a bombing range; 1947/1955—egg yields regularly over 1 million plus extensive conservation and rearing of baby turtles to renew depleted population; 1955/1965—downward trend continues, despite conservation; this is accentuated by the much increased disturbances around the islands, motor boats and large steamers; culminating in the bauxite mining operations at Sematan for 1957. 1966—unprecedented spectacular drop to below 100,000 eggs—or less than 10% of the pre-1955 average yield.

Harrisson (1967) also compares the annual egg lays for 1965—419,066 and 1966—99,307 and concedes "that a drop of 400,000 to 100,000 is unprecedented and runs in contradiction to all known trends and appears to lie outside the range of statistical or logical probability." Chin (1976b) indicates that unconfirmed reports suggest that increasing numbers of turtles nest on the mainland Sematan beaches. The Tanjong Sim-

G. S. DE SILVA

Year 1	Sabah[a] 2	Sarawak 3	Eggs transplanted 4	Percentage of years harvest transplanted 5	Number of hatchlings released 6	Number of eggs sold 7	Total revenue (M$) 8	Expenditure (M$) 9
1964	—	289,691	8,079	2.78	—	243,258	22,074.56	35,567.38
1965	475,450	419,066	8,465	2.01	—	372,768	34,500.26	32,668.85
1966	365,430	98,843	1,554	1.57	—	73,639	8,013.52	28,176.70
1967	677,275	478,622	1,203	0.25	—	431,123	42,370.17	28,314.41
1968	298,797	200,731	707	0.35	—	158,687	31,762.60	20,629.33
1969	650,330	516,581	2,252	0.43	—	469,597	46,749.30	26,841.55
1970	539,593	296,151	2,227	0.75	1,544	227,420	23,278.92	24,348.14
1971	459,700	194,289	180	0.09	127	158,053	17,615.30	24,261.49
1972	406,059	265,525	992	0.37	327	226,109	28,225.24	25,723.77
1973	510,272	323,734	8,535	2.63	8,000	281,551	34,765.88	27,332.20
1974	368,430	204,507	1,191	0.58	785	192,455	29,852.58	32,453.85
1975	380,294	203,140	991	0.48	847	186,249	34,851.26	36,558.53
1976	253,879[b]	299,398	13,159	4.39	12,639	276,578	54,888.52	40,877.18
1977	311,941	159,156	13,134	8.25	11,804	141,600	28,848.10	37,023.50
1978	322,102	253,518	3,003	1.18	14,801	234,331	61,957.75	42,870.60
		4,175,952	80,670			3,673,418	499,753.96	463,647.48

— No data.
a. Sabah figures are given for purposes of comparing harvests.
b. Figure inaccurate due to change of management.

ilajan beaches are also used by turtles, but the nesting density or population is as yet unknown. Eggs are randomly collected or destroyed by predators.

The probable associated reasons for the decline in egg production in Sarawak have been made available by Harrisson (1967). If the collection of eggs is perpetuated by the Sarawak Turtle Board as is now done, then the decline will obviously be accelerated. According to available information, every egg is harvested with only a small percentage left to hatch (Table 7). Chin (1976b) indicates that trawling operations have increased in the South China Sea. Although turtles are not deliberately caught, many are accidentally trapped in the trawling and drowned. He also adds that pollution is becoming a serious problem along the Bornean Coast. Ships of other nations are active in the South China Sea, and they deliberately capture turtles. As the turtles are receiving a very hard battering, it is doubtful if they will survive.

During Harrisson's term as Curator, Sarawak Museum, turtle hatcheries were established on the islands and eggs in varying numbers were transplanted. After his retirement, hatchery work continued under the present Curator Lucas Chin. Hatchery statistics for the years 1964–78 are in Table 7. Chin (personal communication 1979) indicates that the egg decline has been taken seriously, and the Turtle Board of Management has initiated a new conservation program. In 1973, the Fauna Preservation Society of London do-

nated 100 pounds sterling for the purchase of eggs for hatchery purposes. During the period 1976–78 the Turtle Board paid M$2,000 to 3,000 for the purchase of eggs, and decided to improve this conservation program by stages from 1976 to 1980. Unlike in Sabah, the harvesting of turtle eggs within the territorial waters of Sarawak is a government monopoly. A corporate body of trustees manages the industry, and the profits are utilized to finance Malay charities. The legislation will not likely be amended now or in the future.

Philippines

From the Philippines, information regarding turtles is difficult to obtain. The market value of turtles was evident in 1927 but, without the heavy demand in world markets at that time, presumably turtles were not excessively hunted. However, within the last decade Philippine turtle populations appear to bear a heavy burden to meet local and world demands. In 1964, Harrisson indicated that apart from egg collecting, nesting turtles were captured for meat. In 1969, de Silva reported that Filipino fishing vessels hunted turtles in the Sulu and Celebes Seas. Trawler captains informed the writer that their catches could be disposed of within 72 hours in Zamboangao. In 1978, Fontanilla confirmed the plight of turtles in the Philippines and stated that in Zamboangao City, live turtles were sold in 3 sizes—small, medium and large. At Sinanoc, thousands of stuffed turtles were available for export—particularly to Japan. Polunin (1975) quotes Kajihara who estimated that 5,000 large *Chelonia* are captured annually in the Sulu Sea.

Trawlers operating near Pulau Bakkungan Kechil are armed. Crews use swimming or copulating turtles as targets and carcasses with neatly punctured carapaces are occasionally found floating near the island. One carcass carried Tag No. 8199. Polunin (1975) indicated that the main rookeries in the Sulu Sea were Pulau Boaan, Pulau Bagnan, Pulau Taganak and Pulau Bakkungan Besar and situated close to the Sabah border. Fontanilla and de Celis (1978) confirmed this and stated that nesting had declined. The authorities concerned with conservation now appreciate the position and have issued several administrative orders. Due to the turbulent situation in the South, a turtle management program is contemplated on Laubang Island at Barrio Kanaway, Looc. The President of the Philippines has formed a task force to prevent the extinction of turtles in the Sulu Sea.

International Turtle Sanctuary

Close to the Sabah territorial boundary and the Turtle Islands National Park are situated the Philippine Islands of Boaan, Bagnan, Taganak, Lihiman, Langaan, and Bakkungan Besar. Together they form a well-defined group of turtle rookeries. All the islands are heavily exploited for eggs and turtles. Most of the eggs harvested on Taganak and Bakkungan Besar are marketed in Sandakan. In addition, fish blasting occurs. Under these conditions, the conservation work undertaken on the Sabah islands is negated within a distance of about 1.5 to 15 km. Preventive action must be taken in the areas involved. Without being presumptious, it is suggested that the Philippine conservation agencies take cognizance of happenings in the pirate-infested southern Philippine islands and consider converting the islands involved into a turtle sanctuary. If this is done, a first International Turtle Sanctuary could be created. We will cooperate in any way possible or appropriate.

It is unknown whether or not all the Philippine islands involved are inhabited or privately owned. If they are, possibly the islands could be acquired and the inhabitants resettled elsewhere. This will probably be the greatest expense of the exercise. Full-time guards would also be needed on each island.

Acknowledgments

The writer is deeply indebted to Mr. Lucas Chin, curator, Sarawak Museum for making available publications difficult to obtain and research material at short notice; Park Rangers Mr. Edrus Chung, Mr. Richard Yamie, Mr. Alfred Jubilee and Mr. Kassim bin Karim for compiling and checking statistics from numerous files; Mr. George Balazs, University of Hawaii, for sending reference material from time to time; Ms. Anne Feininger and Mr. T. Dachtera of the National Geographic Society, Washington, for furnishing geographical coordinates of little known areas where turtles have been captured and providing straight line distances; and Mrs. Jane Chong, for editing the manuscript. Finally, the writer wishes to express his appreciation to Miss Aini Sloane for patiently deciphering his heavy, illegible hand and typing the drafts and manuscript.

Literature Cited

Carr, A., and S. Stancyk
1975.   Observations on the ecology and survival outlook of the hawksbill turtle. *Biological Conservation* 8:161–71.

Chin, L.
1976a.   Tom Harrisson and the green turtles of Sarawak. *Borneo Research Bulletin* 8:2.
1976b.   Notes on marine turtles (*Chelonia mydas*). *Sarawak Museum Journal* 23:259–65.

de Silva, G. S.
1969a.   Turtle conservation in Sabah. *Sabah Soc. Journal* 5:6–26.
1969b.   Marine turtle conservation in Sabah. *Annual Report, Research Board of the Forest Department* 1969:124–35.
1971.   Marine turtles in the State of Sabah, Malaysia. *IUCN, New Series, Supplement Paper*, 31:47–52.
1978.   *Borneo Research Bulletin* 10:23–24.

Fontanilla, C., and N. de Celis
1978.   The sea turtle that lays the golden egg is vanishing. *Canopy* 4:8–10.

Harrisson, T.
1956.   The edible sea turtle (*Chelonia mydas*) in Borneo. *Sarawak Museum Journal* 7:504–15.
1959.   Notes on the edible green turtle (*Chelonia mydas*), 8. First tag returns outside Sarawak, 1959. *Sarawak Museum Journal* 9:277–78.
1962.   Note II—Notes on the green turtle *Chelonia mydas*. West Borneo numbers, the downward trend. *Sarawak Museum Journal* 10:614–23.
1967.   Notes on marine turtles, 18. A report on the Sarawak turtle industry, with recommendations for the future. *Sarawak Museum Journal* 15:424–36.
1969.   The marine turtle situation in Sarawak. *IUCN Publications, Supplemental Paper*, 20:171–73.

Hendrickson, J. R.
1958.   The green sea turtle (*Chelonia mydas*) in Malaya and Sarawak. *Proceedings of the Zoological Society of London*, 130:456–566.

Hirth, H. F.
1971.   Synopsis of biological data on the green turtle *Chelonia mydas* (Linnaeus). *FAO Fisheries Synopsis*, 85:1:1–8:19.

Polunin, N. V. C.
1975.   Sea turtle reports on Thailand, West Malaysia, and Indonesia with a synopsis of data on the conservation

status in the Indo-West Pacific region. IUCN Report, 40 pp.

Schulz, J. P.
1975. Sea turtles nesting in Suriname. *Zoologische Verhandelingen* 143:1–143.

Uchida, I. U.
1977. Imports of hawksbill turtle shell in Japan. *IUCN/SSC Marine Turtle Newsletter*, January 1977:1.

Siow Kuan Tow
State Director of Fisheries
Kuala Trengganu
West Malaysia

Edward O. Moll
Professor of Zoology
Eastern Illinois University
Charleston, Illinois 61920

Status and Conservation of Estuarine and Sea Turtles in West Malaysian Waters

ABSTRACT

The majority of Malaysians do not eat turtles, yet a thriving industry has grown up around the collection and marketing of their eggs. Four sea turtles (*Dermochelys coriacea, Chelonia mydas, Lepidochelys olivacea, Eretmochelys imbricata*) and 2 estuarine species (*Batagur baska* and *Callagur borneoensis*) figure prominently in this industry. All of these populations are currently declining due to a variety of factors including over exploitation, development of coastal areas, and expansion of fisheries.

Other than *Batagur baska* on the Perak River, large nesting aggregations have ceased to exist on the West Coast; trawling and conversion of nesting beaches into tourist beaches are important causes.

On the less populous East Coast, large nesting aggregations still occur but are rapidly disappearing. Since 1956, egg yields have been nearly halved with *Dermochelys* and *Chelonia* respectively producing only 34 percent and 43 percent of their former levels.

A variety of conservation programs have been started to stem this decline. The Fisheries Department currently operates 5 hatcheries for all of the major species. The Game Department operates an additional 3 hatcheries for *Batagur*. Several turtle sanctuaries have been proposed; an 8-km stretch of *Dermochelys* nesting beach will serve as a prototype of this concept.

Additional actions are recommended, including intensification of hatchery work, expanding sanctuaries, and prohibiting fishing within an 8-km radius of these sanctuaries during nesting seasons.

Introduction

Harvesting of turtle eggs has long provided an important protein resource to the coastal and riverine peoples of West Malaysia. Today, however, the continued existence of the turtle egg industry is being threatened by a multitude of factors including overexploitation of eggs, development of coastal areas, and expansion of

339

fisheries. It is the purpose of this report to discuss the effect of these factors on the turtles and to review conservation efforts that are being taken to maintain the resource.

West Malaysia, or more appropriately Peninsular Malaysia, is situated at the tip of the South East Asian mainland. A central range of high mountains divides the land. The sparsely populated East Coast faces the vast South China Sea and receives a strong northeast monsoon in the months of November to February and a milder southwest monsoon in the months of June to September. As a result of these monsoons, the beach, except at the southern end, is sandy and wide, providing excellent nesting grounds for sea turtles. Whereas, the West Coast, facing the narrow Malacca Strait with its heavy sea traffic, is generally flat, muddy, and has mangrove frontage. Although there are a number of islands and a few short sandy patches on the mainland, the beaches are frequently crowded with holiday makers and, therefore, are rarely visited by nesting females nowadays.

The majority of the population are Malays who profess Muslim religion. Though not strictly forbidden on religious grounds (Hendrickson 1958), Malays refrain from eating turtle meat because, in general, animals of amphibious habitat are considered *haram*. Therefore, turtles have been partially protected for the more than five hundred years since the introduction of Islam to Malaysia. However, the consumption of turtle eggs is allowed; eggs are considered delicacies with aphrodisiacal values, and collection of eggs is nearly 100 percent.

Four species of sea turtles and 2 estuarine species figure prominently in the turtle egg industry:

(at least on the East Coast) usually nests on sea beaches near the river mouth.

Economic Importance

As early as 1937, sea turtles were defined as fish by law (Fisheries Enactment, 1937) and were thus considered a resource to be harvested and managed. Although the majority of the population does not eat turtle meat, and the killing of turtles has long been prohibited on the productive East Coast states, the consumption of eggs is allowed and very much favored.

Hendrickson and Alfred (1961) estimated East Coast egg production at 2 million in 1956 (probably underestimated, see below). Although production has now fallen to just over a million eggs and the price has increased to 5 times the price of a hen's egg, turtle eggs remain a good protein source for the East Coast people. The market value of the eggs amounts to US$240,000. Further, by leasing the egg collecting beaches to the highest bidder, the governments of 3 East Coast states collected another US$98,404 as revenue in 1978 (Kelantan US$985, Trengganu US$96,322, Pahang US$1,097).

The so-called turtle industry is made up of people involved in patrolling beaches to collect eggs, and others involved in transport and marketing of the eggs. An estimated 110 jobs were created by this industry.

Rantau Abang in Trengganu State is a small village, but annually an estimated 50,000 tourists, both local and foreign, flock here in the months of June to August to observe the leathery turtles lay their eggs. The economic activities generated by the influx of tourists benefits many people especially those in the village. To cater to the needs of these tourists, the Tourist Development Corporation has started a US$2.2-million tourist complex project around Rantau Abang. In Pa-

Family	Scientific name	English name(s)	Malay name(s)
Dermochelyidae	*Dermochelys coriacea*	Leathery turtle, leatherback, luth	Penyu Belimbing
Cheloniidae	*Chelonia mydas*	Green turtle	Penyu Agar, Penyu Pulau
	Lepidochelys olivacea	Olive or Pacific ridley	Penyu Lipas, Penyu Rantau
	Eretmochelys imbricata	Hawksbill	Penyu Karah, Penyu Sisek
Emydidae	*Batagur baska*	River terrapin	Tuntong Sungei
	Callagur borneoensis	Painted terrapin	Tuntong Laut, Sutong

The largest species, the leatherback, concentrates its nesting activities on a 19 km stretch of beach at Rantau Abang in Trengganu. The other 3 sea turtle species nest on both coasts but are more concentrated on the East Coast.

River and painted terrapins are estuarine species inhabiting tidal areas of rivers on both coasts. The former nests on sand banks lining the river whereas the latter

hang, a motel at Chendor advertises turtle-watching to attract tourists. Also, pictures of turtles appear on many other travel and hotel brochures.

Other than the *Batagur* population on the Perak River, West Coast turtle populations are too depleted to be of much economic significance. In a good year, the Perak *Batagur* may still lay some 25,000 eggs having a market value upwards to US$4,800 thus providing important income to the local economy. Although Perak licenses egg collectors, the charge is minimal, and state income from this resource is insignificant.

Turtle management in Malaysia may thus be viewed as important not only from the conservation standpoint but also from an economic one.

Legislation

The first legislations concerning Malaysian turtles were promulgated in 1915 in Perak and Pahang. The "River Rights Enactment" of Perak prohibited the killing of turtles of the genera *Orlitia, Callagur, Batagur* and *Hardella* (which is not known to occur in Malaysia) and made collection of turtle eggs on a large portion of the Perak Rver, the perogative of the Sultan.

In the Pahang State Enactment No. 3 or Turtles' Eggs Enactment, the Resident (Chief Administration Officer of the State) was given power to control the collection of turtle eggs, and turtle was defined as any reptile of the genera *Chelone, Thalassochelys, Dermochelys, Orlitia, Callagur, Batagur* or *Hardella*. Subsequently, the whole Pahang beach was controlled, and egg collection was under license.

The legislation was revised by State of Pahang Fisheries Rules 1938 made in 1938 under the Fisheries Enactment, 1937 in which turtle was first classified as fish. Under the Rules, which are still enforced, no person shall capture, kill, injure, sell, or have in his possession any turtle unless authorized, and no person shall in any way prevent or hinder turtles from laying their eggs.

Similar legislation was promulgated in the State of Kelantan under its Turtles and Turtles' Eggs Enactment of 1932 which was amended by the Enactment No. 8 of 1935 to give firmer control.

It is interesting to note that Trengganu, the state with the largest turtle population, waited until 1951 to promulgate the Turtle Enactment of 1951 to prohibit the killing of turtles and to control the collection of turtle eggs. Perhaps before then, the state was too sparsely inhabited.

In 1975, the Fisheries Department reviewed existing legislation and drafted new legislation to: 1) streamline existing legislations into a uniform law for the country; 2) transfer enforcement authority and responsibility to the Fisheries Department, and 3) impose stricter control on the turtle industry to ensure maximum utilization of the resource without endangering the turtles' survival.

This legislation has been submitted to various states and has already been adopted by a few.

Status of Populations

Malaysia achieved independence on 31st August 1957.

Since then, the development both of land and of fishing has been very rapid.

Development Trends

Towns in the coastal region have expanded and continue to grow. More houses are being built on beaches. Due to a better road system, more tourists are visiting once undisturbed beaches. More beach areas are being turned into tourist resorts.

Fish landing has increased from 112,860 metric tons in 1956 (Malayan Fisheries 1957) to 564,898 metric tons (Malaysian Fisheries 1979) in 1978. This gain has resulted from intensification of fishing, improved efficiency due to mechanization, increased use of monofilament drift nets, and the introduction of prawn trawling in 1965 (Siow and Gan 1970). These activities, particularly prawn trawling, have taken a heavy toll on turtles.

The above, in addition to continued exploitation of eggs, have exerted a growing pressure on the turtles of Malaysia, resulting in a general decline of populations.

West Coast

Reports concerning past abundance of sea turtles are few. Cantor (1847) observed that green turtles were "at all seasons plentifully taken in fishing stakes in the straits of Malacca," and that the ridley occurred only rarely on Pinang Island. Flower (1899) thought the ridleys to be less common than either greens or hawksbills on the West Coast. Boulenger (1912) reported green turtles to be very common, especially on the Sembilan Islands, but that hawksbills were rare.

Nesting on the mainland is rare today. On 13 April 1975, a local newspaper published the discovery by the late Governor of Melaka of turtles nesting at Tanjong Kling, a 0.8 km beach 11 km west of Melaka town. On investigation, Kiew (1975) found that both the green turtle and hawksbill nest on the beach and on the nearby islands, Pulau Besar and Pulau Upeh. The number of hawksbill turtles nesting at Tanjong Kling was found to be larger than that of the green turtles which had been more abundant in the past. Kiew encountered 5 hawksbills in 6 nights nesting on the beach during peak season. Unfortunately the beach has since been developed into a well-lighted tourist and industrial area, thus destroying the last known nesting site in the southern part of the West Coast. Kiew (personal communication) considers attempts to save the turtles there to be fruitless.

Pangkor Island like Pulau Sembilan and a few other islands off the Perak coast have small stretches of sandy beaches previously frequented by green turtles, ridleys, and hawksbills. But with more than 1,600 prawn trawlers operating on 171 km of coastline, and with hotels built on the beach, only rarely do turtles nest nowadays. Similarly, the 2 big islands to the north, Pulau Pinang and Pulau Langkawi, have sandy beaches, but due to trawling and a highly developed tourist industry, nesting is sporadic.

The estuarine species, though not uncommon, are believed by West Coast fishermen to be much reduced from former times. The decline of the *Batagur* population on the Perak River has been documented (Loch 1951; Mohamed Khan 1964; Moll 1978). Prior to the second world war, some 375,000 to 525,000 eggs were laid a year. Today only 20,000 to 30,000 eggs are laid in a good year. The decline apparently began during the japanese occupation when large numbers of adults were killed for food. Resurgence of the population has since been inhibited by a variety of factors including heavy egg exploitation, habitat destruction, and poaching.

East Coast

Hendrickson & Alfred (1961) conducted a survey on the nesting populations of sea turtles on the East Coast of Peninsular Malaysia in 1956. They omitted the West Coast because, although turtles were found nesting on the limited sandy beaches and a few islands to the north, the number of eggs marketed was small, compared to the East Coast production. Much of the shore in East Johore, like that of the West Coast, is unsuitable for sea turtle nesting and was also not inlcuded. The remaining 3 states—Pahang, Trengganu, and Kelantan—have individual legislation regulating the collection of the eggs. They assumed, after investigation, that the license fees paid by collectors to the Government was half the total values of the eggs marketed. Therefore, they estimated the egg production of the exclusive leathery turtle beach by dividing the license fees by a constant 0.04 which was half the market price (M$0.08) of the leathery turtle eggs. For other beaches they used the constant of 0.03 being half of the market price (M$0.06) of the green and ridley turtle egg. A predetermined ratio of ridleys to green turtles was used to separate yields contributed by each species. Turtle populations were assumed to have remained fairly constant since the last world war. Figure 1 depicts their findings.

In 1978, the authors together with Dr. Leong Tak Seng of Universiti Sains Malaysia again conducted a survey of the turtle populations on the East Coast. A set format was prepared, interviews were conducted in villages with most of the licensed egg collectors, and government records were examined.

We stress that collection of completely accurate statistics was impossible because: 1. licensed collectors for the main beaches tended to give unreasonably low estimates due to fear of inquiry by tax collectors and to keep potential competition in the dark in the following year's bidding for license areas; 2. most of the collectors on other beaches do not keep careful records. Thus the figures given are often unreliable.

Therefore, in analyzing the data, information such as license fees paid by collectors over the past 5 years, number of egg collectors employed by each licensee, other occupation and income of the licensee, and local price of eggs were taken into account. No single formula was utilized. The results (Table 1 and Figures 2 and 3) are thus no more than the authors' best estimate of the present status of the East Coast nesting populations.

Figure 1. Sea turtle egg production of the East Coast, 1956.

Estimated annual yields of sea turtle eggs

Species	Kelantan	Trengganu	Pahang	Totals
Dermochelys coriacea	Negligible	853,700	Negligible	853,700
Chelonia mydas	40,200	770,200	118,500	928,900
Lepidochelys olivacea	4,500	40,500	13,200	58,200
Totals	44,700	1,664,400	131,700	1,840,800

Source: Reproduced from Hendrickson and Alfred, 1961.

Table 1. Estimate of annual yield of turtle eggs on the East Coast

	Kelantan	Trengganu	Pahang	Johore	Total
Dermochelys coriacea	—	294,000	300	—	294,300
Chelonia mydas	400	298,000	91,000	12,000	401,400
Lepidochelys olivacea	21,000	240,000	34,000	10,000	305,000
Eretmochelys imbricata	—	10,700	5,400	2,500	18,600
Callagur borneoensis	800	13,500	1,200	500	16,000
Total	22,300	856,200	131,900	25,000	1,035,000

Our findings deviate from Hendrickson and Alfred (1961) in that: 1) egg yields of *Dermochelys* and *Chelonia* have been reduced to 34 percent and 43 percent, respectively, of their 1956 levels; 2) conversely, *Lepidochelys* shows a great increase (500 percent) over 1956, and 3) *Eretmochelys* and *Callagur* provided a small but significant contribution to the East Coast egg industry.

We feel the drastic decline in egg yields indicated for *Chelonia* and *Dermochelys* is mostly real (see below). Nearly all those interviewed reported drastic drops in the egg yield over the past 10 years. Increases in fishing activity, especially trawling and drift netting, were blamed for the growing number of dead turtles on the beach every year. One collector reported 25 dead ridleys in a year. In 1973 fishermen operating a bottom long line for rays caused a massive kill of ridleys at Setiu, Trengganu. The senior author, being a fisheries officer, imposed a ban on that type of fishing. But no officer can ban fishing as fish are necessary food for the people.

The increase indicated for ridleys was not supported by our interviews. Again, egg collectors generally reported continual decline in numbers. We feel Hendrickson and Alfred (1961) underestimated ridley numbers (and possibly overestimated green turtles).

Due to time constraints, Hendrickson and Alfred did not do a detailed survey. Rather, they used findings from 5 licensed, major egg-producing areas (unfortunately the names were not given) to estimate yields and species ratios in the remaining areas. Due to the small sample size, it is highly possible that their study areas had atypically large numbers of green turtles and resulted in the skewed ratios (1 ridley to every 10 or 12 green turtles) used in other calculations. This also explains why *Eretmochelys* and *Callagur* egg production was overlooked.

Another source of error resulting from using license fees to calculate area productivity was that accessibility to many coastal areas was difficult at that time. Hence, low tender price was sometimes an indicator of accessibility rather than productivity.

The production in Johore, though not reported by Hendrickson and Alfred, is reported here for reference. It is, as expected by them, smaller than that of Pahang.

Conservation

Hatcheries

The leathery (leatherback) turtle program in Trengganu is the oldest turtle conservation effort in West Malaysia. First proposed by the Malayan Nature Society in 1960 (Wyatt-Smith, 1960), a hatchery was set up at Rantau Abang in 1961. This hatchery, 1 of 5 operated by the Fisheries Department, has operated every year since then (Table 2).

The Fisheries Department next began a green turtle hatchery at Dalam Ru, State of Kelantan, in 1964 (Balasingam 1967a). The hatchery was inactive in some years due to lack of funds or changes in administration (Table 3).

Table 2. Rantau Abang Leathery Turtle Hatchery Results

Year	Number of eggs planted	Number hatched	Percentage hatched
1961	8,366	3,699	44.2
1962	11,654	6,300	54.0
1963	9,956	5,580	56.0
1964	11,535	3,803	32.0
1965	10,071	7,199	71.5
1966	31,250	16,477	52.7
1967	15,650	9,215	58.9
1968	40,000	18,332	45.8
1969	38,008	15,930	41.9
1970	31,050	17,089	55.0
1971	47,391	30,260	63.9
1972	60,000	37,193	62.0
1973	72,260	30,699	42.5
1974	91,147	42,616	46.8
1975	85,922	40,565	47.2
1976	69,480	44,480	64.0
1977	7,803	4,578	58.7
1978	34,391	14,878	43.3
Total	675,934	348,893	51.6

Note: Work on the hatchery in 1979 is in progress.
Source: Fisheries Department, State of Trengganu, Malaysia.

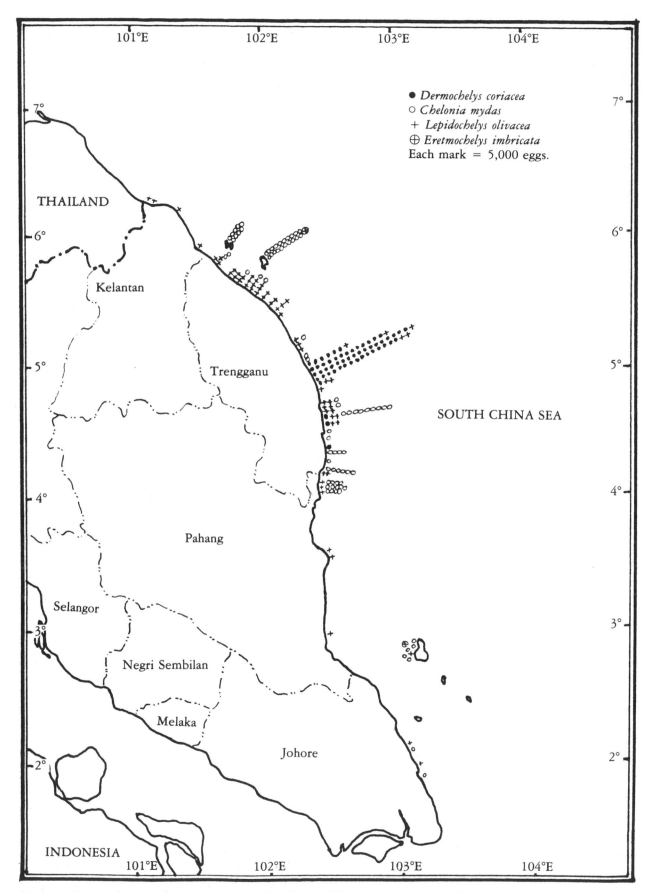

Figure 2. Egg production of sea turtles on the East Coast of Peninsular Malaysia, 1978.

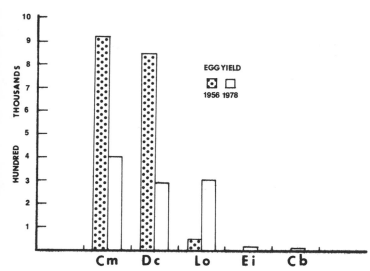

Figure 3. Comparison of estimates for egg yield on the East Coast of Penisular Malaysia in 1956 (Hendrickson and Alfred, 1961) and 1978 (this study). Abbreviations Cm, Dc, Lo, Ei, and Cb stand, respectively, for *Chelonia mydas, Dermochelys coriacea, Lepidochelys olivacea, Eretmochelys imbricata* and *Callagur borneoensis.*

The third hatchery was set up in 1971 at Chendor, State of Pehang for all coastal nesting species. The results are shown on Table 4.

The fourth hatchery was set up in 1978 at Kg. Mangkok, Trengganu chiefly for ridleys and painted terrapins. Out of 5,180 ridley eggs planted, 3,080 hatched out, of which 2,447 were released in Semerak, State of Kelantan at the request of the Director of Fisheries Kelantan. Comparative study on hatchery techniques was conducted by Moll on *Callagur* (Table 5). Work on the hatchery in 1979 is in progress.

This year a fifth hatchery was begun at Pulau Perhentian mainly for green turtles. A total of 10,000 eggs of green turtles have been planted. Hatching has not been completed at the time of this writing.

The Game Department operates conservation programs for *Batagur* in 3 states—Perak, Kedah, and Trengganu. In addition to a hatchery, these programs include "head starting" whereby young turtles are raised for a year in captivity prior to release. Table 6 summarizes these hatchery results.

Attempts were made to operate a hatchery at Tanjong Kling, State of Melaka, by Dr. B. H. Kiew of the University of Malaya in 1976 and another at Muka Head, State of Penang, by Dr. K. H. Khoo of University Sains Malaysia in 1978. Neither materialized due to poor landings of sea turtles. Arrangements are being made to send green turtle eggs by air from Trengganu to Penang for hatching by the Fisheries Department there in 1980.

Sanctuaries

Certain areas of beach are being considered for turtle sanctuaries. Once designated, development of the beach

Table 3. Dalam Ru Hatchery Results

Year	Number planted	Number hatched	Percentage hatched
1964	3,704	1,971	53.6
1965	9,056	4,558	50.3
1966	9,817	4,080	41.6
1967–68	Nil	Nil	—
1969	12,307	2,395	19.5
1970	12,492	5,226	41.8
1971	11,198	5,237	46.8
1972	7,036	2,605	37.0
1973–77	Nil	Nil	—
1978	—	2,447[a]	—
1979	2,080	1,366	65.7
Total	67,690	26,072	

Note: Data include Ridley turtles.
a. Hatchlings supplied from K. Mangkok Hatchery in Trengganu.
Source: Fisheries Department, State of Kelantan, Malaysia.

Table 4. Chendor Turtle Hatchery Results

Year	Number planted	Number hatched	Percentage hatched
1971	4,138	3,514	84.9
1972	14,000	10,619	75.9
1973	6,140	4,341	70.7
1974	5,008	4,511	90.1
1975	Nil	Nil	—
1976	14,595	13,227	90.6
1977	Nil	Nil	—
1978	10,155	8,094	79.7
Total	54,036	44,306	81.9

Note: Work on the hatchery in 1979 is in progress.
Source: Fisheries Department, State of Pahang, Malaysia.

Table 5. *Callagur* hatchery results, 1978

Incubation method	Number planted	Number hatched	Percentage hatched
Outdoor in sand on beach	633	419	66.2
Indoor in sand in plastic buckets	224	195	87.1
Indoor in styrofoam boxes	201	155	77.1
Total	1,058	769	72.6

and intrusion of tourists will be stopped. Licensed egg collection will continue with a cetain proportion being replanted in an artificial hatchery. Predation of eggs and hatchlings is very heavy in natural nests (Hen-

Table 6. *Batagur* **hatchery results for Perak, Kedah, and Trengganu**

Year	Perak R. Number planted	Perak R. Number hatched	Perak R. Percentage hatched	Kedah R. Number planted	Kedah R. Number hatched	Kedah R. Percentage hatched	Trengganu R. Number planted	Trengganu R. Number hatched	Trengganu R. Percentage hatched
1969	500	184	39	—	—	—	—	—	—
1970	1,896	86	05	—	—	—	—	—	—
1971	1,140	218	19	—	—	—	—	—	—
1972	2,940	507	17	—	—	—	—	—	—
1973	1,780	168	09	—	—	—	—	—	—
1974	1,380	395	29	—	—	—	—	—	—
1975	2,420	914	38	—	—	—	—	—	—
1976	2,412	1,245	52	—	—	—	36	36	100
1977	1,953	964	49	—	—	—	141	112	79
1978	3,220	1,412	44	152	114	75	629	429	68
1979	1,440	588	41	325	283	87	1,644	1,232	75
Totals	21,081	6,691	32	477	397	83	2,450	1,809	74

Source: Game Department of West Malaysia.

drickson 1961), and the high market value of the egg encourages heavy poaching. Small clutches of 50 eggs in artificial hatcheries were found to have greater hatching success than larger natural nests (Balasingam 1967b).

A memorandum for setting up a leathery turtle sanctuary and research laboratory at Rantau Abang was submitted to the Government of State of Trengganu in 1975 (Siow 1975). At Rantau Abang, the turtle beach runs parallel to the main trunk road. Fortunately, a 45-m ditch separates the road and the beach, serving as a natural barrier to protect the turtle. It was proposed that 8 of the 19 km of beach be set aside as a turtle sanctuary and fenced in to prevent unauthorized persons from entering. *Casurina* trees were proposed to be planted on the ditch side of the beach to prevent light from shining onto the beach. The project has received financial support from the World Wildlife Fund Malaysia.

A survey of the flora and fauna of Pulau Redang (a major green turtle nesting island off the coast of Trengganu) was completed. Recommendations are that the island be turned into a national park including a marine park. If accepted, the survival of green turtles and hawksbills can be assured (Green 1978).

The Penang State Government is currently studying the proposal of setting up a national park on the northwest side of the island with a fine beach. This would enable the Fisheries Department to conduct a hatchery to regenerate diminishing turtle populations there.

Kuala Baru is a major ridley and *Callagur* rookery in northern Trengganu State. The beach is in fact a long stretch of wide sand dune separating the sea and a large estuary. The senior author will study this area,

and if found feasible for a turtle reserve, will submit a memorandum to the State Government suggesting this designation.

The Outlook

Hendrickson and Alfred (1961) estimated that as little as 2 or 3 percent of the nests, handled properly, could provide sufficient recruitment to the wild population to maintain present levels, provided that other circumstances remain unchanged. Unfortunately, other circumstances, mainly man-made, are changing, and changing rapidly. The intensification of fishing activity has resulted in high mortality of adults which under natural conditions face little predation and hazard. Continuous development of coastal areas has reduced and will further reduce the number and length of beaches suitable for nesting of sea turtles. Such development seems unavoidable; without immediate action, turtle populations will continue to decline.

We suggest the following courses of action as being most practical to conserve turtles in West Malaysia: 1) intensifying hatchery work (i.e. increasing the number and size of present hatcheries with a goal of hatching at least 15 percent of the eggs laid by each species); 2) setting aside a portion (for example 20 percent) of the total beach area as sanctuaries; and 3) prohibiting fishing within an 8 km radius of sanctuaries during breeding seasons.

Acknowledgments

The senior author wishes to thank the Director General of Fisheries Malaysia for approving the presenta-

SIOW KUAN TOW

tion of this paper, Miss Wan Kelsom for typing portions of the manuscript, and officers in the Fisheries Department for their assistance.

The junior author expresses his gratitude to Eastern Illinois University, the New York Zoological Society, the World Wildlife Fund and The Fauna Preservation Society for support of his work in West Malaysia and to M. Buerster for typing portions of the manuscript.

Literature Cited

Balasingam, E.
1967a. Turtle conservation: results of 1965 hatchery programme. *Malayan Nature Journal*, 20:139–141.
1967b. The ecology and conservation of the leathery turtle *Dermochelys coriacea* (Linn) in Malaya. *Micronesia*, 3:37–43.

Boulenger, G. A.
1912. *A Vertebrate Fauna of the Malay Peninsula from the Isthmus of Kra to Singapore, including the Adjacent Islands.* London: Taylor and Francis.

Cantor, T.
1847. *Catalogue of Reptiles Inhabiting the Malay Peninsula and Islands.* J. Asiatic Soc. Bengal. XVI. Reprinted 1966 Amsterdam: A. Asher and Company.

Flower, S.
1899. Notes on a second collection of reptiles made in the Malay Peninsula and Siam. *Proceedings of the Zoological Society of London*, 1899:600–697.

Green, J.
1978. A survey and proposal for the establishment of the P. Redang Archipelago National Park. World Wildlife Fund Malaysia.

Hendrickson, J. R.
1958. The green sea turtle, *Chelonia mydas* (Linn) in Malaya and Sarawak. *Proceedings of the Zoological Society of London*, 130:455–533.
1961. Conservation investigations on Malayan turtles. *Malayan Nature Journal Special Issue*, 214–223.

Hendrickson, J. R., and E. R. Alfred
1961. Nesting populations of sea turtles on the east coast of Malaya. *Bulletin of the Raffles Museum Singapore*, 26:190–196.

Kiew, B. H.
1975. Report on the turtle beach of Tanjong Kling, Melaka. *Malayan Nature Journal*, 29:59–69.

Loch, J.
1951. Notes on the Perak River turtles. *Malayan Nature Journal*, 5:157–160.

Malayan Fisheries
1957. *Annual Fisheries Statistics 1956.* Fisheries Department, Malaya.

Malaysian Fisheries
1979. *Annual Fisheries Statistics 1978.* Fisheries Department, Malaysia.

Mohamed Khan bin Momin Khan.
1964. A note on *Batagur baska* (the river terrapin or tuntong). *Malayan Nature Journal*, 18:184–187.

Moll, E. O.
1978. Drumming along the Perak. *Natural History*, 87:36–43.

Siow, K. T.
1975. *Memorandum bagi mengadakan kawasan simpanan bagi penyu di Rantau Abang Negeri Trengganu.* Fisheries Department, State of Trengganu, Malaysia.

Siow, K. T., and B. H. Gan
1970. *Proposal for the setting up of a national fishing corporation.* Fisheries Department, Malaysia. (restricted distribution)

Wyatt-Smith, J.
1960. The conservation of the leathery turtle, *Dermochelys coriacea. Malayan Nature Journal*, 14:194–199.

Legislation Cited

1) River Rights Enactment, 1915 (Perak)
2) Turtle Eggs Enactment, 1915 (Pahang)
3) The Turtles and Turtle Eggs Enactment, 1932 (Kelantan)
4) Fisheries Enactment, 1937 (Federated Malay States)
5) Fisheries Rules, 1938 (Pahang)
6) Turtles Enactment, 1951 (Trengganu)
7) Fisheries Act, 1963 (Federation of Malaysia)

Ismu Sutanto Suwelo
Direktorat Perlindungan Dan Pengawetan Alam
Jalan Ir. Juanda 9, P.O. Box 133
Bogor, Indonesia

Njoman Sumertha Nuitja
Fakultas Perikanan
Institut Pertanian Bogor
Bogor, Indonesia

Iman Soetrisno
Fakultas Peternakan dan Perikanan
Universitas Brawijaya
Malang, Indonesia

Marine Turtles in Indonesia

ABSTRACT

Indonesia consists of approximately 13,677 islands. Marine turtles can be found along beaches of many of the islands. Five of the 7 known species of marine turtles can be found in the Indonesian Archipelago: *Chelonia mydas, Eretmochelys imbricata, Dermochelys coriacea, Lepidochelys olivacea,* and *Caretta caretta.* Fishermen come to areas where the turtles are abundant, especially on Sukomade (East Java), East Kalimantan, and Flores Sea. The turtles are caught by harpoons and nets. Bajo fishermen are well-known turtle hunters, who build their homes near shallow waters. Other groups of fishermen are Benoa and Buginese. Turtle eggs are liked very much in Indonesa, and they bring higher prices than chicken eggs.

Besides governmental restrictions on turtle catching and egg collection, priority has been given to habitat and population development within the framework of the Indonesian turtle preservation program.

Introduction

Indonesia, an archipelago consisting of 13,677 islands, is situated in the Equatorial Zone. It stretches 5,100 km from east to west and 1,883 km from north to south. Its total area is 1,904,345 km². Indonesia lies between the Pacific and Indian Ocean. Indonesia has a wet tropical climate with a minimum temperature of 18°C and a maximum of 34°C. According to the 1971 census, the population of Indonesia is 130,000,000. About 75 percent of the people live in Java, Madura, and Bali.

Indonesia is rich in natural resources. The diversity of species in Indonesia must be safeguarded, because they are the very elements that build mankind's ecological environment. These natural resources should be utilized for the people's economic development through management based on the principles of nature conservation.

Marine Turtles as an Indonesian Natural Resource

One marine resource that has recently attracted serious attention in Indonesia is the turtles. Of the 7 species of marine turtles in the world, 5 species live in Indonesia. Research into nesting areas in Indonesia has been carried out by Suwelo and Kuncoro (1969), Sumertha Nuitja (1973, 1975 and 1976) and Polunin (1975). Because different names are given for a single species of turtle, a study of what species exist in Indonesia is imperative. For this purpose, Abdurahman, Sumertha Nuitja, and Suwelo (1977) and Sumertha Nuitja (1977) compiled an Indonesian taxonomic list. Since then, more knowledge has gradually come to light concerning the turtle species and their distribution in Indonesia (see Polunin and Sumertha Nuitja, this volume).

The green turtle, *Chelonia mydas* (local names: *penyu sala* in Sumbawa Island, *penyu daging* in Bali, and *penyu nijau* in West Java) is the most commonly caught turtle in Indonesia.

The hawksbill, *Eretmochelys imbricata*, is the second most common turtle. *Penyu sisik* is a local name in Indonesia, and its shell is very popular for ornaments. The storage centers of tortoiseshell are in Palembang, Jakarta, Surabaya, Pontianak, Denpasar, and Ujung Pandang. The carapaces are exported to Singapore, Hong Kong, and Japan, and occasionally to Brussels.

There is also the already rare *Dermochelys coriacea*, popularly known as *penyu belimbing*. This species is legally protected by the Agriculture Minister's Decree No. 327 of 29 May 1978.

The other 2 turtle species are *Caretta caretta* and *Lepidochelys olivacea* (local name *penyu abu-abu*). Their total populations and breeding grounds in Indonesia have not been ascertained. Sumertha Nuitja discovered 2 carapaces of *penyu abu-abu* in Bali in August 1977, so it is probable that this species lives in Nusa Tenggara islands. Although no data are available and their population levels cannot yet be determined, these 2 species need immediate legal protection similar to *D. coriacea*.

Marine Turtle Utilization

Fishermen usually catch turtles with traditional gear such as harpoons and also with modern gill nets. Bajo fishermen know the breeding grounds of the turtles. They catch turtles in Flores Sea, Timor, and Southeast Sulawesi. They build their homes near shallow water. Other groups of fishermen who hunt turtles are Buginese and Benoa. Owing to the vast extent of the Indonesian seas, it is very difficult to control exploitation of marine turtles. Many parts of the turtle are used. This is apparent during visits to Ujung Pandang, Bali, and other places. Turtle eggs are also eaten by the people and can cost more than chicken eggs.

Turtle exports must be authorized. Recent export licenses, issued by the Directorate of Nature Conservation covered: 6,071 carapaces in 1975, 4,870 carapaces in 1976, 6,779 carapaces in 1977, and 6,659 carapaces in 1978.

Population Development Based on Preservation Principles

Because of the growing demand for meat and eggs for consumption within the country and for carapaces for export, the government of Indonesia needs both to utilize and to conserve marine turtles. In addition to the government restricting turtle catching and egg collection, priority has been given to habitat and population development within the framework of a preservation program. Since 1977, the Directorate of Nature Conservation has undertaken the development of turtle populations using the Suwelo method (below). Turtle-raising is also being considered for some locations near marketing centers. Sumertha Nuitja (1970 and 1975), investigated the possibility of using Serangan Island, Bali, and Sumbawa Bay for raising hatchling turtles to maturity. The Governor of Bali supported the idea and asked Sumertha Nuitja to survey Serangan Island for the main site of turtle culture in Indonesia. Other regions also raise turtles, especially on the Seribu islands and near Ujung Pandang. In the rearing facilities on the Seribu islands, however, many young turtles have died of dermatitis, helminthiasis, and tuberculosis.

Efforts like those sponsored by the Governor of Bali should be expanded, particularly the culture of hawksbills, considering the increasing price of tortoiseshell and the difficulty of catching hawksbills. In this way, it will not jeopardize their existence in natural habitats.

Development of Turtle Populations

Taking as an example the development of *Dermochelys coriacea* in Malaysia and taking into account the experience acquired from the turtle egg hatching trials at Sukabumi 10 years ago, techniques can be formulated for developing the turtle population on hatching beaches in Indonesia as follows.

It is recommended that the development efforts be carried out in cooperation with the Provincial Administration and be contracted to a third party.

1. A sufficient number of eggs should be provided by the Provincial Administration free of charge. If this is impossible, eggs should be purchased from the contractor. The number of eggs required is 15 percent of the harvested eggs, although this number is not a necessary condition at the initial stage.

2. The eggs are buried again in the sand. Each hole

is 50 cm deep and filled with 50 turtle eggs. The nests should be made as natural as possible. Nests should be located on sandy beaches in the same nesting area.

3. Each nest should be enclosed with wire netting (30 cm diameter, 30 cm high) and marked with the date of planting and the number of eggs. The wire netting protects the eggs against predators.

4. After incubation in the sun for 50 days (common turtles) or 55 days (*Dermochelys coriacea*), the young turtles emerge.

5. The hatchlings are very active and try to get out of the enclosure and to the sea. They should be counted to determine how many eggs have hatched.

6. The young turtles should be released immediately on the tide line. They will move towards the sea by themselves.

7. They are carried on the waves into the sea and start their lives among the community of turtles and search for their own food.

8. When the turtles have become adults, the females lay eggs on the same beach.

Summary

Indonesia still has very rich populations of sea turtles, particularly *Chelonia mydas* and *Eretmochelys imbricata*. Carapaces are exported to many countries. The eggs and turtle meat are consumed by Indonesians. A well-planned conservation and management program, which would control the exploitation of the turtle populations, would insure that this resource can be exploited continuously in the future.

Literature Cited

Abdurahman, F.; N. Sumertha Nuitja; and I. S. Suwelo
1977. Penyu laut di Indonesia dan usaha budidayanya di Bali. *Seminar Biologi V Malang.*

Polunin, N. V. C.
1975. Sea turtles, report on Thailand, West Malaysia and Indonesia, with a synopsis of data on the conservation status of sea turtles in the Indo-west Pacific Region. Mimeo, Cambridge, England.

Sumertha Nuitja, I. N.
1973. Kemungkinan Budidaya penyu di Pulau Serangan, Bali. LPPL Jakarta.
1975. Beberapa aspek biologi penyu sala, *Chelonia mydas* dan Cara pengelolaannya di Pulau Sumbawa. Fakultas Perikanan, IPB. Bogor.
1976. Studi habitat penyu laut di Pangumbahan, Kabupaten Sukabumi. Kerjasama Fakultas Perikanan IPB dan PPA Pusat Bogor.
1977. Cara mengenal jenis-jenis penyu laut. Fakultas Perikanan, IPB. Bogor.

Suwelo, I. S., and Kuncoro
1969. Penyu laut, produktivitas dan pembinaannya di Indonesia. Rimba Indonesia.

Nicholas V. C. Polunin
World Wildlife Fund
Directorate of Nature Conservation
P.O. Box 133, Bogor, Indonesia

Njoman Sumertha Nuitja
Fakultas Perikanan
Institut Pertanian Bogor
Bogor, Indonesia

Sea Turtle Populations of Indonesia and Thailand

Introduction

Sea turtles and sea turtle products have long been an important commercial resource in many parts of Southeast Asia. In spite of extensive exploitation of both eggs and adults, large populations still survive in several areas. We wish to summarize here what is known of the history and present status of these populations and so to provide a background to the efforts which are being, and must continue to be, made to conserve them.

Thailand and Indonesia are combined in this paper for reasons of space, and because intervening Malaysia is dealt with elsewhere in this volume. In Thailand, surveys—if only somewhat superficial ones—have covered the ground reasonably well, while in Indonesia, a number of important areas such as Irian Jaya, parts of Maluku, and Kalimantan, have barely been looked at. Consequently, our information is at present fragmentary. It is clear, however, that Indonesia, with some 13,500 islands and 81,000 km of coastline, is far richer in sea turtles than is Thailand; most of this account therefore deals with Indonesia.

Historical Background

The first European accounts of sea turtles in the region date back to the mid-sixteenth century, when Portuguese Jesuit missionaries came to the Moluccas (Jacobs 1974); subsequent observations include those of Nieuhoff (1666), Valentyn (1724), Forbes (1885), and Cabaton (1914). However, it is established from early Chinese records (Wheatley 1959; Meilink-Roelofsz 1962) that tortoiseshell was an important trade commodity from much earlier times. Further, the hunting of sea turtles was obviously part of the subsistence of many indigenous peoples (Pelras 1972; Loeb 1972; Polunin 1975), and their expertise was important in the development of the turtle trade then (Fox 1977), as it still is today (cf. Sumertha Nuitja 1974).

Anthropological anecdotes about sea turtles in the region include the description by Loeb (1972) of the tabus surrounding turtle hunting on Siberut in the Mentawei Islands, the account by Covarrubias (1937) of the importance of turtles in Balinese cuisine and cosmogeny, and Kolff's brief mention (1840) of a superstitious aversion to turtle hunting in one of the Moluccan islands. No doubt there are many more such fragments in the region's literature. The discussion of Hendrickson (1958) on the position of turtles in Khoranic law can be taken to apply also to Islamic communities in southern Thailand and Indonesia, but in many areas these rules seem to be loosely adhered to. No such regulations apply to non-Moslem people such as the Balinese, who are predominantly Hindu.

A further element in this background picture is the role played by adat law (cf. Visser 1979) and tabus (cf. Endicott 1970) in traditional Malay and Indonesian life, which often served to regulate use of natural resources. Such regulation probably prevailed among the maritime people who hunted and traded in sea turtles. Notable among these people are both the Buginese of Southwest Sulawesi (Anonymous 1918) and the "sea gypsies," who are variously known as *Moken, Moklen,* and *Urak Lawoi* in Burma and western peninsular Thailand (Hogan 1972), *Orang Laut* in Malaya and the Straits of Malacca (Pelras 1972), and *Bajo* or *Sama-Bajau* in Sulawesi, Nusa Tenggara and the Moluccas (Fox 1977). It is regrettable that so little has been written about these people, especially the sea gypsies, who have for long been involved in the sea turtle trade (Vosmaer 1839; Freijss 1859; Forbes 1885). Although the picture is complicated by the fact that while the Bajo in some areas were almost exclusively engaged in this trade (cf. Crawfurd 1856), in other cases they apparently do not hunt sea turtles at all for semi-religious reasons (J. J. Fox, personal communication, 1978).

Traditionally turtle meat and eggs were used as food, while tortoiseshell, especially from *Eretmochelys imbricata,* has been fashioned into ornaments and utensils (Loebèr 1916; Sumertha Nuitja 1974). Meat from sea turtles, normally *Chelonia mydas,* is consumed heavily in Bali and the Manado area of northern Sulawesi; in Bali particularly, ceremonial uses are common (Suwelo and Kuntjoro 1977). In some places *Dermochelys coriacea*—although the species is now officially protected, (Abdullah and Suwelo 1978) is eaten: Irian Jaya and the Mentawei Islands (Sumertha Nuitja, manuscript); Lembata in East Nusa Tenggara (R. H. Barnes, personal communication, 1979); the Kei Islands (A. Compost, personal communication, 1979). Sometimes *Eretmochelys* is eaten in Irian Jaya (van Hasselt 1922) and Pulau Seribu off Jakarta, from individuals that have been raised on fish (K. Kvalvågnaes, personal communication, 1979). The export market is large, and Indonesia is currently among the world's greatest exporters of tortoiseshell. This trade is of long standing, but is evidently increasing.

Local governments regularly give annual concessions on important nesting beaches to individuals for collecting eggs (Somadikarta 1962; Polunin 1975; Food and Agriculture Organization (FAO) 1977a); often the concessionaires are subject to regulations with respect to seasons and the hatching of some of the eggs collected, but such regulations are hard to enforce. Subadult and adult turtles are traditionally caught in nets, turned on beaches, and harpooned in the sea; in central Indonesia large handmade elastic-powered spear-guns are now widely used. The magnitude of the take incidental to other forms of fishing, notably trawling and long-lining in modern times, has not been quantified. Catch rates for single trawlers in the Java Sea (Losse and Dwiponggo 1977) and southern China Sea (Sudradjat and Beck 1978) appear low, but the effect of the entire fishing effort could be large.

General Ecology

Most data on distribution in Southeast Asia are based on the occurrence of turtles on nesting beaches, and even there the information on the rarer species is usually imprecise. Very little is known about non-nesting animals, with the exception of the data which have been gathered on the main fishing-grounds for *Eretmochelys* in Indonesia (Kajihara 1974). No tagging has yet been carried out in Indonesia, although sea turtles tagged elsewhere have turned up there. These include *Chelonia mydas* from Sabah and Queensland, and *Dermochelys coriacea* from Trengganu. No work has been done on feeding ecology, although it is probable that the algae are as important as seagrasses in the diet of *Chelonia* in central Indonesia.

Five species have been recorded in local seas, namely *Chelonia mydas, Eretmochelys imbricata, Lepidochelys olivacea, Caretta caretta,* and *Dermochelys coriacea.*

The most abundant and best known species is the green turtle, *Chelonia mydas,* which nests intensively at a number of sites (Figure 1); it is probable that several other sites remain to be located. For a few of the documented places some recent data on egg yields are available (Pangumbahan in Java, the turtle islands off the Berau River's mouth in East Kalimantan, and Ko Khram in the Gulf of Thailand), while in another case (Sukamade in Java), the number of turtles nesting has been reported. These data are presented below. In other cases (the Sambas area of West Kalimantan, and the Riau Islands) there are some old data for the early 1960s. For the remaining localities in Figure 1, no information is available. Some data are also presented below for a minor turtle site, Phangnga Province in peninsular Thailand.

As yet *Eretmochelys imbricata,* the hawksbill turtle, has been reported to nest intensively only on the is-

NICHOLAS V. C. POLUNIN

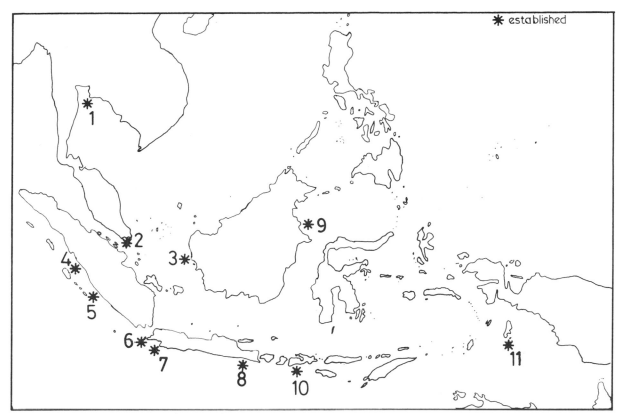

Figure 1. Some major sites of *Chelonia mydas* nesting in Thailand and Indonesia, excluding Malaysia: 1) Ko Khram; 2) Riau Islands; 3) Karimata, West Kalimantan; 4) West Sumatra; 5) Bengkulu; 6) Ujung Kulon, Java; 7) Pangumbahan, Java; 8) Sukamade, Java; 9) Berau, East Kalimantan; 10) Ai-Ketapang, Sumbawa; 11) Pulau Enu, Aru Islands (from Polunin 1975; Sumertha Nuitja 1979).

lands of Nangka and Namperak off Belitung, and on a few beaches in the Ai-Ketapang district of South Sumbawa (Figure 2); however, Kajihara (1974) has also estimated the extent of the main fishing-grounds for this species in Indonesian waters.

Dermochelys coriacea, the leatherback turtle, nests intensively in at least 4 localities: Phuket in Thailand, West Sumatra and Bengkulu Provinces in Sumatra, and on the North coast of the Kepala Burong (Vogelkop) part of Irian Jaya (Figure 3). There may also be some nesting in the Savu Sea area. In addition, a few *Dermochelys* nest occasionally on beaches, such as those on the south coast of Java, where *Chelonia* is dominant. *Dermochelys* nesting occurs typically at localities close to deep ocean.

Apart from sporadic records, almost nothing is known of the status of *Caretta caretta*, the loggerhead turtle, and *Lepidochelys olivacea*, the ridley turtle, in these areas. The former reputedly nests in West Sumatra and occasionally on the Javanese beaches, while the latter probably nests in Nusa Tenggara.

Although all 5 species of sea turtles have been recorded on some nesting beaches, it appears that certain locations are preferred by certain species. Sumertha Nuitja, Eidman, and Aziz (1979) have presented some evidence that species may to some extent be segregated on different nesting beaches in Sumatra, Java and Sumbawa.

Population Trends and Exploitation

It is regrettable that the data on populations, let alone their variations in time, are so scant, for where we have little basis on which to estimate population changes, we have even less chance of establishing a solid connection between any trends and exploitation. Although the reliability of the data may often be in question, an overall downward trend is nevertheless suggested by the egg yields reported for some Thai and Indonesian beaches (Figure 4). The Spearman Rank Correlations are statistically significant at the 5 percent level for Pangumbahan, Berau, Phangnga, and Ko Khram, but not for Sukamade.

That the nesting beach figures indicate declines in breeding populations at least in certain areas is supported by some other circumstantial evidence. It is generally accepted that there has been a reduction in the number of important nesting beaches in Java at least (WWF 1976; FAO 1977b), and the turtle fishery for the Balinese market (although the market is probably expanding) is evidently covering a larger and larger area. Sumertha Nuitja (1974) reports that local turtle populations around Bali were seriously depleted by 1950. In North Sulawesi, turtle nesting is now rare in the Mas Popaya Raja Nature Reserve which was originally established for turtles, and nesting on many other beaches in the area has declined (J. MacKinnon, personal communication, 1979). Turtle nesting seems also

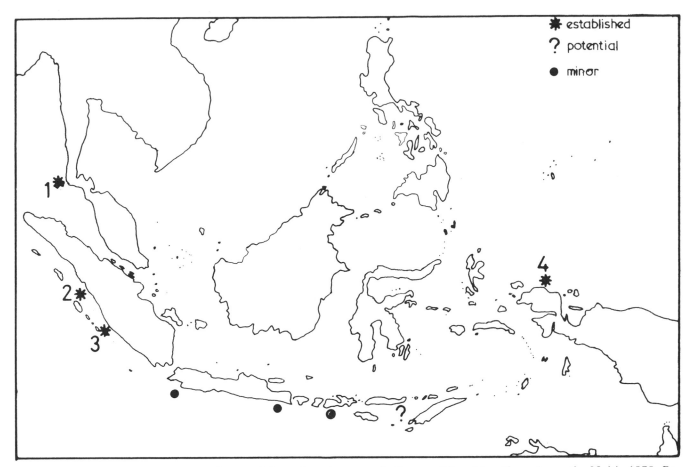

Figure 2. Important nesting sites of *Dermochelys coriacea* in Thailand and Indonesia, excluding Malaysia. 1) Phuket, Thailand; 2) West Sumatra; 3) Bengkulu, Sumatra; 4) North Ke-pala Burong, West Irian (from Sumertha Nuitja 1979; Polunin 1975; FAO/UNDP 1978; Anonymous, 1978b).

to have decreased recently in the Maumere district of Flores (D. Lewis, personal communication, 1979).

The production of turtle products (in particular shell, meat and eggs) for human consumption, both domestic and foreign, is high, and in the case of tortoiseshell is certainly greater than it used to be. Although such data could probably be found in The Hague, in the archives of the Vereinigde Oostindische Compagnie and Dutch governmental departments which subsequently took its place, we have as yet been unable to find any extensive early data on annual tortoiseshell export from Indonesia, apart from occasional mentions in the literature such as Temminck's report of some 2,650 kg of shell exported in 1836 (Temminck 1846). For the early part of the present century, Dammerman (1929) summed up the data for the years 1918–27, while the period 1968–78 is covered by Indonesian Directorate General of Fisheries (Direktorat-Jendral Perikanan) statistics (Anonymous, 1978a).

Figure 5 shows that recorded exports have been higher in recent years than they were previously. The Indonesian fishery statistics (Anonymous, 1978a) also show (Table 1) that between 1971 and 1976 from 348,504 to 1,110,539 kg of sea turtle have been reported caught each year. Of the 999,040 kg for 1976, 95 percent were from 4 areas of the country as follows: East Nusa Tenggara 58.5 percent, North Sulawesi 24.4 percent, Bali 6.9 percent, and Irian Jaya 5.1 percent. Data from export permits issued by the Indonesian Directorate of Nature Conservation (Direktorat Perlindungan dan Pengawetan Alam) show that over 6,000 turtles have been exported annually in the years 1975–78 (Table 1), but these were exclusively from the Medan and Palembang areas of eastern Sumatra. Some 70 percent of the turtles exported in 1978 went to Japan and 21 percent to Singapore. All of the individuals exported live, some 150 in that year, went to the United States. The main tortoiseshell dealers are based in Medan, Palembang, Jakarta, Surabaya and Bali.

Sumertha Nuitja (1974) has reported that in the Kuta and Kesiman districts of Bali 28,800 turtles were consumed in the 3 years 1968–70. In 1973 the Sinar Laut Company of Bali often exported 5,000 to 6,000 stuffed sea turtles and 3,000 sets of turtle leather each month (Polunin 1975). Some 323,509 kg of turtles were reported from the Badung district of South Bali in 1978 (Dinas Perikanan Propinsi Bali, unpublished data). It has been estimated that 2,000 to 3,000 turtles are sold each month in the Denpasar market in Bali (FAO 1977b), while other data indicate that more than 20,000 turtles are consumed annually (M. Halim, personal communication, 1979). It is probable that the

NICHOLAS V. C. POLUNIN

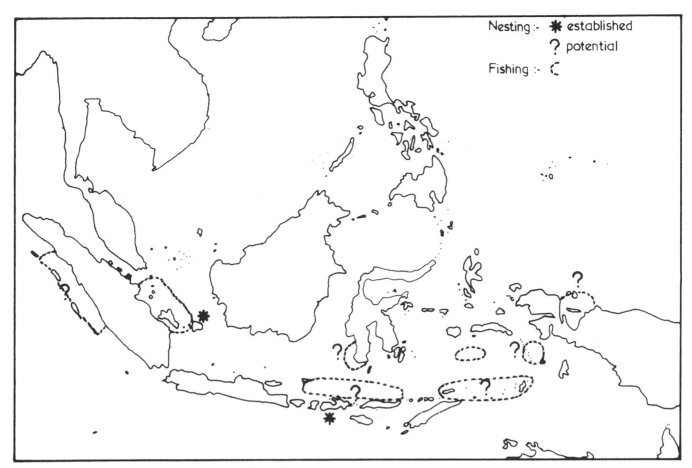

Figure 3. Some important areas for *Eretmochelys imbricata* in Thailand and Indonesia, excluding Malaysia (from Kajihara 1974; Sumertha Nuitja 1979).

Balinese trade is expanding, partly to meet the tourist demand for meat and curios. The number of Bali-based dealers in turtles and turtle products has probably not increased, nor has the number of boats registered in South Bali in the 3 years 1976–78, but participation by boats from other provinces is likely to have increased. In Sumbawa between 5,000 and 9,500 turtles (Table 1) have been reported caught each year (Sumertha Nuitja 1979); most of these were for the Balinese market. The Balinese turtle fishermen include those based at the villages of Tanjung Benoa and Serangan in South Bali, Buginese and Bajos from Sulawesi, and fishermen from Timor, Flores and Sumbawa. The fishery covers areas such as southern and eastern Kalimantan, Sulawesi, Sumbawa, and Flores. In places such as Sumbawa, turtles are often caught by local people, kept alive, and later sold to fishermen traveling to Bali.

Kajihara (1974) has estimated the take of young *Eretmochelys* from the main fishing grounds in Indonesia; some 10,000 and between 15,000 and 20,000 are caught annually in the Ujung Pandang district of South Sulawesi and in Sumatra, respectively. Kajihara (1974) further suggested that the annual take of large *Eretmochelys* from central and eastern Indonesian waters (eastern Java, Flores and Banda Seas) was approximately 5,000 and 30,000 individuals before 1971 and after 1972, respectively.

The above figures clearly indicate very high levels of exploitation of sea turtle stocks. The information on population trends does not overwhelmingly point to consistent decline, but unfortunately we have little data, and it may also be that levels of catch have only recently begun to be critical. Traditionally, at least for *Chelonia* in most of Indonesia, sea turtle utilization was confined to egg collecting and a comparatively low adult take. The conflict between the collection of eggs on nesting beaches and the hunting of adults and subadults is likely to have had a particularly severe impact on sea turtle populations. Further, because of the nature of marine turtle life cycles and our lack of information, it might be awhile before any widespread decline in numbers were detected.

In many places turtle populations have been affected in indirect ways. In several localities in Indonesia (such as the north coast of Java, and the south-facing coasts of Bali) and Thailand (for example Pattaya), beaches that were once turtle nesting sites are now heavily used by people. More subtle forms of nesting habitat alteration may also affect turtles; much of the beach at the small island of Ko Kra in the Gulf of Thailand is now rubble, apparently derived from the blasting of the

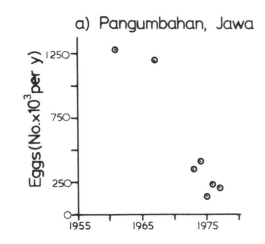

a) Pangumbahan, Jawa

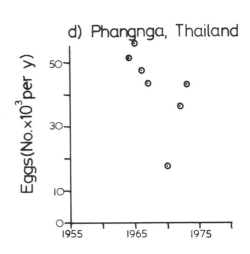

d) Phangnga, Thailand

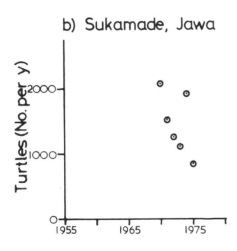

b) Sukamade, Jawa

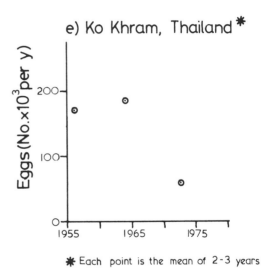

e) Ko Khram, Thailand ✷

✷ Each point is the mean of 2-3 years

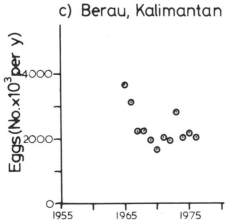

c) Berau, Kalimantan

Figure 4. Data reported from some Thai and Indonesian turtle nesting beaches. Source of data: a) Somadikarta 1962; K. Ikrasaputra, personal communication, 1974; Sumertha Nuitja 1979; b) WWF 1976; c) FAO 1977a; d) Polunin 1975; e) Penyapol 1957; Polunin 1975.

NICHOLAS V. C. POLUNIN

adjacent reef by fishermen.

Estimates of absolute population sizes are bound to be vague, but if we work back from the egg production figures available for some of the important beaches, we conclude that at least 25,000 female *Chelonia* are breeding annually in western Indonesia. For Thailand the figure is a small fraction of this; probably some 1,000 female turtles nest each year (Polunin 1975).

Conservation Measures

Sea turtle conservation measures for the region have been discussed for many years (Anonymous 1919; Meer Mohr 1927; Rappard 1936). It could still be argued that the available data do not show a consistent downward trend and that the populations of *Chelonia* and *Eretmochelys* at least are still sizeable. But it is clear that our data are few and that in some places, such as around Java and Bali, and in the Gulf of Thailand, sea turtle populations have already been seriously depleted, while the pressure of direct and indirect impacts is high and increasing.

Within Indonesia a number of steps can be taken to conserve turtle populations. To begin with, in the critical area of Java, all the main nesting beaches are within (Ujung Kulon and Sukamade) or partly within (Pangumbahan and Citirem) established nature reserves. Efforts must be made to ensure that turtles are fully protected there. In addition, an adequate basis exists for the governmental control of the use of sea turtle resources, both egg collection and the catching of large turtles. Bulk export of turtles and turtle products is officially regulated by permits from the Directorate of Nature Conservation, while, at least in Bali, turtle merchants obtain licenses which are issued by the local fisheries office. Additionally, concessions for egg collecting on most major beaches in Thailand and western Indonesia are dealt with by local governments. In West Nusa Tenggara, because of the impact on nesting turtles, the catching of adults is officially prohibited in the months of January to April.

More effort should be made to monitor present turtle populations and the extent of their exploitation in order to assess more precisely the impact of human activities. Emphasis should be put on a country-wide assessment of sea turtle resources, with detailed studies in problem areas such as Java, Bali, and North Sulawesi. With a better background of appropriate information, management criteria could be decided on; quotas could be set, and other conditions such as closed seasons and size limits could be stipulated in licenses. At Sukamade in Java, turtle egg collecting has been officially halted. The ban is at least partially effective for the present (W. Angst, personal communication, 1979), and the nesting beach is becoming a focus for tourism in the Meru Betiri Nature Reserve. At Ujung

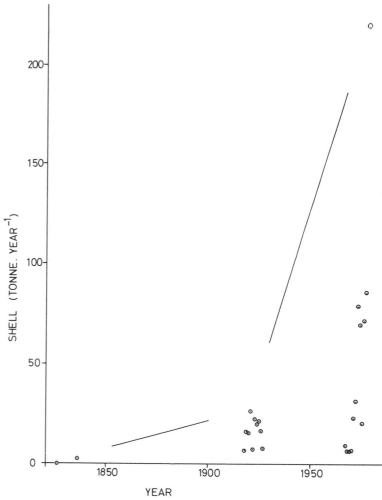

Figure 5. Some data on the export of tortoiseshell from Indonesia (from Temminck 1846; Dammerman 1929; Anonymous, 1978a).

Kulon in Java, nests are being protected from large predators (Anonymous 1979). Both of these are a useful beginning to conservation action but should be more widely extended.

Regulations, if possible, should not seriously restrict traditional subsistence exploitation, but in most areas true subsistence hunting and egg collecting is probably not under governmental control anyway.

Measures must also be taken in importing countries to reduce the international commerce in turtles and turtle products. This would be in line with Indonesia's recent signing of the Convention on International Trade in Endangered Species (CITES).

A study should also be made of the feasibility of turtle rearing, ranching, and farming methods as a possible aid to conservation. Some preliminary work has been carried out in Thailand (S. Rongmuangsart, personal communication, 1977), and in Indonesia (at Pulau Serangan, Bali). Any projects of this kind would initially be dependent on wild populations and should therefore be linked with measures such as protection of nesting beaches. They would also require substantial

Table 1. Some additional data on sea turtle exploitation in Indonesia

Year	1965	1966	1967	1968	1969	1970	1971	1972	1973	1974	1975	1976	1977	1978
Catch of adult and sub-adult turtles reported for Sumbawa Island (number per year)	5,248	6,048	6,820	7,790	8,156	8,059	7,923	9,263	7,946	5,418	—	—	—	—
Catch of sea turtles reported for the Badung district of South Bali (tonnes per year)	—	—	—	—	—	—	—	—	—	—	212.4	30.5	130.3	318.4
Total reported catch for Indonesia (tonnes per year)	—	—	—	—	—	—	464	381	343	1,093	986	446	—	—
Licensed export of turtles and turtle carapaces and skins (number of turtles per year)	—	—	—	—	—	—	—	—	—	—	6,362	7,974	9,129	6,659

— No data.

Sources: Sumertha Nuitja 1979; Reports of the Dinas Perikanan Propinsi Bali, Denpasar; Anonymous 1978a; Unpublished data in the Direktorat Perlindungan dan Pengawetan Alam, Bogor.

investment of capital and time, and should therefore receive guidance from experienced and competent workers.

In the Javanese shadow-puppet theater there is a song about *lamahcai*. In Sudanese language the word merely means "land-water," for these people believe that the terrestrial and aquatic environments are part of a single system, their homeland. Perhaps through the sea turtle, which is unusual among large animals in its dependence on both land and water habitats, this traditional concept can be incorporated into conservation action.

Summary

Indonesia still has large populations of *Eretmochelys* and *Chelonia*, and some important nesting sites for *Dermochelys*. Knowledge of the size of these populations is poor, but exploitation is high and generally increasing. There is evidence of local depletion of turtle stocks. At the same time there is a basis for conservation within the country, which, fortified with more precise information on the populations and their exploitation, would be sufficient to regulate human use of the turtle resource. Outside the country stronger measures can be taken to curb international trade and also provide expertise for turtle management. The region remains rich in turtles and turtle-related lore, and efforts should be made to keep it that way.

Acknowledgments

We thank John MacKinnon, Jan Wind, and Jeff McNeely for their useful comments on this paper, and also Diana Mossman, Sudiarka, Vincent Susanto, and Femme Gaastra for additional information. N.V.C.P. is grateful to the late Tom Harrisson for originally drawing his attention to Southeast Asian sea turtles and to International Union for the Conservation of Nature and Natural Resources (IUCN) and WWF for supporting his work on them. A thank you to Ninta Karina for typing the earlier draft of the manuscript.

Literature Cited

Abdullah, A. S., and I. S. Suwelo
1978. Protected wildlife in Indonesia. In *Wildlife Management in Southeast Asia*, eds. J. A. McNeely, D. S. Rabor, and E. A. Sumardja, pp. 23–39, Bogor: BIOTROP.

Anonymous
1918. *A Manual of Netherlands India (Dutch East Indies)*. London: Naval Staff Intelligence Department.
1919. Schildpad. In *Encyclopaedie van Nederlandsch-Indië*, vol. 3, 3rd ed. The Hague: Martinus Nijhoff.
1978a. *Fisheries Statistics of Indonesia*. Jakarta: Direktorat-Jendral Perikanan.
1978b. *Studi Habitat dan Populasi Penyu Belimbing (Dermochelys coriacea) di Propinsi Bengkulu*. Bogor: Direktorat Perlindungan dan Pengawetan Alam.

NICHOLAS V. C. POLUNIN

1979. Progress in conservation of sea turtles. *Conservation Indonesia* 3:7.

Cabaton, A.
1914. *Java, Sumatra and the Other Islands of the Dutch East Indies.* London: Fisher Unwin.

Covarrubias, M.
1937. *Island of Bali.* London: Cassell.

Crawfurd, J.
1856. *A Descriptive Dictionary of the Indian Islands and Adjacent Countries.* London: Bradbury and Evans.

Dammerman, K. W.
1929. *Preservation of Wild Life and Nature Reserves in the Netherlands Indies.* Bandung: Fourth Pacific Science Congress.

Endicott, K. M.
1970. *An Analysis of Malay Magic.* Oxford: Clarendon.

FAO (Food and Agriculture Organization)

1977a. Proposals for establishment of conservation areas in East Kalimantan. Bogor: FAO (FO/INS/73/013. Field Report 5).
1977b. Nature conservation and wildlife management in Indonesia. Interim Report. Rome: FAO (FO/DP/INS/73/013).

FAO/UNDP
1978. Nature conservation in Irian Jaya. Bogor: FAO (FO/INS/73/013. Field Report 9).

Forbes, H. O.
1885. *A Naturalist's Wanderings in the Eastern Archipelago.* London: Sampson, Low, Marston, Searle and Rovington.

Fox, J. J.
1977. Notes on the southern voyages and settlements of the Sama-Bajau. *Bijdr. Taal-, Land- en Volkenkunde* 133:459–65.

Freijss, J. P.
1859. Reizen naar Mangarai en Lombok in 1854–1856. *Tijdschr. Ind. Taal-, Land- en Volkenkunde* 9:443–530.

Hendrickson, J. R.
1958. The green turtle, *Chelonia mydas* (Linn.) in Malaya and Sarawak. *Proceedings of the Zoological Society of London* 130:455–535.

Hogan, D. W.
1972. Men of the sea—coastal tribes of South Thailand's west coast. *J. Siam Soc.* 60:205–35.

Jacobs, H., editor
1974. *Documenta Malucensia, I (1542–1577).* Rome: Institutum Historicum Societatis IESU.

Kajihara, T.
1974. The hawksbill turtle in Southeast Asia. Japanese Tortoise Shell Association.

Kolff, D. H.
1840. *Voyage of the Dutch Brig of War Dourga, through the Southern and Little-known Parts of the Moluccan Archipelago.* London: James Medden.

Loeb, E. M.
1972. *Sumatra: Its History and People.* Singapore: Oxford University.

Loebèr, J. A.
1916. Been-, hoorn- en schildpadbewerking en het vlechtwerk in Ned-Indie. *Geïllustreerde Beschrijvingen von Indische Kunstrijverheid* 7, 71 pp.

Losse, G. F., and A. Dwiponggo
1977. Report on the Java Sea Southeast Monsoon trawl survey June–December 1976. *Laporan Penelitian Perikanan Laut (Special Report),* 3:1–119.

Meer Mohr, J. C. van der
1927. Aantekeningen betreffende de biologie van *Chelonia mydas. Trop. Natuur* 16:44–53.

Meilink-Roelofsz, M. A. P.
1962. *Asian Trade and European Influence in the Indonesian Archipelago between 1500 and about 1630.* The Hague: Martinus Nijhoff.

Nieuhoff, J.
1666. *Die Gesantschaft der Ost-Indischen Gesellschaft in den Vereinigten Niederlandern, an den Tartarischen Cahm und nunmehr auch Sinischen Kaiser.* Amsterdam: Mors.

Pelras, C.
1972. Notes sur quelques populations aquatiques de l'Archipel Nusantarien. *Archipel* 3:133–68.

Penyapol, A.
1957. A preliminary study of the sea turtles in the Gulf of Thailand. Bangkok: Hydrographic Department, Royal Thai Navy.

Polunin, N. V. C.
1975. Sea turtles: reports on Thailand, West Malaysia and Indonesia, with a synopsis of data on the conservation status of sea turtles in the Indo-West Pacific Region. Morges, Switzerland: IUCN. Mimeo report, 113 pp.

Rappard, F. W.
1936. Een schildpaddenstrand in het wild reservaat Zuid Sumatra. *Trop. Natuur* (Jubileum Uitgave): 124–26.

Somadikarta, S.
1962. Penyu laut di Indonesia. *Buku Laporan Kongres Ilmu Pengetahuan Nasional 2,* 5:593–585.

Sudradjat, A. and U. Beck
1978. Variations in size and composition of demersal trawler catches from the North coast of Java with estimated growth parameters for three important foodfish species. *Laporan Penelitian Perikanan Laut,* 4:1–80.

Sumertha Nuitja, N.
1974. Perikanan penyu dan cara pengelolaannya di Indonesia. *Dokumentasi Komunikasi Institut Pertanian Bogor,* 8:1–18.
1979. Penyu laut dan masalahnya. Manuscript, 67 pp.

Sumertha Nuitja, N.; M. Eidman; and K. A. Aziz
1979. Studi pendahuluan tentang populasi dan habitat peneluran penyu laut di Indonesia. *Perhimpunan Biologi Indonesia, Kongres Nasional Biologi IV,* D1–10:1–17.

Suwelo, I. S., and Kuntjoro
1977. Penyu laut perlu segera diselamatkan. *Oseana* 3–4:7–23.

Temminck, C. J.
1846. *Coup-d'oeil général sur les possessions Néerlandaises dans l'Inde Archipélagique.* Vol. 1. Leiden: Amz.

Valentyn, F.
1724. *Oud en Niew Oost-Indien.* vol. 1b. Amsterdam: de Linden.

van Hasselt, F. J. F.

1922.   Over het eten van Schildpadvleesch. *Trop. Natuur* 11:157–59.

Visser, N. W.

1979.   Nature protection legislation in Indonesia. Morges: WWF Mimeo report, 135 pp.

Vosmaer, J. N.

1839.   Korte beschrijving van het Zuid-Oostelijk schier-eiland van Celebes. *Verh. Bataviaasch Genootsch. Kunst. Wetenschappen*, 17:61–184.

Wheatley, P.

1959.   Geographical notes on some commodities involved in Sung maritime trade. *J. Malay Branch R. Asia. Soc.* 32:1–140.

WWF (World Wildlife Fund)

1976.   *The Javan Tiger and the Meru-Betiri Reserve*. Morges, Switzerland: WWF.

Indian Ocean

C. S. Kar
Gahirmatha Marine Turtle
Research and Conservation Unit
P. O. Satabhaya 754 225
Via. Rajnagar
Cuttack District, Orissa, India

Satish Bhaskar
formerly Field Officer, Madras Snake Park Trust
and Conservation Centre

Status of Sea Turtles in the Eastern Indian Ocean

ABSTRACT

Five sea turtle species have been reported from the eastern Indian Ocean. Their populations are believed to be declining steadily everywhere. Although sea turtles in India and in Sri Lanka have been accorded total legal protection, many difficulties beset enforcement. These result from the remoteness of nesting beaches; the resistance to protective governmental statutes among traditional exploiters of sea turtles; widespread poverty in the region which makes sea turtles, their eggs and derived products an attractive source of income and which precludes the use of adequate staff and facilities to control poaching and illegal trade; and the paucity of knowledge relating to the locations of nesting beaches and feeding areas where detrimental human activity occurs.

One of the largest olive ridley breeding populations in the world is being depleted by the thousands off the coast of Orissa, India, for meat. Human overpopulation has resulted in the colonization of many sea turtle nesting beaches, especially on islands.

Mainland India

The September 1977, Amendments to the Schedules to the Indian Wildlife (Protection) Act, 1972, accord total protection to all sea turtle species excepting the locally unreported *Chelonia depressa*.

The export of sea turtles and derived products was banned in August 1975. India is among the signatories to the Convention on International Trade in Endangered Species.

Between 1963 and 1974 India exported 102,022 kg of sea turtle products valued at roughly $100,880. The products included sea turtle meat, oil, and tortoiseshell (Salm 1976; Murthy and Menon 1976).

The price of 1 kg of tortoiseshell increased from Rs. 0.26 in 1967 to Rs. 20 in 1969 and to Rs. 185.60 in 1975 (Jacob, communication to R. Whitaker, 1977).

Gujerat and the Gulf of Kutch

About 300 km southeast of the Pakistani rookeries at Sandspit and Hawke's Bay lies the Gulf of Kutch (Figure 1) where the green turtle and the olive ridley occur (Bhaskar 1978b). The leatherback is sighted rarely, but unlike the others is not known to nest in the Gulf.

Four fresh olive ridley nests were found on 10 km of beach west of Mandvi in Kutch on 6 July 1978. Suitable nesting beach extends for 60 km. The eggs are preyed upon by jackals, dogs, wild boar, and humans of the Wagir community who sell the excess locally.

Of about 15 islands near Saurashtra's northern coast substantial nesting occurs on uninhabited Bhaidar Island. Nesting also occurs on Beyt, Nora and Chank Islands.

The 3-km nesting beach at Bhaidar was, at places, littered with a profusion of sea turtle egg shells during a survey visit on 15 June 1978. Year-round nesting reportedly occurs on the island and 3 ridleys nested that night. A visiting fisherman also inadvertently netted 3 adult green turtle females whch had entered a creek at flood tide. All were released unharmed as turtle meat is rarely consumed in northern Gujerat. Turtle fat is sometimes used in caulking boats and as protection against marine borers; the salted flipper-hide is on occasion converted into rough shoes for use on jagged coral.

Calcareous, sandy beaches suitable for nesting have, on some islands, been literally trucked away for construction purposes.

Green turtles feed on seagrass pastures between the Gulf Islands. About 10 turtles were sighted at sea in a half-hour near uninhabited Karumbhar Island during a sailboat survey on 15 June 1978.

Green turtles and olive ridleys nest in concentrations still to be determined, but likely to be substantial, along 60 km of beach between Okha and Okha Madhi. Sandy beaches extend far to the south and are likely nesting habitats also.

Turtles were visible at sea close to shore south of Dwarka on 22 June 1978, 10 days after the onset of that year's early monsoon.

Maharashtra

The green turtle and the olive ridley are known to nest but concentrations appear to be low. Recent nesting by ridleys have been recorded at Gorai, Kihim, Manowrie and at Versova. Nesting by green turtles occurs at Back Bay (Mhasawade in Salm 1976) and by unrecorded species at Fort (Mhasawade in Salm 1976) and at busy Chowpatty Beach, but detailed surveys have yet to be undertaken. Nesting turtles and eggs are often taken by humans. Nesting by a green turtle occurred at Salsette Island (Mawson 1921).

Goa

The beaches at Goa today are popular tourist resorts. *Lepidochelys* appears to be the most common turtle nesting on India's west coast south of Goa and a ridley nested at Calangute Beach in February (Salm 1976).

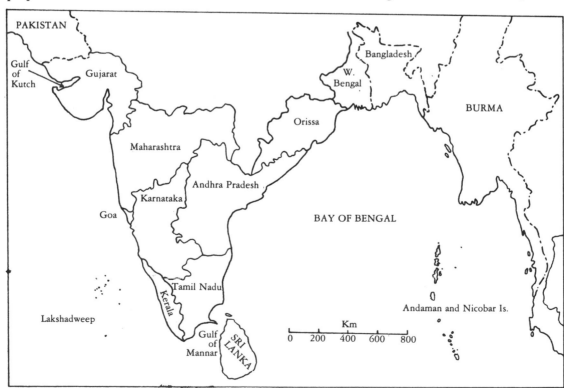

Figure 1. Sri Lanka, Bangladesh, Burma and the maritime states of India.

C. S. KAR

Many turtles of this species are netted by Goa fishermen for consumption. A leatherback was also caught at Baga where infrequent nesting by this species has been reported (Salm 1976). An ovigerous leatherback was taken on 1 November 1933 (Deraniyagala 1939).

Karnataka

Much of the coast is rocky. In the absence of surveys no data are presently available.

Kerala

This state has the highest density of human population in India. Sea walls meant to control erosion preclude nesting along parts of the coast. While only the ridley has been recorded as having nested in Kerala in recent years (Lal Mohan, personal communication, 1979) nesting is reported at Kasargode in the north (Silas, personal communication) and at Kovalam near Trivandrum in the south (Salm 1976). Green turtles occur along the entire coast extending south from Quilon (Shanmugasundaram 1968).

Turtles nest on a 3 km beach at Calicut fishing harbor and at Marad Beach 8 km to the south. Fishermen of the Mukkuvar community collect the eggs for sale and for local consumption. Nesting turtles are also taken. Nesting peaks during September to November, but fishermen report sporadic year-round nesting.

In July 1956, a leatherback nested at Calicut (Jones 1959). Cameron (1923) refers to "giant turtles" that visited the Quilon coast yearly for a period of 1 or 2 months. Fishermen informed him that up to the turn of the century leatherbacks were quite common near Quilon, and that about 40 were caught annually when attempting to come ashore or at sea in special nets. By about 1915 they were seen only occasionally, and about 2 were caught annually. They appeared to frequent the outskirts of Tangasseri reef about 3 km offshore (Smith 1931).

Tamil Nadu

Five species of sea turtle—the ridley, green turtle, hawksbill, loggerhead, and leatherback—are known from the eastern coast of the state. Populations of the first 4 species frequent seagrass and coral reef areas in the Gulf of Mannar and the Palk Bay. Ridleys nest on the predominantly sandy coast north of Point Calimere.

The following data are from Jones and Fernando (1967) and from Valliappan and Pushparaj (1973). Twenty to 25 turtles were sold weekly at the Sunday market at Tuticorin in 1975, but numbers occasionally exceeded 150. Hawksbill meat sometimes poisons those who eat it (Deraniyagala 1939), as happened more recently in August 1979 when 9 persons died at Manapad.

Olive ridleys nest on suitable stretches along the entire east coast of India (Chacko 1942; Jones and Fernando 1967; Daniel and Hussain 1973; Valliappan and Pushparaj 1973; FAO 1974).

Between 4,000 and 5,000 turtles were caught annually in southern Tamil Nadu. The catches were heaviest from October to January. Green turtles constituted about three-quarters of the total catch. Olive ridleys and loggerheads, together formed about one-fifth. The hawksbill was caught occasionally, the leatherback rarely.

Captured green turtles were usually stored alive in pens constructed in shallow water (Kuriyan 1950) to await shipment by rail to coastal markets. Most common at the market were green turtles weighing 50 to 60 kg, although young turtles below 24 cm in plastron width were also caught. The price of live green turtles of various sizes increased twofold to sevenfold between 1967 and 1973.

Recorded ridley nesting sites in Tamil Nadu include 11 of the 21 islands in the Gulf of Mannar (CMFRI 1977) and the coast south of Tuticorin. (The few green turtles occurring off the Tamil Nadu coast do not nest there.) Olive ridleys nest at Point Calimere and on at least 50 km of coast south of Madras. It is the only turtle species known to nest commonly on the Indian coast north of Point Calimere. A proposed Marine National Park, encompassing the islands in the Gulf of Mannar and Nallatanni Island, where nesting is reportedly most common, may help to protect local feeding and breeding sea turtle populations.

A conservative estimate puts the nesting intensity of the ridley along a 50-km stretch south of Madras at 100 nests per km per season, from December to March. Jackals, village dogs, and humans remove about 90 percent of these nests (Whitaker 1977). Itinerant tribals and villagers collect these eggs. Fishermen do not, holding the sea turtle sacred (Valliappan and Whitaker 1974). Nevertheless, turtles drowned offshore in shrimp trawlers sometimes wash ashore.

In 1973–78, the Madras Snake Park Trust and the Central Marine Fisheries Research Institute collected 33,083 eggs for their hatcheries and released 18,475 hatchlings (55.6 percent) into the sea.

Andhra Pradesh

The common species of sea turtle, particularly in the northern half of the state, is the ridley. The green turtle is uncommon, and the leatherback and hawksbill are rare.

While local fishermen on catamarans catch olive ridleys incidentally from about October to February, the proliferation of mechanized fishing trawlers probably was responsible for the larger incidental catch of ridleys in the 1978–79 season than in past years.

The deliberate killing of a leatherback that had come

ashore to nest was reported in 19 May 1979.

At Visakhapatnam, the number of turtles captured per day does not exceed a dozen, although the town was once a trading center where tortoiseshell was processed into ornaments and sold. Turtle eggs are occasionally sold in the town from October to February indicating this to be the nesting season and that nesting must occur nearby (S. Dutt, personal communication, 1979). Some segments of the local population eat turtle meat. Large numbers of sea turtles were transported from Andhra to Calcutta markets during the 1977–78 nesting season (Davis and Bedi 1978), and an unknown volume of this illegal practice continues.

Orissa and West Bengal

Sandy stretches occur along roughly 250 km of coast extending southwest from the Wheeler Islands to the Andhra Pradesh border. Villages are scattered thinly along the coast.

Arribadas of the olive ridley occur along a 35-km stretch called Gahirmatha Beach and possibly elsewhere (Bustard 1976). Personnel of the Gahirmatha Marine Turtle Research and Conservation Unit of the Orissa State Forest Department, who tagged over 1,700 nesting turtles in 1979, estimated the numbers that nested per season as in excess of 150,000 in 1976, 150,000 in 1977, 200,000 in 1978, and 130,000 in 1979.

Prior to 1975, eggs were being taken by private parties from the Gahirmatha Beach under license from the Orissa State Government for sale locally, regionally, and in Calcutta. The estimated legal take in the 1974–75 season was 800,000 eggs (FAO 1975). Following the advice of H. R. Bustard, FAO consultant, the Government ceased issuing licenses to egg collectors in 1975. The Bhitar Kanika Wildlife Sanctuary, which includes the entire 35-km stretch of Gahirmatha Beach, was established that year to protect the nesting turtles, their eggs, the salt-water crocodile, and other wildlife. Since then, nesting turtles and eggs have been effectively protected on shore. However, turtle eggs are still illegally being taken from nesting beaches on the Wheeler Islands and near Astaranga, Chandravhaga, Puri, and Gopalpur-on-Sea.

The gravest threat to the survival of India's east coast olive ridleys is the illegal fishing of turtles off the coast of Orissa (Davis and Bedi 1978) from bases in West Bengal, Orissa, and Andhra Pradesh. Considerable numbers are also taken off the West Bengal and Andhra Pradesh coasts. About 500 ridley carcasses are washed ashore annually on the 20 km of beach being studied at Gahirmatha. The frequency of ridley carcass strandings is about the same along an additional 60-km of beach to the southwest. This number represents a tiny fraction of the illegal offshore catch. Of a random sample of 172 stranded ridley carcasses, 106 (61.6 percent) were females. Adult and hatchling turtles are also caught incidentally in fishing nets.

Digha and Junput on the West Bengal coast are important landing centers from which turtles are transported to the Calcutta market. The numbers of turtles landed at these 2 places alone between 15 October 1978 and 15 January 1979 totalled 21,361 (Biswas, personal communication, 1979). Of late, the increasing numbers of mechanized fishing trawlers result in even higher catches.

Turtles were also being dispatched overland from coastal landing centers in Orissa, West Bengal, and Andhra Pradesh. From November 1974 to January 1975, 6,190 turtles were dispatched alive by rail from Puri to Calcutta (Biswas, personal communication, 1979). The Calcutta market was also being supplied from many other coastal railheads. The transport of turtles by rail from towns in Orissa has largely been controlled since early 1979, but turtles are still being trucked from these and other towns in neighboring Andhra Pradesh and West Bengal.

The first indication that sea turtles occur and are being exploited in the Sunderbans in West Bengal is Biswas' report on the sale of turtle meat at Namkhana.

At Puri, fishermen sell turtles to middlemen at Rs. 20–25 each. These in turn are sold to wholesalers at Calcutta for Rs. 57–60 each. Retailers sell at Rs. 5 to Rs. 6 per kg at Calcutta (Biswas, personal communication, 1979). The turtles weigh 43.4 kg on an average (N = 291).

Turtle meat is also consumed by economically poorer communities in Orissa and is sold in many towns, as at Bhadrakh.

Lakshadweep (India)

Lakshadweep includes the Laccadive, Amindivi and Minicoy Island groups. Four species—the green turtle, hawksbill, olive ridley, and leatherback—nest on the islands, though *Dermochelys* is rare (Gardiner 1906; Bhaskar 1978a).

Nesting is heaviest on 6 seasonally uninhabited islands—Suheli Valiakara, Suheli Cheriakara, Tinnakara, Bangaram, Pitti, and Parali II—where, respectively, 202, 13, 45, 15, 8, and 10 green turtle nests were made during the southwest monsoon (June through September).

Fishermen, coconut harvesters and in the case of Bangaram Island, scuba-diving tourists from abroad, visit the islands during the 6 months between successive monsoon periods. Of the inhabited islands, nests were found on Kadmat (4), Androth (4), Agatti (2) and Minicoy (1) during visits between October 1977 and January 1978 averaging 4 days at each island. Of these 11 nests, 10 had been made by ridleys and hawksbills and 1 by a green turtle.

The ridley, though relatively common near the remaining islands, has not been recorded from the waters around Minicoy. Green turtles and hawksbills are commonly observed feeding close to shore at Minicoy and could be the subject of valuable underwater research.

Sea turtles are avidly harpooned around all the islands but Minicoy for their fat, used in caulking boats. Turtle meat, on rare occasions used as shark bait, is not commonly eaten. More often, the eggs are consumed. Hawksbills are killed and, prior to 1978, tortoiseshell was sold to middlemen on the Indian mainland.

Traditional methods of hunting sea turtles are giving way as motor launches replace sailing craft. Speargun shooting of turtles by tourists around Bangaram Island, though forbidden by law, remains unmonitored.

Enforcement of the law protecting sea turtles is already inadequate, and human population pressures on the islands may soon result in the colonization of the 2 important nesting islands at Suheli.

The Andaman and Nicobar Islands (India)

Four species—the green turtle, olive ridley, hawksbill and leatherback—are known to nest in these islands (Man 1883; Bonington 1931; Davis and Altevogt 1976; Bhaskar 1979b). Sea turtles are killed illicitly for their meat in many parts of the Andamans, and turtle meat is eaten everywhere in the Nicobars. Eggs are also assiduously collected. There was once a regular trade in green turtles between the Andamans and Calcutta (Maxwell 1911). Sea turtles and their eggs are an important traditional item of food for the Onge and Great Andamanese tribals. The influx of refugees and settlers from the Indian subcontinent introduced a new factor in the conservation of turtle stocks in the islands.

Known nesting areas of importance are listed below.

Great Nicobar Island

About 160 old leatherback excavations were found in April 1979, near the mouths of the Alexandria and Dagmar rivers. Nesting by other species also occurs, particularly by hawksbills near Pygmalion Point. Rats prey on hatchling turtles.

Little Andaman Island

About 70 fresh leatherback excavations were present at West Bay in January 1979, and about 10 at South Bay. Above 80 percent of the nests are preyed upon by monitor lizards (*Varanus salvator*). The beaches at Little Andaman and Great Nicobar constitute the only known important leatherback rookeries in India.

The Twin Islands

These 2 uninhabited islets lie off the western coast of Rutland Island. Thirty fresh excavations, probably all made by hawksbills, were found in October 1978 (Bhaskar 1979a).

Rutland Island

Four species nest on the western coast south of Woodmason Bay. Ten recent hawksbill nests were present in October 1978, on the southern coast where monitor predation on the eggs is heavy.

Middle Andaman

Nesting by 4 species occurs on about 10 km of the eastern coast near Betapur where poachers and their dogs take many eggs.

Katchal Island

At least three species—the green turtle, leatherback and either the ridley or the hawksbill or both—nest at South Bay.

South Sentinel Island

Green turtles nest on the island which is uninhabited, rarely visited by humans and is girdled by 6 km of sandy beach. Monitor lizards prey on the eggs (Davis and Altevogt 1976). Other species of sea turtles probably also nest here.

South Reef Island

Year-round nesting occurs perhaps mainly by green turtles.

Treis Island

This island is visited every year by Malays who remain there for six or eight months and collect turtle shells (Government of India 1857).

Many of the islands in North Andaman and in Southern Nicobars have yet to be surveyed for sea turtle habitats.

Turtle hunting by the Great Andamanese and rituals related to sea turtles have been described by E. H. Man (1883).

Sri Lanka

Data from Deraniyagala (1939 and 1953), Salm (1976), Hoffmann (communication to R. Whitaker, 1978) and Jayawardhana (personal communication, 1979).

Five species—the green turtle, olive ridley, logger-head, hawksbill and leatherback—are known from Sri Lanka. In addition, Deraniyagala (1939) suspects that the flatback strays into Sri Lankan waters.

Turtles and their eggs are totally protected by law—*Dermochelys* since 1970, the 4 remaining species since 1972. Sri Lanka is a signatory to CITES.

Lepidochelys olivacea

The olive ridley is the most common Sri Lankan sea turtle both as regards occurrence and nesting. In the Gulf of Mannar, it is commonly captured in nets set for green turtles and is not infrequently liberated.

A large concentration of ridleys in December 1978, apparently migrating to nest on Indian coasts, was reported by Hoffmann. Oliver (1946) and Deraniyagala (1953) reported similar migrating concentrations in the months of September and November, respectively.

The school of sea turtles seen by Deraniyagala occupied about 100 km of sea, the individuals keeping about 200 m apart. Hoffmann writes "the fishing boat I was on went due west from Kalpitiya and then drifted south at the end of a 2-mile long drift net. We crossed another boat coming in which had caught 24 turtles in a similar net. The following day, boats which went due west from Kudirimalai Point caught similar numbers. Two days later, 78 were caught in a single net and the skipper reported the sea as teeming with turtles. In each instance, the turtles were released. Almost all were olive ridleys, the remainder being leatherbacks. The schools were headed north."

Caretta caretta

Fishermen from the Gulf of Mannar where the species is most common quote nesting in June, July, and August. Shells scattered around fishing camps indicated a 1:20 proportion of loggerhead to ridley catches (Deraniyagala 1939).

Chelonia mydas

In the Gulf of Mannar where a green turtle population is actively hunted, the species has been declining steadily. Hundreds of families in the Jaffna district depend on turtle fishing for subsistence. Fishermen using special nets catch the species on both the west and east coasts of Sri Lanka (Salm 1976). Each Sunday, 20 to 30 are slaughtered at Jaffna (Parsons 1962).

Eretmochelys imbricata

Around 1843, hawksbills nested so freely on parts of Sri Lanka's southern coast that the government farmed the right to capture them. Hawksbill scutes were used for ornamental purposes and for inlaying. The flesh was eaten usually by the poorer class of Tamil fishermen in Sri Lanka, and there were occasional cases of poisoning.

Dermochelys coriacea

The leatherback nests on the coast of the Yala National Park in southern Sri Lanka where 3 areas of concentrated nesting totaling 10 km in length were identified by Salm. Both the remoteness of the Yala coast and law enforcement by Park rangers ensure that leatherback eggs remain safe from humans within the Park boundary. Nest predators include wild boar (*Sus cristatus*) and jackals (*Canis aureus lankae*) and possibly leopard and monitor lizards. Salm counted 173 leatherback excavations and 28 nests of smaller turtle species on the 50-km coast of the Park during 9–13 June 1975.

Turtle eggs can be bought from most stalls around the southern part of Sri Lanka and are dug from nests during the day in full view of the public at Bentota, a tourist resort (Salm 1976).

Turtle hatcheries have been established by the government and by the Wildlife and Nature Protection Society.

A proposal to extend the Wilpattu National Park on the northwest coast to include an additional 1,300 km² of coastal waters, if implemented, will benefit turtles (Jayawardhana, personal communication, 1979).

Bangladesh

Four species—the olive ridley, green turtle, hawksbill and leatherback—are reported from Bangladesh (Reza Khan, personal communication, 1979). On the Khulna section of the railways, large numbers of chelonians were seen by Dr. S. L. Hora (Acharji 1950).

Burma

The Burmese exploit turtles for their eggs rather than their meat (Parsons 1962). From Diamond Island alone, the annual take of eggs was 1.6 million or more—about 1.5 million from loggerheads (almost certainly olive ridley—authors) and about 10,000 from hawksbills. The green turtle lays year-round but chiefly from July to November. The loggerhead (olive ridley—authors) nests from September to December, and the hawksbill from June to September. Egg harvesting took place under license from the government but the taking of live turtles was illegal (Maxwell 1911). This ban has apparently been periodically violated as when British ships restocked with nesting turtles (Parsons 1962). The extent of the present egg harvest in Burma, if any, is not known.

"The shores of Little Coco Island swarm with turtles" (Government of India, 1857).

Nesting by leatherbacks has been recorded from the coast of Tenasserim (Pritchard 1963).

Recommendations

1) Detailed surveys to locate sea turtle nesting beaches are urgently required along the coasts of mainland India and nearby islands.

2) The establishment of sanctuaries at nesting and feeding grounds of turtles will help conserve local populations. Suggested locations are: Bhaidar Island in the Gulf of Kutch for nesting olive ridleys; the waters around all Indian islands in the Gulfs of Kutch and Mannar to protect the green turtle feeding populations; Suheli Valiakara, Tinnakara and Minicoy Islands in Lakshadweep, primarily for nesting and feeding green turtles; the Wheeler Islands and the coast extending southwards from the turtle sanctuary of Gahirmatha Beach, to include Hukitola Island and terminating near the port of Paradwip, for nesting olive ridleys; the Twin Islands in the Andamans for nesting hawksbills; the nesting beaches at West Bay on Little Andaman and at the mouths of the Dagmar and Alexandria rivers in Great Nicobar Island to protect the leatherback.

3) A 3 km wide strip of coastal waters extending from Latitude 20°47′N to 20°16′N should be declared off limits to all fishing activity during the months September to March and a speedboat should be acquired by the Orissa State Forest Department to monitor the waters off Gahirmatha Beach and to control the poaching of adults at sea which at present is rampant.

4) The law protecting turtles needs to be rigidly enforced in the state of West Bengal, the main market for turtle meat.

5) Coastal lands need to be initially surveyed for degrees of nesting before being allotted for settlement by humans. Where nesting beaches cannot be adequately protected, settlement should be discouraged. The foregoing has immediate relevance in Suheli, Little Andaman and Great Nicobar Islands.

6) Wide publicity about the plight of sea turtles through the medium of documentary films needs to be taken up.

7) Funds and facilities for conducting field studies and research on sea turtles in India have to be procured. As an example, exact *arribada* populations can only be monitored and tagged with the help of adequate staff and equipment.

Acknowledgments

The authors wish to thank the Orissa State Forest Department and the Madras Snake Park Trust and Conservation Centre for having provided facilities and funds for sea turtle work in India. Special thanks are due to Dr. H. R. Bustard and to Romulus Whitaker for their guidance and advice.

Literature Cited

Acharji, M. N.
1950. Edible chelonians and their products. *Journal of Bombay Natural History Society* 49:529–32.

Bhaskar, S.
1978a. Marine turtles in India's Lakshadweep islands. *IUCN/SSC Marine Turtle Newsletter.*
1978b. Notes from the Gulf of Kutch. *Hamadryad* 3.
1979a. Sea turtles in the South Andaman Islands. *Hamadryad* 4:3.
1979b. Sea turtle survey in the Andaman and Nicobars. *Hamadryad* 4.

Bonington, M. C. C.
1931. Census of India—The Andaman and Nicobar Islands.

Bustard, H. R.
1976. World's largest sea turtle rookery? *Tiger Paper*, 3.

Cameron, T. H.
1923. Notes on turtles. *Journal of Bombay Natural History Society* 29:299–300.

Chacko, P. I.
1942. A note on the nesting habit of the olive loggerhead turtle, *Lepidochelys olivacea* (Eschscholtz) at Krusadai. *Current Science* 12:60–61.

CMFRI
1977. Report on the survey of the islands of Gulf of Mannar for the setting up of a Marine National Park, April 1977.

Daniel, J. C., and S. A. Hussain
1973. The Crocodiles of Bhitar Kanika.

Davis, T. A., and R. Altevogt
1976. Robbers of the South Sentinel. *Yojana*, 15 August 1976.

Davis, T. A., and R. Bedi
1978. The sea turtle rookery of Orissa. *Environmental Awareness*, 1.

Deraniyagala, P. E. P.
1939. The Tetrapod Reptiles of Ceylon. vol. 1.
1953. A coloured atlas of some vertebrates from Ceylon, vol. 2.

FAO (Food and Agriculture Organization)
1974. India: a preliminary survey of the prospects for crocodile farming. By H. R. Bustard.
1975. India: economic potential of gharial and saltwater crocodile schemes in Orissa with notes on the sea turtle industry. By J. M. de Waard.

Gardiner, J. S.
1906. *The Fauna and Geography of the Laccadive and Maldive Archipelagos*, vol. 2. Cambridge University Press.

Government of India
1857. Home Department Public Consultation, no. 33, 2 May 1857.

Jones, S.
1959. A leathery turtle coming ashore for laying eggs during the day. *Journal of the Bombay Natural History Society* 56:137–39.

Jones, S., and A. Bastian Fernando
1967. The present status of the turtle fishery in the Gulf of Mannar and Palk Bay. Proceedings of the Symposium on Taxonomy and Biology of Blue-green Algae.

Kuriyan, G. K.
1950. Turtle fishing in the seas around Krusadai. *Journal of the Bombay Natural History Society* 49:509–12.

Man, E. H.
1883. On the aboriginal inhabitants of the Andaman and Nicobar islands.

Mawson, N.
1921. Breeding habits of the green turtle, *Chelonia mydas*. *Journal of Bombay Natural History Society* 27:956–57.

Maxwell, F. D.
1911. *Report on Inland and Sea Fisheries*. Rangoon: Government Printing Office.

Murthy, T. S. N., and A. G. K. Menon
1976. The turtle resources of India. *Seafood Export Journal* 8.

Oliver, J. A.
1946. An aggregation of Pacific sea turtles. *Copeia* 1946:103.

Parsons, J. J.
1962. *The Green Turtle and Man*. Gainesville: University of Florida Press.

Pritchard, P. C. H.
1963. *Living Turtles of the World*. Jersey City: T. F. H. Publications.
1971. The leatherback or leathery turtle, *Dermochelys coriacea*. IUCN Monograph, Marine Turtle Series, 1:1–39.

Salm, R. V.
1976. Critical marine habitats of the northern Indian Ocean. Contract report to the IUCN, Morges, Switzerland.

Shanmugasundaram, P.
1968. Turtle industry. *Indian Seafoods* 6:18–19.

Smith, M. A.
1931. *The Fauna of British India*, vol. 1.

Valliappan, S., and S. Pushparaj
1973. Sea turtles in Indian waters. *Cheetal* 16.

Valliappan, S., and R. Whitaker
1974. Olive ridleys on the Coromandel coast. Publication of the Madras Snake Trust Park.

Whitaker, R.
1977. A note on sea turtles of Madras. *Indian Forester* 103.

James Perran Ross
Museum of Comparative Zoology
Harvard University
Cambridge, Massachusetts 02138

Mohammed Amour Barwani
Ministry of Agriculture and Fisheries
P.O. Box 467 Muscat
Sultanate of Oman

Review of Sea Turtles in the Arabian Area

ABSTRACT

We review sea turtles and populations in the Arabian Gulf, South Arabian coast, and the Red Sea, noting major nesting and feeding grounds. We estimate population sizes and discuss threats to sea turtles.

The green turtle, *Chelonia mydas*, is the most common turtle in the region. Extensive feeding grounds are in the Arabian Gulf, on the coasts of Oman and the Peoples Democratic Republic of Yemen (PDRY), and in the Red Sea. Large nesting grounds are at Karan Island (Saudi Arabia), Ras al Hadd (Oman) and east of Mukulla (PDRY), each with several thousand turtles nesting each year. However a feature of this area is the large number of small nesting grounds. Small numbers of turtles and turtle eggs are consumed in coastal Oman and in the Gulf, however this exploitation is local and of low intensity. An exception is PDRY where turtle meat has been exported for some years.

Hawksbill turtles, *Eretmochelys imbricata*, generally occur in small numbers throughout the area. Small nesting groups and single nestings often occur but major nesting grounds are found on Hormuz and Larak Island (Iran), Shetvar and Lavan Islands (Iran), Masirah Island (Oman), Jebel Aziz and Perim Island (PDRY) and the Suakin Archipelago (Sudan). Between 100 and 500 turtles nest at each of these locations each year. There is a small trade in hawksbill curios and constant predation of eggs by people, but no systematic exploitation.

A single large nesting ground of loggerhead turtles, *Caretta caretta*, is on Masirah Island, Oman. An estimated 30,000 loggerheads nest each year. Loggerheads are rarely seen elsewhere in the region.

Olive ridley turtles, *Lepidochelys olivacea*, nest in small numbers (150 a year) on Masirah Island. Their distribution may be limited because areas of low salinity and mangroves that this species favors are scarce in the region.

Leatherback turtles, *Dermochelys coriacea*, are seen occasionally but are not known to nest in the area.

Most leatherbacks seen in Oman are subadults. Leatherbacks are caught and rendered for oil on Masirah (Oman) and Larak (Iran) but the level of exploitation is low.

Turtle populations of all species are threatened by the increasing subsistence exploitation that is facilitated by the improved mobility of the people with vehicles and outboard motors. Other threats are coastal development and industrialization which are degrading nesting and feeding habitats.

Further research in the Arabian region should concentrate on coordinated survey and education projects. The opportunity exists to assist local authorities to establish conservation programs before the developing threats reduce the turtle populations from their present healthy levels.

Introduction

This review outlines the current state of knowledge of sea turtles in the area from the Iran-Pakistan border in the east to the horn of Africa (Cape Guardifui) in the west, taking in the shores of the Arabian Gulf and the Red Sea. Information from the region is scarce and fragmentary. Major breeding grounds and feeding aggregations of sea turtles are identified and their population size and current status are evaluated. Present and potential threats to the survival of sea turtles are assessed. A general assessment of sea turtle populations in the area is given to aid in the better understanding of sea turtle populations world wide.

Published works other than taxonomic references on turtles in the Arabian area are restricted to papers by Hirth with various colleagues (Hirth (FAO) 1968; Hirth and Carr 1970; Hirth and Hollingworth 1973; and Hirth, Klikoff and Harper 1973) and the monograph by Basson et al. (1977). We have added various unpublished reports made available by personal communications to the authors. The personal experience of both authors in the area and in particular the extensive survey of sea turtles in the Sultanate of Oman supported by WWF/IUCN has provided additional information. The detailed information on turtle biology obtained over a 3-year interval in Oman has been used to evaluate and extrapolate the information from elsewhere. To augment these sources, letters to the heads of fisheries research organizations in the area were sent with a questionnaire reproduced as Table 1. Answers, received from 4 out of 5 enquiries, are incorporated in the report. Coastline lengths are approximate straight lengths of coast excluding offshore islands.

Turtle Biology in the Arabian Area

The locations of important turtle feeding grounds and nesting beaches are shown in Figure 1. Two species

Table 1. Questionnaire sent to fisheries departments in the Gulf region, spring 1977

1. What species of turtles are seen in your country?
2. What are the common names for turtles in your country?
3. a) Do turtles come ashore to lay eggs in your countries?
   b) Which species?
   c) Where? (Latitude and Longitude, if possible)
   d) How many each year (even rough estimate is useful); e.g. a few, dozens, hundreds, etc.?
   e) Which months of the year do they lay eggs?
4. Are there large numbers of turtles in your countries, even though they do not lay eggs? What sorts; where and how many?
5. a) Are turtles caught by people on nesting beaches in your country?
   b) Are turtles caught in the sea in your country?
6. Do people in your country eat turtle meat?
7. Does your country export any turtles or their parts? If so, where to? How many each year?
8. How many turtles are killed in your country each year?
9. What laws and regulations apply to the catching of turtles?
10. Are the numbers of turtles in your country increasing, decreasing or stable?

predominate in the turtle fauna of the Arabian area, the green turtle and the hawksbill. The loggerhead is known only from the breeding beach on Masirah; its wider distribution remains unknown. Loggerheads are not recorded within the Arabian Gulf, but 1 unconfirmed report suggests they may occur in the Red Sea and nest in the Sinai region. The very small population of ridley turtles nesting on Southern Masirah is the only known occurrence of that species. The relationship of this small population to the much larger nesting colonies of the Indian coast is unknown. Rough estimates of the total population of these species in the region are given in Table 2. Leatherback turtles are reported occasionally throughout the area but are not known to nest there. Data from Oman (Table 3) suggest that many of the leatherbacks observed there are quite small.

The pattern of nesting green turtles and their distribution from nesting grounds to feeding grounds remains incompletely known but appears to be rather complex. There are large concentrations of turtle nesting at Ras al Had in Oman (6,000 ♀/y) and coast of Mukalla in the PDRY (10,000 ♀/y) of the sort seen in the Caribbean. However, there are also widespread small nesting grounds supporting populations of up to

JAMES PERRAN ROSS

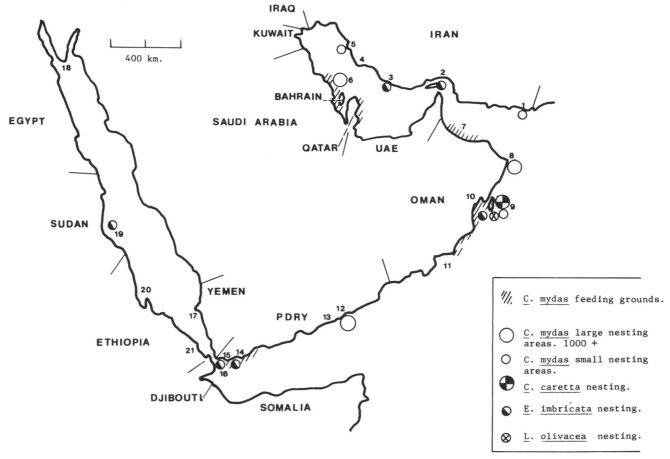

Figure 1. Sea turtle nesting and feeding grounds in the Arabian area, numbered locations: 1) Beris; 2) Hormuz, Larak, Queshm Islands; 3) Lavan, Shetvar Islands; 4) Jabrin Islands; 5) Bandar Bushr; 6) Karan Islands; 7) Batina coast; 8) Ras al Hadd; 9) Masirah Island; 10) Amhawt Island; 11) Kuria Muria Bay; 12) Ithmun, Sharma, Shihr; 13) Mukulla; 14) Jebal Aziz; 15) Khor Umaira; 16) Perim Island; 17) Kamran Island; 18) Sinai; 19) Suakin Archipelago; 20) Dahlak Archipelago; 21) Assab.

Table 2. Annual breeding populations of sea turtles in the Arabian area, rough estimate

Country	Green turtles	Hawksbills	Loggerheads	Ridleys
Iran	500	1,000	—	—
Saudi	500	100	—	—
Qatar	—	100	—	—
Oman	7,000	100	30,000	150
PDRY	10,000	500	—	—
Yemen	200	—	—	—
Total ♀♀/yr.	18,200	1,800	30,000	150
Total population[a]	54,600	5,400	90,000	450

— No data.

a. Using Carr, Carr, and Meylan 1978 estimate that total population (green turtles) = 3 × Annual ♀♀.

a few hundred females each year. For example, in Oman which is well surveyed there are small nesting grounds on almost every isolated beach and island (Figure 2). The movements between these nesting grounds and local or distant feeding grounds are unknown. However, it is notable that in the few records of long distance migrations we do have individual turtles bypass several other nesting and feeding grounds on their migration (Figure 3). A feature of the migrations of green turtles in the area is the regular occurrence of large "fleets" of migrating turtles of the sort described in historical sources in the Caribbean (e.g. Lewis 1940) but not seen there for over 200 years.

Hawksbills are widely but sparsely distributed

Table 3. Records of *Dermochelys coriacea* around Oman

Location		Date		Sex	Straight carapace length (cm)	Notes
Masirah	21°N,59°E	Jun	73	F	126	flotsam[a,b]
Gulf of Oman	24°N,60°E	Apr	77	—	—	trawler[c]
Kuria Muria Bay	17°N,56°E	Mar	78	M	80	trawler
Masirah	21°N,59°E	Sep	78	F	133	flotsam
Batina	24°N,56°E	May	79	—	100	flotsam
Batina	24°N,57°E	Jun	79	F	150+	trawler

— No data.
a. Flotsam = washed up dead on beach.
b. Reported in Hirth and Hollingworth 1974.
c. Trawler = caught by fishing trawler.

throughout the region, often in association with coral reefs. Several large nesting grounds are reported; the Strait of Hormuz and Lavan in Iran, Jabal Aziz and Perim in PDRY, Masirah and possibly the Suakin Archipelago in the Red Sea. In addition, single hawksbills seem to nest on isolated beaches and islands throughout the area, a dispersed pattern common to this species elsewhere. The amount of long distance migration undertaken by hawksbills is unknown.

Many of the populations of sea turtles in the Arabian area are large and have not been reduced in size by exploitation by people. Most people in the region are

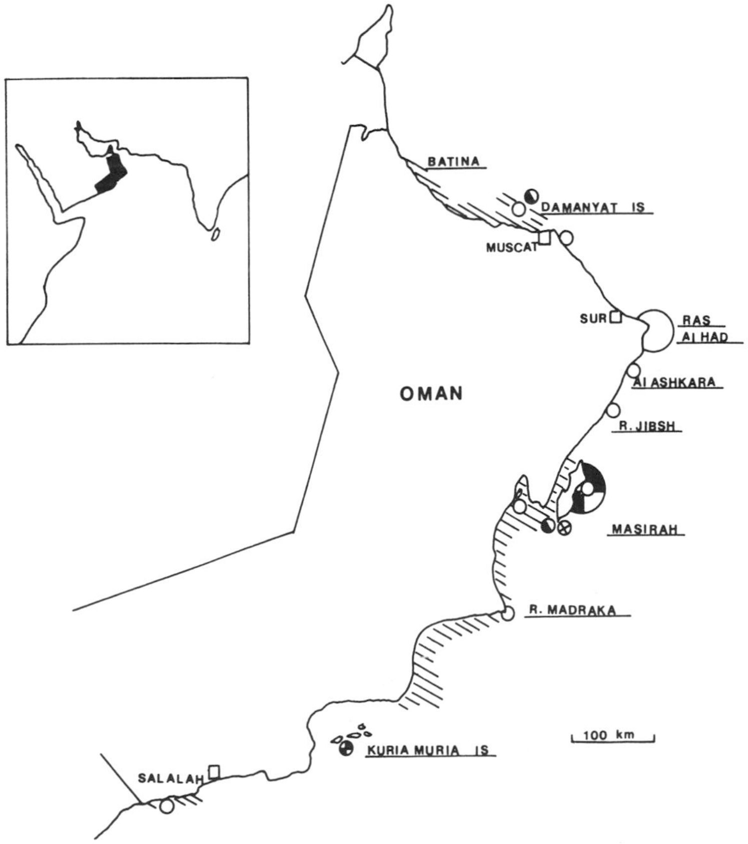

Figure 2. The Sultanate of Oman, showing sea turtle nesting locations. Symbols as in figure 1.

Muslims who have religious prohibitions against eating turtle meat. There are however several areas where this religious prohibition is no longer operative, and a small amount of local consumption of turtle meat is found.

Additional mortality of adult turtles is caused by accidental capture in fishing operations aimed at other species. Turtles are caught and drowned in large pen nets for pelagic fish and in shrimp trawls. There are insufficient records to evaluate the significance of this mortality. Answers to our questionnaire suggest that the number of turtles killed by these means is small. However, Ross (1979b) showed that in the United States shrimp trawling industry the apparently low mortality of 0.03 turtles/trawl/hour results in the death of a significant number of turtles. There are approximately 125 shrimp trawlers operating in the Arabian Gulf (Feidi 1979) and the potential mortality of sea turtles is considerable. Coastal people throughout the region make subsistence use of turtle eggs but there are no reports of organized commerce in eggs. The significance of predation by people must be evaluated in relation to other natural mortalities. Observations at Masirah Island, Oman, show that early in the loggerhead season when nesting density is low, a high proportion of nests are found and robbed. However, when nesting reaches its full density of more than 50 turtles/km/night the proportion of nests robbed by people stabilizes at 6 percent. In comparison the number of nests destroyed by turtles averages between 1 percent and 9 percent of nests laid—for both green turtles and loggerheads in Oman (Table 4). Some authors have argued from computer simulations that the destruction of nests by nesting females acts as a population regulation mechanism (Bustard and Tognetti 1969). It is apparent from the data in Table 4 that at the high nesting densities observed in Oman this effect is small. Most mortality of the loggerhead nests in Oman is due to flooding by the sea (Figure 4). Similar data are shown by Schulz (1975) from Surinam.

Most countries in the Arabian area have numbers of

JAMES PERRAN ROSS

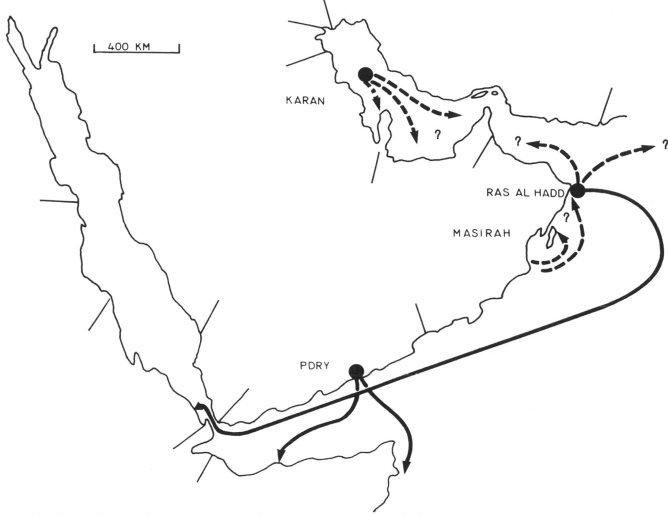

Figure 3. Sea turtle migration routes in the Arabian area. Dotted lines represent speculated or uncertain migration routes. Solid lines are confirmed by tag recoveries.

Table 4. Nest destruction by nesting sea turtles

Species	$\bar{x}$ (percentage)[a]	Range	Number	Nesting density
Ch. mydas Oman	5.6	(3.4–8.5)	246	32 turtles/km night.
C. caretta Oman	3.5	(1.1–7.6)	1,000	50 turtles/km night.

a. The proportion of females nesting that dig up previous nests.

relatively poor people living near the sea. They have hopes of providing protein resources or cash income from the exploitation of sea turtles, particularly green turtles. While the feeding areas for green turtles in the region are extensive in area, the quality and productivity of the feeding grounds are relatively low. Data from Hirth, Klikoff, and Harper (1973), Basson et al. (1977) and Ross (1979a) suggest that the high temperature, high turbidity, and high salinity of the Arabian Gulf waters restrict productivity of the seagrass meadows. The major components of the seagrass meadows are *Halophila* and *Halodule* species which are

hardy pioneer species present in low biomass. Therefore the productivity of green turtles is likely to be low and calculations of the carrying capacity and protein production by green turtles such as those in Basson et al. (1977:107) are overestimates. Ross (in press) has argued that the south Arabian coast is suboptimum habitat for hawksbill turtles due to periodic cold upwelling resulting in smaller size and reduced clutch size in hawksbills from the Oman and PDRY.

Mortality due to human activities is small at the present time in the Arabian region. However the turtle populations there will not remain at their present large

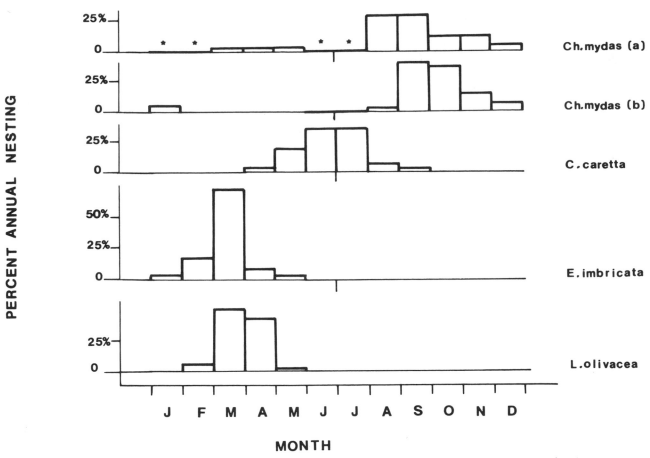

Figure 4. Nesting seasons of sea turtles in the Sultanate of Oman. *Ch. mydas* (a) Ras al Had, *Ch. mydas* (b) Masirah Island. All other species Masirah Island.

size under sustained commercial use for local marketing or export. Conservation strategies in the region must be formulated with this in mind.

National Accounts

Iran

The major source of information from Iran (coastline 2,050 km) are the reports of O. Walczak and W. Kinunen to FAO and the Iran Game and Fish Department. Supplemental information was obtained from J. Frazier (mimeo 1975) and personal communications. Walczak and Kinunen surveyed the coast from Gowater near the Pakistan border to Chah Bahar, from Charak to Bandar Abbas and in the area of Moghum and Bandar Bushr. Isolated nesting grounds are reported in the region of Beris and Bandar Bushr. *Chelonia mydas* (green turtle) and *Eretmochelys imbricata* (hawksbill) are reported to be present. Although the surveyed areas are reported to be "excellent nesting habitat" the number of turtle nests seen was small (<50). The surveys were carried out in spring 1971, and no indication of seasonal variation in numbers of nesting turtles was given.

Anderson (1979) gives records of sightings of *Ch. mydas* and *E. imbricata* and includes precise locations of nesting grounds. Minton (1966) reports green tur-

tles nesting from June to November in nearby Pakistan and Anderson (1979) reports Kinunen-Walczak's observations of hawksbill nesting in April to June.

The main nesting areas reported in Iran occur on islands in the Arabian Gulf. Hormuz, Quesham, and Larak, situated at the eastern end of the Gulf, support nesting populations of hawksbills estimated to be in the order of a hundred females during April and May. A smaller number of *Ch. mydas* nest on the same beaches, and *Ch. mydas* feed offshore. There is predation on turtle eggs by humans, dogs and foxes. On Larak, leatherback turtles are reported to be rendered for oil to treat boat timbers, a common practice throughout the Gulf when the favored source of oil, shark liver, is in short supply.

The islands of Shetvar (= Shotur) and Lavan (= Sheykh Sho'eyb) support a large nesting population of hawksbill turtles estimated at over 500 nests seen in June 1971. Indications from Oman (Ross 1981), and Cousin Island in the Seychelles (Diamond 1976) suggest that this represents at least 300 females nesting, making this a significant nesting ground. A smaller number of green turtles also nest on these islands. Both species are harvested for their shells, and the meat is discarded. Hawksbills are also reported to nest on Jabrin Island (Anderson 1979).

JAMES PERRAN ROSS

Sea turtles of undetermined species are reported from the waters around Sirri Island in the central Gulf.

The beaches in Iran are largely without protection, and exploitation is occurring at an unknown level.

Iraq

There is little information on sea turtles in Iraq (coastline 25 km). The coast is restricted to the estuary of the Tigris and Euphrates and is unlikely to be of importance for sea turtles.

Kuwait

In reply to our questionnaire the director of Fisheries of Kuwait (coastline 200 km) said that 1 green turtle had been caught in 3 years by the Fisheries research vessel, and that no other information was available.

Saudi Arabia Gulf Coast

The only information of the Gulf coast of Saudi Arabia (coastline 500 km) is from Basson et al. (1977). They report *Ch. mydas* nesting on Karan, Jana, Kurayn and Jurayd Islands with 80 percent of the population, estimated at several hundreds, nesting on Karan. Occasional nesting by green turtles on other islands and the mainland are reported but are apparently discouraged by interference from people and other nest predators. Hawksbills are reported to nest infrequently with green turtles from April to July. Green turtles nest in the summer months during May to September. Leathery turtles are recorded as present but uncommon and are not known to nest.

There are extensive feeding areas for green turtles on the seagrass meadows throughout the area. Important areas are near Abu Ali Island, Tarut Bay, between the island of Bahrain and the mainland, and around Jazirat as Samamik. There is in excess of 1,000 km² of suitable feeding area. Three species of seagrass, *Halodule uninervis, Halophila ovalis* and *Halophila stipulacea*, predominate and reach an annual productivity of 128 g dry weight per m². No estimate of the population of green turtles using this resource is given. Nevertheless it is clear that this area represents a major feeding ground within the gulf. Basson et al. (1977) speculate that turtles from Karan may migrate to India and Pakistan, but data are lacking.

No analysis of threats to sea turtles is given, but several can be inferred. There is a large shrimp fishery in the area; accidental mortality of sea turtles is inevitable. A factor currently minimizing this effect is the high productivity of the fishing ground which enables large catches to be made with short trawls around an hour long. However, Basson et al. report a decline in recent catch, and it is likely that as the fishery declines, trawl time will increase and mortality of accidentally caught turtles will increase. A further factor that could threaten these feeding grounds is the rapid establishment of industry along the coast. Increased mobility of local people will lead to more extensive disturbance of nesting turtles. Turtle meat is not known to be consumed in any significant amount in Saudi Arabia, probably owing to religious prohibitions.

Bahrain

Gallagher (1971) reported green turtles feeding near Bahrain (coastline 60 km), and the results from the adjacent coast of Saudi Arabia by Basson et al. (1977) confirm this report. In reply to our questionnaire the Bahrain Fisheries Research Bureau reported that occasional dead green turtles (an estimated 1 or 2 a year) are washed up on Bahrain, but green turtles are not caught or consumed. There are no regulations affecting turtles, and no information on nesting or population sizes.

Qatar

Information from Qatar (coastline 300 km) was obtained from the answer to our questionnaire and from conversation with Peter Hunnam, a marine biologist consultant, who has done underwater survey work in the area. Five species of turtle are reported from Qatar: green turtle, hawksbill turtle, leathery turtle, loggerhead turtle, and flatback turtle. Of these, the identification of loggerheads and flatbacks is doubtful and should be disregarded pending confirmation. The common name for all turtles is *Hemssa* except the leathery turtle which is called *Geldia*.

Nesting of green turtles once occured in small numbers on mainland Qatar at Ras Laffan and Umm Said, but no longer occurs probably due to increased human disturbance. The islands of Sharaawh and Dayinah have nesting hawksbills, and hatchlings are reported during early July. The nesting populations are estimated as "a few only." Aliya and Safaliyah islands are heavily disturbed by people, and no nesting is thought to occur at present. Jazirat Habul has no suitable nesting beaches.

Seagrass flats where green turtles feed are located off the east coast. Turtles of all sizes are caught in trawl nets in this area, but most are said to be returned alive to the sea. Information on the exploitation of sea turtles is conflicting. One informant said turtle is not a common food in Qatar, while the other said green turtles are often seen in fish markets and estimates the annual catch at no more than "a few hundred" per year.

Hawksbill meat and eggs are eaten whenever found. A further threat is the development of a port at Ras abu Khamis which has already led to a degradation of coral reefs in the area and the disappearance of much of their fauna including hawksbill turtles. There is no systematic information on turtles from Qatar and no protective regulations. However, the Department of Fisheries has expressed interest in obtaining more advice and information.

United Arab Emirates

In reply to our questionnaire, the Ministry of Agriculture and Fisheries of the UAE (coastline 600 km) expressed regret that they have not studied their turtle fauna and expressed a desire for more information from us. In the course of the Sultanate of Oman turtle survey we were reliably informed that green turtles from the Ras al Had nesting ground are transported by small truck to Abu Dhabi where they command a high price in the fish market (approximately US $6 a kg). The volume of this trade is small at present, involving at most a dozen vehicles and no more than 100 turtles per year, but it is an ominous beginning. It is likely that there are green turtle feeding grounds along the Emirates coast and green turtles are probably caught for food there but data are lacking.

Oman

A survey of sea turtles was carried out by WWF/IUCN and the Oman government in 1977–79. During the survey, most of Oman's coastline (1,700 km) was surveyed from the air with the help of the Sultan of Oman's Airforce. Further surveys were carried out by boat and Landrover to evaluate remote beaches. The major nesting beaches and feeding areas of sea turtles in Oman have been identified. Local counterparts were hired and trained to carry out intensive studies of turtle biology at 2 main nesting areas: Masirah Island and Ras al Had. From the information obtained from surveys and from the intensive study, recommendations for the conservation of sea turtles in Oman were formulated.

Four species of sea turtles nest regularly in Oman. These are the green turtle, *Chelonia mydas*; the loggerhead, *Caretta caretta*; the hawksbill, *Eretmochelys imbricata*; and the olive ridley, *Lepidochelys olivacea*. Table 5 gives the body sizes and clutch sizes of these species. A fifth species, the leathery turtle, *Dermochelys coriacea*

is encountered occasionally in coastal waters but is not known to nest in Oman.

Two of the species have large nesting grounds in Oman which are of international significance because of their size and because they are relatively undisturbed. The green turtle population at Ras al Had is estimated at a minimum of 6,000 adult females nesting each year. Other smaller aggregations of this species are shown in Figure 2. The loggerhead nesting aggregation on Masirah Island is estimated at a minimum of 30,000 females nesting each year. This is the only large nesting colony of loggerheads in Oman and is the largest known aggregation of this species in the world.

A small population of hawksbills estimated at 90 to 125 females a year nests on the southern tip of Masirah Island, and some hawksbills are resident around Masirah. A small population of olive ridleys, estimated at 150 females nesting each year also nests around the South of Masirah. The seasonal distribution of nesting seasons of these four species is shown in Figure 4.

The only turtle commonly seen away from the nesting beaches in Oman is the green turtle. Feeding areas are in Sawqira Bay, the Gulf of Masirah, and scattered along the Batinah coast. The relationship between the turtles nesting in Oman and those seen on the feeding grounds is not clear. Many of the turtles that nest in Oman may feed as far away as the coasts of Africa and India. A major activity of the survey was to place over 5,000 numbered metal tags on turtles. The only international return to data is a tag put on a green turtle at Ras al Had (Oman) that was returned from Assab (Ethiopia) in the Red Sea. This represents a migration of 2,220 km in 3 months.

The green turtle is eaten by people on Masirah, Amhawt Island and other small villages on the southern coast. A minimum of 1,000 green turtles are captured each year from the feeding grounds and used as food. The value of this resource to the coastal people is greater than US$30,000 per year. All the turtle meat is eaten directly and not sold. The killing of nesting turtles has been banned on Masirah where the excessive killing of nesting turtles has caused the number of turtles to decline.

Table 5. **Body size and clutch size of nesting female sea turtles in Oman**

Species (occurrence)	Mean straight carapace length ± 2 SE	Number	Mean clutch size ± 2 SE	Number
Ch. mydas (Ras al Hadd)	97.1 ± 1.2	(62)	103.5 ± 8.0	(58)
Ch. mydas (Masirah)	93.2 ± 1.4	(90)	97 ± 5.6	(16)
C. caretta	91.2 ± 1.0	(1,378)	107 ± 3.0	(161)
E. imbricata	73.3 ± 1.6	(48)	97[a] ± 16.0	(9)
L. olivacea	71.5 ± 0.6	(100)	118 ± 7.2	(22)

a. Masirah hawksbills also lay an average of 11 small yolkless eggs (range 0–30).

JAMES PERRAN ROSS

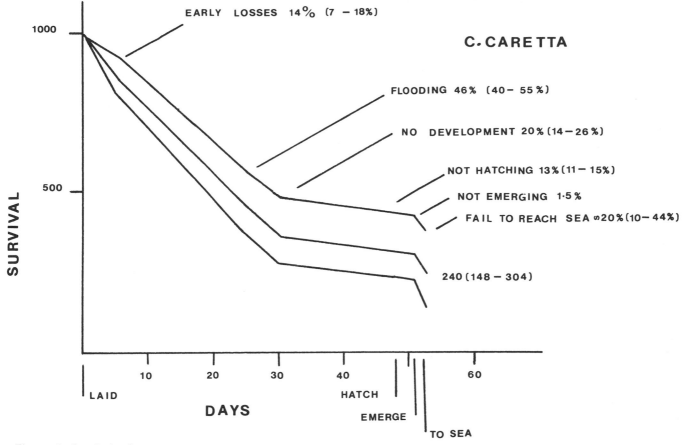

Figure 5. Survival of *C. caretta* eggs and hatchlings from laying to entering the sea. Upper, lower and median estimates are shown.

The major cause of death of loggerhead eggs on Masirah is the sea which unpredictably washes away about 40 percent of the eggs laid each year (Figure 5). People on Masirah take only a small proportion of the eggs to eat. Of greater concern is the rapid development of the village near the main nesting beach. There is some evidence that lights and disturbance from development are causing hatchling turtles to become disoriented and not find the sea. The small populations of hawksbill and olive ridley turtles which nest at the south of Masirah are subject to occasional taking of eggs by people.

Proposals have been put before the Oman Government to protect the nesting beaches in national parks or marine reserves and to control strictly the capture and trade of green turtles. The government has been advised not to promote any commercial development of the turtle resource but to preserve it for controlled traditional use. Further studies are in progress.

People's Democratic Republic of Yemen (PDRY)

All the information on PDRY (coastline 1,200 km) is taken from Hirth and Carr (1970) and Hirth and Hollingworth (1973) in which the results of Hirth's surveys during 1967–72 are presented.

Major nesting grounds of green turtles are reported at Ithmun, Sharma, Musa, Shihr, and Shuhair in the area east of Mukulla. Nesting density is high (reported as approximately 30 turtles/night/km) and the total length of beach available is approximately 13 km. If the nesting parameters of this population are similar to that at Ras al Hadd, an estimated 10,000 females nest in this area during the peak of the season. Other scattered nesting grounds occur along the coast. Hawksbills nest at Jabal Aziz and Perim Island in significant numbers during January–February. The data preclude any estimation but the population is likely to be in the order of several hundred females a year.

Feeding grounds of green turtles are found in the western part of the country. One concentration at Khor Umaira is the site of some useful surveys on seagrasses and their utilization by turtles (Hirth, Klikoff, and Harper 1973).

At the time of Hirth's surveys exploitation of green turtles on both the nesting beaches and feeding grounds was estimated to be between 800–4,000 turtles a year (1964–74). Hirth proposed a limited quota of 1,000 a year but this quota is often exceeded.

No recent figures are available but 4,000 turtles were exported in 1970 and a minimum of 400 to 700 in 1973 (W. King, personal communication). If exploitation has continued at this rate, the nesting population may be depleted.

North Yemen

A partial survey of North Yemen (coastline 500 km) was reported by Walcezak in the early 1970s but it is unpublished and not available. A reliable informant reported (personal communication to Ross) that 3 species of turtle nest at Kamran Island in large numbers. Only green turtles were identified. People do not eat turtles, which they call *Zuqar* in North Yemen, but they do eat turtle eggs, which implies some nesting. No details of turtle numbers or exploitation are available.

Red Sea

We have been unable to discover much useful information about turtles in the northern Red Sea (coastline 2,700 km) except for a personal communication to J. Frazier from Israel (1978). With his permission we briefly report:

Ch. mydas, C. caretta, E. imbricata and *D. coriacea* are all reported from the Sinai region, and there is an unverified report of *L. olivacea. Ch. mydas, C. caretta* and *E. imbricata* are said to breed in the area of Sinai. The whole of the Red Sea is inadequately surveyed but is likely to support both breeding and resident populations of *Ch. mydas* and *E. imbricata.* All species are said to be in decline in the Israeli administered area (1978) due to human disturbance.

Hirth (1980) reports a recent preliminary survey of sea turtles on the Sudan coast and reviews historical sources. Hawksbill turtles nest in good numbers in the Suakin archipelago. Some protection is accorded by the Sudanese Wildlife Conservation Act, but there is concern that tourist development will pose future problems. Hoofien and Yaron (1964) report *Ch. mydas* and *E. imbricata* from the Dahlak Archipelago without giving details of nesting or feeding areas.

Summary and Conclusion

From the rather sparse information we have assembled here it is evident that a significant population of green, loggerhead and hawksbill turtles occurs in the Arabian region. The opportunity to preserve these species at their present high densities is a current challenge. It should be possible to protect the major nesting grounds

particularly as, with the exception of PDRY, there is very little commercial pressure on sea turtles. However the nations throughout the region are well financed and pursuing a course of rapid development. We require a coordinated educational program to ensure that sea turtles are not the victims of this development.

Programs modeled on the WWF-IUCN program in Oman where a survey and scientific study both provided the necessary information and a vehicle for public relations and educational activities should be initiated throughout the area. The present affluence of nations in the regions means that conservation funds invested in this region show a most generous "return" in terms of matching funds and final results.

Literature Cited

Anderson, S. C.
1979. Synopsis of the turtles, crocodiles and amphisbaenians of Iran. *Proceedings of the California Academy of Sciences*, 41:501–28.

Basson, P. W.; J. E. Burchard; J. T. Hardy; and A. R. G. Price
1977. *Biotopes of the Western Arabian Gulf.* Dahran, Saudi Arabia: Arabian American Oil Company.

Bustard, H. R., and K. P. Tognetti
1969. Green sea turtles: a discrete simulation of density dependent population regulation. *Science* 163:939–41.

Carr, A.; M. H. Carr; and A. B. Meylan
1978. The ecology and migrations of sea turtles, 7. The West Caribean green turtle colony. *Bull. Am. Mus. Nat. Hist.* 162:1–46.

Diamond, A. W.
1976. Breeding biology and conservation of hawksbill turtles, *Eretmochelys imbricata*, on Cousin Island, Seychelles. *Biological Conservation* 9:199–215.

Feidi, I.
1979. Regional fishery survey and development project. FAO (R.A.B./71/278). Appendix 12.

Frazier, J.
1975a. Marine turtles of the western Indian Ocean. *Oryx* 13:164–75.
1975b. The status of knowledge on marine turtles in the western Indian Ocean (revised). East African Wildlife Society, Nairobi, Mimeo, 16 pp.

Gallagher, M. D.
1971. *The Amphibians and Reptiles of Bahrain.* Bahrain: private printing.

Hirth, H. F.
1968. Report to the governments of Southern Yemen and the Seychelles Islands on the green turtle. Report FAO/UNDP. CTA 2467, Rome.

Hirth, H. F., and E. M. Abdel Latif
1980. A nesting colony of the hawksbill turtle (*Eretmochelys imbricata*) on Seil Ada Kebir Island, Suakin Archipelago, Sudan. *Biological Conservation* 17:125–30.

Hirth, H. F., and A. F. Carr

1970.   The green turtle in the Gulf of Aden and the Sey-chelles Islands. *Ver der Kon. Ned. Akad. van Wet, Afd. Nat. Tweede Reeks Deel.* 58:1–44.

Hirth, H. F., and S. L. Hollingworth

1973.   Report to the Government of the People's Democratic Republic of Yemen. Report FAO/UNDP. TA 3178, Rome.

Hirth, H. F.; L. G. Klikoff; and K. T. Harper

1973.   Sea grasses at Khor Umaira, People's Democratic Republic of Yemen with reference to their role in the diet of the green turtle, *Chelonia mydas. Fishery Bulletin* 71:1093–97.

Hoofien, J. H., and Z. Yaron

1964.   A collection of reptiles from the Dulak archipelago. *Sea Fisheries Research Station Haifa Bulletin* 35:35–40.

Lewis, C. B.

1940.   The Cayman Islands and marine turtle. *Bulletin of the Institute of Jamaica Science Series* 2:56–65.

Minton, S. A.

1966.   A contribution to the herpetology of West Pakistan. *Bulletin of the American Museum of Natural History* 134:29–194.

Ross, J. P.

1979a. Sea turtles in the Sultanate of Oman. Report IUCN/WWF Manuscript, 53 pp.

1979b. Present status of sea turtles. Report IUCN/WWF Manuscript, 28 pp.

1981.   Hawksbill turtle, *Eretmochelys imbricata*, in the Sultanate of Oman. *Biological Conservation* 19:99–106.

Schulz, J. P.

1975.   Sea turtles nesting in Surinam. *Zoologische Verhandelingen, uitgegeven door het Rijksmuseum van Natuurlijke Historie te Leiden* 143:1–144.

Jack Frazier
Department of Zoological Research
National Zoological Park
and
Division of Reptiles and Amphibians
National Museum of Natural History
Smithsonian Institution
Washington, D.C. 20008

Status of Sea Turtles in the Central Western Indian Ocean

ABSTRACT

The 7 territories of the Central Western Indian Ocean include some of the best known turtle areas in the world. All 5 of the pantropical species are recorded, but only *Chelonia* and *Eretmochelys* are common. Numbers of both species seem to be below former levels, and it is unlikely that marine turtles remain the important resource they once were.

Background

The Central Western Indian Ocean (CWIO) is here defined as: British Indian Ocean Territory (BIOT), Seychelles, Comores, Mayotte, Tanzania, Kenya, and Somalia (Figure 1). Each of these territories has a colonial history, and all are now sovereign states except BIOT, an uninhabited British colony, and Mayotte, a French Department.

This roster includes some of the world's most famous territories for marine turtles. The Seychelles are renowned for their turtles and turtlemen who kept much of Europe supplied with turtle soup and tortoiseshell over the last hundred years. Zanzibar was one of the most important centers in international trade of tortoiseshell. The Bajun of the northern coast of Kenya and southern coast of Somalia are expert turtlemen, with sophisticated techniques for catching turtles at sea, and Comorians, including those from Mayotte, once had similar customs, including the use of sucker fish.

With the exception of Somalia, all of these territories have been surveyed by the author over the past decade. This report is a synopsis of more complete studies, many of which are unpublished (Frazier 1970, 1971, 1975a, 1975b, 1976, 1977, in press) and also complements other reports on adjacent areas of the Indian Ocean (Kar and Bhaskar, this volume; Hughes, this volume; Ross and Barwani, this volume; Sella, this volume). For a discussion of subsistence hunting in the Indian Ocean, see Frazier (1980, this volume).

Note: Revised April 1980.

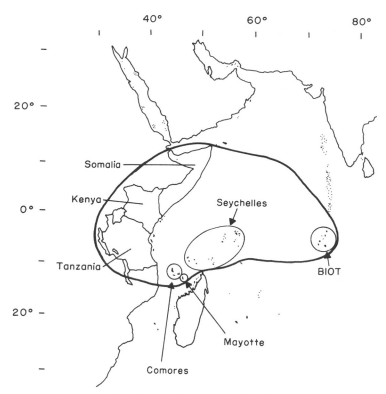

Figure 1. The Central Western Indian Ocean, showing the 7 territories under consideration.

The 5 species of sea turtles that are recorded in the CWIO include all those that are known from the Indian Ocean: green turtle, *Chelonia mydas*; hawksbill, *Eretmochelys imbricata*; loggerhead, *Caretta caretta*; olive ridley, *Lepidochelys olivacea*; and leathery turtle, *Dermochelys coriacea*.

British Indian Ocean Territory

The BIOT now includes only the Chagos Archipelago, the southernmost islands of the Laccadive-Maldive Ridge. There are vast areas of beach and shallow water among the 65 major islands. Possibly because of the Archipelago's remote oceanic position, only *Chelonia* and *Eretmochelys* are recorded, although *Dermochelys* probably pass through deep, offshore water. *Chelonia* and *Eretmochelys* are likely to occur at all the islands and many of the reefs, but documented records are few.

Concentrated nesting occurs at YeYe Island, westernmost and remotest of the Peros Banhos Atoll (Dutton, in press) and Nelson Island. Nests and tracks are large and thought to be from *Chelonia*. This species is reputed to nest from June to September, and the numbers involved are not likely to exceed 300 in a year. The lagoons of some atolls are reputed to have large numbers of immatures, but there are no other reports of nonbreeding or feeding concentrations. Despite the vast shallow water areas, marine pastures are not extensive.

Eretmochelys, as is common in this region, is a dispersed nester, often laying in full daylight. Nesting is reported from only Peros Banhos and Diego Garcia but probably occurs throughout the territory. The main season for nesting is said to be from November to February. Perhaps 300 nest in a year. Rich coral reefs abound and are likely to provide feeding habitat and shelter for sizeable numbers of *Eretmochelys*.

Both species were exploited regularly while the islands were inhabited, and relatively small amounts of tortoiseshell, turtle oil, and meat were exported to Mauritius. The Chagos were known as the "oil islands" from the coconut oil produced and have never been famous for marine turtles. Mating and nesting animals probably took the brunt of exploitation, and reproduction may have been reduced as a result.

Hunting *Chelonia* has been banned completely since 1968, and the inhabitants were evacuated in the early 1970s. No areas are legislated as reserves, but the entire territory is functionally a reserve, for it is uninhabited, except Diego Garcia, now a naval base (Frazier 1977).

Seychelles

The Republic includes a tremendous portion of the western Indian Ocean with 50 major islands. *Caretta* and *Dermochelys* have been reported, but only *Chelonia* and *Eretmochelys* are common, and occur throughout the territory.

Chelonia nests in all island groups but is most numerous on the Aldabras. Nesting is year-round but most active from May to September, during the southeast trade winds. An estimated 1,000 nest annually on Aldabra, and 2,500 are estimated to nest in a year over the Seychelles. Large pastures of algae and phanerograms are in lagoons such as Aldabra and shallow water as on the Seychelles plateau. Feeding animals are common in these areas.

Eretmochelys nest on all island groups, possibly all islands. Cousin Island, in the Granitic Group, has the most concentrated nesting, with 30 to 40 females a year. This is a primary nesting ground for the species in the western Indian Ocean. Nesting, often diurnal, is recorded in all months but June and is most intense from October to January. The numbers estimated to nest annually over the territory are 600. Extensive coral reefs occur throughout Seychelles, providing rich feeding habitats. Some bays in the Granitic Islands have several resident *Eretmochelys*.

Exploitation of sea turtles in Seychelles has been intense for 150 years, since the islands were inhabited. Mating and nesting animals have been especially disturbed, so reproduction has been severely reduced.

JACK FRAZIER

Annual production of *Chelonia* has fallen to a fraction of former levels, indicating that populations have been destroyed. The animal is not as common in the main iwslands as it was once. Curiously, there is little evidence for a decline in *Eretmochelys,* despite ever increasing pressure from soaring prices of tortoiseshell. Increasing human disturbance to beaches and reefs in the Granitic Islands, the center of abundance, adds to problems of increasing direct exploitation. The fate of this turtle is in question.

Chelonia was protected by a total ban from 1968 to 1976, but widespread poaching nullified any chances of populations recovering. This turtle is now insignificant as a resource. Various laws protect the 2 common turtles, but enforcement is rare. There are National Marine Parks and Reserves, at Bai Ternay and St. Anne, but they are not of prime importance to nesting turtles. The future of sea turtles in Seychelles rests on reserves such as Aldabra, Cousin, and Aride Islands (Frazier, in press).

Mayotte

A degraded volcano encircled by barrier reef, this island has the most extensive coral reefs in the region. Though geologically one of the Comoro Islands, its history and political status separate it. Only *Chelonia* and *Eretmochelys* are recorded, but the other 3 species are likely to occur.

There are some 140 beaches on Mayotte and satellite islands, and *Chelonia* nesting spoor was recorded on 17 of these. Most activity is on Saziley in the south and Pamanzi Island in the northeast. Moya and Papani beaches on Pamanzi have an estimated 300 nesting annually; the total for Mayotte is 500. Nesting increases during the southeast trade winds, from June to July, but probably occurs round the year.

Eretmochelys nesting spoor was recorded on one beach only. Perhaps 25 nest in the territory in a year. In contrast to the exposed beaches where *Chelonia* concentrate, this turtle is more commonly on protected beaches of the lagoon. Nesting may occur around the year, but if it is like other territories in CWIO, there is more activity from November to December. Feeding habitat and shelter seem to be abundant with the area of active coral reef that is available.

Human predation concentrates almost exclusively on nesting females, and at Moya beach in 1972 there was little chance that a female could survive the nesting season. Turtle populations seem to be below the carrying capacity of the nesting and feeding habitats, but there are no baseline data for estimating population trends. With increased human density, decreased availability of fish, and increased pressures on food resources, turtles are probably under heavy pressure.

There are no reserves and evidently no protective legislation. An awareness of the need to manage natural resources is not conspicuous in either the populace or the government (Frazier 1972).

Comores

Ngazidia, or Grand Comore, has an active volcano and, like Anjouan, dense human populations. Moheli, oldest and smallest of the 3 islands in the Republic, is the most important for turtles, with its dozen satellite islands, large areas of shallow water, and 89 sandy beaches. *Chelonia* and *Eretmochelys* are recorded from all 3 major islands, and a rare, unidentified turtle has been reported at Moheli. Probably all 5 species occur.

At Moheli, 33 beaches have *Chelonia* nesting, but 81% of nesting is on six beaches. Fourteen females may beach in a night on Mtsanga N'yamba, Itsamia, and the annual total for the island is estimated to be 1,850. Perhaps 50 nest at Grand Comore and Anjouan. Nesting activity increases in June and July, during the southeast trades. Marine pastures are of limited size at the 2 large islands, but Moheli has a considerable area, where even dugongs feed. Immature turtles are regularly caught at the north end of Grand Comore, which may be a nursery area.

Eretmochelys nesting is recorded only from 16 beaches on Moheli. The annual number nesting may be 50, and the main nesting season is probably from November to December. Nesting is dispersed, but may not be diurnal. All islands have active coral reefs, but the small islands south of Moheli have profuse fringe reefs.

Some small turtles are netted, but most human predation concentrates on nesting females. Feral dogs, perhaps more destructive, destroy nests on Moheli. Although females are regularly killed and some meat is traded, turtle populations are large and do not seem to have diminished.

There are no protective laws and no marine reserves. Chissoua Ouenefou, in the south of Moheli, has been suggested as an island reserve. There is little consciousness of resource management, and the more affluent people of Grand Comore and Anjouan are involved in exploitation of food resources on Moheli. The fate of the turtles there is, thus, in question (Frazier 1977).

Tanzania

The marine shore, dissected by several large rivers, is rather sparsely populated. Three inhabited major islands lie offshore: Mafia, Zanzibar, and Pemba. Dozens of small cays and raised reef limestone islands dot the inshore waters. All 5 species of sea turtle are recorded.

Dermochelys is reported infrequently, but may occur regularly in deep water, possibly in migrations to and from the rookery in Natal. *Caretta* has been caught from Zanzibar south. These females were tagged while nesting in South Africa, and some have traveled 2,600 km in 2 months. To date, 4 tagged *Caretta* have been recovered in Tanzania (Hughes, this volume), and the territory seems to be an important nonbreeding area for *Caretta* that nest in Natal.

Lepidochelys has been recorded from much of the coast, although only singly. Several nest yearly on Maziwi Island, and others probably nest in the vicinity of large river mouths farther south, where estuarine habitats may provide requisite feeding habitats.

Chelonia is common, and nesting concentrates on cays around Mafia Island and especially on Maziwi Island. There is also nesting at Ras Dege, one of the more remote headlands on the mainland. Most laying is from June to October. Less than 200 are estimated to nest on Maziwi and the annual total for all of Tanzania is 300. Reef flats provide a large area of shallow water with marine pastures, and there may be resident breeding populations in this territory (Frazier, in prep.). Tanzania may also provide feeding habitat for migrants from the island nesting populations, such as Aldabra and Comores.

Eretmochelys is also common and is thought to nest widely. On Maziwi 20 may nest in a year, and 50, in the Territory. Diurnal nesting is common.

Human exploitation is not organized, but dates back centuries. Turtles are caught on feeding grounds while nesting and are also killed accidentally by dynamite "fishing." Although Zanzibar was a major exporter of tortoiseshell, mainland Tanzania has not been a major supplier, and the shell came from other territories in the region. The populations of all turtles have probably been reduced since prehistory.

The hunting of sea turtles is forbidden in the 7 Marine Fisheries Reserves, most important of which is Maziwi Island. All exploitation of turtles is to be carried out under license, but these laws are seldom enforced. Both the populace and government display vigorous and widespread concern for resource management (Frazier 1976).

Kenya

More arid, with fewer islands, and denser coastal settlements, Kenya also has less marine shore than Tanzania. Some reef flats are so scoured for food that few consumable organisms remain. All species except *Caretta* have been documented, but this turtle is described by Bajun fishermen.

Dermochelys is occasionally seen at some distance offshore. As in Tanzania, these animals may occur regularly in migrations to and from the Natal rookery. *Lepidochelys* are known from Ungwana Bay and near the Tana River, Kenya's largest; sand bars in the area are suspected to have nesting.

Chelonia is common, with nesting concentrated on the near islands of Lamu and Manda, the small island of Tenewi and remote stretches of coast at Ras Biongwe and Ungwana Bay. Most nesting activity occurs after August. An estimated 200 turtles nest annually. Marine pastures grow along the coastal reef flats, and migrants from other nesting populations may feed here.

Eretmochelys is common along the entire coast, nesting on island and mainland beaches, even in populated areas. nesting is not concentrated, 50 turtles may nest in a year. Reefs are not generally rich, but encompass considerable area.

Persistent exploitation over the past 2 millennia and recent increases in coastal development and pollution seem likely to have reduced the numbers of *Chelonia* and *Eretmochelys*. Nest predation on *Chelonia* is severe, and recruitment may be insignificant.

Marine parks are at Malindi, Watamu, and Shimoni, and a reserve is at Kiunga. These are excellent reef areas, but no important turtle beaches are protected. There is a total ban on hunting both *Chelonia* and *Eretmochelys* (Frazier 1979), although tortoiseshell is still exported (Mack, Duplaix, and Wells, this volume).

Somalia

This territory has the longest coastline in the region; it is arid and largely uninhabited. Offshore waters are rich from upwelling. All 5 turtles are known to the Bajun in the south, but only *Chelonia* is apparently common (Grotanelli 1955).

This turtle nests along much of the Indian Ocean coast (Ninni 1937), and there seems to be concentrated nesting (Travis 1967). The numbers involved are unknown, but several thousand a year are likely to nest. Somalia's rich marine pastures support dugongs and *Chelonia* (Travis 1967); there may be resident as well as migratory turtles. Animals that nested in South Yemen have been recaptured in Somalia (Hirth 1978).

Although exploitation for export may have reduced reproduction during the 1960s and early 1970s, these turtles seem to be less disturbed than are many other populations. There are evidently no marine parks or reserves, and no laws protecting turtles.

Summary and Conclusions

Chelonia is the most common turtle in the Central Western Indian Ocean, the total numbers nesting yearly may be about 6,000. *Eretmochelys* is much less numerous, under 2,000 per year. The main nesting season for *Chelonia* is during the southeast trade winds, and *Eretmochelys* nests most actively during the northeast monsoon (Table 1).

Given the number of *Chelonia* that have been exploited annually from just Seychelles, the numbers of

JACK FRAZIER

Table 1. Summary of estimates of annual numbers and main nesting seasons of *Chelonia* and *Eretmochelys* in the Central Western Indian Ocean

| Territory | Number nesting annually | | Main nesting season | |
	Chelonia	*Eretmochelys*	*Chelonia*[a]	*Eretmochelys*[b]
BIOT	300	300	June–Sept	Nov–Jan
Seychelles	2,500	600	May–Sept	Oct–Jan
Mayotte	500	25	June, July?	Nov–Jan?
Comores	1,900	50	June, July?	Nov–Jan?
Tanzania	300	50	June–Oct	Feb–Mar
Kenya	200	50	Aug?	?
Somalia	2,000?	?	?	?
Total	7,700	1,075		

a. *Chelonia* nest during Southeast trades.
b. *Eretmochelys* nest during Northeast monsoon.

animals that nest today appear to be greatly reduced from earlier in the century. The situation with *Eretmochelys* is less clear, but is likely to be similar. It is unlikely that these turtles will be exterminated from the CWIO, but maintenance of breeding reserves, such as Cousin and Aldabra, are imperative for the existence of these animals as significant resources. Nesting reserves and protective legislation are needed in Moheli, Mayotte and Ras Biongwe, Kenya, and effective enforcement is urgently required in all territories.

Acknowledgments

The work was supported by: The Royal Society (London); Natural Research Council (United Kingdom); Fauna Preservation Society; East African Wildlife Society; African Wildlife Leadership Foundation; Department of Agriculture, Government of Seychelles; Government of BIOT; Game and Fisheries Departments of the Governments of Kenya and Tanzania; the Department of Zoological Research, National Zoological Park and the Division of Reptiles and Amphibians, Smithsonian Institution; and a grant from the Smithsonian Scholarly Studies Program to Dr. J. F. Eisenberg. Many colleagues in many countries provided information. The expert assistance of Mrs. Wy Holden and Ms. Virginia Garber in typing and Ms. Sigrid James Bruch in illustrating was invaluable.

Literature Cited

Dutton, R. A.
In press. Observations on turtles in the Chagos Archipelago, and the potential of that area as an important nesting ground for *Eretmochelys imbricata*. *Environmental Conservation*.

Frazier, J.
1970. Report on sea turtles in the Seychelles area. Animal Ecology Research Group, Oxford. Mimeo, 96 pp.

1971. Observations on sea turtles at Aldabra Atoll. *Philosophical Transactions of the Royal Society*, London, B, 260:373–410.

1972. Marine turtles in the Archipel des Comores. Fauna Preservation Society. Typescript, pp. 7 + 12 + 2 + 3.

1975a. Marine turtles of the western Indian Ocean. *Oryx* 13:164–75.

1975b. The status of knowledge on marine turtles in the western Indian Ocean (revised). East African Wildlife Society, Nairobi. Mimeo, 16 pp.

1976. Marine turtles in East Africa and the western Indian Ocean: the less common species. East African Wildlife Society, Nairobi. Mimeo, 80 pp.

1977. Marine turtles in the western Indian Ocean: BIOT and Comores. Fauna Preservation Society. Typescript, 33 + 114 pp.

1979. Marine turtle management in Seychelles: a case study. *Environmental Conservation* 6:225–30.

1980. Exploitation of marine turtles in the Indian Ocean. *Human Ecology*. 8:329–70.

In press. Marine turtles in the Seychelles and adjacent territories, In *Biogeography and Ecology of the Seychelles*, ed. D. R. Stoddart. The Hague: Junk.

Grotanelli, v. L.
1955. *Pescatori Dell'Oceano Indiano*. Rome: Cremonese.

Hirth, H. F.
1978. A model of the evolution of the green turtle (*Chelonia mydas*) remigrations. *Herpetologica* 34:141–47.

Ninni, E.
1937. La pesca indigena. In *Somalia Italiana*, ed. G. Corni, vol. 1, pp. 411–67. Milano: Aste e Stopzia.

Travis, W.
1967. *Voice of the Turtle*. London: George Allen and Unwin.

Jack Frazier
Department of Zoological Research
National Zoological Park
and
Division of Reptiles and Amphibians
National Museum of Natural History
Smithsonian Institution
Washington, D.C. 20008

Subsistence Hunting in the Indian Ocean

ABSTRACT

Sea turtles are exploited by subsistence hunters throughout the Indian Ocean. All 5 species are involved, but *Chelonia* is commonly killed for meat, *Eretmochelys* for tortoiseshell, and *Dermochelys* for oil. Nests of all species are excavated for eggs. International trade in tortoiseshell was well established two millenia ago. Present day coastal populations often lack animal protein, and subsistence-level exploitation concentrates so heavily on reproducing animals and nests that reproduction in many turtle populations has been severely reduced.

Introduction

Subsistence hunting is a difficult topic to describe for various reasons. The term defies unambiguous definition but often connotes traditional forms of exploitation for no monetary reward. Today, the wholly traditional hunter, without concern for money, is a rarity. Noncommercial, subsistence-level hunting is of little importance to governments and so is rarely discussed and almost never documented in detail. Finally, an area as vast as the Indian Ocean (Figure 1) can hardly be treated in a few pages, and it is symptomatic of its isolation and neglect that the situation on over one-seventh of the planet is to be considered in the same space available to a single country. This report is thus a summary of a more thorough study (Frazier, 1980; in prep.).

Sea Turtles Available for Hunting

Five of the 7 species of sea turtle are recorded from the Indian Ocean. Within the region, each species has its areas of abundance, but the most common species in general are green turtles, *Chelonia mydas*, and hawksbills, *Eretmochelys imbricata*. They occur and are hunted in most territories. Loggerheads, *Caretta caretta,* are

Note: Revised April 1980.

391

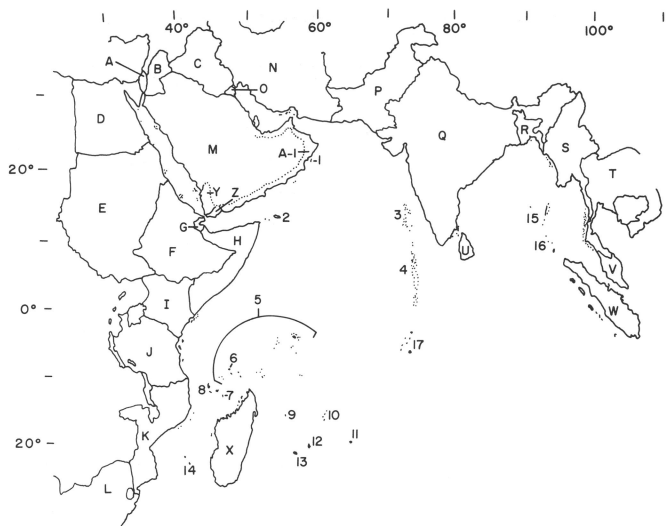

Figure 1. The Indian Ocean. Mainland countries: A) Israel; B) Jordan; C) Iraq; D) Egypt; E) Sudan; F) Ethiopia; G) Djbouti; H) Somalia; I) Kenya; J) Tanzania; K) Mozambique; l) South Africa; M) Saudi Arabia; N) Iran; O) Kuwait; P) Pakistan; Q) India; R) Bangladesh; S) Burma; T) Thailand; V) Malaya; W) Indonesia; X) Madagascar; Y) Yemen Arab Republic; Z) People's Democratic Republic of Yemen; A-1) Oman.   Islands: 1) Masirah (Oman); 2) Socotra (PDRY); 3) Laccadives (India); 4) Maldives; 5) Seychelles; 6) Aldabra (Seychelles); 7) Mayotte (France); 8) Comores; 9) Tromelin (Reunion, France); 10) St. Brandon (Mauritius); 11) Rodriguez (Mauritius); 12) Mauritius; 13) Reunion (France); 14) Europa (Reunion, France); 15) Andamans; 16) Nicobars; 17) BIOT (Chagos).

generally uncommon but abound on Masirah Island, Oman, and are common in South Africa, Mozambique, and Madagascar. Olive ridleys, *Lepidochelys olivacea*, are abundant in the Bay of Bengal and evidently Pakistan. Leathery turtles, *Dermochelys coriacea*, are nowhere common, although one of the largest nesting areas in the world is on the eastern Malaysian peninsula. The only major nesting areas, where they occur regularly in the Indian Ocean, are South Africa, Mozambique, and Sri Lanka. Details of the status of sea turtles in the Indian Ocean region are in Kar and Bhaskar (this volume); Frazier (in press, this volume, in prep.); Hughes (this volume); Ross and Barwani (this volume); and Sella (this volume).

Coastal Peoples that Hunt Turtles

People have occupied the coastal areas of the Indian Ocean basin for hundreds of generations, and many depend on marine resources for food and trade materials. The tortoiseshell trade was well established by the first century A.D. (Freeman-Grenville 1962; Parsons 1972). However, coastal peoples differ tremendously in their expertise as sailors and fishermen. The most adept turtle hunters are the Seychellois, of Seychelles, St. Brandon, and Chagos Islands; Vezo of southwest Madagascar; Bajun of northern Kenya and southern Somalia; Tamils from southern India and northern Sri Lanka; and Selung from the islands of

JACK FRAZIER

Burma and Thailand. Exploitation by these peoples often fits the "subsistence hunter" paradigm, and sea turtles are important in their cultures. However these animals are exploited for nonmonetary reasons throughout the Indian Ocean, although to varying degrees.

General Patterns of Exploitation and Consumption

Meat and eggs are generally consumed, except in some Islamic areas where turtle meat is not allowed for religious reasons. Nutritive products of sea turtles are used by subsistence hunters and are also marketed within the country of origin. *Chelonia* is consumed widely, but other species may also be eaten depending on the area; eggs of all species are commonly eaten. Whole *Chelonia* have been exported for foreign consumption from many countries: Seychelles, Kenya, Somalia, People's Democratic Republic of Yeman (PDRY), and India. In all but PDRY, it has been traditional turtle hunters, using traditional hunting methods, that have provided the turtles for this export trade.

Likewise, *Eretmochelys* is killed by fishermen living at subsistence levels throughout the Indian Ocean region, and the tortoiseshell is exported to oriental and occidental capitals. Turtle oil is produced for medicinal or culinary purposes where *Chelonia* is killed in numbers, and it is popularly used in local communities. This product was also exported in the past.

Trade in turtle leather, or skins, has seen only 2 main centers of development in the region: Pakistan and eastern India. This is neither a traditional activity, nor is it carried out by traditional turtle fishermen. Begun in the last decade, it is one of the fastest growing fisheries of marine turtles and accounts for tremendous numbers, especially of *Lepidochelys*. In other parts of the world, enormous populations of this turtle are being threatened by this type of exploitation (Frazier, ms.) *Lepidochelys* meat, and eggs, have been eaten by local peoples: there is a long-standing trade in eggs from Orissa to Bangladesh (Singh, *in litt.*, 20 July 1976).

Caretta is occasionally killed for meat, but *Dermochelys* is rarely slaughtered, except in certain areas. The oil from the latter is used in boat preservation.

Even in those territories where large numbers of turtles occur and turtle hunters are expert, the main source of protein is fish. Details are not available, but the greatest per capita consumption of turtle was probably in the Aldabra Islands earlier in this century, during the heyday of the *Chelonia* fishery; turtle meat may have been eaten as often as 3 or 4 times a week. This contrasts with the situation in eastern Nicaragua where *Chelonia* traditionally provided 70 percent of the animal protein for Miskito Indians (Nietschmann 1972).

Specific Patterns of Exploitation for Each Territory

South African natives did not traditionally catch turtles (McAllister, Bass, and Van Schoor 1965). Strict laws and efficient enforcement inhibit killing, and no significant exploitation of sea turtles occurs in this territory (Hughes, this volume).

Coastal peoples in Mozambique eat eggs and meat of most sea turtles, even *Dermochelys*. Some turtle products have been marketed locally, but subsistence-level exploitation has been widespread. In the south, women patrol beaches for eggs and nesting turtles; though unorganized, this exploitation has reduced reproduction. Because of the impact of subsistence-level beach predation, the populations of *Dermochelys* and *Caretta* nesting in this territory are thought to be doomed (Hughes 1973). *Eretmochelys* has been exploited for tortoiseshell for at least a century, but this is a small fishery (Frazier, unpubl. data; Hughes 1973).

Madagascar has subsistence hunters all around the coast, but the Vezo and Sakalava in the west are the most experienced as marine exploiters. The latter capture turtles by a variety of techniques, but turning nesting females is not as important as in other territories. This is one of the most important countries for traditional, noncommercial turtle hunting, and turtles are an important part of the culture. However, stuffed turtles and tortoiseshell are prepared for sale to expatriats or for export, and this may account for more turtles than are consumed locally. Annual catches have been estimated to include thousands of each of the 4 species, excluding *Dermochelys* (Hughes 1971).

Reunion has few turtles, hence, little hunting, but its dependencies, Europa and Tromelin Islands in particular, are major nesting areas for *Chelonia*. Because of their remoteness, they have had little concerted exploitation. Passing sailors and temporary inhabitants have caught nesting turtles, but there is no significant hunting today, and the main rookeries are nature reserves (Hughes, this volume).

Turtles on Mauritius were exterminated along with the dodo, but the St. Brandon Islands still have nesting *Chelonia*. Most exploitation here is for export to Mauritius, but the islanders (mainly Seychellois) catch and eat turtle regularly. As the fishing population is transient and small, *Chelonia* exploitation for local consumption is minimal, about 30 per year (Hughes 1975).

The British Indian Ocean Territory (BIOT) now consists of only the Chagos Archipelago, and the Créole des Isles inhabitants, of Mauritian and Seychellois origin, have been evacuated. They hunted turtle, the catch was commonly bought by an island's lessee, and *Chelonia* meat was divided among the islanders. Minimal export occurred (Frazier 1977). Fish was the main source of protein.

The Republic of Seychelles, dispersed over a vast area of the western Indian ocean, is famous for turtles and turtlemen. Harpooning mating animals and turning nesting females are the main techniques of capture. Since its inception, the Seychelles turtle fishery has been geared for export, but turtle products are an important part of the culture. Eating *Chelonia* is presumed to be an inalienable right, and restraint in killing turtles is uncommon. Many gourmet dishes and specialty items have been prepared from turtle products. Some tortoiseshell is crafted locally, but most has been exported (Frazier, in press).

Mayotte, a dependency of France, but geographically one of the Comoro Islands, has no organized turtle fishery. Nesting *Chelonia* are exploited regularly at Pamanzi Island and, although the numbers taken are but several a week, the impact on the nesting population is great. Hunters often cook meat on the beach right after capture and take some choice cuts home in the morning. Waste of eggs and the less choice pieces is tremendous. There may be some local sale of meat, but this is uncommon (Frazier 1972).

Comores are over populated and few turtles occur at the 2 main islands of Ngazidia (Grand Comore) and Anjouan. Before the islands' independence from France, most turtles that were caught at these islands were netted and sold to expatriates and tourists. Moroni Island has thousands of *Chelonia* nesting annually, dispersed over a dozen important beaches. Small numbers are killed while nesting, and occasionally meat is taken to large villages for sale. Most exploitation is for local consumption (Frazier 1977). Although nowadays the people are inexperienced in turtle biology or lore, the fishermen from these islands were once reported to have complicated ceremonies and techniques for hunting turtles and dugongs (Petit 1930).

Tanzania's most able sailors and fishermen are on Zanzibar and Pemba Islands. They catch nesting *Chelonia* happened upon, but, except for sorties to Maziwi Island and other small cays around Mafia Island, turtles are encountered sporadically. Only a few persons are involved with the net fishery in the south. Total catches for local consumption are probably less than 100 per year. *Chelonia* was exported for a few years in the 1960s. Tortoiseshell is, as usual, collected when possible and sold to merchants for export.

Zanzibar was a major clearing house for tortoiseshell from the late 1800s until recently. Subsistence-level fishermen from countries all around the western Indian Ocean supplied the product which was exported to both Europe and the East (Frazier, unpubl. data).

Kenya has few nesting turtles, but the Bajun of the north coast are expert at netting and catching turtles with grapnels and sucker fish. Although *Chelonia* meat and oil are relished, turtle seems less important to their culture than it is to the Vezo. Most coastal people eat meat and eggs if available, but exploitation is sporadic, and now illegal. Bajun fishermen supplied an exporting company from 1952 to 1964. For thousands of years, tortoiseshell has been collected for sale to merchant exporters (Freeman-Grenville 1962).

Somalia's south coast is inhabited by Bajun who are culturally related to Bajun in Kenya, and their relationship with marine turtles is the same. Somalis, although also Moslems, will not eat turtle or other animals from the sea for religious reasons, so exploitation has been concentrated in the south. It has not all been for local consumption; an exporting company was active here about the same time as in Kenya (Travis 1967). The original turtle canning concern was taken over by Russians, but has evidently ceased production with their exodus.

Little is recorded from Djibouti.

Eritrea is Ethiopia's only coastal province. The Islamic people inhabiting the coastal strip occasionally sail to and among the Dahlak Islands where *Chelonia* may be butchered if encountered. What little exploitation occurs is for subsistence (Minot, n.d.).

In Sudan there is also little involvement with turtles. Yemeni fishermen may visit the Suakin Archipelago and catch any nesting animals that they find. There may once have been a large number of turtles slaughtered by passing sailors. *Eretmochelys* is thought to occur in large numbers in the Archipelago (Moore and Balzarotti 1977).

Egypt probably has some nesting turtles, which may be occasionally exploited, but little is known from this territory.

Bedouins traveling along the coast of Sinai may dig up nests and eat eggs. This, the epitome of subsistence hunting, occurs infrequently (Sella, this volume).

Israeli law does not allow exploitation of sea turtles; its tiny coastal strip makes much fishing activity unlikely (Sella, this volume).

Jordan's situation is not known, but the small coastal strip precludes much of a fishery.

Saudi Arabia's Red Sea coast may support sizable turtle populations particularly on the Farasan Islands, but little is known of the situation. Exploitation is likely to be mainly for subsistence, but predation by foreign-owned turtle exporting companies have been rumored.

Fishermen of the Yemen Arab Republic patrol beaches for nests. Male turtles are also eaten, but the numbers killed are small. Nesting beaches on the volcanic offshore islands are less likely to be disturbed than mainland beaches (Walczak 1979).

Yemenis from the People's Democratic Republic of Yemen (PDRY) eat turtle if food is scarce, and some islanders, as on Socotra, eat turtles regularly. Their catching techniques are varied, including the use of sucker fish. *Chelonia* is netted on pastures at Khor Umaira, and a major nesting area has been exploited for export, but this is a nontraditional, commercial fishery (FAO 1973).

Omani Arabs catch *Chelonia* around Kuria Muria and Masirah Islands and dig up nests of any of the 4 nesting species on Masirah. There is some exploitation of nesting *Chelonia* at Ras al Hadd for export, but only a few hundred are caught annually at this major rookery (Ross and Barwani, this volume).

The Persian Gulf states have little documented about their turtle populations. *Chelonia* is imported into Abu Dhabi for consumption. In Qatar both *Eretmochelys* and *Chelonia* are eaten. Turtles are probably not consumed in other states for religious reasons (Ross and Barwani, this volume).

Iran has a large and poorly surveyed coast. Turtle meat is not eaten for religious reasons, but eggs are collected throughout the territory. On Larak Island, oil from *Dermochelys* is prepared for boat preservation (Kinunen and Walczak 1971).

Pakistan has no traditional turtle fishery of significance, but at a recently developed enterprise in the west large numbers are slaughtered and their skins exported (Salm 1976). These may be *Chelonia* or *Lepidochelys*. Eggs and meat may once have been eaten in Karachi (Murray 1884).

India has the largest coastline and the greatest population in the region and turtles are traditionally hunted in many areas. However, the net fishery in the Gulf of Mannar is most developed. Eggs and meat of most species are eaten and sold in local markets. *Chelonia* has been exported to Sri Lanka and elsewhere. Eggs of *Lepidochelys,* and recently whole animals, are taken in large numbers in Orissa and exported to Bangladesh. Possibly many of these turtles are also taken for leather. Oil from *Dermochelys,* and occasionally other species, is used in boat maintenance, and sometimes for medicinal purposes. India has both imported and exported large numbers of turtles and turtle products, but it also has the greatest amount of subsistence hunting in the region (Frazier, 1980).

The Republic of Maldives includes hundreds of islands. Tortoiseshell has been exported for centuries, but as the populace is Moslem, consumption of eggs and meat was banned until recently. In the last decade most exploitation provided stuffed animals for tourists, and subsistence hunters were able to realize the equivalent of a month's income with the sale of 1 turtle (Didi, *in litt.,* 15 June 1976; Colton 1977).

Sri Lanka is also a rich area for sea turtles and turtle culture. According to one estimate, 50,000 people depend on the turtle fishery (Salm 1975), many of them around Jaffna, where there is a net fishery. Formerly artisans of tortoiseshell worked in the south, but *Eretmochelys* there has been extirpated. In the past only *Chelonia* was eaten, and other species were released if caught accidentally. Now all species are eaten, and there

is tremendous pressure on nesting turtles. Protective legislation has been ineffective because of pressures on food resources (Deraniyagala 1939; Hoffmann, in litt., 21 April 1975).

Bangladesh has not been studied, but may support large numbers of *Lepidochelys*. "Hundreds" of eggs are collected from an (?) island off the Sundarban (Choudhury 1968). *Lepidochelys* and eggs are imported from Orissa, India.

Burma is also poorly known, but millions of eggs are reported to have been collected in a year on Diamond Island (Maxwell 1911). Moslems do not eat turtle meat, but most turtles are eaten in the eastern part of the territory. *Lepidochelys* is evidently the main species involved, although *Chelonia* is also exploited (Theobald 1868). The Selung are expert sailors and fishermen, and they hunt turtles for their own consumption (Anderson 1889).

Summary and Conclusions

Marine turtles are hunted by subsistence fishermen throughout the Indian Ocean. Meat, eggs, and oil are valued foods in most territories. These, together with tortoiseshell, have been exploited for millennia.

One of the myths about subsistence hunting, along with the belief that it is completely divorced from monetary considerations, is that it is sustainable and does not damage a resource. However, there are numerous examples in the Indian Ocean where human populations are dense and lacking in protein sources, and their exploitation, concentrated on breeding turtles, has all but stopped the animals' reproduction. The most threatening forms of exploitation are those stemming from commercial enterprises and involve intense, organized exploitation for export. The leather and skin trade is a recent example (Frazier, ms.). In the past, tortoiseshell was collected in a desultory way, but with present-day prices (Mack, Duplaix, and Wells, this volume), *Eretmochelys* is now under tremendous pressure.

Ideally, subsistence exploitation should be uninhibited, but the present situation over much of the Indian Ocean indicates that this is untenable and that unless strict breeding and feeding reserves are established and management policies implemented, it is unlikely that marine turtles will continue to be a significant part of the subsistence hunter's diet.

Acknowledgments

The work was supported by: The Royal Society (London); Natural Research Council (United Kingdom); Fauna Preservation Society; East African Wildlife Society; African Wildlife Leadership Foundation; Department of Agriculture, Government of Seychelles; Government of BIOT; Game and Fisheries Departments of the Governments of Kenya and Tanzania; the Department of Zoological Research, National Zoological Park and the Division of Reptiles and Amphibians, Smithsonian Institution; and a grant from the Smithsonian Scholarly Studies Program to Dr. J. F. Eisenberg. Many colleagues in many countries provided information. The expert assistance of Mrs. Wy Holden and Ms. Virginia Garber in typing and Ms. Sigrid James Bruch in illustrating was invaluable.

Literature Cited

Anderson, J.
1889. Report on the mammals, reptiles and batrachians, chiefly from the Mergui Archipelago, collected for the Trustees of the Indian Museum. *Journal Linnean Society, Zoology* 21:331–50.

Choudhury, A. M.
1968. *Working Plan of Sundarban Forest Division for the Period from 1960–61 to 1979–80.* Government of East Pakistan, Forest Department.

Colton, E. O.
1977. Turtles of the Maldives. *Defenders,* 52:167–70.

Deraniyagala, P. E. P.
1939. *Tetrapod Reptiles of Ceylon.* London: Dulau and Company, Limited.

FAO (Food and Agriculture Organization)
1973. Report to the Government of the People's Democratic Republic of Yemen on marine turtle management, based on the work of H. F. Hirth and S. L. Hollingworth, marine turtle biologists. Report FAO/UNDP (TA) (3178) 51 pp.

Frazier, J.
1972. Marine turtles in the Archipel des Comores. Fauna Preservation Society. Typescript, pp. 7 + 12 + 2 + 3.
1977. Marine turtles in the western Indian Ocean: BIOT and Comores. Fauna Preservation Society. Typescript, 33 + 114 pp.
1980. Exploitation of marine turtles in the Indian Ocean. *Human Ecology* 8:329–70.
In press. Marine turtles in the Seychelles and adjacent territories, In *Biogeography and Ecology of the Seychelles,* ed. D. R. Stoddart. The Hague: Junk.

Freeman-Grenville, G. S. P.
1962. *The East African Coast.* Clarendon: Oxford University Press.

Hughes, G. R.
1971. Sea turtles—a case study for marine conservation in South East Africa. In *Nature Conservation as a Form of Land Use,* pp. 115–23. SARRCUS Symposium, September 1971, Yorongoza National Park.
1973. The sea turtles of Mozambique. Oceanographic Research Institute, Durban. Mimeo, 17 pp.
1975. The St. Brandon turtle fishery. *Proceedings of the Royal Society of Arts and Sciences in Mauritius,* 3:165–89.

Kinunen, W., and P. Walczak
1971. Sport fisheries and marine biology: Persian Gulf sea turtle nesting surveys. Iran Game and Fish Department. Typescript, 12 pp.

Maxwell, F. D.
1911. Government report on the turtle banks of the Irrawaddy Division Rangoon. Cited by M. A. Smith, Reptilia and amphibia, vol. 1, 1931, in *The Fauna of British India, including Ceylon and Burma,* ed. J. Stephenson. London: Taylor and Francis.

McAllitster, H. J.; A. J. Bass; and H. J. Van Schoor
1965. Marine turtles on the coast of Tongaland, Natal. *The Lammergeyer* 3:10–40.

Minot, F., ed.
No date. *Red Sea and Island Resources of Ethiopia.* Nairobi: African Wildlife Leadership Foundation.

Moore, R. J., and M. A. Balzarotti
1977. Report of 1976 expedition to Suakin Archipelago (Sudanese Red Sea); results of marine turtle survey and notes of marine and bird life. Mimeo, 27 pp.

Murray, J. A.
1884. *The Vertebrate Zoology of Sind.* London: Richardson and Company.

Nietschmann, B.
1972. Hunting and fishing focus among the Miskito Indians, eastern Nicaragua. *Human Ecology* 1:41–67.

Parsons, J. J.
1972. The hawksbill turtle and the tortoise shell trade. In *Études de Géographie Tropicale Offertes à Pierre Gourou,* pp. 45–60. Paris: Mouton and Company.

Petit, G.
1930. L'Industrie des Pêches à Madagascar. Paris: Société d'Edit. Géog. Marit. et Colon.

Salm, R. V.
1975. Preliminary report of existing and potential marine park and reserve sites in Sri Lanka, India and Pakistan. IUCN, Morges, Switzerland. Mimeo.
1976. Marine turtle management in Seychelles and Pakistan. *Environmental Conservation* 3:267–68.

Theobald, W.
1868. Catalogue of the reptiles of British Birma, embracing the provinces of Pegu, Martaban and Tenasserim; with descriptions of new or little-known species. *Journal of the Linnean Society, Zoology* 10:4–67.

Travis, W.
1967. *Voice of the Turtle.* London: George Allen and Unwin.

Walczak, P. C.
1979. The status of marine turtles in the waters of the Yemen Arab Republic. *British Journal of Herpetology* 5:851–53.

George R. Hughes
Natal Parks,
Game and Fish Preservation Board
Pietermaritzburg, South Africa

Conservation of Sea Turtles in the Southern Africa Region

ABSTRACT

The Southern Africa region has received uneven attention regarding stock assessment surveys. South Africa, Réunion and its dependencies, and South West Africa-Namibia have few problems. Mozambique, Madagascar, and Angola all have extensive sea turtle populations but require more encouragement and assistance. Mauritius and its dependencies need more dedicated management. None of the 5 species is in immediate danger of extinction, but all would benefit from enforced legislation, and formal protection of breeding grounds.

Introduction

The Southern African region (Figure 1) has always been an area rich in sea turtles. Just how rich is difficult to assess, but intense exploitation once reached substantial proportions (see Hughes 1973, 1974a, 1976a; Lougnon 1970). The extensive exploitation by the coastal peoples of the region has been of such importance that the sea turtle is woven throughout the cultural fabric of some societies, especially in Madagascar (Hughes 1973, 1975; Petit 1930).

In 1963 the Natal Parks Board started a study program in Tongaland, where interest in sea turtles has been increasing. Hughes's major survey (1974a and b) between 1969 and 1973 stimulated workers in other areas to pursue independent studies (Servan 1975; Vergonzanne, Servan, and Batori 1976; Lebeau, Gobert, and Durand 1978; Lebeau et al., in press).

Hughes concluded (1974a and b; 1976b) that 5 species were represented in the area: the leatherback turtle *Dermochelys coriacea,* the green turtle *Chelonia mydas,* the loggerhead turtle *Caretta caretta,* the hawksbill turtle *Eretmochelys imbricata* and the olive ridley turtle *Lepidochelys olivacea,* and that none was in immediate danger of extinction.

For the present review, an attempt was made to obtain first-hand information of the current situation;

varying success was achieved by writing to the countries in the region. Some excellent and most rewarding reports were received. Where no information was forthcoming the latest data available, which may be years old, was used. No apologies are made for using this material because its inclusion will at least present the situation as it was in the recent past.

On the whole the situation remains basically optimistic. Some local populations have increased and others are at least holding steady.

Present Status

It is perhaps unfortunate that this section should be dealt with by country rather than by turtle populations. Regrettably, however, conservation measures are not equally pursued by all countries, and turtles may be safe and threatened over a single kilometer of their feeding range. All the countries reviewed below have at least well motivated legislation in common. In theory the sea turtle is well protected throughout the area; in practice the situation is unsatisfactory.

Angola

During 1971 a series of photographs of nesting leatherback turtles was published in Angola (Anonymous 1971). Regrettably no details were provided but interest in sea turtles was stimulated. Hughes, Huntley, and Wearne (1973) reported that leatherback, olive ridley, green, loggerhead, and hawksbill turtles occurred in Angolan waters, and that the first 3 species were known to nest. The only figures available indicated that in certain areas nesting densities were quite high, for example, 30 tracks on a 500-m stretch of beach. Some turtle beaches were happily situated within the Quiçama National Park.

Huntley (1974) reported that on a flight south of Luanda, he saw 613 nests of green turtles and leatherbacks along 150 km of coastline. Huntley (personal communication) continued aerial surveys north of Luanda and found that sea turtles nest extensively. The distribution of nesting is very uneven: several dense concentrations with extensive, sparsely utilized areas in between. There are far more nesting turtles in the north of Angola, although recent information from South West Africa, indicating that loggerheads may nest in the Skeleton Coast (below), suggests that nesting occurs in the Iona National Park and therefore probably along the entire coastline where suitable beaches occur.

As recently as 1975, exploitation was for domestic use only. Eggs were taken before adult turtles and only occasionally was a turtle killed to sell its carapace to tourists.

There was no official protection until 1973 when a law was proclaimed enforcing vigorous protection and carrying a substantial fine for harming a sea turtle, US$300. But the law was not enforced. No details are available on the recently revised legislation or on the current situation. Despite hopes to the contrary, Angola could not be included in Southern Africa survey. Angola thus remains unsurveyed in detail; a thorough quantitative assessment of the country's large remaining sea turtle populations is highly desirable.

Southwest Africa-Namibia

In 1844 a Captain Morrell reported that turtles nested on Bird Island and at Sandwich Harbour (23° 35′S, 14° 28′E): ". . . green turtles also visit the sandy beaches for the usual purposes" (Morrell 1844).

Those beaches are now protected within the Namibia Desert Park, but there are no recent records of sea turtles nesting in this area. It is also unlikely that the nesters would be green turtles as it would be unheard of for green turtles to nest in such temperate waters. It seems more likely that the turtles were loggerheads. Recently it has been found that loggerheads are occasionally found in the Skeleton Coast Park. The number of nesting turtles has not yet been established, but they are in no danger as the area in which nesting occurs is uninhabited (B. de la Bat, personal communication).

Stranding records indicate that leatherbacks, loggerheads and the occasional green turtle do frequent the waters of Southwest Africa. Coastal surveys are currently being carried out by the Penrith State Museum, Windhoek.

Republic of South Africa

All 5 species are recorded from South Africa (Hughes 1974a), but only 2 species, leatherbacks and loggerheads, are known to nest in the Kwa-Zulu coast, an area known locally as Tongaland (the local black tribe is the aMa-Thonga).

The first laws to protect sea turtles in Natal were passed in 1916. These laws have been progressively updated until the present when sea turtles are protected from all forms of exploitation or interference by Section 101 of Ord. 15 of 1974.

Since 1963–64 a tag-study and protection team has patroled beaches of Tongaland nightly from October through March. Excellent cooperation has been obtained from the Tonga people, and for the past few years staff from the new Kwa-Zulu Conservation Department have joined the Natal Parks Board to share the protection effort.

Results in general have been promising. After a decade of little change both populations appear to be increasing (see Hughes, this volume, Figures 1 and 2).

GEORGE R. HUGHES

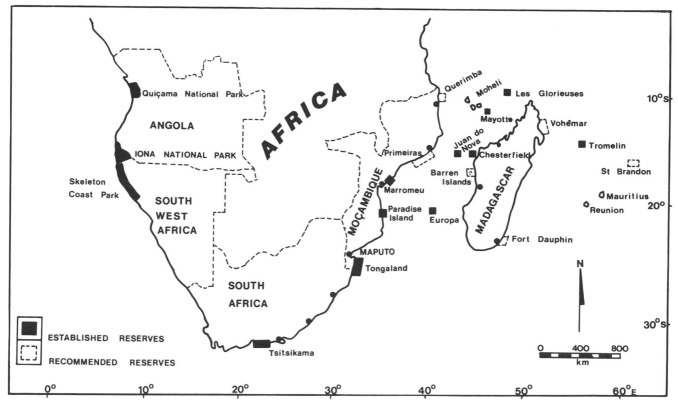

Figure 1. The Southern African Region showing existing marine reserves and parks and some areas recommended for protective status.

Loggerheads have gone from a low of 184 (discounting 1963–64 which was mainly exploratory) to 408 (1978–79) and leatherbacks from 5 (1966–67) to 70 (1977–78).

Considerable attention has been given to hatchling and adult distributions in the sea. Figure 2 illustrates the distributions of sea turtles tagged in Tongaland and recaptured elsewhere. It can be seen that the Tongaland protected beaches are "home" to turtles found along half of the East African coast and the west coast, at least, of Madagascar.

Since 1963 numerous pleas have been made for the total protection of the Tongaland Coast and an offshore zone. On 23 February 1979 the South African government declared an 80-km section of the coast from Cape Vidal to Sodwana Bay a marine reserve. The most important sea turtle breeding beaches were excluded because the area fell under the jurisdiction of the Kwa-Zulu Territorial Authority. On 5 June 1979 the Cabinet of Kwa-Zulu resolved (Resolution 187/79) that steps should be taken to declare a further strip of the Tongaland coast a marine reserve. This strip will stretch from the Mozambique border south for some 61 km, and embracing almost all of the as yet unprotected turtle beaches. Following the next session of Parliament the future of the Tongaland nesting populations should be safeguarded.

Sea turtles are not generally exploited in South Africa and the general attitude of the majority of the populace is conservative. It is the policy of local conservation bodies, however, to protect the interests of local people, when possible from a conservation point of view. Once turtle populations in Tongaland increase to a satisfactory level, it is intended to harvest a minor number. Because the populations have not yet reached satisfactory level, this program is of theoretical interest only.

Mozambique

With 1,300 km of tropical coastline Mozambique is rich in sea turtles with 5 species recorded in the area (Hughes 1971a). All are adequately protected by the Hunting Law (Designation No. 7/78 of 18 April 1978) and Decree No. 117/78 of 18 May 1978 (Anonymous, 1978).

Despite these laws, local authorities have difficulties in controlling the killing of sea turtles or egg collecting. The problem is simply that this new state does not have the personnel or resources to protect adequately or cover the lengthy coastline. There is no legal exploitation, and that exploitation about which something is known is strictly domestic and is believed to result in minor illegal export of tortoiseshell by tourists.

At the 4 marine reserves in Mozambique, sea turtles receive more intensive protection: Reserva Especial do Maputo, Maputo Province; Illia de Inhaca; Parque Na-

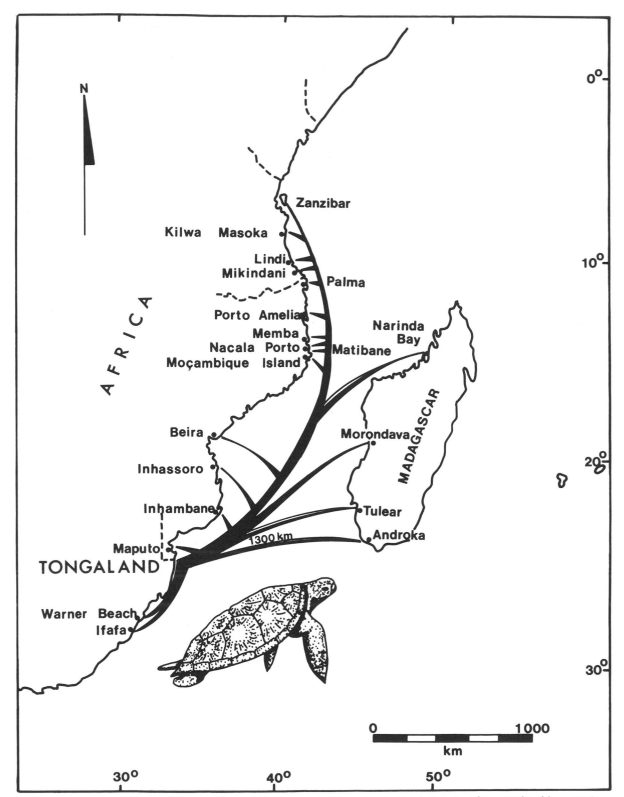

Figure 2. Recovery sites of loggerhead females tagged in Tongaland, Natal, South Africa, 1963–64 to 1979.

cional de Bazaruto, Inhambane Province; and Complexo de Marromeu, Sofala Province.

It is too early to see whether protective efforts, especially in the Maputo Reserve, are having any beneficial effects, but it is suspected that in most areas there is a decline in sea turtle populations.

Madagascar

The coastal waters of Madagascar are host to 5 species

GEORGE R. HUGHES

of sea turtle: green, loggerhead, olive ridley, hawksbill, and leatherback turtles. Figures 2 and 3 indicate quite clearly that protected colonies of sea turtles such as Tongaland, Europa and Tromelin are of great benefit to Madagascar.

This is no minor contribution. Hughes (1971b) during a survey of Madagascar calculated that the sea turtle resource is considerable and extensively exploited. Over 13,000 individual turtles are killed along some 600 km of the southwest coast of Madagascar every year. The past exploitation of hawksbills for the tortoiseshell trade is legendary (Hughes 1973). Even recently, a single taxidermist shop in Diego Saurez had a licence to take 400 hawksbills annually. Throughout the country stuffed juvenile hawksbills are displayed for sale in almost every major market place and general store. All other species are exploited mainly for domestic consumption, but the juvenile hawksbill is an article of commerce (see also Mauritius, below) and is undoubtedly the most threatened species in the territory.

All this exploitation occurs despite the existence of perfectly adequate sea turtle protective legislation.

1) Resolution of 23 May 1923 (J.O. a/vi/23 p. 439) instituted areas reserved for turtles: Nosy Ovambo or Ilot Boise, Diego Province; Nosy Iranja, Nosy Bê Province; Chesterfield Island, Morondava Province; Nosy Trozona, Tulear Province; Nosy Ve, Tulear Province; and Europa, (now under French control, see below). "The protected species are the green turtle (*Chelonia mydas*) and the hawksbill (*Chelonia imbricata*)" [*sic*].

2) Resolution of the 24th October 1923 (J.O. 17/11/23 p. 856) states: "It is forbidden to capture sea turtles; when they are laying; and when the width of the carapace, measured across the plastron, does not exceed 0.50m."

Regrettably surveillance is minimal, and the legislation is ignored by most fishermen. The problems of any law-enforcer, trying to persuade the Sakalava or Vezo people to stop eating sea turtles, are almost insurmountable. Sea turtles are woven into the very cultural fabric of these coastal peoples who practically revere the animal.

There are no coastal marine reserves where the green, hawksbill, loggerhead and, possibly, olive ridley turtle can nest unmolested. A slow and steady disappearance of these species must be expected unless positive action is taken.

The problems facing Madagascar are similar to those of Mozambique. It, too, is a large and poor country that simply cannot afford adequate law enforcement.

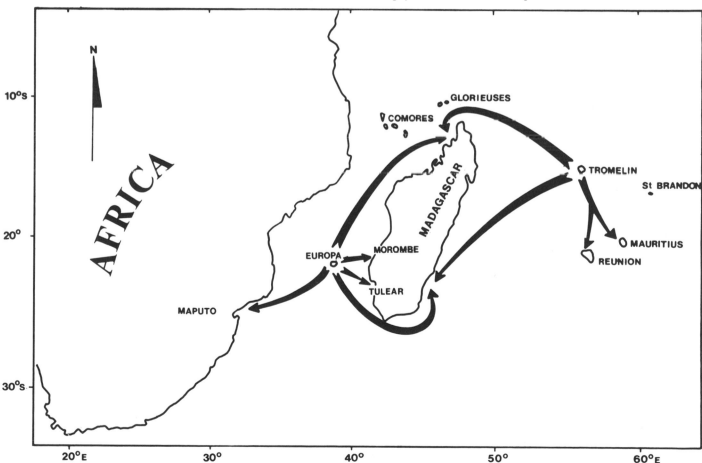

Figure 3. Recovery sites of green turtle females tagged on Europa and Tromelin Islands, 1970–78.

Réunion and Its Dependencies

One of the most rewarding areas for sea turtle study and conservation is Réunion Island with its dependant islands: Europa, Tromelin, Juan do Nova, Les Glorieuses, and Mayotte (Figure 1).

All sea turtles are protected from capture in Réunion itself, and the outer islands have full nature reserve status with complete protection of turtles, nests and hatchlings. The law is not broken on the islands, but a very few turtles are killed by fishermen on Réunion.

The most common species, virtually the only species so far found nesting on the islands, is the green turtle, *Chelonia mydas*. Some hawksbills, *Eretmochelys imbricata*, are seen occasionally on Europa and Tromelin, and loggerheads are infrequently seen in the waters around Réunion but are not resident (Hughes 1974a).

Since 1970 a number of research programs have been conducted on some of the islands. The first, was conducted by Hughes (1970), on Europa. He was followed by Servan (1975); Vergonzanne, Servan, and Batori (1976); and excellent work continues under Dr. Alain Lebeau (Lebeau, Gobert, and Durand 1978, in press). Vergonzanne, Servan, and Batori (1976) and Vergonzanne (in press) have produced reports on the Glorious Islands (Les Glorieuses), and Batori (1974) prepared a report on Tromelin Island. Annual estimates of green turtle females nesting on each island have varied somewhat and are summarized below:

Islands	Number of females	Source and publication date
Europa[a]		
1970–71	4,000–5,000	Hughes 1970
1973–74	3,000	Vergonzanne, Servan, and Batori 1976
1977–78	1,500	Lebeau, Gobert, and Durand 1978
1978–79	9,000–18,000	Lebeau, et al., in press
Tromelin		
1970–71	200–400	Hughes 1974a
1973–74	3,000	Lebeau et al., in press
1973	600	Vergonzanne, Servan, and Batori 1976
1977	1,000	Lebeau, Gobert, and Durand 1978
1978	4,400>	Lebeau et al., in press
Glorious		
1973	70–80	Vergonzanne in press

[a] As regards this past season the number of eggs laid exceeded 3.4 million.

There are no available figures for Juan do Nova, but populations appear to be very limited. Frazier (this volume) deals with Mayotte.

Many hundreds of green turtles have been tagged on Europa and Tromelin, lesser numbers on the Glorious Islands. Figure 3 illustrates the recoveries of turtles from these areas and shows that Madagascar depends heavily for its harvests on the island rookeries protected by the French.

No exploitation of adult green turtles occurs on any of the islands, but 20,000 hatchlings have recently been captured on Europa and transferred to the main island of Réunion for market rearing by a firm called Compagnie Réunionnaise d'Aquaculture et d'Industries Littorales. All collection of hatchlings is undertaken by the Scientific Institute. Every hatchling collected emerges in daylight, a time when it would almost certainly have been taken by frigate birds (*Fregata minor* or *F. ariel*).

It is stressed, therefore, that this "exploitation" program in no way interferes with the natural survival rate or recruitment of green turtles on Europa Island.

The rearing station, situated near the town of St. Leu on the southwest coast of Réunion, intends to sell its products abroad. These exports are expected eventually to reach 100 to 200 tonnes/yr.

Conservation in the areas under French control is admirable, and there is no need for concern.

Mauritius and Its Dependencies

Sea turtles no longer nest on the shores of Mauritius. Its law protecting sea turtles is widely disregarded. For example a recent report stated that "at present *Eretmochelys* are being captured in places like Rodriguez, Madagascar and Agalega, stuffed and sold to tourists for Rs. 500—per turtle." Further, a recent claim for a tag return was submitted by a fisherman who caught the turtle in his net on Mauritius.

The St. Brandon Island fishery for green turtles continues (Hughes 1976a), and the exploitation figures over the past 37 years indicate no sharp decline. Some 10,880 turtles (mean 295 turtles/yr) have been recorded as killed. However, the toll almost certainly has been higher counting the turtles consumed on the islands and thus not on record.

Attempts to obtain more up to date details have not succeeded. It is understood that in addition to North Island, which had always been protected (if not always sacrosanct), Frigate Island has also been declared a reserve.

Hughes (1976a) recommended that, if a local sea turtle farm be established on Mauritius, the harvest of green turtles on St. Brandon should be strictly limited to 150 for at least 5 years; that there should be a closed

GEORGE R. HUGHES

season, and that hawksbill exploitation should cease forthwith. Maurition authorities do not seem to have implemented any protective measures.

Discussion

During the last 5 years, only limited changes have occurred in the conservation situation of sea turtles in the Southern African region. Hughes (1976b) listed 19 areas in the region that, if given Park status with full protection, would assist various species of sea turtle to survive. Some of the 19 have been raised or are about to attain to Park status: 1) South Africa, the Tongaland Coast; 2) Mozambique, Parque Nacional de Bazaruto and the Reserva Especial do Maputo; 3) Réunion, Europa's 4 islands, and Juan do Nova; and 4) Tromelin and Les Glorieuses.

There appears to be an additional protected island in the St. Brandon Archipelago, but this is the work of the private company working on the islands and not the action of the Mauritius government. In Angola it is fortuitous that many turtles nest in the Quiçama National Park.

Even these modest advances are not enough, and all 5 species would benefit from enforced legislation. This is not suggested merely from an esthetic point of view nor from any particular plea for sea turtles as creatures that have a right to exist as does any other creature. This is for the benefit of the peoples of coastal Southern Africa themselves.

There is no doubt that turtle populations respond readily to protection of the nesting grounds provided that the numbers of turtles has not reached too low a level. Two good examples of such responses are the Europa Island green turtles and the Tongaland sea turtles. Both populations have undergone varying degrees of persecution from which the Europa green turtles appear to have completely recovered and loggerheads and leatherbacks in Tongaland are gradually increasing in numbers.

In Southern Africa, where the hunting of sea turtles and the consumption of turtle products are traditionally regarded as a right and not a privilege, turtle conservation must be applied in recognition of the sea turtles' role as a natural resource. Preservation for esthetic value alone in a protein-poor region is hardly justifiable and can expect only limited, if any, support from government bodies incapable of providing substitute sources of protein. On the other hand, governments should not jeopardize the resource as a whole by using such reasoning as an excuse for inaction.

Adequate and convincing proof of the viability of sea turtle populations is available, and the responsibility for the conservation and management of this resource lies in the hands of individual governments.

South Africa, Réunion Island and Namibia do not appear to need any further encouragement. With Mauritius and Réunion, we have 2 opposing situations. Réunion has only limited and well controlled "exploitation" for the benefit of its own people but expends a not inconsiderable amount of time and effort for the benefit of the coastal peoples of Madagascar. Mauritius makes virtually no effort to safeguard an ancillary source of protein for its own people, and seems oblivious to recommendations and appeals. Mozambique, by contrast, is only too well aware of its shortcomings and requests further assistance and guidance.

The economic potential of well managed marine reserves containing sea turtle populations can hardly be measured. Along thousands of kilometers of coastline, sea turtles could range freely and could be harvested by coastal fishermen in their traditional manner. Marine reserves would form the source of stock.

The tourist potential of marine reserves should also be considered and their concomitant revenue-earning potential. Well-targeted educational programs can greatly enhance the value of the reserves and the prestige of the country involved.

Except in cases of dire extremity, all conservation efforts should contribute education, appeal to esthetics, and generate some financial return. Sea turtle conservation need not be an exception. Indeed, few other animal groups can meet all 3 criteria in return for a modest financial expenditure.

The resilient sea turtle has survived, in some cases, centuries of exploitation. As a source of high protein food, they are an important factor in the lives of the developing peoples along the coasts of Southern Africa. Given peace during the nesting season, the sea turtle can continue to be an important food resource and can, with minimal management, increase its contribution to the well-being of the peoples of the Southern African region.

Recommendations

It is recommended that the following regions should be considered for state-established and protected marine parks in order to further improve the survival of sea turtle populations in the region: 1) Mozambique: Primeiras Islands, Querimba Islands; 2) Madagascar: All the areas mentioned above under Madagascar should be redeclared Marine Parks or Nature Reserves and the Barren Islands should be declared a sanctuary. In addition, sanctuaries should be declared for loggerheads at the St. Luce area north of Fort Dauphin and for hawksbills on the Vohemar-Sambara Coast; 3) Mauritius and its Dependencies: The entire St. Brandon Archipelago should be declared a marine reserve albeit with limited exploitation of its natural resources (fish, turtles, rock lobsters) being allowed under state control with limits vigorously enforced.

Quantitative research involving sea turtle experts should be encouraged on the coasts of Angola, Mozambique and Madagascar in order to ascertain the size of those populations identified by the early general surveys.

Acknowledgments

My sincere thanks are expressed to the Director of the Natal Parks Board for his permission to deliver this report. Its compilation would not have been possible without the excellent cooperation of Mr. Paul Dutton, Serviços Veterinario, Maputo, Moçambique; Dr. Alain Lebeau of L'Institut Scientifique et Technique des Pêches Maritimes, Le Port, Réunion; Mr. B. de la Bat, Director of Nature Conservation, Windhoek, South West Africa; and B. J. Huntley, CSIR, Pretoria, South Africa. Attendance of the Conference would not have been possible without the financial assistance of the Southern Africa Nature Foundation and the Natal Parks Board. Finally, my thanks to Maxie Holder and Cindy Pringle for their help.

Literature Cited

Anonymous
1971.   A tartaruga gigante veio desovar a praia da cidade. *Turismo de Angola* 4:25–26.
1978.   Legislaçao sobre a actividade da Caça. *Imprensa Nac. Moçamb.* 1978:1–22.

Batori, G.
1974.   Rapport d'activité. Ile Tromelin. Mimeo, 1–15.

Hughes, G. R.
1970.   The status of sea turtles in South East Africa, 2. Madagascar and the Mascarenes (1) Europa Island. Oceanogr. Res. Inst. Durban, South Africa, Mimeo, 1–47.
1971a.   Preliminary report on the sea turtles and dugongs of Moçambique. *Vet. Moçamb.* 4:45–62.
1971b.   Sea turtles—a case study for marine conservation in South East Africa. In *Proceedings SARCCUS Symposium "Nature Conservation as a Form of Land Use,"* pp. 115–23. Gorongosa National Park, 13–17 September 1971.
1973.   The survival situation of the hawksbill sea turtle (*Eretmochelys imbricata*) in Madagascar. *Biological Conservation* 5:41–45.
1974a.   The sea turtles of South East Africa, 1. Status, morphology and distributions. *Investigational Report of the Oceanographic Research Institute,* Durban, South Africa, 35:1–144.
1974b.   The sea turtles of South East Africa, 2. The biology of the Tongaland loggerhead turtle *Caretta caretta* L. with comments on the leatherback turtle *Dermochelys coriacea* L. and the green turtle *Chelonia mydas* L. in the study region. *Investigational Report of the Oceanographic Research Institute,* Durban, South Africa, 36:1–96.
1975.   *Fauo*—the sea turtle in Madagascar. *Defenders of Wildlife* 50:159–63.
1976a.   The St. Brandon turtle fishery. *Proceedings of the Royal Society of Arts and Science, Mauritius,* III:165–89.
1976b.   Sea turtles in South East Africa. In *Proceedings of the Symposium Endangered Wildlife in Southern Africa,* pp. 81–87. University of Pretoria.

Hughes, G. R.; B. Huntley; and D. Wearne
1973.   Sea turtles in Angola. *Biological Conservation* 5:58.

Huntley, B. J.
1974.   Outlines of wildlife conservation in Angola. *Journal of the Southern Africa Wildlife Management Association* 4:157–66.

Lebeau, A.; B. Gobert; and J-L. Durand
1978.   Rapport sur l'étude de la tortue de mer *Chelonia mydas.* Peuplement, reproduction et biologie des populations de iles Tromelin et Europa. *Inst. Scient. Techn. Pêches Marit.,* La Port, Réunion. Mimeo 1–24.

Lebeau, A. ; G. Biais; J-L. Durand; and B. Gobert
In press.   La tortue verte *Chelonia mydas* (Linne) des iles de Tromelin et d'Europa (Ocean Indien): peuplement et reproduction. *Inst. Scient. Techn. Pêches Marit.,* La Port, Réunion: 1–39.

Lougnon, A.
1970.   *Sous le Signe de la Tortue.* Saint Denis: Libraire Jean Gerard.

Morrell, B.
1844.   *Morrell's Narrative of a Voyage to the South and West Coasts of Africa.* London: Whittaker and Company.

Petit, G.
1930.   L'Industrie des Pêches à Madagascar. Paris: Société d'Editions Géographiques, Maritimes et Coloniales.

Servan, J.
1975.   Etude de la biologie de la tortue verte (*Chelonia mydas*) a l'île Europa. Ph.D. dissertation, University of Paris VI.

Vergonzanne, G.
In press.   Biologie et évaluation du stock de tortues vertes (*Chelonia mydas*) des Iles Glorieuses (nor-ouest de Madagascar). 21 pp.

Vergonzanne, G.; J. Servan; and G. Batori
1976.   Biologie de la tortue verte sur les îles: Glorieuses, Europa et Tromelin. Manuscript.

Atlantic Ocean

L. D. Brongersma
Rijksmuseum van Natuurlijke Historie
Leiden, The Netherlands

Marine Turtles of the Eastern Atlantic Ocean

To facilitate the survey of the marine turtles found in the Eastern Atlantic Ocean, this region may be divided into 3 major areas: European Atlantic Waters; Macaronesian Waters; and African Atlantic Waters.

European Atlantic Waters

European Atlantic Waters (EAW) comprise the eastern part of the North Atlantic Ocean, bordering upon European coasts, together with the seas, bays, channels, and estuaries connected to it, and extending from Iceland in the northwest and North Russia in the northeast to Punta Marroqui, Spain, in the south. In this way EAW include the Barents Sea, North Sea, Baltic Sea, Irish Sea, British Channel, etc.; the Mediterranean is not included (Brongersma 1972:5).

Five species of turtle have been recorded from these waters. In order of frequency these are: the leathery turtle or leatherback, *Dermochelys coriacea* (L.); the loggerhead, *Caretta caretta* (L.); Kemp's ridley, *Lepidochelys kempi* (Garman); the green turtle, *Chelonia mydas* (L.); and the hawksbill, *Eretmochelys imbricata* (L.).

In the past it has been assumed that the turtles found in these waters were just poor waifs that had lost their bearings. Wandering around aimlessly in the ocean, willy-nilly they were borne along by the currents eventually to perish on European coasts. Indeed the climate of the area does not allow breeding there, and in winter it is definitely unfavorable even for the most hardy turtle to stay and survive. An exception may be the seas off the southern part of Portugal and along the south coast of Spain. In summer, turtles may move northwards to fairly high latitudes. The record is held by a young *C. caretta* captured alive at Murmansk (68°55'N) (Konstantinov 1965:111; Brongersma 1972:110); to get there, the turtle must have rounded the North Cape (71°10'N) and Nordkyn (71°05'N). The second best is the *D. coriacea,* taken alive in September 1958 off the Norwegian coast at 69°18'N (Holgersen 1960:135; Brongersma 1972:32).

In years gone by much of our knowledge about the occurrence of turtles along the western coasts of Europe was based upon juvenile *C. caretta* and *L. kempi* being washed ashore in winter, dead or dying. Observations on turtles sighted at sea were scarce at the time, and this is not to be wondered at. Never being numerous, small loggerheads and Kemp's ridleys easily escape notice. In regions, such as Portugal, where loggerheads were stated to be common (Themido 1942:21; except in the North, Ferreira 1893:26), not much attention was paid to them, and the majority of the turtles observed were not recorded individually as to place and time. In fact we do not know how many young turtles actually come to EAW, nor do we know whether any of them succeed in turning back before the cold sets in. From stranded specimens we know that at least some (or perhaps most) die in northern waters; the chances of perishing are greater the farther north the turtles have traveled.

If we still have some doubts about whether *C. caretta* and *L. kempi* succeed in returning to warm waters, there is little doubt that *D. coriacea* does. Once considered extremely rare, the number of observations has grown considerably in the last 3 decennia, and today we know that *D. coriacea* is a regular visitor to the area. Before 1951 only 68 records (since 1729) were available; the years 1951 to 1971 gave 116 records (Brongersma 1972, Table 3a); and since that time many records have been added, for example, from Ireland, the Netherlands, and Portugal.

Duron-Defrenne (1978:68–77) added scores of reports from France. It has been stated (Moulton 1963; Brongersma 1972:101) that *D. coriacea* may travel in groups, and indeed today we have various records of 2, 3, 5, or 6 turtles being sighted together. Once 12 turtles were observed in a restricted area (Duron-Defrenne 1978:79, 83). Thus, *D. coriacea* is a fairly stable element in the fauna of EAW, and especially so in the Gulf of Gascony along the south coast of Britanny and from there southwards to Biarritz (Duron-Defrenne 1978:68–77). The fact that the majority of the records stem from the area of the islands of Ré and Oléron, west of La Rochelle, may be due to the research being organized from La Rochelle. Some of the leathery turtles die in summer from drowning when they become entangled in lines attached to lobster pots, or in nets, or when they are hit by ships' propellers. However, the fact that relatively few specimens that show no wounds wash ashore in winter indicates that the majority of *D. coriacea* visiting the area in summer, leave for warmer waters before winter sets in.

Chelonia mydas, the green turtle, is rarely encountered. The situation is complicated by the fact that before the second world war, green turtles used to be shipped alive to Western Europe to the manufacturers of turtle soup. Turtles that died during the voyage were jettisoned, and among these I count the green turtles washed ashore on the Dutch coast in 1934 (3) and 1937 (1) (Brongersma 1972:180, 181). I know of only 2 records from EAW after the shipping of live turtles had been abandoned: one is that of a young green turtle (carapace 360 mm) stranded alive at Petten, the Netherlands, the other (carapace 362 mm) captured alive in the Ría de Arosa, northwest Spain (Brongersma 1972:180, 186). These 2 I believe to have reached EAW on their own. It may well be that some of the French and Spanish records from the nineteenth century (Brongersma 1972:185–87) belong in the same class, but there is no proof. That *Ch. mydas* is found occasionally on the Portuguese south coast is substantiated by a half-grown specimen from Algarve (Museu do Mar, Cascais).

Eretmochelys imbricata has been mentioned various times as occurring in EAW. As yet I know of only 1 specimen from this area. It was found in the Channel (Brongersma 1972:195), but we do not know the exact locality, whether it was dead or alive, or who found it. In 8 instances specimens reported to be hawksbills proved to have been misidentified: 5 were *C. caretta*, and 3 were *L. kempi* (Brongersma 1972:194). Further proof is needed to show that the hawksbill really does come to EAW on its own.

Summarizing, we may state that *D. coriacea*, *C. caretta*, and *L. kempi* are visitors to EAW, and of these *D. coriacea* is a regular visitor at least to French waters. The question arises as to the origin of the turtles found in EAW. For *L. kempi* the answer seems to be simple as the only known nesting beaches are to be found in Tamaulipas on the Mexican Gulf coast. We must conclude that the juveniles, after passing from the Gulf into the Atlantic Ocean through the Florida Strait, crossed the Atlantic to Europe and, as will be mentioned below, also to the Azores and to Madeira. Most *C. caretta* found in EAW are young specimens, and it seems likely that they too came from the western Atlantic. However, the possibility cannot be excluded that some *C. caretta*, for example, the few adult specimens found in northern EAW, came from populations living farther to the South, or that they came from the Mediterranean (Brongersma 1972:238–39).

Only in Portugal is *C. caretta* exploited but, according to recent information, in a very limited way.

For individual records for each of the species mentioned see Brongersma (1972), where detailed lists are given.

Macaronesian Waters

Macaronesian Waters (MW) comprise the seas around and between the Azores, the Madeira Archipelago, the Selvagens Islands, and the Canary Islands. As in EAW,

L. D. BRONGERSMA

the MW do not harbor breeding populations of turtles. The same five species are involved. *Caretta caretta* is by far the most common species; *Lepidochelys kempi, Chelonia mydas, Eretmochelys imbricata*, and *Dermochelys coriacea* are known from relatively few records.

The Loggerhead Turtle, Caretta caretta *(L.)*

Caretta caretta occurs in large numbers around the Azores, and in the seas north of these islands (to about 42°N) (Brongersma 1971:106; 1972: Chart 5). It has been stated (U.S. Department of Commerce 1978:64) that I estimated 4,000 loggerheads were slaughtered annually in the Azores, but this is a misunderstanding. The estimate of 4,000 is that of Mr. Dalberto Teixeira Pombo (of Sta. Maria, Azores) for the number of loggerheads killed annually in Madeira (*in litt.,* 16.vi. 1969). However, in this letter Mr. Pombo indicates that in the Azores the demand for the loggerhead for human consumption was increasing. Large numbers of *C. caretta* are also found around Madeira. Following a visit to Madeira in 1967, I estimated the number of loggerheads slaughtered annually to be 1,000 (and probably more); Mr. Pombo arrived at a number of 4,000. Recently, when I again visited Madeira, I found the exploitation still going on, in a somewhat different manner. During my 1967 visit, I got the impression that loggerheads were slaughtered for human consumption and that as a side-line the cleaned-out turtles were stuffed to be sold to tourists. The slaughtering was more or less a home industry at the time. During my visit in July 1979, it appeared that the turtles were not slaughtered primarily for human consumption, but for the tourist trade. In 2 souvenir shops, 1 and 2 loggerheads, respectively, were on sale for 800 escudos each (about US$20). In another shop (annex workshop), 29 loggerheads were drying on the roof. As the number of cruise ships visiting Madeira had declined since 1967, local sales have decreased; this is compensated, however, by exporting stuffed turtles to the Canary Islands where they can be sold to tourists. The turtles vary in carapace length from about 25 to 50 cm. Larger loggerheads may occasionally be slaughtered locally for human consumption, but these are not stuffed, as tourists will not take home cumbersome adult loggerheads. The turtles are caught by fishermen at night from fairly small boats while fishing with hook and line for the *espada preta* or scabbard fish (*Aphanopus carbo* Lowe). Should they chance upon loggerheads during the night, they are captured whenever possible. Slaughtering and stuffing is not a home industry any more. Behind the Funchal fish market only 1 man was at work on turtles. On 9, 11, and 17 July, I saw 9, 3, and 8 loggerheads, respectively; I was told that none was obtained on 10 July. Within 2 months (May and June) the man had dealt with about 200 turtles (all

loggerheads); in winter he gets fewer specimens. I do not know whether there are other slaughterers around Funchal. My informants believed that 4,000 loggerheads annually would be too high a current estimate, but it was believed that 2,000 a year might be correct.

Caretta caretta is also known from the Selvagens Islands (carapaces found on the island in the Rijksmuseum van Natuurlijke Historie [RMNH], Leiden, the Netherlands). The species is found in the Canary Islands. Steindachner (1890:305) recorded the presence of *C. caretta* off Gran Canaria and Tenerife, and he stated that it was somewhat more common along the coasts of Lanzarote. I saw 2 small specimens taken in the waters off Gran Canaria in 1966. On a visit to Tenerife in 1975, I learned from a fisherman that *C. caretta* was more or less regularly taken in summer, all small specimens. He showed me a juvenile (carapace about 23 cm) which he had at home in a bathtub. On 26 March 1975, while on a ferry from Los Christianos, Tenerife, to San Sebastian, Gomera, I saw a small *C. caretta* off the Tenerife coast. The Naturmuseum Senckenberg, Frankfurt a. Main, Federal Republic of Germany, has a loggerhead shell from Punta Retinga, Island of Hierro (Brongersma 1968:129). In an advertisement in the periodical "Delphin, Revue der Unterwasserwelt" (August 1975:26) to recruit scuba divers to come to Hierro, it is stated "dort wo noch Seeschildkröten zu sehen sind" [where one still can see marine turtles]. In a swimming pool near Jaméo del Agua, Lanzarote, there were 6 turtles, all loggerheads as far as I could see.

Whether there is any exploitation of loggerheads in the Canary Islands, I do not know. Considering the small size of the specimens found in the area, exploitation for human consumption does not seem likely. The fact that small specimens are captured fairly regularly in summer may have led to some exploitation for the tourist trade. However, since stuffed turtles are imported from Madeira for the souvenir trade, it is unlikely that large numbers are captured in the Canary Islands.

The Kemp's Ridley Turtle, Lepidochelys kempi *(Garman)*

A very young specimen (carapace 99.7 mm) of *Lepidochelys kempi* was found on Corvo, Azores, in 1913 by Col. F. A. Chaves (Mus. Monaco, no. 2660; Deraniyagala 1938:540; 1939:1–4, 2 figures; Brongersma 1972:265, Figure 30; Pritchard and Marquez 1973:26); 3 specimens have been recorded from Madeira (summer 1949, Museu Municipal, Funchal, Madeira, no. 3978; May 1950, Naturmus. Senckenberg, no. 41057; summer 1950, Mus. Mun., no. 3194; Brongersma 1968:133; 1972:266, Figure 31). The species has not been recorded from the Selvagens and Canary Islands.

The Green Turtle, Chelonia mydas (L.)

The green turtle has been found in the Azores and in Madeira, but there are very few records. They are represented in the Museu Machado, Ponte Delgado, São Miguel, Azores, and in the Museu Municipal, Funchal, Madeira. Old records, like the one by Drouët (1861:129) who states that *Chelonia mydas* [sic] is not rare in the Azores where it provides the inhabitants and the seafarers with a healthy and rather agreeable food, may have been based on a misidentification of *C. caretta*, probably because most people considered the green turtle the only edible one. The green turtle has been recorded from Madeira, by Mertens (1935:89), Sarmento (1948:262), Maul (1948:295), and Brongersma (1968:134, Museu Municipal, Funchal, no. 22242).

There is as yet no record of *Ch. mydas* from the Selvagens Islands, and but a single record from the Canary Islands (Tenerife; Duméril, and Bibron 1835:544).

The Hawksbill Turtle, Eretmochelys imbricata (L.)

This species has been found in the Azores and in Madeira, and it is represented in the collections of the Museu Machado, Ponte Delgado, and of the Museu Municipal, Funchal, but it must be considered very rare in the area. From Madeira it has been mentioned by Mertens (1935:89), Maul (1948:295), and Brongersma (1968:135). It has been found in the Selvagens Islands (specimen in RMNH, Leiden); I do not know of any record from the Canary Islands.

The Leatherback Turtle, Dermochelys coriacea (L.)

Dermochelys coriacea has been observed in the Azores and in Madeira, but there are few records, for example, 1 female captured off Vila Franco do Campo, São Miguel, Azores, 31.v.1966 (Brongersma 1970:333), and 1 taken off Mosteiros, west coast of São Miguel, 9.ix.1977, Museu Machado, Ponte Delgado, São Miguel. From Madeira, the leatherback or leathery turtle was recorded by Sarmento (1948:264), Maul (1948:295), and by Brongersma (1968:135, female, south coast of Madeira, 19.vii.1955, Museu Municipal, Funchal, 5952; Pritchard 1971:32). Sarmento (1948:262) refers to 6 leathery turtles having been seen near the coasts of Madeira in the preceding 10 years. As yet there are no records from either the Selvagens or the Canary Islands, but it is very likely that occasionally or rarely the species appears in these islands.

West African Waters

West African Waters (WAW) comprise the waters along the west coast of Africa, from the Straits of Gibraltar to Cape Agulhas (South Africa), including the Cape Verde Islands, the Bissagos Islands, and other islands close to the African coast, as well as the islands in the Gulf of Guinea, such as Fernando Póo, São Thomé, Principe and Rolas, and St. Helena. But little reference will be made to the turtles of Angola, Southwest Africa, and South Africa, as these countries will be dealt with at more length in the contribution of the Southwest Indian and Southeast Atlantic Oceans by Hughes (this volume).

Recently, Ross et al. (1978) reviewed the present status of sea turtles. The maps indicating the principal nesting beaches indicate 2 nesting sites on the west coast of Africa: those of *Lepidochelys olivacea* in Senegal and Angola. No nesting is indicated on the west coast of the African continent, or on the islands for *Caretta caretta, Chelonia mydas, Eretmochelys imbricata,* and *Dermochelys coriacea*. Indeed recent information about the presence and nesting of the various species along the west coast of Africa is very scarce. Where such information is lacking, it may be of use to point to information obtained in the past. It is hoped that this may stimulate people to obtain information about the presence of turtles, their nesting, and also about the exploitation and other dangers that threaten various populations of turtles in their countries. Of course, records do exist (for example, those of Loveridge and Williams 1957:484, 489, 494, 497, 502), but sometimes the author and date of the records are hard to trace from lists. Therefore, I give fairly detailed lists of records for the 5 species found in WAW with references to literature (if possible to the original reference; sometimes later references are given).

There was a time when *Lepidochelys olivacea* was not recognized as a distinct species; the specimens were placed with *Caretta caretta*. One of the best known examples is Gadow's paper (1899) in which he developed a theory about the reduction of the number of scutes in the carapace, during an individual's growth based upon a comparison of hatchlings, which in fact were *L. olivacea,* and halfgrown and adult specimens, which were *C. caretta*. Accepting *C. caretta* and *L. olivacea* as distinct species, the following records must be transferred from *C. caretta* to *L. olivacea*: San Pedro, Ivory Coast, by Loveridge and Williams (1957:494), and that from Cameroon by Nieden (1910:3, 5). Boettger (1888:18) recorded a specimen from Banana, Zaïre as *Thalassochelys olivacea*, and Bocage (1895:6–7) transferred it to *Thalassochelys caretta* (i.e., *C. caretta*), but it should be returned to *L. olivacea*. A hatchling figured by me as *C. caretta* (Brongersma 1941: Figure 5c) is in fact a hatchling of *L. olivacea* (from Liberia). It may well be that the record of *C. caretta* from Cameroon (Tornier 1902:665) was also based upon *L. olivacea*.

Observations of turtles in WAW at some distance from the coast were made by the Guinean Trawling Survey (GTS); I have indicated as best I could, the

L. D. BRONGERSMA

positions of the various stations.

Lepidochelys kempi has not been recorded from WAW. Pasteur and Bons (1960:101) consider it possible that the species might accidentally reach the Moroccan coast. As the species is known to enter the Mediterranean (Brongersma and Carr, in prep.) it might strand on the Moroccan coasts bordering the approaches to the Straits of Gibraltar. Bons (1972:120) lists *Lepidochelys olivacea kempii* under the species that probably are to be found in Morocco, or which should be looked for.

The Loggerhead Turtle, Caretta caretta (L.)

The following localities have been recorded. Morocco (Bons 1972:10); Menasra (north of Kenitra, about 34.36°N, Pasteur and Bons 1960:27); Mogador (Pellegrin 1912:256); plage Blanche (the coast between the former Spanish possession of Ifni and Rio de Oro, Pasteur and Bons 1960:100). Rio de Oro (Carr 1952:382). Mauritania: western part of the Baie du Lévrier (about 20.40 to 21.10°N); Banc d'Arguin (Maigret 1975:118–119 not seen), Maigret 1977:11–12; Trotignon and Maigret 1977:27–28, (not seen). Cape Verde Islands (Bertin 1946:91, 105; Bannerman and Bannerman 1968:54); São Vicente (Bocage 1896:66, a very young specimen); São Vicente and Sal (Angel 1937:1696); Boa Vista (Schleich 1979:12). Senegal: Hann, Joal, Fadiouth (Cadenat 1949:19); Dakar, plage des Almadies, Gorée, Kayar (Cadenat 1957: 1371, 1373); Cap Vert (Dakar, Gorée, Kayar) (Maigret, 1977:11, 12); îles de la Madeleine, off Dakar (Dupuy and Maigret 1979:4). Guiné-Bissao; GTS II, LR, Sta. 2/2, 11.37°N, 17.01.5°W (Williams 1968:98). Sierra Leone: GTS I, LR, Sta. 9/2, 07.54°N, 11.37°W (Williams 1968:98). Ghana: GTS I, LR, Sta. 30/2a, and 32/1b (Williams 1968:98). Gabon (Loveridge and Williams 1957:494, perhaps *L. olivacea?*). Congo: Pointe Noire (A. Crosnier, *in litt.,* 16.iii.1968). Its presence in Angolan waters has been reported by Hughes, Huntley, and Wearne (1973:58).

J. Blache (*in litt.,* 22.ii.1968) states that turtles are rare around the island of Gorée, and A. Blanc (*in litt.,* 9.iii.1968) also is of the opinion that *C. caretta* is rare, in Senegal.

The species has been reported by Pasteur and Bons (1960:27) to breed on the Moroccan coast as far north as Menasra, and these authors (1960:101) infer that it also breeds on the plage Blanche. Bocage (1896) accepts the very young specimen from São Vicente, in the Cape Verde Islands, as proof of the species breeding there. In Senegal, *C. caretta* nests or used to nest on Gorée Island, on the beach of Almadies (near Dakar), and it was found to nest on 1 of the very small beaches of the îles de la Madeleine (off Dakar). This is the only nesting recently observed (Dupuy and Maigret 1979:4).

The Olive Ridley Turtle, Lepidochelys olivacea (Eschscholtz)

As mentioned above various old records of *Caretta caretta* in reality were based upon *Lepidochelys olivacea.*

The species has been recorded from the following localities: Mauritania: Port Etienne (21°N) (Carr 1957: 48, 49: Villiers 1958:186; Pasteur and Bons 1960:101). Senegal: Dakar, Hann, Gorée, Guet N'Dar, N'gaparo, Joal (Cadenat 1949:17, 19, 33, 35; 1957:1370, 1374; Carr 1957:48, 49, Figure 1, hatchling from Dakar; Loveridge and Williams 1957:497; Maigret 1977:12). Liberia: 2 hatchlings (Brongersma 1961:27); Grand Cape Mt. (Brongersma 1961:27); north of Point Marshall (Carr 1957:49, 50). Ivory Coast: San Pedro, hatchlings (Deraniyagala 1943:82, 92; Carr 1957:49; Loveridge and Williams 1957:497). Ghana: Tema (leg. Irvine, Brit. Mus. (N.H.) 1940.2.23.3; Carr 1957:49; Loveridge and Williams 1957:497: Tenia). Cameroon: Victoria (Nieden 1910:5, *C. caretta*; Carr 1957: 48, 49, 50). Gabon (Duméril, 1860:170; Carr 1957:49). Congo: Pointe Noire (A. Crosnier, *in litt.,* 16.iii.1968). Zaïre: Banana (Boettger 1888:18; Bocage 1895: 6, *Thalassochelys caretta*); Banana and Moanda (Carr 1957: 48, 49); "Congo" (Brongersma 1961:27). Angola: Ambriz (Brongersma, 1961:27), Luanda (Hughes, Huntley, and Wearne 1973:58).

Maigret (1977:12) writes that probably the species breeds on the Senegal coast, but that there is no confirmation of this. The photograph of a hatchling from Dakar (Carr 1957, Figure 1) definitely points to nesting in the area. Likewise the hatchlings from Liberia, Ivory Coast, Ghana, Cameroon, Zaïre, and Angola point to nesting taking place along the coastline from Senegal to Angola; Hughes, Huntley, and Wearne (1973) suggest that nesting takes place near Luanda, Angola. Deraniyagala (1943:92) writes "Probéguin Côte d'Ivoire," thus suggesting that Probéguin is the locality; a list received from the Paris Museum, gives San Pedro as the locality and Probéguin as the collector.

The Green Turtle, Chelonia mydas (L.)

The species has been recorded from the following localities: Morocco: Pasteur and Bons (1960:99) are convinced that incidentally the green turtle will be found along the Moroccan coast. They add that it may nest from the "plage Blanche" (the coast between the former Spanish possessions of Ifni and Rio de Oro) southwards; Bons (1972:120) considers it very probable that the species occurs on the coast of Tarfaya (the area of Cape Juby) and even farther to the north. Mauritania: Gulf of Arguin (Pasteur and Bons 1960:99; Parsons 1962:45); eastern part of the Baie du Lévrier, Banc d'Arguin, breeding on Pointe d'Arguin (Maigret 1975, not seen; Trotignon and Maigret 1977, not seen; Maigret 1977:10, 11, and 1978). Cape Verde Islands: São Vicente (Loveridge and Williams 1957:484); for records by Parsons (1962:43–45), see discussion below. Senegal: Langue de Barbarie (nesting), Saint Louis, Gandiole, Cap Vert, Kayar, plage des Almadies (nesting), Dakar, Hann, Somone (nesting), M'Bour, Joal (nesting), Tare (Cadenat 1949:17, 22, and 1957:1369, 1371, 1374; Maigret 1977:10; 1978), Guiné-Bissao (Bocage 1866:5; 1896:74; Monard 1940:147); GTS II, LR, Sta. 4/5, 10.15°N, 16.34°W (Williams 1968:98). Sierra Leone: Turtle Island off Sierra Leone (J. Tomlinson, in: Parsons 1962:45); Sussex (Phaff 1964:15; 1967:49, "soepschildpad", figures). Liberia (Büttikofer 1884:31; 1890, vol. 2:147, 438, 439; Johnston 1906:819, 833); near Robertsport (nesting) (Büttikofer 1890, vol. 1:266, 267, 302); Monrovia, Robertsport (Loveridge and Williams 1957:484). Ghana (Irvine 1947:309; Parsons 1962:45); GTS I, LR, Sta. 32/1a, 05.38°N, 00.07°W (Williams 1968:98); a hatchling from Tema, Ghana, leg. Irvine, Brit. Mus. (N.H.). Togo (Villiers 1958:331, lists a vernacular name from Togo). Fernando Póo: south coast (Eisentraut, 1964:472–74, nesting). São Thomé (Greef 1885:49, nesting). Ilha do Principe (Bocage 1903:52, specimen collected in 1881); Ilheo das Rolas (Greef 1885:49, nesting). Congo: Loango (Pechuël-Loesche 1882:277, nesting). Cabinda (Angola): Chinchoxo (Loveridge and Williams 1957:484). Zaïre: Banana (Boettger 1888:17). Angola: Luanda (Bocage 1866:5, 1895:6); Bahia dos Tigres (Monard 1937:146, formerly frequently nesting, but became rare, A. J. Vilela 1923); Hughes, Huntley, and Wearne (1973:58, nesting south of Luanda). Southwest Africa: a specimen caught locally is in the Museum at Swakopmund. South Africa, St. Helena (A. Loveridge, *in litt.*, 1.xi.1968, sometimes nesting).

In Senegal, *Ch. mydas* is the commonest of the sea turtle species (Cadenat 1949:17, 22). The turtles are captured more or less accidentally when they become entangled in nets used to catch sharks. The remark by Parsons (1962:45) that "A small turtle industry is said to have existed in recent years at Requins in Senegal" is erroneous in so far as "Requins" does not refer to a locality, but to the "filets à requins" (nets to catch sharks). However, the figure of about 70 turtle shells in the yard of a sharks-fishery at Joal (Cadenat 1949, Figure 1; Villiers 1958, Figure 57) shows that there was some exploitation. Maigret (1978) remarks that according to fishermen of Glandiole (St. Louis region) 20 years ago there were about 300 nests each year on the beaches at the Langue de Barbarie; in 1975 only 3 nestings were observed.

Parsons (1962:43–45) mentions the islands of Sal, Boavista, Maio, Fogo, and São Tiago as localities, and he adds that the green turtle nests (nested) on Sal, Boavista, and Maio. However, Angel (1937:1696), Bertin (1946:91, 105); Bannerman and Bannerman

(1968:54), and Schleich (1979) mention only *Caretta caretta* and *Eretmochelys imbricata* from the Cape Verde Islands. Cadenat (in Parsons, 1962:45) states that on several trips to the islands he had never seen a green turtle. Parsons based his statement on the evidence of early voyagers visiting the Cape Verde Islands, for example, Dampier (1697, ed. of 1968:60; Bannerman and Bannerman 1968:14), but Dampier stated only that in the months of May to August "a sort of small Sea-Tortoise came hither to lay their Eggs." The small size of the nesting turtles makes it unlikely that Dampier saw green turtles nesting in the Cape Verde Islands. The scope of the present survey does not allow a complete discussion of this matter. However, it seems more probable that Dampier saw hawksbills (known to occur in the Cape Verdes) or olive ridleys (not yet recorded from the islands). The fact that the turtles were collected as a source of meat may have led to misidentifying them as green turtles. A careful study of the original sources is necessary before a definite conclusion can be reached about *Ch. mydas* once nesting in great numbers in the Cape Verde Islands.

From the list of records it is clear that *Ch. mydas* breeds in many places: perhaps in southernmost Morocco, but definitely in Mauritania, Senegal, Sierra Leone, Liberia, Ghana, Fernando Póo, São Thomé, Principe, Rolas, Congo, Zaïre, Angola, and on rare occasions on St. Helena.

The Hawksbill Turtle, *Eretmochelys imbricata (L.)*

The following localities have been recorded: Morocco? (Pasteur and Bons 1960:100, very probably as an accidental visitor; Bons 1972:120). Mauritania: Banc d'Arguin (Loveridge and Williams 1957:489); between Cape Timiris (19.23°N) and St. Louis in Senegal (16.16°N) (Maigret 1977:11). Cape Verde Islands (Schleich 1979:12); Togo (Boulenger 1905:197; Loveridge and Williams 1957:489); Senegal: Hann, Joal (Cadenat 1949:22; Loveridge and Williams 1957:489); St. Louis, Hann, Casamance (Cap Skirring), Betenti (Parc national du delta du Saloum) (Maigret 1977:11); îles de la Madeleine (Maigret 1978:4). Gambia, Brit. Mus. (N.H.), leg. Tucker, no. 45.12.29.12. Sierra Leone: Bonthé (Loveridge and Williams 1957:489); Sussex (Phaff 1964:15, 16, figure; 1967:49, figure, "havikssnavel"). Liberia (Büttikofer 1884:31; 1890, vol. 2:438; Johnston 1906:833); Angel River (Loveridge and Williams 1957:488, 489); Liberia, 1884, leg. Büttikofer and Sala, RMNH 8104, 3 hatchlings. Ghana: Anamabu (Irvine 1947:311); Tenia (= Tema) (Loveridge and Williams 1957:489); Gold Coast, leg. Irvine, Brit. Mus. (N.H.), 1930.6.9.38 (head); G.T.S. I, TH, Sta. 32/a, 5.40°N 0.13°E (Williams 1968:98). Togo (Villiers 1958:329, vernacular name). Cameroon: Longji (Longuy) (Nieden 1910:5). Gabon (Loveridge and Williams

L. D. BRONGERSMA

1957:489). Fernando Póo (south coast, Eisentraut 1964:471). São Thomé (Greef 1885:49; Mus. nat. Hist. nat., Paris, leg. Almada Negreiros; Loveridge and Williams 1957: 489). Rolas (Greef 1885:49; Loveridge and Williams 1957:489). Angola (Hughes, Huntley, and Wearne 1973:58). South Africa (Loveridge and Williams 1957:489). St. Helena (A. Loveridge, *in litt.*, 1.xi.1968).

Maigret (1977:11) believes that *E. imbricata* nests in Mauritania and Senegal, but there are as yet no exact observations. Büttikofer (1884:31; 1890, vol. 1:267) mentions its breeding in Liberia, and the 3 hatchlings in the Leiden Museum support this statement. Similarly we may accept the breeding records for Fernando Póo (Eisentraut 1964:471), São Thomé and Rolas (Greef 1885:49).

Maigret (1977:14) and Schleich (1979:12) mention the exploitation of *E. imbricata* in the Cape Verde Islands. F. Reiner, Museu do Mar, Cascais, recently told me that it is heavily exploited in São Thomé (in a way similar to *C. caretta*'s exploitation in Madeira) for the souvenir trade.

The Leatherback Turtle, Dermochelys coriacea *(L.)*

The following localities have been recorded: Morocco (Bons 1972:110): off Casablanca, Cap Cantin (= Meddouza) (Pasteur and Bons 1960:29). Senegal: Hann (Cadenat 1949:17, 35 Figure 1); Hann, Rufisque (Loveridge and Williams 1957:513); Hann, Joal (Villiers 1958:191, Figure 164); Rufisque, Bargny, Joal, Langue de Barbarie, Pointe de Sangomar (nesting), south of Palmarin (nesting) (Maigret 1977:12); coast and peninsula of Sangomar (Dupuy and Maigret 1979:5, nesting); A. Blanc (*in litt.*, 9.iii.1968). Liberia: (Büttikofer 1884:31; 1890, vol. 1:267; Johnston 1906:819, 833; Brongersma 1970:332, pl. xi, hatchling, 13.iv.1893); Mahfa River (Büttikofer, 1890, vol. 2:438, nesting; Loveridge and Williams 1957:501, 503). Ivory Coast (Villiers 1958:192, Figures 169, 170, hatchling). Ghana: (Irvine 1947:312); Tenia (= Tema) (Loveridge and Williams 1957:502, 503); [St. George d'] Elmina, Tema, "Gold Coast" (Brongersma 1970:332, pl. xii, hatchlings); Salt Pond (Pritchard 1972:148; 1976:752, Figure 3, specimen tagged at Bigisanti, Surinam). Togo: (Matschie 1893:208); Sebbe (Sebe or Zebe) (Tornier 1901:66, 3 specimens still within the egg membranes; Loveridge and Williams 1957:503). Gabon (Loveridge and Williams 1957:503). Zaïre: "Congo" (leg. Kamerman, viii, 1883, RMNH, 5477, taken from the egg or just hatched, remnants of yolk sac present). Angola: Rio Dende (hatchling, leg. F. Reiner, 1970, Museu do Mar, Cascais, Portugal); some 200 km of coast S. of Luanda (Hughes, Huntley, and Wearne 1973:58, nesting; Huntley 1978:1374).

By direct observations or by the presence in collections of hatchlings (or specimens taken from the egg) nesting has been demonstrated to take place in Senegal, Liberia, Ivory Coast, Ghana, Togo, Zaïre, and Angola.

Species Unknown

J. A. Sayer of FAO reported nesting activity in the area of Ouidah, Benin; the species was not identified (Anne Meylan, *in litt.*, 12.x.1979).

Conclusions

European Atlantic Waters have a population of turtles that move in when the water temperature has risen in summer, to leave again (or perish) when the cold sets in. A female *Dermochelys coriacea* came ashore near the Pointe d'Arçay, Vendée, France, on 17 August 1978, at 18 hrs; it was disturbed by the crowd that had assembled around it, and it returned to the water. Also, in the summer of 1938, at twilight, with rising tide, a *D. coriacea* came ashore at the beach of Vert-Bois, Island of Oléron, Charente-Maritime, France (Duron-Defrenne 1978:75, 83). Interesting as these observations are, there is no reason to assume that successful nesting can take place on the French Atlantic coast.

The population in Macaronesian Waters consists of turtles that stay in the area for a part of their life; turtles do not breed in the area, but they do get there when they are young, moving away again when they become adult. For both areas (EAW, MW) the question arises: whence do these turtles come, and where do they go? The occasional presence of *Lepidochelys kempi* in EAW and MW (as well as the Mediterranean) can only be explained by assuming a migration from the Gulf of Mexico through the Florida Strait into the Atlantic, where they may travel with the currents to Europe, to the Azores, and to Madeira. Likewise, young *Caretta caretta* might come from American beaches. A chart showing the positions of turtle sightings on the high seas (Brongersma 1972, Chart 5) shows a large concentration of records in the Azores area, around Madeira and between these islands and the approaches to the Straits of Gibraltar. The fact that sightings are much less common between 30° and 50° W and are lacking between 50° and 60°W may seem to plead against the assumption that young turtles move in a more or less steady flow from the Western to the Eastern Atlantic. However, one must take into account that the majority of records stem from merchantmen, who follow more or less fixed shipping lanes; the waters of the Azores and Madeira carry heavy shipping traffic, and correspondingly more records are received from that area.

That turtles can and do cross the Atlantic from west to east is shown by a *C. caretta* tagged and released (head-started) by Ross Witham on Hutchinson Island

(Florida, United States) and recently recaptured off Porto Moniz, Madeira; also by a head-started *Ch. mydas* (from the same source) having been washed ashore on the island of Flores in the Azores (Maul, Witham, and Brongersma, in prep.). Still, one cannot ignore that *C. caretta* breeds (or used to breed) on the Moroccan west coast northwards to Menasra (about 34°36′N), and that turtles are known to move through the Straits of Gibraltar in both directions (east to west and west to east). In September they are abundant on the Atlantic coast of Southern Spain; in July, August, and September many are captured on the Mediterranean coast near the Straits of Gibraltar (Dr. Julio Rodríguez-Roda, Instituto do Investigaciones Pesqueras, Cadiz, Spain, *in litt.* 23.ii.1968). Some of these turtles might succeed in reaching the Azores, Madeira, and the Canary Islands.

To solve these problems, it is necessary to start tagging programs. The difficulty is that in this case one should try to obtain turtles from fishermen, and then tag and release the turtles. I understand that some tagging is done by Mr. Dalberto Teixeira Pombo on Santa Maria Island, Azores. It would be of great value if tagging could be extended to other islands in the Azores and to Madeira. One would also like to know to what extent, in winter, the turtles of MW may move to warmer parts of the ocean.

Although there is a heavy exploitation of *C. caretta* in Madeira, it seems that the numbers occurring in the area have not diminished, and apparently the breeding stock from which they stem has not yet been affected.

West African waters harbor breeding populations of turtles. Our knowledge of nesting in this area is limited mostly to observations on individual nestings, or on hatchlings having been preserved in collections. Very little is known about the numbers of females that come to nest; the exceptions are the observations by J. Maigret in Senegal, and those by G. R. Hughes in Angola. Large stretches of coastline remain unexplored. On his search for the possible occurrence of ridleys in northwest Africa, Carr (1957:54) wrote: "But a flight along the coast from Mauritania to Morocco showed the whole shore there to be practically uninhabited for hundreds of miles and a place where ridleys might abound, with nobody the wiser"; substituting "turtles" for "ridleys" the statement still holds. Not only would one like to know much more about nesting sites, the species nesting there, and the number of females that come ashore but also whether migrations take place and, if so, over what distances. A remarkable record is that of a female *D. coriacea* which had been tagged in Surinam, and which was recaptured in Ghana (Pritchard 1972:148; 1976:752).

A decline in the number of turtles may be due in part to over-exploitation in past centuries. Today intensive exploitation for human consumption does not seem to take place in the Eastern Atlantic. Of course, people living on the coast will use some turtles and their eggs but the turtles can withstand this sort of exploitation. Exploitation increases when people move into coastal areas after continuous drought makes food scarce in the hinterland; this has been observed around St. Louis in Senegal (Maigret 1977:13). The exploitation of *E. imbricata* for tortoiseshell and for turtles stuffed as souvenirs for tourists takes place in the Cape Verde Islands (Maigret 1977:14), and very heavily in São Thomé (F. Reiner, personal communication). Maigret (1977:14) considers the exploitation of shells of the various species of turtles for the souvenir trade as a danger, which may have already been realized in the Dakar area and at Cap Skirring. A very serious danger is the urbanization of coastal areas and the use of the beaches by tourists: the nesting sites are destroyed, the eggs are collected for consumption, and the souvenir trade is stimulated (Maigret 1977:13; Dupuy and Maigret 1979:2).

To obtain a better insight into the turtle populations along the west coast of Africa, still existing nesting beaches will have to be located and a record made of the species nesting there, the numbers of females that come to lay, and the results of the nesting. Threats to the survival of these populations must also be identified. A tagging program may help to get some idea of possible migrations.

Literature Cited

Angel, F.
1937. Sur la faune herpétologique de l'archipel du Cap-Vert. *C. R. XIIᶜ Congr. Int. Zool., Lisbonne* 1935:1693–700.

Bannerman, D. A., and W. M. Bannerman
1968. *History of the Birds of the Cape Verde Islands: Birds of the Atlantic Islands. 4.* Edinburgh: Oliver and Boyd.

Bertin, L.
1946. Le peuplement des îles atlantides en vertèbres hétérothermes. *Mem. Soc. Biogeogr.* 8:87–107.

Bocage, J. V. Barboza du
1866. Lista dos reptis das possessoes portuguezas d'Africa occidental que existen no Museu de Lisboa. *Jorn. Sc. math. phys. nat., Ac. Sci. Lisboa* 1:1–48.
1895. *Herpétologie d'Angola et du Congo.* Lisbonne: Impr. Nat.
1896. Reptis de algunas possessoes portuguezas d'Africa que existen no Museu de Lisboa. *Jorn. Sc. math. phys. nat., Ac. Sci. Lisboa* 4:65–104.
1903. Contribution à la faune des quatres îles du Golfe de Guinée. *Jorn. Sc. math. phys. nat., Ac. Sci. Lisboa* 7:25–59.

Boettger, O.
1888. Materialien zur fauna des untern Congo II. Reptilien und batrachier. *Ber. Senckenb. naturf. Ges. Frankfurt a.M.* 1888:3–108.

Bons, J.
1972. Herpétologie Marocaine. I. Liste commentee des amphibiens et reptiles du Maroc. *Bull. Soc. Sc. Nat. Phys. Maroc* 52:107–26.

Boulenger, G. A.
1905. Report on the reptiles collected by the late L. Fea in West Africa. *Ann. Mus. Civ. Stor. nat., Genova.* 2:196–216.

Brongersma, L. D.
1941. De Huid en de Huidspieren. In *Leerboek der Vergelijkende Ontleedkunde van de Vertebraten* (2d ed.), J. E. W. Ihle, ed. pp. 27–94. Utrecht: A. Oosthoek.

1961. Notes upon some sea turtles. *Zool. Verh. Leiden* 51:1–46.

1968. Notes upon some turtles from the Canary Islands and from Madeira. *Proc. Kon. Nederl. Akad. Wet., Amsterdam, C.* 71:128–36.

1970. Miscellaneous notes on turtles. III. *Proc. Kon. Nederl. Akad. Wet., Amsterdam, C.* 73:323–35.

1971. Ocean records of turtles (North Atlantic Ocean). *IUCN Publications, New Series, Supplemental Paper,* 31:103–8.

1972. European Atlantic turtles. *Zool. Verhand. Leiden,* 121:1–318.

Büttikofer, J.
1884. Mededeelingen over Liberia. Resultaten van eene onderzoekingsreis door J. Büttikofer and C. F. Sala in de jaren 1879–1880. *Bijbl. Tijdschr. Aardrijksk. Gen.* 1884:1–147.

1890. *Reisebilder aus Liberia.* 1. Leiden: E. J. Brill.

Cadenat, J.
1949. Notes sur les tortues marines des côtes du Sénégal. *Bull. Inst. Fr. Afr. Noire* 11:16–35.

1957. Observations de cétaces, siréniens, chéloniens et sauriens en 1955–1956. *Bull. Inst. Fr. Afr. Noire* 19:1358–75.

Carr, A.
1952. *Handbook of Turtles.* Ithaca, New York: Cornell University Press.

1957. Notes on the zoogeography of the Atlantic sea turtles of the genus *Lepidochelys. Rev. Biol. Trop.* 5:45–61.

Dampier, W.
1968. *A New Voyage Round the World, with an Introduction by Sir Albert Gray. Reprint 1697 ed. with New Introduction by P. G. Adams.* New York: Dover Publications.

Deraniyagala, P. E. P.
1938. The Mexican loggerhead turtle in Europe. *Nature* (London) 142:540.

1939. The distribution of the Mexican loggerhead turtle, *Colpochelys kempi* Garman. *Bull. Inst. Ocean. Monaco* 772:1–4.

1943. Subspecies formation in loggerhead turtles (Carettidae). *Spolia Zeyl., Geol. Zool. Anthro.* 23:79–92.

Drouët, H.
1861. *Eléments de la Faune Acorienne.* Paris: J. B. Baillière et Fils, J. Rothschild.

Dumeril, A.
1860. Reptiles et poissons de l'Afrique Occidentale. Etude précédée de considerations générales sur leur dis-tribution géographique. *Arch. Mus. nat. Hist. nat.. Paris* 10:137–268.

Duméril, A., and G. Bibron
1835. *Erpétologie Générale ou Histoire Naturelle Complète des Reptiles, 2.* Paris: Roret.

Dupuy, A. R., and J. Maigret
1979. La protection des écosystèmes cotiers, un exemple concret: les parcs nationaux du Sénégal. Seminaire UNESCO sur les écosystèmes cotiers en particulier lagunes cotières et estuaires de la côte ouest de l'Afrique. Mimeo, 6 pp. Dakar, Senegal, 11–15 July 1979.

Duron-Defrenne, M.
1978. Contribution à l'étude de la biologie de *Dermochelys coriacea* (Linne) dans les Pertuis Charentais. Thesis, Université de Bordeaux.

Eisentraut, M.
1964. Meeresschildkroten an der kuste von Fernando Póo. *Natur u. Museum* 94:471–75.

Ferreira, J. Bettencourt
1893. Revisao dos reptis e batrachios de Portugal. *Jorn. Sci. Math. Phys. Nat., Acad. Sci. Lisboa* 8:19–27.

Gadow, H.
1899. Orthogenetic variation in the shells of chelonia. *A. Willey's Zool. Res.* 3:207–22.

Greef, S. R.
1885. Ueber die fauna der Guinea-Inseln S. Thome und Rolas. *Sitz. ber. Ges. Beford. ges. Naturwiss. Marburg* 1884:41–79.

Holgersen, H.
1960. Laerskilpadde sed Karmoy. *Stavanger Mus. Arb.* 1959:131–38.

Hughes, G. R.; B. Huntley; and D. Wearne
1973. Sea turtles in Angola. *Biological Conservation* 5:58–59.

Huntley, B. J.
1978. Ecosystem conservation in southern Africa. In *Biogeography and Ecology of Southern Africa,* M. J. A. Werger and A. C. van Bruggen, eds. pp. 1333–84. Monographiae Biologicae, 31. The Hague: W. Junk.

Irvine, F. R.
1947. *The Fishes and Fisheries of the Gold Coast.* London: Crown Agents for the Colonies.

Johnston, H.
1906. *Liberia,* vol. 2. London: Hutchinson and Company.

Konstantinov, K.
1965. [Turtle in Barents Sea.] *Priroda* 3:111.

Loveridge, A., and E. E. Williams
1957. Revision of the African tortoises and turtles of the suborder Cryptodira. *Bulletin of the Museum of Comparative Zoology* 115:163–557.

Maigret, J.
1975. Notes et informations: les tortues de mer du Banc d'Arquin. *Bull. Lab. Peches Nouadhibon* (Mauritanie) 4:118–19 (not seen).

1977. Les tortues de mer au Sénégal. *Bull. AASNS* 59:7–14.

1978. Sea turtles nesting on the coast of Senegal. *IUCN/SSC Marine Turtle Newsletter,* 8:4.

Matschie, P.
1893. Die reptilien und amphibien des Togogebietes. *Mitth. Forschungsreisenden u. Gelehrten a. d. deutsch. Schutz-

geb., (*Wiss. Beih. D. Kolonialbl.*), 6:207–15.

Maul, G. E.
1948. Lista sistematica dos mamiferos, aves, repteis e batraquios assinalados no Arquipelago da Madeira. In
Vertebrados da Madeira, 1, (2d ed.), A. A. Sarmento,
ed. pp. 275–96. Funchal, Madeira.

Mertens, R.
1935. Zoologische Eindrucke von einer atlantischen In-
selfahrt. *Bl. Aq. Terrk.* 46:82–89.

Monard, A.
1937. Contribution à l'herpétologie d'Angola. *Arq. Mus.
Bocage* 8:19–153.
1940. Résultats de la mission scientifique du Dr. Monard
en Guinée Portugaise 1937–1938, VIII. Reptiles.
Arq. Mus. Bocage 11:147–80.

Moulton, J. M.
1963. The recapture of a marked leatherback in Casco Bay,
Maine. *Copeia* 1963:434–35.

Nieden, F.
1910. Die reptilien (ausser den schlangen) und amphibien.
In *Die Fauna d. deutsch. Kolonien*, pp. 1–74. Berlin:
Zool. Mus., Reihe I, Kamerun, H.

Parsons, J. J.
1962. *The Green Turtle and Man.* Gainesville: University
of Florida Press.

Pasteur, G., and J. Bons
1960. Catalogue des reptiles actuels du Maroc, revisions
de formes d'Afrique, d'Europe et d'Asie. *Trav. Inst.
Scient. Cherifien, Ser. Zool.* 21:1–132.

Pechuël-Loesche
1882. Die thierwelt. In *Die Loanga-Expedition, 3*, pp. 199–
316. Abth: Halfte.

Pellegrin, J.
1912. Reptiles, batraciens et poissons du Maroc (Mission
de Mme Camille du Gast). *Bull. Soc. Zool. Fr.* 37:255–
62.

Phaff, P.
1964. P. Phaff en zonen, Sierra Leone: in zeeschildpadden
en barracuda's. *Vers van 't Vat (Heinekens Brouwerijen
Nederland)*, 15th year, 107:14–17.
1967. De wonderlijkste vangst van mijn leven. *Tussen de
Rails, Maandbl. Nederl. Spoorwegen* 16:48–49.

Pritchard, P. C. H.
1971. The leatherback or leathery turtle, *Dermochelys coriacea. IUCN Monographs Marine Turtle Series*, 1:1–
39.
1972. Project 690. Coordination of marine turtle conservation. *World Wildlife Yearbook 1971–1972*:147–49.
1976. Post-nesting movements of marine turtles (Cheloniidae and Dermochelyidae) tagged in the Guianas.
Copeia 1976:749–54.

Pritchard, P. C. H., and R. Marquez M.
1973. Kemp's ridley turtle or Atlantic ridley, *Lepidochelys
kempi. IUCN Monographs, Sea Turtle Series*, 2:1–30.

Ross, J. P., and S. S. C. Marine Turtle Group
1978. Present status of sea turtles, a summary of recent
information and conservation principles. Morges:
IUCN/WWF. Mimeo, 42 pp.

Sarmento, A. A.
1948. *Vertebrados da Madeira, 1. Mamiferos, Aves, Repteis,
Batraquios*, 2d ed. Junta Geral do Distrito Autonomo

do Funchal.

Schleich, H. H.
1979. Sea turtle protection needed at the Cape Verde Islands. *IUCN/SSC Marine Turtle Newsletter*, 12:12.

Steindachner, F.
1890. Ueber die reptilien und batrachier der westlichen
und ostlichen gruppe der Canarischen Inseln. *Ann.
Naturh. Hofmus., Wien* 6:287–306.

Themido, A. A.
1942. Anfibios e repteis de Portugal (Catalogo das Colec-
coes do Museum Zoologico de Coimbra). *Mem. Es-
tud. Mus. Zool. Univ. Coimbra* 133:1–49.

Tornier, G.
1901. Die crocodile, schildkroten und eidechsen in Togo.
Arch. Natg., Beih.: 65–88.
1902. Die crocodile, schildkroten und eidechsen in Kamerun. *Zool. Jahrb., Syst.* 15:163–677.

Trotignon, J., and J. Maigret
1977. Les tortues de mer du Banc d'Arguin. *Asoe Soutien
du P. N. B. A.* 27:28. (not seen).

U.S. Department of Commerce
1978. Final Environmental Impact Statement, Listing and
Protecting the Green Sea Turtle (*Chelonia mydas*),
Loggerhead Sea Turtle (*Caretta caretta*), and Pacific
Ridley Sea Turtle (*Lepidochelys olivacea*) under the
Endangered Species Act of 1973. 144 pp.

Vilela, A. J.
1923. A pesca e industrias derivadas no Distrito de Mos-
amedes. (not seen, quoted from Monard 1937:146).

Villiers, A.
1958. Tortues et crocodiles de l'Afrique Noire Française.
Initiations Africaines 15: 1–354.

Williams, F.
1968. General Report on the Guinean Trawling Survey,
vol. 1. Lagos, Nigeria, 828 pp.

L. D. BRONGERSMA

Igal Sella
Bustan Hagalil
Akko, Israel

Sea Turtles in the Eastern Mediterranean and Northern Red Sea

ABSTRACT

The ecology of the sea turtle in the Mediterranean Sea was a subject unfamiliar to science until the 1950s, even though in the first half of the century they were hunted indiscriminately and on a very large scale in Israel and Turkey.

This hunting did not bring the population to the brink of extermination. However, in Israel, a serious additional hazard developed in the 1950s, due to sand excavations. Since then the annual number of specimens in the sea, as well as the number of nests, has decreased. In 1979 only 2 nestings were recorded in Israel along 250 km of shore, as compared with 15 per km per year at the beginning of the 1950s.

There seems to be no chance of a natural revival even after the total prohibition of fishing, declaration of nesting preserves, and a slight improvement in the condition of sandy beaches. This year the Nature Reserves Authority in Israel began to collect eggs for the purpose of raising and freeing 1-year-old turtles into the sea. Activities of this type should be undertaken in cooperation with the other countries concerned, namely Turkey and Egypt.

Introduction

The sea turtle has been known as an economic factor in the eastern Mediterranean since before the beginning of the century. Gruvel (1931) reports on turtles off the shores of Syria and Turkey, and on trade in turtles with England and Egypt. A report on the fisheries of Palestine (Hornell 1934) describes the export of 2,000 turtles a year from Palestine to Egypt.

Lortet (1883) mentions sea turtles on the shores of Syria, Lebanon, and Israel (Haifa). In Haifa he saw "several hundreds which were washed up onto the shores." It is of course possible that these were females that had gone ashore to nest. In the 1920s Haifa children were accustomed to such sights; it is likely that these were turtles concentrated on shore for the purpose of being sent abroad.

At this point, all interest and recorded information ceased until our time. Nor did anyone foresee the almost total destruction of the turtle population in Israel and in Turkey until 1963.

Recent interest in sea turtles began in Israel in 1954, but not to the extent of developing serious research on the subject. Most of the observations until 1958 were made by amateurs. In the same year, organized recording began, though not on the level required by the subject. The destruction of the shores, which increased towards the end of the 1950s, convinced the Society for the Protection of Nature that serious action was required. This approach led to the first nesting research in Israel (1964) and to preliminary research in Turkey (1965) and Sinai (1968).

The basic purpose of the study was to define the problem of the survival of sea turtles in the Mediterranean. Thus, a partial study was undertaken of such aspects as incubation conditions on various shores, species composition of the population, and the size of nesting specimens. However, the information collected was insufficient, because only very small remnants of the population could still be found.

A most important source of information was an aged fisherman from Acre, the late Abu Hanafi, who had organized turtle hunting in the 1920s and the beginning of the 1930s. The data he gave us were accurate and should be treated accordingly. This conclusion is important in order to estimate correctly the extent of destruction of the turtle population in such a short time.

Relying on this and other sources, it is possible to estimate that between the end of the first world war and the end of the 1930s, at least 30,000 sea turtles were caught in systematic fishing off the shores of northern Israel by Abu Hanafi's crews. At the same time, other fishermen were also active in this field, but we have no definite information on them.

Similar numbers were caught in Turkey, off the coast of Mersin and Adana, mainly in the 1960s. Fishing in these areas continues today. Additional damage through occasional fishing, egg collecting, accidental destruction of clutches, pollution of the shores by crude oil, underwater explosions, and other disturbances also continues. As a cumulative result, the turtle population in the eastern Mediterranean has been thinned out alarmingly, especially in Israel.

Distribution of the Species

The following species have been found in the eastern Mediterranean: *Dermochelys coriacea*, *Chelonia mydas*, *Caretta caretta*, and *Trionyx triunguis* (see Appendix). *Eretmochelys imbricata* is mentioned by Gruvel (1931) and by Wermuth and Mertens (1961), but its occur-

rence has not been substantiated by our study in the eastern Mediterranean.

Dermochelys coriacea is rather rare, but we have some proof of possible nestings. On 30 June 1963 trails were found on the beach at Palmachim (south of Tel-Aviv, Figure 1), but the trails did not end in any nests. The width of the tracks, 1.10 m, and the incomplete excavation, about 2 m in diameter, indicate that these were tracks of *Dermochelys*.

Chelonia mydas, as related by fishermen, now appears at least singly in the eastern Mediterranean between Turkey and the Nile Delta.

Nesting shores in the past (as told by Abu Hanafi) were found on all sandy beaches in north Israel without any distinct relation to the size of the grains of sand. Grain size varies from a minimum of 0.065 mm on Acre beach to 1.7 mm on Nahariya beach. Abu Hanafi did not know of nesting on the shores of Syria and Lebanon, but he did know about the spring migration of the species to the shores of Turkey, and he assumed that nesting also occurred there.

In the course of our research, scattered nests were found in Israel in the following localities: the beaches of Netanya, Caesarea, Atlit, Nahariya, and Rosh Haniqra. As related by fishermen in Turkey, in 1965–67, there were nesting beaches at Viransehil, Kazanli, Tuzla, Karatas, and Yumurtalik. In these places, according to the same sources, large numbers of turtles were caught, and there was also much nesting activity.

Smaller concentrations which were not hunted, and for which we have no estimates of quantities, are known at Tasucu, Silifke, Chahenem, and Side. According to information we received at Yumurtalik, nesting also occurs at Samandagy, but we found no on-site evidence.

Caretta caretta is known all along the shores of the eastern Mediterranean from Turkey to Egypt. Gruvel (1931) indicated that this was the most common species in the Bay of Iskenderun, whereas today *Chelonia* clearly is the most prevalent.

During the 1950s, I found some 15 nests per km a year on the stretch of coast between Nahariya and Rosh Haniqra (5 km). A similar number of nests were found in 1958 on the beach of Atlit (8 km). On the rest of the shores of Israel and northern Sinai, a length of about 400 km, we may find occasional nests. No accurate counts were undertaken, but a rough survey by aircraft counted 100–150 nests in 1968.

Since the beginning of our study, not a single emergence has been recorded on the shores of Haifa Bay (22 km), indicating a clear preference for the coarse-grained beaches. However, we have to note that the number of specimens caught before and during the 1960s off this coast, was large (Table 1).

IGAL SELLA

Table 1. Quantity of turtles caught in Haifa Bay and north of Acre, and brought to the Acre market

Year	Chelonia	Caretta
1963	1	15
1964	2	16
1965	8	5
1966	7	4
1967	11	15
1968	0	0
1969	24	54
1970	0	0
1971	0	0
(All were tagged and released)		
Average weight of male (kg)	61.5	37.5
Average weight of female (kg)	45.2	27.7
Maximum weight (kg)	100	65

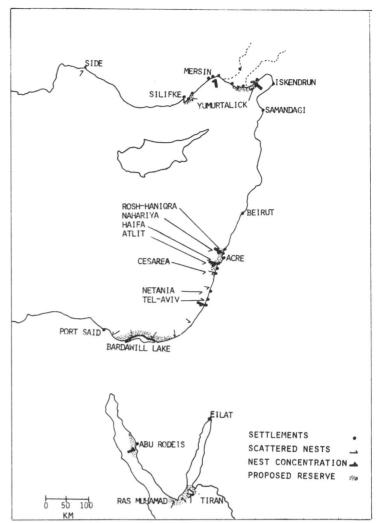

Figure 1. Eastern Mediterranean nesting sites and proposed turtle nature reserves.

Fishing

Both common species of sea turtles, and sometimes their eggs, are eaten by Moslems and Christians in Israel and Egypt. In Turkey this is not customary.

Today in Israel there is no systematic fishing, and even collecting of eggs along the shores of northern Sinai is only incidental. We know about systematic fishing from various sources, but full and accurate data were furnished to us by Abu Hanafi.

Massive fishing in Israel started immediately after the first world war, and reached a peak in the middle of the 1930s off the shores of Nahariya, Haifa Bay, and Atlit. Abu Hanafi alone employed up to 12 crews of 2 boats each during the above period, April to July, the nesting and mating season. The size of the mesh in the nets was about 40 cm, and every specimen caught in these nets was taken.

At the height of the season, some 600 specimens were caught a day, 90 percent of them *Chelonia*. Hanafi estimated that during these years some 30,000 turtles of both species had been caught. The normal weight of *Chelonia* in those days was 100 to 150 kg, and *Caretta* weighed no more than 60 to 80 kg.

Systematic fishing was carried on into the 1960s but on a much reduced scale. In the second world war it stopped altogether due to the thinning out of the population and decreasing profit. From then until the 1960s, fishing continued based on occasional catches, but not for export. The quantities that reached Acre market, which has always been the center of turtle fishing and commerce, are shown in Table 1. Turtle fishing is prohibited by Israeli law. For the purpose of our research we encouraged fishermen to bring and sell us their entire catch, but sometimes turtles were slaughtered, and we were informed only afterwards or not at all. Therefore our data are not entirely complete.

Since 1970 trade in turtles in Acre has stopped altogether. Even turtles caught by chance are returned to the sea, owing more to lack of profit than prohibition by law.

Events in Turkey have followed a similar path since the 1950s. In May 1965, we made a trip to that country to locate the fishing and nesting shores. In 1967 we were given an additional chance to visit these shores and to meet the people who are actually involved in fishing and commerce.

Official records of turtle commerce before 1967 do not exist, and local people refrained from speaking for fear that the information would reach undesirable addresses. Nevertheless, in our opinion, the following information is reliable.

A fishing company from Iskenderun began to buy turtles from fishermen on the shores of Mersin and its surroundings. The slaughter house at Iskenderun could absorb a good number of turtles, and at the end of the 1960s a number of groups specialized in this field. This slaughter house's entire production was destined for Europe.

During the main hunting season, from April to June, 200 turtles and more were brought to the slaughter house each day. Usually they weighed 120 to 150 kg, but 15-kg juveniles were not returned to the sea (M. Swartz, personal communication).

Between 1952 and 1965, up to 15,000 specimens were taken from the shores of Mersin alone. Toward the mid-1960s, the turtle population thinned out considerably, and the center moved to the estuary of the Seyhan and Ceyhan rivers, south of Adana. In May 1965, 100 specimens or more were caught each day in this new hunting area, all *Chelonia*. In this single area by May 1965, apparently more than 10,000 turtles had been captured.

Dr. U. Hiersch observed turtle fishing off the shores of Yumurtalik in April 1972, and he was informed that the seasonal catch reached approximately 1,200 turtles (from a letter to Prof. Mendelssohn, Tel-Aviv University).

Excavations

Additional severe damage to the turtle population in Israel was brought about by the excavation of sand for the production of concrete in 1954–63 (Niv and Nir 1969). In these years the nesting beaches of Nahariya, Rosh Haniqra, and Atlit, which were previously the main and almost the only nesting beaches, were severely damaged. The strip of beach between Rosh Haniqra and Nahariya was destroyed down to the beach-rock layer. At Atlit a strip of beach 80 m wide was removed from the original 120 m (Figure 2). Other beaches were also badly damaged.

At the same time, increasing numbers of tracks ended without any nesting, clutches rotted, and embryos developed abnormally (Table 2).

The prerequisites for normal nesting and incubation are a stable temperature of ± 28°C at a depth of more than 30 cm, and no flooding by waves. These 2 conditions became disrupted as soon as the excavations passed the natural line of the wave flow.

As compared with about 15 successful nestings per

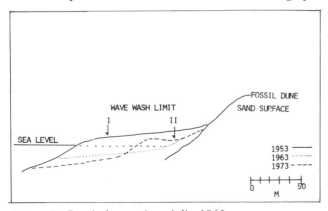

Figure 2. Beach destruction, Atlit, 1963.

Table 2. Nesting success, 1964

Category	No. on Nahariya – Rosh Haniqra 5 km	No. on Atlit 8 km
Nests	16	10
Barren emergences	10	18
Spoiled clutches	3	3
Abnormal hatchings	11	3
Normal hatchings	2	4

km per year until the 1960s, in 1964 the number of nests had decreased, as shown in Table 2.

All the defective or partially defective nestings were found within the wash line and were flooded at least once or as many as four times during the season. Some were found to be very near to the surface, and as a result underwent extreme temperature fluctuations of 18–35°C per day, resulting in spoilage. One nest was found in a concentration of gravel, and the young were not able to emerge to the surface.

At the urging of the Society for the Protection of Nature in Israel, a state committee was set up to examine the problems caused by the excavations. It recommended a halt to all the excavations on all the shores.

The recommendation was adopted, and within 5 years an improvement was apparent. However, this improvement has not yet brought the shores back to their original state. The destruction of the beaches, together with the extreme thinning out of the turtle population, seem to have reduced the number of turtles below the minimum necessary for natural survival of the species in Israel.

Results of the 1979 nesting survey by the Nature Reserves Authority show the steady decline in nest numbers. This year only 2 nests and 7 non-nesting emergences were encountered along the Israeli coast (250 km).

In Turkey in 1965, at least in the vicinity of Mersin, there were excavations on the nesting shores. We have no information as to what is happening there today.

Unnatural and Premature Mortality

An estimated 20 to 30 dead turtles are cast onto the shore every year between Nahariya and Ashqelon (200 km). Information nearly always reaches us too late to establish the cause of death or the date, or sometimes due to the disintegration of the corpse, the exact species. Each year, our count of dead turtles adds up to a similar figure. This is not logical because the number should decrease every year, in relation to the decrease in the number of nests and living turtles in the sea.

IGAL SELLA

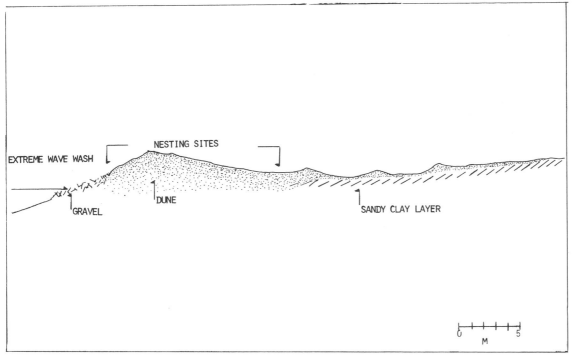

Figure 3. Cross section, Ras Shartibe nesting beach.

Summary and Conclusions

At the beginning of the century, 30,000 to 40,000 turtles lived off the northern shore of Israel (a length of 35 km). Some of the turtles hunted in Israel may have belonged to the Turkish population and may have been caught during their spring migration northwards. In Turkey, over about 100 km of shoreline, the numbers were similar. In both areas, the turtle populations have come close to extinction due to a similar process and over a similar length of time.

We have no knowledge of any other nesting grounds in the Eastern Mediterranean. Because of this and the very low potential rate of natural increase, we cannot foresee the rehabilitation of the species in the near future, if at all, without man's active intervention. Aid could come through preservation and rehabilitation.

In the framework of preservation, all sea turtles should be declared protected species (Israel has such a law) and hunting should be prohibited, at least for a limited period, pending the development of a method of artificial propagation. At the same time, international control of trade in turtles should be initiated. Nature reserves should be established with the main purpose of protecting the nesting beaches and mating area, irrespective of whether or not these two overlap (Figure 1).

Rehabilitation by artificial methods should be tested locally and in minimum quantities to ensure the survival of the 2 species. However, it is worthwhile to consider rehabilitation also for commercial purposes.

A common plan for the countries of the Eastern Mediterranean would be more economical than separate local plans, due to the migratory character of the turtles in this area.

In Israel the following plans are being implemented: besides 2 nature reserves in Atlit and Rosh Haniqra, an artificial raising system is being undertaken to raise 1-year-old turtles in the maximum number available.

The current state of research on turtles in the area is far from satisfactory. Therefore, before, and parallel with, any action for preservation and rehabilitation, research on a suitable level must be completed in the 3 main countries concerned: Turkey, Israel, and Egypt.

Literature Cited

Hornell, J.
1934.   Report on the Fisheries of Palestine. Manuscript, 65 pp.
Gruvel, A.
1931.   Les Atats de Syrie. Paris.
Lortet, L.
1883.   Etudes Zoologiques sur la Faune du Lac de Tibériade Suivies d'un Aperçu sur la Faune des Lacs d'Antioche et de Homs—I. Poissons et Reptiles du Lac de Tibériade et de Quelques Autres Parties de la Syrie. Lyon.
Niv, D., and I. Nir
1964.   Shore excavation report. Ministry of Development, Jerusalem.

Sne, A., and T. Wisebrod
1969.   Discovery of the Felosium Arm of the Nile Delta. *Teva Vearetz* 16:219.
Wermuth, R., and R. Mertens
1961.   Schildkroten, Krokodile und Bruckenechsen.

Appendix 1. Sea Turtles in Sinai

Up to 1967 we gathered very little information on sea turtles in the Red Sea, or to be more precise, the tip of the Bay of Eilat. The information we possess today was gathered from Israeli sources.

On the shores of Sinai and the Island of Tiran the following species are known today:

1.   *Caretta caretta*. Some bones (of a few specimens) were found in a small cave on the beach of Ras Muhammed (identification by Prof. A. Carr). None has yet been caught alive.

2.   *Lepidochelys olivacea*. Two specimens have been identified south of the Peninsula (Prof. A. Ben-Tuvia, University of Jerusalem, personal communication).

3.   *Dermochelys coriacea*. This species is very rarely seen and caught in Eilat Bay. A few were observed by a helicopter pilot from the air throughout the month of July 1969 off the nesting shore of *Chelonia* (see below) south of Abu Rodeis. No nests have been found (D. Ron, personal communication).

4.   *Eretmochelys imbricata*. This species is occasionally seen and caught in various places along the shores of Sinai. No nests have been found to date.

5.   *Chelonia mydas*. The green turtle is seen and caught more than any other species along the shores of Sinai. Nesting activity is known around the southern point of Sinai and Tiran.

Chance collecting of eggs and fishing are known in Sinai, but the Bedouin population on the Sinai shores is so sparse that they have no effect on the existence of the species there. Nevertheless, we can point to a few facts which jeopardize the existence of the only colony known to date.

The nesting areas of *Chelonia* were identified in a survey undertaken in 1968 (Figure 1). The entire shores of Sinai and the island of Tiran were examined during flight from a height of 100 m and driving by jeep.

In all cases except 1, nests were found scattered singly or in small groups. Only in 1 locality, Ras Shartib on the Bay of Suez south of Abu Rodeis, was a comparatively high concentration of nests found. In October 1967 we found no fewer than 40 nests, or what appeared to be nests. In July to September 1969, we counted 37 nests in a stretch of 200 m. In an aerial observation at the end of September, 30 more nests were observed (D. Ron, personal communication).

Congestion of the nests is very great here. Most are dug one on top of the other in a limited strip between the wash line and the end of the beach dunes, which are about 2 m above the regular boundary of the waves and no wider than 15 m. Beyond the belt of beach dunes, tiny sand mounds are scattered 30 to 50 cm high on a hard layer of sandy clay (Figure 3). In this section we found dozens of trial diggings but not a single nest. South and north of this section there are no beach dunes, and the waves wash up to the area of the small mounds. We found no additional nests 40 km north and 30 km south.

The coastal belt seems to have declined and destruction of the beach dunes to have advanced, leading to a constant reduction of the stretches suitable for nesting.

Due to lack of time and the great difficulties in finding nests dug one into the other, incubation conditions were not properly examined. However, from the small number we did find, the percentage of successful nesting is clearly very small. The general failure is increased by 2 new factors. An oil tank farm has been put up on the border of the nesting strip, and the shore is polluted by crude oil. Development of the oil industry naturally draws people and their dogs, which rove all over the area and dig up some of the nests.

From all of the above, it appears likely that the only proper colony known in Sinai is being destroyed.

Appendix 2. *Trionyx triunguis* in the Mediterranean Sea

This tropical fresh water species was once common in Israel in every stream and small river flowing into the Mediterranean. Today, because of pumping and pollution, they have become rare. They are also known in Lebanon, Syria and Turkey (related by fishermen).

In our study we found that this species appears regularly in the Mediterranean Sea. Gruvel (1931) gives evidence of finding this species as an unusual phenomenon in the Bay of Iskenderun at 30-m depths. We found soft-shelled turtles along the shores of the Eastern Mediterranean as shown below:

There is, therefore, no reason to think that their appearance in the sea is accidental, or that their penetration into the sea is caused by floods. In experiments carried out in the physiological laboratory of the Tel-Aviv University, Prof. A. Shkolnik and his student tried to "acclimatize" these turtles to sea water, but without success. This interesting phenomenon should be included in the framework of research and preservation plans for the turtles of the Mediterranean Sea.

Place (from south to north)	Date		Dead or alive	Depth of sea (m)	Distance from shore (km)	Distance from fresh water (km)
Bardawil Lake	Sept	1979	Disintegrated skeleton	On shore	—	150.0[a]
Tel-Aviv	June	1978	Alive in net	?	?	6.0
Haifa Bay	Oct	1963	Alive on rod	6	2.5	3.0
Haifa Bay	June	1972	Alive in net	6	2.0	2.0
Haifa Bay	Sept	1972	Alive on rod	4	0.5	0.5
Iskenderun Bay	May	1965	Alive in net	10	12.0	25.0
Karatas Lagoon	June	1967	Disintegrated skeleton	On shore	—	20.0
Side Lagoon	May	1965	Disintegrated skeleton	On shore	—	12.0
Side Lagoon	May	1965	Alive in net	5	0.5	12.0

— No data.

a. This specimen undoubtedly died within historic times (according to the state of the skeleton) 150 km away from the nearest fresh water. Even the old eastern arm of the Nile Delta (Sne and Wisebrod, 1969), which dried up in the first century is 60 km away.

Remzi Geldiay
Tufan Koray
Süleyman Balik
Department of Biological Oceanography and Institute of
Hydrobiology
Faculty of Science
Ege University
Bornova-Izmir, Turkey

Status of Sea Turtle Populations (*Caretta c. caretta* and *Chelonia m. mydas*) in the Northern Mediterranean Sea, Turkey

ABSTRACT

Observations on the species, distribution, mating behavior, nest making, the relationship between marginal shield distributions and the size of subadult individuals and young were made on the sea turtles nesting on the Mediterranean coastal beaches of Turkey. Nearly 2,000 km of the Mediterranean coastline is composed of sandy beaches. For investigation of the mean clutch size and the annual mean egg production, 5 stations were chosen, covering a total of almost 100 km.

Caretta caretta nested at all 5 stations without differentiation. *Chelonia mydas* made its nests only at Belek, Side, and Alanya. Also, *Trionyx triunguis*, which lives in freshwater, at times in lagoons, and sometimes even in the sea (Atatür 1978) was found in large numbers in our study regions. These carnivorous animals are disliked by fishermen.

The mating of loggerheads began in mid-April and lasted until the last week of May. Mating occurred just opposite the shore where the nests were made. The first report of a loggerhead nest came on 8 May and the nest making season lasted until the end of August. But almost 50 percent of the nests had been made by 20 June.

The total number of eggs laid at the research station areas during the egg laying season was 140,000. On the average, 47.0 percent of the eggs were destroyed by predators or the tide.

In the distribution of marginal shields in sub-adult and young *Caretta caretta* the 11 to 11 and 12 to 12 distribution was frequently observed; important correlations were found in some of the relationships in size.

Introduction

Some species of sea turtle that are widespread in the Pacific and Atlantic Oceans reach as far as the eastern Mediterranean and compose large populations in specific areas along this region, but scientific knowledge about them remains scarce.

The purpose of this study is to investigate the sea turtles visiting Turkey's Aegean and Mediterranean coasts (36°N, 26°E and 42°N, 36°E) and to help preserve these populations. To do this, representative species must be examined, their inter-relationships investigated, their population size recorded, and their ecology understood.

The good flavor and the high protein content of their meat has increased the turtle's commercial importance in many regions. Also, these animals are important in areas where their carapaces are sold as decorative articles to tourists, or where superstitions state that their meat, fat, and blood cure several illnesses (such as breathlessness and hemorrhoids). Many factors, including pollution and tourist traffic force the turtles to leave their eggs haphazardly or to look for a new territory in order to lay eggs. In this way, they might lose continuity between generations and they, like other species before them, would become extinct. Many dangers await young hatching, including the numerous predators. As long as predation does not become overwhelming, it can be tolerated as a requirement of the food chain needed to maintain the natural balance.

Methods

The Aegean and Mediterranean coasts of Turkey include long, sandy beaches totaling more than 2,000 km. For this reason investigations were carried out at 5 stations chosen during the first, inventory-taking survey, trip. The stations chosen are shown in Figure 1:

- Köycegiz Dalyanköy Bogaz Beach, 7-km long, 36°47′N, 28°38′E;
- Kumluca, 10-km long, 36°22′N, 30°19′E;
- Belek, 40-km long, 36°50′N, 30°58′E;
- Side-Nigit, 30-km long, 36°46′N, 31°28′E; and
- Alanya, 12-km long, 36°36′N, 32°05′E.

The research groups (composed usually of 3 persons) were sent to these regions at 1- or 2-week intervals (leaving 1 observer in the area during the intervals) and spent the same length of time at each station. These groups made observations either on foot or by jeep from 2100 hr to 0400 or 0600 hr. Track countings and nest evaluations were done in daytime. Investigation of false emergences was also emphasized.

The behavior of individual females during the sequence of emergence on land, nest digging, and egg laying were observed. They were captured during their return to the sea and measured for the length and width of the curved and straight carapace, length and width of the plastron, width of the nose shield and lengths of the head with a steel yardstick and a compass suited for morphometry. Also, the major shields of the carapace and plastron were counted, and the weights were determined with a portable scale.

By day, the number of emergences onto land and the number of nests were counted. The distance to the shoreline, depth, inner temperature, relative humidity, and the number of eggs were determined for each of the 50 nests opened. The amount of destruction caused by people (tractors, etc.), predators (crabs, dogs, foxes, pigs, jackals, and some birds), and tides was calculated for the marked nests.

The biometric measurements taken on the 15 captured subadult *C. caretta* were statistically examined.

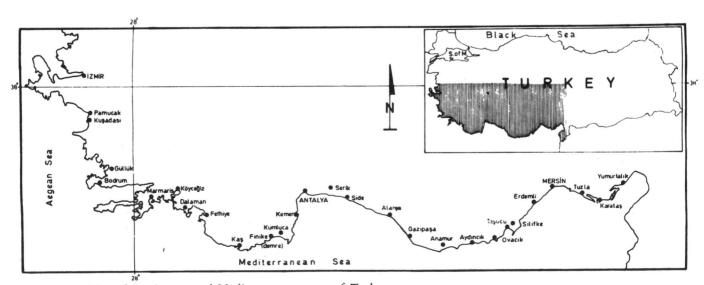

Figure 1. Map of the Aegean and Mediterranean coasts of Turkey.

REMZI GELDIAY

Sea Turtles Visiting and Making Nests on the Northeastern Mediterranean

The first report directed seriously towards the subject of sea turtles in Turkey outside of routine observations was Hathaway (1972). Hathaway (1972) stated that there are more turtles in the Mediterranean than in any other sea. However, in the absence of any definitive report on the species and number of these animals, some facts are known about their feeding habits, but almost nothing about their migrations, mating behavior, and the role of instinct during migration, emergence, and nest making. So far, our only knowledge about turtles visiting Turkey and their numbers comes from the statistics: the year and weight (in kg) of the fish and turtle catch in Turkish waters. These statistics are difficult to read for turtles. The number of turtles cannot be calculated from total catch weights because individual weights vary enormously from juvenile to adult. These statistics, for example, report a drop in the turtle catch from 286,505 kg in 1968 to 52,355 kg in 1969. Why did the catch decline in 1969? Did fewer turtles visit Turkish waters, or did a drop in price discourage turtle hunters, or were fishermen careless about reporting their turtle catches (Hathaway 1972)? The explanation for the sudden decrease of 1969 is not known. Additionally, no differentiation is made between species, another possible source of miscalculation of numbers.

For these reasons, work on population numbers (with population dynamics research), along with differentiation of species, presents a difficulty in the investigation of the large groups on Turkish coasts. For now this work has been done by following the emergences onto the beaches or by examining the tracks.

The turtles' destination after leaving the Turkish coast is unknown. Although this research is a part of our program, we had to leave this inquiry for future seasons due to circumstances not under our control. The tags (monel metal tags) that we had ordered were not sent to us on time.

Where do these animals go after they have laid their eggs? Do they, as many people state, roam the Mediterranean (Brongersma 1972) or do they pass to the western Indian Ocean (Indowest Pacific Ocean) through the Suez Canal, or do they come from the Atlantic and return to this ocean? There is no doubt that these questions will be answered easily and with certainty after the marking operations are completed. The other remaining questions (age, maturation, underlying reason for false crawls, etc.) can be answered with some careful scientific observations.

In general, sea turtles are found mostly in tropical seas, but they are also found in considerable numbers in subtropical and temperate regions. In the Mediterranean there are 5 species: *Caretta caretta, Chelonia mydas, Dermochelys coriacea, Eretmochelys imbricata, Lepidochelys olivacea* (Basoglu 1973). But all of these species lay eggs on the Mediterranean coast.

The species of sea turtles known to nest along the coasts of the Turkish seas [Mediterranean, Aegean, Marmara (inner sea), and Black seas] can be summarized as follows.

Family: Dermochelydae

SUBSPECIES: *Dermochelys c. coriacea*

Found in small numbers in Turkey only on the Mediterranean coast, this species also occurs in the Greek Mediterranean (Ondrias 1968), and even though specific regions are not given, in the Mediterranean and Aqaba Gulf around Israel (Hoofien 1972). Also, the subspecies *Dermochelys c. schlegelii* should not have been published (Mertens and Wermuth 1960).

Family: Cheloniidae

SUBSPECIES: *Caretta c. caretta*

Though found in large numbers along the Black, Marmara, Aegean, and Mediterranean seas of Turkey, this turtle is known to nest only on the Aegean and Mediterranean shores. It also occurs along the Greek Aegean and Mediterranean coasts (Ondrias 1968), the Mediterranean shores of Israel (Hoofien 1972), and the western portion of the Bulgarian Black Sea (Beskov and Beron 1964). There is no evidence that they nest in these areas. Russian literature reports that they are also present on the Black Sea (Terentjev and Chernov 1975).

SUBSPECIES: *Chelonia m. mydas*

In Turkey the green turtle is distributed along the Black, Marmara, and Aegean Seas but particularly large populations form on the Mediterranean. It is not known whether green turtles nest on the Aegean and Mediterranean coasts of Greece where their presence has been reported (Ondrias 1968). They have also been reported from the Mediterranean and Aqaba Gulf of Israel (Hoofien 1972). Found on the western Black Sea region of Bulgaria (Beskov and Beron 1964), they are not known to nest there.

SPECIES: *Eretmochelys imbricata*

Hawksbills are probably found along the southern shores of Turkey and Greece (Mertens and Wermuth 1969). They have been reported from the Mediterranean and Aqaba Gulf of Israel (Hoofien 1972).

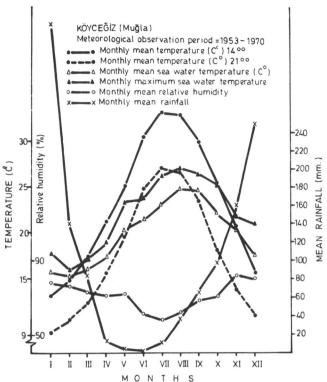

Figure 2. The average annual temperature, sea water temperature, relative humidity and rainfall for Köycegiz over a 17-year period.

SPECIES: *Lepidochelys olivacea*

Olive ridleys are rare along the Mediterranean coasts but resemble *Caretta caretta*.

Work at Köycegiz

Köycegiz-Dalyanköy Beach is in the region between the Aegean and the Mediterranean and east of Marmaris Bay (36°47′N, 28°38′E). It has a typical Mediterranean climate.

Figure 2 summarizes meteorological observations at Köycegiz for 17 years. According to these data, the highest average temperatures are recorded for July (27.8°C) and August (27.3°C). The lowest average temperature of the year is in January (9.5°C). August (25°C) and September (24.5°C) have the highest average sea water temperatures; February (15.3°C) the lowest. The most rain falls in December (245.5 mm), the least in August (1.8 mm).

The Dalyanköy Beach stretches around the entire mouth of the canal joining the sea with the lake of Köycegiz. Mollusc shells are always strewn on this beach. The littoral region is rocky without forming a set.

The pale yellow sand, when wet, turns a yellowish-tan. *Euphorbia paralias* is frequent at a distance of 10 m from the shoreline. Almost 20 m from the shoreline *Nerium* sp. and from place to place *Scirpus* sp. can be seen.

Many *Ocypode cursor* (Order Decapoda) are found on the Dalyanköy Beach. The forests of plane trees at the junction between lake and sea are hideways for foxes (*Vulpes vulpes anatolica*). Only from May to September, when people visit the beaches, are dogs seen and many of them have learned to open sea turtle nests.

In this region observations made during the summer months of May to August show that only *Caretta caretta* nests here. Two males were caught by fishermen in the sea with a net. During work in the egg laying season, 330 nests were observed. But, almost 43.6 percent of these were destroyed by predators and tides (Figure 3). Only *Chelonibia* sp. was noted as the characteristic epizoic organism. *Trionyx triunguis* is found in large groups in Köycegiz lake, canal and lagoons. It is understood from the complaints of the fishermen that they pass into the sea from time to time.

Although one of the aims was to determine the population of *Caretta caretta*, the re-nesting interval could not be calculated because the monel metal tags arrived too late. Thus, Hughes's equations to estimate population size (1974) could not be used. The total number of nests was not used to reach an estimate because of the probability of large error.

The mating season begins in the middle of April and lasts until the last week of May according to Dalyanköyü coastal fishermen and our own observations. A mating pair was observed at a distance of 300 m from the shore on 14 May. As far as we know, supported by outside reports, mating occurs on sunny days and lasts 3 hours. Research in 1978–79 showed that the nesting season began in mid-May and lasted until the end of August along the Turkish Mediterranean coast (Figure 4).

Our first report of a nest in Dalyanköy was 8 May. In April and September, the observed tracks did not lead to nests.

The frequency of the "false crawl" (Carr, Carr, and Meylan 1978) attracts attention in this region. Of the 11 emergences onto land discovered the night of 12 June, only 2 ended with the production of a successful nest.

For the duration of the nesting season, the emergences onto land occurred between 2200 hr and 0400 hr. Emergences were not observed during the day or towards evening. Once a *Caretta caretta* at 0600 hr was observed. Two individuals were seen making nests on the Dalyanköy Beach at 2200 hr but emergences become more frequent between 0100 hr and 0300 hr. This agrees with the suggestions of Schulz (1975) that *Chelonia mydas* emergences are more frequent in a rising sea.

An animal reaching the shoreline remains there a little while, though completely out of the water. After this rest, the emergence begins with continuous movements. Figure 5 summarizes the distance of nests from

428   REMZI GELDIAY

shoreline at the research stations. On the average, 50 percent of the nests were made 15 to 20 m from the shoreline.

The Dalyanköy nest-making females and males caught in the sea had a minimum straight carapace length of 55 cm and a maximum of 74.6 cm. They probably reached sexual maturity in this area. The minimum weight of the females was 40.0 kg, and the maximum was 75 kg with an average of 57.5 kg. As the number of times these animals laid eggs was not determined, these values include an error relative to the egg weight in the ovaries.

Evaluation of the 50 nests we opened indicated a likely clutch size of 70 to 120 eggs. The average egg number was 93, the minimum 23 and the maximum number 134. Diameter measurements of the top 10 eggs of each nest gave the minimum value as 3.7 cm and the maximum as 4.2 cm with an average diameter of 3.9 cm. The average weight of an egg was 20.3 g, lower than reports by other workers (Caldwell 1959; Hughes 1974) and probably due to the small size of the animals.

Incubation period is defined as the time from egg laying to hatchling emergence (Caldwell 1959). Data from marked nests and from the laboratory incubator indicate a minimum 50-day and a maximum 64-day duration with an average of 57 days.

The temperature inside the nest and the relative humidity may affect incubation period (Yntema and Mrosovsky 1979). Inner nest temperatures at Dalyanköy ranged from 24°C to 28°C with the average at 26°C.

Measurements of straight carapace length of 50 hatchlings captured upon leaving the nest were a minimum of 37 mm and a maximum of 42 mm with the average at 39.9 mm. Straight carapace width varied between 30 mm and 35 mm with an average of 32.3 mm. Weights of the same animals ranged from 12.7 g to 18.3 g with 16.1 g as the average.

Relationships between lengths of subadult individuals captured at the 5 stations were investigated within the 95 percent confidence limit (Figures 6, 7, and 8). Important correlations were found in measurements of weight and carapace length, and carapace length and carapace width for 50 C. caretta young (Figures 9 and 10).

The majority of subadult males and females examined on the Dalyanköy Beach showed marginal shield distributions of 11 R, 11 L; 11 R, 12 L; 12 R, 11 L; and 12 R, 12 L. Only 1 individual showed the 12 R, 13 L distribution.

The 50 C. caretta hatchlings mentioned above had mostly the 11 R, 11 L, and 12 R, 12 L marginal shield distributions. Some showed the 12 R, 11 L and 11 R, 12 L distributions.

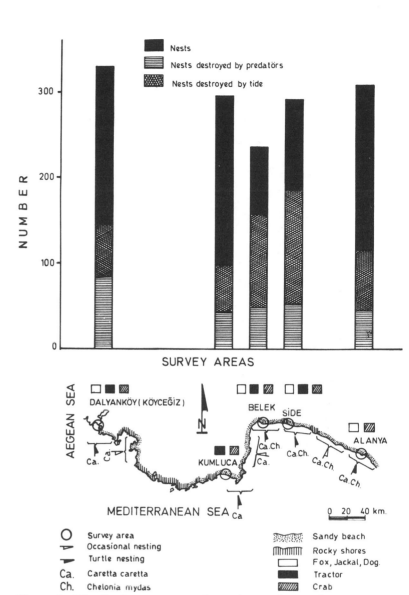

Figure 3. Distribution, nest capacity and nest threatening factors for *Caretta caretta* and *Chelonia mydas* at the research stations.

Other Beaches

Kumluca Beach

Kumluca Beach is located on the inner curve of the Bay of Finike (36°47′N, 28°38′E). It has the characteristic Mediterranean climate and plant cover.

Belek and Side Nigit Beaches

Belek and Side Nigit Beaches extend the length of Antalya Bay (Belek 36°50′N, 30°58′E; Side 36°51′N, 31°28′E) with a Mediterranean climate.

Alanya Beach

Alanya Beach is located in Antalya Bay (36°36′N, 32°05′E) with a Mediterranean climate.

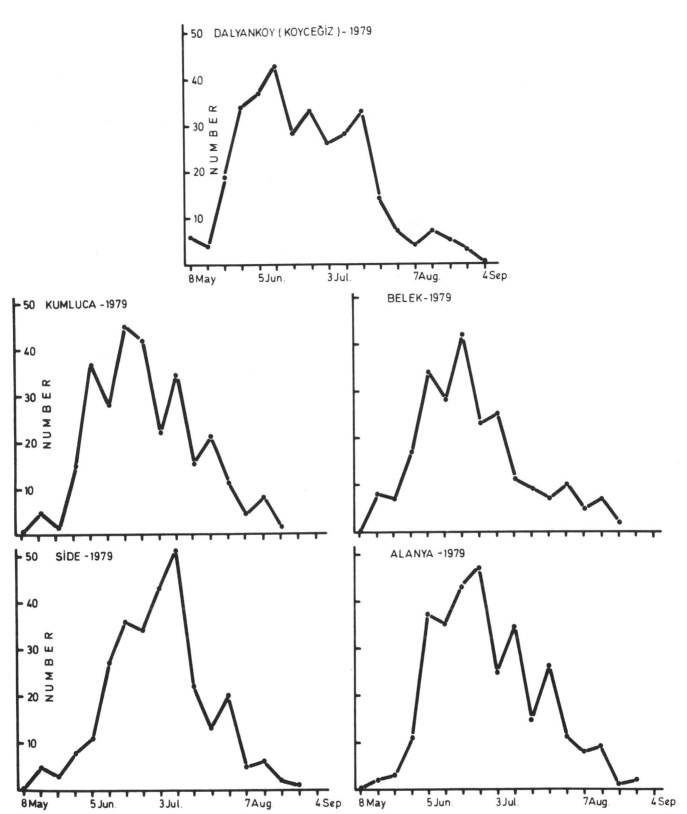

Figure 4. Nesting activity at the research stations during the 1979 egg laying season.

Discussion

Distribution of Caretta caretta *and* Chelonia mydas

Hildebrant and Hatsel (1937:35) described *C. caretta* as a cold sensitive species, but it reaches more north-

erly than other Atlantic species (Brongersma 1972). It nests up to 35°N on the eastern North American coast (Coles 1914, ibid.), reaches higher latitudes in Virginia (Carr 1952, ibid.), goes up to 35°N on the Atlantic shores of Morocco (Pasteur and Bons 1960, ibid.) and up to 43°N on the western Mediterranean shores of

REMZI GELDIAY

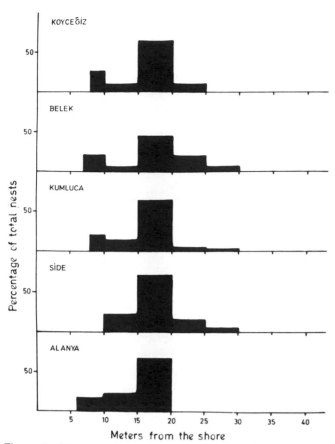

Figure 5. Distance of the nests from the shoreline.

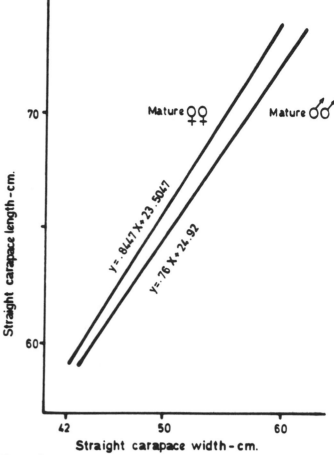

Figure 6. Relationship between straight carapace length and width of *Caretta caretta* subadult males and females.

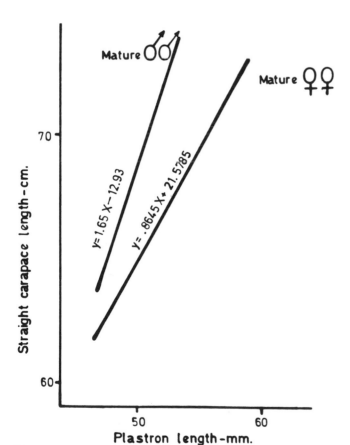

Figure 7. Relationship between plastron length and straight carapace length in *Caretta caretta* subadult males and females.

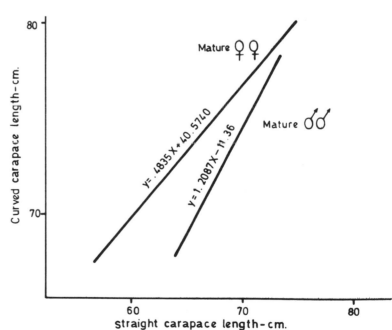

Figure 8. Relationship between straight carapace length and curved carapace length of subadult male and female *Caretta caretta*.

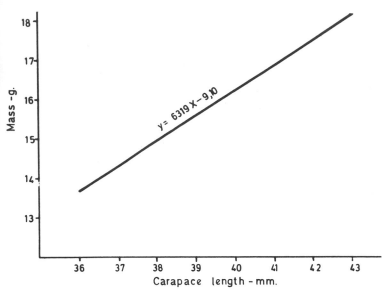

Figure 9. Relationship between carapace length and weight of *Caretta caretta* hatchlings.

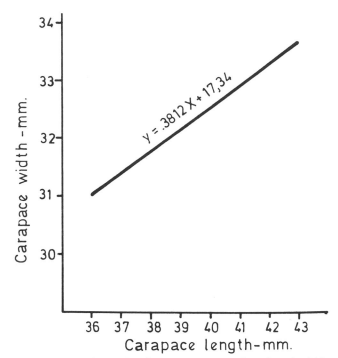

Figure 10. Relationship between carapace length and width of *Caretta caretta* hatchlings.

Italy (Bruno 1969, ibid.). It is also known to nest at 35°N on the Japanese coast (Nishimura 1967, in Hughes 1974).

The Mediterranean coast of Turkey (36°N) presents a major nesting region for *C. caretta*. Loggerheads are observed in the Aegean Sea and even reach the Black Sea (41°N) through the Sea of Marmara. We do not have any information about the animals' passage into the Black Sea.

Brongersma (1972) stated that *C. caretta* in northern

and central latitudes were small or half-grown specimens. This interpretation is supported by our observations that all of the animals captured were subadults.

Hughes (1974) reported that distribution of *C. caretta* was bounded by the 25°C summer isotherm. This is verified by nest making on Turkish Mediterranean shores which show a 24–26°C summer isotherm.

Chelonia mydas is found only at Belek, Side, and Alanya. It does not move farther west. This may be due to the warmer temperatures in the east (Alanya, 28°C) as compared to the western waters (Dalyanköy, 25°C).

Hatchling Numbers on the Turkish Mediterranean Coast

The number of *Caretta caretta* nests around the research stations were counted as number of nests per kilometer for the nesting season. The beaches with the highest numbers of nests to the lowest are as follows: Dalyanköy, 47 nests/km; Alanya, 30 nests/km; Kumluca, 29.4 nests/km, and Belek and Side, 10 nests/km. *Chelonia mydas* nests may have been included in the calculations for the beaches of Belek, Side, and Alanya. A mathematical estimate, based on 93 eggs for each nest, puts the total number of eggs at 135,000; 47.0 percent damage leaves 71,550 healthy eggs. We should emphasize that our investigative regions cover only 100 km of the almost 2,000-km Mediterranean coastline. Actual production rate must be much more than our calculations.

Variations in Marginal Shields of Young and Mature Animals, and Linear Relationships in Length

Deraniyagala (1953, in Hughes 1974) separated Atlantic and Indo-Pacific *Caretta caretta* stocks by marginal shield numbers, vertebrae, and their curvation. This classification was believed to be incomplete in later years (Caldwell, Carr, and Ogren 1959; Hughes 1974), but it still remains useful.

Marginal shield counts of 50 hatchlings at the end of the incubation period showed 58.0 percent had an 11-11 distribution, 20 percent 12-12, and the remainder mixed.

Measurements of 15 subadults showed 46.1 percent had an 11-11 distribution, 15.3 percent had 12-12, and the remainder mixed. The 11-11 and 12-12 distribution percentages in hatchlings and subadults resemble one another. The 11-11 distribution shows a higher percentage than the 12-12. These results are markedly different from those of other workers, but the sampling size was too small to permit any definite conclusions.

There were no important correlations found between lengths of Atlantic loggerheads by Caldwell, Carr, and Ogren (1959) and Gallagher *et al.* (1972). Hughes (1974) showed linear relationships between lengths in the southeast African *C. caretta*.

Statistical analyses on basic measurements taken from subadult *C. caretta* captured on eastern Mediterranean shores suggested the existence of important correlations between some of them.

This situation falls into opposition with the views of Caldwell, Carr, and Ogren (1959) and Gallagher *et al.* (1972). As far as we know from publications, their smallest turtles were longer than our biggest specimens. In other words, they investigated adult animals. Hughes (1974) worked both with mature animals and subadults.

The relationship between curved carapace length and straight carapace length of subadult females and males is shown in Figure 8. As we can see females are more curved than males. Figure 6 shows the relationship between straight carapace length and straight carapace width of subadult males and females. Males are somewhat wider than females and this fact is true in larger size classes also. The relationship between plastron length and straight carapace length is summarized in Figure 8. Males from the same length group have shorter plastron measurements than the females.

There was no correlation found between curved carapace length and width of head and between straight carapace length and head width in subadult males and females.

Summaries of relationships among weights and carapace lengths and widths of 50 *C. caretta* hatchlings are shown in Figure 9 and Figure 10.

The first developmental steps of the young occur with increase in carapace length versus increase in weight. Increase in width of the carapace is slower when compared to the increase in length.

Acknowledgments

We wish to express our deepest gratitude to all the officials on the IUCN/WWF Joint Project Office and all the interested parties for their support of this research in the northern Mediterranean (Turkey). We express our great appreciation to Dr. Archie Carr, Dr. L. D. Brongersma and Dr. N. Mrosovsky for their kind assistance in sending the literature that we had difficulty in securing. Also, we wish to thank our department members, Zehra Ünver, Yilmaz Karacalioglu, Ethem Atalik and Osman Şimşek, for their unending energy and efforts during the preparation of this work.

Literature Cited

Atatür, M.
1978. *Trionyx triunguis* 'in morfoloji ve osteolojisi, ana-dolu'daki biotop ve dagilisi uzerinde arastirmalar ve biyolojisine ait bazi gozlemler. Ph.D. dissertation.

Basoglu, M.
1973. Diniz kaplumbagalari ve komsu memleketlerin sahillerinde kaydedilen turler. *Turk Biyoloji Dergisi. Cilt* 23:1–124.

Beskov, V., and P. Beron
1964. *Catalogue et Bibliographie des Amphibiens et des Reptiles en Bulgarie.* Sofia: Editions de l'académie Bulgare des sciences.

Brongersma, L. D.
1972. European Atlantic turtles. *Zool. Verhand. Leiden* 121:1–318.

Caldwell, D. K.
1959. The loggerhead turtles of Cape Romain, South Carolina. *Bulletin of the Florida State Museum* 4:319–48.

Caldwell, D. K.; A. Carr; and L. Ogren
1959. Nesting and migration of the Atlantic loggerhead turtle. *Bulletin of the Florida State Museum* 4:295–308.

Carr, A.; M. H. Carr; and A. B. Meylan
1978. The ecology and migrations of sea turtles, 7. The West Caribbean green turtle colony. *Bulletin of the American Museum of Natural History* 162:1–46.

Gallagher, R. M.; L. M. Hollinger; R. M. Ingle; and C. R. Futch
1972. Marine turtle nesting on Hutchinson Island Florida in 1971. *Florida Department of Natural Resources, Marine Research Laboratory Special Scientific Report* 37:1–11.

Hathaway, R. R.
1972. Sea turtles: unanswered questions about sea turtles in Turkey. *Balik ve Balikcilik* 20:1–8.

Hoofien, J. H.
1972. *Reptiles of Israel.* Tel-Aviv: Department of Zoology, Tel-Aviv University.

Hughes, G. R.
1974a. The sea turtles of South-East Africa. I. Status, morphology and distributions. *Investigational Report of the Oceanographic Research Institute*, Durban, South Africa, 35:1–144.

1974b. The sea turtles of South-East Africa. 2. The biology of the Tongaland loggerhead turtle *Caretta caretta* L. with comments on the leatherback turtle *Dermochelys coriacea* L. and the green turtle *Chelonia mydas* L. in the study region. *Investigational Report of the Oceanographic Research Institute*, Durban, South Africa, 36:1–96.

Mertens, R. U., and H. Wermuth
1960. Die Amphibien un Reptilien Europas. 3. Liste. Senckenberg. Buch 38. Frankfurt am M, Federal Republic of Germany.

Ondrias, J. C.
1968. Liste des Amphibiens et des Reptilies de Grèce. *Biologia Gallo-Hellenica* 1:111.

Schulz, J. P.
1975. Sea turtles nesting in Suriname. *Zoologische Verhandelingen, uitgegeven door het Rijksmuseum van Natuurlicke Historie te Leiden* 143:1–144.

Terentjev, P. V., and S. A. Chernov
1965. *Key to Amphibians and Reptiles.* Jerusalem: Israel Pro-

gram for Scientific Translations.

Yntema, C. L., and N. Mrosovsky

1979. Incubation temperature and sex ratio in hatchling
loggerhead turtles: a preliminary report. *IUCN/SSC
Marine Turtle Newsletter* 11:9–10.

REMZI GELDIAY

Joop P. Schulz
Foundation for Nature Preservation (STINASU)
Paramaribo, Surinam

Status of Sea Turtle Populations Nesting in Surinam with Notes on Sea Turtles Nesting in Guyana and French Guiana

Four species of sea turtles nest frequently on the Suriname beaches: *Chelonia mydas, Lepidochelys olivacea, Eretmochelys imbricata,* and *Dermochelys coriacea.* The loggerhead, *Caretta caretta,* has been reported nesting only once.

There are 2 main nesting areas. Galibi (Galibi-Baboensanti-Eilanti) is located at the mouth of the Marowijne River. It is a nature reserve. At Matapica and Krofajapasi (formerly called Wia-Wia or Bigisanti when it was situated in the Wia-Wia Nature Reserve), nests are protected by special decree. A few small beaches are situated between Matapica and the mouth of the Surinam River.

For a detailed description of the beaches, see Schulz (1975). The most characteristic feature of the nesting beaches is the continuous alteration of the shoreline and the westward movement (± 1.5 km/yr) of the Wia-Wia beach.

Surinam Populations

The Green Turtle, Chelonia mydas

There is no explanation for the considerable fluctuations recorded in the yearly number of green turtle nests.

Population size is based on an estimated average of 3 nests laid per female during a nesting season (Schulz 1975, pp. 71–73) and an average interbreeding period of 2.3 years (Schulz 1975, pp. 62–68). It is estimated that in the period 1976–79 roughly 5,000 females made up the female population of green turtles nesting in Surinam. The population feeds off the coast of the states of Maranhão (?), Piauí, Ceará, Rio Grande do Norte, Paraiba, Pernambuco and Alagoãs in Brazil (Schulz 1975, pp. 104–10; unpublished data, Schulz and Reichart).

Year	1968	1969	1970	1971	1972	1973	1974	1975	1976	1977	1978	1979
Number of nests	±5,000	2,495	3,115	5,755	6,885	6,600	7,465	3,610	8,080	4,955	8,465	4,330

The Olive Ridley, Lepidochelys olivacea

The startling decline in numbers of nesting olive ridleys is probably caused by the numbers of turtles that are caught in the nets of the shrimp trawlers that operate off the coasts of Surinam, French Guiana and Venezuela.

Guyana

The only published information is in Pritchard (1969). Based on his observations in 1964–65, he concluded that "Shell Beach seems to be a site of considerable ridley nesting activity." He found hawksbill turtles and green turtles nesting in fair numbers on Shell Beach.

Year	1967	1968	1969	1970	1971	1972	1973	1974	1975	1976	1977	1978	1979
Number of nests	2,875	3,290	1,665	1,750	1,595	1,270	890	1,080	1,070	1,160	1,030	870	795

For a discussion of the dispersion of the population between breeding periods, see Schulz (1975, pp. 111–14).

The ridley population nesting in Surinam, estimated at 2,100 to 3,000 females in 1967–68, has dropped to an estimated 550 to 800 in the last 2 years. Estimates are based on the assumption of an average of between 1.4 and 2 nests/female/season and an interbreeding period of 1.4 years (Schulz 1975, p. 87).

The Leatherback Turtle, Dermochelys coriacea

The rise in the number of leatherbacks nesting in Surinam is probably largely caused by the turtles shifting from the deteriorating French Guiana beach at the mouth of the Marowijne River to the Surinam beaches.

Nesting of green turtles also occurred on some less important beaches east of Shell Beach.

In 1976, I was informed in Georgetown, Guyana, that turtle slaughtering continued unabated on Shell Beach and other sites, which corroborates Pritchard's comment that "parties . . . on the beach were slaughtering virtually every turtle that nested."

In June 1976 I flew over all the nesting beaches (except Punta Plaia on the border with Venezuela) without seeing a single turtle track. I fear that decades of unrestricted slaughtering of turtles has resulted in the extirpation of the Guyanese nesting populations. This predicates the desirability to do a survey of the Guyana beaches.

Year	1964	1967	1968	1969	1970	1971	1972	1973	1974	1975	1976	1977	1978	1979
Number of nests	95	90	200	305	255	285	380	900	785	1,625	670	5,565	2,160	3,900

As little data are available on the number of nests a female lays during one season and on the length of the interbreeding period, an estimate of the population size seems very risky.

Protection in Surinam

Laws give total protection to all marine turtles plus nests in Galibi Sanctuary and on the Matapica-Krofajapasi beaches. The latter have been protected by special decree since the beaches moved outside the borders of Wia-Wia Nature Reserve.

The collecting of eggs is allowed on the small beach west of Matapica. An annual quota of approximately 300,000 to 400,000 green turtle eggs is allowed to be taken from Galibi beaches. These eggs are sold on the local market thorugh STINASU.

From 1968 to 1978, some 300,000 eggs have been sold to the Cayman Turtle Farm. Since 1978, approximately 12,000 eggs/yr have gone to the Surinam Turtle Farm run by STINASU.

French Guiana

The nesting beach in western French Guiana, called Organabo Beach when it was situated near the mouth of Organabo Creek in the 1960s, is by far the world's most important nesting ground of the leatherback, and the population nesting in French Guiana the world's largest. In the 1970s the beach had moved so far westward that the western boundary reached the confluence of the Marowijne and the Mana rivers. By 1979 this beach had virtually disappeared, leaving only a sandspit, washed over by high tides.

The only suitable nesting ground remaining is the approximately 1.5-km-long stretch of sand at the junction of the mouths of the Marowijne and Mana rivers (Les Hattes-Aouarra). Nests are so crowded that a considerable number are destroyed by later arriving females. Moreover, every night spectators drive their cars onto the beach and create massive disturbances among the nesting turtles.

Based on 3 counts during the 1978 and 1979 seasons

JOOP P. SCHULZ

(May to July), I estimated, very roughly, that between 10,000 and 15,000 nests are laid between Aouarra and les Hattes during a season. Fretey, who did extensive tagging in 1978–79, certainly has more precise data. Pritchard (1969) estimated that during a "good mid-season night" 300 leatherbacks nested on the beach that still existed at that time. During any night in July he ". . . would tag well over 100 leatherbacks on a single walk to the end of the beach and back."

Where did all these turtles go? The increase of leatherbacks nesting on the Surinam beaches does not account for the decrease in the number of nests in French Guiana.

Literature Cited

Pritchard, P. C. H.
1969.   Sea turtles of the Guianas. *Bulletin of the Florida State Museum* 13:85–140.
Schulz, J. P.
1975.   Sea turtles nesting in Surinam. *Zoologische Verhandelingen, uitgegeven door het Rijksmuseum van Natuurlijke Historie te Leiden* 143:1–144.

Bernard Nietschmann
Department of Geography
University of California
Berkeley, California 94720

The Cultural Context of Sea Turtle Subsistence Hunting in the Caribbean and Problems Caused by Commercial Exploitation

Changing Patterns

In the Caribbean, the diverse cultural patterns that once linked many indigenous coastal and island societies to sea turtles are mostly gone now, having declined with the demise of the turtles and with the catastrophic loss of human population that occurred after the coming of the Europeans. The varied indigenous patterns of local subsistence use of sea turtles were replaced largely with intensive methods introduced and organized principally by the English and other north Europeans to feed their mariners, colonists, and slaves, and later to supply an export commodity for markets abroad. Yet in spite of almost five centuries of intensive pressure on sea turtles and indigenous peoples, vestiges and amalgams of former patterns persist. Here and there, in isolated backwaters, in corners of the Caribbean too far or too poor or too formidable to attract large-scale foreign colonization, indigenous societies survived as did reduced populations of the formerly abundant green turtles. These patterns, among the last of their kind, are now changing due to internal and external pressures and constraints.

In the big and small islands of the Caribbean, where so much was lost long ago but where also some of the old and introduced cultural traits and lifeways have been maintained in peasant fishing-communities, highly specialized livelihoods are threatened by marine resource depletion and demographic and economic change.

And on the horizon, pressures build for another pattern of change; this one guided by the desire to protect and conserve vanishing species. But what will happen to the indigenous peoples whose cultures are adapted in part to those animals? And what of the peasant peoples whose fishing skills have long provided a measure of economic and social independence and an important means of subsistence? What do these peoples stand to lose if sea turtles are lost through intensive exploitation or if they are protected by prohibitive legislation?

Ecological and Cultural Transformations

The intrusion and spread of European colonization in the Caribbean area meant rapid and widespread changes in biota and environments, the displacement and loss of native peoples and cultures, and the introduction of new peoples, lifeways and approaches to resource use. The decline of native societies in the West Indies was quick and final. Disease, slavery, loss of food supply, and cultural dislocation reduced populations estimated to have numbered several million to a few hundred surviving Amerindians on Dominica and St. Vincent by the end of the 1600s. Along with much of the native island fauna, green turtle populations were severely reduced.

The Island Arawaks and Caribs were skilled in obtaining resources from their marine environment, including sea turtles (Breton 1665; Dutertre 1667, II; Labat 1742, I; Price 1962). Unlike the Island Arawak, the Island Carib were never enslaved on a large scale and resisted European intrusions. They survived for a longer period than did the Island Arawak and "the prolonged and intimate interaction between these Island Carib fishermen and the French settlers played a major role in shaping future local fishing habits" (Price 1962:1370). In the Lesser Antilles French traders introduced large-meshed turtle nets and metal harpoon points to the Island Carib—as did the English in other parts of the Caribbean—and later these materials were sold and traded to African fishing slaves and to communities of freed and runaway slaves (Moreau de Saint-Méry 1797:873; Price 1962:1374, 1376). On the edge of the plantation system, a distinct and continuing fishing subculture evolved based first on African slaves, who provided food for the plantation staff and guests. Although some slaves from the Gold and Ivory Coasts may already have been experienced fishermen, "most slaves clearly learned to fish in their new home, and the techniques they practiced indicate that both the French and the Island Caribs served as teachers" (Price 1962:1371). These people and freed and runaway slaves served as the nucleus around which developed the unique and self-perpetuating fishing subculture in many parts of the Caribbean. "It should be clear that an early synthesis of Island Carib, French and Negro techniques, completed by the mid-eighteenth century, appears almost totally unchanged in fishing villages throughout the islands today" (Price 1962: 1376–77).

Thus, even though populations of Caribbean Amerindians were lost, much of their fishing technology and knowledge was maintained within the economic and later emerging cultural amalgam of African and European influences. The independent and distinctive means of livelihood in West Indian fishing communities are based in large measure on the persistence of Amerindian fishing knowledge and technology and continuing access to marine resources. Although I know of no West Indian society or community that is heavily dependent on sea turtles for subsistence or livelihood—other than the Cayman Islanders, on many islands these animals do contribute a significant source of food and small but important amounts of money.

Uninhabited at the time of European discovery, the Cayman Islands later became an important turtling ground. As the early settlers' dependency on turtles and the sea grew, socioeconomic patterns developed which relied heavily on the commercial and subsistence exploitation of green and hawksbill turtles. After 200 years of intensive turtling in local and foreign waters, large-scale exploitation came to an end when the Nicaraguan government closed the turtle grounds to the Cayman Islanders in the 1960s and markets were lost in the United States and Europe as the result of national and international legislation (Hirst 1910; Lewis 1940; Carr 1956; Parsons 1962; Nietschmann 1979b).

On the mainland margins of the Caribbean, several indigenous societies are culturally and nutritionally dependent upon green turtles whose survival has been threatened by commercial overexploitation.

Types and Scale of Exploitation

As a result of antecedent and introduced cultural and

Table 1. Types of sea turtle exploitation

Subsistence

1. exploitation, exchange and consumption are socially and nutritionally important and part of an indigenous culture complex
2. exploitation and consumption primarily by individual households in peasant fishing communities
3. opportunistic catches by subsistence fishermen
4. exchange and socially-based sale of meat within communities provide meat and small sums of money

Market

5. sale of live turtles, meat and eggs to regional markets
6. sale of live turtles to packing companies for export
7. sale of turtle products (calipee, shell, skins, oil) for export
8. opportunistic catches (primarily of hawksbill) by lobster divers and fishermen for market sale

Incidental

9. incidental catches by shrimp trawlers

economic systems, there exists in the Caribbean region a variety of contexts within which sea turtles have been and are being exploited.

In general, exploitation of sea turtles for subsistence is less of a threat to sea turtle survival than is market exploitation. Subsistence-related activities are dependent on the size of local human populations, their degree of reliance on the animals, and the society's cultural needs and exploitation controls. Exploitation for market, both regional and export, is much more open-ended, dependent on large, external populations with the power to purchase more than can be obtained; hence, the drain on sea turtles that provide meat and luxury by-products would rapidly escalate without legislative controls.

The Cultural Context of Turtling and Turtles

Among the traditional indigenous peoples who still make their own living rather than earning it and who still rely on turtling and turtles for part of their subsistence, the procurement, distribution and consumption of green turtles remain important to their culture. More than any other Amerindian society, the coastal Miskito Indians of Nicaragua and Honduras are culturally dependent on green turtles, as will be seen in the following material. Although other coastal Amerindians are less reliant upon sea turtles, the cultural and social context of resource acquisition and distribution are generally similar in structure and content.

Subsistence provisioning involves production for immediate use, distribution within a discrete social unit and area, and consumption by the producers and their dependents. In subsistence, resource exploitation is organized and internally regulated within kin-based networks to satisfy biological and cultural needs.

Turtling is more than a means to get meat, turtles are more than simply a source of meat, and turtle meat is more than just another meat. For several surviving coastal Caribbean Amerindian societies, turtling and turtles are part of a way of life, not merely a means of livelihood. The activity and the product are not elements that can be simply lost or substituted without consequent deep change in cultural patterns.

Turtling is part of a cultural complex that links people, society, environment, and biota. Rooted in cultural history and followed for generations, it is one of the principal means through which knowledge of the sea and marine life is passed on, technological patterns are maintained, and sea and resource procurement skills are socially rewarded. For many males it is an important if not a major activity and it supplies significant often substantial amounts of protein.

Chelonia mydas is the most important sea turtle exploited for meat by Amerindian peoples. Green turtle meat constitutes a significant source of protein and

item for social exchange. In coastal Miskito villages subsistence turtling supplied up to 70 percent of animal protein in the diet prior to intensive commercial exploitation (Nietschmann 1973). Among these people turtle meat transcends all other food in esteem and social significance.

Miskito society is structured by kinship relationships and all subsistence activities have a social context. Individuals are obliged to freely share subsistence resources, especially if it is turtle meat. Socially based exchanges of resources between kin honor consanguineal and affinal relationships, spread meat distribution through the community, and insure that differential procurement is evened out so that many households share the results of an individual's skills. The giving of meat is as important as its receipt. Without these exchanges, social relationships and the quality of diet in the communities would decline.

But turtle meat also has a symbolic as well as social and nutritional value. Between the sea and the table turtle meat moves along a chain of cultural levels where each transformation increases its symbolic value. Males obtain and butcher the turtles, whose meat is distributed by women to kin and friends. As the meat moves from animals to hunter, from males to females, from individuals to kin and to the larger society, it is increasingly imbued with symbolic significance. In the end, the item is no longer simply meat but a cumulative symbolic record of relationships between nature and people, men and women, and the individual and society.

Thus, turtling and turtle meat are for the Miskito a means by which part of the structure and organization of society is maintained and their place in their world is defined.

Surviving Traditional Societies and Turtles

Because of the worldwide demise of green turtles[1] and traditional turtling societies, few situations remain where native peoples depend on *Chelonia*. In recent and contemporary times, the most important traditional turtling people have been the Seri in Mexico, the Marshall Islanders in Micronesia, the Torres Strait Islanders in Australia, the Andaman, Nicobar and Maldive Islanders of the Indian Ocean, the Vezo and Sakalava of Madagascar, the Bajuni of Kenya and Somalia, and the Miskito of Nicaragua and Honduras.

In the Caribbean, the Miskito are the foremost Amerindian turtling society. Other native groups that still exploit turtles are the Rama of Nicaragua (Nietschmann and Nietschmann 1974); Guaymí (Gordon 1969)

---

1. A survey of subsistence hunting of sea turtles is beyond the range of this short paper and my personal field experience. I thank Bill Rainey for sharing this information on sea turtle distributions and exploitation in the Caribbean.

Table 2. International recoveries of tags placed on nesting green turtles at Tortuguero and Aves Island

From turtles tagged at Tortuguero, 1956–76			From turtles tagged at Aves Island, 1971–76		
Place	Number	Percentage	Place	Number	Percentage
Nicaragua	957	86.2	Lesser Antilles	5	50
Colombia	45	4.1	Dominican Republic	2	20
Panama	28	2.5	Mexico	1	10
Mexico	28	2.5	Nicaragua	1	10
Venezuela	24	2.1	Venezuela	1	10
Cuba	13	1.2		10	
All others	15	1.4			
	1,110				

Source: Carr, Carr, and Meylan 1978:9

and San Blas Cuna (Stier 1976:45–46) of Panama; some Guajiro in Colombia (Kaufmann 1971:76; Rebel 1974:136); Maya along the northeastern coast of Yucatán; and Black Carib (Garifuna) in the Honduras coast–Bay Island area.

Many other peoples elsewhere in the Caribbean also exploit turtles for subsistence and small-scale market sales: for example, the Cayman Islanders; Creoles in Bluefields, San Juan del Norte (Greytown), Limón, Bocas del Toro area and Colón; Colombians in the Golfo de Urabá and waters off Cartagena, Barranquilla and Ríohacha; Venezuelans in the Golfo de Venezuela and off the Los Roques Islands and on Aves Island;[2] various groups throughout the Lesser Antilles; American and British Virgin Islanders; Dominicans in the Cabo Samaná and Puerto Plata areas; Cubans along the southern coast; Jamaicans, especially off the Morant and Pedro Cays; and people from San Andrés and Providence Islands who journey to Serrana Bank, Roncador Cay, and the Albuquerque Cays.

Hawksbills have scattered distributions yet are heavily exploited for their valuable shell. Hawksbill meat is also eaten by some peoples. Areas where hawksbill exploitation is heavy include the Gulf of Honduras (Cabo de Tres Puntas); Bay Islands; Miskito and Set Net Cays, Corn Islands and Cocal Beach in Nicaragua; Almirante Bay, Chiriquí Beach, and San Blas Islands in Panama; Roncador Cay and Serrana Bank in Colombian waters; Los Roques Islands in Venezuela; Huevos Island off Trinidad; the British Virgin Islands; Isla Mona off Puerto Rico; the Pedro Cays; and Cayman Brac, BWI.

The areas where green turtles are most heavily exploited in the Caribbean can be located by using the tag return data compiled by Archie Carr and Bill Rainey (Table 2).

2. Aves Island is the major nesting beach for green turtles in the eastern Caribbean. In 1978 the Venezuelan government established a military outpost there.

The overwhelming number of tags returned from Nicaraguan waters reflects the localized concentration of green turtles and highly focused exploitation for subsistence and market by the Miskito Indians. All other relationships between a native society and sea turtles in the Caribbean fade when compared with the Miskito's dependence on green turtles and the scale of subsistence and recent commercial exploitation.

The Cultural Impact of Commercialized Turtling on the Miskito and Nicaraguan Green Turtle Population

In 1969, on the heels of Nicaragua's failure to accept the Conferencia Tripartita agreement with Costa Rica and Panama for a three-year moratorium on green turtle exploitation, the first of three turtle processing companies began purchasing and exporting turtle meat and calipee. The Miskito had long been involved in a series of export market resource exploitation cycles (rubber, lumber, gold, bananas), each one of which made them increasingly dependent upon purchased goods and wage labor. Economic conditions were depressed in Miskito communities when the turtle companies started up.

The Miskito were the world's best turtle hunters, and the last large green turtle population in the western Caribbean inhabited the nearby shallow waters off eastern Nicaragua. The new companies provided a year-round connective link between local supply and distant demand. The Miskito started to sell large numbers of green turtles to the companies that exported the meat, oil, and calipee to foreign countries. Whereas they were once the central focus of the Miskito's subsistence system, green turtles now became the primary means to secure money to purchase, and the only surviving green turtle refuge came under severe exploitation pressure. From 1969 through 1976, up to 10,000 green turtles were exported every year. Already depleted by Cayman Island turtlers on the feeding ground at Miskito Bank and by Costa Ricans on the nesting beach at Tortuguero, the most

BERNARD NIETSCHMANN

significant West Caribbean green turtle population was subjected to a sudden and intensive rise in human predation. The resulting reduction of the population soon became evident. The average amount of time it took to capture one turtle went up from two man-days in 1971 to six man-days in 1975. Even though hunting was less efficient, more turtles were taken from the depleted herds because more Miskito were hunting and were doing so almost year-round. Furthermore, tags placed on nesting turtles at Tortuguero to study migration and life cycle patterns began to be returned from Nicaraguan waters in unprecedented numbers, indicating a massive upward change in the scale of exploitation (Nietschmann 1979a:8–9).

For the Miskito the commercialization of turtling was different from previous economic booms in that it threatened the internal core of their subsistence and associated social relationships. Economic entanglement led to social quicksand as they became overly dependent on intensive exploitation for sale, rather than their former moderate exploitation for provisioning and social exchanges. Miskito society became as threatened as the species they hunted and netted for market sales. Decline of the turtle populations created the need to further intensify and both resource decline and intensification of turtling led to shortages of meat and agricultural foods and to social conflicts in Miskito communities.

The turtle companies extended credit and advanced foodstuffs to the Miskito so that they could stay on the turtle grounds for long periods, year-round rather than their normal pattern of short, seasonal trips. Simply to pay off the credit extended by the companies and to purchase food for families to live on while the turtlemen were away from the villages meant intensive exploitation. Labor had to be further diverted from subsistence activities to commercial turtling as the resource declined. The quality and variety of the diet fell as money obtained from the export of protein was used to purchase imported white flour, rice, beans and sugar that replaced traditional foods. As the resource further dwindled, and as labor was further overextended, sufficient yields could be maintained only by diverting turtles from subsistence consumption and social distribution to market sale. By 1972, the downward spiraling pattern of market exploitation had already passed the economic threshold where money earned from turtling could not buy the equivalent amount of food formerly produced in subsistence. During a year period between 1972–73, 913 turtles were obtained by Little Sandy Bay Miskito. Of these, 743 (81 percent) were sold to the companies and only 170 consumed in the village (Weiss 1977). By 1975, the resource and subsistence drain hit social rockbottom. Rather than incur the wrath of disgruntled kin by returning to the villages with only a turtle or two, many turtlemen preferred to sell all to the companies as there were no social pressures to distribute money. Social relationships became strained, turtlemen and their families were accused of selfish interests and denial of Miskito traditions, and members of the many households which were without turtlemen became nutritionally impoverished and felt socially abandoned.

In traditional subsistence, the risk of failure was ameliorated by access to a wide range of plants and animals and by generalized sharing of the yields between kin. Differential procurement was common but differential receipt was rare. As the Miskito became further involved in market activities by channeling more and more labor to acquire the culturally most important item to sell, economic risk was added to subsistence risk and magnified through the individualization of production efforts. Where all households once operated within a social network that saved them from a possible economic mistake or plain bad luck, some individual households began to have to secure resources, income, and food and run the risk of possible economic shortfall.

As the economy became increasingly monetized, a few households in every village became financially better off than the majority. Alterations in the focus of production and the individualization of production created economic differences between individuals and households. In order to secure a surplus to sell rather than to share, labor and materials that once helped support the less able, the elderly, widows, widowers, the sick, and the injured were diverted to market.

Until the Miskito began to sell a subsistence resource, they were able to keep distinct the two economic systems: one based on generosity and sharing between kin without expectation of return, and the other based on individual receipt of goods and materials with no expectation to share. But when labor and materials that were once exchanged between kin were channeled into market sales, social devaluation resulted. To sell a subsistence resource is a social contradiction. If a household produces food, it is obliged to share; if it purchases food, its members are under no such social obligation; if they sell what should be shared, they bring into conflict the opposing rules in the two economic systems (Nietschmann 1979a:12–13).

Between 1969 and 1977 intensive commercial turtle exploitation for export markets began to erode the ecological and social heart of Miskito subsistence and culture. Many other factors also contributed: increasing inflation in the price of purchased goods, depletion of other resources, wage labor migration from villages, and the younger Miskito's increasing dissatisfaction with the nonmaterial rewards of a subsistence-based way of life. In many other areas of the Caribbean migration, inflation, and changing lifestyles are reducing the attractiveness of turtling whether for subsistence or for small-scale market sales.

National and international conservation led to the closing of many foreign markets that trafficked in endangered and threatened species. The 1973 Convention on International Trade in Endangered Species of Wild Fauna and Flora (CITES) and U.S. endangered species legislation (1973) were the two most important of many conservation measures enacted that included the protection of sea turtles. In addition, the West Caribbean population of green turtles received a reprieve from impending disaster when the Tortuguero nesting beach was declared a National Park by the Costa Rican government in 1975, and when the Nicaraguan government closed the turtle companies in early 1977 (the Miskito were permitted to continue subsistence turtling).

The loss of the market for green turtles created economic stress in Miskito communities but the decline of the resource would have led to the same result. Their response was to expand once again subsistence agriculture and hunting and fishing which had an immediate benefit in 1978 when the shipment and distribution of food supplies and goods to the east coast were interrupted by the Nicaraguan Revolution. The Miskito still have sufficient land and a relatively rich resource base and stock of green turtles which will provide for both subsistence and social well-being. Yet because of long exposure to outside goods and markets they will continue to feel economically deprived.

The new Nicaraguan government has major problems of economic and social reconstruction in ravaged western Nicaragua. When the government turns its attention to eastern Nicaragua, they may find that some of the social and ecological wounds caused from bleeding resources away from local societies have already begun to heal. May they be sensitive and wise.

If the Tortuguero nesting beach continues to be protected as a National Park and if the Nicaraguan green turtle population and those of other countries can be protected from commercial exploitation, I see no reason why subsistence turtling by local peoples should not continue, especially by Amerindian peoples such as the Miskito, Rama, Guaymí, and San Blas Cuna.

Elsewhere in the Caribbean, curtailment of market exploitation of sea turtles might cause economic hardship but I know of no island or mainland society whose culture is dependent on tortoiseshell; turtle calipee, meat, eggs, oil, skin; or stuffed turtle curios and shellacked carapaces.

Effective management and restocking of Caribbean sea turtle populations can be achieved by national and international programs to protect nesting beaches and marine habitats, and to prohibit commercial exploitation, export and import of sea turtle products (meat, eggs, oil, calipee, leather, shell, and curios). Just as

Miskito society was not saved by selling a subsistence resource, endangered and threatened species cannot be saved by selling them.

There is a possibility that the Nicaraguan government will establish a marine reserve or park in the Miskito Cays area. This zone contains some of the best seagrass and reef habitat and largest number of green turtles to be found anywhere in the Caribbean. Hawksbills may still be fairly abundant. If within a large enough area both sea turtles and the marine habitat were to be protected from disturbance by turtlers, fishermen, lobster divers, and shrimp trawlers then these species would have an optimum chance for increasing their populations. The Miskito Cays-Miskito Bank region is large enough so that subsistence turtling could still be done in areas well outside a designated reserve.

The exploitation of sea turtles for subsistence by Caribbean mainland and island peoples often provides an important source of meat in local diets subject to nutritional decline caused by high-priced purchased foods. Furthermore, the cultural context of turtling and turtle meat in Amerindian societies must not be threatened by market exploitation or by overly prohibitive legislation. If Caribbean peoples once more have use of sea turtle populations, then management and conservation efforts will be well rewarded.

Literature Cited

Breton, R. P. R.
1665. *Dictionnaire Caraïbe-Français.* Réimprimé par Jules Platzmann, Edition Facsimile, Leipzig, 1892.

Carr, A.
1956. *The Windward Road.* New York: Knopf.

Carr, A.; M. H. Carr; and A. B. Meylan
1978. The ecology and migrations of sea turtles, 7. The West Caribbean green turtle colony. *Bulletin of the American Museum of Natural History* 162:1–42.

Dutertre, R. P. J.
1667. *Histoire Générale des Antilles Habitées par les François.* Paris: Thomas Iolly.

Gordon, B. L.
1969. Anthropogeography and rainforest ecology in Bocas del Toro Province, Panama. Department of Geography, University of California, Berkeley.

Hirst, G. S. S.
1910. *Notes on the History of the Cayman Islands.* Kingston: P. A. Benjamin Manufacturing Company.

Kaufmann, R.
1971. Die Lederschildkroete *Dermochelys coriacea* L. in Kolumbien. *Mitt. Inst. Colombo-Aleman Invest. Cient.* 5:87–94.

Labat, R. P. J.
1742. *Nouveau Voyage aux Isles de L'Amérique.* Paris: T. Le Gras.

Lewis, B. C.
1940. The Cayman Islands and marine turtle. *Bulletin, Institute of Jamaica, Science Series* 2:56–65.

Moreau de Saint-Méry, M.

1797.   *Description Topographique, Physique, Civile, Politique, et Historique de la Partie Française de l'Isle Saint-Domingue,* 3 vols. nouvelle edition, Paris: Société de l'histoire des Colonies Françaises, 1958.

Nietschmann, B.

1973.   *Between Land and Water: The Subsistence Ecology of the Miskito Indians.* New York: Seminar Press.

1979a.   Ecological change, inflation and migration in the far western Caribbean. *The Geographical Review* 69:1–24.

1979b.   *Caribbean Edge: The Coming of Modern Times to Isolated People and Wildlife.* New York: Bobbs-Merrill.

Nietschmann, B., and J. Nietschmann

1974.   Cambio y continuidad: Los Indígenas Rama de Nicaragua. *América Indígena* 34:905–18.

Parsons, J. J.

1962.   *The Green Turtle and Man.* Gainesville: University of Florida Press.

Price, R.

1962.   Caribbean fishing and fishermen: A historical sketch. *American Anthropologist* 68:1363–83.

Rebel, T., ed.

1974.   *Sea Turtles and the Turtle Industry of the West Indies, Florida and the Gulf of Mexico.* Coral Gables, Florida: Unversity of Miami Press.

Stier, F.

1976.   Bananas: An account of environmental and subsistence on Playa Chico, San Blas. Working Papers on Peoples and Cultures of Central America, no. 2.

Weiss, B.

1977.   Economia del tortuguero: En cada venta una perdida. In *Memorias de Arrecife Tortuga,* ed. B. Nietschmann, pp. 161–82. Managua: Banco de América.

Henry H. Hildebrand
413 Millbrook
Corpus Christi, Texas 78418

A Historical Review of the Status of Sea Turtle Populations in the Western Gulf of Mexico

ABSTRACT

All species of sea turtles in the western Gulf of Mexico have declined in abundance in every state from Louisiana to Yucatan with the possible exception of the leatherback, *Dermochelys coriacea*, because of an intense exploitation by man of the eggs, subadults and adult turtles.

The major feeding grounds of the Kemp's ridley, *Lepidochelys olivacea*, are identified as the crustacean-rich grounds of Louisiana and the Tabasco-Campeche area of the southern Gulf. Thus, both the feeding grounds and the nesting grounds are in the western Gulf of Mexico.

The historical decline of the green turtle (*Chelonia mydas*) fishery in Texas is partially charted. At its peak it was in excess of 230,000 kg/yr, but it was virtually nothing when the catch of turtles was outlawed in 1963. Evidence was presented that climatic conditions as well as overfishing played a role in this decline.

The small population of loggerhead turtles, *Caretta caretta*, in the waters of Texas is being further depleted by incidental catch in shrimp trawls.

The state of Tamaulipas is not an important feeding grounds for turtles, but it has extremely important nesting grounds for the Kemp's ridley and, in the past, for other species as well.

The state of Campeche has the greatest landings of any state in the western Gulf, but the catch of all 5 species has been declining for more than a decade.

Introduction

The data base for this survey leaves much to be desired. In Texas, the turtle fishery began about the middle of the last century, and by 1900 the catch had declined to insignificance. In other words, one is left with the chore of writing the obituary of a fishery which disappeared before any scientific gathering of data was attempted. In Louisiana, on the other hand, an organized fishery never existed. However, an unknown but

447

sizeable number of turtles became a delectable part of the shrimpers' diet while at sea. Again this type of exploitation by excellent Cajun cooks leaves few records. On the Mexican Gulf Coast until the middle of this century, turtles were regularly taken in a subsistence fishery and very few were sold outside the local village. Turtle statistics gathered from tax records are available from 1948 to the present; however, there is no breakdown by species. Since 1966 data have accumulated from a good scientific program developed by the Mexican fisheries department. René Márquez has contributed greatly to these investigations.

Laud or Leatherback Turtle, *Dermochelys coriacea*

Information on the leatherback is very sparse in the western Gulf of Mexico. It is occasionally taken by shrimp boats in offshore waters, but the numbers are so small that this mortality is not significant. Furthermore there seem to be no data that can be used to assess changes in abundance of the leatherback over the years.

Leary (1957) observed on 17 December 1956 an estimated 100 leatherbacks, ranging in length from 1 to 2 m, along a 50 km line extending north from Port Aransas, Texas. The turtles were apparently feeding on the dense aggregations of cabbage head jellyfish, *Stomolophus meleagris*. This jellyfish is carried out of the bays by the strong currents produced by the northers. This concentration of feeding leatherbacks so near the beach is unusual, but undoubtedly they were attracted by these jellyfish concentrations which occur each year with the onset of winter. After the cold wave of 31 January to 2 February 1951 I observed 2 dead leatherbacks 1.2 to 1.5 m in length, adrift in the channel at Port Aransas. The potential for heavy mortality in winter schools of leatherbacks is certainly present, but no records of such a catastrophe have been unearthed in interviews or in the literature.

The leatherback is certainly widely distributed in Mexican waters, but only in Tabasco (Márquez 1976) have small concentrations been reported. These aggregations are found particularly off Barra de San Pedro from August to November.

Nesting records are rare. Hildebrand (1963) reported the leatherback nesting at Little Shell on Padre Island. One report was based on the record of a taxidermist who had mounted one in 1928. It was turned on the beach, and he still had an egg preserved in a mason jar in his shop from this turtle. The other authority for nesting leatherbacks was Lewis Rawalt, an Audubon warden and a recognized authority on Padre Island. He had observed the species nesting in the mid-1930s, and he had taken a picture of 1 occurrence. I have no records of leatherbacks nesting on Padre Is-

land during the last 40 years.

The nesting of the leatherback along the east coast of Mexico is based on informed individuals. The species has nested in the past at Rancho Nuevo, Tamaulipas according to residents of this rancherria. A few times in the past, they had rendered a leatherback for oil. A leatherback will yeild up to 30 liters of oil according to my informant, and it is reputed to have medicinal value particularly for skin and lung disorders. Fishermen at Anton Lizardo, Veracruz have told me that the species occasionally nests on a nearby island. The turtle is not common there, but when it is encountered it is killed for its meat and oil. I observed the head and entrails of a leatherback that had been butchered near Veracruz on 24 February 1980. Carranza (1959) states that the leatherback is used for shark bait in Yucatan and that it also nests on Alacranes Reef.

Carey or Hawksbill Turtle, *Eretmochelys imbricata*

According to the available information the hawksbill is the rarest marine turtle in the Gulf. However, despite the gathering of eggs and the capturing of turtles at all sizes for the curio trade, a small population has persisted at Veracruz during the 25 years that I have known the area. According to Veracruz fishermen, the species nests from Isla Lobos to Anton Lizardo. They also state that the species prefers to nest on stormy nights. Perhaps this explains its persistence at Veracruz. Although the species is not readily taken with ordinary commercial gear, it is easily captured by skin divers. According to Carranza (1959) and Fuentes (1967) the species is taken in small numbers in both Campeche and Yucatan, but its greatest abundance is in Quintana Roo on the Caribbean coast of the peninsula, a region outside my purview.

There are few records of the hawksbill from Texas. Most specimens I have seen are small individuals (in their first year of life) that wash up on the beach in a moribund condition. Some of these have survived when cared for in aquaria. In 25 years of observations on the Texas coast I have seen only 2 hawksbill which were captured in a healthy condition. One was caught along the jetty at Port Mansfield and the other at an offshore oil rig. Both of these specimens were small, possibly in their second year of life. There are no records for adult hawksbills from the coast of Texas or Louisiana.

Cahuama or Loggerhead Turtle, *Caretta caretta*

In the western Atlantic the greatest concentration of loggerheads is found along the Atlantic coast from North Carolina to southern Florida. Smaller concentrations

HENRY H. HILDEBRAND

are found along the entire Gulf coast from the Florida Keys to Quintana Roo. Lund (1974) estimates the eastern United States population at 25,000 to 50,000 adult loggerheads. There is no reliable count for loggerheads in the Gulf. It is certainly an ubiquitous species, but its total number is small. The species is regularly fished in the waters of the Yucatan peninsula where its center of abundance is on the Caribbean side. There is a secondary but much smaller concentration in the state of Campeche. Fuentes (1967) gives a catch of approximately 365 cahuama for Campeche during a year. Since the fishermen will capture every one they can, this catch is indicative of a small population.

In Texas information gathered by me indicates that loggerheads occur throughout the summer as isolated individuals around oil field platforms, rock reefs and obstructions. Sports fishermen may, at times, see large adults feeding on Portuguese man-of-war at considerable distance from the coast. There may be a southward migration in the fall and a northward one in the spring, but this has not been confirmed by tagging data.

There is no fishery for the loggerhead in Texas. It was utilized before the first world war by the inhabitants of the coastal villages, and a few were marketed. Rabalais and Rabalais (in press) have studied the strandings of marine turtles on the coast of south Texas from Cedar Bayou to Brazos Santiago. They logged 202 dead loggerheads during the period from September 1976 to 1 October 1979. Most of these turtles were subadult (98 percent were smaller than 76 cm). Strandings were greatest during the fall and spring, and in this sense they were correlated with peak inshore shrimping for white shrimp. A few loggerheads are also snagged by sports fishermen, but most of them are released unharmed because it is illegal to harm turtles in Texas. There has been no organized tagging effort on the Texas coast because of the small and scattered population. However, in the early 1960s my students tagged 8 loggerheads. Four of these tags—half—were recovered within a year. These results support a high mortality rate due to the activities of man. Dead turtles on the beaches of South Texas are not a new phenomenon, but no counts were made during the early 1950s when I frequently traveled the beach.

Nearly all substantiated records for nesting loggerheads are in the eastern Gulf of Mexico. There is a record of a loggerhead nesting on Chandeleur Island (Ogren 1977) and many reports of turtle tracks on the same islands. B. Melancon (personal communication) gathered loggerhead eggs on Grand Isle in the 1930s. Texas is usually placed in the nesting range of the loggerhead mainly on the basis of informed observers. Reports of the nesting of large, unidentified turtles must surely be this species. Indeed, the now nearly 300-year-old account of gathering of sea turtle eggs by members of La Salle's ill-fated expedition may refer to

the loggerhead. However, I have been able to document fully only 2 nests—one in 1977 and the other in 1979 on south Padre Island. Hatchlings from these nests were identified by me.

In Mexico I have gathered numerous reports of the tracks of large turtles on the beaches of Tamaulipas north of La Pesca and in Veracruz from Tampachichi to Barra de Corazones. These may prove to be loggerhead nests. Almost all of them are soon despoiled by man. Isolated individuals do nest in some years at Rancho Nuevo, Tamaulipas. Loggerheads also nest around the peninsula of Yucatan, but most of the known sites are on the Caribbean side.

Lora or Kemp's Ridley, *Lepidochelys kempi*

Although the Kemp's ridley was described a century ago by Garman, most of the information on this species has been gathered during the past 25 years. There was no organized fishery for the species anywhere except as a by-catch of a green turtle fishery near Cedar Key, Florida. It is probable that the gray loggerhead eaten occasionally by Port Aransans prior to the first world war was a ridley.

Hildebrand (1963) estimated the nesting population at Rancho Nuevo in 1947 at 42,000 ridleys based on a statement of Andres Herrera and an analysis of the concentration of the turtles in the film. Exploitation of the population has been high for an undetermined number of years. In fact, the colony was located by following up a chance remark of a friend that an Arab trader packed out jute sacks full of eggs on a pack train of 40 or 50 burros. At the start of the investigations in the late 1960s, 5,000 to 6,000 nested there, but by 1975 the population has declined to a few hundred. However, since that time there has been a very slight increase in the number of nesters.

Exploitation of this rookery prior to its official protection in 1966 was confined almost entirely to the collection of eggs. This activity was so widespread in the 1950s and the early 1960s that the demise of the species was predicted (Hildebrand 1963) if it was not stopped. I am not personally aware of any large scale butchering of turtles for their meat or skin, although I have heard and seen such reports. A permit to slaughter ridleys was granted in 1970, but fortunately the arribada did not arrive on schedule and only 5 turtles were killed according to official accounts.

One can ask whether the present rookery at Rancho Nuevo is the sole survivor of many former nesting colonies or whether it is unique. There is no clear evidence of the species nesting in arribadas elsewhere along the Gulf Coast. There are records of scattered nesting from Port Aransas to Alvarado, Veracruz. Fishermen from Rancho Cruz told me that the lora nested by the thousands at Rancho Nuevo and by the hundreds in the area from Punta Jerez to Barra del Tordo. E. Liner (personal communication) informs me that this species probably nested on Ile Derniere in Louisiana prior to the destructive hurricane of 1856. Ogren (1977) saw a small turtle which may have been this species crawling up a beach in the same general area more than a century later. Positive identification could not be made from the air, but daytime emergence and small size indicate a ridley. Percy Vioscai (1961) stated that the ridley nested on the Chandeleur Islands. This record apparently was based solely on tracks seen from the air.

Many inferences concerning the life history of this species have been made on scanty data. The following hypothesis best explains the existing data on the ridley, in my opinion. The Kemp's ridley nests in arribadas at Rancho Nuevo, Tamaulipas and the feeding grounds for the subadults and adults are the highly productive white shrimp-portunid crab beds of Louisiana (Marsh Island to the Mississippi Delta) and Tabasco-Campeche (Chupilco, Tabasco to Champoton, Campeche). As in the case of the nesting grounds, there is some leakage of individuals and small groups to other areas away from the primary sites.

This hypothesis is supported by the following information. First, the ridley feeds primarily on decapod crustaceans, particularly 2 genera (*Ovalipes* and *Callinectes*) of the swimming or portunid crabs. Portunid crabs occur in the greatest concentration in Louisiana and the Tabasco-Campeche area of Mexico.

Secondly, current patterns—either the loop current for northward transport or an eddy for southward transport—favor the distribution of the ridley to the 2 areas. Because of the variability of these currents, large numbers of young turtles in some years could be transported through the Florida Straits via the Gulf Stream. The sporadic nature of the records from the Atlantic Coast certainly support this contention. Most of the turtles which pass through the straits are probably lost to the population. Some may make it back to the nesting grounds. This seems possible if we consider the long migrations from the crab-rich grounds off Ecuador to nesting grounds in Mexico made by *L. olivacea*.

Thirdly, temperatures are favorable in both areas for the ridley. The Mississippi River has built a ramp across the shelf, and the water temperatures are not subject to as much fluctuation as elsewhere in the western Gulf. These waters support a winter population of king mackerel, *Scomberomorus cavalla*, so feeding ridleys can live there.

Fourthly, the offshore drift lines of logs and other debris from the rivers should provide hiding and feeding places for the young juveniles. There are accounts of large numbers of small turtles in such areas. These small turtles have not been caught and identified, but their location points to the ridley.

Fifthly, many young ridleys were formerly taken in shrimp trawls in shallow water near the outer Gulf beaches. Liner (1954) did not give catch rates for the 11 ridleys caught off Terrebonne Parish in 1952 in 4 m of water. However, these data suggest a catch rate of 1 ridley per trawler per day during the spring of the year. His turtles ranged from 3 to 26 kg; many of the ridleys caught off Louisiana, now and in the past, weigh under ten pounds. The presence of immature ridleys in the Tabasco-Campeche white shrimp grounds is supported by my observations and fishermen's statements at Dos Bocas, Tabasco and by Fuentes (1967).

Sixthly, the use of both areas as feeding grounds by mature adults has been demonstrated by tag returns from nesting females (Chavez 1969). The existing data from tag returns and incidental capture support the concept that the species migrates primarily along shore rather than in the open Gulf. Mature turtles probably frequent deeper water than the immatures. Carr and Caldwell (1958) reported a mature female from 26 m off Terrebonne Parish.

New information will be difficult to obtain because the present law on endangered species has virtually dried-up all information from the fishermen who accidently catch the species, and this makes tagging programs hardly worthwhile in elucidating migrations and mortality.

Tortuga Blanca or Green Turtle, *Chelonia mydas*

The history of the green turtle fishery in the western Gulf of Mexico still remains to be written. The fishery was the first to develop in the new state of Texas, and it was the first to disappear. It was little more than a memory by 1900. The abundance of large green turtles in the bays was commented upon by the first European sttlers. When trade started with markets outside the State of Texas is uncertain, but it must have been shortly after independence from Mexico. Turtles could be shipped alive by schooner to New York, or they could go by steamer to New Orleans and then be transhipped to other points. The other method was to process the turtles into meat and soup which could be transported as a canned product. Again there is some question when the first cannery was established. One was in operation in 1859 when $15,000-worth of turtle soup was canned at Indianola. However, there is an indication that the first cannery was established in 1849. The cannery ran an ad in the Corpus Christi Ranchero for 4 months in 1860 and paid $4.00 for every turtle delivered to Indianola. No mention was made of size or species. This cannery disappeared early in the Civil War years. After the war and the yellow fever epidemic of 1867, there was again interest in food processing in Texas. Turtle canning came to be closely associated with the beef packeries. Ray Stephens (personal communication) informed me that the waste from the packing plant was deposited in the bay, and it attracted large numbers of turtles. I assume that the packers thought if they were going to fatten the turtles they might as well can them for a profit. Turtle canneries were reborn at Fulton, Texas, probably in 1872. In 1877 the cannery was buying turtles at $0.02 a pound. A catch of 11 turtles in 1 week at Shamrock Point during the last half of September 1877 was recorded in Ed Mercer's diary (Local History Collection, Corpus Christi Library).

Charles Stevenson (1893) surveyed the Texas fisheries for the U.S. Commission of Fish and Fisheries. He reported a catch of 265,000 kg of green turtle for the entire coast of Texas. Nearly all the turtles were caught in Aransas Bay, Matagorda Bay, and the lower Laguna Madre. Most of the turtles were caught in turtle nets—222,000 kg in Aransas Bay and 22,000 kg in the lower Laguna Madre. In addition 20,000 kg were caught in fish seines in Aransas Bay and 900 kg in Galveston Bay. The main fishing town on Matagorda Bay, Indianola, was destroyed by a hurricane in 1886, and the city was not rebuilt. Consequently the Matagorda Bay statistics for 1890 are included in the figures given above for Aransas Bay, the point where the landings took place.

The scattered accounts of the fishery show turtles were caught primarily from April to November. The turtles were said to be in poor condition in the early part of the year, but fat from August to November. Whether the turtles migrated southward in the winter is not clear. John Priour (personal communication) told me that after severe cold waves a few people would take wagons along the south shore of Corpus Christi turning moribund turtles on the way out and loading them into the wagons on the way back. The turtles were carted to John Superach, a local restaurant owner, who would buy them. One might wonder why the fishery was so intense when the price was never more than $0.04 a kg, and the average catch was less than 1 turtle a day. According to John Priour in other types of work you worked from sunup to sundown for $0.50 a day, so turtle fishing, even if only one 180 kg turtle was caught a week, was attractive to a number of people.

Mostly the turtle nets were set near the major passes but within the bay. The majority of the turtles were caught as they returned to their feeding grounds from their nightly resting places in the deeper water of the bay. In the lower Laguna according to Viktor Delgado (personal communication) they used a different method. A man was placed in the mast of a Port Isabel scow to find the turtles and determine the direction they were moving in the clear water. They would sail around the turtle and set their nets across its path.

The turtle cannery which packed 900 green turtles weighing 110,000 kg in 1890 (Stevenson 1893) moved to Tampico in 1896 apparently because of a greater supply of turtles in that area. The fishery deteriorated badly in the last half of the 1890s. The severe freeze of 1894–95 was a disaster, and the much more severe freeze of 1899 dealt the coup de grace. Some blamed overfishing, particularly on the nesting grounds, for the failure of the fishery to recover from the freeze. Still others placed the blame on the jettying of unjettied passes or the shoaling of passes like Cedar Bayou after 1899. Nevertheless, a few people continued to turtle in a desultory fashion. Harry Mills (personal communication) hung up his turtle nets for the last time in 1935. He told me that he thought he was the last turtler on the coast. A few were marketed as a bycatch by fishing and shrimping crews until it became unlawful in 1963.

Green turtles still inhabit the same meadows they did before the turn of the century but in greatly reduced numbers. The greatest concentration is in the lower Laguna Madre near Port Isabel. A fisherman with a trot line and good location might catch 5 in a year, all small. They are usually foul hooked in the flipper and can be readily released.

The nesting grounds of the green turtle are not well known in the Gulf of Mexico and the nesting grounds for the greens which supported the fishery in the nineteenth century in Texas were never identified. Some still nest on Cayo Arcas, Cayo Arenas, Arrecife Alacranes and Arrecife Triangulos on Campeche Bank. An occasional green still nests at Rancho Nuevo. Repeatedly, individuals have told me that the species regularly nested at Playa Washington prior to the second world war. This location is approximately 19 km south of the mouth of the Rio Grande. Neck (1978) reported green turtles nesting at the mouth of the Rio Grande in 1889. I believe this is in error and that these turtles were ridleys. The most probable nesting place for Texas green turtles are the beaches between Boca Jesus Maria and Tuxpan. If the movement of a turtle cannery from Fulton, Texas to Tampico in 1896 made sense, it was because it was near the nesting ground for the green turtle. No significant feeding grounds are found there.

Status and Outlook

In this section I will review the turtle population in each state in the western Gulf of Mexico. Again, I must point out as in the species accounts that there is a paucity of data.

Louisiana

Kemp's ridley could be logically labeled the Louisiana

turtle because its greatest abundance is found there. It is beyond a doubt the commonest marine turtle in the state. However, it is concentrated in the shallow water from Marsh Island to the Mississippi Delta. Based on information supplied by shrimpers it has declined greatly in abundance during the past 25 years. There appears to be little prospect for a quick improvement in its survival status.

One would expect the highly productive waters of Louisiana to support a significant number of leatherbacks at least seasonally, but there are no supporting data.

Louisiana is not now and probably never was a major nesting area for marine turtles.

Texas

The most important species in Texas was and is the green turtle which fed in the seagrass meadows from Matagorda Bay to the lower Laguna Madre. It is probable that the population is now increasing, but this is based solely on the number of reports we receive concerning the species. It is possible that the increased population and head-starting in Isla Mujeres has resulted in emigration to Texas. The greatest threat to an improved situation in Texas may be a reduction in the acres of grass beds by dredging.

The subadult loggerhead population is obviously under stress because of heavy mortalities due to trawling. Under present fishing pressure along the coast, one would expect the mortality to increase.

The ridley occurs in Texas in small numbers, and it is probable that all adult ridleys moving from Louisiana to the nesting grounds stay near the coast. As in Louisiana and Tamaulipas, they are subject to loss in shrimp trawls.

The leatherback feeds on cabbage head jellyfish, and although it may be locally abundant, it is not greatly affected by human activity in Texas.

The sand beaches of Texas are now and apparently always were of minor importance as nesting beaches. Two, *Lepidochelys kempi* and *Caretta caretta*, of the 3 species that formerly nested in the state still do, but in smaller numbers.

Tamaulipas

Tamaulipas has been known for years as an important nesting area. The only major rookery for Kemp's ridley occurs at Rancho Nuevo. Its beaches were also nesting sites for the loggerhead and the green, although today few use them. The commercial landings of turtles in the state have apparently always been small. On the other hand, the eggs have been heavily exploited. Very few nests are overlooked by man outside the heavily protected area at Rancho Nuevo.

Veracruz

In contrast to the desolate coast of Tamaulipas, many people live near the coast in the state of Veracruz. All 5 species nested along the coast, and the eggs, juveniles and adults were subject to intense exploitation by man. The northern part of the state around Cabo Rojo was probably an important nesting site for the green turtle and possibly for the loggerhead. The area from Isla Lobos to Alvarado was a good nesting and feeding ground for the hawksbill.

The catch, according to official records, has always been small. However, the fishery, although of a subsistance nature, has always been very intense. Turtle nets are still used near the city of Veracruz although the expectation of the fishermen is very low. For example in 1 village 5 turtles were caught during the season—2 green turtles while I was in the area at the end of October 1979.

Tabasco

The fishery in Tabasco has always been small. The principal species is Kemp's ridley. No major nesting beaches are known in the state.

Campeche

Campeche has for a number of years been the most productive state in harvested turtle resources in the Gulf of Mexico. In part, this is due to 2 major fishing ports and more qualified fishermen than any other state on the east coast of Mexico, but it is also due to more productive waters. In addition to nesting beaches at Isla Aguada and Isla Carmen, there are extensive feeding grounds for green turtles, loggerheads and ridleys. The catch has declined considerably in recent years, and there seems little prospect for improvement in the near future.

Yucatan

The catch has always been small in the state and the few active nesting grounds are located on offshore cays.

Literature Cited

Carr, A. F., and D. K. Caldwell
1958. The problem of the Atlantic ridley turtle in 1958. *Rev. Biol. Trop.* 6:245–62.
Carranza, J.
1959. Los recursos naturales del sureste y su approvechamiento. *La Pesca.* Ediciones Inst. Mex. de recursos naturales renovables, A. C. Mexico. 238 pp.
Chavez, H.
1969. Tagging and recapture of the lora turtle (*Lepidochelys kempi*). *International Turtle and Tortoise Society Journal* 3:14–19, 32–36.

Fuentes, C. D.

1967. Perspectivas de cultivo de tortugas marinas en el Caribe Mexicano (1). *Inst. Nal. Inv. Biol. Pesq. Prog. Marcado Tortugas Marinas* 1:1–9.

Hildebrand, H. H.

1963. Hallazgo del area de anidacion de la tortuga marina "lora," *Lepidochelys kempi* (Garman), en la costa occidental del Golfo de Mexico. *Ciencia. Mexico* 22:105–12.

Leary, T. R.

1957. A schooling of leatherback turtles, *Dermochelys coriacea*, on the Texas coast. *Copeia* 1957:232.

Liner, E. A.

1954. The herpetofauna of Lafayette, Terrebonne and Vermilion parishes, Louisiana. *Louisiana Academy of Sciences* 17:65–85.

Lund, F.

1974. Marine turtle nesting in the United States. Report to the United States Fish and Wildlife Survice, 39 pp.

Márquez, R.

1976. Estado actual de la pesqueria de tortugas marinas en Mexico, 1974. Inst. Nal. de Pesca. INP/SI:1–46.

Neck, R. W.

1978. occurrence of marine turtles in the Lower Rio Grande of South Texas (Reptilia, Testudines). *Journal of Herpetology* 12:422–27.

Ogren, L. H.

1977. Survey and reconnaissance of sea turtles in the northern Gulf of Mexico. National Marine Fisheries Service, Panama City, Florida. Mimeo, 8 pp.

Rabalais, S. C., and N. N. Rabalais

1979. The occurrence of sea turtles on South Texas beaches. Manuscript.

Stevenson, C. H.

1893. Report on the coast fisheries of Texas. Report of the Commissioner, United States Commerce of Fish and Fisheries, pp. 373–420.

Vioscai, P.

1961. Turtles, tame and truculent. *Louisiana Conservationist* 13:5–8.

Conservation Theory, Techniques, and Law

David Ehrenfeld
Cook College, Rutgers University
New Brunswick, New Jersey 08903

Options and Limitations in the Conservation of Sea Turtles

ABSTRACT

The options for conserving sea turtles are limited by many things, but especially by the biology of the animals, themselves, and by our inadequate knowledge of them. These limiting factors include mysterious life cycles and obscure ecological relationships, long migrations across international boundaries, unknown population dynamics, unknown taxonomic relationships of different populations, nesting cycles of highly variable length, and an exceedingly long maturation time.

The combination of our incomplete knowledge about sea turtles and the numerous constraints imposed by their biology dictates a very conservative conservation strategy. Many of these limiting factors will not change markedly in the future. I conclude that the best we can do is to concentrate on the protection of existing wild populations, using the simplest and least risky techniques of conservation.

Fortunately, the techniques with the lowest risk and greatest promise are also those with the lowest cost and requiring the least elaborate technologies. (This is also true of conservation-related research.) Highest conservation priority should be given to the following items (listed in no special order): 1) protection of nesting grounds and aquatic habitats, including minimization of environmental disruption at these sites; 2) use of hatcheries and short-range transplantation of nests to protect eggs at the nesting beach; 3) conservation education; 4) control of international trade; 5) national and international coordination of conservation strategies; and 6) dissemination of improved fishing trawls (when available).

I accord lower priority to: 1) long-range transplantation of nests; 2) headstarting; 3) fisheries-type management of the turtle catch; 4) manipulation of sex ratios; 5) cottage industry turtle ranching; and 6) non-commercial captive breeding to maintain gene pools. Commercial ranching and farming cause a net drain on wild populations of sea turtles, and do not belong in a conservation strategy.

It is no coincidence that the conservation methods that have the greatest potential for saving wild sea turtles are those not limited by the biology of these animals or by our ignorance of it.

Introduction

In the preamble to the Draft Conservation Strategy we wrote that "of the . . . factors . . . that determine the fate of sea turtles, only one, the biological factor, is nonnegotiable in a conservation strategy." This idea is of critical importance. No matter what we decide at this conference, and no matter what conservation measures are adopted later, if they are not in accord with the biological facts of life of sea turtles, they will not work. In other words, the options for conserving sea turtles are limited by the animals, themselves, and by our inadequate knowledge of them—limited to an extent rarely encountered in conservation.

The biological limiting factors include: mysterious and inaccessible life cycles for all species, with many of their ecological relationships totally obscure; long migrations that take turtles across international boundaries; unknown population dynamics; equally unknown taxonomic relationships of different populations; nesting cycles of variable length, which make yearly census data difficult to interpret, especially for green turtles; and an exceedingly long maturation time, which makes it likely that many of us will be in our graves before it is possible to know whether our conservation policies have done any good. Of course research will erase some of our ignorance, but most of these limitations are not going to change very much in the near future, and some are fixed in the genes of the turtles and will not change at all.

To see where the limitations apply, and to determine which conservation options have the greatest potential for success, it is necessary to examine critically the conservation techniques that have been suggested for sea turtles. Therefore, I will list many of them, as follows, with some of the technologically simpler and less expensive methods first.

Protection of Nesting Beaches

One of the simplest ways to conserve sea turtles is to make their nesting beaches sanctuaries, either by law or by official regulation. The effectiveness of this procedure depends mostly upon the local traditions of respect or disregard for laws and regulations, and upon the degree to which they are enforced. Another factor that can be important is the size of the reserve—whether it has enough depth to maintain the ecological integrity of the beach, itself, and whether it is wide enough to include the major turtle nesting areas. If conditions are favorable, it is sometimes possible to achieve considerable success in conserving sea turtles by using this technique, without the need to worry too much about the many subtleties of the biology of the turtles.

The simple protection of nesting beaches can be supplemented by a variety of practices designed to protect the area from destructive development, especially development related to tourism and recreation. Sella (this volume) has described how the removal of sand from Israeli beaches for construction purposes destroyed the nesting habitat there; and Witham (this volume) has listed the ways in which it is possible to lessen the human impact upon nesting beaches. This needs no further comment.

Another conservation technique that can greatly enhance the protection of nesting beaches (and sea turtles in general) is local conservation education. This technique is uncomplicated and relatively inexpensive, but it has been little used to date.

Protection of Feeding Grounds and Other Aquatic Habitats

Here the principle is the same as in the protection of nesting beaches, but the application is more difficult. It is far harder to delineate and patrol several thousand hectares of open water than to do the same for a few kilometers of linear beachfront. Nevertheless it can be done, as has been shown by the United States in protecting the turtle hibernaculum sites in the waters off of Cape Canaveral, Florida. Again, once it has been determined that turtles are using a particular area intensively for some purpose, it is not necessary to know too much more about them to effect simple conservation.

Sometimes it may be necessary to protect the aquatic ecosystem from damage by various kinds of human activities: destruction of reefs or reef faunas, and pollution by chemicals, silt, and other contaminants. Petroleum and related compounds are especially significant. Research concerning the responses of turtles to pollutants is lacking and would be interesting, but we do not need research to tell us that a 20-km oil slick is going to be bad for the turtles that get in its way, or, for that matter, that a reef that is repeatedly dynamited is not going to support a large population of hawksbills. As far as habitat degradation is concerned, it is important from the standpoint of management to remember that what happens upcurrent may determine what happens inside the reserve itself.

Management of Turtle Catch for Maximum Sustained Yield

One kind of active manipulation of populations in their aquatic ecosystems is the application of modern fish-

DAVID EHRENFELD

eries management techniques to sea turtles. Here I think we come to the first serious limitation imposed by sea turtle biology on a conservation option. I am not referring to the general criticism of the maximum sustained yield concept; this has already been well-covered by Dodd (this volume), and I agree with both him and Larkin (1977) in their conclusions. What concerns me are the specific problems caused by applying the methodologies developed for fish to the catching of sea turtles. We simply do not have the kind of long-term data on population dynamics and catch per unit effort that are necessary for even rudimentary fisheries management.

In addition, the slow growth rates of sea turtles make fisheries management of them especially difficult. Pritchard (this volume) reports extremely slow maturation rates for *Lepidochelys*, *Caretta*, and *Chelonia*. Balazs (this volume) has data that show that green turtles probably take several decades or more to reach sexual maturity. The danger of any fisheries management models applied to turtles is that the long lag time between turtle hatching and maturity will prevent managers from seeing the effects of their miscalculations during their tenure in the job, or during their lifetimes. Sea turtles are not like most commercial fish species, which mature much more quickly. What happens "now" to a managed turtle population is largely the result of past history, not current management practices, and this is very misleading. Our knowledge is such that sea turtle populations are not yet ready for fisheries management practices aimed at regulating the catch; if they ever are, it will probably be with dynamic pool models that take such variables as age structure of the population, growth rates, and mortality into account. But this kind of information may continue to elude us.

Manipulation of Eggs and Hatchlings at the Nesting Beach

There are 3 kinds of biological management at nesting beaches, each of which involves some interference with eggs and occasionally with hatchlings. Perhaps the least intrusive of these is the local nest transplantation method described by Stancyk (this volume). If this practice does reduce nest predation significantly, it may prove to be a boon to conservation. First, however, some fairly easy questions need to be answered. Will predators learn, in the course of a few nesting seasons, to find the artificial nests? Stancyk indicates that this is a possibility. Will nest transplantation fool predators other than raccoons in other parts of the world? Will the hatching rate be reduced in some places where workers may be badly trained and supervision is lax? Are the artificial nests being dug to the proper depth so that incubation temperatures and other microenvironmental factors are as natural as possible?

The removal of eggs to hatcheries is a more manipulative technique than short-range nest transplantation, and there is evidently enough of a difference to have resulted in a lower hatching rate under experimental conditions. Nevertheless, on beaches where natural hatching is low or nonexistent because of predation, hatcheries are clearly necessary. The Suwelo method (this volume) of protected incubation under natural conditions, coupled with immediate release of hatchlings after emergence is a safe and effective conservation technique, which also has the great advantages of minimal technology requirements and low cost. The principal danger is that the effort may be wasted if too small a percentage of local eggs are used; there is no guarantee that "15 percent of the harvested eggs," a figure cited by both Suwelo and his coworkers and by Siow Kuan Tow and Moll (this volume), will be enough to keep the populations in long-term equilibrium.

The discovery, described by Mrosovsky and Yntema (this volume), that incubation temperature can affect the sex of hatchlings, shows us that it is important to keep incubation conditions as natural as possible in the hatchery. Beyond this, the development of the elegant sexing method described by Owens (this volume) may tempt hatchery managers to use incubation temperatures to alter population sex ratios in some direction that is judged likely to increase fertility rates in the wild. I would caution against this. Our physiological understanding of sea turtles, primitive as it is, is far advanced over our genetic, evolutionary, and ecological knowledge. We have no way of knowing what deliberate manipulation of sex ratios will do to a population over the course of many years, thus there is a great potential for damage from such well-intentioned management schemes.

A final word about hatcheries: in looking at the data reported from Malaysia (Siow and Moll, this volume), I note that the different turtle hatcheries had markedly different annual rates of hatching (20 to 53 percent, 32 to 71 percent, 70 to 90 percent). Unless there is some trivial explanation for this, it might be worthwhile to find out what caused the differences, which are likely to transcend differences related to the species of eggs that are incubated.

The third and least natural method of manipulating eggs at nesting beaches for conservation purposes is to combine the use of hatcheries with headstarting programs, in which the hatchlings are raised to a size at which they are deemed to be less vulnerable to predation before they are released. Headstarting has become a common practice, and the existence of a headstarting program is often used to justify the removal of eggs from nesting beaches for other purposes such as commercial ranching and farming, or long-range transplantation efforts. I want to emphasize, however, that there has not been a proven return of an adult headstarted turtle to its nesting beach. This does not mean that headstarting does not work. But headstarting does involve removing a turtle from a complex and totally unknown sequence of experiences that it would have had in its natural environment, and that may play a necessary role in its development. Everything we know about development in other vertebrates indicates that the genetically programmed sequence of developmental events is distorted in an aberrant environment. The early life histories of sea turtles appear to be very elaborate and take place in a sequence of different environments; there is no reason to believe that environment is less important to them than to other vertebrates.

As Pritchard (1979) has said, the "captive rearing of hatchling sea turtles for release is an experimental procedure, and should never be used as a justification for higher levels of harvest of wild turtle populations, or conducted to the exclusion of direct release of hatchling turtles." There is nothing wrong with headstarting *as an experiment*, provided that it, together with all other uses of eggs and hatchlings, remains an insignificant percentage of the reproductive effort at a given beach. We often hear that survival of hatchlings in the first year of life is only 1 to 2 percent, and that therefore headstarting programs should receive high priority. Yet we should consider that the figure of 1 to 2 percent survival is pure conjecture, not based on one shred of evidence, and that the survival and reproductive success of headstarted turtles after release is also unknown.

Efforts to Establish New Nesting Beaches

There have been a number of these efforts, from the massive Operation Green Turtle, in the 1960s, to the present heroic attempt to give *Lepidochelys kempi* a new chance for survival at Padre Island. The latter program has the benefit of accumulated knowledge, and has been carefully thought out in most respects. Klima and McVey (this volume) identify 4 factors that are considered to be the minimum necessary for potential success of a long-distance transplantation program. These are: 1) natural incubation conditions and orientation exposure for hatchlings on the new beach; 2) headstarting of released turtles; 3) an adequate marking technique; and 4) biologically appropriate release conditions. Of these, only the second is questionable (although the third may be hard to achieve). Headstarting is questionable because, unlike the other "minimum" conditions, it offers the very real possibility of lowering rather than raising the chances for success of the transplantation program. By combining headstarting and long-distance transplantation in the same experiment, 2 sets of independent variables are mixed together. Should the effort to establish a new ridley nesting beach at Padre Island fail or achieve only limited success, we may never know whether it was the headstarting or the transplantation that did not work. It might be better to release the majority of transplanted hatchlings directly upon emergence from the nest, reserving the minority for headstarting. I would certainly advise that this be done in Israel, to maximize the chances for success of their transplantation effort.

In their paper, Klima and McVey give 5 reasons for headstarting *L. kempi.* I have discussed, without citing them, some of these reasons in the section on headstarting above, and in the "noncommercial captive breeding" section, below. But one of these reasons deserves comment here. Klima and McVey state, "to verify the establishment of a second nesting beach at Padre island a 'headstarting' program is required, to produce turtles which can be tagged to provide later identification." But because the tags are unlikely to last until the turtles reach maturity, this is a very weak justification for headstarting.

I want to make one more observation that concerns both headstarting and long-distance transplantation. These experiments are all designed with the idea in mind that hatchling—or even embryonic—turtles may be "imprinted" with the odor or taste of chemicals released by the sand of their natal beaches. This is certainly a possibility, and it costs very little to take it into account. But even though I am one of the originators of the beach imprinting idea, I still must agree with Hendrickson (this volume) that the hypothesis is totally unproven. It may be that other characteristics of the beach environment are more important: infrasound, magnetic field characteristics, nature of the offshore waters, and so forth. If this is so, then headstarting may produce defective animals unable to respond to the cues from their own, or any, nesting beach. Again, this reinforces my warning that headstarting should never be used as a complete substitute for natural nest emergence of hatchlings on their natal or adopted beaches.

Technology to Reduce Incidental Take

The development of this technology may prove indispensable for the conservation of many populations of sea turtles, and it should be pursued energetically. The existence of this research program, however, should not prevent us from recommending that certain critical sea turtle habitats be closed to shrimp and other fishing until trawls that exclude sea turtles are commercially available.

DAVID EHRENFELD

"Cottage Industry" Programs to Raise Sea Turtles for Subsistence, Cash Income, and Release

After the spectacular failure of the Torres Straits Islands turtle farming scheme, it is unlikely that this type of technique will receive widespread support from either conservationists or government officials. But ideas of this sort never seem to die, as we have seen at this conference, so some remarks are appropriate. First, as Nietschmann has repeatedly and lucidly explained, the introduction of cash payments for resources into a subsistence culture, the act of coupling such a culture to the world or regional economy, destroys both the resource and the culture. Cottage industry headstarting or ranching programs turn a subsistence resource into a market commodity. (They also emphasize the value of luxury goods, such as tortoise shell and turtle leather, which rightly have little worth in a subsistence culture.) Even the much simpler policy of buying eggs from native peoples for resale and for conservation purposes is fraught with some of the same risks, although the damage can be intangible and may not appear for a number of years. I accept Dr. Siow's statement (conference discussion) that egg purchase is sometimes necessary. But there is a danger in teaching people that conservation is always accompanied by a cash profit, and a certain danger in running conservation programs on the proceeds from the sale of a resource. We have discovered this in the United States, where state fish and game departments are supported by hunting fees—often with most unsatisfactory results. I think it is very wrong also to assume the superior attitude that peoples in poor countries are incapable of having or acquiring moral feelings of conservation.

Second, it is worth noting that both headstarting and farming are techniques that require sophisticated technology and a high level of scientific control. These features are not available to peoples emerging from subsistence cultures.

Insofar as cottage industry farming or the sale of a part of the egg harvest are based on the assumption that headstarting works, then they are even more risky as conservation ventures.

Control of International Trade

The control of international trade in turtles and turtle products cannot be faulted as a conservation tool from a biological point of view. More will be said about it in conjunction with the discussion of commercial farming, below.

Commercial Ranching, Plus Headstarting to Augment Wild Populations

Apart from any benefits associated with headstarting and the release of a small percentage of their captive turtles, turtle ranches are entirely detrimental to conservtion. The value of headstarting is sufficiently unproven so that it is not enough to justify any commercial ranching operation. In many other respects, ranching is similar to farming, discussed below.

Commercial Farming

I have written about commercial farming elsewhere (Ehrenfeld 1974, 1980) and have concluded that it is detrimental to the conservation of sea turtles for a number of reasons. I see no need either to repeat or to modify my argument now. We have heard some people reject the conservation premises upon which turtle farming is based, and we have heard others defend them. How does one decide between them? Remembering my purpose here to describe the nonnegotiable constraints that sea turtle biology places on various conservation techniques, I will limit my discussion of this controversy to a single table. I have based this table on a paper by Webber and Riordan (1976). The paper was entitled, "Criteria for candidate species for aquaculture," and in my table I have simply evaluated sea turtles according to the criteria of suitability that they list (see Table 1).

What this table says to me is that because of intractable, unchangeable limitations imposed by the biology of the sea turtles, ranching and farming will remain practicable only while international demand for all turtle products, especially the luxury ones of shell, leather, and stuffed animals, remains high, and while the prices of these products also remain very high. It will therefore be necessary for the industry to seek ever wider markets and higher prices if they wish to survive in an inflationary world. According to Mack, Duplaix, and Wells (this volume), the sea turtle is now "the most profitable wild animal in large scale international trade." This explains the survival, even expansion, of the biologically absurd turtle farming industry. We are told that Cayman Turtle Farm and other farms will saturate the markets for turtle products, while continuing to expand these markets. If turtles were gold, and someone found a complex, capital-intensive way to farm gold and make a modest profit, is it likely that the wild gold mines would be abandoned?

Noncommercial Captive Breeding: Preservation of Sea Turtles in Zoos and Aquaria

It has been suggested that we preserve the gene pools of *L. kempi* and possibly other endangered sea turtles by maintaining them in captivity in selected zoos, aquaria, or special breeding ponds. Insofar as this involves the use of a few (perhaps 50) captive-raised individuals, I can see little harm to the idea. But any significant use

Table 1. Suitability of sea turtles as candidates for aquaculture (closed-system farming): evaluation of biological and economic characteristics

Characteristics	Suitable	Marginal or questionable	Unsuitable
Growth rate			X
Ability to take advantage of natural food production in captivity			X
Ability to feed with inexpensive processed foods or waste products			X
Suitability for polyculture			X
Tolerance to crowding		X	
Easy access to unlimited supply of wild juveniles, or complete control of reproductive cycle (including economic control)			X
Short reproductive cycling time			X
Potential for genetic improvement		X	
Hardiness		X	
Initial capital requirements			X
Water purification and waste management costs		X	
Market demand	X		
Price of products	X		

of wild-caught *L. kempi* for captive propagation seems to me to be totally unwarranted, for at least 4 reasons. First, captive breeding programs to save endangered species have been notoriously unsuccessful in the case of species whose biology is complex and badly understood. The biology of sea turtles is complex and badly understood.

Second, in the absence of natural selection pressures, the gene pools of captive animals often undergo rapid and destructive change. The great zookeeper, Hediger (1955), vividly describes how wild animals in zoos or under domestication lose, after a number of generations, both their special sensory abilities and many of their special behaviors associated with reproduction. The difficulty of reintroducing captive-reared Hawaiian geese into their native habitat is but one of many examples.

Third, if we start preserving "gene pools" in captivity, where do we draw the line? Do we keep *L. kempi*, because it is a named species, but discard the Aves Island green turtles because they are considered just a subspecies? There are not enough facilities to save every endangered gene pool.

And fourth, there is always the possibility that showy and popular efforts to create captive breeding populations of sea turtles will drain away efforts and funds from conservation activities that deserve a much higher priority.

This ends my survey of options and limitations in the conservation of sea turtles. In looking over the list, I believe some principles emerge. Most important is that a combination of our incomplete knowledge about sea turtles and the numerous constraints imposed by their biology dictates a very conservative conservation strategy. I conclude that the best we can do is to concentrate on the protection of existing wild populations, using the simplest and least risky techniques of conservation. Fortunately, the techniques with the lowest risk and greatest promise are also those with the lowest cost and requiring the least elaborate technologies. This is also true of much of the research related to sea turtle conservation. And fortunately, most of the conservation techniques are not mutually exclusive and can be applied simultaneously. Finally, conservationists must remember that the results of their efforts, good or bad, are most likely to be seen by their children.

In Table 2, I have given my personal list of priorities for research and techniques of conservation. I have not included commercial ranching and farming in the list, because I believe that they have only a negative impact on conservation. While reading this table, I urge the reader to remember that, as Hendrickson (this volume) has clearly shown, the options and limitations for conservation of sea turtles vary markedly from species to species (see Table 2).

I do not mean to imply that the items in this table with medium or low priority should not be done, rather that they should be done only when we are sure that the effort will not divert needed workers or funds from more important kinds of conservation activity.

It is no coincidence that the conservation methods that have the greatest potential for saving wild sea turtles are those not limited by the biology of these animals or by our ignorance of it—namely, control of international trade, widespread conservation education, coordination of conservation efforts, and the simpler kinds of habitat protection. The greatest irony of this convention may well be that some of the most

DAVID EHRENFELD

Table 2. Priorities in the conservation of sea turtles

Priority	Research	Conservation methods
High	Life histories, especially migrations and the non-nesting portions; population dynamics; critical habitats; effects of egg manipulation (including temperature) and other hatchery-related research; taxonomy and related population genetics; simple, inexpensive, effective tagging methods; improved fishing trawls; effects of nesting beach alterations; turtle product species identification methods	Protection of nesting grounds and aquatic habitats, including minimization of environmental disruption at these sites and designation of critical habitats; short-range transplantation of nests, use of egg hatcheries; conservation education; control of international trade; intergovernmental and interorganizational coordination of conservation strategies; dissemination of improved trawls (when available)
Medium	Control of infectious diseases and parasites in captive animals, especially juveniles; study of biological effects of pollutants; nutritional research; fisheries management research; effects of headstarting and long-distance transplantation	Long range transplantation of nests; headstarting
Low	Effects of manipulations in closed-cycle breeding systems; some high-technology research (endocrinology, sensory physiology, etc.)	Fisheries management of turtle catch; manipulation of sex ratios away from the population norm; cottage industry ranching; noncommercial captive breeding to maintain gene pools

effective conservation actions we can take are not strongly dependent on any further increase in our knowledge of sea turtles.

Acknowledgments

I thank my wife, Joan, for her usual invaluable comments on the manuscript. Dr. Churchill Grimes provided helpful information on the models used by fisheries managers. Dr. Karen Bjorndal gave useful criticism of the final draft and corrected several errors. During the past few years, I have had many conversations about the problems of sea turtles with Dr. Archie Carr, Wayne King, and Peter Pritchard; some of the better ideas in this paper are undoubtedly theirs, but I have forgotten which ones.

Literature Cited

Ehrenfeld, D.
1974. Conserving the edible sea turtle: Can mariculture help? *American Scientist* 62:23–31.
1980. Commercial breeding of captive sea turtles: status and prospects, In *Reproductive Biology and Diseases of Captive Reptiles*, edited by J.B. Murphy and J.T. Collins, pp. 93–96. Lawrence, Kansas: Society for the Study of Amphibians and Reptiles.

Hediger, H.
1955. *Studies of the Psychology and Behaviour of Captive Animals in Zoos and Circuses*. London: Butterworths Scientific Publications.

Larkin, P.
1977. An epitaph for the concept of maximum sustained yield. *American Fisheries Society Transactions* 106:1–11.

Pritchard, P.
1979. "Head-starting" and other conservation techniques for marine turtles Cheloniidae and Dermochelyidae. *International Zoo Yearbook*, 19:38–42.

Webber, H., and P. Riordan
1976. Criteria for candidate species for aquaculture. *Aquaculture* 7:107–23.

Henry A. Reichart
Staff Member, Surinam Forest Service
Director, Foundation for
Nature Preservation in Surinam (STINASU)

Farming and Ranching as a Strategy for Sea Turtle Conservation

ABSTRACT

Regional socio-economic conditions should be included in the formulation of a world strategy for marine turtle conservation. Current conservation principles are based on standards developed by affluent societies and are not necessarily effective in lesser developed countries. A rational approach to conservation of sea turtles in those areas would be to make it economically attractive. This can be accomplished by allowing captive-rearing of turtles, in particular the green turtle (*Chelonia mydas*).

Such projects can benefit conservation in 2 ways: first, by providing the economic incentive to protect the wild populations, which are the source of hatchlings for the project; second, by making it compulsory for the projects to include headstart programs, in which predetermined numbers of captive-reared animals are released to the sea at regular intervals. Marine turtles are a valuable natural resource with considerable economic benefit. International trade in marine turtle products cannot be stopped by resolutions or punitive measures; nor can the wild populations be rebuilt to optimal densities by hands-off policies. Taking advantage of the considerable reproductive potential, and by reducing drastically the high, early-age natural mortality through captive rearing, the wild populations may be increased by headstarting. The Convention for International Trade in Endangered Species (CITES) should make provisions to allow for ranching schemes, and for trade in captive-raised marine turtle products only so that developing countries will be provided with economic motivation to conserve their wild turtle populations.

Introduction

The world strategy for sea turtle conservation must consist of a mosaic of complementary courses of action which aim at the global goal of preserving marine turtle species. Conservation should be their only unifying feature.

The strategy developed during this conference ought to provide for adaptability to local socio-economic conditions, which can vary significantly from country to country. These conditions will greatly influence any conservation attempts. For example, wildlife conservation in North America and wildlife conservation in Latin America serve the same purpose—to protect the well-being of the species. The approach to that goal, if the strategy developed here is to be effective, will have to be quite different in the 2 regions. Where the halo effect of a show-business celebrity kissing a baby seal on some icefloe in Canada may serve conservation efforts in North America well, it will only provoke ridicule in lesser developed countries, and may even be counterproductive. It must therefore be strongly recommended that any world strategy agreed upon include heavy reliance on cultural and economic factors of the particular region involved. Such a strategy master plan will be infinitely more effective than a conservation ethic originating in the affluent countries, that is meant to be applied worldwide.

Conservation principles are based on a valid concept: respect for the rights of wildlife. At no time should this motivation for wildlife conservation be ignored. In fact, it must be included as an important guiding principle in the formulation of a world strategy. Unfortunately, this basic tenet of good conservation is not a salable item in developing countries, and we must adjust our strategy accordingly.

Most marine turtle habitats are located near the lesser developed countries of the world. The main concern of these countries is not the conservation of wildlife, but rather the exploitation of any ready resource in order to improve the standard of living for their citizens. Many examples show that only short-term benefits are being considered in resource-use decisions, while the long-term aspects are conveniently ignored—to be dealt with by future generations. This irrationality is by no means restricted to the developing nations; it has become a fine art in industrial countries as well. Signs, however, point to the growing awareness in some countries that natural resource conservation is not only compatible with economic well-being but is also essential to it. A case in point is the sudden concern of some East African nations for their wildlife resources. This concern is by no means motivated from compassion for the animals themselves but rather by the income derived from the tourists who come to see these animals. Marine turtles do not provide the spectacle the East African wildlife does to attract large-scale tourism. Instead, the strategy for their conservation should be based on a different, but nevertheless economic value.

Discussion

Sea turtles are well known as a commercially valuable natural resource. Management of some of their populations and controlled trade in their products should thus be considered a potentially constructive mechanism in the strategy for marine turtle conservation.

To some people these activities—trade and conservation—appear to conflict, but—if properly applied—they could well be the key to saving the species. Trade and conservation have for too long been automatically regarded as a dichotomy by many conservationists, but a reappraisal of the underlying philosophy may show that these two activities need not be mutually exclusive. International trade in sea turtle products will continue unabated and will probably increase as human populations increase. Unless we change tactics, and use trade to our advantage, wild turtle populations will become irreversibly depleted, maybe before the end of this century.

A few populations seem to be recovering as the result of protective measures (Pritchard, this volume) but it is highly questionable whether the legislative measures of the past decades have made any improvement in the status of marine turtle populations; continuation of current attitudes will be fruitless. The language of the developing world is one of economics. This suggests the inclusion of the following in the preamble of a rational strategy for turtle conservation in those countries: to conserve marine turtle populations it must be made economically atttractive to do so.

The Convention on International Trade in Endangered Species of Wild Fauna and Flora (CITES) is the proper forum to deal with these aspects. CITES provides a major step in a global effort to conserve endangered wildlife, where trade is an important factor in its endangered status. The problem with CITES, however, is that no provisions exist to take advantage of biological properties, such as reproductive potential, of certain animal taxa in order to boost their chances in the race against extinction.

Most presently endangered species have attained this dubious honor through man's activities, be it habitat destruction or excessive harvesting. For instance, no amount of conservation effort will restore the spotted cats to anywhere near their original numbers. They have been overhunted, their habitat is practically gone, and they are not particularly prolific. At best, we can hope to establish sanctuary ecosystems where these species can maintain themselves at current or slightly greater densities. On the other hand, some endangered species are reproductively prolific and the bulk of their natural mortality occurs in early life. Their chances for rapid population recovery are decreased if only natural processes are allowed to run their course. Marine turtles especially fit in this category, most of all the green turtle (*Chelonia mydas*), often called man's most valuable reptile. Utilizing the high reproductive potential of these animals, and simultaneously slashing their early

HENRY A. REICHART

natural mortality through captive rearing of hatchlings, offers a powerful tool to help rebuild and conserve these endangered species.

Despite the destruction of some sea turtle nesting beaches, by and large the feeding grounds are still intact. It would take two major courses of action to rebuild the wild populations: 1) reduce man's incidental and intentional catch of wild turtles; 2) restock the wild populations with captive-reared animals.

Reduction of man's incidental catch will necessitate closer cooperation with fishermen, whether local operators or international fishing fleets. Experience has indicated, at least in the Guianas, that fishermen do not like to catch marine turtles in their nets. An enmeshed mature turtle not only causes considerable damage to expensive nets, but it also reduces the target fish catch. To untangle a mature turtle is such an arduous task that most fishermen hack off the flippers and set the carcass adrift. Each year several such mutilated animals are stranded on Surinam beaches. When extrapolated over the western Atlantic, this number suggests a senseless waste of mature turtles at a stage in their lives when we can least afford to lose them. It is not uncommon for a single fisherman in the Guianas to set 1 kilometer of nets at a time. Multiply this by the number of local fishermen, then add the cumulative lengths of nets set or dragged by fishing fleets of several nations operating further offshore, and it becomes obvious that a lethal barrier is stretching across turtle travel routes in the southern Caribbean. Fishermen would be quite cooperative in accepting alternate net designs which would decrease incidental turtle catch. Antagonistic interaction with the fishing industry should be avoided; it resolves nothing, and the loss of their cooperation will also be a loss for marine turtle conservation.

To reduce the intentional catch of wild sea turtles, where such practice is illegal is, in general, a matter of wishful thinking. Anyone inclined to go out and catch a turtle will do so. Legislation or international resolutions will merely be an exercise in frustration. The only way to reduce this type of exploitation is to substitute an economically attractive alternative. This leads to the second course of action in conserving sea turtles: stocking the wild populations with captive-reared animals from commercial operations.

It was for this reason that during the second meeting of CITES, in San Jose, Costa Rica, in March 1979, some countries argued against tightening the definition for the term "bred in captivity." In spite of this, the U.S. proposed definition of the term was adopted by the convention in virtually original form. This definition should be considered a grave threat to conservation attempts based on captive-breeding schemes. The ramifications are that to initiate captive-breeding projects for marine turtles will no longer be economically

attractive. A farm conforming to the definition requirements would need 15 to 20 years before it could engage in international trade of farmed turtle products. It would require several million dollars-worth of investment with no return to be expected for that period. Any plans for such a project by a developing country will be abandoned. As a result, the hunting pressure on wild turtles will increase.

A turtle ranch would be an economic asset for a developing country. It can be the source of much needed foreign exchange, and it would provide employment for some of the chronically unemployed people in those areas. These 2 aspects alone would make governments and local people aware that conservation of wild turtles can be economically attractive, because these animals form the source of supply for their captive-rearing operations. Developing countries would then have the motivation to protect the wild populations living in their territorial waters.

Preservationists in the industrial nations should clearly understand that without an economic incentive, the people of lesser developed areas in the world cannot be persuaded to care about conservation. Instead of imposing rigorous bans on international trade in captive-reared turtle products, a certain amount of flexibility should be maintained in the decision-making procedures to allow for it.

Based on this concept, and at the request of several member states to the Convention, the Secretariat of CITES was instructed to investigate the possibility of allowing ranching schemes for some marine turtles. This would eliminate the necessity of establishing and maintaining breeding stock for the operation in countries which have a viable green turtle population near their shores. CITES approval would enable some developing countries to utilize this valuable resource with a minimal investment, and give them the incentive to contribute to the conservation of endangered sea turtle species.

Captive-rearing of sea turtles for economic benefit alone does not implicitly aid conservation, but it can open the way for supplemental conservation efforts that local authorities will support and, much more importantly, that the local people will accept. As an additional conservation measure an internationally recognized turtle ranch should be required to include, among other things, a headstart program of releasing some of its turtles to the wild at regular intervals.

Captive-rearing solely for the purpose of release to the wild is a noble goal, one that only affluent societies can afford. Even if such projects were financed by them in a developing country, its rationale would escape the local people, and the entire project would likely end up as a showcase for abuse and corruption.

Allowing subsistence hunting of marine turtles by selected indigenous people (Pritchard 1977) is merely an expedient attempt at turtle conservation by pacifying the natives' demand for unlimited access to the turtle resource. Even though the meat is sought after by many, there are also many natives who do not care to eat turtle meat; some favor only the eggs. Much of this has to do with cultural tabus.

What most of these people do have in common is the awareness of the value of turtle products on the international market, and they would like to capitalize on it. Given the choice, they would rather have the security of a regular income from employment at a turtle ranch than to be relegated to a steady diet of turtle meat.

International trade in turtle products will continue regardless of restrictions or resolutions. The most sensible thing to do is to acknowledge this fact of life and try to develop rational guidelines for this trade by using captive-reared green turtles so it will benefit conservation and the economics of developing nations. The use of prohibition-type tactics in trade restrictions will only encourage unscrupulous people to exploit, and thus destroy, the wild populations because the demand for sea turtle products cannot be legislated out of existence. Instead, national and international regulatory bodies should be empowered to provide controls for trade in farmed and ranched turtle products, with very stiff penalties to violators of established codes of operations. While the search for a practical labeling technique to distinguish captive-reared turtles from wild ones continues, rearing facilities, processing sites, and export and import stations must be rigorously inspected. Any violation must be severely punished by immediate cancellation of the internationally sanctioned operating permit. The cost and responsibility of such controls should be borne proportionally by the participants in each phase of the trade.

For conservationists the key attraction of commercial captive-rearing schemes, is of course, the headstart program. Headstarting should not merely be a secondary function of the facilities, but must be made an integral part of the overall management plan for the project. In fact, an internationally recognized permit

Table 1. Surinam marine turtle conservation program (*Chelonia mydas*)

Annual cycle

Natural (hands-off policy)		Managed (Headstart)	
Start.................	1,000,000 eggs	Start.................	1,000,000 eggs
Doomed	250,000 eggs	Ranch project and market.................	300,000 eggs
	750,000 eggs		700,000 eggs
Hatch failure .... (33 percent)	250,000 eggs	Hatch failure .... (28 percent)	200,000 eggs
To the sea.............	500,000 hatchlings	To the sea.............	500,000 hatchlings
Mortality to age 1 year... (99 percent)	495,000	Mortality to age 1 year... (99 percent)	495,000
	5,000 yearlings		5,000 yearlings
		Release captive-reared turtles.................	1,000 yearlings
			6,000 yearlings

Negative aspects

• No eggs for market will result in extensive poaching and destruction of nests of all marine turtle species nesting in Surinam.

• No eggs for market means loss of funds to hire beach personnel.

• No eggs for market will result in loss of public support on conservation measures and will almost certainly cause poaching of mature nesting females.

Positive aspects

• Sale of eggs provides funds to protect nests of all other marine turtle species in Surinam.

• Hatch failure of natural nests reduced through better predator control.

• Market availability of eggs has reduced egg poaching to almost nil.

• Management program enjoys full support of the Surinam people and there is no poaching of sub-adult or adult marine turtles.

HENRY A. REICHART

to operate a turtle ranch or farm must have this aspect as an obligatory feature.

The Surinam green turtle ranching project, which at this time is only a feasibility study, may serve to illustrate the concept. Of the 4 species of sea turtles nesting on Surinam beaches, the green turtle alone lays about one million eggs each year. A quarter of a million of these eggs would be destroyed by spring tides, if left unattended (Schulz 1975). This constitutes a great loss of a valuable natural resource, both in economic and conservation terms. Through the efforts of Surinam conservation organizations most of the so-called doomed eggs are harvested and sold under close supervision on local markets. Hatchlings for the ranching project also come from these eggs. Proceeds of the egg sales are used to finance conservation work in Surinam on all marine turtles nesting there. The rationale for the controlled egg harvest is shown in Table 1.

For commercial purposes, the ranch turtles are to be raised to an age of 4 years, when they will be harvested. This 4-year cycle is based on the results of a high protein diet. Experiments are in progress in Surinam with alternate local food sources but too little is known at this time to give specific growth rate data. Meanwhile, each year a certain number of turtles in each year class will be marked and released to the sea. The general program pattern is shown in Table 2.

As stated earlier, most natural mortality for sea turtles occurs during the egg and hatchling stage. It is a reasonable, and possibly optimistic, estimate that under natural conditions 1 out of every 100 eggs laid will become a turtle that reaches the age of 1 year. Or thinking in terms of eggs harvested for the project: with the Surinam headstart program, when 1,000 yearlings are released to the sea (as was done last year), the potential of 100,000 eggs is put back for the 10,000 hatchlings taken yearly for the project. The possible benefits to conservation of such headstart programs

Table 2. Turtle ranch pilot project with headstart program, Foundation for Nature Preservation, Surinam (STINASU)

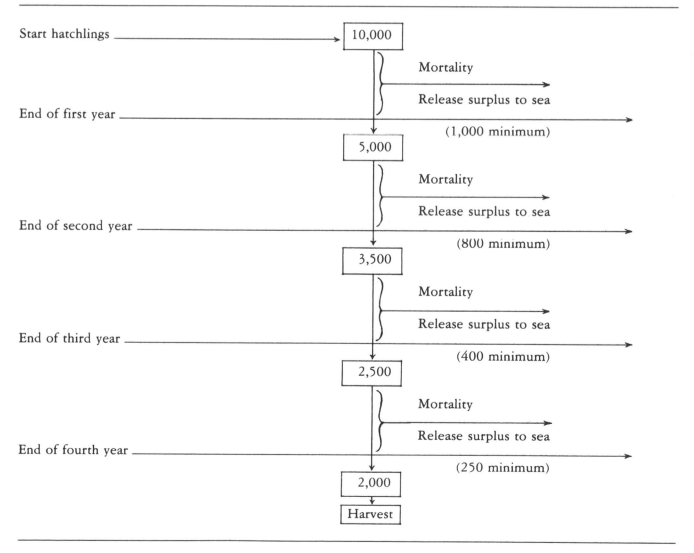

are obvious. There are unknown aspects of headstarting, and the questions being raised are: 1) Will they survive in the wild, after having been raised in captivity? 2) Will they follow species-specific travel routes? 3) Will they find the feeding grounds of the parental stock? 4) Will they be accepted by the wild population, and be allowed to mate? 5) Will they find their nesting beach again as breeding adults?

As a result of tag returns and occasional sightings some of these questions can already provisionally be answered. Headstart turtles are surviving in the wild. Some tag returns which included weight and size information indicate healthy growth. From what is known of the behavior of wild green turtles in this area, the animals are dispersing in the anticipated directions. Some of the turtles released are already on the feeding grounds of the parental stock off the Brazilian coast. They have successfully negotiated the 2,000-odd km voyage from Surinam.

Whether the released turtles will eventually find their beach of origin again after they are mature is largely academic, especially for Surinam. The beach where they emerged as hatchlings will no longer be there—it will have shifted westward by 15 to 20 km, leaving inaccessible mudflats where once the original rookery was. Beach imprinting is still a theory and too much emphasis is placed on its significance in headstart programs. It is more important to provide recruits for the wild breeding stock than to speculate on a homing mechanism for some specific beach.

The merits of headstarting is the subject of intense debate. It has been stated that none of the headstart projects attempted so far has helped increase the numbers of wild turtles in the sea (Ross 1978). A negative appraisal of headstarting on these aspects often indicates preconceived bias and is not based on accumulated data or experience. Current protectionist strategy has not improved the lot of wild turtle populations, and it is time that philosophical discussions on the merits of headstarting make way for the practical application of such programs. Time is running out for the species. Only empirically derived data will provide the answers. With the current Surinam project, a start has been made for a long-term mark and release program of captive-reared green turtles. With subsequent releases of various age classes, emerging patterns may suggest testable hypotheses.

Ranching is a very attractive turtle utilization scheme for developing countries which have the resources, and the inclusion of a compulsory headstart program will be an acceptable feature to them. Additional scientific benefits derived from such projects should not be underestimated. Farming has already provided a wealth of data on various aspects of turtle biology. In addition, it has pioneered the development of techniques in using artificial nest boxes for the hatching of turtle eggs.

The use of this tool alone has enabled Surinam conservation organizations to put more hatchlings in the sea than all other well-meant conservation efforts in Surinam put together. Captive-rearing projects provide an opportunity to study the turtles at close range. Results of such studies can form the basis on which to continue work with wild populations. The facilities can also be used to provide breeding stock of all sea turtle species for select conservation projects.

Too much publicity has been given to the disadvantages of commercial captive-rearing schemes for marine turtles. These claims are the opinion of some, but by no means all, turtle experts. Their views, however, are echoed *ad infinitum* by laymen. As a result, the negative publicity is widespread. The fact is that there are no hard data to show that commercial turtle farms or ranches are detrimental to the wild populations; it is pure conjecture. Equally positive arguments could well be provided, but these are usually suppressed in public interviews.

The data on increased trade in turtle products are undoubtedly valid (Mack, Duplaix, and Wells, this volume), but it must be kept in mind that as human populations are expanding, the use of natural resources will increase. There is hardly a commodity in the world whose trade aspects have not increased. Therefore, no number of trade restrictions will decrease the demand for turtle products. Pressure on wild sea turtle populations will thus persist unless an alternate source for the same product is provided.

Conclusion

A captive-raised green turtle can reach maturity in 8 to 12 years, considerably sooner than a wild one. By releasing captive-reared turtles to the sea at a time when they may reach reproductive condition sooner than wild ones, we accelerate recruitment to the population and thus improve natality, which is an important parameter of population increase. At this critical stage of marine turtle conservation the focus of efforts should be on increasing the numbers of sexually mature animals in the sea. By the largely passive action of letting the turtle's natural reproductive cycle take care of hoped-for population growth, the adherence to current conservation measures will invite disaster for the wild populations.

Through the millennia, a balance has emerged between natality and mortality factors for sea turtles. Man, in just a short time, has been able to upset this balance. Natural reproductive processes will be unable to adjust and compensate for this disturbance, and the species will lose. Nature needs an assist; man, despite his negative effects on the species so far, has the technique to build up the populations. Headstarting is a viable conservation tool, but it will not be feasible unless it

HENRY A. REICHART

is made economically attractive, which farming or ranching does.

Literature Cited

Pritchard, P. C. H.
1977. *Marine Turtles of Micronesia.* San Francisco: Chelonia Press.
Ross, J. P.
1978. Present status of sea turtles. A summary of recent information and conservation priorities. Report to IUCN/WWF.
Schulz, J. P.
1975. Sea turtles nesting in Suriname. *Stichting Natuurbehoud Suriname (STINASU) Verhandeling* 3:1–143.

C. Kenneth Dodd, Jr.
Office of Endangered Species
U.S. Fish and Wildlife Service
Washington, D.C. 20240

Does Sea Turtle Aquaculture Benefit Conservation?

ABSTRACT

As populations of sea turtles continue to decline, the aquaculture of various species, particularly the green, hawksbill, and olive ridley, has been proposed as a method whereby present population levels can be maintained while satisfying the demand for products. In this paper, I examine aquaculture as it may affect conservation of these species. The following topics are discussed: flaws underlying sea turtle aquaculture theory, the chances of increasing survivorship by aquaculture, relative advantages and disadvantages of aquaculture products to wild products, world wide food potential of aquaculture-derived meat, possible conservation benefits of aquaculture research, and other questions. I conclude that aquaculture could stimulate markets thus leading to a proliferation of farms and ranches with no conservation outlook, that it encourages the marketing of luxury products which then undermines local conservation laws and makes enforcement difficult if not impossible, and that farms may spawn ill-conceived yet well-publicized "conservation" activities that could mislead the consuming public into believing that sea turtles are recovering significantly. All could lead to the continued exploitation of wild populations. In addition, aquaculture is based on false premises about the extent of biological knowledge of the species and the applicability of the maximum sustainable yield concept to sea turtle harvest. The proliferation of turtle farms and ranches cannot be justified on conservation grounds.

RESUMEN

A medida que las poblaciones de las tortugas de mar continuan declinando, el aquacultura de varias especies, particularmente el de la tortuga verde, el carey, y la golfina, ha sido propuesto como un método por medio del cual los niveles de poblaciones presentes pueden ser mantenidos mientras se satisface la demanda de sus productos. En este estudio examino cómo el aquacultura de las tortugas de mar puede afectar la conservación de estas especies. Los siguientes tópicos son discutidos: defectos fundamentales en la teoria del aquacultura de las tortugas de mar; ¿incrementará el aquacultura la sobrevivencia de estas especies?; ¿son los productos derivados de aquacultura superiores a aquéllos obtenidos de las especies salvajes?; ¿alimentará a un mundo hambriento la carne derivada de aquacultura?; ¿beneficiará a la conservación la investigación de aquacultura? Otros problemas son también analizados. Concluyo que aquacultura podría estimular los mercados, de este modo conduciendo a la proliferación de criaderos sin perspectivas de conservación, que favorece la venta de productos de lujo lo cual detrimenta las leyes locales de conservación y hace difícil y a veces imposible el enforzamiento de las mismas, y que los criaderos pueden generar ideas erróneas pero bien difundidas de que realizan actividades de "conservación" lo cual alucina al público consumidor con la creencia de que las tortugas de mar están recuperándose significativamente. Todo esto puede determinar la continua explotación de las poblaciones salvajes. Además, aquacultura se basa en falsas premisas acerca de la extensíon del conocimiento biológico de las especies y la aplicabilidad del concepto de máximo rendimiento sostenible en la cosecha de tortuga de mar. La proliferación de criaderos de tortugas no puede ser justificado en bases de conservación.

There is no doubt among sea turtle biologists that the green turtle, *Chelonia mydas,* is threatened with extinction throughout large portions of its circumglobal range, nor is there doubt that many populations have been severely depleted through exploitation in past times by colonial powers (Parsons 1962), incidental take, subsistence take, habitat destruction, and through the demand of luxury markets in the modern world for turtle meat, eggs, soup, and shell. A valuable species often inhabiting places with little or no effective protection lends itself to exploitation. Turtle biologists throughout the world have voiced concern at this situation, and some have suggested ranching or farming of sea turtles as a way to maintain present population levels and yet satisfy the demand for products.

The idea of turtle farming or ranching is not new (Brongersma 1978a). Indeed, in some areas it developed into a local cottage industry although true captive culture through several generations was never achieved (Le Poulain 1941). At present, turtle farms and ranches appear to be proliferating. Perhaps the best known of these operations, and the closest to becoming a closed cycle farm, is Cayman Turtle Farm in the Grand Cayman Islands (see Hendrickson 1974 and Simon, Ulrich, and Parkes 1975 for an overview of this farm) which began operation in 1968 as Mariculture Ltd. Other farms and ranches are or have been located in the Torres Straits in Australia (Bustard 1972; Carr and Main 1973; Applied Ecology Pty, Ltd., 1978), Corail on Reunion Island (Hughes, personal communication), the Seychelles, South Yemen (Leitzell 1978), Mexico, Malaysia, the Philippines (Fontanilla and Carrascal-de Celis 1978), Indonesia (Suwelo, 1973), Suriname (Reichart, this volume), and perhaps additional areas. While conservation of the green turtle is not the primary motivation behind these operations, it is often raised as a justification for their establishment and continuance (see for instance Cayman Turtle Farm, Ltd., 1978). The question therefore arises about the validity of these organizations' conservation claims and techniques.

Proponents of sea turtle farming and ranching cite a number of arguments which they believe support their position; I will outline these briefly. I refer those who desire to review these arguments in greater detail to the following references: Hendrickson (1974), Reiger (1975), Ehrenfeld (1974), Brongersma (1978a, b), Reichart (this volume). These arguments are basically 4.

1) Farming will drive illegal turtle fishermen out of business by offering better and more uniform goods in large quantity thus satisfying market demand. By cornering the market, farms will thus eliminate poaching and result in the protection of beaches which otherwise might not receive any protection. In addition, the elaborate packaging will allow better enforcement of both local laws and international agreements, thus reducing illegal trade.

2) Turtle farms will supply a high source of protein in many areas suffering a protein deficient diet.

3) Turtle survivorship will be increased through the headstarting of young, the salvaging of doomed eggs, and the nonreliance on wild populations of adults, except occasionally to increase genetic variability.

4) Useful research will result, leading eventually to closed-cycle farming techniques that can be adopted throughout the world, to increased knowledge of basic turtle biology which, in turn, will further the conservation of wild populations, and to the development of methods for rearing species or populations verging on extinction. None of these arguments can be justified if subjected to careful scrutiny.

I will restrict my discussion to those operations whose purpose is to raise sea turtles, either by ranching or farming, to provide products for consumption or use to people other than themselves. This would include not only the wholesale commercial marketing of turtle products, such as practiced by Cayman Turtle Farm, but also such activities as the sale of shells or stuffed turtles to tourists even if the edible parts were consumed by the turtle raisers themselves. A ranch is defined as any operation that relies on wild-caught animals or eggs taken from natural beaches and which rears them to an appropriate slaughter size. A farm is a true closed-cycle system or one at least approaching

C. KENNETH DODD, JR.

a closed cycle, that is, it does not in any way rely on wild populations of turtles for present or future marketing. This definition of closed cycle is in accord with that adopted at the meeting of the nations that belong to the Convention on International Trade in Endangered Species of Wild Fauna and Flora (CITES) in San Jose, Costa Rica (19–30 March 1979). The discussion centers on the green turtle because it is most likely to be involved in farming or ranching; however, the arguments apply equally to other schemes advanced for the hawksbill (*Eretmochelys imbricata*) and olive ridley (*Lepidochelys olivacea*). Many of the ideas presented were initially set forth by Ehrenfeld (1974, 1980); his arguments are as apropos now as when first published.

Flaws in Sea Turtle Aquaculture Theory

Perhaps the major flaw underlying marine turtle aquaculture lies in the allegation that too little is known about the basic biology of the various species to insure that actions taken to set up a farm are not detrimental to wild populations, whether by removal of eggs or adults from their habitats for large scale breeding stock. As reviewed by Bustard (1979), we know virtually nothing of turtles during the "lost year," we know little about longevity, sex ratios, population size, reproduction, growth in the wild, recruitment, mortality, etc. Yet if marine turtle aquaculture is developed to a large extent, as its proponents advocate, it would be essential to know these characteristics to insure that during the initial development of closed-cycle operations, perhaps as long as 10 or more years, wild populations were not being severely harmed by removal of eggs and adults.

The argument is often raised that many of the eggs removed for aquaculture purposes are doomed eggs, eggs that would be lost through beach erosion, predation, or other natural causes. While some eggs are certainly doomed, these eggs can just as easily be moved to safe locations as be used for commercial purposes (Ehrenfeld 1980). Indeed, if they are truly doomed, would it not be better to allow them to be used by native peoples in need of high quality protein instead of being taken to be raised as adults to become gourmet items on the well-stocked tables of America and Europe? Ehrenfeld (1980) has also pointed out that there will always be more and more pressure to locate additional doomed eggs to supply various demands. This leads to another problem not only with doomed egg quotas, but any part of the population removed for commercial purposes; how are quotas to be set?

Because so little is known about the recruitment of turtles to the population, in reality we can only guess how many doomed eggs, nondoomed eggs, and adults can be removed while maintaining a viable wild population. Yet in most instances, egg quotas are set in advance based on limited past experience. While this

is usually done carefully, the serious risks involved could lead to the extirpation of populations. For instance, the numbers of individuals nesting on a particular beach may fluctuate drastically. Thus, what was an acceptable level of egg removal 1 year would not necessarily be useful in predicting the next year's removal, or the removal of eggs several years hence. Only by long term investigations of nesting populations prior to exploitation could predictions be made with any degree of accuracy. Unfortunately, very few such studies have been undertaken. Schulz (1975), working in Surinam, has stated the problem clearly:

It is evident that the weak point in the scheme (removal of eggs) is the arbitrariness of the fixation of the quota. We know the feeding grounds and the migration route of the green turtle population nesting in Surinam. We have a good estimate of the average number of eggs produced per annum by a mature female and of the number of hatchlings that every year crawl to the sea. But actually we have no idea about the age distribution, the total reproduction of the population, the duration of life, the dispersion and other information on which an optimal-yield exploitation of the eggs should be based. And even if we could establish how many hatchlings are required to produce one mature female and how many eggs one female produces during her lifetime, there are no quantitative data available on mortality (including the catches in Brazilian waters).

Schulz also pinpoints another problem in both the conservation and exploitation of sea turtles in this passage, namely, that the exploitation or conservation activities of one country may be at odds with the activities in an adjacent country. How, for instance, is it possible to fix exploitation quotas for eggs in one country if all the adults go to another country where they are intensively harvested? Before accurately assigning quotas to insure adequate recruitment, it would be necessary to know where the adults go and the threats that they face. Yet such basic information is almost universally lacking or ignored, especially in many countries where aquaculture is being considered. Schulz (1975) continues:

So there is no answer to this basic problem of conservation, the problem of optimal yield: to what extent can the population nesting in Surinam be exploited for its eggs, maintain itself within a certain size range and at the same time yield a reasonably high production of eggs. The problem becomes practically unsolvable due to the fact that the population is also subjected to the capture of mature turtles in Brazilian waters (on which we are unable to exert even the slightest influence). There being no possibility of predicting the effect upon the population of the harvest of a certain amount of eggs, we can but do our best to continue with closely following the annual number of green turtle nests and to try to adjust the annual egg quota to this trend.

The future will learn whether in this it will be possible to balance the number of young turtles drowned in the nets of the rapidly growing number of shrimp trawlers and the catch of adults by Brazilian turtle catchers.

In this context, it would be well to note that even with the best of care in setting egg quotas for exploitation, such as in Sarawak, severe declines have still occurred (Harrisson 1976).

Schulz (1975) touches another topical question: whether or not sea turtles are proper subjects for aquaculture—the idea that they and their eggs are resources that can be harvested on a maximum sustainable yield (MSY) basis, at least during the initial development of a closed-cycle system. Proponents of aquaculture believe they can; I would disagree. The concept of maximum sustainable yield has been well regarded in the management of species, but as Holt and Talbot (1978) point out, "The embodiment of simplistic formulations in legislation has reinforced a belief that hypotheses, such as that the size of a stock essentially determines the yield it can sustain in perpetuity, have in fact been validated, and that the desirable state of a resource system can be exactly specified in terms of a single criterion. That belief does not survive scrutiny . . ." These authors discuss in detail the MSY concept and conclude that it.:

1) *focuses attention on the dynamics of particular species or stocks without explicit regard to the interactions between those species or stocks and other components of the ecosystem;*
2) *concerns only the quantity and not the quality of potential yield or other value from the resource;*
3) *depends on a degree of stability and resilience of the resource that may not exist;*
4) *focuses attention on the output from resource use, without regard to the input of energy, of other natural resources, and of human skill and labor required to secure the output;*
5) *may admit, and even encourage, overexploitation.*

Each of these criticisms applies to obtaining stock for sea turtle farms. How long would eggs need to be removed from beaches? Cayman Turtle Farm and its predecessor obtained eggs for over 10 years, yet have still not developed a true closed-cycle operation (Dodd 1978; Leitzell 1978). Eggs have been taken on a MSY basis, a basis which in fact has no support in the case of sea turtles, and which, as pointed out by Holt and Talbot (1978), is not a conservation technique by itself (also see Larkin 1977, and Ehrenfeld, this volume, for a discussion of the MSY concept). How long would adults need to be removed from wild populations? This is still not clear. Cayman Turtle Farm and its predecessor have reportedly spent over $17,000,000 to date, yet have had only limited success in developing a closed-cycle system. During over 10 years of operation, they have continued to rely on eggs and adults from wild

populations. It is this developmental period, when wild stocks could be most severely affected, that aquaculture proponents so often overlook. I do not mean to single out Cayman Turtle Farm as a culprit; however, anyone wishing to set up turtle farms should consider its experience and examine carefully the concepts underlying development of initial stock as well as the need for profit while attempting a closed-cycle system.

The arguments enumerated above are directed at the establishment of farms, but they apply equally to ranches since ranches make no pretense about ever ending their reliance on wild populations. The underlying assumptions remain that turtles can be harvested on a MSY basis (not valid) and that enough is known about sea turtle biology and population characteristics to set take quotas (also not valid). Proponents may allege that ranching may divert continuous take of adults but, as will be discussed later, this argument is weak.

Will Aquaculture Increase Survivorship?

Proponents of aquaculture claim that increased survivorship will result if a proportion of young from doomed eggs or eggs layed from captive adults are raised to juvenile size and then returned to the sea. Unfortunately, while there has been much speculation about rates of survivorship and recruitment, as Ehrenfeld (1974) and Bustard (1979) point out, nothing is known about these factors in wild populations. Therefore, our best guesses about the number of juveniles to be returned to maintain a population are just that—guesses. While it is indeed better to release some young instead of allowing whole clutches to wash into the sea, the wholesale removal of eggs from good beaches cannot be justified by saying that only 1 percent would survive anyway. The assumption of a solid survivorship value is not based in fact (Hirth and Schaffer 1974). It is also not certain that all aquaculture operations would be willing to return a proportion of juvenile turtles to the ocean.

When, where, and how would juveniles be released? The green turtle is very site-specific when nesting (Carr 1973; Carr, Carr, and Meylan 1978), and Rainey (personal communication) has shown that as many as 5 genetically identifiable populations may dwell in the Caribbean alone. Before a release could be successful, the genetic identity of each individual and each wild cohort's approximate location would have to be known. Each turtle would have to be taken to that area and released. Many operations may not be willing or able to shoulder the high expenses and careful recordkeeping this would entail. And still the problem of imprinting and other aspects of a young turtle's life history would have been ignored (Ehrenfeld 1974, 1980, and Pritchard 1979).

C. KENNETH DODD, JR.

Finally, it is argued that survivorship will increase should local people become accustomed to ranching instead of harvesting wild eggs and adults. It is difficult to understand how this would be so unless 1) *every* individual in an area was involved in the ranch and thus had a stake in its operation, and 2) there was no outside market for products. Simultaneous fulfillment of both conditions is unlikely (Applied Ecology Pty, Ltd., 1978). Altruism is unlikely (for example, stopping harvest of an available exploitable commodity, Hardin 1977), and worldwide demand for turtle products reaches every sea turtle location (Balazs and Nozoe 1978). Again, how many eggs and adults could be ranched without hurting the wild population?

Are Aquaculture Products Superior to Wild Products?

Proponents of aquaculture claim that products of farm-raised sea turtles, except oil or calipee, are superior in quality (thick, clear, less-scarred scutes; softer and more flexible leather; better tasting meat) to products of wild populations (Leitzell 1978). While the subjective nature of this claim has been pointed out (Ehrenfeld 1974), the difficulty of insuring that wild turtle products do not enter commerce make the claim irrelevant. For instance, how does a customs agent decide the relative beauty of a tortoise shell comb? Hendrickson (1979) has shown that biochemical analysis cannot now distinguish wild turtle meat, leather, or shell products from those derived from aquaculture, even if each article carried across a border by tourists or each shipment could be analyzed. Thus, the wholesale marketing of such items could undermine local laws designed to protect wild populations. Anyone could label a can of meat as being from a turtle farm; customs inspectors could not tell the difference. Leitzell (1978) provides additional evidence of the impossibility of enforcing exemptions for aquaculture products from trade restrictions.

Dodd (1978) has also questioned the assumption that superior products will drive cheaper ones off the market. The assumption is wishful thinking and totally ignores worldwide business and marketing practices. If a demand is created, as indeed any business must do to show a profit and satisfy shareholders, products will be marketed to satisfy the demand. In fact, "less superior" products might be cheaper and find a greater market than those of aquaculture farms. The marketing of "superior" products could thus seriously harm wild populations indirectly and lead to further exploitation of depleted stocks. There is no reason to believe that a person of modest means would not be as happy with a wild sea turtle product as a wealthy person with an aquaculture product.

Proponents of aquaculture claim that no new markets would be stimulated by "superior" aquaculture products. However, both Leitzell (1978) and Ehrenfeld (1974) have convincingly argued that such claims are unfounded. An industry that does not cultivate both existing and new markets is an industry with a short future. The intentions of Cayman Turtle Farm and its predecessor in this matter are clear from the advertising accompanying promotions (Leitzell 1978). The advertising attending turtle product marketing would undoubtedly stimulate demand which would be filled from wild populations. The interest generated by a single large commercial farm should be more than enough evidence of the potential for large markets throughout the world (Mack, Duplaix, and Wells, this volume). It is folly to believe that illegal poaching and the proliferation of farms, often without any attempt at conservation, would not try to satisfy this market. It is also unlikely that farms would refrain from selling their products until they achieve closed-cycle status (Dodd 1978).

Will Aquaculture Meat Feed a Hungry World?

The notion that the green turtle will provide a valuable source of protein for hungry people may be the weakest argument to justify aquaculture. Turtle farms are capital-intensive operations and will likely remain so for some time to come (Ehrenfeld 1974). They entail large holdings, advertising, and marketing networks that can be financed only by pricing products to cover costs. The main markets for such products are Europe, Japan, and America, areas not suffering from protein want. Instead, turtle soup and meat are sold as luxuries in gourmet restaurants and stores. Before being banned in the United States, soup sold at $1.85 the 13-ounce can, hardly fare for the poor. Even Cayman Turtle Farm has abandoned the protein argument, admitting that the real market is for polished shell, not meat (P.C.H. Pritchard, personal communication.)

As noted, the large scale removal of eggs to supply a large, and probably foreign-financed company may rob local people of a limited protein source. The market stimulation caused by the proliferation of turtle farms could also lead to the conversion from a protein-sufficient, subsistence economy to a protein-deficient, cash economy, as happened in Nicaragua (Nietschmann 1972; 1979; this volume), as people exploit their resources at the urging of foreign buyers.

Could turtle ranches supply needed protein within a limited geographical area? Perhaps. However, it would depend on whether the turtles are allowed to exploit their proper trophic level (Ehrenfeld 1974) or are fed and raised on high quality protein, as has often been done on turtle farms and ranches (Hendrickson 1974).

For anything but natural-diet feeding, a source of income will have to be derived to offset feeding costs, thus raising questions of production for a market and outside financing (Applied Ecology Pty, Ltd., 1978). In this case, the main purpose of the ranch now becomes production, not feeding local people.

What if turtles are fed on turtle grass (or equivalent natural diet)? In this case, a ranch would become unnecessary since turtles could be caught for subsistence use without the bother of penning them. The Miskito did this for centuries without apparent harm; only when an outlet to supply the world market became available did their way of life change radically (Nietschmann 1972). With healthy populations of turtles, ranches become unnecessary to serve the needs of local coastal people.

Will Aquaculture Research Benefit Conservation?

Proponents of aquaculture contend that valuable research is performed which will eventually benefit the conservation of wild populations (Cayman Turtle Farm 1978). Undoubtedly some beneficial research will result which may have application to rearing turtles in captive propagation, such as the critically endangered Kemp's ridley (but see Leitzell 1978) or as offshoots of other investigations (Applied Ecology Pty, Ltd., 1978). However, the vast majority of research projects will of necessity be concentrated in areas dealing with raising sea turtles for market, such as nutrition and disease control in crowded conditions. For instance, of the 61 research projects cited by Cayman Turtle Farm in their lawsuit concerning U.S. import prohibition of sea turtle aquaculture products, only 4 related directly to the natural history and conservation of *Chelonia mydas* (Cayman Turtle Farm 1978; Dodd 1978). The rest dealt with various aspects of turtle farming or laboratory-related studies of physiology and endocrinology. While interesting, such studies are often not vital to the continued existence of the species and cannot be used to justify continued exploitation of this species (but see Owens, this volume). Operations not as well funded as Cayman Turtle Farm could hardly be expected to finance much conservation-related activities without commercial application. Commercial farms and ranches are, after all, just that—commercial. They do not raise turtles for conservation but for profit. Research derived from aquaculture ventures is not worth the risk to wild populations.

Other Problems

There are a number of other problems with the aquaculture of sea turtles, such as disease and infection control, physical housing and maintenance, costs of upkeep and marketing, and the potential release of individuals via natural disaster, thus introducing what may be ill-adapted individuals into local gene pools. Ehrenfeld (1974) discusses these in detail. While many problems, such as disease, may be overcome in time, the solutions to problems associated with housing and so on are capital- and labor-intensive thus insuring that sea turtle products will remain luxuries. Indeed, economic pressures have recently forced farm closings in Australia (Cooper, personal communication) and Indonesia (Polunin and Nuitja, this volume). This further weakens the conservation argument of aquaculture proponents.

Why is interest in sea turtle aquaculture today so prevalent? For one reason—it could be profitable because of the potential luxury market. All other arguments in its favor are secondary to this although I do not mean to imply that all proponents favor trade. Do we need luxuries at the expense of declining populations? Brongersma (1978b) states, "Although one may consider turtle products as luxury, one does not give up luxury to which one has become accustomed, especially not at a time when the standard of living has become very high." This is a surprising sentiment and should be examined closely by all countries where aquaculture is being considered. It is also debatable. For instance, many countries have curtailed trade in crocodilians and spotted cats without hurting their economy or standard of living. It is important that the markets are not likely to be in the countries where the turtles are raised, nor in most cases is the capital locally available. Aquaculture encourages the exploitation of species universally recognized as endangered or threatened, usually as a result of exploitation in world trade. Sea turtle aquaculture encourages this continued exploitation and encourages native peoples to rely on dwindling resources. As King (1978) has pointed out, the only way to protect commercialized species is by a universal ban on such trade.

Conclusion

In the brief amount of space available, I have set forth the major evidence against the aquaculture of sea turtles. Aquaculture: 1) stimulates markets which further stimulates the proliferation of farms and ranches; 2) encourages the marketing of farmed products for the luxury trade which then undermines local conservation laws; 3) farms may spawn ill-conceived yet well publicized "conservation" activities. All of the above lead to the exploitation of wild populations which may then decrease local standards of living over a long period of time. Sea turtle farms and ranches are based on false premises about the extent of biological knowledge of the species and about the applicability of the maximum sustainable yield concept. Are aquaculture ventures

C. KENNETH DODD, JR.

justified in the future? Perhaps, but only after populations are allowed to rebuild to the point where they are no longer endangered or threatened. At present, sea turtle aquaculture can only be considered a threat to the survival of these endangered species.

Acknowledgments

I would like to thank C. J. Limpus, G. R. Hughes, J. R. Hendrickson, N. Mrosovsky, L. D. Brongersma, and R. D. Cooper for providing information for this paper, although by no means do they necessarily agree with the conclusions. Archie Carr, Jack Frazier, David Ehrenfeld, F. Wayne King, and P. C. H. Pritchard provided valuable comments and criticism of the manuscript.

Literature Cited

Applied Ecology Pty. Ltd.
1978. Fifth annual report 1977–78. Australian Government Publ. Serv. Canberra, 28 pp.

Balazs, G. H., and M. Nozoe
1978. Preliminary report on the hawksbill turtle (*Eretmochelys imbricata*) in Indonesia, Philippines, Malaysia, and Singapore. 1973 report of the Japanese Tortoise Shell Association, English version, 73 pp.

Brongersma, L. D.
1978a. Schildpaddenfarms. *Verslag gew Verg. Afd. Natuurk., Kon. Ned. Akad. Wetensch.* 87:136–40.
1978b. De bedreigde Zeeschildpadden. *Panda* (March): 35–39.

Bustard, H. R.
1972. *Australian Sea Turtles.* Glasgow: Wm. Collins Sons and Company.
1979. Population dynamics of sea turtles. In *Turtles: Perspectives and Research,* eds. M. Harless and H. Morlock, pp. 523—40. New York: John Wiley and Sons.

Carr, A.
1973. *So Excellent a Fishe.* Garden City, New York: Anchor Natural History Books.

Carr, A., and A. R. Main
1973. Turtle farming project in northern Australia, a report on an inquiry into ecological implications of a turtle farming project. Canberra, Australia: Union Offset Company Party Limited.

Carr, A.; M. H. Carr; and A. B. Meylan
1978. The ecology and migrations of sea turtles, 7. The West Caribbean green turtle colony. *Bulletin of the American Museum of Natural History* 162:1–46.

Cayman Turtle Farm, Limited
1978. Comments of Cayman Turtle Farm, Limited, on reconsideration of the prohibitions prescribed in 50 C. F. R. 17.42(b)(1) and 227.71 (July 28, 1978) insofar as those prohibitions apply to farmed turtle products derived from the mariculture operations of Cayman Turtle Farm, Limited. Shaw, Pittman, Potts and Trowbridge (1800 M Street, N.W.) Washington, D.C., 34 pp.

Dodd, C. K., Jr.
1978. Memorandum–Response to comments of Cayman Turtle Farm, Limited, on the final rules regarding mariculture exemptions for the green sea turtle, *Chelonia mydas.* 15 pp. (available from author).

Ehrenfeld, D. W.
1974. Conserving the edible sea turtle: can mariculture help? *American Scientist* 62:23–31.
1980. Commercial breeding of captive sea turtles: status and prospects. In *Reproductive Biology and Diseases of Captive Reptiles,* eds. J. B. Murphy and J. T. Collins, SSAR Contributions to Herpetology Number 1.

Fontanilla, C., and N. Carrascal de Celis
1978. The sea turtle that lays the golden egg is vanishing. *Canopy* 4:8–9, 12.

Hardin, G.
1977. *The Limits of Altruism: An Ecologist's View of Survival.* Bloomington: Indiana University Press.

Harrisson, T.
1976. Green turtles in Borneo. *Brunei Museum Journal* 3:196–98.

Hendrickson, J. R.
1974. Marine turtle culture—an overview. Proceedings of the 5th Annual Meeting, World Mariculture Society, pp. 167–81.
1979. Chemical discrimination of tortoiseshell materials and reptilian leathers. Report to U.S. Fish and Wildlife Service, Albuquerque, N.M., 54 pp.

Hirth, H. F., and W. M. Schaffer
1974. Survival rate of the green turtle, *Chelonia mydas,* necessary to maintain stable populations. *Copeia* 1974:544–46.

Holt, S. J., and L. M. Talbot
1978. New principles for the conservation of wild living resources. *Wildlife Monographs* 59:1–33.

King, F. W.
1978. The wildlife trade. In *Wildlife and America, ed. H. P. Brokaw, pp. 253–71. Washington, D.C.: U.S. Government Printing Office.*

Larkin, P. A.
1977. An epitaph for the concept of maximum sustained yield. *Transactions of the American Fisheries Society* 106:1–11.

Leitzell, T. L.
1978. Decision memorandum—Whether the sea turtle regulations should be amended to allow importation of sea turtles. National Oceanic and Atmospheric Administration, U.S. Dept. of Commerce, Washington, D.C. 31 pp.

Le Poulain, F.
1941. Note sur les tortues de mer du Golfe de Siam. In *Les Tortues de l'Indochine,* ed. R. Bourret, pp. 215–18. Nhatrang: Inst. Oceanographique de l'Indochine.

Nietschmann, B.
1972. Hunting and fishing focus among the Miskito Indians, eastern Nicaragua. *Human Ecology* 1:41–67.
1979. Ecological change, inflation and migration in the far western Caribbean. *Geographical Review* 69:1–24.

Parsons, J. J.
1962. *The Green Turtle and Man.* Gainesville: University

of Florida Press.

Pritchard, P. C. H.
1979.   "Head-starting" and other conservation techniques for marine turtles Cheloniidae and Dermochelidae. *International Zoo Yearbook* 19:38–42.

Reiger, G.
1975.   Green turtle farming, a growing debate. *Sea Frontiers*, 215–23.

Schulz, J. P.
1975.   Sea turtles nesting in Suriname. *Zoologische Verhandelingen* 143:1–143.

Simon, M. H.; G. F. Ulrich; and A. S. Parkes
1975.   The green sea turtle (*Chelonia mydas*): mating, nesting and hatching on a farm. *Journal of Zoology*, London, 177:411–23.

Suwelo, I. S.
1973.   Notes on turtle ranching at Tidung Island. Universitas Nasional, Jakarta. Manuscript, 3 pp.

C. KENNETH DODD, JR.

Edward F. Klima
James P. McVey
Department of Commerce, NOAA
National Marine Fisheries Service
Galveston Laboratory
4700 Avenue U
Galveston, Texas 77550

Headstarting the Kemp's Ridley Turtle, *Lepidochelys kempi*

ABSTRACT

A summary of the international program to restore and preserve the Kemp's ridley turtle is provided. The program can be divided into 3 main parts: 1) enhancement of nesting success and survival at Rancho Nuevo, Tamaulipas, Mexico; 2) establishment of a second breeding population at Padre Island National Seashore in Texas, and 3) an experimental study to evaluate the concept of headstarting.

In 1978 the Galveston headstart turtle program obtained a 68 percent survival rate and released approximately 2,000 juvenile turtles. The Kemp's ridley turtle is best reared in individual containers to avoid aggressive behavior. Immediate treatment of damaged or ill turtles with antibiotics provided a 95 percent recovery rate. The turtles were released at 3 locations in the Gulf of Mexico. All turtles received flipper tags, and selected turtles released at Everglades National Park and Homosassa, Florida, were equipped with radio transmitters to allow radio-tracking by plane and boat. The yearling turtles did not remain in the areas where released and exhibited pelagic behavior rather than benthic orientation. One animal recovered at Jekyll Island, Georgia, 8 months after release showed an increase of 2,700–4,000 g in weight and provides evidence that 1 of the project objectives, to demonstrate survival after release, may have been met.

Introduction

The Kemp's ridley turtle, *Lepidochelys kempi,* is an endangered species that nests primarily on a single beach in the western hemisphere—Rancho Nuevo, in the state of Tamaulipas, Mexico. In 1947, over 40,000 nesting females used this isolated beach; however, the nesting population in recent years has ranged between only 200 to 500 females a year. Unless positive steps are taken to protect the nesting beach and improve recruitment, the species is threatened with extinction.

Representatives of the U.S. Fish and Wildlife Serv-

Table 1. Summary of nesting and hatching success of Kemp's ridley turtle

Year	Number of nests collected	Estimated number of nesting females	Total number of eggs	Number of eggs in corral	Number of eggs in styrofoam containers (Mexico)	Number of eggs in styrofoam containers (Padre Island)	Hatching rate in corral (percentage)	Hatching rate in styrofoam containers (Mexico) (percentage)	Hatching rate in styrofoam containers (Padre Island) (percentage)
1978	711	450	85,000	65,000	18,000	2,000	57	64	88.1
1979	950	500	97,600	89,000	6,500	2,100	68	80	85.6

Source: U.S. Fish and Wildlife Service.

ice and the National Park Service presented an action plan for the restoration and enhancement of the Kemp's ridley turtle to representatives from the Texas Parks and Wildlife, National Marine Fisheries Service and the Instituto Nacional de Pesca who met in January 1977 in Austin, Texas. This group of state and federal scientists agreed to the proposed plan, which provides for: 1) enhancement of nesting success and survival at Rancho Nuevo, Tamaulipas, Mexico; 2) establishment of a second breeding population at Padre Island National Seashore, Texas[1]; and 3) an experimental study to evaluate the concept of headstarting.

An international program to implement the plan was begun in 1978, and this paper discusses the initial results of the cooperative effort to save the Kemp's ridley from extinction with special emphasis on the headstarting aspects of the program being conducted at the National Marine Fisheries Service, Southeast Fisheries Center's Galveston Laboratory.

Enhancement of Nesting Success and Survival, Rancho Nuevo, Mexico

The Instituto Nacional de Pesca and the Fish and Wildlife Service joined forces on the beach at Rancho Nuevo to protect the eggs and nesting adults and to document the present nesting intensity. Mexican marines patrol the beach to keep predators and poachers away, and Mexican and U.S. biologists record the number of turtles, nests and eggs. Nests are marked at the time of nesting, and the eggs removed and placed in man-made nests within a fenced corral to minimize predation by man and wildlife. A small number of eggs are placed in styrofoam hatching chests for protection and for movement to the United States as part of the establishment of a second breeding population and the headstarting program. In 1978 over 85,000 eggs were collected and protected and in 1979 almost 100,000 eggs were incubated (Table 1). The program is considered successful because more hatchlings have gone to sea than in the years immediately preceding the cooper-

ative international program. We are unaware of the hatching rate before the collection and protected incubation program was initiated in 1978, but we assume because of reduced predation by wildlife and man that the present hatching success is significantly greater now than before.

Establishment of a Second Breeding Population at Padre Island

Not until 1961 was Rancho Nuevo identified as the prime nesting area for Kemp's ridley turtles (Carr 1963; Hildebrand 1963). Small numbers of Kemp's ridley turtles have nested periodically along the lower Texas coast during recent years. The National Park Service requested the Fish and Wildlife Service to conduct a study of the feasibility of establishing a second Kemp's ridley nesting population at Padre Island National Seashore as part of the restoration plan. The study showed that nests laid on Padre Island had been fertile, and that the beach slope and profile and sand grain size at Padre Island were similar to those at Rancho Nuevo. Some differences were noted between air and water temperatures but these were considered insignificant, especially during the nesting season.

The cooperating agencies agreed to attempt the establishment of a second nesting population at Padre Island National Seashore. The mechanical and biological problems associated with transplanting sea turtle eggs have been resolved over many years of effort, and the process is now routine for experienced personnel. However, the mechanisms of imprinting hatchling sea turtles to a given beach are not understood. Factors complicating evaluation of transplanting programs are the enormous mortality of hatchlings in their first year of life and the lack of suitable tagging methods for new hatchlings.

The agencies identified the following factors as the minimum necessary for the potential success of a transplant program to establish a second nesting colony.

1. A natural orientation exposure for hatchlings on the proposed natal beach and near shore waters. Incubation should occur in the sand from the proposed natal beach to ensure proper chemical imprinting dur-

---

1. 1978-ABC-IV-0751, No. 27611-8786- (Mexican Permit); 1979-ABC-IV1258, Exp. 4287- (Mexican Permit).

EDWARD F. KLIMA

ing the incubation period.

2. A captive rearing program of 6 months to 1 year to bring the hatchlings up to a size where, presumably, predator mortality will be reduced and the turtles can be tagged.

3. An adequate technique for marking juvenile turtles to allow recognition as adults.

4. A release program that places the young in the proper area and habitat so they enter the environment at an appropriate place and time in association with naturally occurring young of the same year class.

There was also concern that the low populations remaining in Mexico could not support any removal of eggs for such a program. It was decided, however, to limit the removal of eggs for a transplant program to a small number (less than 5 percent) and that the number of yearling turtles supplied by the headstart program would outweigh any losses of eggs because of the natural high mortality rate during the first year.

In 1978 and again in 1979, approximately 2,000 eggs were obtained from egg laying females at Rancho Nuevo. The eggs were not allowed to touch Rancho Nuevo sand, but were caught and placed in styrofoam containers containing Padre Island sand and then flown to Padre Island for incubation and imprinting. The hatchling turtles were allowed to walk down the beach, from what biologists considered the probable nesting area, to the water where they were allowed to swim a few minutes before being collected and transported to Galveston. Our educated guess is that imprinting on natal beaches occurs during incubation and during the walk down the beach into the water and the swim away from the beach. The study's experimental design provides the hatchlings with this imprinting potential.

Experimental Headstart Program

The culture and later release of turtles in the sea as a means of increasing turtle populations, headstarting, is an unproven management concept. The technique, though practiced by commercial turtle farmers and some government conservation agencies, has never been scientifically tested to determine the degree of reliability as an acceptable management technique. The headstarting program will provide answers to questions raised by researchers concerning the fate of cultured turtles in the sea, such as: 1) Do they survive after release? 2) Do they breed and do they breed where released or on natal beaches? 3) What is the optimum marine habitat to release post hatchlings or juvenile turtles?

Headstarting Kemp's ridley turtles was identified as a major component of the overall recovery plan for this species because: 1) the population has seriously declined to a level that might prevent natural recovery

unless recruitment is improved by assisting hatchlings through the first year; 2) in order to verify the establishment of a second nesting beach at Padre Island a headstarting program is required to produce turtles which can be tagged to provide later identification; 3) the project lends itself to scientific evaluation of the headstart technique for turtle management; 4) the headstart period can be used to provide valuable information on the life history of the species; and 5) maintaining hatchlings in captivity provides a possible brood stock should the species face immediate extinction because of an environmental disaster.

The decision to involve the Galveston Laboratory in the Kemp's ridley turtle experiment headstarting program was unanimously approved by U.S. Fish and Wildlife, Texas Parks and Wildlife, U.S. National Parks Service, Instituto Nacional de Pesca, and the National Marine Fisheries Service in the multiagency action plan of January 1978. The rationale for the decision was that the Galveston Laboratory had the necessary physical plant to support the program. The laboratory is the one closest to the natural nesting population and has extensive expertise in aquaculture and has had experience rearing loggerhead turtles, *Caretta caretta*.

Turtle Culture

The Galveston Laboratory has reared turtles for the past 2 years utilizing commercial feeds, semiclosed raceways, and individual containers. Growth and survival rates are carefully recorded and techniques have been developed to control disease and to minimize aggressive behavior between turtles. The ultimate objective is to develop optimal culture techniques and to obtain information on early life history of marine turtles. During July and August 1978, 3,081 Kemp's ridley turtle hatchlings were brought to Galveston and placed in a facility designed to provide optimum water quality and disease control. These turtles had come from 2 incubation and imprinting sites: 1,226 were hatchlings from Rancho Nuevo, Mexico; 1,855 had been incubated and allowed to go to sea at Padre Island National Seashore in the hope of imprinting them to a new beach so that a second nesting population could be established (Table 2).

Continuous modifications of the holding systems and disease treatments have led to increased survival and disease control. The survival until 9 May 1979, the time of the final release, was 68 percent. Several individuals had reached 1,200 g, but the average size at both release times was about 600 g.

The present holding system contains 15 raceways each with 106 buckets with perforated bottoms, 9 tanks 2-m in diameter also containing perforated buckets and 210 individual basins. Four 24,000-liter insulated reservoir tanks are equipped with immersion heaters to

Table 2. Headstarted Kemp's ridley hatchlings received and released in 1978 and 1979

Site of imprinting	Arrival date	Number	Average weight (g)	Number released	Survival released (percentage)
Padre Island	3–8 August 1978	1,855	17	1,321	71
Rancho Nuevo	11 August 1978	1,266	17	749	61
Padre Island	7–24 July 1979	1,658	14.5	—	—
Rancho Nuevo	26 June 1979	188	14.5	—	—

warm the water in winter. This system allows for the individual maintenance of 2,000 turtles. Two 40,000-liter waste treatment tanks process turtle wastewater before the effluent is released from the facility.

Results of experiments to determine optimum foods and feeding rates disclosed no significant differences between combinations of fresh foods and turtle pellets, and turtle pellets alone. No difference in growth rates was observed between single or multiple daily feedings. Pelleted turtle feed was chosen for its convenience and good growth results, but we do feel that feeding live shrimps, crabs and fish before release helps prepare young captive turtles to feed in the wild. The turtles did not hesitate to feed on natural foods when presented with live foods. Figure 1 shows no difference in the average growth rate between Rancho Nuevo and Padre Island imprinted turtles fed pelleted food. Turtles reached an average size of 153 g, 336 g, and 587 g in 3, 6, and 8 months, respectively (Wheeler, NMFS, personal communication).

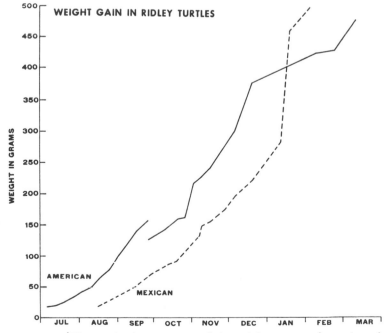

Figure 1. Average growth-in-weight curves for cultured Kemp's ridley sea turtles hatched at Rancho Nuevo, Mexico, and Padre Island, Texas.

Disease Control and Behavior

Aggressive behavior between turtles was the greatest problem in holding Kemp's ridley turtles. The physical damage caused by biting opened the way for secondary infections, which would cause death if not treated in time. Early detection of damage and immediate treatment with ampicillin and other antibiotics resulted in recovery of 95 percent of the damaged turtles. Healing was facilitated by isolating damaged turtles in individual buckets within a raceway. This was adopted as the best way of preventing the damage that leads to infection and mortality. Also, labor was reduced once turtles were placed in buckets within the raceway.

Behavioral experiments to determine methods of controlling aggressive behavior have been started and preliminary results show there is a hierarchy within groups of turtles; certain turtles are more aggressive than others regardless of hierarchy; and high temperature and corresponding higher activity lead to more aggression (Howe, University of Houston, personal communication). This work is continuing, and we will use the information obtained to design better holding facilities for turtles, in the hope of enabling us to culture the majority in groups for easier maintenance.

Disease is a major problem in the mass culture of Kemp's ridley turtles. At least 16 kinds of disease conditions have been observed in the headstarting program, and some have been significant causes of mortality (Leong, NMFS, personal communication): eyelid infection, emaciation syndrome, fungal infection of the lung, peritonitis, and intestinal obstruction. These diseases are particularly noticeable in group-held turtles, which are under more stress than individually held turtles. Techniques to improve diagnostic capabilities, i.e., X-ray and hemotological analyses, are being developed.

Release of Cultured Turtles

The release and later nesting of cultured turtles is the aim of the program. The release location of cultured turtles is extremely important in that Drs. Carr, Hildebrand, Márquez, and Pritchard, and our staff, have agreed to select sites that place young turtles in the habitat they would normally encounter in the wild.

EDWARD F. KLIMA

Unfortunately, there is little information available concerning distribution of juvenile ridley turtles. After searching the literature and reviewing unpublished data, we have concluded that south Florida and Homosassa, Florida, are suitable habitats for releasing cultured juvenile turtles. In 1979, we planned to release 400 g cultured turtles as soon as enough of the 1978 year class achieved this size. By February, several hundred turtles were ready for release. South Florida was selected as the best location for the mid-winter release because of the warmer water temperatures and the natural occurrence of the species in the area. A total of 525 Padre Island imprinted turtles were released between 22 February and 5 March 1979 at Everglades National Park. A second site was selected for a spring release off Homosassa, which appears to provide an ideal habitat for green turtle populations. Ridley turtles had historically made some use of this area. A total of 1,368 turtles, of both Rancho Nuevo and Padre Island imprinted turtles were released off Homossassa on 8

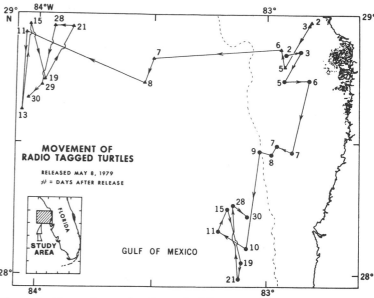

Figure 2. Chart of tagged radio tracked Kemp's ridley turtles released 8 May 1979 off Homosassa, Florida.

Table 3. Recovery of Kemp's headstarted turtles

Released	Recovered	Days out	Condition	Rereleased
1. Cape Sable	Florida Keys	49	Healthy[1]	Yes
2. Cape Sable	Del Ray Beach, Fla	21	Injured[1]	Yes
3. Cape Sable	Florida Keys	14	Healthy	Yes
4. Cape Sable	Florida Keys	25	Healthy	Yes
5. Cape Sable	Florida Keys	32	Healthy	Yes
6. Cape Sable	Miami, Fla.	47	Weak	Yes
7. Cape Sable	Pompano, Fla.	66	Healthy	Yes
8. Cape Sable	Key Biscayne	26	Thin[2]	Yes
9. Cape Sable	Florida Keys	32	Healthy[1]	Yes
10. Cape Sable	Florida Keys	40	Feeding	Yes
11. Cape Sable	Florida Keys	31	Slow	Yes
12. Cape Sable	Miami, Fla.	54	Tar	Yes
13. Cape Sable	Florida Keys	55	Healthy	Yes
14. Cape Sable	Florida Keys	17	Healthy[1]	Yes
15. Cape Sable	Florida Keys	28	Dead	No
16. Cape Sable	Florida Keys	31	Poor	Died
17. Cape Sable	Jekyll Island, Ga.	234	Excellent[3]	Yes
18. Homosassa, Fla.	Mississippi Sound	51	Healthy[4]	Yes
19. Homosassa, Fla.	Port Everglades	120	Healthy	Yes
20. Homosassa, Fla.	Homosassa, Fla.	1[5]	Healthy	Yes
21. Homosassa, Fla.	Homosassa, Fla.	1[5]	Healthy[1]	Yes
22. Homosassa, Fla.	Clearwater, Fla.	19	Healthy	Yes
23. Homosassa, Fla.	Homosassa, Fla.	1[5]	Healthy	Yes
24. Homosassa, Fla.	Weeki-Wachee Springs	48	Healthy[1]	Yes
25. Homosassa, Fla.	Port Richie, Fla.	42	Healthy	Yes
26. Homosassa, Fla.	Homosassa, Fla.	1[5]	Healthy[1]	Yes

1. Flipper injured or gone.
2. Found in parking lot.
3. Gained 2,700–4,000 g.
4. Increase of 394 g in weight.
5. Easy to catch.

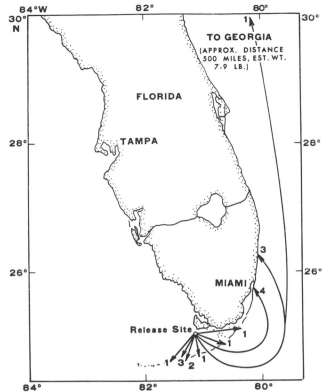

Figure 3. Recoveries of flipper tagged turtles released 22 February 1979 to 5 March 1979 at Everglades National Park.

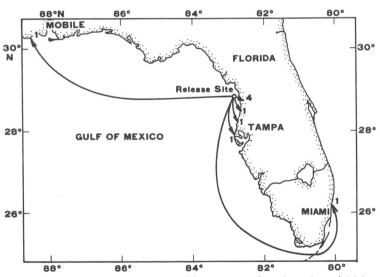

Figure 4. Recoveries of flipper tagged turtles released 9 May 1979 at Homosassa, Florida.

transmitters and were followed by plane and boat. Some turtles were tracked for as long as 30 days; diving behavior and movement were observed.

Preliminary analysis shows that the transmitters were essential in determining probable movement of the released group. Many of the radio-tracked turtles were observed diving and behaving normally during the 30-day tracking period. Several transmitters became detached during tracking, indicating that a better means of attachment is necessary. Because of possible detachment of transmitters we visually verified attachment of the transmitters to the turtles by locating the transmitters from planes and directing a boat to the site for location using a hand-held receiver and visual verification. Using this method we were able to find several transmitters that had broken away. We were also able to verify the attachment of a transmitter to a turtle after one week thus validating all earlier plane observations. The movements of 2 turtles after the Homosassa release are plotted in Figure 2. These turtles were representative of 2 trends of movement observed in the 10 turtles with radio-transmitters. One group of radio-tagged turtles tended to move offshore (west) and another group moved along shore (south). They remained in the immediate area of the release for 5 to 6 days and then a significant movement of 80 to 160 km occurred, either west or south. The turtles stayed in the same general area until the completion of the 30-day tracking period. We are now trying to relate the movement observed to wind and wave conditions recorded during the tracking period.

The recovery of flipper-tagged turtles through September 1979 has been surprising; thus far, 27 headstarted tagged turtles have been recovered—17 from the South Florida release, 9 from the Homosassa release (Table 3). Twenty-five recovered turtles were captured alive and released; most appeared active and in good health. The turtles recovered from the Everglades Park release were found in the Florida Keys, Biscayne Bay and up the east coast to Delray Beach, Florida. Eight months after release, the weight of 1 turtle recaptured off Jekyll Island, Georgia, had increased 2,700 to 4,000 g. The turtles recovered from the Homosassa release were found south of the release point to Tampa, but there were recoveries from Biloxi, Mississippi, and Fort Lauderdale, Florida (Figures 3 and 4). Several recoveries have occurred after 6–8 weeks.

Many of the recoveries occurred within estuary systems or inside of barrier islands, indicating a possible orientation to brackish-water conditions. We feel, however, that it is still too soon to make any conclusions regarding yearling turtle habitat preference.

These results tend to confirm that headstarted Kemp's ridley turtles survive in the wild and that a major question concerning the effectiveness of the program can be answered in the affirmative.

and 9 May 1979. A third area was Padre Island's National Seashore Park, because of the attempt to establish a second nesting beach at that location, at which 98 turtles were released in July.

An integral part of this year's program was to determine the movement and survival of the young Kemp's ridley turtle after release. All released turtles were tagged with monel flipper tags, and 10 to 12 turtles in the first 2 releases were equipped with small radio-

EDWARD F. KLIMA

Future Plans for Kemp's Ridley Headstart Program

In July 1979, 1,846 hatchling Kemp's ridley turtles were received in Galveston. This year the turtle research program will emphasize studies of the early life history requirements of marine turtles. Behavioral studies to help in the modification of aggressive behavior and to determine orientation to chemical and physical parameters will be conducted. Further work will be done on developing systems for holding the turtles in groups and on the development of semiclosed systems, which will reduce the need to heat large volumes of water during the cold winter months. Special attention will be given to improving disease diagnosis and control. We also hope to consolidate the information gained thus far into a manual on turtle diseases and cures.

If labor and space permit, other species of turtles will be maintained so that early life history requirements between species can be compared. Tagging and release studies will be continued to obtain additional information on survival and movement after release.

The final evaluation of the program will take many years as we must wait for the effects of our work to appear on nesting beaches, either at Rancho Nuevo, Padre Island, or elsewhere. The actual age to sexual maturity is not known, but estimates range from 5 to 10 years and older.

Literature Cited

Carr, A.
1963.   Pan-specific reproductive convergence in *Lepidochelys kempi* (Garman). *Ergebnisse der Biologie* 26:298–303.

Hildebrand, H. H.
1963.   Hallazgo del area de anidacion de la tortuga marina "lora," *Lepidochelys kempi* (Garman) en la costa occidental del Golfo de Mexico. *Ciencia* 22:105–12.

Hilburn O. Hillestad
J. I. Richardson
Southeastern Wildlife Services, Inc.
113 Hoyt Street
Athens, Georgia 30601

Charles McVea, Jr.
J. M. Watson, Jr.
Pascagoula Laboratory
Southeast Fisheries Center
Natinal Marine Fisheries Servce, NOAA
P.O. Drawer 1207, Pascagoula, Miss. 39567

Worldwide Incidental Capture of Sea Turtles

ABSTRACT

Incidental capture and mortality resulting from capture are currently recognized as a threat to the survival of certain species of sea turtles. Evaluation of the effects of mortality resulting from incidental capture in some species has received increased scrutiny. This paper presents a review of incidental capture of sea turtles resulting primarily from commercial fishing activities, presents estimates of capture and mortality rates for turtles in North American waters, and discusses implications of capture and mortality. Encirclement nets, set nets, longlines, seines, and shrimp trawlers are responsible for varying rates of capture and mortality. Shrimp trawlers are considered to capture and drown more sea turtles worldwide than any other form of incidental capture. An excluder panel designed for otter trawls may potentially reduce incidental mortality rates of some species to tolerable levels.

Introduction

Most species of sea turtles have suffered an alarming worldwide decline in numbers. Although none have become extinct, some species are now clearly endangered; others are threatened in portions of their ranges.

Historically, marine turtle conservation has consisted of preservation by reducing, managing, or eliminating the direct take of animals. We have focused on the establishment of nesting sanctuaries and the elimination or control of legal and illegal harvest of eggs and individuals in order to perpetuate populations. These efforts have been partially successful; most governments have made progress in establishing nesting sanctuaries and enacting laws regulating harvest (Carr, Carr, and Meylan 1978). Enforcement of these laws still leaves much to be desired in many areas of the world, but the trend is positive.

Some species of sea turtles are affected by stresses other than direct take or nesting area disturbances in various areas of the world. These stresses are associated with the incidental or accidental capture of these animals by commercial and noncommercial interests. Hillestad, Richardson, and Williamson (1978), Pritchard (1976), and others have raised the question of the impact incidental capture may have upon certain species. Carr, Carr, and Meylan (1978) summarizes this concern and state that incidental capture may administer the coup de grace to some.

Our purpose here is to review, in summary form, the worldwide incidental capture of sea turtles and to discuss the implications of incidental capture upon various species.

Incidental Capture

Incidental capture of sea turtles is primarily related to the commercial fishing industry. However, turtles are captured as nontarget species as a result of other activities in many parts of the world.

Shoreline Set Nets

These nets, which include those used for commercial fishing and for reducing shark populations near bathing beaches, usually are set near shore and remain stationary for certain time periods.

Our knowledge concerning the regular capture of turtles by sturgeon nets is centered on the South Carolina coast in the southeastern United States. These nets are set perpendicular to the shoreline near river entrances in the early spring. The nets entangle sturgeon ascending rivers on spawning runs; these nets also entangle and capture sea turtles (primarily *Caretta*). Historical data on turtle captures by sturgeon nets are lacking, but mortality apparently is high.

Sturgeon fishing has declined in recent years in the Southeast; consequently, turtle captures have also declined. The state of South Carolina regulates this fishery and limits the sizes of nets and dates of their use to a period in the early summer prior to the arrival of large numbers of turtles. Incidental capture of sea turtles in sturgeon nets now appears to be low in this area and probably constitutes an insignificant threat to turtles in the coastal waters of the Carolinas.

In Natal and northeastern Australia, set nets are used to capture sharks near bathing beaches. Thirty-one *Caretta caretta* (63 per cent of total catch) were caught in shark nets in Natal in 1968 (Hughes 1969). Green turtles accounted for 35 percent (17 animals) of the capture. One hundred seventy-six (majority *Caretta*) turtles were captured between 1965 and 1968; 85 *Caretta* and 49 green turtles were captured after 1968 (Hughes 1974). Most of the animals were subadults; only one female was fully mature. Hughes (1974) suggests that the larger turtles may migrate in deeper water outside of the littoral zone where nets are set.

Fifty-seven turtles were captured in Cairns Inlet shark nets in northeastern Australia between 1971 and 1974 (Limpus 1975). Most were *Chelonia mydas* and *C. depressa*; 1 *Lepidochelys olivacea* and 1 *Dermochelys coriacea* were captured.

The potential impacts, if any, of incidental capture of sea turtles in Natal and Australian waters have not been evaluated.

Fisheries Set Nets

The fisheries industry employs numerous types of set nets throughout the world; those capturing sea turtles probably are encirclement nets. Drownings of turtles in shrimp and menhaden nets are increasing (Carr 1972). Although we have been unable to locate other published references relating to the capture of sea turtles by menhaden nets, we know that sea turtles have been observed in these nets in southeastern U.S. waters (D. Harrington, Marine Extension Service, Georgia, pers. comm.). Most likely other fisheries employing encirclement nets capture sea turtles also, but nothing is known of the rates of the spatial and temporal aspects of this incidental capture. Mortality rates also are unknown, but, since these nets (especially menhaden) allow turtles to surface while entangled, mortality may be low.

Longlines

Longlines (48 km in length) are used by fisheries worldwide, and turtles are occasionally taken by this method. Few references to incidental capture on longlines exist; however, a *D. coriacea* has been reported captured on a shark longline off Ireland (Atkins 1960). Cato, Prochaska, and Pritchard (1978) report that 230 kg of *C. mydas* were taken by longline for commercial purposes in Florida in 1962. Observers from the National Marine Fisheries Service presently are aboard Japanese longline fishing vessels in the Gulf of Mexico and the Atlantic. Fishing is conducted primarily for tuna with the incidental catch of various sharks and billfish. Although representing thousands of hours of fishing effort, few turtles have been captured by these vessels. Thus longline fishing probably catch and injure few turtles.

Seines

Seines, especially when pulled from shore, occasionally capture sea turtles. Six hundred and eighty kg of *C. caretta* were captured by haul seines in Florida in 1962 (Cato, Prochaska, and Pritchard 1978). Two *D. coriacea* were captured by seines near Panacea, on the Gulf coast of Florida, in March and April of 1962 (Yeager 1965). Hillestad captured a yearling *C. mydas* with a

HILBURN O. HILLESTAD

beach seine off Blackbeard Island, Georgia, in 1972. Although some sea turtles are captured by seining, this form of incidental capture is probably insignificant.

Shrimp Trawlers

Worldwide, the shrimp trawling industry seems to capture more sea turtles than any other commercial fishery. Shrimp, distributed throughout the world, are in constant demand. Trawling is concentrated primarily in the relatively shallow waters near shore in both temperate and tropical zones. Many of the most intensively trawled waters are adjacent to major sea turtle nesting beaches or feeding grounds.

All species of sea turtles are captured by shrimp trawlers. The majority of captures appear to consist of *C. caretta*, *C. mydas*, *C. depressa,* and *L. kempi*. *L. olivacea* are captured in certain areas but fewer than *L. kempi*; *D. coriacea* and *Eretmochelys imbricata* are least caught.

In the following discussion we review the worldwide incidental capture of sea turtles by shrimp trawlers geographically and, when possible, by species from a spatial and temporal perspective.

AUSTRALIAN WATERS

Cogger and Lindner (1969) discuss sea turtles of northern Australia and report that *E. imbricata* and *C. depressa* are taken by Australian-Japanese prawn trawlers. No quantification of the *E. imbricata* catch was given, but most animals were subadults. Apparently most size classes were reflected in the catches.

Limpus (1973) reports on 3 female *C. caretta* that drowned in trawl nets of shrimpers off Queensland. He states that female loggerheads were caught by trawlers 0.5 to 4 km offshore only during the nesting season. Also in this region, otter trawls in 10 m of water caught up to 3 females per trawl per hour. Males rarely occurred in the same area during the nesting season. Limpus (*op. cit.*) did not have sufficient data to evaluate *Caretta* captures and mortality.

AFRICAN AND INDIAN OCEAN WATERS

Very little has been reported on the incidental capture of sea turtles by trawlers in this area.

SOUTH AMERICAN WATERS

Sea turtles have been captured in fair numbers by trawlers off French Guiana; these animals were mostly *L. olivacea* (Pritchard 1969). Fourty-four green turtles tagged in Surinam during their peak nesting period (March to May) were recovered by trawlers the following October through March (Pritchard 1973). Pritchard submits that green turtles "... are more likely

to be caught when actually enroute from the nesting ground; exhausted and spent after months of reproductive activity, they lack the ability to avoid nets ..." However, the number captured may have been a function of varying trawler activity; no data were given on the trawling season.

Thirty-nine *L. olivacea* tagged in Surinam were recovered by shrimp trawlers in 1 year (Pritchard 1973). This finding led Pritchard to speculate that the carnivorous ridley may enhance its capture by continually searching for prey and running a higher risk of capture.

Three male and 3 female *L. olivacea* were captured in experimental trawls by the RN *Calamar* off Surinam in 1967–68. All were adults; most were captured in the fall.

As the result of a tagging study of *C. caretta* on the Caribbean coast of Colombia, Kaufman (1975) suggests that shrimp trawlers operating off the nesting beach are a significant cause of mortality for turtles in that area. *C. caretta* numbers have been depressed there for some time.

CENTRAL AMERICAN WATERS

All shrimp grounds in Central America are trawled, but precise data are lacking to evaluate the capture and mortality of sea turtles in these waters. Carr, Carr, and Meylan (1978), in discussing the West Caribbean green turtle colony, comment on trawler captures of sea turtles "... trawlers have moved into new ground: the trawls now used are much larger than they once were, and the usual haul time nowadays is long enough to drown many turtles caught." These factors increase the chance of turtle captures, and the latter increases the possibility of death.

NORTH AMERICAN WATERS

Commercial trawling on the South Atlantic and Gulf of Mexico coasts of the United States and Gulf coast of Mexico probably accounts for a large percentage of worldwide trawler-related captures and mortality of sea turtles.

In temperate waters, loggerheads and Kemp's ridleys are frequently captured. *C. caretta* have comprised the bulk of the catch in South Carolina and Georgia (Hillestad, Richardson, and Williamson 1978; Ulrich 1978), although immature greens and ridleys occur infrequently. Approximately 500 ridleys have been captured annually by shrimp trawlers from Cuba, Mexico, and the United States (Márquez 1976). Ridleys have also been captured off Louisiana (Liner 1954; Dobie, Ogren, and Fitzpatrick 1961), off south Texas (Carr 1961), and in the Florida Keys (Sweat 1968). Ridleys have been killed in trawls off Veracruz, Tabasco, and Campeche, Mexico (Pritchard 1976).

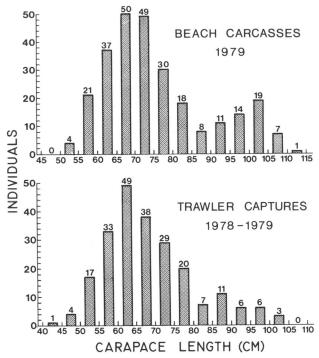

Figure 1. Comparison of size distribution of 269 *Caretta caretta* stranded on Georgia beaches, 1979 (top), and 274 turtles captured by shrimp trawlers, Georgia and South Carolina, 1978–79 (bottom).

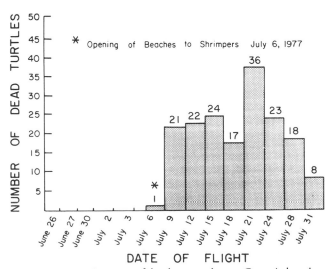

Figure 2. Aerial survey of dead sea turtles on Georgia beaches 26 June to 31 July 1977.

Infrequently, shrimpers have captured leatherbacks along the Georgia and South Carolina coasts (Schwartz 1954). Hillestad has observed 2 dead leatherbacks in Georgia and has interviewed 2 shrimpers who have caught them off shore of Georgia. At least 2 leatherbacks have been captured by shrimpers off Louisiana (Dunlap 1955).

Quantification of Incidental Capture and Mortality in U.S. Waters

Only in recent years have data become available to quantify incidental capture in U.S. waters. Initially, data sets were based on interviews. Among these were Cox and Mauerman (1976), Ulrich (1978), and Hillestad, Richardson, and Williamson (1978). Cox and Mauerman (1976) and Anonymous (1976, 1977) present interview-based summaries of incidental capture and mortality of turtles by shrimpers in Texas, Louisiana, Alabama, and western Florida. The highest reported capture rate was 1 turtle every 27 fishing days in Florida (6 turtles a season). Louisiana shrimpers captured 1 turtle in 53 fishing days (4 a season), Texas shrimpers about 5 turtles a season, and Alabama shrimpers 1 turtle in 72 fishing days (about 2 a season).

Mortality estimates from Alabama, Florida, and Louisiana shrimpers ranged from 21 to 25 percent. Estimates from Texas were 16 percent. In 1978 observers were placed on board Texas shrimp boats to record incidental captures as part of a National Marine Fisheries Service program to evaluate experimental trawling. Their observations generally confirmed capture and mortality rates for Texas waters as reported above. In 1976 Hillestad, Richardson, and Williamson (1978) interviewed the captains of the 321 vessels of the Georgia fleet and observed incidental captures aboard selected vessels. The data indicated capture of 0.09 turtles per hour of trawl for a 5.7 month season and mortality of 7.9 percent. Based on these data and on the resident fleet size of 321 vessels, 30.7 turtles were captured per vessel-year in Georgia and 778 sea turtles were drowned. (These data do not include out-of-state shrimpers who fish in Georgia and unload their catches elsewhere.)

Onboard observation of captured sea turtles in Georgia indicate that most are subadults (Figure 1). Beach strandings of carcasses reflect the same relationship: 88 percent subadults, 12 percent adults (Figure 1). Beach strandings are highly correlated with shrimping activity (Figure 2) and provide an index of shrimper-induced mortality.

South Carolina shrimpers captured 1 to 3 turtles a week, and mortality was estimated at 18.2 percent in 1976 and 43.3 percent in 1977 (Ulrich 1978). During 1978 and 1979, observers were placed on board commercial shrimp vessels during the brown and white shrimp season of South Carolina, Georgia, and northeastern Florida. Observers were placed aboard vessels in Key West during the pink shrimp season there.

Figure 3 shows the size distribution of 224 turtles captured in Georgia and South Carolina in 1978–79. These turtles were primarily subadults, supporting interview-based data presented earlier. Figure 4 summarizes the temporal distribution of this catch. Peak capture occurred in July, which coincided with peak nesting in Georgia and South Carolina.

Size distribution and total number of both trawler captures and beach strandings were similar in the subadult size classes. Beach strandings of large animals exceeded trawler capture, however.

Discussion and Implications

Data from various studies presented earlier quantify captures and mortality of sea turtles in certain North American waters. Clearly, captures are high in certain areas and at certain times, such as in the temperate waters of South Carolina and Georgia during mid-summer shrimp season when inshore shrimping is permitted. Incidental capture is low in the Key West–Dry Tortugas area during the winter pink shrimp season. Capture rates are fairly low in most reaches of the Gulf of Mexico, compared to captures in the Georgia Bight.

Mortality rates vary widely in North American waters. Mortality apparently is influenced by many factors such as turtle size and condition at capture and trawl duration. Evidence indicates that smaller turtles drown more quickly than larger turtles, i.e., they are less capable of enduring long trawl periods. This fact may bias the observed recorded size distribution of captured subadult animals. (Subadult animals should be more prevalent in the populations and they should be encountered more commonly by the trawls.) There is insufficient evidence to indicate whether small animals are capable of evading trawls. Ogren, Watson, and Wickham (1977) report that 2 of 3 adults failed to evade the trawl during experiments in the Gulf.

Within the nearshore waters of Georgia, it is common to recapture the same turtle in the same day, often immediately following the initial capture. Such turtles are probably more prone to drowning than turtles not previously captured. The probability of recapture is high in Georgia and South Carolina waters due to the very large trawling effort expended in a small area. For example, the Georgia fleet of 321 vessels annually covers an area equivalent to 75,061 km² within a 860 km² coastal area. A typical Georgia commercial shrimper trawls an average of 23,383 ha per day.

In a recent study (Hillestad, Richardson, and Williamson 1978), most Georgia and South Carolina shrimpers trawled for approximately 2 hours with nets

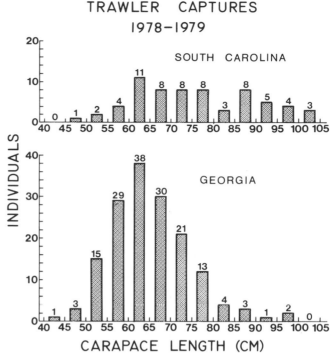

Figure 3. Size distribution of 224 sea turtles incidentally captured by shrimp trawlers, Georgia and South Carolina, 1978–79.

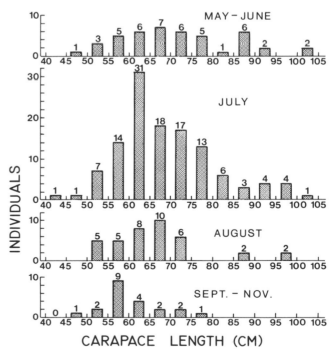

Figure 4. Temporal distribution of 224 sea turtles incidentally captured by shrimp trawlers, Georgia and South Carolina, 1978–79.

averaging 17.2 m in width. A highly significant correlation existed between numbers of turtles captured and net widths greater than 12.4 m. A similar correlation (p < .01) existed between capture rates and net widths for Alabama and Louisiana shrimpers in certain waters.

The foregoing data do not permit the evaluation of incidental capture, i.e., capture and mortality data must be analyzed in relation to population levels, recruitment into the population, and other factors. The fact of capture and mortality must be separated from the effect of mortality.

It is intuitive that any mortality incurred by Kemp's ridley is significant since stocks are currently very depressed, perhaps beyond their biological threshold to recover. In North American waters, capture of leatherbacks by trawlermen is low. Therefore, this species probably suffers little mortality from this cause. Most green turtles captured in North American waters are subadults. Capture rates of green turtles are low, apparently lower than captures of ridleys which probably have lower total population levels. The effect of North American incidental capture and mortality of green turtles is probably less significant than direct harvesting (Carr, Carr, and Meylan 1978).

The loggerhead is the most commonly captured sea turtle in North American waters and presently is probably more capable of sustaining trawler-induced mortality than other species. The species has large, protected nesting beaches in northeastern Florida, Georgia, and South Carolina, and large numbers of this species occur in these waters each spring and summer. Due to its relatively high abundance, *Caretta* is well suited for field testing and development of an excluder panel for otter trawls (Seidel and McVea, this volume). This vanguard effort by the National Marine Fisheries Service should provide the basic solution to reducing incidental capture and subsequent mortality of sea turtles. The reduction of incidental capture and mortality, the continuation of successful efforts to protect nesting beaches, and the control of directed take of sea turtles throughout the world will, it is hoped, enable sea turtle populations to recover from their threatened and endangered conditions.

Literature Cited

Anonymous
1976. Incidental capture of sea turtles by shrimp fishermen in Florida: Preliminary report of the Florida West coast survey. University of Florida Marine Advisory Program. Mimeo, 3 pp.
1977. Alabama shrimp fishermen interviews for 1977–78. Marine Resources Office. Alabama Cooperative Extension Service. Mimeo, 1 p.

Atkins, L. S.
1960. The leathery turtle, or luth, *Dermochelys coriacea* (L.) in Co. Cork. *Irish Natural. J.* 13:189.

Carr, A.
1961. The ridley mystery today. *Animal Kingdom* 64:7–12.
1972. Great reptiles, great enigmas. *Audubon* 74:24–35.

Carr, A.; M. H. Carr; and A. B. Meylan
1978. The ecology and migrations of sea turtles, 7. The West Caribbean green turtle colony. *Bulletin of the American Museum of Natural History* 162:1–46.

Cato, J. C.; F. J. Prochaska; and P. C. H. Pritchard
1978. An analysis of the capture, marketing and utilization of marine turtles. National Marine Fisheries Service. Contract 01-7-042-11283. Mimeo, 119 pp.

Cogger, H. G., and D. A. Lindner
1969. Marine turtles in northern Australia. *Australian Zoologist* 15:150–59.

Cox, B. A., and R. G. Mauerman
1976. Incidental catch and disposition by the Brownsville-Port Isabel gulf shrimp fleet. Cameron Cooperative Extension Service, San Benito, Texas, and Texas Shrimp Association, Brownsville, Texas. Mimeo, 5 pp.

Dobie, J. L.; L. H. Ogren; and J. F. Fitzpatrick, Jr.
1961. Food notes and records of the Atlantic ridley turtle (*Lepidochelys kempi*) from Louisiana. *Copeia* 1961:109–10.

Dunlap, C. E.
1955. Notes on the visceral anatomy of the giant leatherback turtle (*Dermochelys coriacea*). *Tulane Medical Fac.* 14:55–69.

Hillestad, H. O.; J. I. Richardson; and G. K. Williamson
1978. Incidental capture of sea turtles by shrimp trawlermen in Georgia. *Proceedings of the Annual Conference of the Southeast Association of Fish and Wildlife Agencies*, 23.

Hughes, G. R.
1969. Report to the survival service committee on marine turtles in southern Africa. *Marine Turtles, IUCN Publication, New Series, Supplemental Paper*, 20:56–66.
1974. The sea turtles of Southeast Africa. II. The biology of Tongaland loggerhead turtle *Caretta caretta* L. with comments on the leatherback turtle *Dermochelys coriacea* L. and the green turtle *Chelonia mydas* L. in the study region. *Investigational Report of the Oceanographic Research Institute*, Durban, South Africa, 36:1–96.

Kaufman, R.
1975. Studies on the loggerhead sea turtle, *Caretta caretta caretta* L. in Colombia, South America. *Herpetologica* 31:323–26.

Limpus, C. J.
1973. Loggerhead turtles (*Caretta caretta*) in Australia: food sources while nesting. *Herpetologica* 29:42–45.
1975. The Pacific ridley, *Lepidochelys olivacea* (Eschscholtz), and other sea turtles in northeastern Australia. *Herpetologica* 31:444–45.

Liner, E. A.
1954. The herpetofauna of Lafayette, Terrebonne, and Vermilion Parishes, Louisiana. *Lafayette Academy of Sciences* 17:65–85.

Marquez, R.

1976. Natural reserves for the conservation of marine turtles of Mexico. *Florida Marine Research Publications* 33:56–69.

Ogren, L. H.; J. W. Watson, Jr.; and D. A. Wickham

1977. Loggerhead sea turtles, *Caretta caretta*, encountering shrimp trawls. *Marine Fisheries Review Paper* 1270:15–17.

Pritchard, P. C. H.

1969. Sea turtles of the Guianas. *Bulletin of the Florida State Museum* 13:85–140.

1973. International migrations of South American sea turtles (Cheloniidae and Dermochelidae). *Animal Behavior* 21:18–27.

1976. Endangered species: Kemp's ridley turtle. *Florida Naturalist* 49:15–19.

Schwartz, A.

1954. A record of the Atlantic leatherback turtle (*Dermochelys coriacea*) in South Carolina. *Herpetologica* 10:7.

Sweat, D. E.

1968. Capture of a tagged ridley turtle. *Quarterly Journal of the Florida Academy of Sciences* 31:47–48.

Ulrich, G. G.

1978. Incidental catch of loggerhead turtles by South Carolina commercial fisheries. National Marine Fisheries Service. Contract 03-7-042-35121. Mimeo, 10 pp.

Yerger, R. W.

1965. The leatherback turtle on the Gulf Coast of Florida. *Copeia* 1965: 365–66.

Wilber R. Seidel
Charles McVea, Jr.
National Marine Fisheries Service, NOAA,
Southeast Fisheries Center,
Pascagoula, Mississippi 39567

Development of a Sea Turtle Excluder Shrimp Trawl for the Southeast U.S. Penaeid Shrimp Fishery

ABSTRACT

In 1978, a 3-year project was initiated to develop a sea turtle excluder trawl for use in the Penaeid shrimp fishery of the southeastern United States. The National Marine Fisheries Service (NMFS) with cooperation of the commercial shrimp industry is presently testing several basic trawl modifications to reduce incidental sea turtle capture without significantly reducing shrimp catch efficiency. Initial design concepts, prototype selective trawl methodology, and summarized field evaluation data for sea turtle excluder trawls are discussed.

The decline of many sea turtle populations throughout their ranges has been well documented during the last 2 decades (Carr 1952; Pritchard 1969; Rainey and Pritchard, no date). Factors contributing to reduced turtle stocks include: 1) overexploitation for a variety of turtle products including human food, 2) destruction of nesting habitat due to coastal development, 3) predation, 4) inadequate protection, 5) natural mortality, and 6) other man-induced mortality including incidental capture in demersal trawls.

Although migration and aggregate nesting of sea turtles undoubtedly have evolutionary significance, these predictable behavior patterns have resulted in serious predation by man and beast. Unfortunately, certain areas of seasonal turtle concentration overlap with productive demersal trawl fisheries. Incidental turtle capture and mortality by trawlermen alone might not significantly affect population stability, but in addition to other known mortality sources, undirected take could have significant impact on already reduced sea turtle stocks.

The Penaeid shrimp fishery of the Gulf of Mexico and midwestern Atlantic represents one of the most important commercial fishery industries of the world. In general, shrimp seasons in the southeast fishery region open in mid-June and continue through December with the pink shrimp (*Penaeus duorarum*) season occurring in the Dry Tortugas from January through April. Intensive shrimping occurs during the opening weeks of each season with boats from neighboring coastal states participating in areas of predicted high shrimp density. The highly productive near shore waters of South Carolina, Georgia, and eastern Florida, though limited in fishing area, receive fishing pressure from these representative states as well as pressure from coastal states of the northern Gulf of Mexico. This fishing effort early in the shrimp season coincides with peak sea turtle concentrations in limited areas primarily adjacent to the major nesting beaches of the loggerhead turtle (*Caretta caretta*) along the continental United States. Other sea turtles including Kemp's ridley (*Lepidochelys kempi*) are occasional visitors in these waters and are accessible to trawling activities. Sea turtle incidental capture by demersal trawls in these areas has been known for some time, but only recently have incidental capture rates been determined for these areas. An incidental sea turtle capture rate of 0.09/hr during a 6.7 month, 1976–77 Georgia shrimp season indicates the severity of the problem (Hillestad, Richardson, and Williamson 1978).

As the need to protect declining sea turtle stocks became more apparent, some protection was provided by the Endangered Species Act of 1973, regulations, and intensified enforcement. The Act, however, has no mechanism which specifically addresses the inadvertent incidental capture of listed endangered animals; therefore, a direct conflict with standard shrimping techniques became inevitable.

In an effort to mediate the problem, protect threatened and endangered sea turtle species, and insure the viability of the shrimp industry activities in high turtle density areas, the National Marine Fisheries Service (NMFS) initiated a gear development project to reduce significantly capture and mortality of these animals in shrimp trawls. If successful, the selective shrimp trawl should allow the coexistence of sea turtles and shrimp harvest in areas of seasonal sea turtle abundance. The project was assigned to the Harvesting Technology Branch of the National Marine Fisheries Service, Southeast Fisheries Center Laboratory, Pascagoula, Mississippi. The project goal is to develop and introduce a shrimp trawl to the commercial shrimp industry which will greatly reduce sea turtle captures without significantly reducing shrimp production.

Attempts to develop selective shrimp trawls which can reduce the amount of fish bycatch in the northeast Pandalid shrimp fishery (High, Ellis, and Lusz 1969) and the southeast Penaeid shrimp fishery (Watson and McVea 1977) have been relatively successful depending primarily on the bycatch species composition. These designs utilize water flow patterns within the trawl to mechanically separate target species from nontarget

species through strategically placed webbing barriers. Underwater observations of sea turtle interaction with towed demersal trawls (Ogren, Watson, and Wickham 1977) suggests an alternate approach for separating sea turtles from the shrimp catch. Observed animals tend to outswim the trawl, but once overtaken, scutes and claws become entangled in the webbing. It was apparent from these observations that conventional "within trawl" separation concepts would not effectively separate turtles, and a barrier placed ahead of the trawl was conceived to deflect the turtles up and safely over the headrope. The forward excluder barrier (Figure 1) is hung from the headrope of the trawl to a separate groundline connected between the trawl doors. Additional floats are added to the headrope to maintain the desired trawl height above bottom and provide a measure of stability to the barrier.

In addition to reducing sea turtle captures, an excluder trawl must also maintain effective shrimp production. Unless this capability can be demonstrated, the shrimp industry will not accept use of the trawl without legislation and enforcement. It is hoped that the development of the sea turtle excluder trawl will be successful and that use of the trawl can be initiated in problem areas without legal conflicts developing between the shrimp industry and other environmental concerns.

Maintaining shrimp catch efficiency of selective trawls in the various areas of the United States where turtles are incidentally captured is complicated by behavioral differences among the 3 species of commercially important Penaeid shrimp. White shrimp (*Penaeus setiferus*) tend to disperse in the water column in response to tidal flow and shrimping intensity, resulting in the use of high-opening trawls when white shrimp increase in the catch composition. On the other hand, brown shrimp (*Penaeus aztecus*) and pink shrimp (*Penaeus duorarum*) tend to remain close to the bottom even when stimulated by intensive fishing pressure. Therefore, fishermen want trawls with maximum spread and low height opening to cover as large a bottom area as possible with each drag. This behavioral difference between the brown and white Penaeid shrimp necessitates 2 completely different net configurations. The situation is further complicated by the U.S. shrimp fishery's use of a considerable number of different net designs.

In November 1977, a feasibility study was conducted to determine methods for sea turtle separation as well as possible reductions in shrimp catch efficiency using a forward barrier design concept. During 100 hours of comparative double-rigged trawling, a standard trawl to excluder trawl turtle capture ratio of 9:1 was obtained with a shrimp loss of less than 30 percent. With these encouraging results, plans were made to conduct 1978–79 field evaluation studies of various

WILBUR R. SEIDEL

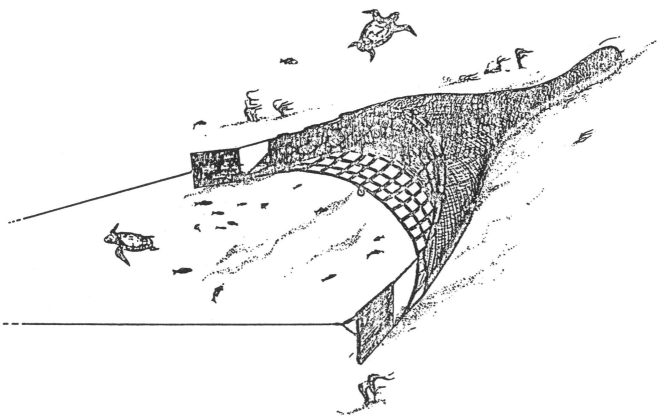

Figure 1. Forward panel sea turtle "excluder" shrimp trawl. Courtesy Southeast Fisheries Center, Harvesting Technology Unit.

excluder trawl designs on selected shrimp grounds of the coastal United States.

Development

Environmental groups want to reduce incidental capture of sea turtles, even to the point of prohibiting shrimping in some areas. This attitude stems partly from decreasing turtle populations in general with particular concern over protecting the remnants of the Kemp's ridley population, and the somewhat hesitant approach various agencies have demonstrated while addressing this critical problem. Such fishing restrictions, however, could be disastrous to the shrimp industry. Since the excluder trawl offers an acceptable solution to all interested concerns, a 3-year schedule was accepted as a reasonable time period in which to accomplish development of the trawl. Three years, however, is not much time to establish a design concept, develop prototype trawls, sufficiently field test the trawls, and demonstrate the trawls' feasibility to the industry. Development is further restricted by the relatively short seasonal concentration-overlap of shrimp and turtles in a given area. In addition, even in areas of large turtle concentrations, captures occur infrequently aboard individual vessels; therefore, many hours of test dragging are required to produce enough turtle captures to establish test trawl efficiencies. Owing to

these experimental restraints, Government-owned or chartered research vessels alone could not develop the trawl and establish its efficiency in a 3-year period. The active participation of shrimp industry vessels was required to continuously evaluate prototype designs and modifications under actual commercial conditions.

Representatives of the shrimp industry and progressive vessel captains agreed to participate in the project and cooperate in the trawls' development. A relatively large number of participating commercial shrimp vessels were enlisted to test the prototype nets and to evaluate the many different net characteristics under different commercial conditions. In a show of good faith, the commercial vessel captains agreed to provide their assistance basically at "no gain." A cooperating vessel's only consideration realized was protection against reduced shrimp production associated with extreme prototype excluder trawl designs. All participating shrimp vessels are double-rigged trawlers. The test net is installed on 1 outrigger and towed simultaneously in direct comparison against the same size standard net. If a shrimp loss does occur in the test net, the vessel is reimbursed for this loss. Although participation and cooperation is voluntary, at the end of the test season (3 to 4 months) the vessel owner is given a trawl of his choice, essentially in repayment for the use of his standard trawl during the test. In addition, a full-time observer is placed aboard each vessel to record test data. During the 3- to 4-month evaluation period, each design of the excluder trawl is fished commercially 10 to 12 hr/day (usually 6 day/week). Without the cooperative industry-government project and active participation of the commercial shrimp vessels, an adequate evaluation of the many different net characteristics that must be tested under actual fishing conditions to produce an effective trawl could not be accomplished.

After establishing the basic design concept of a forward barrier, prototype nets with various design characteristics being used industry-wide had to be developed for field testing. Since shrimp boats tow many different net types and sizes, many different excluder trawls had to be built and prepared for direct comparison against each cooperating vessel's standard trawl. Development of efficient net designs in which to install excluder panels properly, fitting excluder panels to the different net sizes, and pretest evaluation and tuning of each trawl required extensive use of scuba divers. Each design characteristic was prepared and evaluated at commercial towing speeds by scuba divers working on nets in approximately 8 m of water on hard sand bottom off Panama City Beach, Florida. A commercial shrimp vessel was obtained under continuous charter to support diving operations and prototype net development. To date, the project diving activity has been a major effort to evaluate new trawl designs and prepare excluder trawls properly for testing on the large number of commercial shrimp vessels in the project.

Results

The first test season began in June 1978 in the mid-western Atlantic and continued through the summer in the northern Gulf of Mexico off Texas and Louisiana. It was terminated in March 1979 after 3 months of trawl evaluation on the Dry Tortugas pink shrimp grounds. A total of 27 shrimp vessels participated in these studies. Results for this effort are presented in Table 1. The forward excluder panel was initially tested because it seemed to offer the best potential for reducing turtle captures based upon behavioral observations and results of the 1977 feasibility study. Although data indicate the forward barrier reduces turtle captures relatively well, certain problems related to the weighting system made the design difficult to handle on deck particularly in rough seas. In addition, excessive shrimp loss during seasonal white shrimp concentrations was inherent in the forward barrier design. It also became apparent by the end of the first test season that it would be inappropriate to continue attempts to develop barrier patterns for the almost exhausting variety of trawl designs. Instead, one efficient basic trawl design was required that could 1) fish competitively with the different standard trawl types in use and 2)

provide the construction flexibility necessary for proper excluder panel installation.

Difficult handling associated with accessory hardware on the excluder trawl and tangling sometimes when putting the net overboard were particularly irritating to those shrimp fishermen unwilling to regiment the deployment of their gear. At least 22 kg of weight were required to hold the barrier groundline in contact with the sea bottom, and the weighting system was difficult to keep from tangling in the large mesh webbing barrier. In addition, during very shallow water shrimping, sea turtles did not respond to the repelling nature of the forward barrier, but rather sounded in an attempt to escape the trawl. When this reaction occurred, it increased the possibility that the turtle would pass under the barrier groundline and be captured. Very likely, turtles in an area of intensive shrimping are also caught more than once a day. When this occurs, an alive but stunned turtle reacting abnormally to an approaching trawl could easily trip under the groundline of the forward barrier and become entangled in the net. To overcome these problems with the forward barrier, development and evaluation of a reverse barrier excluder trawl was initiated late in 1978.

The reverse barrier design (Figure 2) provides several positive options for solving problems encountered in previous excluder designs. Late 1978 results with this design indicated a significant increase in shrimp catch efficiency (Table 1) and a decrease in previous operational handling and tangling problems. Reverse barrier trawls have considerably more overhang with the barrier laced from the headrope to the footrope, and white shrimp in the water column tend to be trapped under the trawl headrope before encountering the barrier. Only turtles small enough to pass through the barrier mesh can be captured since the entire mouth of the trawl is covered by the barrier. The excluder panel mesh size can also be selected to be as restrictive as possible as long as it does not reduce shrimp catch efficiency.

Preliminary 1979 reverse barrier fishing results presented in Table 2 summarize drags completed during the East Coast brown shrimp season. Of the 2 barrier mesh sizes being tested, data clearly indicate the 66-cm stretch mesh superior to that of the 81-cm barrier. Twenty-four turtles were captured in standard trawls pulled simultaneously against the 66-cm excluder trawl which caught only 3 turtles for a separation rate of 87 percent. Shrimp loss associated with the 66-cm barrier ranged from 2 percent to 36 percent during tests with the Super X-3 Tongue Trawl. These results include data from the entire test period, although early recorded shrimp loss rates were significantly reduced aboard some vessels during the study as indicated by Table 3. Removal of floats and the addition of weight to the tongue improved contact with the sea floor, and

WILBUR R. SEIDEL

Table 1. Trawl catch results

Time	Area	Number of vessels	Greatest percentage turtle separation	Percentage shrimp loss	Number of tows	Hours fished
Forward panel						
26 Jun–15 Dec 1978	East coast	15	75	25–45	1,424	3,438
15 July–15 Dec 1978	North Gulf of Mexico	6	Insufficient turtles	17–35	985	4,723
6 Jan–31 Mar 1979	Florida–Tortugas Gulf of Mexico	6	Insufficient turtles	25–45	195	901
Reverse panel						
21 Jun–25 Jul 1979	East coast	7	100	0–25	390	882

Table 2. 1979 Turtle excluder fishing summary

Trawl type	Barrier type	Turtle reduction (percentage)	Shrimp loss (percentage)
Flat	Forward 51 cm mesh	—	53
Jib 2-seam	Forward 81 cm mesh	—	42
Flat, 2-seam	Reverse 51 cm mesh	—	29
Jib	Reverse 81 cm mesh	—	23
Super X-3	Reverse 66 cm mesh	86	28
Super X-3	Reverse 81 cm mesh	0	33
Super X-3 "Tongue"	Reverse 66 cm mesh	89	17
Super X-3 "Tongue"	Reverse 81 cm mesh	0	25

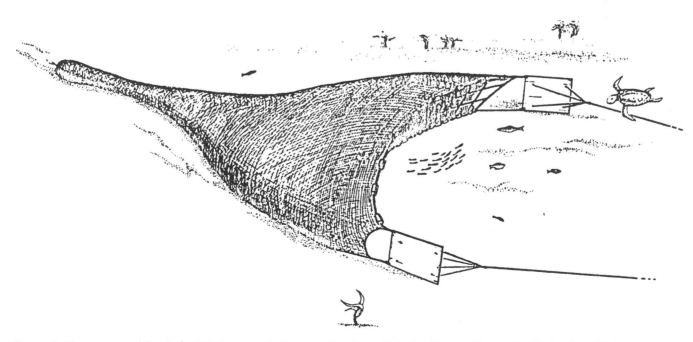

Figure 2. Reverse panel "excluder" shrimp trawl. Courtesy Southeast Fisheries Center, Harvesting Technology Unit.

Table 3. Gear modification log, Vessel Code "A"

Rigging description for excluder trawl	Shrimp loss percentage
1) Initial Rigging	67
12 (2.7 kg) floats	
8 mm Loop chain	
91 cm Tickler chain	
2) Removed 6 floats	37
3) Add 2 floats, add 11.3 kg to tongue	8
4) Two floats removed, tickler chain −122 cm	24
5) All floats removed, tickler chain −91 cm	8
6) Replaced 11.3 kg weight with 6.8 kg weight, added 4 floats	19
7) Barrier removed	9[a]

Note: Standard trawl = 15 m semiballoon trawl. Excluder trawl = 15 m super X-3 tongue trawl, barrier code RN26.

a. Increase.

shrimp loss was reduced to 8 percent aboard Vessel A. With the barrier completely removed, the Super X-3 Tongue Trawl was 9 percent more efficient than the standard 15-m semiballoon (control trawl).

Once the gear technical team tunes the trawl and leaves the vessel, some captains in their honest attempt to improve trawl efficiency continued to make changes to the rig. These changes, although sometimes beneficial, are more often detrimental to the gear performance. The single most difficult problem associated with the reverse barrier, however, is related to the vertical lift created as the barrier moves through the water. This lift causes the trawl to fish light on the bottom and can result in an increased shrimp loss. To overcome the problem, project personnel are presently working on a weighting system that will give continuous positive bottom contact. Once accomplished, the reverse barrier excluder trawl will be able to produce shrimp catches very nearly equal to that of a standard trawl.

The 3-year project for development of a sea turtle excluder trawl is on schedule. First-year objectives were successfully accomplished, changes indicated by field test results were incorporated into the trawl design for 1979, and 1979 results are being analyzed to provide direction for final year test and evaluation during 1980. Significantly improved catch rates of shrimp and very good reduction in turtle captures are indicated with the reverse barrier excluder trawl. Differences in fishing vessels, catch composition, topography, animal behavior on the various shrimp grounds, the broad spectrum of trawl types in use by the commercial shrimp industry and the logistical problems of maintaining gear and observers aboard a large number of vessels in different locations have presented difficult problems. Accomplishments to date, however, show considerable promise that a successful sea turtle excluder trawl will be available for introduction to the U.S. shrimp industry by the end of 1980 field testing.

Literature Cited

Carr, A.
1952. *Handbook of Turtles.* Ithaca, New York: Cornell University Press.

High, W. L.; I. E. Ellis; and D. L. Lusz
1969. A progress report on the development of a shrimp trawl to separate shrimp from fish and bottom-dwelling animals. *Commercial Fisheries Review* 31:20–33.

Hillestad, H. O.; J. I. Richardson; and G. K. Williamson
1978. Incidental capture of sea turtles by shrimp trawlermen in Georgia. Presented at 23rd Annual Conference of the Southeastern Association of Fish and Wildlife Agencies, 5–8 November 1978, Hot Springs, Virginia.

Ogren, L. H.; J. W. Watson, Jr.; and D. R. Wickham
1977. Loggerhead sea turtles, *Caretta caretta,* encountering trawls. *Marine Fisheries Review* 39:15–17.

Pritchard, P. C. H.
1969. The survival status of ridley sea turtles in American waters. *Biological Conservation* 2:13–17.

Rainey, W. E., and P. C. H. Pritchard
Undated. Distribution and management of Caribbean sea turtles. Virgin Islands Ecological Research Station, Caribbean Research Institute. St. Thomas: College of the Virgin Islands. Manuscript, 21 pp.

Watson, J. W., Jr., and C. McVea, Jr.
1977. Development of a selective shrimp trawl for the southeastern United States Penaeid shrimp fishery. *Marine Fisheries Review* 39:18–24.

WILBUR R. SEIDEL

Peter C. H. Pritchard
Florida Audubon Society
1101 Audubon Way
Maitland, Florida 32751

Recovered Sea Turtle Populations and U.S. Recovery Team Efforts

ABSTRACT

A few sea turtle populations, once depleted by heavy human exploitation, are known to have recovered to some degree. They include the loggerhead and leatherback populations in Natal, South Africa; the green turtle populations of Europa Island; and possibly the green turtle populations of Florida and Mussau, Papua New Guinea. Nevertheless, a recovery is always highly protracted and requires virtually complete protection in order to take place. Turtle populations that have been subject to massive egg exploitation may continue to decline for a decade or more following complete protection, as adults die of natural causes and are not replaced because of reduced or absent age-classes corresponding to the years of egg exploitation. On the other hand, populations subject to heavy harvesting of female turtles but that were not exposed to intensive egg collection may start to show recovery as soon as the population is protected; but such recovery is still very slow and in the short term may be masked by normal fluctuations in numbers of nesting turtles.

A false impression of recovery can be given either by unknown conditions prompting an abnormal fraction of the nesting population to nest in a given year, or by deteriorating conditions in adjacent nesting areas forcing ever-increasing numbers of turtles to nest on the remaining intact beaches.

The functions of the U.S. National Marine Fisheries Service-Fish and Wildlife Service Southeastern Region Sea Turtle Recovery Team are described. The team is appointed to write a recovery plan to delineate the tasks necessary for the restoration of all sea turtle populations in the southeastern region to a status that would not require their being listed as endangered or threatened. No recovery team is currently planned for the western region (U.S. Pacific Territories), but data-gathering in that area has been initiated.

In recent years it has been recognized that many, indeed most, sea turtle populations in the world are depleted to a greater or lesser extent, and responsible governments are undertaking efforts to protect and restore such populations so that they may once again become abundant, stable and of potential resource value. However, the slow maturation time of sea turtles, their migratory habits, and the heavy odds against a given hatchling reaching maturity combine to make the results of any turtle conservation plan less than immediately obvious. The data base is still too incomplete for me to be able to give a clear prediction as to what can be expected when one gives a previously overexploited turtle population either partial or complete protection, but the question is still worthy of discussion. Certain phenomena can give the impression that a turtle population is recovering when it is not in fact doing so, and these need to be considered. On the other hand, some turtle populations have been protected for several or even many years and have still failed to recover. Indeed, many turtle biologists and conservationists have become so discouraged by the nontangible results of their protective efforts that they are beginning to question quite seriously whether a depleted turtle population has the potential to recover, however thorough the protection. These ideas shall be discussed in turn.

For most species of sea turtle, gradual recovery—or depletion—of a population is masked by year-to-year variations in numbers of turtles that arrive on the nesting beach, since turtle populations are usually evaluated in terms of the number of nesting females. At other phases of their life cycle, reliable means of population census have not been developed. Factors whose nature is still not understood bring a much higher proportion of adult females into breeding condition in some years than in others. Carr, Carr and Meylan (1978) show that at Tortuguero, Costa Rica, an estimated 15,426 turtles nested in 1972, 5,723 in 1974, and 23,142 in 1976; but despite this variation there is no evidence that the turtle population suffered a marked decline from 1972 to 1974 from which it more than recovered by 1976. Both predation and conservation efforts were of comparable magnitude throughout this period, and no natural castastrophes were reported; it is simply a case of subtle climatic, feeding, and other factors having brought a great number of turtles into breeding condition in 1972 and 1976, with concomitant reduced nesting in adjacent years since green turtles almost never nest in successive years.

Similarly, Limpus (1978) reports that, throughout Queensland, Australia, green turtle nesting almost failed to take place at all during the 1975–76 nesting season, yet in the following season nesting was back up to almost typical levels. As an example of extreme fluctuation, 1,100 green turtles nested on Heron Island in 1974–75, but only 19 turtles nested there in 1975–76, even though the population remained unexploited. Clearly, under circumstances such as these, overall changes in the turtle population are extremely hard to detect, and must be evaluated either by careful transects of feeding areas or by taking multiyear averages of numbers of turtles seen nesting.

The ridley turtles of the genus *Lepidochelys* frequently if not typically nest annually; consequently, they are less susceptible to apparent fluctuations such as occur with the green turtle, and counts of nesting females provide a more reliable guide to the overall status of the population. However, it is not yet known whether ridleys show a labile 1- or 2-year cycle; but annual nesting may well be the norm—Pritchard (1969) reports that, of 445 nesting olive ridleys tagged in Surinam in 1967, 214 were found renesting in 1968. Allowing for mortality, tag loss, and turtles missed, this is an impressive recovery percentage.

Another circumstance that may give an unwarranted impression of a recovering population occurs when turtles are observed to nest in steadily increasing numbers in a protected area because they are progressively displaced from neighboring beaches by natural or artificial disruptive or distructive forces. As an example, the loggerhead turtles nesting at Cape Romain, South Carolina, have shown an impressive increase in numbers over the last 3 decades (Baldwin and Lofton, 1959; Hopkins et al., in press). But this may be attributable not so much to a recovering population as to human development or disruption of adjacent beaches forcing more and more turtles to nest within the undisturbed beach area in the Cape Romain Wildlife Refuge.

Indeed, predation on the abundant nests at Cape Romain by raccoons is so intense that recruitment is almost zero, and a population collapse can be envisioned in the near future.

Similarly, Schulz (1975, this volume) has documented the increasing numbers of leatherback turtles nesting in Surinam, especially on the beaches in the Marowijne River; the increasing numbers are as follows:

Year	Number of nests
1964	95
1967	90
1968	200
1969	305
1970	255
1971	285
1972	380
1973	900
1974	785
1975	1,625
1976	670

PETER C. H. PRITCHARD

Year	Number of nests
1977	5,565
1978	2,160
1979	3,900

Nevertheless, it would be wrong to conclude that overall the leatherback population nesting in the Guianas is increasing. The real explanation is that the neighboring French Guiana beaches, which provide nesting grounds for the world's largest known colony of the species, were undergoing progressive erosion in 1964–75, and increasing numbers of the breeding leatherbacks were displaced toward the still intact or accreting beaches of Surinam.

Some turtle populations have not only failed to recover, but have also undergone conspicuous further decline even though they have been offered strong protection. For example, the only significant colony of the olive ridley turtle (*Lepidochelys olivacea*) in the western Atlantic nests on a small beach called Eilanti in Surinam. Protection of this colony from egg collectors was instigated in 1967 and 1968, when the Surinam Forest Service, the World Wildlife Fund, and I cooperated in a program of egg purchases from the local Carib Indians. The entire season's production—nearly 300,000 eggs a season—was thus purchased, the eggs were incubated, and the hatchlings released. In subsequent years, it has been illegal to exploit ridley eggs in Surinam, and this edict has been enforced; protection of the nesting beach has been effectively 100 percent from 1967 to the present. Yet the number of ridleys nesting at Eilanti decreased as follows during those years:

Year	Number of nests
1967	2,455
1968	2,598
1969	1,074
1970	1,266
1971	1,249
1972	1,051
1973	690
1974	638
1975	531

The numbers of ridleys nesting on the neighboring Marowijne beaches dropped similarly during this period, from 465 nests in 1968 to 201 in 1975. A modest increase was recorded, from 95 to 236, at the other Surinam nesting beach (Bigisanti), but this increase was far from sufficient to account for the drop reported from the other beaches. Schulz (this volume) report that the decline has continued in the last few years;

1,070 nests were made in the whole of Surinam in 1975, 1,160 in 1976, but only 1,030 in 1977, 870 in 1978 and 795 in 1979.

Similarly, the only known population of the Kemp's ridley (*Lepidochelys kempi*) whose sole nesting ground is in the vicinity of Rancho Nuevo, Tamaulipas, Mexico, has undergone a marked drop despite the presence of vigorous beach patrols and a hatchery since 1966. In 1966, the largest *arribada* or nesting aggregation of the season included about 1,338 females, while other arribadas the same season were estimated or counted at 200, 150 to 200, 98, 20, 200, and 25, giving a total of around 2,100 nests, not counting the nests made by sporadic individuals on intervening days. In 1977, the twelfth year of good protection, arribadas of the following size were recorded: 45; 200; 50; 40; 60; 20; 170; 6; 15; and 15 totaling 621. Slight increases were logged in the next 2 years—924 nests in 1978 and 1,013 nests in 1979.

Two factors probably caused the turtles to continue to diminish despite protection of the beach:

1) Both species of ridley are very subject to accidental mortality in shrimp trawlers. Most of the tag recoveries from these 2 populations were reported by shrimpers, and in many cases the turtles were not returned alive to the sea. This factor alone may be sufficient to prevent the recovery of a highly depleted population even if beach protection is good. However, it is also probable that recruitment of young adults into the population during the years in question has been extremely depressed. For many years prior to the instigation of beach protection in 1967, the Carib Indians collected every single egg laid on the beach at Eilanti. It is not entirely clear how long this had been the case, but certainly the habit was well established when I first visited the beach in 1966; each Indian had rights to eggs in a certain section of the beach, and they knew well when to expect an arribada. Consequently, almost no hatchlings were produced by this population for many years prior to 1967, and thus a drop in population would be inevitable as adult turtles died off. The population would be expected to level out and start rising again as hatchlings from the 1967 and subsequent seasons started to reach maturity, but this reversal of trend was not apparent by 1979. Márquez, Villanueva, and Peñaflores (1976) calculate that *L. olivacea* reaches sexual maturity at an age between 7 and 9 years, but these data were based on specimens in semicaptivity in an estuary in Jalisco, Mexico.

I want to analyze the case of the Kemp's ridley population in some depth, not because it is a recovered population, but because it has failed to recover even though many observers would have expected or hoped for recovery already in view of the considerable resources that have been and are being invested in this population, the only population of the species in existence. Perhaps we can, by analyzing the situation and bringing together available knowledge, initiate some predictions of when recovery may start to become visible.

The history of the Kemp's ridley population prior to 1966 is poorly documented. The largest arribada in 1947 was estimated from a film to have numbered about 40,000 nesting turtles, but no real data exist on the size of subsequent arribadas until 1966. However, Hildebrand (1963) gathered some fragmentary information on populations—and exploitation levels—during these intervening years. Hildebrand was informed in 1958 that each year an Arab came to the beach and departed with a mule train numbering 40 or 50 animals each loaded with sacks of ridley eggs to sell on the Tampico market. Another informant reported that 4 or 5 turtles had nested on the beach on 13 May 1960, and that the following day, from an aircraft, "turtles could be seen all along the beach."

From information such as the above, it appears likely that, for a number of years preceding the installation of the conservation camp in 1966, the eggs of Kemp's ridley were subject to methodical and nearly complete exploitation by man. Further evidence is provided by Adams (1966), who reported meeting an egg collector in 1963 who had sold over 20,000 eggs, and who reported that many others were doing likewise. Adams also reported from his interviews with the head of the Rancho Nuevo *ejido* in 1964: "We were told many different stories of how the number of turtles had decreased in the last few years. How buyers from Tampico and Mexico City would come every year and leave with truckloads of eggs for ready market in the cities. How boats would lay offshore and scoop up thousands of live turtles and cut them open just for the eggs."

The breeding population of Kemp's ridleys may have collapsed to close to the present levels between 1969 and 1970. Vargas (1973) illustrates by means of histograms the estimated sizes of arribadas of *kempi* during 1966 to 1970. From these charts, the total numbers of turtles in arribadas of these years were:

Year	Number of turtles
1966	1,800
1967	5,000
1968	7,000
1969	10,000
1970	900

I have no explanation for the rapid build-up during 1966–69, nor the collapse in 1969–70. However, Casas-Andreu (1978) reports 6 arribadas during the 1970 season, on 17 April, 1–3 May, 16–17 May, 27 May, 2 June, and 10 June. Vargas indicates no arribadas at all during April 1970 in his histograms, and his estimate

for the total numbers nesting in arribadas in 1970 (900) is less than half that of Casas-Andreu (2,000). It is possible that the arribadas of earlier years were only approximately estimated and rounded off to the nearest thousand, so it is quite possible that the rise and collapse of the population as indicated by Vargas' figures were more dramatic than justified.

In 1973, I was informed by villagers at Rancho Nuevo that about 1,000 turtles had nested on 28 April and that all the eggs had been taken. Subsequent monitored nesting was as follows: 3 May, 21; 13 May, 133; 14 May, 8; 15 May, 4; 18 May, 7; 19 May, 57; 22 May, 2; 28 May, 50 to 75; 30 May, 1; 2 June, 4; 8 June, 5; 11 June, 9; 12 June, ca. 36, and 19 June, 3. By this time, the fragmentation of the arribadas into small groups or solitary individuals was becoming apparent.

It is fair to assume that recruitment of hatchling Kemp's ridleys into the population was extremely low for some years preceding protection in 1966. This, together with the omnipresent shrimp trawlers, may account for the progressively declining numbers on the nesting beach subsequent to 1970. It is within our estimates of the maturation time of Kemp's ridley (Márquez 1972) calculated that Kemp's ridley takes 5.5 years to reach minimum breeding size, and 7 years to reach average breeding size) to suggest that the slight up-turn in numbers of breeding ridleys in 1977–79 indeed reflects the "new wave" of recruitment of 1966–69 and subsequent years. However, more years of data will be necessary to confirm this.

A few data are available on the ability of populations of the loggerhead turtle (*Caretta caretta*) to recover following protection of eggs and nesting adults. Richardson et al. (1978) initiated a tagging and hatchery program for this species at Little Cumberland Island, Georgia, in 1965. Beach patrols were not so intensive that all turtles nesting were tagged, but each year a substantial proportion of turtles was tagged, and the ratio of tagged individuals to neophytes was calculated. The proportion of neophyte individuals progressively decreased from 100 percent in 1965 and 97 percent in 1966 (some turtles had been tagged in 1964) to 30 percent in 1975 and 27 percent in 1976. The regression curve was linear, indicating an apparent complete lack of recruitment of new animals into the population. Up to 1964, predation on eggs by raccoons had probably been extremely intense, but subsequent to that time 6,000 to 10,000 hatchlings were released annually. It appears likely that the subsequent 12 years of beach patrols were insufficient for even the earliest of these hatchlings to reach maturity. The only alternative explanations are that 100 percent of the hatchlings failed to survive (since survival of even 6 to 10 animals would have been revealed in the analysis of the neophyte ratio), or that the turtles nested on other beaches. The latter explanation would require as a corollary that turtles hatched on other beaches would sometimes nest on Little Cumberland, but the evidence suggests otherwise. However, Ehrhart (1979) found that turtles tagged while nesting elsewhere in Florida and even in Georgia nested at Merritt Island in 1978.

When nesting turtles are subject to regular killing by man, the nesting colony is likely to show progressive and often rapid depletion. Bass et al. (1965) report on the first year of work (1963–64) on the loggerhead colony at Tongaland, South Africa, where the nesting females had been subject to a significant level of predation by local people for several seasons, though the practice was thought to be "fairly new," and loss of eggs had apparently not been serious. For each season since 1963–64 beach patrols were conducted throughout the nesting time, and predation by man on eggs and turtles was largely brought under control.

As Hughes (this volume) has shown, the number of loggerheads nesting at Tongaland showed a steady rise in subsequent years. The number of years (16) is sufficient for the overall population trend, unmasked by year-to-year fluctuations, to be revealed. With a progressive increase from fewer than 90 nesting turtles in 1963–64 to approximately 400 in 1978–79, this appears to be our first example of a recovered, or at least recovering, population. The history of the leatherback population in Tongaland has been identical, and the species has shown a corresponding recovery under protection, from fewer than 25 nesting females in each season from 1963–64 to 1970–71, to more than 50 nesting turtles each season from 1973–74 to 1978–79 (Hughes this volume).

Certain populations of the green turtle (*Chelonia mydas*) can also be considered possible candidates for "recovered" or "recovering" status. To demonstrate recovery and determine the rate at which it can occur, we need to select situations where a population has been subject to heavy exploitation of breeding adults, followed by a protracted period of protection. Europa Island, in the Mozambique Channel, may be a suitable example, although unfortunately the former exploitation there, though undoubtedly heavy, is inadequately documented. However, Frazier (this volume) reports that organized turtle exploitation may have begun on Europa as early as 1860, but was probably not consistent until 1903 when the first settlement was established. Between 1903 and the passage of protective legislation in 1933 exploitation was probably heavy. Paulian (1950) reported large numbers of turtle bones still to be seen in the northern part of the island. However, recovery now appears to be complete, and recent visitors report a nesting green turtle population so large that it is probably self-limiting (by nest destruction by later turtles nesting in the same spot). Even so, numbers of turtles nesting each year appear to be strikingly variable. These are reviewed by Hughes (this volume), who quotes 4,000 to 5,000 turtles nesting in 1970–71, 3,000 in 1973–74, 1,500 in 1977–78 and 9,000 to 18,000 in 1978–79.

Another population that may have some claim to be in at least an early stage of the recovery phase is the green turtle in Florida. However, a difficulty in the argument is that, although numerous texts make casual mention of green turtles' once having nested in numbers in Florida, there is extremely little available in the way of first-hand accounts to document this. As Carr and Ingle (1959) observed, "Other writers apparently assumed that breeding occurred in Florida because the grazing flats on the Gulf Coast and in the Indian River were so heavily populated with turtles. This does not necessarily follow. It is possible that nesting loggerheads, and the feeding aggregations of green turtles in adjacent waters may have been incorrectly assumed to be aspects of one phenomenon."

At any rate, the Dry Tortugas definitely were once an important green turtle nesting ground, though the greens there were exterminated many years ago and today apparently only loggerheads nest there.

In recent years, there has been an increasing trickle of records of green turtles nesting on the Atlantic coast of the Florida Peninsula; these may represent the remnant of a once major nesting colony. Carr (1952) wrote that there had been no report of a green turtle nest on the Florida coast for 40 years, and while there was far less activity by herpetologists looking for nesting turtles during those years than there is today, it seems surely true that there could not have been much nesting by the species in Florida during those years. Carr and Ingle (1959) reported the first 2 definite breeding records for the Florida coast—1 on 11 July 1957 2 miles north of Vero Beach, Indian River County, and 1 on 27 June 1958 at Hutchinson Island, Martin County. In subsequent and recent years, Ross Witham working on Merritt Island, Brevard County, have reported regular, though not abundant nesting by green turtles. Ehrhart (1979) reports as follows regarding *Chelonia mydas* nesting on Merritt Island: "In five previous summers we had seen a total of only 11 individuals of this species and never more than 3 per year at KSC. In 1978, no less than 14 were encountered there and we have definite records of 23 nesting emergences . . . Only one of the *Chelonia* was a recapture from a previous year.

"There is good evidence that greater numbers of green turtles nested elsewhere on the east coast of Florida as well and that a larger than usual contingent nested at Tortuguero, Costa Rica, the major Western Hemisphere Rookery. As a result, some sea turtle biologists now feel that there may be 300 to 400 adult females left in the Florida population, rather than the 100 to 200 formerly proposed."

Similarly, Witham has found that green turtle nests on Hutchinson Island increased erratically but steadily from 2 in 1967 to 45 in 1975 and to 61 in 1978. Therefore, while the data base on the original population, its exploitation, and subsequent protection is defective, it seems a reasonable hypothesis to suggest that a once great nesting population of green turtles in Florida that was virtually wiped out by intemperate exploitation is now showing some initial but definite signs of a comeback.

One other green turtle population may be in a recovering or recovered phase. This population both grazes and nests around Mussau Island, north of New Ireland in the Bismarck Sea. Visiting this island in 1978, I was informed by the local people that green turtles had, in former times, been heavily exploited and had become scarce. However, in the early 1930s, the entire island population had converted to the Seventh Day Adventist Faith, and as a tenet of this belief had ceased to eat sea turtles. Occasional animals were taken by outsiders in subsequent years, but not many. I was impressed to see large numbers of surprisingly tame green turtles of both sexes and both juvenile and adult sizes on the reefs and in the shallows off Mussau. The green turtles there were far more conspicuous and seemingly abundant than anywhere else in Papua New Guinea, and while the past status of the species is based purely on the memory of village elders, it seems reasonable to consider the Mussau green turtle population has recovered after about 45 years of conscientious protection.

Rapid recovery of a green turtle population that has been subject to complete or nearly complete egg exploitation is probably impossible. Data now being gathered in several parts of the world suggest that green turtles, although having the potential to reach maturity in about 8 years under captive conditions and on a high protein diet, grow extremely slowly in the wild. For example, the average growth rate of immature green turtles in the 46 to 59 cm range in the Galápagos Islands was determined to be only 0.53 cm/yr by Green and Ortiz (this volume). Slightly higher rates were found in the southern Great Barrier Reef of Australia by Limpus 1979, namely an average of 1.31 cm/yr for turtles in the 60 to 90 cm range. Only slightly greater growth rates have been found in Mosquito Lagoon, Florida by Mendonça (personal communication). Doubtless the precise growth rate depends upon the feeding environment in which the animal finds itself, and Balazs (this volume) has documented that, in the Hawaiian archipelago, growth rates are much more rapid around the southernmost, biggest island, Hawaii itself, than in the smaller, more northern islands.

The accumulating data suggest that green turtles typically take several decades to reach maturity under natural conditions. Thus, a population that has been heavily overharvested for eggs, such as the Sarawak Turtle Islands population, may have to be completely protected for several decades before the nesting population "bottoms out" and finally starts to rise again. Apparent increases before such an interval for any green turtle population are almost certainly the result of the normal year-to-year fluctuations discussed earlier in this paper.

Recovery Teams and Recovery Plans

The concept of the recovery team was conceived by Earl Baysinger of the United States Fish and Wildlife Service, and was developed by the Office of Endangered Species. The intention of the concept is that the Service should designate a small working group of experts on a particular endangered species. These experts are usually drawn primarily from government agencies but also include representatives of the academic and private sectors, whose duty is to draw up a multistep plan that, if conscientiously executed, would bring about the recovery of the species to the point where it could be taken off the Endangered Species List. Team members are instructed to concern themselves purely with biological considerations rather than attempting to interpret political considerations. The plan finally agreed upon by a recovery team is then submitted to the regional director (FWS or NMFS as appropriate), who distributes it to interested agencies and other parties for comment. After modification, if necessary, the plan is sent to the director for final approval. The Endangered Species Act also provides that "critical habitat" may be declared for an endangered species, and once so declared, projects within that area that have federal funding or require a federal permit must demonstrate themselves to be free of adverse impact upon that critical habitat. Teams may make their views known as to what they recommend to be the critical habitat for their species, though the designation of the habitat is not one of the team's responsibilities.

The essentially volunteer nature of recovery teams has resulted in some recovery plans' taking a long time to be finished, and recently Congress has made its intentions clear that teams should receive "adequate funding." Recently, some recovery plans have been drawn up by a single individual working under contract to the Fish and Wildlife Service.

Three sea turtle species—the hawksbill, leatherback, and Kemp's ridley—were added to the U.S. Endangered Species List in 1973, but the question of agency jurisdiction, necessary for the establishment of recovery teams, was not resolved until about 5 years later. The eventual memorandum of understanding between FWS and NMFS established that the Department of the Interior should have jurisdiction over turtles on land and that the Department of Commerce should

PETER C. H. PRITCHARD

have jurisdiction over turtles while they were in the water. This resulted in such allocations of duties as Interior's having responsibility for controlling the importation of illegal turtle products, and Commerce having responsibility for the development of devices to minimize incidental catch of turtles by shrimp trawlers and others. It was also decided that the recovery team should be a joint responsibility, but with National Marine Fisheries Service (Department of Commerce) having the lead role and the responsibility of providing team funding.

A Southeast Region Sea Turtle Recovery Team has been formed, accountable primarily to the NMFS Southeastern Regional Director. It has responsibility for all 6 species of sea turtles found in the southeast region and has a larger budget than Department of Interior Teams ($10,000 annually rather than $2,000). The team is unusually large—12 members—who represent the National Marine Fisheries Service (2 members); Fish and Wildlife Service; Natural Resource Departments of South Carolina, Georgia and Florida; the Caribbean Fishery Management Council; the academic world; the citizen conservation movement; the commercial trawling industry; the Florida Audubon Society; and the South Carolina Wildlife and Marine Resources Department. The team also has approximately 30 formal consultants.

The team is considering not only species whose range extends within the United States but also specific populations whose members migrate between U.S. and foreign waters. The plan thus will present numerous recommendations to foreign governments. The team was directed by the regional director to include in the recovery plan relevant social, economic, and political information.

Because of the international nature of sea turtle conservation and the recognition that a population or a species can be properly protected or conserved only if the management program extends over all phases of the life cycle, the team gave considerable thought to the delineation of the populations over which it would deliberate. It was decided that all populations would be included that ever resided within the waters of the U.S. Southeastern Region, which as defined by the National Marine Fisheries Service includes the coast and adjacent waters of the entire U.S. Gulf Coast, the Atlantic Coast as far north as Virginia, the U.S. Virgin Islands, and Puerto Rico. This definition, it was felt, was compatible with both the NMFS directives and biological reality.

The team is thus interested in Kemp's ridley, whenever and wherever it is to be found. It will also include consideration of the olive ridley. Although this species is only known to nest in South America in the West Atlantic, individuals occasionally reach the Greater Antilles, including Puerto Rico and are frequently caught accidentally by U.S.-based shrimp trawlers.

Green turtles nesting in Florida are of special interest to the team which also proposes to follow all Caribbean populations of this species as well as the population nesting in Guyana. The Surinam population not only appears to be under good management already but also has been shown to migrate only to the east, towards Brazil, after nesting, so it was felt there was no need to include it in the plan. Neither did it meet the criteria for inclusion.

The plan will devote major attention to the loggerhead in the United States. Not only does this species comprise over 98 percent of all marine turtle nestings on mainland United States, but it is also subject to 2 pressures—excessively high egg predation by raccoons and heavy losses of subadults and adults in shrimp trawls—that may well cause its reasonably abundant populations to collapse in the next 2 decades.

Caribbean populations of the loggerhead will also be included in the plan, but not those of Brazil or the eastern Atlantic.

The nesting and reef-dwelling hawksbills of the U.S. Virgin Islands, Puerto Rico, and Mona will be of special interest to the team. However, it will address itself to this species generally throughout the Caribbean.

Recovery teams have been criticized as being excessively academic, merely formalizing already obvious species survival strategies, and increasingly becoming an end in themselves rather than the first step in saving threatened and endangered species. Another criticism is that the formation of a recovery team and its subsequent 1 or 2 years of deliberations over a recovery plan dangerously delays the instigation of necessary protective measures. These criticisms can be answered.

Without a recovery plan, a species may receive attention only if it happens to catch the imagination of an authority on the species who can devote the necessary time and attract the necessary resources to instigate recovery procedures. Moreover, such a haphazard, piecemeal approach may result in neglect of vital aspects of the recovery procedure. For some highly localized species, a recovery plan may be as simple as a single action or recommendation; but for a group as complex and far-flung as sea turtles, the formation of a recovery team ensures the logical definition of the recovery process, the identification of all its necessary components and the designation of agency responsibility.

The second criticism also does not apply in the case of the Sea Turtle Recovery Team. In no way are protective measures, recovery procedures, or relevant research being delayed until the work of the team is finished. While private interests and individual states carried most of the burden of responsibility for sea turtle conservation for several decades, the federal agencies are now making up for lost time most commendably. The stage was set by the passage of the Endangered Species Act and the listing of the various species as threatened or endangered; currently, only *Chelonia depressa*, endemic to northern Australia, is not listed. Subsequently, the National Marine Fisheries Service started a major research program to solve the problem of incidental capture of sea turtles in shrimp trawls, and is currently involved with a 10-year head-start program for the critically endangered Kemp's ridley and a wide variety of sea turtle research programs, being conducted mostly by the private sector under contract. The U.S. Fish and Wildlife Service has also been active, and, among other things, has the primary responsibility for the U.S. contribution to the beach protection effort for the Kemp's ridley nesting ground at Rancho Nuevo, Mexico. Other federal agencies involved in one way or another with sea turtle research include the National Park Service and the National Aeronautics and Space Administration.

As already mentioned, the existing Sea Turtle Recovery Team has responsibility for the southeastern region only. For the present time, no team is proposed for the western region. Nevertheless, NMFS Hawaii is in the early stages of planning a major research and data-gathering program in the western region. Even though there are very few sea turtles on the West Coast of the U.S. mainland, Hawaii has a small but important population of green turtles, and various populations of green turtles and hawksbills are scattered through the endless islands of the Trust Territory (Micronesia) and other U.S. Pacific Islands, including Guam and Samoa. Knowledge of these populations is at a much earlier stage than that of the turtles of the Southeastern Region; the need for a recovery team will be addressed once the problems have been defined.

If marine turtles are to be saved from further depletion and ultimate extinction, strong action by many governments and concerned conservationists will be necessary in many countries. But I have confidence that the recovery team will play more than a parochial role. All 6 threatened or endangered species of sea turtle—the green turtle, leatherback, loggerhead, hawksbill, Kemp's ridley, and olive ridley—are found in the southeastern region, at least marginally, and thus fall within the sphere of interest of the recovery plan. Moreover, even if the U.S. initiative fails to achieve any significant action by other nations, we should not forget that the marine turtle populations over which the United States has technical jurisdiction are spread over far more of the globe than just U.S. mainland waters. This jurisdiction includes the thousands of islands and 3 million square miles of ocean in the Pacific Trust Territory; the 1,000-mile long Hawaiian chain and adjacent waters; Samoa; Guam; and, in the Caribbean, Puerto Rico, and the U.S. Virgin Islands.

But it is our hope that the recovery plan will, in fact, receive the approval of other nations so that integrated and comprehensive plans for the restoration of all populations of sea turtles will be established and agreed upon in a true spirit of international cooperation for a common cause.

Finally, I should mention that even though I happen to be co-leader of the Southeastern Sea Turtle Recovery Team, I have presented this summary of information on the team and its functions purely as a private individual, and opinions expressed are mine alone.

Acknowledgments

My field work on Kemps's ridley in Mexico was supported by the World Wildlife Fund (U.S. National Appeal) in 1967, 1970 and 1973, and by the U.S. Fish and Wildlife Service (Albuquerque office) from 1978–80. The work in Mussau Island was conducted in the course of field work in Papua New Guinea entirely sponsored by the PNG Wildlife Department. The Kemp's ridley studies were conducted in connection with the Mexican Instituto Nacional de Pesca (René Márquez M., chief of sea turtle programs). Tags for the work on Kemp's ridley up to 1973 were generously supplied by Dr. Archie Carr, who also provided clearing-house service for tag returns. To all these institutions and individuals I express my warm gratitude.

Literature Cited

Adams, D. E.
1966.   More about the ridley. Operation: Padre Island. Egg transplanting. *International Turtle and Tortoise Society Journal* 1:18–20.

Baldwin, W. P., and J. P. Lofton
1959.   The loggerhead turtles of Cape Romain, South Carolina. Abridged and annotated by D. K. Caldwell. *Bulletin of the Florida State Museum* 4:309–48.

Bass, A. J.; H. J. McAllister; and H. J. Van Schoor
1965.   Marine turtles of the coast of Tongaland, Natal. *Lammergeyer* 3:12–40.

Carr, A. F.
1952.   *Handbook of Turtles*. Ithaca, New York: Cornell University Press.

Carr, A. F.; M. H. Carr; and A. B. Meylan
        The ecology and migrations of sea turtles, 7. The West Caribbean green turtle colony. *Bulletin of the American Museum of Natural History* 162:1–46.

Carr, A. F., and R. M. Ingle
1959.   The green turtle (*Chelonia mydas*) in Florida. *Bulletin of Marine Science in the Gulf and Caribbean* 9:315–20.

Casas-Andreu, G.
1978.   Analisis de la anidacion de las tortugas marinas del genero *Lepidochelys* en Mexico. *An. Centro Cienc. del Mar y Limnol. Univ. Nal. Auton. Mexico* 5:141–58.

Ehrhart, L. M.

1979.   Threatened and endangered species of the Kennedy Space Center. Semi-annual Report to NASA, volume V. Contract NAS 10-8986. Mimeo, 214 pp.

Hildebrand, H. H.

1963.   Hallazgo del area de anidacion de la tortuga marina lora, *Lepidochelys kempi* (Garman), en la costa occidental del Golfo de Mexico. *Ciencia* 22:105–12.

Hopkins, S. R.; T. M. Murphy; K. B. Stansell; and P. M. Wilkinson

In press.   Biotic and abiotic factors affecting nest mortality in the Atlantic loggerhead turtle. Contribution 89, South Carolina Marine Resources Center.

Limpus, C. J.

1978.   The reef: Uncertain land of plenty. In *Exploration North: Australia's Wildlife from Desert to Reef*, ed. H. J. Lavery, pp. 187–222. Richmond, Australia: Richmond Hill Press.

1979.   Notes on growth rates of wild turtles. *Marine Turtle Newsletter* 10:3–5.

Márquez, R.

1972.   Resultados preliminares sobre edad y crecimiento de la tortuga lora, *Lepidochelys kempi* (Garman). Memorias IV Congreso Nacional de Oceanografia, 17–19 November 1969, Mexico D.F., pp. 419–27.

Márquez, R.; A. Villanueva; and C. Peñaflores

1976.   Sinopsis de datos biologicos sobre la tortuga golfina *Lepidochelys olivacea. Inst. Nat. Pesca., Mexico.*, pp. 1–61.

Paulian, R.

1950.   L'Ile Europa, une dependance de Madagascar. *Le Naturiste Malgache* 2:77–85.

Pritchard, P. C. H.

1969.   Sea turtles of the Guianas. *Bulletin of the Florida State Museum* 13:85–140.

Richardson, T. H.; J. I. Richardson; C. Ruckdeschel; and M. W. Dix

1978.   Remigration patterns of loggerhead sea turtles (*Caretta caretta*) nesting on Little Cumberland and Cumberland Islands, Georgia. *Florida Marine Research Publications* 33:39–44.

Schulz, J. P.

1975.   Sea turtles nesting in Suriname. *Zoologische Verhandelingen, uitgegeven door het Rijksmuseum van Natuurlicke Historie te Leiden* 143:1–144.

Vargas, T. P. E.

1973.   Resultados preliminares del marcado de tortugas marinas en aguas Mexicanas (1966–1970). *Inst. Nac. Pesca, Serie Informativa*, pp. 1–27.

Stephen V. Shabica
National Park Service
Gulf Islands National Seashore
Coastal Field Research Laboratory
NSTL Station, Mississippi 39529

Planning for Protection of Sea Turtle Habitat

Introduction

Planning for habitat protection involves 2 steps: the identification of the habitat upon which the resource is dependent, and the elaboration of criteria and implementation of regulations to insure that habitat degradation will be prevented and that habitat improvement will occur. In the case of sea turtles, habitat has generally been thought of as the nesting beaches. However, sea turtle habitat goes far beyond simply the nesting beaches. Marine turtle habitat characteristics have been described by Rebel (1974) and more recently by Schwartz (1978).

Threats to Habitats

Man's ability to produce environmental and ecological perturbation and his inability to perceive or predict the environmental and ecological consequences of this perturbation have led to what Bella and Overton (1972) term an "environmental predicament: man's ability to modify the environment will increase faster than his ability to foresee the effects of his activities." Man has been modifying the environment throughout history. Yet, it is only in the last half century that these perturbations have begun to stress the integrity of the earth's ecosystems. Man is not only a part of the environment, but he is also inextricably linked with all ecosystems. We are only now beginning to appreciate the effects of our technological society on the resources and organisms on which we depend for sustenance and survival. We are also increasingly aware of our effects on species such as marine turtles.

Witham (this volume) as well as others have demonstrated that human encroachment of sea turtle nesting habitat is perhaps the primary cause of the decline of sea turtle activity in many areas. Declines in nesting populations of green and loggerhead turtles over a 3-year period on Hutchinson Island, Florida are attributed by Worth and Smith (1976) to expanding urban development including reduced background vegeta-

513

tion, artificial lights, human activity, and increased predation by raccoons brought about by the displacement of the raccoon population as a result of construction activities. On Kiawah Island, South Carolina, Dean and Talbert (1975) observed that nesting activity was lowest in areas where beach homes were present and where restriction on lighting and beach traffic were lacking, even if the beach appeared ideal for nesting. Additionally, they found that areas which appeared geomorphically unsuitable to nesting were "increasingly utilized by turtles due to the areas' isolation and lack of human activity." They conclude that "only because of highly limited development in the past have the Kiawah loggerheads not encountered the human pressures evident on other Charleston beaches." Lund (1974) suggests that the consequences of coastal strand development may be more profound in northern states than in southern states such as Florida. The Carolina seashores attract people during the summer nesting season whereas Florida beaches are primarily utilized in the winter. This effectively "increases the chances of disrupting nesting turtles with bonfires, vehicles, lights, etc." Coastal development and construction in nesting areas are considered the greatest threats to sea turtles in Queensland, Australia by Bustard (1972). He also cites oil exploration and mineral sands mining as posing threats to sea turtle nesting grounds. Dean and Talbert (1975) observed that increasing development along South Carolina beaches has resulted in the displacement of loggerhead females to more protected nesting grounds (Cape Romain National Wildlife Refuge, Belle W. Baruch Institute for Marine Biology, and the Santee Coastal Reserve of South Carolina's Heritage Trust Program). The increase in loggerhead turtle nesting in the Cape Sable area of Everglades National Park is thought to be related to the park's establishment in 1947. Davis and Whiting (1977) suggest that increasing developmental pressure surrounding the park resulted in the displacement of turtles to the park beaches from pressured areas. Caldwell (1962) suggests that the abandonment of the Jekyll Island rookery in favor of the beaches of the Cumberland Islands in Georgia was due to human and developmental encroachment on Jekyll Island.

The blowout on 3 June 1979 of the oil drilling rig IXTOC-I in the Bay of Campeche has resulted in the release of crude oil into the Gulf of Mexico at reported rates of 15,000 and 30,000 barrels per day. Drilling of a relief well or capping of the well is not expected until mid-November. Of prime consideration is the potential impact of this oil on the nesting habitat of the Kemp's ridley turtle and on the long-term effects from oil fouling of both adults and hatchlings. Fortunately, more than two-thirds of the 1979 hatchlings had entered the Gulf of Mexico before the spill. The remainder were released off the coast of Florida and

within *Sargassum* patches offshore of Rancho Nuevo. As of this writing, 1 dead ridley, 6 dead green turtles, and 5 oil-fouled but recovered green turtles have been associated with this spill in U.S. waters. This oil spill has demonstrated that planning for habitat protection of sea turtles must include the oceanic environment. Even after surface-oiled beaches have been cleaned, one must wonder about the fate of buried oil. How will an oil layer within the nesting beach affect the eggs or the hatchlings? Would oil, either on the surface or buried, be cause for the female to reject a nesting beach? The head start program (Wauer 1978) carried out by Mexico and the United States at Padre Island National Seashore and the Galveston Laboratory of the National Marine Fisheries Service may prove futile if nesting beaches, migration routes, and feeding grounds become fouled by oil. Thus, habitat protection for sea turtles involves not only the protection of nesting areas from development and intense human presence, but also the protection of nursery habitats, hibernation areas and dispersal and migratory routes from pollution.

The habitats within which sea turtles hibernate (Felger, Cliffton, and Regal 1976; Ogren and McVea, this volume) must be included in habitat protection. Data must be obtained on these populations so that activities such as fishing, dredging, or construction can be scheduled to prevent interference with the populations.

Lund (1974) suggests that fluctuations and movements of nesting populations may be due to the temporary disruption of nesting beaches caused by beach erosion and stresses "the importance of federal wildlife refuges, national parks and national seashores, state parks and even public beaches as possible reservoirs of turtle nesting in the future." In the United States, the largest and most significant nesting populations of sea turtles occur within federal wildlife refuges, national parks and national monuments. The loss of turtle nesting habitat outside these areas is thought to be due primarily to the urbanization and development along active or potential nesting beaches (Lund 1974). In Australia, Bustard (1972) sees the establishment of national parks as a means of protecting sea turtle nesting grounds from habitat degradation, although this is not without its problems since aboriginal populations are granted the right to exploit wild populations within these areas.

Many prime turtle nesting habitats are found on barrier islands and beaches. The very dynamic nature of these systems requires jurisdictional and legal planning in order to maintain the continuity of the area if the habitat moves due to physical or successional changes. In this regard, buffer areas must be sufficiently large to provide for long-term variations of the critical area. Clark (1976) states, "tragically, the barrier islands are increasingly the focus of intense real estate speculation and development activity, setting up a strong conflict.

STEPHEN V. SHABICA

Natural values and public access problems are rapidly lost in the face of the seashore building boom. More than half of the major barrier islands and beaches are already fully committed to private housing and commercial enterprise."

Habitat Protection

We must approach the problems of sea turtle habitat protection in an integrated way:

The establishment of a few parks and reserves is a hopeful sign but is also only a partial approach. There is no alternative to massive, integrated effort of scientists, social scientists, lawyers, politicians and public relations personnel in a regional, worldwide attack on the problem. Ecological science must take a lead since the primary need is to modify man's lifestyle to the realities of ecosystems which he is far from understanding or controlling in the sea and coastal zone, but which he is presently destroying (Ray 1976).

How can we plan for the protection of sea turtle habitat when we know very little about their critical habitats? How are sea turtles able to return to their natal beaches? What sensory cues, if any, does the female receive from the beach that allow her to choose or reject a particular area? How important are migration routes, and growing and hibernation bays and lagoons? We do know that human encroachment in the form of developments, industry, lights, vehicles, and activity nearby and within nesting areas have profound consequences to the sea turtle population; that onshore and offshore mining and accidents associated with petroleum exploration and exploitation may adversely affect sea turtle habitats. Research is beginning to provide answers to some of these questions.

One planning method, termed the diversity approach (Bella and Overton 1972), calls for the uneven distribution of man's environmental perturbations. "Developmental efforts are confined to a number of selected systems and regions while specific steps are taken to prevent and even reduce development in others." This approach has been in progress along the southeastern coast of the United States, although it does not appear to be by design. Rather, intense development has not been permitted to occur along some turtle nesting beaches and on barrier islands as a result of the establishment of federal wildlife refuges and national seashores, and state parks, and the participation of local communities as on Little Cumberland Island, Georgia, and private organizations as on St. Catherines Island, Georgia. These activities remove the pressure of development from sea turtle nesting habitat, yet intense peripheral development occurs nearby. Recognizing the limitations of present ecological knowledge when considering whether or not to preserve an area, the environmental predicament leaves

open the option of including the area without the traditional accompanying specific proof.

Ray (1976) proposed a planning methodology in which "ecosystems science" served as the basis for marine conservation. The framework he developed is equally applicable to sea turtle habitat protection. In this methodology he suggests that ". . . First we must work toward the identification of critical areas and the buffer zones upon which the critical areas depend." The U.S. Fish and Wildlife Service and the National Marine Fisheries Service have proposed (44 Federal Register 159: 47863-64. 15 August 1979) a legal definition of "critical habitat":

(1) the specific areas within the geographical area occupied by a species, at the time it is listed in accordance with section 4 of the Act, [Act refers to the Endangered Species Act of 1973, 16 U.S.C. 1531 et seq.] *on which are found those physical or biological features (i) essential to the conservation of the species and (ii) which may require special management considerations or protection, and (2) specific areas outside the geographical area occupied by a species at the time it is listed in accordance with the provisions of section 4 of the Act upon a determination by the Secretary that such areas are essential for the conservation of the species.*

The surrounding areas on which the critical habitat depends are defined as buffer zones. The designation of a critical habitat may prove difficult in some instances due to the sea turtle's mobile nature. However, the environmental predicament provides us with the ability to designate habitat as critical.

Reserves and Parks

Sea turtles are endangered and threatened by man, and by setting aside habitats as either reserves, preserves, or parks, we protect not only the habitat but also the species. However, as Ray (1976) points out, ". . . what . . . reserves cannot do is survive intact outside . . . the ecosystems of which they are only a part." A buffer zone, therefore, "must be established to include the support systems which usually derive largely from outside . . ." As part of any plan to preserve critical habitats, long-term research is required not only of the habitat but also of the species itself. The greater our understanding of the sea turtle in terms of reproductive biology, hatchling survival, dispersal patterns, and growth, the better equipped we will be to protect its habitat. The problems associated with the establishment of parks and reserves in Third World nations have been addressed by Western and Henry (1979).

Buffer zones, as discussed by Ray (1976), ". . . are created to protect the core . . . buffers must accommodate the shift of the core in cases of biological, ecological, or geomorphological change, for example,

... the movement of beaches. ... Buffer zones may differ fundamentally from the core by not being under the direct ownership or jurisdiction of the agency managing the core area. Therefore, control of human activity within the buffer may be through administrative action, easements, or by other means ..." In identifying or establishing boundaries for core areas "entire ecological units (habitats and communities)" should be included.

The planning process and management only begins with the designation of a habitat vital to the survival of the sea turtle, and core and buffer zones protected from human encroachment. In planning habitat protection, whether by designation as a critical habitat within a reserve, refuge, or park, activities outside of the buffer area or habitat boundary must be addressed. The National Parks and Conservation Association (1979) found that out of 203 national park system superintendents surveyed in the United States, nearly two-thirds stated that their areas suffer from incompatible adjacent activities including residential, commercial, industrial, and road development, grazing, logging, agriculture, energy extraction and production, mining, and recreation. These activities were cited as having an adverse effect on water, air and terrestrial resources. Activities hundreds or thousands of miles removed from the habitat sites must also be considered. The placement of shoreline protection structures to reduce erosion updrift of a nesting beach may prove helpful to the area being protected, but the reduction of sediment transport to the rookery area could cause erosion and eventual loss of the habitat. Of a more recent nature, the IXTOC-I oil well failure in the Bay of Campeche suggests that we must be prepared to deal with and protect areas from oil impact not only at the state level but also internationally. International cooperation is essential. Planning must be achieved prior to such events so that the resource is not jeopardized by improperly applied recovery or rehabilitation methods.

International Reserves for Protection

The creation of international wildlife refuges or protected regions in which resource management would be coordinated internationally rather than locally, would be a step toward solutions. Iran set the precedent for this form of cooperation at the United Nations Conference on the Human Environment in Stockholm in 1972, when its government offered to cede its sovereignty in the National Park Sashte Argan in favor of international control (Kopp and Yachkaschi 1978). However, this may no longer be recognized. Since nesting and nursery areas, migratory routes and other such areas are critical to the propagation and survival of sea turtles, these areas could be included in Inter-

national Protected Areas so that all aspects of the turtles' life history would occur within an area cooperatively managed by a single entity. Most protection of widely ranging turtles should occur at such a level, for the best efforts by one nation or state could be negated by a laissez-faire attitude of another. The listing of such regions in the World Heritage Trust Program (UNESCO, undated) would bring recognition not only to endangered and threatened turtles but also to their life history requirements which depend on ecological rather than on political boundaries. The State of Massachusetts has provided for resource preservation by designation of areas of critical environmental concern (Massachusetts Coastal Zone Management, undated). In California, Areas of Special Biological Significance have been identified for protection (J. W. Hedgeth, personal communication).

Planning Strategy

Planning for the protection and management of sea turtle habitat should include the following procedural considerations and should take place both nationally and internationally.

1) Coordinated surveys should be made of sea turtle nesting, growing and hibernation areas, and migration routes. These surveys should include known and suspected sea turtle areas. The habitats and community structure, as well as the type and intensity of sea turtle activities, within each area should be summarized. A listing of natural and human-related perturbations occurring within or near each habitat should be compiled. The surveys do not need excessive detail and can be based on literature reviews and secondary sources. They should, however, be comprehensive and "should attempt to identify natural units both irrespective of political or legal boundaries" (Ray 1976). By identifying the habitats, the nature of sea turtle activity, and perturbations, priority areas can be selected for protection.

This is not simply a call for another survey. Previous surveys have been site- or nation-specific and have not provided the comprehensive data base needed for the worldwide protection of marine turtles and their habitats. This is a request, however, that we begin to assemble and update continuously, under 1 organization, and in 1 format, all the information now available on sea turtles in general and their specific habitats.

2) High priority areas would be considered for inclusion as critical sea turtle habitat.

3) Methods of protecting critical sea turtle habitat would be appropriate to its status as public or private property. Habitats within refuges, reserves, or parks would be assessed to determine the degree of protection afforded the habitat by enabling legislation and the management policies in existence. In some in-

STEPHEN V. SHABICA

stances, management practices might be suitable. Many significant sea turtle habitats located within the United States and its territories already have some protection by virtue of their inclusion in, for example, national parks: Canaveral National Seashore, Everglades National Park and Buck Island Reef National Monument; wildlife refuges: Cape Romain National Wildlife Refuge; federal reserves: Kennedy Space Center; private ownership: Little Cumberland Island and St. Catherines Island, Georgia; and critical habitats: Sandy Point, St. Croix, U.S. Virgin Islands. In those areas where the habitat is already under protection, planning should include the identification of core areas, possible expansion or enlargement of buffer zones or restrictions of human and vehicular access, or fishing activities during nesting or hibernating seasons. Critical sea turtle habitat which is unprotected or not directly under the control and protection of public agencies or private organizations would require "executive action for the immediate protection of certain most critical areas or to cause cessation of harmful practices" (Ray 1976) until ecological research demonstrated otherwise. This is especially clear in view of the environmental predicament which suggests that our ignorance is large. A sea turtle would be better served if such areas were identified and protected. This action would be followed by legislation (Environmental Law Institute 1977) at either national or international levels directed at the establishment of a park, reserve, or refuge encompassing the habitat and a large buffer zone. This is not to imply that, once identified, all critical areas and buffer zones must be protected, for it is clear that these areas and zones will include large portions of the world's oceans. Our ability to protect these areas through restrictions on fisheries, shipping, and coastal development would be difficult at best.

4) In high priority habitats the planning process must include research and management programs tailored to individual sites to be protected. Research efforts directed toward non-nesting biology and ecology of those species most threatened by human activity are essential. These would provide the information necessary to make trade-offs between preserving pristine oceanic ecosystems containing marine turtles and man's exploitation of the sea in such a way as to protect enough habitat with sufficient safeguards to ensure the survival of sea turtles. Management must recognize the socioeconomic issues within and surrounding habitats but not allow existing issues, policies, and practices to deter it from taking firm steps to protect the habitat. This action would include the identification of the "legal and financial means by which preservation may be achieved" (Ray 1976).

Following the general worldwide survey of sea turtle nesting, growing and hibernation areas, and migration routes, each habitat must be described and all available information compiled in an internationally accepted computer-compatible format. The following types of information should be acquired: name of area; geographical location; latitude and longitude; surface area; physical features including beach slope and texture; ecologically dominant biota including sea turtle species and their relative abundances, predators, and vegetation lines; special values (scientific, recreational or other interest); conservation status, degree of naturalness, degree and nature of any threats; present ownership or jurisdiction; character and use of contiguous land and sea areas emphasizing effectiveness as buffer zones; proposed purpose or present use of area; knowledgeable contacts; and references to both scientific and popular literature.

By programming the above information for computer storage and retrieval, and by updating the information periodically, the status of all worldwide sea turtle habitat can be ascertained. This system would permit us to, among others, determine those habitats most threatened and needing immediate action.

In order to accomplish this, an international steering committee should be formed and protective action begun. This committee, formed of ecologists, public interest groups, planners, social scientists, and environmental lawyers would coordinate the acquisition of the data and recommend habitats needing immediate protection to the international community. In consultation with the involved states, programs would be developed to protect high priority sea turtle habitats without regard to political boundaries. The United Nations Educational, Scientific, and Cultural Organization (UNESCO) or the International Union for Conservation of Nature and Natural Resources (IUCN) might possibly serve as the administrative organization for the establishment of such a steering committee.

The task before us is not simple. Our goal is high, nothing less than the protection of the very habitats upon which marine turtles depend for survival. It can be achieved, but it will require cooperation and hard work at national and international levels.

Acknowledgments

I wish to thank the following individuals for their critical reviews, additions, and viewpoints: Mr. Gary E. Davis, Dr. Joel W. Hedgpeth, Dr. A. R. Weisbrod, Mr. Jim Wood, and Mr. Jim C. Woods, and above all Mrs. June Erickson for her expert editing and manuscript preparation.

Literature Cited

Bella, D. A., and W. S. Overton
1972. Environmental planning and ecological possibilities. *Journal of Sanitary Engineering Division, ASCE* 98:579–92.

Bustard, R.

1972.   *Sea Turtles: Their Natural History and Conservation.* New York: Taplinger Publishing Company.

Caldwell, D. K.

1962.   Comments on the nesting behavior of Atlantic loggerhead sea turtles, based primarily on tagging returns. *Journal of the Florida Academy of Sciences* 25:287–302.

Clark, J., ed.

1976.   Barrier islands and beaches. Technical proceedings of the 1976 Barrier Islands Workshop. The Conservation Foundation, Washington, D.C.

Davis, G. E., and M. C. Whiting

1977.   Loggerhead sea turtle nesting in Everglades National Park, Florida, USA. *Herpetologica* 33:18–28.

Dean, J. M., and O. R. Talbert

1975.   The loggerhead turtles of Kiawah Island. *Environmental Inventory of Kiawah Island*, pp. T-1 to T-19. Columbia, South Carolina: Environmental Research Center, Inc.

Environmental Law Institute

1977.   The evolution of national wildlife law. U.S. Government Printing Office.

Felger, R. S.; K. Cliffton; and P. J. Regal

1976.   Winter dormancy in sea turtles: independent discovery and exploitation in the Gulf of California by two local cultures. *Science* 191:283–85.

Kopp, H., and A. Yachkaschi

1978.   Development and status of protected areas in Iran. *Parks* 2:11–14.

Lund, F.

1974.   Marine turtle nesting in the United States. Report to United States Fish and Wildlife Service. Manuscript, 39 pp.

Massachusetts Coastal Zone Management

Undated.   Designating areas of critical environmental concern. Boston: Coastal Zone Management Office.

National Parks and Conservation Association

1979.   *National Parks and Conservation Magazine* 53:4–9.

Ray, G. C.

1976.   Critical marine habitats. *IUCN Publication. New Series* 37:15–59.

Rebel, T. P.

1974.   *Sea Turtles and the Turtle Industry of the West Indies, Florida, and the Gulf of Mexico.* Coral Gables, Florida: University of Miami Press.

Schwartz, F. J.

1978.   Sea turtles—distribution and needs. In *North Carolina Workshop on Sea Turtles*, ed. F. Barick, pp. 6–13. Raleigh: North Carolina Wildlife Resources Commission.

UNESCO (United Nations Educational, Scientific, and Cultural Organization)

Undated.   *Operational Guidelines for the World Heritage Committee.* Paris: Division of Cultural Heritage.

Wauer, R. H.

1978.   "Head start" for an endangered turtle. *National Parks and Conservation Magazine* 52:16–20.

Western, D., and W. Henry

1979.   Economics and conservation in third world national parks. *Bioscience* 29:414–18.

Worth, D. F., and J. B. Smith

1976.   Marine turtle nesting on Hutchinson Island, Florida, in 1973. *Florida Marine Research Publication* 18:1–17.

STEPHEN V. SHABICA

Ross Witham
Florida Department of Natural Resources
Marine Research Laboratory
100 Eighth Avenue SE
St. Petersburg, Florida 33701

Disruption of Sea Turtle Habitat with Emphasis on Human Influence

ABSTRACT

The effects of human activity on sea turtle nesting habitats were assessed by reviewing published and unpublished Florida nesting data. Disruptions caused by residential, recreational, and business use of beach fronts include impacts from artificial lights, physical barriers such as jetties, groins, and sea walls, and vehicular traffic on beaches. Negative effects associated with petroleum production are also presented. Management techniques for mitigating these impacts are recommended.

Introduction

Disruptive alterations of sea turtle habitat result from both short-term and long-term human activities, as well as from natural causes such as erosion, accretion, and high tide flooding. Hirth (1971), while mentioning that nocturnally nesting sea turtles could be scared by "... barking dogs, lights, etc.," discussed primarily natural disruptions to their beach rookeries. Man's most apparent disruptive activities are related to beach front development which increases artificial lights, structures, and human activity.

LeBuff (1973) stated, without citing references, "The negative effect that development and increased public use has on marine turtle nesting has been well documented in the southeastern United States." While not firmly established, it seem probable that until rather recently most sea turtle rookeries have been in coastal areas where development was minimal or nonexistent. Lund (personal communication) mentioned a lack of suitable nesting beach and an apparent increase in false crawls in the seawalled area of Jupiter Island. Conversely, Mann (1977) noted that nesting females on Florida's southeast coast did not avoid lighted, developed beaches or favor undeveloped beaches.

Disruption of nesting habitat due to development has been assessed by reviewing published and unpublished Florida data. While in many areas, developed and undeveloped, have not been systematically studied, existing data yield information useful in evaluating artificial disruption of nesting and other nearshore habitats. A review of some available data follows.

Discussion

The most obvious and best studied impact of beach development is artificial lighting. Daniel and Smith (1947) reported that emergent hatchlings could ". . . sometimes be stimulated to crawl back onto the sand . . . ," in response to artificial light. Hendrickson (1958), McFarlane (1963), Ehrenfeld (1968), Mrosovsky and Carr (1967), Mrosovsky and Shettleworth (1968), Mrosovsky (1972), and Philibosian (1976) documented positive phototactic responses. Mann (1977) noted up to 100 percent landward movement where landward light intensity exceeded seaward intensity. He further reported post-emergent hatchling death rates from 0.02 to 95.96 percent, with lower percentages occurring where barriers, natural or man made, prevented hatchlings from crawling to the landward horizon. High relief barriers, by disrupting the landward movement of hatchlings, apparently permitted hatchlings to reorient toward the sea when dawn's light became stronger than landward artificial light. Condition and fate were not determined for misoriented hatchlings that eventually reached the sea after extensive terrestrial meanderings.

The light orientation propensity of young sea turtles has suggested the need to study the possible effects of lighted, permanent structures at sea. These structures attract sizeable populations of pelagic fishes (Mertens 1976; Hastings, Ogren, and Mabry 1976). Since fish predation is considered to be a major cause of hatchling mortality (Hirth 1971), young turtles attracted to lighted, offshore structures might be subjected to increased predation.

Nesting sites have been degraded or lost because erosion control structures have made unsuitable topographic changes at several beaches within sea turtle nesting ranges. Jetties at inlets are man's earliest physical beach alterations. Erosion and accretion patterns around inlet jetties are determined by sand transporting currents; erosion occurs on the down current side and accretion on the up current side (Anonymous, 1955). Inlet jetties without sand-transfer systems result in long-term erosion and accretion. Eroded areas frequently become unsuitable for nesting when high tides cover beaches to the vegetation line. Accreting areas continue to provide suitable nesting habitat.

Groins (short jetties at noninlet areas of beaches) are erected in an attempt to control beach erosion; their effects are similar to those of inlet jetties.

More recently, rip rap, a wall made up of loose stones, has been used along the surf zone of existing groins in an attempt to overcome their ineffectiveness. Rip rap alone is used as a stop-gap to protect upland structures. Groins and rip rap have dubious effectiveness in protecting or restoring eroded beaches. Use of rip rap is unsatisfactory when it hampers turtle access to suitable nesting beaches.

Seawalls, in several configurations, have been used in attempts to control erosion. They are made of concrete, steel, or a combination of both; some have vertical walls, others have curves designed to impart a recurving motion to incoming waves. One type of seawall is made of waffle-like elements built along an upward slope under the existing beach. Because seawalls fail to prevent beach erosion, the present trend is to use rip rap as a stop-gap measure and sand nourishment to rebuild eroded beaches.

Beach nourishment is usually done by dredging suitable offshore sand and pumping it onto the beach. Sand shifting and settling after deposition on beaches may limit initial use by turtles for nesting. Sand pumped onto an eroded beach with nests may trap hatchlings (Mann 1977); therefore nourishment should be limited if possible to non-nesting seasons. Dredging during non-nesting months may not be possible due to winter sea conditions, however endangered eggs can be safely moved for protected incubation (Witham and Futch 1977).

Recreational, residential, and business use of beach fronts has resulted in construction of condominiums, hotels, motels, restaurants, fishing piers, and electric generation facilities. In some areas, extensive development occurred before there was any interest in assessing its effects. In other areas, development proceeded after the problems were recognized, but before management plans were implemented.

Florida's coastal construction management program began in 1970 with the enactment of Sec. 161.052 Florida Statutes (F.S.). This law required a 50-ft setback from mean high water. Difficult to enforce, this law was amended (Sec. 161.053 F.S.) to establish a surveyed set-back line in cooperation with individual counties. An example of a development prior to the set-back law is a motel built on former dunes resulting in a lighted beach which encourages nightly guest activity. Turtle nesting appears to decrease within the limits of increased human activity. False-crawling females that deposit eggs elsewhere could benefit hatchlings which would otherwise be disoriented by the motel's lights. A compatible development having its buildings well behind the dune line does not interfere with either nesting females or emergent hatchlings. However, problems resulting from increased beach activities could be a factor during nesting seasons (Mann 1977).

Vehicular traffic on beaches can affect rookeries in several ways. Some densely populated areas use heavy

ROSS WITHAM

beach cleaning equipment which is capable of digging up or collapsing nests (Mann 1977). In other areas, recreational off-road vehicles are sometimes used extensively and make ruts deep enough to dig up or collapse nests. Recreational use of horses causes similar problems. Depressions created by these activities can entrap seaward bound hatchlings.

Tar balls, which result from oil spills (Dedera 1977), soil beaches. One beach cleaning contractor (J. Peart, personal communication) estimated a tar content ranging from 5 to 50 percent in the approximately 180 metric tons of seaweed and other debris he removes annually. In addition to its frequent appearance on beaches, tar is widely distributed in oceanic environments (Wade et al., 1976). Accumulating observations (Kleerekoper and Bennett 1976; Witham 1978, unpublished data) suggest adverse effects upon small sea turtles by oil and tar. These effects require quantitative and qualitative investigations, and, in the interim, efforts must be made to reduce the amount of hydrocarbons entering the seas.

The cooling water taken from the ocean by electric generation facilities entrain larger turtles in cooling canals. The offshore intake structure of Florida Power and Light Company's St. Lucie Plant may look like a reef to some turtles, suitable for resting, and these turtles are subsequently drawn into the cooling system. Actively feeding leatherbacks probably follow large numbers of jellyfish into the intake. More than 130 turtles of 3 species—loggerhead, *Caretta caretta*, green turtle, *Chelonia mydas*, and leatherback, *Dermochelys coriacea*—were removed from the St. Lucie intake canal in 1 year (D. Worth, personal communication). About 16 percent of loggerhead and green turtles were dead when removed from the system; the remainder were tagged and released. Three leatherbacks were released alive, but at least 1 had severe skin abrasions from being caught in a net.

Evidence from Florida demonstrates continued use of nesting habitat by turtles in developed and developing areas. Mann (1977) reported nesting densities ranging from 2 to 80/km from Delray Beach to Lauderdale-by-the-Sea. He commented that "Many females did nest on highly developed beaches, with bright background lighting, tall buildings, human activity, and dredged sands very different from those which would naturally be present, even though some of the more natural beaches available nearby were only lightly nested." Fletemeyer, (personal communication 1978) reported 12 nests/km on the 8.4 km of brightly lighted, highly developed Ft. Lauderdale beach. Some of his data suggest that human activity, rather than lighting alone, reduces nesting activity. Wagner (personal communication 1978) reported 352 nests on 4.2 km of beach at Boca Raton. More than 16,000 loggerhead and 1,500 green turtle hatchlings were released as part of the city's nest protection work. Hutchinson Island nesting surveys (Gallagher et al., 1972; Worth and Smith 1976; Henderson, personal communication) used nest counts on 9 1.25-km beach sections to estimate nesting populations. These surveys reported total nest counts of 1,412 in 1971, 1,263 in 1973, and 1,446 in 1975, although during this time there was continued coastal development. Where development or use is detrimental, damage can be mitigated by management techniques. These techniques include transplanting endangered eggs, collecting hatchlings, and nourishing beaches as well as restricting beach use, shielding lights, and removing barriers.

Literature Cited

Anonymous
1955. Studies and recommendations for the control of beach erosion in Florida. Coastal Engineering Staff, College of Engineering, University of Florida, Gainesville, Florida. Special Report to Florida State Legislature, 40 pp.

Daniel, R. S., and K. U. Smith
1947. The sea-approach behavior of the neonate loggerhead turtle. *Journal of Comparative Physiol. Psychology* 40:413–20.

Dedera, D.
1977. The disasters that didn't. *Exxon, USA* Third Quarter:11–15.

Ehrenfeld, D. W.
1968. The role of vision in the sea-finding orientation of the green turtle (*Chelonia mydas*). 2. Orientation mechanism and range of spectral sensitivity. *Animal Behavior* 16:281–87.

Gallagher, R. M.; M. L. Hollinger; R. M. Ingle; and C. R. Futch
1972. Marine turtle nesting on Hutchinson Island, Florida in 1971. *Florida Department of Natural Resources Marine Research Laboratory Special Scientific Report*, 37:1–11.

Hastings, R. W.; L. Ogren; and M. T. Mabry
1976. Observations on the fish fauna associated with offshore platforms in the northeastern Gulf of Mexico. *Fishery Bulletin* 74:382–402.

Hirth, H. F.
1971. Synopsis of biological data on the green turtle *Chelonia mydas* (Linnaeus) 1758. *FAO Fisheries Synopsis*, no. 85.

Hendrickson, J. R.
1958. The green sea turtle, *Chelonia mydas* (Linn.), in Malaya and Sarawak. Proceedings of the Zoological Society of London 130:455–535.

Kleerekoper, H., and J. Bennett
1976. Some effects of the water soluble fraction of Louisiana crude on the locomotor behavior of juvenile green turtles (*Chelonia mydas*) and sea catfish (*Arius felis*) Preliminary results. NOAA Symposium Fate and Effects of Petroleum Hydrocarbons in Marine Ecolsystems, 10–12 November 1976, Seattle, Washington.

LeBuff, C. R.
1973. Extension of the Caretta Research Field Program in Lee County—1973. Caretta Research Project Annual Report.

Mann, T. M.
1977. Impact of developed coastline on nesting and hatchling sea turtles in southeastern Florida. Master's thesis, Florida Atlantic University.

McFarlane, R. W.
1963. Disorientation of loggerhead hatchlings by artificial road lighting. *Copeia* 1963:153.

Mertens, E. W.
1976. The impact of oil on marine life: a summary of field studies. Sources, effects and sinks of hydrocarbons in the aquatic environment. Proceedings of a Symposium, 9–11 August 1976, at American University, Washington, D.C., AIBS:507–14.

Mrosovsky, N.
1972. The water-finding ability of sea turtles: behavioral studies and physiological speculations. *Brain Behavior and Evolution* 5:202–25.

Mrosovsky, N., and A. Carr
1967. Preference for light of short wavelengths in hatchling green turtles, *Chelonia mydas*, tested on their natural nesting beaches. *Behaviour* 28:217–31.

Mrosovsky, N., and S. J. Shettleworth
1968. Wavelength preferences and brightness cues in the water-finding behavior of sea turtles. *Behaviour* 32:211–57.

Philibosian, R.
1976. Disorientation of hawksbill turtle hatchlings, *Eretmochelys imbricata*, by stadium lights. *Copeia* 1976:824.

Wade, T. L.; J. G. Quinn; W. T. Lee; and C. W. Brown
1976. Source and distribution of hydrocarbons in surface of Sargasso Sea. Sources, effects, and sinks of hydrocarbons in the aquatic environment. Proceedings of a Symposium, 9–11 August 1976, at American University, Washington, D.C., AIBS:270–81.

Witham, R.
1978. Does a problem exist relative to small sea turtles and oil spills. In *Proceedings Conference on Assessment of Ecological Impacts of Oil Spills*, 14–17 June 1978, at Keystone, Colorado, AIBS:629–32.

Witham, R., and C. R. Futch
1977. Early growth and oceanic survival of pen-reared sea turtles. *Herpetologica* 33:404–9.

Worth, D., and J. B. Smith
1976. Marine turtle nesting on Hutchinson Island, Florida in 1973. *Florida Marine Resources Publication* 18:1–17.

ROSS WITHAM

Daniel Navid
Executive Officer
IUCN Commission on Environmental Policy,
Law and Administration
Bonn, Federal Republic of Germany

Conservation and Management of Sea Turtles: A Legal Overview

ABSTRACT

This paper provides an overview of existing national legislation and international conventions pertaining to the conservation of sea turtles. The impact that the draft Law of the Sea Treaty might have on sea turtle conservation in its elaboration of new state rights and responsibilities over ocean space and marine resources is also considered. Two reference appendices are provided to applicable legal materials.

The present legal situation is unsatisfactory as national legislation has been piecemeal and uncoordinated throughout the ranges of the sea turtles, and insufficient use has been made of international conventions to develop common conservation approaches. Enforcement difficulties are also touched upon. Future legal requirements to better conserve sea turtles, their eggs and their habitats are suggested. However, legal initiatives are dependent upon advances in knowledge of the ecological requirements of sea turtles.

The long term conservation of migratory species requires international cooperation. The newly adopted Convention on the Conservation of Migratory Species of Wild Animals, which covers all sea turtles, is suggested as a possible vehicle for this cooperation, while the increased utilization of regional conservation conventions is also promoted.

As a matter of priority, attention is urged to the improvement of national legal measures. The expected delay until the Migratory Species Convention can come into force and have sufficient parties to become an effective instrument, the time necessary to either develop new or to improve existing regional conservation conventions, the advances in scientific knowledge that are required for framing effective international conservation and management programs, and the importance that individual state action can have in conserving particular sea turtle populations are all cited as reasons for devoting immediate attention to improving the situation at national level.

Finally, two legal approaches that could be taken to

control the specific problem of incidental take of sea turtles in fishery operations are suggested.

Introduction

In this paper a brief overview is provided of the existing legal situation pertaining to the conservation of sea turtles, followed by observations for future legal initiatives at both national and international level to better control the exploitation of sea turtles and to conserve their habitats.

.The legal situation to date has not been fully satisfactory. Conservation measures have been taken predominately at national level, and these appear to have been haphazard and insufficient when the total range of the various turtle species is concerned. Only recently have there been significant international developments: at global level, the Convention on International Trade in Endangered Species of Wild Fauna and Flora, which can effectively control international trade in live specimens and products, and the newly adopted Convention on the Conservation of Migratory Species of Wild Animals, which could provide a global framework for comprehensive conservation and management measures for sea turtles. At regional level there are also promising initiatives to either strengthen existing general conservation conventions or to elaborate new ones.

Ultimately the conservation (which means sustained use as well as protection) of marine migratory species which pass into and out of areas under national jurisdiction, requires a coordinated international approach. However, this cooperation will no doubt take time, and so improved national provisions must be promoted as the immediate means for the conservation of sea turtles and their habitats.

A few caveats should be mentioned about the role of law in the conservation process.

First of all, law should be seen as just one piece of the puzzle for conservation. It can reinforce and strengthen all other aspects, but requires the right circumstances to be effective. This means fitting legal measures within the applicable social, political and economic framework.

Second, in the field of nature conservation, meaningful legal advances are possible only if predicated upon advances in scientific knowledge (for sea turtles on such factors as range and migration routes).

Third and finally, even the most sophisticated and well-intentioned law is worthless if it is not effectively enforced. Enforcement problems can stem from a number of sources: lack of political will; inadequate personnel, finance or equipment; lack of public awareness; social and economic problems; or legal obstacles such as lack of agency cooperation within a government. Specific enforcement difficulties in regard to sea

turtle conservation cannot be considered in this overview paper but enforcement is a very serious problem which must be addressed if conservation programs for sea turtles are to be effective. More will be said on this point below.

Current National and Provincial Legislation

Appendix I provides citations to some 70 current conservation laws and regulations directly applicable to sea turtles. In analyzing these enactments, it is evident that present legislation has been piecemeal (for example, most only deal with one aspect of the problem, such as the collection of eggs), and legal provisions have not been coordinated between jurisdictions (for example, open season dates set in various countries).

Current International Conventions

Cooperative efforts to conserve sea turtles are addressed by one regional (the African Convention on the Conservation of Nature and Natural Resources), and two global conventions (the Convention on International Trade in Endangered Species of Wild Fauna and Flora, and the Convention on the Conservation of Migratory Species of Wild Animals). The first two are in force; the last remains open for signature until 22 June 1980. Data concerning these conventions may be found in Appendix II.

Two other regional conventions, the Convention on Nature Protection and Wildlife Preservation in the Western Hemisphere and the Convention on the Conservation of Nature in the South Pacific, are also structured to directly deal with the conservation of selected species as both provide that lists of taxa that merit special importance be maintained by their contracting parties. Unfortunately, concrete actions for cooperative conservation programs that might benefit sea turtles have not yet been achieved under either, although it is being increasingly recognized that the Western Hemisphere Convention could provide a valuable mechanism for regional conservation in the Americas for migratory wildlife including local populations of sea turtles. The main problem with this latter convention is the lack of an administrative mechanism to facilitate implementation. In addition unlike other international conservation conventions, it does not contain an agreed list of species that all parties undertake to conserve, but rather provides for individual national listings of species to be protected (whose effect to other parties is very unclear). Also, the section of the convention calling for the conservation of migratory species is limited to birds.

The possibility of developing new regional conservation conventions for West Africa and Southeast Asia is now being explored within UNEP and ESCAP re-

DANIEL NAVID

spectively. Although it may be several years before anything comes of these, efforts should be made to have these agreements comprehensively and directly cover the conservation of living resources, including sea turtles.

Many existing international agreements have an indirect impact upon sea turtle conservation. These include conventions dealing with protected areas, pollution control, and fishery operations. It is not possible to consider each in this brief overview paper. Instead, a few are cited as conventions which *could* prove useful for sea turtle conservation, and the special problem under fishery conventions—incidental take of sea turtles—is alluded to below.

Examples at regional level include the Convention for the Protection of the Mediterranean Sea Against Pollution and the Kuwait Regional Convention for Cooperation on the Protection of the Marine Environment from Pollution, and at international level include the Convention on Wetlands of International Importance Especially as Waterfowl Habitat and the Convention Concerning the Preservation of the World Cultural and Natural Heritage. It must be underlined, however, that these conventions are all either too specialized, or suffer from too many other inherent weaknesses to provide substantial contributions to sea turtle conservation.

Finally, attention needs to be given to the draft Law of the Sea Treaty as that instrument will have great impact upon the framework for both national and international measures for sea turtle conservation. The following is a brief description of the three existing conventions of direct relevance and the Law of the Sea draft text.

African Convention on the Conservation of Nature and Natural Resources, 1968

This convention, in force since 1969, was developed to provide a comprehensive basis for the conservation of natural renewable resources (soil, water, flora and fauna) throughout Africa. All marine turtles are covered by the convention, being listed as Class A protected species which are totally protected throughout the entire territory of the parties, with taking allowed only under special circumstances.

The convention needs to be more fully implemented throughout Africa. This could be best fostered through the functioning of an active secretariat. However, in relation to the conservation of sea turtles the convention will still be somewhat inadequate. Eggs are apparently excluded from coverage, nationals of parties are not covered by the provisions of the convention when they operate outside national territory, and most seriously, the regional application of the convention may preclude coverage of the full range of some sea

turtle populations involved.

Convention on International Trade in Endangered Species of Wild Fauna and Flora (CITES), 1973

This convention was designed to prevent the extinction of species and to assist in the improvement of their status by controlling strictly international commerce in listed species, their products and derivative parts. Since entry into force in 1975, CITES has shown that it can be an effective mechanism for international cooperation.

Three Appendices to the Convention list the species covered. Appendix I includes species that are threatened with extinction. Consequently, international trade in these species and their products and derivatives is subject to very strict regulation, being authorized only under exceptional circumstances. Appendix II species are those which either may become threatened with extinction if international trade is not regulated, or species which must be subject to regulation in order that trade in these first species may be controlled. Appendix III is of less significance, including species which a particular party identifies as being subject to regulation within its jurisdiction and for which cooperation is requested from the other parties to control trade in the species.

Sea turtles have been subject to much discussion and consideration by the parties to CITES. The parties have for the most part been able to agree upon stringent trade controls for all sea turtles with the following presently included on the convention appendices:

	Appendix I	Appendix II
Cheloniidae	Caretta caretta Chelonia mydas	Cheloniidae spp. (all species)
	Eretmochelys imbricata Lepidochelys kempii Lepidochelys olivacea	
Dermochelyidae	Dermochelys coriacea	

Two parties maintain reservations in regard to Appendix I listings for particular sea turtles—France for *Chelonia mydas* and *Eretmochelys imbricata* and Italy for *Chelonia mydas*. This means that for those listings, France and Italy respectively are not considered to be party states and may therefore engage in international trade in the species without observing the Appendix I provisions.

Convention on the Conservation of Migratory Species of Wild Animals, 1979

The Migratory Species Convention was developed in response to the need for a global system for the conservation of migratory animals. It recognizes that migratory species, a term very broadly defined in the convention, are a resource to be conserved and managed by all states that exercise jurisdiction over any part of the range of the species. It is a framework convention which is intended to provide direction and guidelines for further conservation agreements for migratory animals as well as to provide directly a mechanism for individual states to unilaterally conserve endangered migratory species.

Two appendices to the convention list the species selected for coverage. Species may be listed in either or both appendices. Appendix I includes endangered species for which immediate and stringent conservation measures by the party states along the range of listed species ("Range States") are required. Appendix II species are those recommended to be the subject of agreements by their Range States. This appendix is in fact the heart of the convention, as it implicitly recognizes that agreements between states (particularly for groups of migratory species) are essential to conserve and manage migratory species. The basic elements for the agreements are set forth in the convention in the form of guidelines. Agreements are to be comprehensive, with consideration given to research requirements, exchange of information, conservation of habitats, removal of or compensation for obstacles to migration, procedures for coordinating action to suppress illegal taking and so forth.

The appendices developed at the Bonn Diplomatic Conference at which the convention was concluded were purposely kept short, and were adopted as initial expressions of the types of species to be covered by the convention. Nevertheless, sea turtles are well represented, with *Lepidochelys kempii* and *Dermochelys coriacea* entered onto Appendix I and all species of Cheloniidae and Dermochelyidae entered onto Appendix II.

Draft Law of the Sea Treaty

The Third United Nations Conference on the Law of the Sea has had as its task the elaboration of a new treaty that will set out the rights and obligations of states for the future use of the oceans and their resources. The conference apparently is winding down to a conclusion, with a consensus already finding its way into international law on many of the crucial issues (as may be seen in the current draft text[1]). Under these circumstances some aspects of the future legal regime of the seas can be evaluated with a degree of certainty.

One portion of the draft treaty that seems to be firmly in place is the treatment of jurisdictional limits and the related rights and responsibilities for coastal states regarding the conservation of living resources in areas under their jurisdiction.[2] From the previous situation of 3- and 12-mile limits, coastal states are now able to claim jurisdiction over resources in areas that might extend beyond 200 miles from their shores. However, the Law of the Sea Conference has never had the conservation of marine species as its major focus, and conservationists repeatedly have pointed out gaps in coverage in the draft texts. Much future attention in other fora will need to be given to remedy the more important conservation deficiencies of the treaty.[3]

A brief review of the new jurisdictional framework under the draft treaty text is useful for an indication of how this new legal regime might affect the conservation of sea turtles.[4]

INTERNAL WATERS

These are the waters on the landward side of a baseline which is normally the low-water line along the coast (Articles 5–8). They are not covered by the Law of the Sea text as they are legally considered to be part of the coastal "land." Such waters are of tremendous ecological importance since they include estuaries, lagoons and other inshore areas. In view of the text's silence, coastal states are free to conserve, or may legitimately exploit, living resources in these areas without obligation under this treaty. It may also be interpreted that the silence of the treaty means that coastal states are free to control all forms of pollution, or conversely to pollute, with impunity in these waters.

---

1. Informal Composite Negotiating Text/Rev. 1, UNCLOS document A/Conf.62/WP10/Rev.1 28 April 1979. Article numbers referred to below concerning the Law of the Sea refer to the numbers in this text.
2. FAO reports in document COFI/78/Inf.9 that as of 1 April 1978 out of some 130 coastal states, 85 already claim fishery jurisdiction beyond 12 miles, with 67 of these claiming limits of 200 miles either as exclusive economic zones, fishery zones, extended territorial seas or in other terms.
3. See *Conservation and the New Law of the Sea—IUCN Statement on the Informal Composite Negotiating Text UNCLOS III*, February 1978 Morges, Switzerland, which was widely distributed within and outside of the Law of the Sea Conference. This Statement pointed out IUCN's evaluation of the conservation deficiencies of the draft treaty and provided proposals in the form of draft articles for improving those aspects of the text.
4. For a more detailed treatment of the new jurisdictional situation, see de Klemm, *Conservation and the New Informal Composite Negotiating Text of the Law of the Sea Conference*, Environmental Policy and Law, vol. 4, no. 1, Lausanne, April 1976. This article refers to the previous version of the Law of the Sea text, but it is still relevant as the text has not been changed regarding these jurisdictional points.

DANIEL NAVID

TERRITORIAL SEA

Each coastal state is given the right in the treaty to establish the breadth of its territorial sea up to a limit of 12 nautical miles measured from its baseline (Article 3). The coastal states have the same authority for conservation or exploitation in the territorial sea as in internal waters. However, coastal states cannot prohibit innocent passage of foreign ships in these waters, although they may establish shipping lanes (Article 17). This exception has relevance for pollution control and establishment of protected areas.

EXCLUSIVE ECONOMIC ZONE

EEZs are areas that coastal states are free to establish under the treaty which can extend up to a maximum distance of 200 nautical miles to sea measured from their baseline (Article 57). In their EEZ, coastal states have sovereign rights for the purpose of the preservation of the marine environment and for the exploitation of both living and nonliving natural resources (Article 56). Coastal states are charged with the determination of their capacity to harvest living resources in this area, and subsequently with the setting of an allowable catch figure, along with the obligation to avoid overexploitation (Articles 61 and 62). Consequently coastal states may regulate or prohibit resource exploitation by other states within these zones.

Nevertheless, all states enjoy the freedom of navigation in the EEZ area (Article 58) and although this right is qualified later [Article 211 (6)], it is a point which again has implications for marine pollution and for the establishment of protected areas. Other states are obliged to conform to the coastal state's laws in regard to vessel source pollution within EEZs, but such laws must accord with international practice (Article 211).

HIGH SEAS

All remaining waters comprise the high seas (Article 86). Therein states enjoy the right of freedom of fishing and navigation (Article 87), subject, of course, to any international agreements the states have undertaken on the subjects. Pollution on the high seas by ships is regulated by a few provisions of the treaty, but as there is no global enforcement agency, enforcement is left to the flag states or to port states (Articles 94, 210, 211, 216–220).

Special jurisdictional limits also apply to the continental shelf regime but those need not concern us here.

These new jurisdictional limits, already unilaterally adopted by many states, greatly alter the traditional legal regime of the oceans. For example, the former principle of freedom of the seas for fishing has been seriously eroded since the vast majority of fishing operations are carried out in areas which will now come under coastal state jurisdiction.[5]

Sea turtles pass through all of the jurisdictional zones mentioned above. Because of extensions in national jurisdiction, certain coastal states shall largely be responsible for the conservation of sea turtles during much of their life cycles. This could be a welcome turn of events as coastal states having greater authority can undertake more comprehensive conservation measures than in the past. Nevertheless, this could be anything but an improvement should some of these coastal states prove unable or unwilling to take needed conservation measures in their extensive areas, as they alone have the authority to so act therein.

The migration routes of many sea turtles cross the waters of several coastal states. Thus, the failure of some states to take proper conservation measures could easily negate the efforts of other states along the range of the species. Several articles of the draft Law of the Sea Treaty advocate cooperation by states directly or through appropriate international bodies for conservation, management or optimum utilization of migratory species (Articles 63–67), but these are unfortunately only general principles for cooperation and do not point the way for concrete results.

In sum, the new Law of the Sea significantly alters the "rules" for state responsibility and rights over marine species, but it does not provide proper guidelines for conservation or sustained exploitation of these species. Furthermore, the division of ocean space into national areas on a greater scale than in the past, although potentially advantageous for conservation, can be an unwelcome development as far as migratory marine species are concerned since a precise system for international cooperation is not provided. A serious gap for the conservation of these species may result, and it must be faced in other international arenas.

Future Legal Requirements

In the foregoing many inadequacies in present legal provisions for the conservation of sea turtles were noted. On the national level, existing legislation apparently has not been adequate to maintain or improve the status of the species. At international level insufficient attention has been given to the problem of sea turtle conservation with only control of international trade in sea turtles and their products now apparently well in hand. Furthermore, in the elaboration of the new Law of the Sea Treaty, conservationists have been dismayed by the omission of provisions for the framing

---

5. *Supra*, note 2: FAO estimates that as much as 90 percent of future fishery operations will take place in areas under national jurisdiction.

of conservation responsibilities for coastal states while these states are given new authority over the resources in tremendous areas of ocean space. Under these circumstances, urgent attention must be given to improving legal provisions if sea turtle populations are to be maintained or enhanced.

Requirements at National Level

As mentioned above, a threshold problem for improving legal measures is one of scientific information. Law cannot be created in a vacuum. Data must be provided as to the range, migration routes, critical habitats and the ecological requirements of sea turtles so that appropriate conservation legislation can be framed. However, the premier problem for sea turtle conservation will certainly be that of political will at national level. From the legal point of view, assuming that scientific input is satisfactory, there are already plenty of tools available for conservation, but these must be seized by governments. Two necessary types of legal enactments at national level are legislation that comprehensively considers sea turtles as far as exploitation is concerned (addressing local use as well as international trade), and legislation that preserves sea turtle habitat. For the former, there is the precedent of comprehensive action being taken by 1 country (the United States) for particular marine species (marine mammals) with a great success.[6] For the latter, there again are precedents for the establishment of marine (and of course land) sanctuaries.[7]

Requirements at International Level

Even with the appropriate political will for enlightened national measures, the conservation and management of migratory species require cooperation between states. One framework for a common international approach now exists in the new Migratory Species Convention, and so it is timely to consider in more detail the treatment that this convention gives to the various requirements and problems of sea turtle conservation.

All sea turtles are included under Appendix II to the Migratory Species Convention, being recommended to be the subject of conservation agreements, and in addition, two species, *Lepidochelys kempii* and *Dermochelys coriacea* have been included in Appendix I to be the subject of stringent unilateral measures. The convention is predicated upon the principle that all the states along the range of a migratory species must participate in its conservation and management.

Scientific judgement will determine the states which will constitute the range states for given species of sea turtles. Regional arrangements for identified distinct populations (which can be considered separate "migratory species" under the convention) will provide certain political and practical advantages. However, if migration routes go beyond regional areas, a more global approach will be required (see the discussion above on the African Convention). One attempt was made in 1969 to establish a regional agreement for *Chelonia mydas* by Costa Rica, Nicaragua, and Panama. That convention was not enacted and the attempt has since been abandoned by the three countries.[8] Such efforts should be encouraged in the future, if they are based upon the conservation of distinct populations.

By mandating measures for the conservation of Appendix I species, and by setting guidelines for the provisions of agreements on Appendix II species, the Migratory Species Convention should ensure that a large measure of harmonization will exist in the legal provisions of the states along the range of marine turtle species.

The convention contains quite clear provisions concerning the exploitation of listed species. For Appendix I (endangered migratory species) exploitation will be allowed only under exceptional circumstances and if the exploitation does not operate to the disadvantage of the species. In the case of Appendix II species, the convention recognizes that exploitation can be an integral part of conservation and management but that it must be controlled to ensure sustainable benefits.

The convention contains several provisions concerning habitat conservation including directions for the restoration of habitats that have been previously degraded and the establishment of reserve areas for both Appendix I and II species.

The convention acknowledges the crucial role that scientific research must play in the determination of conservation measures. A Scientific Council is established that will provide scientific advice to the parties and promote the undertaking of research. Furthermore, the guidelines for agreements include reference to the need for provisions to be made within the agreements to facilitate research work.

As noted earlier, difficulties in the enforcement of national conservation enactments have often been cited as a major problem in efforts to conserve sea turtles. For developing countries, primary enforcement difficulties can often be traced to lack of resources and

---

6. Marine Mammal Protection Act of 1972, Pub.L.92–522. 21 Oct. 1972.

7. See, for example, for marine areas, the U.S. Marine Protection, Research and Sanctuaries Act of 1972, Pub.L.92–532, 23 Oct. 1972 which provides for the establishment of marine sanctuaries necessary for the purpose of preserving or restoring areas (which may extend out to the edge of the continental shelf) selected for their conservation, recreational, ecological or aesthetic values.

8. Letter from J. de J. Coneja, Subdirector de Coordinación Económica, Ministerio de Relaciones Exteriors y Culto, San José Costa Rica, 2 November 1978.

DANIEL NAVID

trained personnel as well as the underlying economic aspects of the problem. The Migratory Species Convention implicitly recognizes that certain parties will require assistance in the form of financial aid and expertise to allow them to fully implement the convention and leaves the details to be worked out in the various agreements under the convention. A special resolution was included in the Final Act to the convention that urged the parties promote financial, technical, and training assistance for developing countries to enable them to fully implement the convention and also urged international and national aid organizations to give priority in their programs to assist developing countries under the convention.

Experience has shown that international conventions can expedite and guide national legislation. Two reasons have been cited for this: 1) states having assumed an international responsibility to take an action are bound to act effectively to meet this responsibility; and 2) certain states, because of their internal political systems, are able to go farther with their national legislation if it can be based upon an international convention than would be politically feasible without it.

The Migratory Species Convention would thus seem to be a promising means to facilitate international cooperation for the conservation of sea turtles. The reaction of governments shows clearly that there is support for the concept of a migratory species convention. However, several of the most influential states were not able to support the convention at the Bonn Conference, principally because these countries could not accept the application of the convention to marine species. Fortunately, the controversy over marine species was largely confined to marine fish, with all but one delegation favoring the inclusion of sea turtles under the ambit of the convention.

A frequently cited worry of the states that could not support the inclusion of marine species in the convention was that the Migratory Species Convention might conflict with future obligations being developed at the Law of the Sea negotiations despite the fact that a provision was made in the Migratory Species Convention for the Law of the Sea text to prevail should there be any conflict between the two. In any event, there should not be any contradictions as the Migratory Species Convention serves to provide specific rules for the principles of cooperation set forth in the Law of the

Sea text. It is hoped that objections to the Migratory Species Convention based upon potential incompatibility with the new Law of the Sea will soon be overcome with the adoption of that treaty.

Required Immediate Legal Action

Even if concerned states take concerted international action to conserve sea turtles by means of the Migratory Species Convention or other international instruments, it may be some time before such agreements either come into force and have a sufficient number of parties to make international cooperation meaningful, or, in the case of certain existing regional agreements, are adequately strengthened to provide significant benefits to sea turtles.[9] In the meantime, the conservation situation for sea turtle species cannot be allowed to further deteriorate, and so there is a need for states to gear up in anticipation of strong formalized agreements. Cooperation, for example, could begin among interested states in joint research and management efforts.

International agreements, although ultimately necessary to conserve migratory species, should not be solely relied upon as the answer for the future conservation of sea turtles. In addition to the problem of delay until they might provide benefits mentioned above, there are several reasons to give precedence to national legislation for sea turtle conservation. For one, national measures will be required to implement the provisions of international agreements. Also, until sufficient knowledge is gained as to the range and migration routes of sea turtle populations, common international management action might not be fully effective, forcing attention to continue to be focussed at national programs. Finally, for certain sea turtle populations conservation measures may only be needed from individual states (for example, some sea turtle populations may not migrate, or may only be subject to exploitation within the jurisdictional boundaries of 1 state).

Finally, incidental take of sea turtles in fishery operations deserves special concern as a matter of priority. The Migratory Species Convention treats a country whose ships incidentally take a particular species of sea turtle as a range state for that species, and that country will be expected to join in a common conservation program with the other range states of the species.[10] Still, as it may be quite a while before the convention can be brought into play, two other legal actions could be taken.

1) Include provisions to regulate incidental take operations in national legislation: These should encompass efforts to regulate alterations in fishery gear, to

---

9. For example, Article XVIII of the Migratory Species Convention provides that it will enter into force on the first day of the third month following the deposit of the fifteenth instrument of ratification, acceptance, approval or accession. This might be compared with the Endangered Species Convention which only required 10 instruments, yet took over 2 years to enter into force, and perhaps only now 6 years later is it really becoming effective. The amendment of existing regional conventions would similarly be a long-term process.

10. See Migratory Species Convention Article I (h) and (i) on this point.

close fishing in certain areas and at certain times, and possibly to follow the approach taken in the U.S. Marine Mammal Act[11], which regulates the importation of commercial fish caught using methods that resulted in the incidental killing or injury of marine mammals.

2) Include provisions to regulate incidental take operations in international fishery agreements: At international level, fishery commissions should be urged to regulate the taking of nontarget species by also mandating appropriate gear and closing areas and seasons as necessary when sea turtles congregate in large numbers. In several cases amendments might be required for international fishery conventions to provide jurisdiction over nontarget species, but as many of these conventions are being renegotiated in view of the new Law of the Sea, the time might now be propitious for raising such amendments.

Conclusion

The conservation situation for sea turtles can be improved if increased attention is given to legal steps at both national and international level through:

1. immediate attention to the strengthening and/or develoment and enforcement of comprehensive national legislation which regulates the exploitation of sea turtles and their eggs, and protects sea turtle habitat, and

2. the coordination of national measures through international cooperation. The new Migratory Species Convention appears to provide the best framework for this cooperation, and it deserves support from all those interested in the conservation of migratory species in general and sea turtles in particular. To complement this global effort, strong regional conservation conventions should be promoted for distinct populations of sea turtles. Several existing conservation conventions require scrutiny and strengthening before they can provide significant benefits for sea turtles.

Finally, in order to determine gaps in existing legal coverage, and to set priorities for future legal action, it will be extremely useful to maintain a systematic inventory of relevant national legislation and international conventions concerning sea turtles.

---

11. *Supra*, note 6.

Appendix 1. National legislation directly concerning sea turtles

The following is a listing of the national and provincial legal instruments available at IUCN's Environmental Law Center in Bonn which specifically deals with sea turtles. Within IUCN's collection of some 18,000 legal texts from over 130 States, a great number of enactments can be found which, although not specifically mentioning sea turtles, can be extremely important for sea turtle conservation (for example, texts for protected areas, marine pollution, fishery conservation, etc.). Furthermore, this appendix does not contain references to legislation that is only used to implement the Endangered Species Convention, it being considered sufficient to list the parties to that convention in Appendix II recognizing that those states have taken the obligation to enact national regulations for international trade in sea turtles and their products.

Given the difficulties involved in acquiring legal texts throughout the world, this appendix is necessarily incomplete even for enactments which directly concern sea turtles. IUCN would consequently be most grateful for information as to additional relevant laws in force as well as assistance in their acquisition.

Australia

Fisheries Act, 1952–68
IUCN no. 805680 (J–952031300)

Australia–Northern Territory

Fisheries Ordinance 1965–67
IUCN no. 805860 (J–965000000)

Australia–Queensland

Fisheries Act, 1976
IUCN no. 805730 (J–976121600)

Bahamas

The Fisheries Act, 1969
IUCN no. 807580 (J–969052800)

The Marine Products (Fisheries) Rules (1954)
IUCN no. 805730 (J–954092300)

Belize (UK)

Fisheries Regulations (1967)
IUCN no. 811990 (J–967000000)

Bermuda

The Fisheries Act (1972)
IUCN no. 808700 (J–972051200)

Brazil

Law No. 5197 of 3 January 1967
Protection of the Fauna
IUCN no. 810920 (H–967010300)

Portaria No. 3 481 – DN (Lista Oficial de Espécies
Animais Ameacada de Extincao da Fauna Indigena)
(1973)
IUCN no. 810920 (H–973053100)

Cayman Islands (UK)

The Marine Conservation (Turtle Protection) Regulations, 1978
IUCN no. 817300 (H–978091900)

The Marine Conservation Law (1978)
IUCN no. 817300 (H–978091401)

The Endangered Species Protection and Propagation
Law (1978)
IUCN no. 817300 (H–978091400)

Costa Rica

Ley de Pesca y Caza Maritimas (1949)
IUCN no. 826150 (I–949011100)

Decreto Ejecutivo No. 9 del 24 Mayo de 1963
IUCN no. 826150 (I–963052400)

Decreto No. 5680 Parque Nacional de Tortuguero
(1975)
IUCN no. 826150 (R–975110300)

Cyprus

Regulation 8A – Fisheries Law Ch. 35 1971
IUCN no. 827770 (J–971000000)

Dominica

Forestry and Wildlife Act, 1976
IUCN no. 943370 (H–976062200)

Ecuador

Ley de Proteccion de la Fauna Silvestre y de los Recursos Ictiologicos (1970)
IUCN no. 832300 (H–970111700)

Fiji

Fisheries Ordinance (1942)
IUCN no. 833400 (J–942010100)

Fisheries Regulations (1965)
IUCN no. 833400 (J–965020600)

Gabon

Décret-Loi no. 22/PM du 30 décembre 1960, les taux
de permis de chasse, taxes d'abattage, taxes cynégétiques et licences de chasseurs professionels
IUCN no. 837900 (I–960123000)

Gambia

Fisheries Act, 1971
IUCN no. 838640 (J–971000000)

Ghana

Wildlife Conservation Regulations, 1971
IUCN no. 841340 (H–971043000)

Gilbert and Ellice Islands
(now Kiribati and Tuvalu)

Wildlife Conservation Ordinance, 1975
IUCN no. 841800 (H–975052800)

Grenada

Birds and other Wild Life (Protection of) Ordinance
(1957) (as amended)
IUCN no. 848540 (H–957012600)

Guadeloupe

Arrêté portant réglementation de l'exercise de la
pêche marine côtière dans les eaux du département
de la Guadeloupe (1979) No 79–6 AO/3/3
IUCN no. 836050 (J–979032600)

Guyana

Fisheries (Aquatic Wild Life Control) Regulations (1966)
IUCN no. 850650 (J–966000000)

Guyane (France)

Arrêté no. 69 239–ID/2BAG réglement la capture
des tortues de mer, la récolte de leurs oeufs et l'utilisations de ces produits dans le Département de la
Guyane (1969)
IUCN no. 835060 (J–969022600)

Honduras

La Ley de Pesca (1959)
IUCN no. 851700 (J–959051900)

Iran

Game and Fish Law
Khordad 2526 (June 1967) as amended Esfand 2533 (March 1975)
IUCN no. 857400 (I–967000000)

Ivory Coast

Loi no. 65–255 du août 1965, relative à la protection de la Faune et à l'exercise de la chasse
IUCN no. 864040 (H–965080400)

Kenya

The Wildlife (Conservation and Management) Act (1976)
IUCN no. 869160 (H–976021300)

Madagascar

Ordonnance no. 62–079 établissant un droit de sortis sur les animaux sauvages (1962)
IUCN no. 878030 (H–962092900)

Malaysia – Sabah

The Fauna Conservation (Turtle Farms) Regulations, 1964
IUCN no. 879870 (H–964062600)

Malaysia – Sarawak

Wild Life Protection Ordinance (1957)
IUCN no. 879900 (H–957000000)

Malaysia – Trengganu

Turtles Enactment, 1952
IUCN no. 879960 (H–952012000)

Mauritius

Fisheries Ordinance, 1948
IUCN no. 883000 (J–948021700)

Mexico

Disposiciones sobre la captura, approvechamiento y comercializacion de la tortuga marina (1968)
IUCN no. 883680 (H–968090000)

Montserrat (UK)

Turtle Ordinance (1951)
IUCN no. 885760 (H–951000000)

New Caledonia (France)

Dispositions de l'Arrêté rendant exécutoire la délibération no. 220 du 3 août 1977
IUCN no. 836980 (H–977080300)

Nicaragua

Decreto No. 937 (Reformas a la Ley Especial sobre Pesca) (1964)
IUCN no. 891690 (J–964101100)

Decreto (Isla del Venado; 1960)
IUCN no. 891690 (I–960101100)

Decreto reglemento la explotacion y prohibe la destruccion de las tortugas (1958)
IUCN no. 891690 (I–958082000)

Pakistan–Baluchistan

Baluchistan Wildlife Protection Act, 1974
IUCN no. 869130 (H–974072200)

Pakistan–Sind

The Sind Wildlife Protection Ordinance, 1972
IUCN no. 896120 (H–972040600)

Panama

Decreto Ejecutivo no. 104 (Por el cual se adiciona el Decreto Ejecutivo no. 23; 1974)
IUCN no. 897020 (H–974090400)

Saint Lucia

Turtle, Lobster and Fish Protection Act, 1971
IUCN no. 943395 (J–971060700)

Saint Vincent (UK)

Birds and Fish Protection Ordinance (1901)
IUCN no. 943390 (H–901000000)

Senegal

Décret no. 67–510 du 30 mai 1967 portant Code de la Chasse et de la Protection de la Faune
IUCN no. 911600 (I–967053000)

DANIEL NAVID

Seychelles

Turtles Ordinance (1925)
IUCN no. 912000 (H–925000000)

Turtles Regulations (1929)
IUCN no. 912000 (H–968071700)

The Green Turtles Protection Regulations (1968)
IUCN no. 912000 (H–96807100)

Solomon Islands (UK)

The Fisheries Ordinance, 1972
IUCN no. 812020 (J–972000000)

The Fisheries Regulations, 1972
IUCN no. 812020 (J–972000001)

Sri Lanka

Fauna and Flora Protection Ordinance (1938)
IUCN no. 933000 (H–938030100)

Surinam

Hunting Act (1954)
IUCN no. 887590 (I–954000000)

Trinidad and Tobago

Fisheries Ordinance (1916 as amended to 1975)
IUCN no. 926880 (J–916000000)

Protection of Turtle and Turtle Eggs Regulations, 1975
IUCN no. 926880 (H–975090800)

Thailand

Fisheries Act (1947)
IUCN no. 923780 (J–947011300)

Turks and Caicos Islands (UK)

Fisheries Protection Ordinance, 1949
IUCN no. 929720 (J–94908300)

Fisheries Protection Regulations, 1976
IUCN no. 929720 (J–976000000)

United States

Endangered Species Act of 1973
IUCN no. 933000 (H–973122800)

Florida

Florida Statutes Section 370.12(1) Chapter 74 20
IUCN no. 933500 (H–000000000)

Hawaii

Regulation 36: Relating to the protection of marine turtles (1974)
IUCN no. 933600 (H–974041100)

New York

Conservation Law § 11–0536–6
IUCN no. 934650 (H–00000000)

North Carolina

General Statutes – Subchapter 101 – Endangered and threatened species
IUCN no. 934700 (H–979072800)

Texas

Parks and Wildlife Laws, § 978d–1, P.C. (1971)
IUCN no. 935200 (H–97100000)

Regulations for Non-game species
IUCN no. 935200 (H–000000002)

Regulation for Endangered species
IUCN no. 935200 (H–000000001)

Virgin Islands (UK)

Endangered Animals and Plants Ordinance, 1976
IUCN no. 942900 (H–976052800)

Appendix 2. International conventions providing for cooperative conservation measures for sea turtles

African Convention on the Conservation of Nature and Natural Resources

Date of adoption	15/9/1968	Date of entry into force	9/10/1969
Place of adoption	Algiers	Depositary	Organization for African Unity

Parties	Entry into force (day/month/year)	Parties	Entry into force (day/month/year)
Central African Emp.	16/4/1970	Niger	26/2/1970
Djibouti	7/5/1978	Nigeria	7/5/1974
Egypt	12/5/1972	Senegal	24/2/1972
Ghana	9/10/1969	Seychelles	14/11/1977
Ivory Coast	9/10/1969	Sudan	30/11/1973
Kenya	9/10/1969	Swaziland	9/10/1969
Madagascar	23/10/1971	Tanzania	22/12/1974
Malawi	6/4/1973	Uganda	30/12/1977
Mali	3/7/1974	Upper Volta	9/10/1969
Morocco	11/12/1977	Zaire	13/11/1976

Convention on International Trade in Endangered Species of Wild Fauna and Flora

Date of adoption	3/3/1973	Date of entry into force	1/7/1975
Place of adoption	Washington	Depositary	Switzerland

Parties	Entry into force (day/month/year)	Parties	Entry into force (day/month/year)
Australia	27/10/1976	Mauritius	27/07/1975
Bahamas	20/06/1979	Monaco	18/07/1978
Bolivia	04/10/1979	Morocco	14/01/1976
Botswana	12/02/1978	Nepal	16/09/1975
Brazil	04/11/1975	Nicaragua	04/11/1977
Canada	09/07/1975	Niger	07/12/1975
Chile	01/07/1975	Nigeria	01/07/1975
Costa Rica	28/09/1975	Norway	25/10/1976
Cyprus	01/07/1975	Pakistan	19/07/1976
Denmark	24/10/1977	Panama	15/11/1978
Ecuador	01/07/1975	Papua New Guinea	11/03/1976
Egypt	04/04/1978	Paraguay	13/02/1977
Finland	08/08/1976	Peru	25/09/1975
France	09/08/1978	Senegal	03/11/1977
Gambia	26/11/1977	Seychelles	09/05/1977
German Dem. Rep.	07/01/1976	South Africa	13/10/1975
Germany, Fed. Rep.	20/06/1976	Sri Lanka	04/05/1979
Ghana	12/02/1976	Sweden	01/07/1975
Guatemala	05/02/1980	Switzerland	01/07/1975
Guyana	25/08/1977	Tanzania	26/02/1980
India	18/10/1976	Togo	23/01/1979
Indonesia	26/03/1979	Tunisia	01/07/1975
Iran	01/11/1976	United Arab Emirates	01/07/1975
Israel	16/03/1980	United Kingdom	31/10/1976
Italy	31/12/1979	United States	01/07/1975
Jordan	14/03/1979	U.S.S.R.	08/12/1976
Kenya	13/03/1979	Uruguay	01/07/1975
Liechtenstein	28/02/1980	Venezuela	22/01/1978
Madagascar	18/11/1975	Zaire	18/10/1976
Malaysia	18/01/1978		

Convention on the Conservation of Migratory Species of Wild Animals

Date of adoption	23/6/1979
Place of adoption	Bonn
Date of entry into force	—
Depositary	Federal Republic of Germany
Open for signature	Bonn, to 22/6/1980

Edgardo D. Gomez
Director, Marine Sciences Center
University of the Philippines
Diliman, Quezon City

Consultant, Task Force Pawikan Working Group
Ministry of Natural Resources
Visayas Avenue, Quezon City

Problems of Enforcing Sea Turtle Conservation Laws in Developing Countries

ABSTRACT

Sea turtle conservation laws should cover the turtles themselves, their eggs, and their habitats with special reference to nesting beaches and forage grounds. In some countries, the problems start here. There may be no specific conservation laws for sea turtles, or the laws may be inadequate in terms of scope or penalties. This situation may be a result of jurisdictional ambiguities. It is not always clear which governmental department or bureau has jurisdiction over sea turtles.

The actual implementation or enforcement of the laws may be regarded from various aspects. The geography of island nations and archipelagoes makes the task difficult. The majority of these are developing countries in the tropics.

When the inhabitants of these countries are considered, several points come to light. By definition, the economies of these countries are smaller than those of the developed nations. Consequently, economic pressures, if not actual survival needs, may make some members of the population oblivious of conservation laws. Sometimes, the problem is plain ignorance of the law. Some of the people have not received adequate education and hence are unaware of conservation concepts. In other situations local populations may have traditional rights and practices which are not superseded by the national or general laws. The continued exercise of these traditional practices would not be problematical except that the populations of these countries have increased so much that the exploitation of sea turtles may have reached dangerous levels.

Enforcement agencies in some developing countries may be hampered in their conservation law activities. Too often there are not enough enforcers or wardens. Similarly, their equipment (such as patrol craft) may be insufficient, as may be the operational budget. Then, there is always the problem of the laxity for one reason or another of a certain percentage of the police force. In some countries, all this may be further complicated by a widespread or regional degeneration of the peace and order.

Other factors such as the external world demand for turtles and turtle products, the profit motive of the middleman, and the inadequacy of public information contribute to the difficulty of conservation law enforcement.

Introduction

This paper draws heavily on the author's familiarity with problems of enforcing marine resource conservation laws, especially with respect to corals. Difficulties in enforcement are quite similar for the various marine resources and for most developing countries.

The treatment of the topic is from a general viewpoint without indicating specific countries. Time constraints in the preparation of the manuscript imposed this format. The reader is referred to the papers of Navid and Bavin in this volume for some of the specific laws concerning sea turtle conservation. As may be noted, these laws may refer to the turtles themselves, to their eggs, or to their habitats.

The Inadequacy of Laws

Not all countries with sea turtles have conservation laws to protect these resources. In some countries, the laws concerning sea turtles may be inadequate in scope or in strength. For example, some laws protect the turtles but do not cover the eggs. Few countries have laws protecting turtle habitats. If there are protective regulations covering rookeries, there may be none protecting the feeding grounds, or vice versa. In some situations, laws protecting habitats may have as their primary object some other commodity and apply only incidentally to turtles. An example of this are protected coral reef areas where hawksbills are known to forage.

If the penalties for the violation of conservation laws are light, unscrupulous traders may disregard the laws and pay the fines from the profit of their harvest.

Jurisdictional Ambiguity

Jurisdictional ambiguity may appear trivial on the surface, but in some countries it may be a real issue. Jurisdiction over sea turtles may be ambiguous or overlap with respect to agencies concerned with sea turtles. This usually results in poor enforcement especially if the wrong agency is given the jurisdiction. In most countries marine turtles fall under fisheries departments while in others they are under the wildlife or forestry departments, or both. The rationale here is that while they are in the water, they are fisheries resources; when on land, they are wildlife species. Needless to say, the departments or agencies concerned do not have equal law enforcement capabilities.

Problems in Implementation

Even where laws leave no jurisdictional ambiguities, problems of enforcement may arise due to several other factors.

The first difficulty may be one of geography or physiography. Many developing countries are island nations or archipelagoes. One needs only to consider the South Pacific island countries and such archipelagoes as the Philippines with some seven-thousand islands and Indonesia with nearly twice that number. It is physically impossible to watch every island visited by marine turtles. Dangerous reefs or weather conditions make many of these islands inaccessible by sea. As for land travel, many islands have poor or no roads to turtle beaches. These constraints make it very difficult, if not impossible, for law enforcers to cover much of the area where turtles are found. On the other hand, hunters and traders who are familiar with local conditions or who employ local inhabitants often gain access to turtles in these areas.

The second difficulty arises from the subject people. In developing countries, many coastal people are subsistance fishermen or hunters. Economic pressures to support families, usually large, may make some members of the population oblivious of conservation laws. When the dictum is "Anything that moves is fair game," turtles will not be spared. When fishing is poor, a large piece of protein that swims by is not going to be protected by its carapace and plastron. If it is a question of food on the table or obeying the law, one can guess what the choice will be. Then, too, many rural peoples have never even heard of the laws promulgated in the city.

Another scenario may be that of simple, uneducated folk who cannot understand new conservation regulations promulgated by their government as a result of the recommendations of a distinguished group of turtle biologists and other authorities gathered in Washington, D.C. to save the turtles from extinction. They may never even have heard of Washington, much less of "missing arribadas." And what they cannot understand, they cannot accept and follow.

Mention has been made in this conference of traditional rights and practices that are not superseded by national legislation. The exercise of such traditional rights would not normally be problematical except for the fact that the populations have outstripped the traditions. Because of large human populations in many countries, some traditional practices need to be modified.

The third difficulty in enforcement pertains to the law enforcement agencies. Too often there are not enough police or wardens. Even if the personnel is sufficient, the equipment may be insufficient. Patrol craft and land vehicles are often scarce. Should there

EDGARDO D. GOMEZ

be alternate modes of transport, operational funds may be inadequate. If there is only a small budget for travel, the enforcers cannot go to the field.

The financial inadequacies may manifest themselves in other ways. A poorly paid warden may not do his job conscientiously especially if it entails risks. Unscrupulous traders may offer the poorly paid civil servant with incentives to apply the law more strictly in some areas while leaving other areas unpatrolled. Related to this is the problem of favoritism or selective enforcement because of bonds of blood or friendship. Extended families are commonplace in developing countries.

Banditry and the breakdown of peace and order are conditions more likely to prevail in developing countries than in developed areas. An unstable region makes the enforcement of conservation measures more difficult.

Other Problems

Conservation problems in developing countries often arise or are aggravated because of external factors. Poaching by foreign vessels in the territorial waters of developing countries is one example. More often, it is world demand for turtles and turtle products that leads to the depletion of resources in developing countries because some of the local inhabitants are only too eager to earn foreign exchange. In these conditions, profit seeking middlemen can prosper by putting conservation low in their list of priorities, and subsistence fishermen and hunters are often exploited by their own countrymen.

Related to this is an inadequate public information system that leaves even conscientious citizens ignorant about the issues. This ignorance may extend to government officials in different parts of the administration who may unwittingly contribute to the negation of conservation measures. Thus, while the departments involved in conservation may be trying their best, other departments dealing with trade, cottage industries, or export promotion may continue to follow policies that conflict with conservation.

Clark R. Bavin
Chief, Division of Law Enforcement
U.S. Fish and Wildlife Service
Washington, D.C.

Enforcement of Restrictions on Importation of Sea Turtle Products

Without a doubt commercial trade has a devastating effect on sea turtle populations. We know that sea turtle products remain in great demand in the international market-place. I believe that controlling their importation and exportation throughout the world is one of the most effective methods of sea turtle conservation. In order to control this commercial trade, we have established an elaborate import control system.

Before discussing these control mechanisms let me emphasize that I do not suggest that our procedures should be implemented by other nations. However, these procedures work for us, and there may be some techniques that would be helpful to other nations depending on their location, capabilities, and problems.

First, let me briefly explain the federal laws which give us the power to protect sea turtles. For many years our states have enforced laws which protect nesting sea turtles and their eggs. These laws restrict the taking, possessing, and selling of turtles, their eggs and products as well as destruction of their nests. In addition federal law protected sea turtles on federal refuges for many years. But it was not until the Lacey Act was amended in 1969 that the Federal Government was authorized to enforce restrictions on sea turtle importations. The Lacey Act is a law originally enacted in 1900 which makes it a federal crime to import into the United States any wildlife that has been taken, transported, possessed or sold in violation of the law of a state or a foreign country. The 1969 amendments made this statute applicable to sea turtles for the first time.

Also in 1969 the Endangered Species Act was passed which imposed importation prohibitions on species listed by the Secretary of the Interior as endangered. This law was replaced in 1973 by a new act which allows the Secretary to compile a list of endangered species, that is those in danger of extinction, and threatened species, defined as species which if not protected may become endangered within the foreseeable future.

541

In 1970 the hawksbill, leatherback and Kemp's ridley were listed as endangered. In 1978 certain populations of the green turtle and olive ridley were listed as endangered and all other populations listed as threatened. Also in 1978 the loggerhead turtle was listed as threatened. All of that may sound very complicated to someone who is trying to differentiate between the threatened and endangered status of certain populations. Nevertheless, from an enforcement standpoint, importation of such turtles or their parts or products is prohibited without a proper permit.

In September of 1978 Cayman Turtle Farm filed suit to prevent us from enforcing such restrictions against their products. Pending the outcome of this review the regulations were not applied to products from their operation. In May of 1979 the U.S. District Court for the District of Columbia ruled that these restrictions do apply and importation was prohibited. Presently all U.S. restrictions apply to all such turtles and their products.

The United States is also a party to the Convention on International Trade in Endangered Species of Wild Fauna and Flora (CITES), and the provisions of this Convention apply to the importation and exportation of sea turtles and their products.

In summary, without proper scientific permits it is unlawful for any person to import into the United States any of the six sea turtles which are listed as threatened or endangered. In addition, it is unlawful to sell any of these sea turtles or their products in interstate or foreign commerce, or to transport them in interstate or foreign commerce in the course of a commercial activity.

Now let us look at how the United States enforces these sea turtle restrictions. There are over 300 Customs ports of entry where persons and cargo can enter the United States legally. All of these ports of entry are staffed by officers of the U.S. Customs Service, and they are authorized to enforce laws and regulations concerning fish and wildlife importations, including sea turtles and their products. The U.S. Customs Service is the initial screen, and we rely heavily on them to interdict illegal wildlife importations. Backing up the Customs Service are Special Agents and Wildlife Inspectors of the U.S. Fish and Wildlife Service. In addition, the National Marine Fisheries Service has enforcement officers who work closely with us to provide federal protection to sea turtles.

One of our basic enforcement tools is the "designated port of entry." There are only eight of the over 300 Customs ports designated for general wildlife importations. These are: New York, Miami, New Orleans, Chicago, Los Angeles, San Francisco, Seattle, and Honolulu. All wildlife importations, with certain exceptions, must enter the United States through one of these 8 designated ports. One exception is for certain personal effects that accompany the traveler. Another exception is by permit which allows importations at specific nondesignated ports during specified periods of time. These are subject to inspection so that control is still maintained. There are also exceptions to the designated port requirement along the Canadian and Mexican borders for importation and exportation of wildlife which originate in either country. Basically, all wildlife entering the United States through commercial channels from other than Canada or Mexico must enter at one of the eight designated ports. Thus, we funnel most wildlife importations into a few major ports where close screening can be performed. If wildlife arrives at a nondesignated port, the Customs officer simply redirects the shipment to proceed in bond to the nearest designated port.

At each of the designated ports the U.S. Fish and Wildlife Service employs Wildlife Inspectors. These inspectors are trained in wildlife identification and the intricacies of wildlife law which relate to import restrictions on all fish and wildlife and their parts or products. When Customs officers discover a wildlife shipment, they refer it to our Wildlife Inspectors who determine whether the shipment may legally enter the country.

To assist these Wildlife Inspectors in performing their job we have a special wildlife importation and exportation declaration form which must be filled out before most wildlife can be legally imported into the United States. These declarations include the name and address of the consignee and consignor, a description of the wildlife by common and scientific name, quantity involved, declared value, and country of origin. The importer must also declare that any foreign wildlife permits required have been obtained and are accompanying the shipment and that the shipment is otherwise legal under all foreign and domestic law. It is also the responsibility of the importer to properly identify and declare the wildlife which is being presented for entry.

We also have regulations which require that all containers and packages containing wildlife and wildlife products be conspicuously marked on the exterior of the package with the name and address of the shipper and the consignee and a complete description of the number and kinds of wildlife included in the package or container. For extremely valuable shipments a special permit can be issued to allow a symbol to be marked on the container in lieu of the other information. Marking of containers is important because it alerts Customs officers and Wildlife Inspectors that wildlife is contained in a particular shipment and thus requires special scrutiny. Often unscrupulous importers and exporters move wildlife in packages labeled machine parts, old clothing, leather goods or under some other descrip-

CLARK R. BAVIN

tion which does not come close to wildlife. It is obvious to us that this is an area where wildlife smugglers feel threatened and it provides a valuable enforcement tool.

Wildlife Inspectors make physical inspections of wildlife shipments as an additional means of gaining compliance. It is impossible for us to inspect physically all wildlife shipments; so our inspectors spot-check shipments or portions of shipments. They inspect the contents of the packages and containers as well as all accompanying documents, permits, and other papers. We attempt to do a paper inspection of every shipment and a physical inspection of as many shipments as is possible given the manpower resources and the volume of shipments at the particular time.

Each wildlife importation is subject to special clearance procedures by the U.S. Fish and Wildlife Service. Before the Customs Service allows a wildlife shipment to enter domestic commerce and "clear Customs," it requires our Wildlife Inspectors or agents to stamp the wildlife declaration and other import documents cleared for entry.

On many occasions wildlife shipments arrive with incomplete foreign documents or without the requisite importation permits. These shipments are held in Customs custody or released under bond pending the receipt of the required documents. If the appropriate documents are not forthcoming in a reasonable period, the matter is turned over to a Special Agent of the U.S. Fish and Wildlife Service for investigation. In addition, even after clearance has been granted by a Wildlife Inspector and the shipment has entered the country, we can use a procedure called "post clearance investigation." That system is invoked where we have released a shipment erroneously and discovered additional information that indicates the shipment is illegal. Clearance by a Wildlife Inspector is not a certificate of legitimacy and post clearance investigations are a possibility.

Under our new system of licensing, all persons in the business of importing or exporting wildlife or their parts or products will have to obtain a specific license from the U.S. Fish and Wildlife Service to engage in such business. This licensing system will not only furnish names and addresses of all people in the wildlife importing and exporting business so that we may communicate with them regarding laws, regulations and procedures, but it will also require them to keep books and records and make reports, as we deem appropriate, concerning their import and export activities. It will also allow us to inspect their facilities to verify their compliance with wildlife statutes. We expect this new enforcement tool to be very productive in bringing about compliance.

All of these tools give us an effective system for enforcing importation laws. We know, however, that there is still illegal traffic in wildlife. We are doing our best to interdict these illegal shipments, but there are several problem areas.

First of all, the sheer volume of wildlife importations is staggering. In 1978 alone over 13.1 million wildlife hides and skins were imported, along with 368,000 live birds, 2.5 million live reptiles and amphibians, 152,000 game trophies, 260 million tropical fish, and over 187 million individual products manufactured from wildlife. These figures are particularly staggering when you realize that they include only declared and documented shipments which are ostensibly legal. Smuggled wildlife, of course, is not declared and therefore not included in these figures. The exact extent of illegal traffic is unknown, but based on actual seizures and intelligence we believe it may run as high as 10 to 25 percent of total wildlife shipments depending on the species involved.

Another major problem is the tourist traffic, that is, the individual traveler who goes abroad and buys wildlife items, including sea turtle products, and brings them back to the United States. Under our domestic legislation there is no exception, and a tourist cannot bring into the country a sea turtle product that he has purchased abroad. The problem is obviously the same as for declared shipments, that is, the sheer volume of passengers entering the United States every year makes it impossible for either Customs or the Fish and Wildlife Service to adequately inspect for such items. Tourists bringing back wildlife items cannot be required to come through specific designated ports. Therefore, they often escape undetected unless a sharp Customs inspector recognizes a declared item as coming from sea turtle or some other prohibited species.

To assist tourists and others in understanding the restrictions on wildlife importations, the Fish and Wildlife Service has conducted and will continue to conduct a public relations and education program. We have been quite successful during the past year in interesting numerous newspaper reporters and magazine writers in the problem of illegal wildlife commercialization. A number of major metropolitan daily newspapers and national magazines have printed articles on wildlife smuggling and the effect illegal trade has on wildlife populations. By encouraging reporters covering these stories, we have reached millions of Americans throughout the nation. We have also worked with television and radio. Two of the 3 major networks, ABC and NBC, did segments on wildlife smuggling on evening and weekend news earlier this year. Several members of our enforcement staff have participated in radio talk shows and were interviewed on local stations and National Public Radio.

Another valuable and extremely effective tool has been our series of widely used TV and radio spots. The 3 television spots were distributed to 450 television stations—or 60 percent of the U.S. television market—in early 1975 and were used by some stations as long as 4 years. We conservatively estimate that over 60 million people saw one of the spots the first year it was aired.

Another part of our public relations effort is the distribution of fact sheets and a brochure aimed primarily at tourists, called "Facts About Federal Wildlife Laws." The Fish and Wildlife Service has distributed about 35,000 copies of this brochure annually, and expects to distribute another 50,000 copies in 1980. We also distribute the brochure in bulk to U.S. Customs facilities at our major ports of entry, and to U.S. Embassies and Consulates abroad for direct distribution to tourists preparing to enter the United States.

The forensic identification of sea turtle parts and products presents another major problem. Often, turtles seized as evidence have been processed into frozen meat, turtle oil, and tortoiseshell products. At this point, the identification of the species-origin of the turtle product is by no means a trivial project. The forensic examiner not only has to determine the species of the turtle involved, but must also demonstrate to a scientific certainty that the meat or oil is of turtle origin and not of any other processed animal species. Clearly, the examiner testifying in such cases needs an extensive background in the identification of animal tissues in general as well as specific expertise in identifying the physiological characteristics that distinguish individual turtle species.

Today, the Fish and Wildlife Service is in a position to coordinate the diversity of wildlife identification expertise in the scientific community in order to prosecute federal wildlife violators. Our Forensic Science Branch is contacting experts in turtle-product identification for the purpose of unifying and coordinating our ability to identify commercial wildlife products. For those of you who have not been contacted yet, and who are interested in becoming involved in our efforts to coordinate and resolve some of the existing problems in forensic identification of turtle products, we would be most happy to work with you.

In summary, commercial trade most certainly has had a devastating effect on sea turtle populations. There continues to be great demand in the international marketplace for sea turtle products. It has been our experience that so long as the demand exists, an illicit supply will emerge to meet that demand. U.S. law enforcement agencies intend to enforce our import controls aggressively, and we expect our efforts to take U.S. nationals out of the marketplace and remove the U.S. dollar as an incentive to the commercialization of sea turtles. We also stand ready to cooperate to the best of our ability with other nations similarly interested in the future of sea turtles.

CLARK R. BAVIN

David Mack
Nicole Duplaix
TRAFFIC (U.S.A.)
1601 Connecticut Avenue, N.W.
Washington, D.C. 20009

Susan Wells
Species Conservation Monitoring Unit
219(C) Huntington Road
Cambridge CB3 ODL, U.K.

Sea Turtles, Animals of Divisible Parts: International Trade in Sea Turtle Products

Introduction

Most parts of sea turtles are of potential commercial value; the shell can be used for jewelry and ornaments; the skin of the flippers and neck can be tanned and used for leather articles; the meat is consumed; the offal is used in soup; the oil is used as a cosmetic base.

There are 7 species of sea turtle, but only 3 are heavily exploited for trade: the green turtle, *Chelonia mydas*, the olive ridley turtle, *Lepidochelys olivacea*, and the hawksbill turtle, *Eretmochelys imbricata*. The green turtle is taken for its meat, and its calipee/calipash (belly cartilage), neck, and tail bones are used in manufacturing turtle soup. The olive ridley is captured mainly for its skin and secondarily for its meat and oil. Due to the thickness and color pattern of the scute, the hawksbill has the most valuable shell.

In this report, we present world trade data on sea turtle products and identify importing and exporting countries and trade routes. The impact of the Convention on International Trade in Endangered Species of Wild Fauna and Flora (CITES) on trade is assessed. All sea turtles (except some Australian populations) were on Appendix I of CITES by February 1977, thus prohibiting commercial trade in their products by party states.

Most of the data in this paper have been obtained from trade records published by government statistical offices and from CITES annual reports compiled by party states. Unfortunately, many countries do not list the import-export of turtle products under separate tariff headings, and these were inferred whenever possible from other countries' trade statistics. For example, Cuba does not list export trade statistics for turtle shell by volume, but this can be deduced, in part, from the amount of shell that Japan reports importing from Cuba.

Some countries separate tortoiseshell into a number of categories; most commonly "tortoiseshell" (bekko in Japan) is listed separately from "claws and waste" of tortoiseshell. To simplify the data, all categories of raw tortoiseshell (BTN tariff heading 05.11 and SITC tariff heading 291.11), and all categories of worked tortoiseshell (BTN tariff heading 95.01 and SITC tariff heading 899.110) have been added together.

Government statistics are well known for being inaccurate, and many discrepancies exist between the volume and value of imports and exports recorded by pairs of countries. These are due to a number of factors, including variations in classification of commodities, methods of recording (e.g. imports may be recorded with country of "origin" or of "consignment"), methods of valuation and exchange conversion, time lags between departure and arrival at country of destination, transportation charges and import duties. Where discrepancies are large, there is good reason to believe that illegal transactions may be taking place (Bhagwati 1974).

International Import and Export: Trade Statistics and Routes

Tortoiseshell Trade

Most of the tortoiseshell trade is believed to be from the hawksbill sea turtle. Wild green and olive ridley turtle scute is approximately 0.5-mm thick, and, while too brittle for jewelry or ornaments, it can be used as delicate inlays and veneer on furniture (King, personal communication, 1979). The intricate design and distinct gradations in color make the carapace the choice part of the hawksbill shell. The edge of the shell (referred to as hooves or claws) is thick and irregular, and both this and the plastron are less valuable than the carapace.

Since 1976, 45 countries within the range of wild populations are known to have exported raw (unworked) tortoiseshell (Table 1). According to government statistics, the major exporters of raw tortoiseshell between 1976 and 1978 were Indonesia, Thailand, the Philippines, India, and Fiji (Table 1). Export statistics are available for few tropical countries, but the figures estimated from importing countries' data show that the Caribbean and Central America are important sources, as is, to a lesser extent, East Africa (Table 1).

The volume of raw tortoiseshell involved in international trade has increased dramatically since the early 1970s. Between 1967 and 1970 Indonesia exported less than 10,000 kg of raw tortoiseshell annually. In 1978, it exported 219,585 kg, more than double the exports in 1977. Over 95 percent of Indonesia's exports are destined for Hong Kong, Japan, and Singapore (Tables 2a and 2b), although Hong Kong and Japan record much lower figures for their imports from Indonesia. Thailand exported between 10,000 kg and 15,000 kg of raw tortoiseshell annually between 1973 and 1975, but by 1978, exports had increased to 53,618 kg, going mainly to Hong Kong and Taiwan (Tables 2a and 2b). Indian exports rose from under 3,000 kg annually prior to 1975 to 82,855 kg in 1977 and 11,918 kg in the first 2 months of 1979 (Tables 2a and 2b). During the 1960s, the Philippines exported less than 5,000 kg annually (JTSA, 1973), but exports have risen steadily since 1975, reaching 38,145 kg in 1978. Although in the past Philippine exports of raw tortoiseshell were destined mainly for Japan, they have increasingly gone to other countries as well, particularly to Taiwan (Tables 2a and 2b). Although no country records large imports of raw tortoiseshell from Fiji (Japan records imports of about 300 kg annually in 1978 and 1979 (Table 3)), Fijian statistics show exports of large quantities since the mid-1970s (except in 1977), most of which have been destined for Japan and Europe (Tables 2a and 2b). Ecuador exported large quantities of raw tortoiseshell in 1976 and 1977 at a very low export value. This was probably olive ridley shell, a secondary product from the skin and leather trade.

The main importers of raw tortoiseshell from 1976 to 1978 were Taiwan, Hong Kong, and Japan (Table 4). Imports into Taiwan and Hong Kong have increased considerably over these three years, mainly from Indonesia and Thailand. Imports into Japan increased in 1979 (Table 3), having remained stable since 1974. Japan's imports come mainly from Asia and the Caribbean. In Europe, the main importer is West Germany (Table 4), although imports into this country have decreased since 1974 when over 28,000 kg were imported. Unfortunately, in 1978, many European countries ceased listing tortoiseshell under a separate tariff heading in their trade statistics, which makes it very difficult to monitor their trade. In the past, there have been considerable discrepancies between customs figures and figures in annual CITES reports for some countries. For example, in its annual CITES report for 1976, the UK recorded no imports of raw tortoiseshell and re-exports of 850 kg. UK customs statistics for 1976 recorded the import of 320 kg and the export of 1,742 kg. The 1977 West Germany CITES report recorded imports of 55 kg of "turtle shell," whereas West German customs recorded the import of 8,281 kg of raw tortoiseshell.

Singapore, Malaysia, and Hong Kong are the main re-exporters of raw tortoiseshell (Table 5), and all have increased their re-exports over the last few years. Malaysian re-exports leave from Sabah and are almost certainly of Philippine origin; domestic exports from Malaysia leave from Peninsular Malaya.

The data available for worked tortoiseshell in trade are more difficult to interpret; a number of countries record only values of exports and imports, and, where volumes are recorded, these may include the weights of materials such as metal and wood which go to make up the articles. However, the Far East (Indonesia, the Philippines, and Taiwan in particular) is clearly the

DAVID MACK

Table 1. Domestic exports of raw tortoiseshell (kg)

Country	1976	1977	1978
Asia			
Indonesia	71,373	85,577	219,585
Thailand	23,859	37,941	53,618
India	21,460	82,855	11,918[a]
Philippines	15,607	27,905	38,145
Malaysia	7,253	8,879	(9,311)[b]
Singapore	370	2,501	230
Pakistan	(745)	—	—
Maldives	(625)	(317)	(567)
Sri Lanka	2	—	—
Burma	—	(1,100)	(500)
Bangladesh	—	(4,960)	(4,150)
Vietnam	—	(1,854)	—
Laos	—	—	(781)
Indian Ocean	—	(68)	—
Total	141,294	253,957	338,805
Oceania/Pacific Islands			
Fiji	53,587	362	35,243
Solomon Isl.	(873)	(756)	(528)
Australia	(1,087)	(192)	—
Total	55,547	1,310	35,771
Central and South America			
Ecuador	12,323	37,423	—
Mexico	6,334	—	—
Panama	(5,885)	(4,450)	(6,505)
Nicaragua	(1,446)	(2,573)	(1,014)
Costa Rica	1,390	(260)	(47)
Belize	(12)	(40)	—
Honduras	—	(71)	(9)
Venezuela	(1,000)	—	—
Total	28,390	44,817	7,575
Africa			
Somalia	(5,099)	(236)	(30)
Tanzania	1,813	1,836	1,625
Kenya	1,661	872	761
Mozambique	(463)	(290)	—
Madagascar	(164)	—	—
Seychelles	(106)	(577)	(1,198)
Mauritius	(55)	—	—
Reunion	(377)	—	(46)
Cape Verde	(63)	—	—
Total	9,801	3,811	3,660
Caribbean			
Barbados	22	—	(23)
Cuba	(6,985)	(3,984)	(6,600)
Haiti	(1,219)	(1,173)	(1,004)
Cayman Isl.	(4,002)	(3,875)	(7,500)
Bahamas	(532)	(922)	(1,018)
Dominican Rep.	(367)	(1,000)	(62)

Table 1.

Country	1976	1977	1978
Jamaica	(343)	(1,136)	(128)
Puerto Rico	(262)	(264)	(25)
Fr. W. Indies	(152)	(236)	(276)
St. Vincent	(130)	(230)	(144)
Brit. Dominica	(126)	(507)	—
St. Lucia	—	(489)	(349)
Grenada	—	(59)	—
Total	14,140	13,875	17,129
World Total	249,172	317,770	402,940

Key
— Not available.
( ) Figures estimated from importing countries' data.
a. January and February only.
b. May include re-exports.
Source: Published government statistics.

Table 2. Exports and countries of destination for major raw tortoiseshell exporters (kg)

a. 1977

Country of destination	Exporting country					
	Indonesia	India	Thailand	Philippines	Ecuador	Fiji
Japan	55,442	6,000	—	26,259	17,038	—
Hong Kong	1,127	1,134	28,031	—	—	—
Singapore	27,920	—	5,000	—	—	—
Taiwan	—	—	4,910	1,269	—	—
Kuwait	—	50,050	—	—	—	—
Italy	95	1,699	—	25	19,861	—
Fed. Rep. Germany	—	20,816	—	—	—	—
United States	—	1,656	—	—	524	—
Other countries	993	1,500	—	352	—	362
Total	85,577	82,855	37,941	27,905	37,423	362

b. 1978

Country of destination	Exporting country				
	Indonesia	India Jan–Feb	Thailand	Philippines	Fiji
Japan	40,368	2,245	—	29,847	16,803
Hong Kong	125,008	—	26,990	—	—
Singapore	52,313	—	5,628	—	—
Taiwan	—	—	20,500	7,600	—
Italy	400	—	—	384	—
Fed. Rep. Germany	—	—	—	—	9,144
Spain	—	—	—	—	9,144
United States	—	9,673	—	164	—
Other countries	1,496	—	500	150	152
Total	219,585	11,918	53,618	38,145	35,343

— Not available.
Source: Published government statistics.

DAVID MACK

Table 3. Japanese imports of raw tortoiseshell (kg)

Country of origin	1976	1977	1978	1979 Jan–Oct
Asia				
Indonesia	6,464	10,114	5,735	19,068
Philippines	3,160	3,313	1,439	3,399
Singapore	3,129	4,080	1,844	2,413
Thailand	0	200	1,550	1,380
Maldives	485	317	567	1,470
Other countries	2,861	1,696	499	3,758
Subtotal	16,099	19,720	11,634	31,488
Africa				
Kenya	2,712	2,655	2,850	2,051
Tanzania	2,152	1,474	1,410	5,824
Seychelles	106	577	1,066	1,054
Other countries	777	0	46	67
Subtotal	5,747	4,706	5,372	8,996
Caribbean				
Cuba	6,985	3,984	6,600	4,475
Cayman Islands	4,002	3,863	7,500	6,312
Haiti	1,094	1,173	1,004	1,351
Bahamas	532	922	1,018	1,332
Jamaica	343	1,136	128	474
Other countries	796	1,785	879	615
Subtotal	13,752	12,863	17,129	14,559
The Americas				
Panama	5,885	4,450	6,505	4,589
Nicaragua	1,446	1,573	1,014	949
Other countries	182	371	122	412
Subtotal	7,513	6,394	7,641	5,950
Pacific				
Australia	1,087	192	6	0
Fiji	189	82	399	463
Solomon Is.	873	756	528	799
Other countries	0	0	42	0
Subtotal	2,149	1,030	975	1,262
European countries	800	1,105	1,288	3,040
Total	46,060	45,818	44,039	65,295

Source: Published government statistics.

main region exporting carved tortoiseshell (Table 6), and exports have increased from less than 24,000 kg in 1975 to over 92,000 kg in 1978. Although still destined mainly for Japan, Hong Kong, and Singapore, exports now also go directly to European countries (Table 7). Philippine exports of worked tortoiseshell reached a peak in 1976 (in 1974, only 425 pieces were exported), but they have declined slightly since then (Table 6). In the past, most went to Japan, but they are now going increasingly to European countries (Table 7), as are Taiwanese exports of worked tortoiseshell.

Japan, Italy, and West Germany have been the main importers of worked tortoiseshell since 1976 (Table 8). Japan imports primarily from Indonesia, Singapore, the Philippines, and Taiwan, although it also has its own traditional carving industry. Fiji appears to be a

Table 4. Imports of raw tortoiseshell (kg)

Country	1976	1977	1978
Asia			
Taiwan	52,427	37,704	128,846
Japan	46,060	45,818	44,039
Hong Kong	26,620	42,788	102,275
Malaysia	9,133	30,060	—
Singapore	4,140	21,002	18,469
South Korea	6,100	6,100	7,333
Mainland China	(3,911)	(3,381)	(3,827)
Vietnam	(2,700)	(647)	—
Thailand	1,238	2,231	2,622
Nepal	—	(1,699)	—
Kuwait	—	(50,000)	—
Total	152,329	241,430	307,411
Europe			
Federal Republic of Germany	3,937	8,281	(9,309)
Netherlands	3,000	3,000	—
Italy	2,500	3,000	(784)
Spain	1,531	824	1,080
France	1,000	1,000	(240)
Belgium	400	100	—
United Kingdom	320	26	—
Switzerland	126	39	—
Total	12,814	16,270	11,413
Americas and the Caribbean			
United States	(5,160)	(11,853)	(164)
Mexico	18,021	—	—
Canada	—	—	(50)
Barbados	—	(22)	—
Total	23,181	11,875	214
Pacific			
Fr. Pac. Isl.	(425)	(352)	(150)
New Hebrides	—	—	(102)
New Caledonia	—	(302)	—
Australia	(975)	(60)	—
Total	14,000	714	252
World total	202,324	270,289	319,290

— Not available.

( ) Figures estimated from exporting countries' data.

Source: Published government statistics.

major center for the worked tortoiseshell in the Pacific; imports have risen from US$6,605 in 1972 to US$62,718 in 1978, mainly from the Philippines and a smaller proportion from India. Fiji exports worked tortoiseshell to American Samoa, Western Samoa, and the United States.

From February 1977 until the ban on imports of farmed turtle products in June, 1979 (see later), im-ports of worked tortoiseshell into the United States have been limited to farmed green turtle shell from the Cayman Turtle Farm (CTF). However, a number of countries continued to report exports to the US (Table 7), and, although over this period U.S. customs officials seized approximately 1,000 hawksbill shell ar-ticles from tourists returning to the United States, this does not account for the figures recorded in other

DAVID MACK

Table 5. Re-exports of raw tortoiseshell (kg)

Country	1976	1977	1978
Asia			
Singapore	20,026	30,014	45,578
Malaysia	5,587	46,212	—
Hong Kong	7,497	6,471	10,128
Taiwan	2,376	338	2,233
Japan	24	274	2,258
Total	35,510	83,309	60,197
Europe			
Netherlands	64,000	2,000	—
United Kingdom	1,742	—	—
Portugal	400	200	—
Federal Republic of Germany	47	73	—
Italy	58	—	—
Total	66,247	2,273	—
World total	101,757	85,582	60,197

— Not available.

Source: Published government statistics.

countries' export statistics. Large quantities of worked tortoiseshell are sold as souvenirs in many countries; this trade goes unrecorded, and it is impossible to estimate its volume.

Turtle Skin and Leather Trade

The recent growth of the turtle skin trade is a typical example of international trade's turning to a new species for a product when the traditional source is depleted through overexploitation. In this case, turtle skin became important when the leather trade found it increasingly difficult to obtain traditional reptiles (such as crocodiles) due to scarcity, bans, and better protection in the early 1960s and 1970s. Since very few countries record turtle skin or leather under a separate tariff heading, the extent of the trade may be larger and involve more countries than we have recorded.

At present, olive ridley turtles from Mexico and Ecuador are the main source of skins. In Mexico, sea turtles are captured by fishermen who are part of fishery cooperatives. Antonio Suarez purchases most of these at present and processes them at his 3 plants on the Pacific coast of Mexico. Table 9 provides the volume and number of olive ridley turtles captured by fishermen. Each year, there is a large discrepancy between the number of turtles captured as reported by the legal fishery cooperatives to the Mexican Department of fishery and the number of turtles caught as estimated by Antonio Suarez. According to Suarez's estimates, from 1966 to 1977, an average of 130,000 olive ridleys was taken annually in the states of Oaxaca, Michoacan, and Jalisco.

Since 1975, it has been illegal to export raw turtle skin from Mexico (Márquez, in litt., 1979), and, as a result, most exports are now in the form of leather. In 1976, 22 kg of turtle skins were exported from Mexico, whereas exports of turtle leather rose from 10,041 kg in 1974 to 23,787 kg in 1976, with most of it destined for the United States and Japan. However, there appears to be a large scale illegal trade in raw skins, since Japan records imports of over 50,000 kg of raw turtle skins from Mexico since 1976 (Table 10).

The trade in olive ridley skins in Ecuador is discussed in detail in the paper by Green and Ortiz in this volume. Three companies are still exporting turtle skins, mainly to Japan and Italy. 161,070 kg of skins were exported in 1978, and 139,900 kg were exported in the first 6 months of 1979.

A third important source of skins is from the farmed green turtles from the Cayman Turtle Farm (CTF). Japan imported 23,514 kg of skin in 1978 and 6,988 kg of skin in the first 10 months of 1979 from CTF. The United States imported 14,000 pieces of skin from CTF in the first 5 months of 1979, before the ban on the import of farmed products came into force, and, in 1978, West Germany imported 2,603 skins from CTF, according to its CITES report.

Japan imports large quantities of skins from Asian countries, in particular, the Philippines, Singapore, Indonesia, and Pakistan; these may be from wild green or olive ridley turtles (Table 10). Japan has been recording imports of turtle skin and leather since 1976 (Tables 10 and 11) and is probably the largest consumer of skins in the world. From 1976 to 1978, Japan

Table 6. Exports of worked tortoiseshell (kg and US dollars)

	1976		1977		1978	
Country	Volume	Value	Volume	Value	Volume	Value
Asia						
Indonesia	69,065	396,629	90,792	531,813	92,099	340,533
Philippines	24,330[a]	23,630	11,615[a]	64,306	7,835[a]	95,524
Taiwan	6,044	49,868	2,984	48,000	2,218	100,259
Japan	40	20,081	91	25,511	37	19,057
South Korea	62	13,027	85	9,138	—	—
Thailand	27	5,672	3	673	3	2,439
India	249	4,431	349	2,477	—	—
Singapore[b]	—	2,092	—	10,356	—	23,063
Malaysia[b]	—	39	—	2,033	—	—
Total	75,487	515,469	94,304	694,307	94,357	580,875
Europe						
Italy	700	41,363	1,400	30,382		
United Kingdom	1,725	26,555	525	7,475	—	—
Spain	1,000	4,977	1,000	4,265	—	—
Federal Republic of Germany	18	2,383	4,700	1,723	—	—
Belgium	500	11,579	0	0	—	—
Switzerland	54	1,440	7	1,004	—	—
Netherlands	0	0	—	1,630	—	—
France	1,157	11,507	1,071	27,069	—	—
Total	5,154	99,804	8,703	73,548	—	—
Other countries						
Mexico	76	1,763	—	—	—	—
Fiji[b]	—	4,711	—	9,498	—	8,052
Total	76	6,474	—	9,498	—	8,052
World total	80,717	621,747	103,007	777,353	94,357	588,927

— Not available.
a. Number of pieces (not included in totals).
b. Only values available for these countries.
Source: Published government statistics.

imported an average 10,061 kg of leather (98 percent from Mexico) and 92,198 kg of skins yearly (over 50 percent from Ecuador). In the first 10 months of 1979, Japan imported 159,728 kg of turtle skin and 19,274 kg of turtle leather, which is a considerable increase on previous years.

There is now a considerable market for turtle leather products in a number of European countries. Italy may be the main importer, importing large numbers of skins from Ecuador (see Green and Ortiz, this volume), and Spain and France may also be important centers for the turtle leather trade. West Germany imported CTF skins in 1978 and re-exported a number to Switzerland; turtle leather products are seen on sale in the United Kingdom.

The United States was a major consumer of skin and leather before CITES and U.S. legislation were introduced. From January to May 1977, the United States imported over 31,000 pieces of olive ridley skins and almost 3,000 turtle skin leather articles (shoes) from Mexico (Table 12). In May 1977, the United States banned the import of wild turtle products, but, in 1978, import permits show that 5,706 pieces of wild olive ridley skin from Mexico were illegally imported. The 1978 United States CITES report records the import of 2,000 olive ridley skins from Mexico. In 1979, imports were recorded only from CTF. Only a few shipments of wild turtle leather articles have been seized since 1977: 448 items in 10 shipments (Table 12).

Table 7. Exports and countries of destination for major worked tortoiseshell exporting countries, 1978

| Country of destination | Exporting countries | | | |
	Indonesia kg	Philippines pieces	Taiwan kg	Fiji US dollars
Japan	47,150	6,344	515	—
Hong Kong	25,369	—	—	—
Singapore	16,575	—	—	—
Mainland China	2,200	—	—	—
Federal Republic of Germany	7	1,176	9	—
Italy	200	155	18	—
Spain	—	—	440	—
France	66	—	146	—
Belgium	269	—	—	—
Netherlands	30	—	—	—
United Kingdom	—	100	1	—
Australia	233	60	—	—
United States	—	—	337	477
American Samoa	—	—	—	7,386
Western Samoa	—	—	—	189
Other countries	—	—	752	—
Total	92,099	7,835	2,218	8,052

— Not available.
Source: Published government statistics of exporting countries.

Sea Turtle Meat Trade

Most green turtles that are taken for their meat are consumed locally, and, where they are exploited for international trade, the use of their meat may be secondary only to the use of their skin. In the past, calipee was the main export product since it could be dried and easily transported. Many New World countries killed turtles for meat in the early 1970s (Rebel 1974), and Nicaragua, Mexico, Ecuador, and Costa Rica were the largest exporters according to U.S. import statistics (Table 13). Both Nicaragua and Costa Rica have since curbed or stopped this trade, and Mexico and Ecuador have few buyers for their meat, although illegal shipments and smuggling of Mexican and Ecuadorian meat are known to occur. Crates of olive ridley meat have been labelled as green turtle for export from Ecuador (Ortiz, personal communication); further information on the Ecuadorian meat trade is to be found in the paper by Green and Ortiz in this volume.

Today, CTF is probably the largest exporter of green turtle meat in the world, since many countries are now restricted to imports of "farmed" turtle meat due to CITES regulations on the trade of wild sea turtle products. Prior to the U.S. ban on the import of farmed turtle products, all CTF meat was exported via the United States through the port of Miami. Large quantities were imported for domestic consumption as well as for re-export (Table 14). However, considerable discrepancies exist between the available figures for imports up to the time of the ban (Table 15). The reasons for these discrepancies have not yet been ascertained, but they emphasize the problems of obtaining reliable information in order to monitor the wildlife trade.

Since the U.S. ban on farmed products, West Germany and the United Kingdom have become the main importers of turtle meat, calipee/calipash, etc., from CTF. A number of firms in Europe manufacture turtle soup, and turtle steak is seen increasingly on restaurant menus. The 1977 West German CITES report recorded the import of 31,819 kg of turtle meat from CTF and 18,568 kg from Somalia. The latter import was prior to April 1977, when green turtles (other than farmed ones) were added to CITES Appendix I; the 1978 report recorded the import of 2,370 kg of meat from CTF and the re-export of over 17,000 kg to the United Kingdom and smaller quantities to Denmark and France. The 1977, U.K. CITES report records the import of 3,072 kg of tail and neck bone and 907 kg of calipee from CTF; in 1978, 1,088 kg of tail and neck bone were imported. Between January and November 1979, U.K. licenses had been taken up for the import of 375 kg of steak, 4,082 kg of neck and tail bone, 6,019 kg of calipee/calipash, and 2,718 kg of skinned flipper. This is a considerable increase from the 1978 imports. Switzerland imports meat and calipash from West Germany and, in 1978, re-exported 318 cartons of soup to Canada. Japan may also be a major consumer of turtle meat, but no data are available.

Other Turtle Products

Although turtle eggs are collected in huge numbers for food in many countries, they do not play a large role in international trade. The only data available are for Malaysia: Sarawak imported 334,600 eggs in 1976 and 99,800 eggs in 1977 from Indonesia (species not specified), and Sabah imported 80,800 eggs from the Philippines in 1976. Malaysia also re-exports eggs to Brunei. Turtle egg trade in Central America is discussed by Cornelius in this volume.

Stuffed turtles have become a very important product for the tourist souvenir trade. Large numbers of adult and juvenile hawksbills and green turtles are stuffed or freeze-dried in countries such as Indonesia, the Philippines, the Maldives, the Seychelles, Thailand, Madagascar, and Panama. Japan is probably the main consumer, but tourists from European countries such as the United Kingdom continue to buy them. As with

Table 8. Imports of worked tortoiseshell (kg–US dollars)

Country	1976 Volume	1976 Value	1977 Volume	1977 Value	1978 Volume	1978 Value
Asia						
Japan	113,286	874,507	101,674	757,462	97,605	847,422
Singapore[a]	—	15,884	—	13,980	—	17,306
Malaysia[a]	—	5,793	—	8,024	—	—
South Korea	9	1,926	0	0	—	—
Thailand	905	1,796	0	0	2,100	3,603
Taiwan	0	0	0	0	15	486
Indonesia	0	0	11	50	0	0
India	—	—	801	2,104	—	—
Total	114,200	899,906	102,486	781,620	99,720	868,817
Europe						
Italy	109,300	19,135	50,200	20,169	—	—
France	8,125	197,715	2,963	93,418	—	—
Federal Republic of Germany	1,058	26,211	22,434	252,390	—	—
Belgium	3,100	24,997	500	7,729	—	—
United Kingdom	112	19,872	777	21,100	—	—
Spain	25	10,463	1,000	39,175	—	—
Switzerland	359	6,270	51	5,249	—	—
Netherlands	2,000	5,295	2,000	6,519	—	—
Malta[a]	—	614	—	1,334	—	—
Total	124,079	310,572	79,925	447,083	—	—
Other						
Fiji[a]	—	29,189	—	44,815	—	62,718
Mexico	2	257	—	—	—	—
Total	2	29,446	—	44,815	—	62,718
World total	238,281	1,239,924	182,411	1,273,518	99,720	931,535

— Not available.
a. Only values available.
Source: Published government statistics.

tortoiseshell souvenirs, most of this trade goes unrecorded.

There are also very few data available on trade in turtle oil. This is processed in Mexico and at CTF and may be exported to a number of countries for use in beauty creams. During the first quarter of 1979, U.S. Customs at New Orleans confiscated 107 turtle products from U.S. tourists; over half of the products confiscated were turtle creams mainly from Mexico.

Turtle Farming

Although turtle farming has been considered for many years as a practical method of harvesting turtles commercially, it was not until 1968 that the first carefully planned sea turtle farm was established by American and British interests under the name of Mariculture,

Ltd., on Grand Cayman Island. As a "seed" stock, several thousand green turtles were imported and raised during the first year of operation (Rebel 1974).

Following their first year, Mariculture imported approximately 60,000 doomed eggs annually from the beaches of Surinam and hatched them in their facilities. These young turtles were either used to supplement the breeding stock or were killed and sold as turtle products.

In 1975, Mariculture, Ltd., was liquidated, sold to a consortium of British and West German industrialists and the Cayman Island government, and renamed the Cayman Turtle Farm (CTF). In March 1978, it ceased obtaining eggs or adults from the wild (Johnson, personal communication, 1979).

During the March 1979 CITES meeting in Costa Rica, the parties defined "farmed animals" as those

DAVID MACK

Table 9. Olive ridley turtles captured in Mexico by fishery cooperatives

| Year | Take reported to government by cooperatives | | Total take[a] | |
	Number of turtles	Thousands of kg	Number of turtles	Thousands of kg
Oaxaca				
1966	2,737	104	60,000	2,280
1967	84,368	3,206	120,000	4,560
1968	9,053	344	65,000	2,470
1969	53,131	2,019	60,000	2,280
1970	41,053	1,560	50,000	1,900
1971	—	—	25,000	950
1972	—	—	30,000	1,140
1973	53,046	2,015.74	90,000	3,420
1974	25,493	968.73	60,000	2,280
1975	58,575	2,225.84	70,000	2,660
1976	40,407	1,535.46	55,000	2,090
1977	56,706	2,154.85	75,000	2,850
Total	424,569	16,133.62	760,000	28,880
Av/yr	35,381	1,344.47	63,333	2,406.67
Michoacan				
1965	447	17	15,000	570
1966	26	1	15,000	570
1967	1,447	55	25,000	950
1968	1,526	58	30,000	1,140
1969	684	26	5,000	190
1970	474	18	5,000	190
1971	—	—	15,000	570
1972	—	—	10,000	380
1973	—	—	15,000	570
1974	987	37.5	10,000	380
1975	889	33.79	10,000	380
1976	1,819	69.11	5,000	190
1977	575	21.86	5,000	190
Total	8,874	337.26	165,000	6,270
Av/yr	683	25.941	12,692	482.3
Jalisco				
1968	16,687	634.11	150,000	5,700
1969	1,037	39.42	10,000	380
1970	1,055	40.08	20,000	760
1971	—	—	40,000	1,520
1972	—	—	40,000	1,520
1973	16,947	643.97	100,000	3,800
1974	19,830	753.54	40,000	1,520
1975	10,896	414.05	40,000	1,520
1976	20,057	762.16	40,000	1,520
Total	86,509	3,287.33	480,000	18,240
Av/yr	9,612	365.26	53,333	2,206.67

Note: The conversion factor from volume to number of turtles is 38kg/1 turtle.
— Not available.
a. Report by Antonio Suarez to Departamiento de Pesca, Mexico.

Table 10. Japan: Imports of turtle skin

Country	1976 (kg)	1977 (kg)	1978 (kg)	1979 (Jan–Oct) (kg)
Ecuador	40,275	62,073	40,807	120,599
Mexico	35,231	5,244	1,061	9,075
Cayman Is.	—	36	23,514	6,988
Nicaragua	883	2,322	640	—
Panama	—	—	2,546	—
United States	1,676	—	—	—
Philippines	18,610	6,408	3,857	3,600
Singapore	—	—	9,673	12,261
Indonesia	—	145	6,261	3,477
Pakistan	4,648	1,016	5,360	3,248
Taiwan	—	—	726	—
Belgium	3,283	—	—	—
France	—	—	—	480
Total	104,606	77,244	94,445	159,728

— Not available.
Source: Published government statistics.

Table 11. Japan: Imports of turtle leather

Country	1976 (kg)	1977 (kg)	1978 (kg)	1979 (Jan–Oct) (kg)
Mexico	11,065	6,835	11,646	18,256
Singapore	186	145	154	143
Belgium	—	—	—	875
Federal Republic of Germany	120	—	—	—
Netherlands	—	28	—	—
Italy	—	—	3	—
Total	11,371	7,008	11,803	19,274

— Not available.
Source: Published government statistics.

born (hatched) from parents which had mated in a captive environment. Under this definition, many sea turtle products presently sold by the Cayman Turtle Farm are not farmed, since the eggs were taken from wild populations and only hatched in their facilities; however, West Germany, the United Kingdom, and Switzerland, all CITES members, still import green turtle products originating from CTF for their luxury soup industry, in the belief that CTF will eventually become self-supporting. The United States is the only country so far to have banned the import of farmed products.

The large investment involved in raising captive turtles to a size suitable for export means that the retail products are going to be expensive; hence, they are restricted to a luxury market. At present, the main criticism of turtle farms is that they encourage and maintain a market for turtle products and, in some cases, cause wild turtle products to be sold under the guise of farmed ones at a time when many populations are seriously threatened by commercial exploitation.

Numbers of Turtles Involved in Trade

Given the inaccuracy of trade statistics and the fact that in many countries turtle products such as tortoiseshell may be stored for long periods before being exported, it is not possible to estimate actual catch numbers from trade data. However, the statistics presented in this paper point to an ever increasing quantity of turtle products on the world market at the same time as

DAVID MACK

Table 12. United States: Imports and seizures of turtle skins and leather from major ports of entry

Port of entry	1977			1978			1979 (Jan–May)		
	Imports	Seizures No.	Quan.	Imports	Seizures No.	Quan.	Imports	Seizures No.	Quan
Miami	—	—	—	—	—	—	8,000 pc (CTF) 1 ar (CTF)	1	18 ar
Los Angeles	—	2	88 ar	—	2	2 ar	—	1	1 ar
New York	—	1	132 ar	—	1	188 ar	—	1	18 ar
SW Region (Brownsville, Laredo and El Paso)	21,732 pc* (Mexico) 10,150 sq* (Mexico) 2,987 ar (Mexico)	1	1 ar	5,706 pc* (Mexico) 4,000 pc (CTF)	—	—	6,000 pc (CTF)	—	—
Total	21,732 pc 10,150 sq	4	221 ar	9,706 pc	3	190 ar	14,000 pc 1 ar	3	37 ar

Key: * = from olive ridley turtles; pc = pieces; ar = articles; sq = squares; CTF = Cayman Turtle Farm (country of origin).
Source: 3–177 Declaration of Import Permits, courtesy of Law Enforcement Division, Fish and Wildlife Service, U.S. Department of the Interior.

Table 13. United States: Imports of turtle products by country of origin, 1966–76 (thousands of kg)

Country of origin	1966	1967	1968	1969	1970	1971	1972	1973	1974	1975	1976
Mexico	7.0	4.6	—	4.8	13.2	11.2	2.5	10.0	6.9	6.3	—
Nicaragua	38.3	35.5	—	—	—	—	84.4	72.2	135.9	76.5	1.2
Costa Rica	11.0	—	4.7	42.7	2.0	—	12.3	—	—	—	13.6
Ecuador	—	—	—	1.3	—	—	—	12.3	28.3	22.0	19.1
Grand Cayman	—	—	—	—	—	—	—	—	—	9.0	17.6
West Indies	—	—	—	—	—	—	.3	1.4	—	3.6	—
Jamaica	—	—	—	—	—	—	.1	5.8	—	—	—
Bahamas	—	—	—	—	—	—	.4	1.9	—	—	—
Colombia	—	—	—	3.2	—	.9	—	—	—	—	—
Guatemala	—	—	—	—	—	—	3.2	—	—	—	—
Honduras	—	—	—	—	—	13.6	—	—	—	—	—
Dominican Rep.	—	—	—	—	—	—	—	.2	—	—	—
Venezuela	—	—	—	—	—	—	—	.6	—	—	—
Yemen	—	—	—	—	—	—	—	—	10.9	—	—
The Netherlands	2.0	—	—	—	—	—	—	—	—	—	—
Total	58.3	40.1	4.7	52.0	15.2	25.7	103.2	104.4	182.0	117.4	51.5

— Not available.
Source: Derived from the National Marine Fisheries Service, NOAA, Unpublished Statistics. From Peter Pritchard with permission (in Cato, Prochaska and Pritchard, Unpublished Report, December, 1978).

scientists report ever declining numbers in most turtle populations. This suggests that the turtle catch is, in fact, increasing in many areas, although, as discussed below, the increase in volume on the world market may be due to dealers' getting rid of stocks or stockpiling in anticipation of protective legislation.

Table 14. United States: Imports of turtle meat from major ports of entry

Port of entry	1977 Quantity	1977 Country of origin	1978 Quantity	1978 Country of origin	1979 (Jan–May) Quantity	1979 (Jan–May) Country of origin
Miami	6,609 kg 2,230 ct[1] (50,682 kg)	CTF	8,909 kg 1,765 ct[1] (40,114 kg)	CTF	65,498 kg 150 ct[1] (3,409 kg)	CTF
	18,800 kg 1,783 ct[2] (32,418–40,523 kg)	Ecuador				
	1,477 kg 400 ct[2] (7,273–9,091 kg)	Costa Rica Fr. W. Ind.				
West Palm Beach	21,560 kg 114 ct[2] (2,073–3,182 kg)	CTF[3] Ecuador				
Total	140,892– 151,924 kg		49,023 kg		68,907 kg[3]	

Note: New York, Los Angeles and the Southwest Region (Brownsville, Laredo, El Paso) did not report turtle meat imports during these years.
1. A carton (ct) of meat from CTF (Cayman Turtle Farm) contains 50 lbs. (22.7 kg).
2. A carton (ct) of meat from Ecuador and the French West Indies was estimated to contain between 40 and 50 lbs. (18.2–22.7 kg).
3. Of this total, 9,318 kg was re-exported to West Germany and 4,546 kg was re-exported to the United Kingdom.
Source: 3–177 Declaration of Import Permits, courtesy of Law Enforcement Division, Fish and Wildlife Service, U.S. Department of the Interior.

Table 15. Imports of CTF meat into the United States according to different sources

Source of data	1978	1979 (Jan–May)
U.S. Customs statistics	120,874 kg	106,157 kg
U.S. Declaration of Import Permits	49,023 kg	69,361 kg
National Marine Fisheries Service	84,950 kg	44,100 kg
U.S. CITES Report	4 kg (stew)	—

— Not available.
Sources: U.S. Customs statistics compiled by Bureau of Census, Department of Commerce. U.S. Declaration of Import Permits, Fish and Wildlife Service, Department of the Interior. National Marine Fisheries Service, Department of Commerce. U.S. CITES Report compiled by the Wildlife Permit Office, Fish and Wildlife Service, Department of the Interior.

Hawksbill Turtles

There is increasing evidence that species other than hawksbill may now be used in the tortoiseshell trade. CTF exports polished green turtle shell, and Ecuador may be exporting shell from olive ridleys. Philippine tortoiseshell exports inlcude a large number of green turtle carapaces and scutes (Alvarez, in litt., 1979). However, most tortoiseshell exports are probably still hawksbill, and the following calculation of the numbers represented by the world trade is based on this assumption.

There is considerable variation in the estimates available for the yield of tortoiseshell from a hawksbill. A further complication is that the tariff heading for raw tortoiseshell, which in many countries covers only the scutes, also covers whole carapaces in some countries (e.g. Thailand and the Philippines). The following estimates have been obtained:

Average Weight of Scutes	Weight of Carapace
.68 kg (JTSA 1973)	3.64 kg (Rebel 1974)
.91 kg (Uchida 1977)	(maximum commercial
1.6 kg (Parsons 1972)	yield)

These maximum and minimum estimates were used to calculate the numbers of hawksbills involved in the world trade (Table 16). These numbers must be interpreted with great care. *They are not estimates of annual catches.* Exports from a number of countries such as India, Indonesia, and the Philippines may well be tortoiseshell that has been stored. The figures in Table

DAVID MACK

Table 16. Number of hawksbills involved in world trade

Year	Weight of raw tortoiseshell exported (excluding CTF and Ecuador)—Table 1	Numbers of turtles using maximum and minimum scute weight	Numbers of turtles using carapace weight
1976	249,172	155,000–367,000	68,000
1977	317,770	198,000–468,000	87,000
1978	402,940	251,000–593,000	111,000

16 may possibly be overestimated, since re-exported tortoiseshell may be included in the total if a country does not list it separately from domestic exports of tortoiseshell.

The tortoiseshell export figures for many countries have been calculated from other countries' import data, and, therefore, the size of the export figure is to some extent dependent on the number of countries for which import figures were available. At the time of writing, 1978 statistics for a few countries were still not available, and, as mentioned earlier, some European countries did not record tortoiseshell separately in their trade statistics in 1978. As a result, the 1978 total may actually be underestimated. The number of hawksbills involved in trade may be even larger if worked tortoiseshell and stuffed turtles for the souvenir trade are taken into account.

Olive Ridley Turtles

From 1970 to 1977, an estimated one million ridleys were taken on the eastern Pacific coast by Mexico and Ecuador to supply the skin and leather trade (Table 9; Green and Ortiz, this volume). Skins from an estimated 85,000 ridleys were exported in 1978 by Ecuador (Green and Ortiz, this volume), and, during the same year, Mexico captured 50,000 ridley turtles (mainly gravid females) in the state of Oaxaca alone (Cliffton, in litt., 1979).

Green Turtles

In addition to the large number of olive ridleys captured in Mexico, green turtles are also taken; in 1978, fishermen in Mexico took 5,000 of them, mainly males (Cliffton, in litt., 1979).

According to Japan's import data, a number of Southeast Asian countries export turtle skins (Table 10). These skins are likely to be from green turtles, since imports of skins from most of these countries have a much lower value than Ecuadorian and Mexican ridley skins and an even lower value than the farmed green turtle skins from CTF. A wild turtle skin weighs 3.0 kg to 3.4 kg (Hirth and Hollingworth, 1973); using this estimate, between 13,833 and 15,677 green turtles were killed in Southeast Asia (excluding Singapore)

between 1976 and 1978, and the skins were imported by Japan. As mentioned earlier, skin may be the secondary product of the green turtle; the turtles may be killed primarily for meat which is consumed locally.

According to a representative of CTF at the World Conference on Sea Turtle Conservation, 12,000 green turtles are being killed yearly at CTF. This could account for the volume of meat, leather, and shell from CTF observed in trade in 1978. The highest figure for the import of CTF meat into the United States in 1978 is that provided by U.S. Customs: 120,874 kg. Since an average of 16.8 kg of meat, calipee, and steak is obtained from a CTF turtle (Rebel 1974), a minimum of 4,510 turtles (depending on whether all meat and calipee is exported from one turtle) were killed for the U.S. meat trade in 1978. Japan imported 23,514 kg of turtle skin from CTF in 1978; Japan's imports represent 9,406 turtles since raw skin of a CTF turtle weighs approximately 2.5 kg (Johnson, personal communication). Of the 7,500 kg of raw tortoiseshell imported into Japan from CTF in 1978, 6,321 kg were bekko (probably scutes), and 1,179 kg were claws and waste. Since a CTF turtle yields .45 kg of scute (Johnson, personal communication), the bekko imports represent 14,047 turtles.

The Effect of CITES on Sea Turtle Trade

The Convention on International Trade of Endangered Species of Wild Fauna and Flora (CITES) came into force on 1 July 1975. By the end of the first meeting of parties in Berne, Switzerland, in 1976, all marine turtles (except the flatback, *Chelonia depressa*, and the Australian population of green turtles) were listed on Appendix I of the convention. These amendments came into effect on 2 February 1977, thus prohibiting commercial export and import of sea turtle products (except products from farmed stocks) by party states.

By 1 January 1980, 59 countries had ratified CITES. Currently, France has a reservation on the green and hawksbill turtles, and Italy has one on the green turtle. However, none of the exporting countries which are parties to CITES has reservations.

At least 23 CITES party states have recently been involved in trade in sea turtle products. Many countries have legislation covering marine turtles (see Navid, this volume), but data from official trade statistics and other sources show that domestic or CITES regulations are being enforced in very few countries. This is partly due to the fact that even though a country may be party to CITES, its own legislation may not necessarily fulfill the requirements of the convention, or that some of its trade is being justified on the grounds that the stocks are "pre-convention."

A number of nonparty countries heavily involved in the trade have recently ratified (for example Indonesia, Kenya, Italy, and Tanzania) or shown their intention to ratify CITES. This may cause panic buying and selling before legislation comes into effect and could account for the big increase in tortoiseshell on the world market in the last 2 years, as shown in our analysis of official statistics. While some countries have curbed their exports or imports since ratifying the convention, other countries have not.

Laxity in Enforcement?

There is evidence that some of the following parties may not be effectively enforcing CITES. Parentheses refer to date that CITES came into force:

INDONESIA (March 1979)

Indonesia's ratification in December 1978 may have accounted for the huge exports that year as traders unloaded their stocks of raw tortoiseshell before CITES came into force (Table 1). However, Japan imported 4,911 kg of raw shell and 459 kg of turtle skin from Indonesia between July and October 1979 (Tables 3 and 10).

MALAYSIA (January 1978)

Singapore and Hong Kong reported imports of over 9,000 kg of raw tortoiseshell from Malaysia in 1978.

PAKISTAN (July 1976)

Japan imported over 3,000 kg of turtle skin from Pakistan in the first 10 months of 1979; it is not known which species is involved (Table 10).

INDIA (October 1976)

In 1977, 82,855 kg of raw tortoiseshell were recorded in India's export figures, and, in January and February 1978, 11,918 kg were exported (Table 1), in spite of the fact that exports of tortoiseshell have been banned since 1975 (Karand Bhaskar, this volume).

KENYA (March 1979)

It is still too early to tell if the convention is being effectively enforced in Kenya. Prior to 1979, Japan imported approximately 2,500 kg of tortoiseshell annually from this country; between July and October 1979, Japan imported 896 kg from Kenya (Table 3).

TANZANIA (February 1980)

Japan's imports of raw tortoiseshell from Tanzania increased dramatically in 1979, probably in anticipation of the ratification in November 1979 (Table 3).

SEYCHELLES (May 1977)

Japan has increased imports of tortoiseshell from the Seychelles since 1976 and, in 1979, had already imported over 1,000 kg by October (Table 3).

BAHAMAS (September 1979)

Japan doubled the amount of raw tortoiseshell it imported from the Bahamas between 1976 and 1978; imports rose again in the first 9 months of 1979 to 1,332 kg (Table 3). This increased export to Japan may be the result of unloading stocks before CITES legislation came into force.

PANAMA (November 1978)

Panama has been a regular supplier of raw tortoiseshell to Japan in the past; in 1978, Japan imported 6,505 kg of raw shell from this country. In 1979, tortoiseshell continued to be exported, however, and Japan imported 4,589 kg between January and October (Table 3).

ECUADOR (July 1975)

Ecuador is the world's largest exporter of olive ridley skins. In the first half of 1979, about 140,000 kg of skins were exported (Green and Ortiz, this volume). Sea turtles are classified as fish in Ecuador and come under the jurisdiction of the Fishery Department. This Department does not feel bound by the rules of the Convention (Ortiz and Cantos 1978; 1978 Ecuador CITES Report). The CITES Management Authority in Ecuador is currently working to solve this problem.

NICARAGUA (November 1977)

Between January 1978 and October 1979, Japan imported about 2,000 kg of raw shell and over 600 kg of turtle skins from Nicaragua (Table 3).

DAVID MACK

WEST GERMANY (June 1976)

According to customs statistics, West Germany imported over 8,000 kg of raw tortoiseshell in 1977, although this was not recorded in the 1977 annual CITES Report (Table 4). India recorded exports of over 20,000 kg to West Germany in 1977 (Table 2a). In 1978, Fiji recorded exports of over 9,000 kg to West Germany, which was also not recorded in the annual West German CITES Report (Table 2a). As mentioned above, West Germany still imports large quantities of meat from CTF.

UNITED KINGDOM (October 1976)

The U.K. ratification of CITES affected a number of dependent territories including Hong Kong and Belize, both of which are involved in the turtle product trade. Between January and October 1979, Japan imported 314 kg of raw tortoiseshell from Belize. In 1978, Hong Kong was the second largest importer of raw tortoiseshell, importing over 100,000 kg that year (Table 4). Between January and October 1979, Japan imported 1,976 kg of raw shell from Hong Kong. As mentioned above, the United Kingdom imports large quantities of meat from CTF.

FRANCE (August 1978)

Prior to its ratification, France was a major importer of worked tortoiseshell and regularly imported raw shell (Table 4). To ensure that this trade could continue, France placed a reservation on both green and hawksbill turtles. There are reports from Mexico that turtle leather is still exported to France; these skins are probably from olive ridleys, a species on which France did not place a reservation.

ITALY (December 1979)

Italy is a major center for the worked tortoiseshell trade and regularly imports raw tortoiseshell (Tables 4 and 6). Italy is also the European center of the turtle skin and leather trade, most of which is imported from Ecuador. Italy ratified CITES with a reservation on the green turtle but they import skins mainly from the olive ridley and tortoiseshell from the hawksbill.

UNITED STATES (July 1975)

The United States has been trying to control imports of tortoiseshell since 1970, when the hawksbill turtle was placed on the U.S. Endangered Species List, but it appears that some products are still getting into the country. India exported almost 3,000 kg of raw tor-

toiseshell to the United States in 1976, 1,656 kg in 1977, and 9,673 kg in January and February 1978 (Tables 2a and 2b). Taiwan exported worked tortoiseshell to the United States in both 1977 and 1978. Several shipments of turtle shell products were seized during these years, but the amounts do not appear sufficient to account for the above volume.

Other Possible Infringements

Over the last 3 years, other parties (Costa Rica, Brazil, Canada, Sri Lanka, Madagascar, Australia, and Papua New Guinea) have either exported or imported small quantities of raw or worked tortoiseshell. In general, these amounts are insignificant compared to the volume of the international tortoiseshell trade. Thailand and the Peoples' Republic of China have recently expressed their intention to ratify CITES. Thailand is one of the major exporters of raw tortoiseshell; if it ratifies, much pressure could be removed from the hawksbill turtles of Southeast Asia, especially if Indonesia and India, both CITES members, were to fully enforce CITES also. Mainland China is involved primarily in the worked tortoiseshell trade, although it regularly supplies Japan with small quantities of raw tortoiseshell.

Much could be done to reduce the commercial exploitation of marine turtles by improving the enforcement of legislation in countries which are parties to CITES. In 1977, 6 party states or governed territories (India, Ecuador, Australia, Costa Rica, Belize, and Puerto Rico), where CITES had already come into force, exported an estimated 121,034 kg or 38 percent of world exports of raw tortoiseshell (Table 1). Seven party states or governed territories (Hong Kong, the United States, West Germany, Nepal, Australia, Switzerland, and the United Kingdom) imported an estimated 64,746 kg or 24 percent of world imports of raw tortoiseshell (Table 4). Data for fewer countries are available for 1978, but at least 6 party states (India, Seychelles, Nicaragua, Malaysia, Costa Rica, and Puerto Rico) exported over 22,000 kg of raw tortoiseshell. In the first 10 months of 1979, at least 13 percent (8,989 kg) of Japanese imports of raw tortoiseshell came from CITES parties, and over 70 percent of Japanese raw turtle skins were from CITES countries (Ecuador and Pakistan).

Conclusions

The present size of the international tortoiseshell trade gives considerable cause for alarm. Although European countries may still play an important role in the worked tortoiseshell trade, the Far Eastern countries, particularly Japan, Hong Kong, and Taiwan, are the main consumers. The Japan Tortoise Shell Association (JTSA) concluded in its report (1973) that if Japan reduced its volume of imports, exporting countries would automatically lower their catch of hawksbills. Unfortunately, no attention has been paid to this recommendation. The status of nesting populations of hawksbills in Southeastern Asia is virtually unknown (Ross 1979), but, in 1973, the JTSA reported that people involved in the trade claimed signs of hawksbill depletion.

The hawksbill has probably never maintained a high density, and few large populations are known (Ross 1979). Ross lists a number of priority areas for this species, many of which correspond to the regions of greatest exploitation: India, Thailand, the Philippines, the Seychelles, and the Caribbean. The large exports of tortoiseshell from Fiji suggest that the Pacific should also be made a priority region. In 1972, Bustard reported that hawksbills were already depleted in Fiji.

Data in this paper further stress the importance of the recommendations made by Ross (1979) for the hawksbill: "An effective ban on the international trade in this species is an absolute necessity for its survival. Immediate surveys and rescue programs are needed for the areas of major exploitation."

A second major threat to the hawksbill is the flourishing souvenir trade in many tropical regions. More effort should be made to inform tourists from developed countries of the detrimental effects of buying sea turtle articles. Controls on the import of such souvenirs should be strictly enforced.

Olive ridleys are seriously threatened throughout their range. The main pressures from trade are on the eastern Pacific populations on the west coast of Central America. Their habit of nesting in huge concentrations means that they are particularly vulnerable to intense commercial exploitation.

Mexico has the largest populations of olive ridleys—an estimated 485,000 adults in 1978 (Márquez, Villanueva, and Peñaflores 1976)—but, due to exploitation for commercial trade, the population is decreasing at an alarming rate. It has been estimated that, if this take continues unabated, the olive ridley rookery would cease to exist on the Pacific coast of Mexico by 1985 (Felger and Cliffton 1978).

In Ecuador, the CITES Management Authority is at odds with the Fisheries Department which controls legislation relating to sea turtles. The companies processing turtles feel that the ridleys are a migratory species, and, if they do not utilize their products, other neighboring countries will (Green and Ortiz, this volume). While most of the world's attention is focused on Mexico and their large-scale killing of olive ridleys, Ecuador actually currently takes more than twice as many.

The major pressure on green turtles may be the take of adults and eggs for local consumption. Although little data are available on international trade in green turtles products, the exploitation of this species in Mexico gives cause for concern. If the commercial take continues unabated, the green turtle rookery nesting on the Pacific coast of Mexico is likely to be extinct by 1980 (Felger and Cliffton, 1978). In Southeast Asia, green turtles may be threatened by the market in Japan for skin and possibly for meat. If other areas exporting green turtle products were to close as a result of stronger controls and new legislation, Southeast Asia could become a more important supplier for this trade.

Acknowledgments

A review of the international trade in sea turtles demands an effort of international dimensions. Many people were contacted in our quest for data and yet more data—scientists, government officials, and conservationists. This paper could not have been completed without the help and patience of all concerned. We would like to thank Dr. F. Wayne King, who suggested we undertake this task. We are also particularly grateful to Clark Bavin, Chief of the Law Enforcement Section of the U.S. Fish and Wildlife Service, and to Peter Sand, former Secretary General of CITES. Professor Archie Carr, Peter Pritchard, George Balazs, Richard Felger, Kim Cliffton, Henry Reichart, Joop Schulz, René Honegger, R. Márquez, F. Ortiz Crespo, Ken Dodd, J. Mortimer, P. Ross, J. Frazier, N. Mrosovsky, I. Uchida, and many others generously shared their field data and reviewed the TRAFFIC (U.S.A.) first drafts. A special thanks goes to John Burton, who edited the final draft of the paper and coordinated the TRAFFIC (International) contribution. The personnel of both TRAFFIC offices provided invaluable help in typing and preparation of the manuscript: Shirley Bennett and Hilly Boorer in London and Cara Worthington, Laura Grady, and Zoe Combs in Washington. To all a very grateful thank you.

Copies of original data deposited in offices of IUCN Wildlife Trade Monitoring Unit, Cambridge, U.K.
A more detailed, 84 page sea turtle trade report is available from TRAFFIC(U.S.A.).

Literature Cited

Bhagwati, J.
1974.   *Illegal Transactions in International Trade.* Amsterdam/Oxford: North Holland Publishing Company.
Cato, J. C.; F. J. Prochaska; and P. C. H. Pritchard
1978.   An analysis of the capture, marketing and utilization of marine turtles. Manuscript, 119 pp.
Felger, R. S., and K. Cliffton
1978.   Conservation of the sea turtles of the Pacific Coast of Mexico. IUCN/WWF Report no. 1471. Manuscript.

DAVID MACK

Sea Turtle Conservation Strategy, Action Plan, and Action Projects

Sea Turtle Conservation Strategy

The first draft of the Conservation Strategy *was written by David Ehrenfeld. The draft was revised during the Conference to the version presented here.*

Situation and Objectives

Few groups of animals are more valuable and magnificent and at the same time more misused than sea turtles. Able to serve as a source of protein for coastal peoples in the tropics, they have been overexploited most frequently to feed, clothe and adorn the wealthy in Europe, North America, and eastern Asia. Populations are being lost through land development that destroys nesting beaches, through reef destruction, through the accidental drowning of turtles in trawl nets, and through the failure of states to join together to protect species that migrate from areas under one coastal jurisdiction to others. Even states intent on managing the resource wisely have destroyed sea turtle populations by developing management plans that ignored the biological needs of the species. Very few populations of sea turtles remain undiminished. The majority are depleted. Many are extinct. Six of the seven species are endangered.

The objective of this strategy is to develop conservation action based on the biology of the species that will return sea turtles to former abundance while allowing controlled exploitation for the benefit of generations of humans yet to come.

The Problem

The fate of sea turtles in the modern world is being determined by the interaction of many factors. These include: 1) the use of sea turtles as food by peoples who live where sea turtles are found; 2) the use of sea turtle products in local commerce (for example, sea turtle eggs sent to local markets); 3) the international trade in sea turtle products; 4) the differing attitudes toward conservation in different countries; 5) the incidental destruction of sea turtles that occurs during the fishing of other species; 6) the effects of nesting beach alteration or destruction; 7) the effects of marine and land-based pollution; and 8) the natural recovery rates of the various sea turtle populations under different conditions of exploitation and incidental stress. The biological constraint (8) is in turn determined by such variables as growth rate, food resources, migratory habits, the fixity of nesting behaviors (including preference for certain nesting sites) and others.

Of these eight factors (there may be more) that determine the fate of sea turtles, only one, the biological factor, is non-negotiable in a conservation strategy. Sea turtles, even the most resilient of the species, are neither shrimp nor herring. They mature very slowly compared with most commercially important species, and when mature their reproduction is vulnerable to disruption by many kinds of human activity in addition to ordinary turtle fishing. Among other widely exploited marine species, only the great whales, and possibly the sturgeons, show similar biological constraints on exploitation. In determining a conservation strategy, this ultimate limitation must be kept constantly in mind.

Sea Turtle Conservation Policy

This document sets forth, in outline format, policy considerations for the conservation of sea turtles.

I. *Habitat Protection*
   Habitat conservation can be achieved through a variety of management techniques. These may include the creation of protected areas such as national parks or reserves, management efforts, or simple limitation of access or activities in specific areas at specific times. Management techniques need to be carefully evaluated for particular areas so the measures selected are most appropriate. Habitats that should be protected are:
   A. Terrestrial Habitats
      1. concentrated nesting beaches
      2. diffuse nesting beaches
      3. basking sites
   B. Aquatic Habitats
      1. internesting areas
      2. migration routes
      3. feeding grounds
      4. hibernacula
II. *Management* [Considerations under Eggs, Hatchlings, Adults and Subadults listed in order of priority or preference.]
   A. Eggs[1]
      1. No intervention other than protection.
      2. Criteria for intervention—intervention is justifiable when hatching rate is reduced by

         a. heavy predation
         b. heavy human exploitation
         c. physical damage to nesting beach
      3. Types of intervention—the least manipulative techniques should be used.
         a. protect eggs *in situ*, control of predation
         b. transplant to adjacent hatch sites
         c. remove to hatcheries
   B. Hatchlings[1]
      1. Protection of *in situ* nests—limit beach traffic and disturbance at vulnerable preemergence and emergence stage
      2. Immediate release of hatchery hatchlings
      3. Retention of hatchlings for headstarting
      4. Removal to safe habitat (e.g., airlifting beyond oil spills)
   C. Adults and Subadults
      1. Complete protection and prevention of interference with reproductive activities on nesting beaches, and in internesting habitats (see also sections VI. Conservation Education and VII. Legislation, below)
      2. Prevention, reduction and control of exploitation[2] in
         a. migratory routes
         b. feeding grounds
         c. hibernacula
III. *Control of Exploitation*
   One goal of conservation is the rational sustained use of wildlife for the greatest benefit of humans now and in the future. Since overexploitation is responsible for the endangerment and extinction of many populations of marine turtles, maximum control of exploitation is mandated.
   A. Commercial
      1. As long as sea turtles remain endangered, the ending of commercial exploitation of all sea turtle products is a long-range goal or ideal of the conservation strategy. We do not anticipate,

---

1. Although they may be of great value, the more manipulative techniques (removal of eggs or hatchlings, colony transplantation, headstarting) are unproven techniques and should not be applied to a substantial portion of the eggs and hatchlings of a given colony. Tests of the success of manipulative efforts should be a part of every operation. In all manipulations, efforts should be made to keep conditions as natural as possible (e.g., natural temperature regimes for eggs, exposure of newly emerged hatchlings to natural sensory imprints from the beach).

2. Endangered and declining populations deserve complete protection through prevention of exploitation in all habitats.

however, that this goal will be achieved quickly. An end to international trade in all sea turtles and their products was mandated by placement of the species (with the exception of some Australian-Papua New Guinea populations) on Appendix I of the Convention on International Trade in Endangered Species of Wild Fauna and Flora (CITES) in 1973 and 1976. But because many of the principal international trading nations are not Party to the Treaty or even though they are Party to it, have taken reservations for sea turtles, or do not adequately implement it, the Convention has had only limited effect. At the moment, the highest priority should be given to ending:

a. The leather trade. This is a new industry whose demise would not have any major undesirable cultural or economic side effects. The present leather trade constitutes an intolerable drain on the sea turtle populations, especially those of *Lepidochelys olivacea* and *Chelonia mydas*. Current world trade should be terminated and all measures taken to achieve this end.

b. The trade in tortoiseshell. The trade in tortoiseshell should cease in those countries where it has no special traditional cultural significance. Those countries where tortoiseshell has a cultural value (e.g., in marriage ceremonies) should be encouraged to preserve and recycle antique supplies, to promote the use of synthetic substances, and with all dispatch to phase out the importation of new material.

c. Eggs collected for sale in distant markets. Eggs should be collected only for noncommerical consumption—and then only in those cases where a program is in effect to ensure that the great majority of eggs from that beach will be left to hatch, and that hatching will be under conditions as natural as possible. Conservation Education (see VI, below) should be used to counter the myths about special properties of eggs, in those countries where these superstitions are a cause of high commercial demand.

d. Trade in stuffed juvenile sea turtles. This totally unnecessary luxury trade is having a serious impact on populations of *Eretmochelys imbricata*. It should cease and all measures should be taken to achieve this end.

After the demonstrated recovery of abundance of sea turtles, some level of exploitation may be possible. However, any exploitation program must be based on the best available biological information and must be in accordance with national and international law.

B. Noncommercial Hunting

1. Noncommercial hunting is defined as a traditional way of obtaining food practiced by aboriginal peoples who are not yet part of a cash economy or technological society. In this context, noncommercial hunting can be a valid activity, especially when it is carried out so as to have a minimum impact on turtle populations. Nevertheless, there are some turtle populations that are endangered even by legitimate noncommercial hunting, and in those cases techniques of self-regulation and biologically sound conservation practices should be encouraged (see I. Habitat Protection, II. Management, and V. Research and Population Assessment sections). Where the noncommercial hunting of sea turtles is valid, subsistence users have first right to the resource.

C. Farming

In addition to the prime objective of marketing sea turtles raised under artificial and/or semiartificial conditions, farming has been claimed by some to provide incidental conservation benefit by relieving the commercial pressure on wild animals. Others feel that such operations create the risk of increasing pressures on wild populations.

1. Before the benefits and risks of commercial turtle culture can be fully evaluated, more data are needed, as follows:

a. The feasibility of complete, closed-cycle farming, with no dependence on wild populations (either eggs or adult breeders) should be studied. "Feasibility" refers to both biological and economic factors.

b. The considerations that determine the minimum (and possibly maximum) sized operation that is commercially feasible ought to be ascertained.

c. The impact of commercial turtle culture (farming and ranching) on prices of turtle products, on the creation of new markets, on the capture of turtles from wild populations, and on the trade in products derived from wild-caught sea turtles should be evaluated.

2. In the absence of definitive answers from the above inquiries, the following cautions are necessary:

a. Commercial mariculture must be in conformity with all applicable conservation regulations and laws, whether local, national, regional or international.

b. Care should be taken that special legal provisions and exemptions for farmed products are not misused by importers and exporters of wild turtle products.

c. Any effort by commercial mariculture interests to develop markets for new turtle products or to create demand for turtle products where it did not previously exist is insupportable.

d. The establishment of new commercial turtle "farms" must be discouraged until it is certain that such operations will not cause, directly or indirectly, a further decline in turtle populations.

IV. *Incidental Catch*

Incidental catch is a major threat to many sea turtle populations and must be eliminated or reduced to very low levels.

A. All countries should be prepared to establish restricted fishing zones in areas of high turtle concentration (as has been done by Mexico, near Rancho Nuevo, and by the United States, near Cape Canaveral).

B. The development of fishing techniques and equipment that preclude the incidental take of sea turtles should be given high priority. This technology should be made freely available to all states.

C. Information concerning the magnitude of the incidental catch of sea turtles is sorely needed. The industries involved in this incidental catch should be encouraged to assist in the gathering of information.

D. International fisheries commissions should address the problem of incidental catch in the framing of their regulations. If necessary, amendments should be promoted for international fishery conventions to give specific jurisdiction to fishery commissions over nontarget species.

E. Turtles which remain alive after being incidentally captured in fishing nets should be resuscitated and released.

F. The deliberate mutilation and killing of sea turtles during commercial fishing for other species must be ended.

V. *Research and Population Assessment*

A. Data on the location, and estimated or census-determined sizes of all populations of sea turtles is needed. Except in the case of *Lepidochelys kempi*, which exists as a single population, it is not obvious that there is any value in devoting time to estimating the population sizes of entire species.

B. Information on all aspects of the basic biology of sea turtles is needed. Of special relevance to conservation is information about growth rates, complete life histories, population dynamics (reproductive rate, mortality rate, and age at sexual maturity), phylogenetic and taxonomic relationships of different populations, and effective tagging methods.

C. Important issues of management techniques include testing the biological effectiveness of restocking, transplanting, and headstarting programs, and studying the effects of incubation temperatures and other environmental conditions on sex determination.

VI. *Conservation Education*

Conservation education in different countries will be enhanced through cooperation of local conservation organizations and agencies. Provisions should be made to supply them with information about sea turtles so that they can:

A. Organize their own political action and educational campaigns

B. Perform market surveys and gather information about trade in sea turtle products as well as local consumption of these products

C. Organize tagging programs and make surveys of activity at nesting beaches

D. Educate coastal people to identify the different kinds of sea turtles and to aid in the gathering of information about them

E. Develop recommendations for children's books (including parts of school texts), comic strips, and posters in various languages, on the subject of the plight of local sea turtles, and the value of a wildlife heritage

F. Develop survey teams that would census and salvage turtles that had washed up on the beach and, when possible, determine the cause of death

G. Maintain records about sea turtle populations and trade in sea turtle products, which would facilitate year-to-year comparisons

VII. *Legislation*

A. National

1. A worldwide systematic inventory of turtle conservation laws is needed to determine where gaps in coverage exist and what the priorities for action should be.

2. Where gaps exist, comprehensive conservation legislation (dealing with exploitation and habitat protection) should be enacted and implemented.

3. Effective mechanisms for enforcement of legislation should be developed. These should emphasize the development of strong enforcement techniques, and the training of effective conservation officers, drawn from the people among whom they would work. To facilitate control of international commerce, points of entry for such commerce into a country should be limited to those which can be staffed with trained officers.

4. Attention should be given to the strengthening of penalties for the breach of national legislation to reflect the severity of violations.

B. International

1. All states that have not already done so should become Party to the Convention on the International Trade in Endangered Species of Wild Fauna and Flora (CITES) without reservation. States Party to CITES which have taken reservations for sea turtles should withdraw those reservations. All Parties to CITES should fully implement their obligations with vigor.

2. All states under whose jurisdiction sea turtles pass any part of their life cycle and any states that exploit sea turtles on the high seas should enter into cooperative conservation programs for turtles, and in particular those states should become party to regional or umbrella conventions as the framework for development of necessary international cooperation. The Convention for the Conservation of Migratory Species of Wild Animals represents a useful effort to develop an umbrella convention applicable to sea turtles.

3. Existing regional conservation conventions should be strengthened and implemented (e.g., Western Hemisphere, African, South Pacific Conventions).

VIII. *Cooperative Efforts*

The exchange of information and the development of joint conservation programs among the many disparate and often isolated organizations and states (e.g., governmental agencies and nongovernmental organizations and adjacent range states) should occur.

Implementation of the Strategy

A Standing Committee should be established to monitor and facilitate the further development and the implementation of the Sea Turtle Conservation Strategy.

This Committee should be associated with the Marine Turtle Specialist Group of the Survival Service Commission of the International Union for Conservation of Nature and Natural Resources (IUCN), and should include representatives from the various regions of the world. The IUCN and the World Wildlife Fund are requested to accept responsibility for the overall coordination of this Standing Committee and the active cooperation of the various elements of the IUCN, including the Traffic Specialist Group, the Commission on National Parks and Protected Areas, and the Commission on Environmental Policy, Law, and Adminstration is essential.

International and national nongovernmental organizations should assist with implementing the Strategy, as appropriate, and especially with public information and education and with the promotion of necessary governmental action.

Participation in the Action Plan by governmental agencies, and particularly those involved with marine turtle research and conservation, is requested, because such participation is essential to the successful implementation of the Action Plan. The United Nations Environment Program and the United Nations Food and Agriculture Organization are encouraged to provide financial and programmatic support to this global conservation program.

For the purpose of preparing a report assessing the progress made in implementing the Strategy, the Standing Committee should meet with the IUCN Survival Service Commission at its meeting immediately prior to the 3rd Conference of the Parties to CITES, in the first quarter of 1981.

Recent Advances in Sea Turtle Biology and Conservation, 1995

J. D. Miller
Queensland Department of Environment and Heritage
P.O. Box 5391
Townsville
Queensland 4810 Australia

Nesting Biology of Sea Turtles

The literature on the reproductive biology of marine turtles is immense (e.g., Hirth 1971; Pritchard 1971; Anon. 1980; Bjorndal 1982; Witzell 1983; Miller 1985; Dodd 1988; Groombridge and Luxmoore 1989; Márquez M. 1990; National Research Council 1990). Much of the recent literature is contained in the Proceedings of the Annual Symposia on Sea Turtle Biology and Conservation (NOAA Technical Memoranda).

Marine turtles share a generalized life cycle that includes iteroparous reproduction (Hirth 1980; National Research Council 1990) with the possible exception of the Kemp's ridley, stereotyped nest-building behavior (Hendrickson, this volume), laying of relatively large numbers of eggs several times during the reproductive period (Hirth 1980; Van Buskirk and Crowder 1994), and relatively strong attachment to particular locations for nesting (e.g., Bjorndal et al. 1985; Limpus et al. 1992), but inter- and intra-specific variation exists (Van Buskirk and Crowder 1994).

Determining the details of the nesting biology of sea turtles requires the identification of individual turtles, usually through tagging. The position of application on the flipper, habitat (i.e., nesting area/foraging area), type of tag (monel, inconel, titanium), and the species of turtle tagged affect the rate of tag loss (Limpus 1992). With proper assessment of tag loss, the probability of recapturing tagged turtles increases (Limpus et al. 1994).

Nesting Biology

The nesting biology of marine turtles is deceptively simple. Mating occurs during a relatively short female receptive period in the vicinity of the nesting beach (Owens 1980; National Research Council 1990); the individual's mating season is completed before egg laying commences.

The nesting process is similar among all species of sea turtles. Several behavioral sequences have been described but the real differences are small (Hendrickson, this volume). The reproductive and ovipositional cycles are regulated by changes in specific serum gonadotropins and gonadal steroids (Guillette et al. 1991; Wibbels et al. 1992).

Reproductive Output

All sea turtles lay several clutches during a nesting season (Hirth 1980; Van Buskirk and Crowder 1994). Determining the number of times a turtle nests during a season is important, particularly if such data are averaged and used in calculations to estimate the number of female turtles in the population (Pritchard 1990). However, the number of clutches laid is often derived from incomplete coverage of the nesting season or nesting area, or loss of the individual from the nesting group (Bjorndal et al. 1985).

The number of eggs laid per clutch varies among the species (Hirth 1980; Van Buskirk and Crowder 1994). Smaller species typically lay more, smaller eggs than larger species, with two exceptions: flatback turtles typically lay about 50 large eggs and leatherback turtles lay only about 100 large eggs.

The embryological development of Cheloniidae and Dermochelyidae sea turtles (based on green, hawksbill, loggerhead, flatback and leatherback turtles) is similar in detail up to stage 22, after which generic and species specific characters become increasingly evident (Miller 1985).

Clutches of sea turtle eggs typically have a high hatching success (80% or more) unless external factors (e.g., predation, environmental change, microbial infection) intervene (National Research Council 1990). To assess the reproductive output of a nesting population, the emergence success should be calculated using a large number of nests including nests that do not produce hatchlings.

Periodicity

The mean remigration interval reported for sea turtles is 2, 3, 4, or 5 years, with a range of from 1 to 9 or more years (Dodd 1988; Limpus et al. 1994; Van Buskirk and Crowder 1994) depending on the species. The actual remigration interval for the population may be longer; the discrepancy may result from tag loss, incomplete survey coverage, too short of a study period, or some as yet undocumented aspect of their biology.

Philopatry

Genetic studies and tagging studies provide evidence that sea turtles exhibit a high degree of philopatry in subsequent nesting seasons (Limpus et al. 1984, 1994; Bjorndal et al. 1985; Bowen et al. 1992; Norman et al. 1994). However, not all turtles return to the nesting site in subsequent seasons (Limpus et al. 1994); some utilize other nesting sites in the general area (Dodd 1988). The determination of philopatry depends on identification of the individual and the coverage of the nesting season and nesting beach.

Site Fixity

Within a season, a turtle will tend to renest in relatively close proximity (0–5 km) during subsequent nesting attempts, although a small percentage may utilize more distant nesting sites in the general area (Limpus et al. 1984; Bjorndal et al. 1985; Dodd 1988; Eckert et al. 1989). The ecological consequences of this behavior require elucidation.

Literature Cited

Anonymous
1980. Behavioral and reproductive biology of sea turtles. *American Zoologist* 20:485–617.

Bjorndal, K. A., editor
1982. *Biology and Conservation of Sea Turtles*. Washington D.C.: Smithsonian Institution Press.

Bjorndal, K. A.; A. Carr; A. B. Meylan; and J. A. Mortimer
1985. Reproductive biology of the hawksbill turtle *Eretmochelys imbricata* at Tortuguero, Costa Rica, with notes on the ecology of the species in the Caribbean. *Biological Conservation* 34:353–368.

Bowen, B. W.; A. B. Meylan; P. Ross; C. J. Limpus; G. H. Balazs; and J. C. Avise
1992. Global population structure and natural history of the green turtle (*Chelonia mydas*) in terms of matriarchal phylogeny. *Evolution* 46:865–881.

Dodd, C. K.
1988. Synopsis of the Biological Data on the Loggerhead Sea Turtle *Caretta caretta* (Linnaeus 1758). U.S. Fish and Wildlife Service, Biological Report 88(14). 110 pp.

Eckert, K. L.; S. A. Eckert; T. W. Adams; and A. D. Tucker
1989. Internesting migrations by leatherback sea turtles (*Dermochelys coriacea*) in the West Indies. *Herpetologica* 45:190–194.

Groombridge, B., and R. Luxmoore
1989. *The Green Turtle and Hawksbill (Reptilia: Cheloniidae) World Status, Exploitation and Trade*. Lausanne, Switzerland: CITES Secretariat.

Guillette, L.; K. Bjorndal; A. Bolten; T. Gross; B. Palmer; B. Witherington; and J. Matter
1991. Plasma estradiol-17b, progesterone, prostaglandin F, and prostaglandin E2 concentrations during natural oviposition in the loggerhead turtle (*Caretta caretta*). *General and Comparative Endocrinology* 82:121–130.

J. D. MILLER

Hirth, H. F.

1971. Synopsis of biological data on the green turtle *Chelonia mydas* (Linnaeus) 1758. *FAO Fisheries Synopsis* FIRM/S85:75p.

1980. Some aspects of the nesting behavior and reproductive biology of sea turtles. *American Zoologist* 20:507–524.

Limpus, C. J.

1992. Estimation of tag loss in marine turtle research. *Wildlife Research* 19:457–469.

Limpus, C. J.; P. Eggler; and J. D. Miller

1994. Long interval remigration in eastern Australian *Chelonia*. In *Proceedings of the Thirteenth Annual Sea Turtle Symposium*, compilers B. A. Schroeder and B. E. Witherington, pp. 85–88. NOAA Technical Memorandum NMFS-SEFSC-341.

Limpus, C. J.; A. Fleay; and M. Guinea

1984. Sea turtles of the Capricornia Section, Great Barrier Reef. In *The Capricornia Section of the Great Barrier Reef: Past, Present, and Future*, eds. W. T. Ward and P. Saenger, pp. 61–78. Brisbane: Royal Society of Queensland and Australian Coral Reef Society.

Limpus, C. J.; J. D. Miller; C. J. Parmenter; D. Reimer; N. Mclachlan; and R. Webb

1992. Migration of green (*Chelonia mydas*) and loggerhead (*Caretta caretta*) turtles to and from eastern Australian rookeries. *Wildlife Research* 19:347–358.

Márquez M., R.

1990. FAO Catalogue. Sea turtles of the world. *FAO Fisheries Synopsis* 125:1–81.

Miller, J. D.

1985. Embryology of marine turtles. In *Biology of the Reptilia, Vol. 14 (Development A)*, eds. C. Gans, F. Billett and P. F. A. Maderson, pp. 270–328. New York: John Wiley and Sons.

National Research Council

1990. *Decline of the Sea Turtles: Causes and Prevention*. Washington, D.C.: National Academy Press.

Norman, J.; C. Moritz; C. Limpus; and R. Prince

1994. Population genetics as a tool for managing marine turtle populations. In *Proceedings of the Australian Marine Turtle Conservation Workshop*, compiler R. James, pp. 92–105. Canberra: Australian National Parks and Wildlife Service.

Owens, D.

1980. The comparative reproductive physiology of sea turtles. *American Zoologist* 20:549–563.

Pritchard, P. C. H.

1971. The leatherback or leathery turtle, *Dermochelys coriacea*. *IUCN Monograph Marine Turtle Series* 1:1–39.

1990. Kemp's ridleys are rarer than we thought. *Marine Turtle Newsletter* 49:1–3.

Van Buskirk, J., and L. Crowder

1994. Life-history variation in marine turtles. *Copeia* 1994:66–81.

Wibbels, T.; D. Owens; P. Licht; C. Limpus; P. Reed; and M. Amoss

1992. Serum gonadotropins and gonadal steroids associated with ovulation and egg production in sea turtles. *General and Comparative Endocrinology* 87:71–78.

Witzell, W. N.

1983. Synopsis of biological data on the hawksbill turtle, *Eretmochelys imbricata* (Linnaeus, 1766). *FAO Fisheries Synopsis* 137:1–78.

Blair E. Witherington
Florida Marine Research Institute
Tequesta Field Laboratory
19100 Southeast Federal Highway
Tequesta, Florida 33469 USA

Hatchling Orientation

As hatchlings, sea turtles begin a life of astonishing migratory feats. When the first edition of this volume was published, movements of sea turtles at many life-history stages were well documented, but an understanding of orientation cues and guidance mechanisms was just beginning. At that time, Archie Carr described the lack of tenable theories to explain migratory abilities as "an embarrassment to science" (this volume). In the years since this prompting, substantial progress has been made toward redemption.

Sea-finding Orientation

The orientation of hatchling sea turtles from nest to sea is a robust behavior that lends itself well to experimentation. Capitalizing on the strong propensity of hatchlings to move in the brightest direction, N. Mrosovsky and colleagues conducted numerous studies to develop a model describing how orienting hatchlings use light cues (Mrosovsky and Kingsmill 1985). Their experiments with partially blindfolded hatchlings showed that their turning toward brightness is initiated by light angle and intensity differences that are "measured" at the retina. In this measurement, a hatchling's eyes seem to function as if each were a complex array of light detectors. With this orientation mechanism, hatchlings average brightness input over a wide horizontal range (~180°) and a narrow vertical range (~10°) (Witherington 1992).

The perception of brightness is dependent upon light wavelength and intensity. Experiments have shown that independent of intensity, short-wavelength light (near-ultraviolet, blue, green) is more attractive to hatchlings than long-wavelength light (yellow, red) (Witherington and Bjorndal 1991a). Unlike other species, loggerhead (*Caretta caretta*) hatchlings have a unique tendency to orient away from monochromatic yellow light (Witherington and Bjorndal 1991b).

577

In addition to brightness, horizon silhouette and/or shapes associated with the horizon also influence hatchling orientation (Salmon et al. 1992; Witherington 1992). Hatchlings orient away from vertical stripes and elevated silhouettes, generally irrespective of brightest direction. In highly directed light fields, however, with brightness in one direction far exceeding that of competing directions, hatchlings move toward brightness. Beach conditions under which hatchlings orient seaward without the seaward direction being brightest are common. Under these conditions, hatchlings deprived of form vision by waxed paper cannot orient seaward (Witherington 1992).

Artificial sources near nesting beaches produce highly directed light fields that misdirect hatchlings (Verheijen 1985). Mortality from this behavioral disruption remains a substantial conservation problem. Although darkening beaches is the only complete solution, replacing problem light sources with low-pressure-sodium-vapor sources, which emit monochromatic yellow light, reduces effects on orienting hatchlings (Witherington and Bjorndal 1991b).

Orientation at Sea

After they enter the sea, hatchlings depend on cues other than light to lead them away from land (Salmon and Wyneken 1994). At least two sets of cues, wave direction and the geomagnetic field, direct swimming orientation. Swimming hatchlings orient into oncoming waves, which establishes seaward movement in the near-shore zone (Wyneken et al. 1990). Geomagnetic cues seem to be responsible for guiding a seaward course as hatchlings travel offshore (Lohmann and Lohmann 1994).

Laboratory studies show that hatchlings are guided by an inclination compass, like that of birds, rather than a polar compass (Light et al. 1993). An inclination compass functions by distinguishing the angle of the Earth's magnetic field lines, which vary with latitude, and can provide information both on poleward direction and degree of latitude.

Hatchlings acquire a magnetic directional preference during their initial swimming (Lohmann and Lohmann 1994). Once this preference is established in the laboratory, hatchlings will adjust their bearing predictably when the magnetic field around them is altered to imitate inclination angles at different latitudes (Lohmann and Lohmann, in press). Sea turtles are the first animals to show an ability to determine latitude magnetically. The magnetic sense in hatchlings has the potential to indicate both direction and position and may be the key to a guidance system used by sea turtles at all life-history stages.

Literature Cited

Light, P.; M. Salmon; and K. J. Lohmann
1993. Geomagnetic orientation of loggerhead sea turtles: evidence for an inclination compass. *Journal of Experimental Biology* 182:1–10.

Lohmann, K. J., and C. M. F. Lohmann
1994. Acquisition of magnetic directional preference in loggerhead sea turtle hatchlings. *Journal of Experimental Biology* 190:1–8.

In press. Detection of magnetic inclination angle by sea turtles: a possible mechanism for determining latitude. *Journal of Experimental Biology.*

Mrosovsky, N., and S. F. Kingsmill
1985. How turtles find the sea. *Zeitschrift fuer Tierpsychologie* 67:237–256.

Salmon, M., and J. Wyneken
1994. Orientation by hatchling sea turtles: mechanisms and implications. *Herpetological Natural History* 2:13–24.

Salmon, M.; J. Wyneken; E. Fritz; and M. Lucas
1992. Seafinding by hatchling sea turtles: role of brightness, silhouette and beach slope as orientation cues. *Behavior* 122:56–77.

Verheijen, F. J.
1985. Photopollution: artificial light optic spatial control systems fail to cope with. Incidents, causations, remedies. *Experimental Biology* 44:1–18.

Witherington, B. E.
1992. Sea-finding behavior and the use of photic orientation cues by hatchling sea turtles. Ph.D. dissertation, University of Florida.

Witherington, B. E., and K. A. Bjorndal
1991a. Influences of wavelength and intensity on hatchling sea turtle phototaxis: implications for sea-finding behavior. *Copeia* 1991:1060–1069.

1991b. Influences of artificial lighting on the seaward orientation of hatchling loggerhead turtles (*Caretta caretta*). *Biological Conservation* 55:139–149.

Wyneken, J.; M. Salmon; and K. J. Lohmann
1990. Orientation by hatchling loggerhead sea turtles *Caretta caretta* L. in a wave tank. *Journal of Experimental Marine Biology and Ecology* 139:43–50.

Alan B. Bolten
Center for Sea Turtle Research
Bartram Hall
University of Florida
Gainesville, Florida 32611 USA

George H. Balazs
National Marine Fisheries Service
2570 Dole Street
Honolulu, Hawaii 96822-2396 USA

Biology of the Early Pelagic Stage— The "Lost Year"

The life history of sea turtle hatchlings from the time they leave the nesting beach, enter the sea, and become part of the pelagic community until they return to coastal, benthic foraging habitats as juveniles, has been known as the "mystery of the lost year" (Carr, this volume). Research efforts to unravel the mystery of this early pelagic stage have focused on the North Atlantic and North Pacific loggerhead populations (*Caretta caretta*). Despite considerable effort, researchers have not found areas where pelagic-stage turtles can be consistently found for any other population or species. The flatback turtle (*Natator depressus*) may not have a true pelagic stage (Walker and Parmenter 1990).

The orientation behavior and sensory cues used by hatchlings first to find the sea and then to maintain direction once in the sea are reviewed by Witherington (this volume). Wyneken and Salmon (1992) have analyzed the swimming frenzy that occurs once the hatchlings have entered the sea. Variation in the Earth's magnetic field may provide the cues for the post-hatchlings to orient while in the pelagic current systems (Lohmann and Lohmann 1994, in press).

Archie Carr (1986, 1987a,b) hypothesized that the lost year turtles were associated with driftlines, convergences, and rips in the North Atlantic Gyre system. Carr (1986) suggested that loggerhead hatchlings from the southeastern USA rookeries become incorporated into the Gulf Stream Current and from there, those post-hatchlings that are in the eastern portion of the Gulf Stream become incorporated into the "Azorean" Current that carries them past the Azores, Madeira, Canary Islands and back again to the western Atlantic. The size frequency distribution of loggerheads in the eastern Atlantic complements the "missing" size classes in the western Atlantic (Carr 1986; Bolten et al. 1993) and was the first line of evidence that the turtles in the two regions belong to the same population. Mitochondrial DNA sequence patterns are being analyzed to confirm this relationship. In addition, the movement patterns that

Carr (1986) hypothesized within the North Atlantic Gyre system have been documented (Eckert and Martins 1989; Bolten and Martins 1990; Bolten et al. 1992a,b; Bjorndal et al. 1994). Satellite telemetry is being used to document specific movement patterns within the North Atlantic Gyre (Bolten et al., unpublished data). Movement of Atlantic turtles into the Mediterranean has been documented by tag returns (Manzella et al. 1988; Bolten et al. 1992a) and confirmed by genetic analyses (Laurent et al. 1993). Transoceanic movements for loggerheads in the North Pacific have been documented using genetic analyses (Bowen et al., in press).

Duration of the pelagic stage has important demographic implications. Preliminary results comparing growth rates estimated from recaptures and those estimated from length-frequency analysis suggest the "lost year" for the Atlantic loggerheads is more likely a "lost decade" (Bolten et al., in press). Zug et al. (in press) using skeletochronology, report a similar time period for the North Pacific loggerheads.

Jellyfish (e.g., *Pelagia noctiluca*) are the major natural diet component of pelagic loggerheads in the North Atlantic; in the North Pacific, the principal food sources are the neustonic coelenterate *Velella velella* and the gastropod *Janthina* sp. Young posthatchlings feed on a variety of invertebrates (including insects) that are associated with the *Sargassum* ecosystem (Richardson and McGillivary 1991; Witherington 1994).

Ingestion of and entanglement in marine debris (e.g., plastics, tar, and discarded fishing gear) have an impact on survivorship of pelagic populations (Balazs 1985; Carr 1987a,b; Witherington 1994). Incidental take in commercial fisheries (e.g., driftnets and longline fisheries) poses another major threat to pelagic turtles (Aguilar et al. 1992 and 1993 as summarized in Balazs and Pooley 1994; Wetherall et al. 1993; Balazs and Pooley 1994; Bolten et al. 1994).

Genetic markers may provide the necessary tools to link pelagic populations with specific rookeries. From this linkage, and through collaborative efforts with oceanographers, factors affecting distribution and movement patterns may be elucidated in the future.

Literature Cited

Balazs, G. H.
1985. Impact of ocean debris on marine turtles: entanglement and ingestion. In *Proceedings of the Workshop on the Fate and Impact of Marine Debris*, eds. R. S. Shomura and H. O. Yoshida, pp. 387–429. NOAA Technical Memorandum NOAA-TM-NMFS-SWFS-54.

Balazs, G. H.; S. Pooley; and 14 collaborators
1994. *Research Plan to Assess Marine Turtle Hooking Mortality: Results of an Expert Workshop Held in Honolulu, Hawaii, November 16–18, 1993.* NOAA Technical Memorandum NMFS-SWFSC-201.

Bjorndal, K. A.; A. B. Bolten; J. Gordon; and J. A. Camiñas
1994. *Caretta caretta* (loggerhead) growth and pelagic movement. *Herpetological Review* 25:23–24.

Bolten, A. B.; K. A. Bjorndal; and H. R. Martins
1994. Life history model for the loggerhead sea turtle (*Caretta caretta*) population in the Atlantic: Potential impacts of a longline fishery. In *Research Plan to Assess Marine Turtle Hooking Mortality: Results of an Expert Workshop Held in Honolulu, Hawaii, November 16–18, 1993*, pp. 48–55. NOAA Technical Memorandum NMFS-SWFSC-201.

In press. Life history model for the loggerhead sea turtle (*Caretta caretta*) in the Atlantic. *Proceedings of the First Symposium on the Fauna and Flora of the Atlantic Islands.* Madeira, Portugal: Museu do Funchal.

Bolten, A. B., and H. R. Martins
1990. Kemp's ridley captured in the Azores. *Marine Turtle Newsletter* 48:23.

Bolten, A. B.; H. R. Martins; K. A. Bjorndal; M. Cocco; and G. Gerosa
1992a. *Caretta caretta* (loggerhead) pelagic movement and growth. *Herpetological Review* 23:116.

Bolten, A. B.; H. R. Martins; K. A. Bjorndal; and J. Gordon
1993. Size distribution of pelagic-stage loggerhead sea turtles (*Caretta caretta*) in the waters around the Azores and Madeira. *Arquipélago* 11A:49–54.

Bolten, A. B.; C. Santana; and K. A. Bjorndal
1992b. Transatlantic crossing by a loggerhead turtle. *Marine Turtle Newsletter* 59:7–8.

Bowen, B. W.; F. A. Abreu-Grobois; G. H. Balazs; N. Kamezaki; C. J. Limpus; and R. J. Ferl
In press. Trans-Pacific migrations of the loggerhead turtle (*Caretta caretta*) demonstrated with mitochondrial DNA markers. *Proceedings National Academy of Sciences, USA.*

Carr, A.
1986. Rips, FADS, and little loggerheads. *BioScience* 36:92–100.

1987a. New perspectives on the pelagic stage of sea turtle development. *Conservation Biology* 1:103–121.

1987b. Impact of nondegradable marine debris on the ecology and survival outlook of sea turtles. *Marine Pollution Bulletin* 18:352–356.

Eckert, S. A., and H. R. Martins
1989. Transatlantic travel by a juvenile loggerhead turtle. *Marine Turtle Newsletter* 45:15.

Laurent, L.; J. Lescure; L. Excoffier; B. Bowen; M. Domingo; M. Hadjichristophorou; L. Kornaraki; and G. Trabuchet
1993. Étude génétique des relations entre les populations méditerranéenne et atlantique d'une tortue marine (*Caretta caretta*) à l'aide d'un marqueur mitochondrial. *Comptes Rendus de l'Académie des Sciences, Sciences de la vie* (Paris) 316:1233–1239.

Lohmann, K. J., and C. M. F. Lohmann

1994. Acquisition of magnetic directional preference in loggerhead sea turtle hatchlings. *Journal of Experimental Biology* 190:1–8.

In press. Detection of magnetic inclination angle by sea turtles: a possible mechanism for determining latitude. *Journal of Experimental Biology.*

Manzella, S. A.; C. T. Fontaine; and B. A. Schroeder

1988. Loggerhead sea turtle travels from Padre Islands, Texas, to the mouth of the Adriatic Sea. *Marine Turtle Newsletter* 42:7.

Richardson, J. I., and P. McGillivary

1991. Post-hatchling loggerhead turtles eat insects in *Sargassum* community. *Marine Turtle Newsletter* 55: 2–5.

Walker, T. A., and C. J. Parmenter

1990. Absence of a pelagic phase in the life cycle of the flatback turtle, *Natator depressa* (Garman). *Journal of Biogeography* 17:275–278.

Wetherall, J. A.; G. H. Balazs; R. A. Tokunaga; and M. Y. Y. Yong

1993. Bycatch of marine turtles in North Pacific high-seas driftnet fisheries and impacts on the stocks. *North Pacific Commission Bulletin* 53(III):519–538.

Witherington, B. E.

1994. Flotsam, jetsam, post-hatchling loggerheads, and the advecting surface smorgasbord. In *Proceedings of the Fourteenth Annual Symposium on Sea Turtle Biology and Conservation*, compilers K. A. Bjorndal, A. B. Bolten, D. A. Johnson, and P. J. Eliazar, pp. 166–167. NOAA Technical Memorandum NMFS-SEFSC-351.

Wyneken, J., and M. Salmon

1992. Frenzy and postfrenzy swimming activity in loggerhead, green, and leatherback hatchling sea turtles. *Copeia* 1992:478–484.

Zug, G. R.; G. H. Balazs; and J. A. Wetherall

In press. Growth in juvenile loggerhead seaturtles (*Caretta caretta*) in the North Pacific pelagic habitat. *Copeia.*

Scott A. Eckert
Hubbs-Sea World Research Institute
2595 Ingraham Street
San Diego, California 92109 USA

Telemetry and the Behavior of Sea Turtles

The modern era of microelectronics has provided unique insight into the movements of sea turtles. In this rapidly evolving field, many studies are aired only in symposia proceedings, and techniques are often obsolete at the time of publication. Prospective investigators should contact their active colleagues. White and Garrott (1990) is an indispensable reference.

Early attempts to electronically monitor movement used VHF radio transmitters which emit a signal at a preprogrammed rate (Murphy 1979; Chan et al. 1991; Liew and Chan 1993). Individual transmitters are identified by frequency or pulse rate and located using a receiver and directional antennae; transmitters can be modified for data transmission. Active investigators include myself (Hubbs-Sea World Research Institute; leatherbacks), Javier Alvarado (University of Michoacán, Mexico; East Pacific green turtles), and Steve Morreale (Cornell University) and André Landry (Texas A&M University) (Kemp's ridleys).

Instead of a radio signal, sonic transponders emit a moderate to high frequency sonic pulse detectable using a directional hydrophone (Ogden et al. 1983; Standora et al. 1984; Yano and Tanaka 1991). Transponders can also be modified to transmit non-locational data (e.g., temperature, depth). Caveats include short range and battery life; background noise can confuse interpretation. Active investigators include Ed Standora (State University College of Buffalo; loggerheads), Liew Hock Chark and Chan Eng Heng (Universiti Pertanian Malaysia; green turtles), and Joanne Braun and Sheryan Epperly (National Marine Fisheries Service, Beaufort; technique development).

Platform transmitter terminals (PTT) transmit data to an orbiting satellite. Life expectancy is highly variable depending on battery configuration. Loggerhead and leatherback studies have been published by Hays et al. (1991) and Keinath and Musick (1993), respectively. Active investigators include myself (leatherbacks), Steve Morreale (leatherbacks, Kemp's ridleys), Pamela Plotkin (Texas A&M University; olive ridleys), George Balazs (National Marine Fisheries

Service, Honolulu; green turtles), and Richard Byles (U.S. Fish and Wildlife Service, Albuquerque; several species).

Microprocessor data-loggers sample and record sensor data (e.g., velocity turbines, pressure transducers, electrical signals, ambient light, or thermocouples) (Eckert et al. 1989; Sakamoto et al. 1993). Multiple variables can be measured without compromising behavior, and resolution (accuracy) exceeds that provided by direct observation techniques. The disadvantage is that instruments must be recovered for data retrieval. Active investigators include myself (leatherbacks), Yasuhiko Naito (Japanese National Institute of Polar Research; loggerheads), and Robert Van Dam (Scripps Institution of Oceanography; hawksbills).

Literature Cited

Chan, E. H.; S. A. Eckert; H. C. Liew; and K. L. Eckert
1991. Locating the internesting habitats of leatherback turtles (*Dermochelys coriacea*) in Malaysian waters using radiotelemetry. In *Proceedings of the 11th International Symposium on Biotelemetry, Yokohoma, Japan*, eds. A. Uchiyama and C. J. Amlaner, Jr., pp. 133–138. Tokyo: Waseda University Press.

Eckert, S. A.; K. L. Eckert; P. Ponganis; and G. L. Kooyman
1989. Diving and foraging behavior of leatherback sea turtles (*Dermochelys coriacea*). *Canadian Journal of Zoology* 67:2834–2840.

Hays, G. C.; P. I. Webb; J. P. Hayes; I. G. Priede; and J. French
1991. Satellite tracking of a loggerhead turtle (*Caretta caretta*) in the Mediterranean. *Journal of Marine Biological Association of the United Kingdom* 71:743–746.

Keinath, J. A., and J. Musick
1993. Movements and diving behavior of a leatherback turtle (*Dermochelys coriacea*). *Copeia* 1993:1010–1017.

Liew, H. C., and E. H. Chan
1993. Biotelemetry of green turtles (*Chelonia mydas*) in Pulau Redang, Malaysia, during the internesting period. In *Biotelemetry XII: Proceedings of the 12th International Symposium on Biotelemetry, Ancona, Italy*, eds. P. Mancini, S. Fioretti, C. Cristalli and R. Bedini, pp. 157–163. Pisa, Italy: Editrice Universitaria Litografia Felici.

Murphy, T. M., Jr.
1979. Sonic and radio tracking of loggerhead turtles. *Marine Turtle Newsletter* 11:5.

Ogden, J. C.; L. Robinson; K. Whitlock; H. Daganhardt; and R. Cebula
1983. Diel foraging patterns in juvenile green turtles (*Chelonia mydas* L.) in St. Croix, United States Virgin Islands. *Journal of Experimental Marine Biology and Ecology* 66:199–205.

Sakamoto, W.; K. Sato; H. Tanaka; and Y. Naito
1993. Diving patterns and swimming environment of two loggerhead turtles during internesting. *Nippon Suisan Gakkaishi* 59:1129–1137.

Standora, E.; J. R. Spotila; J. A. Keinath; and C. R. Shoop
1984. Body temperatures, diving cycles and movement of a subadult leatherback sea turtle, *Dermochelys coriacea*. *Herpetologica* 40:169–176.

White, G. C., and R. A. Garrott
1990. *Analysis of Wildlife Radio Tracking Data*. San Diego: Academic Press Inc.

Yano, K., and S. Tanaka
1991. Diurnal swimming patterns of loggerhead turtle during their breeding period as observed by ultrasonic telemetry. *Nippon Suisan Gakkaishi* 57:1669–1678.

SCOTT A. ECKERT

Brian W. Bowen
BEECS Genetic Analysis Core
P.O. Box 110699
University of Florida
Gainesville, Florida 32611 USA

Molecular Genetic Studies of Marine Turtles

Molecular genetic analyses of marine turtles have provided much information over the last five years, and this body of conservation-oriented research will expand substantially over the next decade. Most of the research accomplished to date has utilized mitochondrial DNA (mtDNA). The idiosyncrasies of this maternally inherited genome may limit utility for some aspects of population genetics and molecular evolution, but these same features provide powerful applications in cases where maternal behaviors (such as natal homing) define population structure (Avise, in press). Nuclear DNA analyses are likely to yield many revelations in the decade to come, as indicated by Karl et al. (1992) and FitzSimmons et al. (1995). For a more comprehensive review of the papers summarized here, the reader is referred to Bowen and Avise (in press).

Natal Homing

Based on the nest site fidelity of adult females, Carr (1967) and others suggested that marine turtles nest on their natal beach. Hendrickson (1958) proposed an alternative scenario in which neophyte nesters follow experienced breeders to a nesting site and subsequently use this site for nesting effort. These alternate scenarios have been tested for three species with mtDNA analyses. Evidence from green turtles (Meylan et al. 1990; Bowen et al. 1992; Allard et al. 1994; Norman et al. 1994; Encalada et al., in review), loggerhead turtles (Bowen et al. 1993a), and hawksbill turtles (Broderick et al. 1994; Bass et al., in review) are concordant in supporting the natal homing hypothesis for these species. An important corollary of these findings is that nesting populations are distinct demographic units. Depletion of nesting aggregates will not be compensated by recruitment from other populations.

585

Phylogeography and Molecular Evolution

The mtDNA data indicate that green turtle rookeries within an ocean basin are demographically independent over ecological timescales but closely related in an evolutionary sense (Bowen et al. 1989; Encalada et al., in review). Subsequent analyses have extended this qualitative conclusion to loggerhead turtles (Bowen et al. 1993a) and hawksbill turtles (Bass et al., in review). Over intervals of thousands of years, changes in climate and coastal geography probably drive an ongoing process of rookery extinction and colonization. Furthermore, occasional lapses in natal homing must occur to allow the colonization of new nesting habitat. These processes may limit the development of deep evolutionary separations within each ocean basin. Despite this mixing, patterns of regional colonization may be inferred from mtDNA sequences (Encalada et al., in review).

In regard to the global phylogeography of cheloniid species, evidence from ridleys (Bowen et al. 1991), green turtles (Bowen et al. 1992) and loggerheads (Bowen et al., in press a) indicate that continental barriers have been of overriding importance. Deeper evolutionary history of marine turtles has also been analyzed with mtDNA sequence data. Findings of relevance to marine turtle biology include confirmation of a distant relationship between *Natator depressus* and other cheloniid species, the affiliation of *Eretmochelys* with the tribe Carettini rather than Chelonini, the genetic distinctiveness of *Lepidochelys kempi* from *L. olivacea*, and the paraphyly of *Chelonia mydas* with respect to the putative *C. agassizi* (Bowen et al. 1991, 1993b). Nuclear DNA and mtDNA analyses do not support the taxonomic distinction of *C. agassizi* (Bowen et al. 1992; Karl et al. 1992). Sequence divergences at intergeneric and interfamilial levels, when assessed against fossil-based separation times, indicate that marine turtle mtDNA evolves more slowly than under the "conventional" vertebrate molecular clock (Avise et al. 1992).

Mixed Stock Analysis of Feeding Populations

One outcome of the mtDNA analyses reviewed here is the discovery that many nesting populations of green, loggerhead, and hawksbill turtles contain unique haplotypes (Norman et al. 1994; Lahanas et al. 1994). These "endemic" nucleotide sequences, and corresponding haplotype frequency shifts between nesting populations, may be used to estimate the contribution of reproductive aggregates to feeding grounds or migratory corridors (Avise and Bowen 1994; Bowen et al., in press b).

In an analysis of mtDNA polymorphisms in juvenile green turtles, Lahanas et al. (in review) demonstrate that nesting colonies contribute to a Caribbean feeding habitat in cohorts which are roughly proportional to the size of source (breeding) populations and suggest that this feeding population represents a random mix of individuals from regional nesting populations. In contrast, Sears et al. (in press) conclude that loggerheads in a South Carolina feeding aggregate are not a random mix of individuals from regional nesting populations, but appear to be drawn at higher than expected frequency from the Georgia/South Carolina nesting population relative to the larger Florida nesting population. Broderick et al. (1994) demonstrate that two hawksbill foraging areas in northern Australian waters are genetically distinct from proximal nesting beaches, consistent with the hypothesis that hawksbill nesting cohorts overlap extensively in feeding areas (see also Bowen et al., in press c).

Molecular genetic markers have also been used to document unusually long migrations in juvenile loggerhead turtles. Laurent et al. (1993) demonstrated that mtDNA genotypes endemic to West Atlantic nesting colonies occur in Mediterranean feeding populations. Bowen et al. (in press b) demonstrate that haplotypes endemic to Japanese and Australian nesting populations occur in juvenile turtles off Baja California, indicating a developmental migration of over 10,000 km.

Literature Cited

Allard, M. W.; M. M. Miyamoto; K. A. Bjorndal; A. B. Bolten; and B. W. Bowen
1994. Support for natal homing in green turtles from mitochondrial DNA sequences. *Copeia* 1994:34–41.

Avise, J. C.
In press. Mitochondrial DNA polymorphisms and a genetics-demography connection of conservation relevance. *Conservation Biology.*

Avise, J. C., and B. W. Bowen
1994. Investigating sea turtle migration using DNA markers. *Current Opinion in Genetics and Development* 4:882–886.

Avise, J. C.; B. W. Bowen; E. Bermingham; A. B. Meylan; and T. Lamb
1992. Mitochondrial DNA evolution at a turtle's pace: evidence for low genetic variability and reduced microevolutionary rate in the Testudines. *Molecular Biology and Evolution* 9:457–473.

Bass, A. L.; D. A. Good; K. A. Bjorndal; J. I. Richardson; Z.-M. Hillis; J. Horrocks; and B. W. Bowen
In review. Testing models of female migratory behavior and population structure in the Caribbean hawksbill turtle, *Eretmochelys imbricata*, with mtDNA control region sequences.

BRIAN W. BOWEN

Bowen, B. W.; F. A. Abreu-Grobois; G. H. Balazs; N. Kamezaki; C. J. Limpus; and R. J. Ferl
In press b. Trans-Pacific migrations of the loggerhead sea turtle demonstrated with mitochondrial DNA markers. *Proceedings of the National Academy of Sciences, USA.*

Bowen, B. W., and J. C. Avise
In press. Conservation genetics of marine turtles. In *Conservation Genetics: Case Histories from Nature,* eds. J. C. Avise and J. Hamrick. New York: Chapman and Hall.

Bowen, B. W.; J. C. Avise; J. I. Richardson; A. B. Meylan; D. Margaritoulis; and S. Hopkins-Murphy
1993a. Population structure of the loggerhead turtle (*Caretta caretta*) in the northwest Atlantic Ocean and Mediterranean Sea. *Conservation Biology* 7: 834–844.

Bowen, B. W.; A. L. Bass; A. Garcia-Rodriguez; C. E. Diez; R. van Dam; A. B. Bolten; K. A. Bjorndal; M. M. Miyamoto; and R. J. Ferl
In press c. Origin of hawksbill turtles in a Caribbean feeding area as indicated by mtDNA sequence analysis. *Ecological Applications.*

Bowen, B. W.; N. Kamezaki; C. J. Limpus; G. H. Hughes; A. B. Meylan; and J. C. Avise
In press a. Global phylogeography of the loggerhead turtle (*Caretta caretta*) as indicated by mitochondrial DNA haplotypes. *Evolution.*

Bowen, B. W.; A. B. Meylan; and J. C. Avise
1989. An odyssey of the green sea turtle, *Chelonia mydas*: Ascension Island revisited. *Proceedings of the National Academy of Sciences, USA* 86:573–576.
1991. Evolutionary distinctiveness of the endangered Kemp's ridley. *Nature* 352:709–711.

Bowen, B. W.; A. B. Meylan; J. P. Ross; C. J. Limpus; G. H. Balazs; and J. C. Avise
1992. Global population structure and natural history of the green turtle (*Chelonia mydas*) in terms of matriarchal phylogeny. *Evolution* 46:865–881.

Bowen, B. W.; W. S. Nelson; and J. C. Avise
1993b. A molecular phylogeny for marine turtles: trait mapping, rate assessment, and conservation relevance. *Proceedings of the National Academy of Sciences, USA* 90:5574–5577.

Broderick, D.; C. Moritz; J. D. Miller; M. Guinea; R. J. Prince; and C. J. Limpus
1994. Genetic studies of the hawksbill turtle: Evidence for multiple stocks and mixed feeding grounds in Australian waters. *Pacific Conservation Biology* 1:121–131.

Carr, A.
1967. *So Excellent a Fishe: A Natural History of Sea Turtles.* New York: Scribner.

Encalada, S. E.; P. N. Lahanas; K. A. Bjorndal; A. B. Bolten; M. M. Miyamoto; and B. W. Bowen
In review. Phylogeography and conservation genetics of the green turtle (*Chelonia mydas*) in the Atlantic Ocean and Mediterranean Sea: A mitochondrial DNA control region sequence assessment.

FitzSimmons, N. N.; C. Moritz; and S. S. Moore
1995. Conservation and dynamics of microsatellite loci over 300 million years of marine turtle evolution. *Molecular Biology and Evolution* 12:432–440.

Hendrickson, J. R.
1958. The green turtle, *Chelonia mydas* (Linn.) in Malaya and Sarawak. *Proceedings of the Zoological Society of London* 130:455–535.

Karl, S. A.; B. W. Bowen; and J. C. Avise
1992. Global population structure and male-mediated gene flow in the green turtle (*Chelonia mydas*): RFLP analysis of anonymous nuclear DNA regions. *Genetics* 131:163–173.

Lahanas, P. N.; K. A. Bjorndal; A. B. Bolten; S. E. Encalada; M. M. Miyamoto; and B. W. Bowen
In review. Genetic composition of a green turtle feeding ground population: Evidence for multiple origins.

Lahanas, P. N.; M. M. Miyamoto; K. A. Bjorndal; and A. B. Bolten
1994. Molecular evolution and population genetics of Greater Caribbean green turtles (*Chelonia mydas*) as inferred from mitochondrial DNA control region sequences. *Genetica* 94:57–67.

Laurent, L.; J. Lescure; L. Excoffier; B. W. Bowen; M. Domingo; M. Hadjichristophorou; L. Kornaraki; and G. Trabuchet
1993. Genetic relationships between Mediterranean and Atlantic populations of loggerhead turtle *Caretta caretta* with a mitochondrial marker. *Comptes Rendus de l'Academie des Sciences,* Paris, Sciences de la vie 316:1233–1239.

Meylan, A. B.; B. W. Bowen; and J. C. Avise
1990. A genetic test of the natal homing versus social facilitation models for green turtle migration. *Science* 248:724–727.

Norman, J. A.; C. Moritz; and C. J. Limpus
1994. Mitochondrial DNA control region polymorphisms: Genetic markers for ecological studies of marine turtles. *Molecular Ecology* 3:363–373.

Sears, C. J.; B. W. Bowen; R. W. Chapman; S. B. Galloway; S. R. Hopkins-Murphy; and C. M. Woodley
In press. Demographic composition of the juvenile loggerhead sea turtle (*Caretta caretta*) feeding population off Charleston, South Carolina: Evidence from mitochondrial DNA markers. *Marine Biology.*

David Wm. Owens
Department of Biology
Texas A&M University
College Station, Texas 77843-3258 USA

Applied and Behavioral Endocrinology

Behavioral endocrinology studies, initiated at Cayman Turtle Farm, first on *Chelonia mydas* (Owens and Morris 1985) and more recently on *Lepidochelys kempi* (Rostal 1991), have now been expanded to wild populations using several different species (Wibbels et al. 1990; Rostal 1991).

The preliminary gonadotropin work of Licht (1984) has been significantly modified with the realization that what was originally thought to be an assay for follicle stimulating hormone (FSH) was in actuality an assay for neurophysin, which is a carrier molecule for arginine vasotocin (AVT) (Licht et al. 1984). The AVT and neurophysin are secreted at the same time and in a very dynamic way during oviposition (Figler et al. 1989), as are prostaglandins (Guillette et al. 1991). Using a new FSH assay in *Caretta caretta* and *C. mydas*, Wibbels et al. (1992) found FSH and LH secreted simultaneously at ovulation. Thyroid hormone levels have also been correlated to nesting, growth, migration and hibernation (Moon 1992).

In wild turtles progesterone levels peak with ovulation (Wibbels et al. 1992), and the ovulatory process may be impacted by stress as correlated to corticosterone (B) (Valverde et al. 1994). There are several additional stress/B studies in progress at various labs. Estrogen (E) is significantly elevated during the follicular development period prior to migration, while testosterone (T) peaks dramatically and E drops as migration begins (Wibbels et al. 1990). The ratio of E to T in the allantoic fluid of hatchlings (Crain et al. 1994) has shown promise as a hatchling sexing technique.

The T-based sex determination (TSD) technique for immatures has been validated and used in several studies (Wibbels et al. 1991). The method has proven useful where the effects of TSD could be important to their conservation (Bolten et al. 1992). We have recently been able to estimate the number of nests that a female has deposited by the changes in her T levels. The technique has proven useful when researchers

589

could not actually monitor each female for each of her nests (Rostal 1991). In addition, the circulating steroids in an adult can now be used to predict their reproductive status (Wibbels et al. 1990, 1992).

Literature Cited

Bolten, A. B.; K. A. Bjorndal; J. Grumbles; and D. W. Owens
1992. Sex ratio and sex-specific growth rates in immature green turtles, *Chelonia mydas*, in the southern Bahamas. *Copeia* 1992:1098–1103.

Crain, D. A.; T. S. Gross; A. Bolten; K. Bjorndal; R. Carthy; and L. Guillette, Jr.
1994. Development of a non-invasive sexing technique for hatchling loggerhead turtles (*Caretta caretta*). In *Proceedings of the Fourteenth Annual Symposium on Sea Turtle Biology and Conservation*, compilers K. A. Bjorndal, A. B. Bolten, D. A. Johnson, and P. J. Eliazar, p. 30. NOAA Technical Memorandum NMFS-SEFSC-351.

Figler, R. A.; D. S. MacKenzie; D. W. Owens; P. Licht; and M. S. Amoss
1989. Increased levels of arginine vasotocin and neurophysin during nesting in sea turtles. *General and Comparative Endocrinology* 73:223–232.

Guillette, L. J., Jr.; K. A. Bjorndal; A. Bolten; T. Gross; B. Palmer; B. Witherington; and J. Matter
1991. Plasma estradiol-17B, progesterone, prostaglandin F, and prostaglandin E2 concentrations during natural oviposition in the loggerhead turtle (*Caretta caretta*). *General and Comparative Endocrinology* 82:121–130.

Licht, P.
1984. Chapter 3: Reptiles. In *Marshall's Physiology of Reproduction, Vol. 1*, ed. G. E. Lamming, pp. 206–282. New York: Churchill Livingstone.

Licht, P.; B. T. Pickering; H. Papkoff; A. Pearson; and A. Bona-Gallo
1984. Presence of a neurophysin-like precursor in the green turtle (*Chelonia mydas*). *Journal of Endocrinology* 103:97–106.

Moon, Dae-Yeon
1992. The responses of sea turtles to temperature changes: Behavior, metabolism, and thyroid hormones. Ph.D. dissertation, Texas A&M University.

Owens, D. W., and Y. A. Morris
1985. The comparative endocrinology of sea turtles. *Copeia* 1985:723–735.

Rostal, David C.
1991. The reproductive behavior and physiology of the Kemp's ridley sea turtle, *Lepidochelys kempi* (Garman, 1880). Ph.D. dissertation, Texas A&M University.

Valverde, R. A.; D. W. Owens; D. S. MacKenzie; and R. Patterson
1994. Hormone levels and ovulation correlates in the olive ridley sea turtle. In *Proceedings of the Fourteenth Annual Symposium on Sea Turtle Biology and Conservation*, compilers K. A. Bjorndal, A. B.

Bolten, D. A. Johnson, and P. J. Eliazar, p. 156. NOAA Technical Memorandum NMFS-SEFSC-351.

Wibbels, T.; R. E. Martin; D. W. Owens; and M. S. Amoss, Jr.
1991. Female-biased sex ratio of immature sea turtles inhabiting the Atlantic coastal waters of Florida. *Canadian Journal of Zoology* 69:2973–2977.

Wibbels, T.; D. W. Owens; P. Licht; C. Limpus; P. Reed; and M. S. Amoss, Jr.
1992. Serum gonadotropins and gonadal steroids associated with ovulation and egg production in sea turtles. *General and Comparative Endocrinology* 87: 71–78.

Wibbels, T.; D. Owens; C. Limpus; P. Reed; and M. Amoss
1990. Seasonal changes in gonadal steroid concentrations associated with migration, mating, and nesting in loggerhead sea turtles. *General and Comparative Endocrinology* 79:154–164.

James R. Spotila
Department of Bioscience and Biotechnology
Drexel University
Philadelphia, Pennsylvania 19104 USA

Metabolism, Physiology, and Thermoregulation

Considerable progress has been made in the last 15 years in elucidating the physiological mechanisms of sea turtles. This has been accomplished by laboratory studies and by the application of sophisticated physiological techniques to field conditions. Radio, sonic, and satellite telemetry have demonstrated that sea turtles range widely in the oceans and dive to great depths: as much as 233 m in loggerheads and 1000 m in leatherbacks (Eckert et al. 1986, 1989; Sakamoto et al. 1990). Sea turtles have complexly subdivided multicameral lungs with small terminal air sacs through which a high flux of oxygen can occur (Jackson 1985; Lutcavage et al. 1987). Cartilaginous support, and perhaps the smooth muscles distributed throughout the lungs, allows gas exchange at rates up to 12 L sec^{-1}. This is equivalent to one vital capacity per second and approximates the flow velocity of adult humans during a maximal forced expiration. This support apparently allows nearly complete lung collapse which prevents gas trapping when turtles dive to great depths (Jackson 1985). Loggerhead and green turtles use their lungs as the principal O_2 store to support aerobic diving to moderate depths because their blood and tissue O_2 stores are similar to those of their terrestrial relatives (Wood et al. 1984; Lutcavage et al. 1987, 1989, 1991).

Leatherback turtles are unique because of their ability to dive to great depths, their long distance migrations into cold temperate waters, their diving physiology, and their metabolic adaptations (Paladino et al. 1990). Deep dives by leatherbacks appear to be supported by an increased O_2 carrying capacity of blood and tissue because the lungs undoubtedly collapse owing to increased hydrostatic pressure. Hematocrits and hemoglobin and myoglobin concentrations are among the highest recorded in reptiles, and approach levels found in diving mammals. However, blood O_2 affinity, Hill coefficient, and Bohr effect are similar to other turtles (Lutcavage et al. 1990). Resting metabolic rates of leatherbacks are three times those predicted from allometric relationships for green turtles and other reptiles scaled to leatherback size (0.387 W kg^{-1}, 1.15 ml O$_2$ kg^{-1} min^{-1}), but half the values predicted for mammals of this size (Paladino et al. 1990). Lutcavage et al. (1992) reported a resting metabolic rate of 1.1 ml kg^{-1} min^{-1} and calculated a range of dive times of 5 to 70 min that could be supported aerobically. The lower value for metabolic rate reported by Lutcavage et al. (1990) appeared to be due to a reduced tidal volume caused by an increased resistance to air flow inherent in their gas collection system. They used a small Hans Rudolph valve (no. 2600) and relatively small (34 L) Douglas bag as compared to the giant (7200) Hans Rudolph valve and 200 L Douglas bag used by Paladino et al. (1990). The smaller apparatus is sufficient for most sea turtles but it is inadequate for the leatherback.

The thermal biology of sea turtles depends upon the heat transfer properties of their environment as well as the physical and physiological characteristics of the turtles (Spotila and Standora 1985). Large sea turtles use a unique combination of thermoregulatory adaptations to control body temperature. Green turtles display regional endothermy. Warm pectoral muscles probably increase this turtle's swimming ability and may facilitate long-distance migrations (Standora et al. 1982). A resting leatherback has internal temperatures higher than carapace and ambient temperatures indicating that heat is generated internally and not absorbed from the environment (Standora et al. 1984). Mathematical modelling indicates that leatherbacks can use large body size, peripheral tissues as insulation, and circulatory changes to maintain warm temperatures in cold water and to avoid overheating in warm water (gigantothermy) (Paladino et al. 1990). Future research is needed to elucidate the cellular basis for these adaptations and to determine the effect of body size on physiological thermoregulation.

Literature Cited

Eckert, S. A.; K. L. Eckert; P. Ponganis; and G. L. Kooyman
1989. Diving and foraging behavior of leatherback sea turtles (*Dermochelys coriacea*). *Canadian Journal of Zoology* 67:2834–2840.

Eckert, S. A.; D. W. Nellis; K. L. Eckert; and G. L. Kooyman
1986. Diving patterns of two leatherback sea turtles (*Dermochelys coriacea*) during internesting intervals at Sandy Point, St. Croix, U.S. Virgin Islands. *Herpetologica* 42:381–388.

Jackson, D. C.
1985. Respiration and respiratory control in the green turtle, *Chelonia mydas*. *Copeia* 1985:664–671.

Lutcavage, M. E.; P. G. Bushnell; and D. R. Jones
1990. Oxygen transport in the leatherback sea turtle *Der-mochelys coriacea*. *Physiological Zoology* 63: 1012–1024.

1992. Oxygen stores and aerobic metabolism in the leatherback sea turtle. *Canadian Journal of Zoology* 70:348–351.

Lutcavage, M. E., and P. L. Lutz
1991. Voluntary diving metabolism and ventilation in the loggerhead sea turtle. *Journal of Experimental Marine Biology and Ecology* 147:287–296.

Lutcavage, M. E.; P. L. Lutz; and H. Baier
1987. Gas exchange in the loggerhead sea turtle *Caretta caretta*. *Journal of Experimental Biology* 131:365–372.

1989. Respiratory mechanics in the loggerhead sea turtle, *Caretta caretta*. *Respiration Physiology* 76:13–24.

Paladino, F. V.; M. P. O'Connor; and J. R. Spotila
1990. Metabolism of leatherback turtles, gigantothermy, and thermoregulation of dinosaurs. *Nature* 344: 858–860.

Sakamoto, W.; I. Uchida; Y. Naito; K. Kureha; M. Tujimura; and K. Sato
1990. Deep diving behavior of the loggerhead turtle near the frontal zone. *Nippon Suisan Gakkaishi* 56: 1435–1443.

Spotila, J. R., and E. A. Standora
1985. Environmental constraints on the thermal energetics of sea turtles. *Copeia* 1985:694–702.

Standora, E. A.; J. R. Spotila; and R. E. Foley
1982. Regional endothermy in the sea turtle, *Chelonia mydas*. *Journal of Thermal Biology* 7:159–165.

Standora, E. A.; J. R. Spotila; J. A. Keinath; and C. R. Shoop
1984. Body temperatures, diving cycles, and movement of a subadult leatherback turtle, *Dermochelys coriacea*. *Herpetologica* 40:169–176.

Wood, S. C.; R. N. Gatz; and M. L. Glass
1984. Oxygen transport in the green sea turtle. *Journal of Comparative Physiology* 154:275–280.

Lawrence H. Herbst
Elliott R. Jacobson
College of Veterinary Medicine
University of Florida
Gainesville, Florida 32610 USA

Diseases of Marine Turtles

This synopsis of marine turtle diseases is presented to raise awareness about their potential role as direct causes of morbidity and mortality in marine turtles.

Viral

Gray-patch disease.

ETIOLOGY: Herpesvirus (Rebell et al. 1975).

PATHOLOGY: Focal papules or spreading plaques of epidermal necrosis.

SIGNIFICANCE: High morbidity and mortality in post-hatchling mariculture-reared green turtles. Stress may be a factor in outbreaks (Kleese 1984). While latent infection may be ubiquitous, disease in wild turtles has not been documented (Haines 1978).

Lung-eye-trachea disease (LETD).

ETIOLOGY: Herpesvirus (Jacobson et al. 1986).

PATHOLOGY: Conjunctivitis, pharyngitis, tracheitis, and bronchopneumonia.

SIGNIFICANCE: Only reported in mariculture-reared green turtles.

Bacterial

Various bacterial infections.

ETIOLOGY: Variety of Gram positive and Gram negative bacteria (reviewed by Lauckner 1985; Glazebrook and Campbell 1990a,b; Aguirre et al. 1994). *Vibrio alginolyticus*, *Aeromonas hydrophila*, and *Flavobacterium* sp. are associated with ulcerative stomatitis–obstructive rhinitis–pneumonia complex (Glazebrook et al. 1993).

593

PATHOLOGY: Abscesses and granulomas, septicemia, toxemia.

SIGNIFICANCE: Many are opportunistic pathogens requiring predisposing factors. Several species (e.g., *Mycobacterium*, *Vibrio*, *Salmonella* sp.) may be pathogenic to humans.

Chlamydiosis.

ETIOLOGY: *Chlamydia* sp. (Homer et al. 1994).

PATHOLOGY: Myocarditis.

SIGNIFICANCE: Mortality in mariculture-reared green turtles. Significance to wild populations is unclear.

Fungal

ETIOLOGY: Various species including *Sporotrichium*, *Cladosporium*, *Paecilomyces* spp., and *Penicillium* (Jacobson et al. 1979; Lauckner 1985; Glazebrook et al. 1993).

PATHOLOGY: Fungal dermatitis and fungal bronchopneumonia.

SIGNIFICANCE: All reports are from captive green turtles and loggerheads. Most fungi require predisposing factors to cause disease.

Protozoal

Amebiasis.

ETIOLOGY: *Entamoeba invadens* (Lauckner 1985).

PATHOLOGY: Enteritis and hepatitis.

SIGNIFICANCE: Mortality reported in captive marine turtles only.

Coccidiosis.

ETIOLOGY: *Caryospora cheloniae* (Leibovitz et al. 1978).

PATHOLOGY: Intestinal mucosal epithelial necrosis and encephalitis.

SIGNIFICANCE: Mass mortality in captive hatchling and juvenile green turtles and in free-ranging green turtles from Australia (Gordon et al. 1993).

Endoparasites

Trematodiasis.

ETIOLOGY: Diverse fauna of digenean trematodes in gastrointestinal and cardiovascular systems (Lauckner 1985).

PATHOLOGY: Usually inapparent infections. Gastrointestinal species may cause enteritis and cardiovascular trematodes may cause systemic vascular and perivascular inflammation. One species associated with gallbladder papillomatous hyperplasia in green turtles (Lauckner 1985).

SIGNIFICANCE: Gastrointestinal trematodiasis contributes to debilitation in severe cases. Cardiovascular trematodiasis may cause sporadic mortality.

Cestodiasis.

ETIOLOGY: Cestodes (tapeworms).

PATHOLOGY: Adult pseudophyllidean tapeworms reported in the gastrointestinal tract are probably nonpathogenic. Tetrarhynchid cestode cysts on serosa of gastrointestinal tract and in other viscera.

SIGNIFICANCE: Rare. Reported only in loggerheads which may be an aberrant host (Lauckner 1985).

Nematodiasis.

ETIOLOGY: Variety of gastrointestinal nematode species have been described (Lauckner 1985). *Sulcascaris sulcata* (Ascaridoidae) commonly affects *C. mydas* and *C. caretta*.

PATHOLOGY: Enteritis, gastrointestinal obstruction.

SIGNIFICANCE: Heavy infections may cause debilitation and death.

Ectoparasites

Leeches (Hirudinea).

ETIOLOGY: *Ozobranchus* species.

PATHOLOGY: Erosive skin lesions at attachment sites. Heavy infestations associated with severe anemia.

SIGNIFICANCE: May infest all marine turtle species although not reported from *Dermochelys*. Mortality in captive green turtles and loggerheads attributed to heavy infestation (Schwartz 1974). Potential vectors of bloodborne disease although none reported.

Crustacea (barnacles).

ETIOLOGY: Encrusting and burrowing barnacle species (Lauckner 1985).

PATHOLOGY: Shell lesions, oropharyngeal obstruction, visceral and joint lesions in burrowing species.

SIGNIFICANCE: Most are innocuous epibionts. Severely debilitated turtles may have increased epibiont loads. Heavy loads may increase the costs of locomotion. Burrowing species (e.g., *Tubicinella* sp.) may penetrate the body cavity and kill their host.

Neoplastic

Green turtle fibropapillomatosis (GTFP).

ETIOLOGY: Probably viral (Herbst et al., in press). Role of herpesvirus demonstrated in tumors is unproven.

PATHOLOGY: Cutaneous fibropapillomas and visceral fibromas.

SIGNIFICANCE: Cosmopolitan; high prevalence among wild juvenile and adult green turtles in some areas. Reported in captive green turtles. Similar condition reported in loggerheads, flatbacks, and olive ridleys (Herbst 1994).

Conclusion

The causes of individual strandings and mass mortalities among free-ranging marine turtles are poorly documented, especially when not self-evident (entanglement, trauma). Although some non-infectious disease problems are well documented in free-ranging populations, we have a better understanding of infectious diseases in captive turtles. A fuller appreciation of the significance of disease in the ecology of free-ranging sea turtles will require prompt, carefully conducted necropsies of stranded turtles and the development and application of diagnostic reagents specific for marine turtle diseases (Herbst and Klein, in press).

Literature Cited

Aguirre, A. A.; G. H. Balazs; B. Zimmerman; and T. R. Spraker
1994. Evaluation of Hawaiian green turtles (*Chelonia mydas*) for potential pathogens associated with fibropapillomas. *Journal of Wildlife Diseases* 30:8–15.

Glazebrook, J. S., and R. S. F. Campbell
1990a. A survey of the diseases of marine turtles in northern Australia. I. Farmed turtles. *Diseases of Aquatic Organisms* 9:83–95.
1990b. A survey of the diseases of marine turtles in northern Australia. II. Oceanarium-reared and wild turtles. *Diseases of Aquatic Organisms* 9:97–104.

Glazebrook, J. S.; R. S. F. Campbell; and A. T. Thomas
1993. Studies on an ulcerative stomatitis–obstructive rhinitis–pneumonia disease complex in hatchling and juvenile sea turtles *Chelonia mydas* and *Caretta caretta*. *Diseases of Aquatic Organisms* 16:133–147.

Gordon, A. N.; W. R. Kelly; and R. J. G. Lester
1993. Epizootic mortality in free-living green turtles *Chelonia mydas*, due to coccidiosis. *Journal of Wildlife Diseases* 29:490–494.

Haines, H.
1978. A herpesvirus disease of green sea turtles in aquaculture. *Marine Fisheries Review* 40:33–37.

Herbst, L. H.
1994. Fibropapillomatosis of marine turtles. *Annual Review of Fish Diseases* 4:389–425.

Herbst, L. H.; E. R. Jacobson; R. Moretti; T. Brown; J. P. Sundberg; and P. A. Klein
In press. Transmission of green turtle fibropapillomatosis with filtered cell-free tumor extracts. *Diseases of Aquatic Organisms*.

Herbst, L. H., and P. A. Klein
In press. Monoclonal antibodies for the measurement of class-specific antibody responses in the green turtle, *Chelonia mydas*. *Veterinary Immunology and Immunopathology*.

Homer, B. L.; E. R. Jacobson; J. Schumacher; and G. Sherba
1994. Chlamydiosis in mariculture-reared green sea turtles (*Chelonia mydas*). *Veterinary Pathology* 31:1–7.

Jacobson, E. R.; J. M. Gaskin; M. Roelke; E. C. Greiner; and J. Allen
1986. Conjunctivitis, tracheitis, and pneumonia associated with herpes virus infection in green sea turtles. *Journal of the American Veterinary Medical Association* 189:1020–1023.

Jacobson, E. R.; J. M. Gaskin; R. P. Shields; and F. H. White
1979. Mycotic pneumonia in mariculture-reared green sea turtles. *Journal of the American Veterinary Medical Association* 175:929–932.

Kleese, W. C.
1984. Environmental effects upon herpesvirus infections in captive green sea turtles. In *Diseases of Amphibians and Reptiles*, eds. G. L. Hoff, F. L. Frye, and E. R. Jacobson, pp. 203–210. New York: Plenum Press.

Lauckner, G.
1985. Diseases of reptilia. In *Diseases of Marine Animals, Vol. 4, Part 2*, ed. O. Kinne, pp. 551–626. Hamburg: Biologische Anstalt Helgoland.

Leibovitz, L.; G. Rebell; and G. C. Boucher
1978. *Caryospora cheloniae* sp. n.: A coccidial pathogen of mariculture-reared green sea turtles (*Chelonia mydas mydas*). *Journal of Wildlife Diseases* 14: 269–275.

Rebell, G.; A. Rywlin; and H. Haines
1975. A herpesvirus-type agent associated with skin lesions of green sea turtles in aquaculture. *American Journal of Veterinary Research* 36:1221–1224.

Schwartz, F. J.
1974. The marine leech *Ozobranchus margoi* (Hirudinea: Pisciocolidae), epizootic on *Chelonia* and *Caretta* sea turtles from North Carolina. *Journal of Parasitology* 60:889–890.

N. Mrosovsky
Department of Zoology
University of Toronto
Toronto, Ontario, Canada M5S 1A1

Temperature and Sex Ratio

In the 1979 meeting (Mrosovsky and Yntema, this volume), we raised three priorities for those managing sea turtles: 1) learning about the effects of fluctuating temperatures, including those in styrofoam boxes; 2) studying within-species variation in pivotal temperatures; 3) locating thermosensitive periods within incubation. To these I now add another: discovering what sex ratios are in nature. This could provide a benchmark against which to assess changes resulting from global warming or from conservation practices (this is not the same as saying that one should always aim to maintain natural ratios). Apart from demonstrations that incubation in styrofoam boxes does indeed introduce masculinizing biases (e.g., Dutton et al. 1985), data on the priorities raised in 1979 remain sketchy. By default, practice or theory is sometimes based on pivotals from only one or two clutches, not a safe way to estimate population means. Knowledge of variance in pivotals is important for assessing how turtles might fare during global warming.

Nevertheless, much has been learned about thermal influences on sexual differentiation in reptiles. The proceedings of a recent symposium (Lance 1994) provide access to this literature. Unfortunately a gap still exists between the variable temperatures in nature and the constant temperatures in most laboratory work. This makes it difficult to predict sex ratio accurately from nest temperatures. One relevant finding, with freshwater turtles, is that excursions above the pivotal are more influential than similar-sized excursions below the pivotal (Bull 1985). For definitions of terminology, see Mrosovsky and Pieau (1991).

Another problem in estimating sex ratios from nest temperature is lack of detailed information on thermosensitive periods, except for a study on leatherbacks (Desvages et al. 1993). The reason for wanting to estimate sex ratios is to avoid having to sacrifice hatchlings and the associated histology. The glycerine method (van der Heiden et al. 1985) for sexing has not been adequately validated for sea turtles, and two failures to do this have been reported (Jackson et al. 1987; Mrosovsky and Benabib 1990). No endorsement of the glycerine method for sea turtles by histology has been published as far as I am aware. Choosing an appropriate method of sexing is basic for any study. Regrettably, fixing the gonad and subsequent histology is still probably the most feasible reliable method to use in remote areas. It is hoped that this may soon change.

Since 1979, the idea that temperature directs sexual differentiation has moved from fringy to orthodox, so much so that temperatures are now being recorded on beaches in many parts of the world. In some cases inattention to methodology reduces the value of the data. Even a few tenths of a degree can make a difference to sex ratio. It is not possible to be too obsessional about calibration; for this a good quality Hg thermometer is advised. Paying one or two thousand dollars for dataloggers does not guarantee that circuitry will remain in unblemished perfection in the wet salty sand of a tropical beach. For those using thermocouples, the how-to-do-it manual by Spotila et al. (1983) is invaluable.

Clearer formulation of aims might improve some studies. It is often appropriate to measure temperatures within the egg mass. However, if the aim is to compare thermal profiles in one situation to those in another (northern vs. southern rookery, renourished vs. natural beach, hatchery site vs. nesting area), then burying some temperature probes in the sand rather than in nests themselves should be considered—unless one can arrange for turtles to nest on the same date in the different areas, depositing standard numbers of eggs at standard depths, with nests disposed in representative transects.

Finally, on methods, nest temperatures change over the course of a day. There are various solutions to this problem (Godfrey and Mrosovsky 1994). In case it is not feasible to take readings at appropriate times, I place some probes 60 cm below the surface; at this depth temperature variation is reduced.

Literature Cited

Bull, J. J.
1985. Sex ratio and nest temperature in turtles: comparing field and laboratory data. *Ecology* 66:1115–1122.

Desvages, G.; M. Girondot; and C. Pieau
1993. Sensitive stages for the effects of temperature on gonadal aromatase activity in embryos of the marine turtle *Dermochelys coriacea*. *General and Comparative Endocrinology* 92:54–61.

Dutton, P. H.; C. P. Whitmore; and N. Mrosovsky
1985. Masculinisation of leatherback turtle *Dermochelys coriacea* hatchlings from eggs incubated in styrofoam boxes. *Biological Conservation* 31:249–264.

Godfrey, M. H., and N. Mrosovsky
1994. Simple method of estimating mean incubation temperatures on sea turtle beaches. *Copeia* 1994: 808–811.

Jackson, M. E.; L. U. Williamson; and J. R. Spotila
1987. Gross morphology vs. histology: sex determination of hatchling sea turtles. *Marine Turtle Newsletter* 40:10–11.

Lance, V. A., editor
1994. Environmental sex determination in reptiles: patterns and processes. *Journal of Experimental Zoology* 270(1).

Mrosovsky, N., and M. Benabib
1990. An assessment of two methods of sexing hatchling sea turtles. *Copeia* 1990:589–591.

Mrosovsky, N., and C. Pieau
1991. Transitional range of temperature, pivotal temperatures and thermosensitive stages for sex determination in reptiles. *Amphibia-Reptilia* 12:169–179.

Spotila, J. R.; E. A. Standora; S. J. Morreale; G. J. Ruiz; and C. Puccia
1983. Methodology for the study of temperature related phenomena affecting sea turtle eggs. U.S. Fish and Wildlife Service Endangered Species Report 11.

van der Heiden, A. M.; R. Briseño-Dueñas; and D. Rios-Olmeda
1985. A simplified method for determining sex in hatchling sea turtles. *Copeia* 1985:779–782.

Karen A. Bjorndal
Center for Sea Turtle Research
Department of Zoology
University of Florida
Gainesville, Florida 32611 USA

George R. Zug
Division of Amphibians and Reptiles
National Museum of Natural History
Washington, D.C. 20560 USA

Growth and Age of Sea Turtles

During the 1979 conference, the very slow growth rates of sea turtles under natural conditions were revealed (Balazs, this volume, and references cited therein). Previously, sea turtle growth rates had been assumed to be moderately rapid with most cheloniids attaining sexual maturity in five to six years. The new estimates of as many as 40–60 years to attain sexual maturity revolutionized our view of the life history of sea turtles and of the time frame required for conservation actions. Considerable effort has been invested in the studies of growth and age in wild sea turtles because both parameters are critical for demographic studies and development of appropriate management programs. These studies have taken four basic approaches.

First, growth models (e.g., von Bertalanffy, logistic, Gompertz) have been applied to mark-recapture data (Frazer and Ehrhart 1985; Frazer and Ladner 1986; Bjorndal and Bolten 1988; Boulon and Frazer 1990). The von Bertalanffy model usually has the best fit for turtle data (Frazer et al. 1990). Most studies have used the models to extrapolate beyond the studied size range to estimate age at sexual maturity. Such extrapolation requires caution, although tests on a data set of known-aged freshwater turtles (Frazer et al. 1990) revealed that the removal of smaller individuals from the data caused little change in the asymptote of the growth curve and thus in the maturity estimate. Removal of large individuals caused significant changes in the asymptote, which would affect estimates of maturity. Maturity estimates depend upon the body size selected to represent maturity; mean size of nesting females is considered the best index (Frazer and Ehrhart 1985).

Second, some studies have calculated growth rates based on mark-recapture data and presented mean growth rates for 5- or 10-cm size classes of turtles (Bjorndal and Bolten 1988; Collazo et al. 1992; Limpus 1992; Boulon 1994). Years to grow through each size class are calculated and summed to estimate the number of years necessary to grow through the size

599

interval of the study population. This approach offers conservative growth estimates, particularly if there is no extrapolation beyond the size range of the study population.

Third, skeletochronology—based on periosteal layers in the humerus—provides estimates of age and growth rates without the time investment required for mark-recapture studies (Frazier 1985; Zug et al. 1986; Zug 1991). The technique is handicapped by the natural process of remodeling that progressively eliminates the inner (younger) growth layers as the turtle grows (Zug 1990). Protocols are available for estimating the number of lost layers and ages, but researchers must be aware of the assumptions and bias of each protocol (Parham and Zug, unpublished manuscript). Periosteal layers represent annual growth layers in juvenile loggerheads in temperate waters (Klinger and Musick 1992).

Fourth, length-frequency analysis, in which estimates of von Bertalanffy growth parameters are generated from length distributions of sample populations, does not require long time investments to study growth and age. This technique successfully estimated growth rates for a population of immature green turtles of known growth rates based on a mark-recapture study (Bjorndal and Bolten 1995; Bjorndal et al. 1995).

Many questions remain unanswered. Research effort has been uneven among species of sea turtles—the pelagic leatherback and olive ridley are, not surprisingly, the poorest known. From the studies cited above and others, we know that there is substantial variation among species, and among populations and geographic regions for any species. However, at least for immature green turtles, growth rates apparently do not vary between sexes in a population (Bolten et al. 1992). Growth rates and age estimates based on captive-raised specimens have proved unreliable (Bjorndal 1985) and should be avoided.

Literature Cited

Bjorndal, K. A.
1985. Nutritional ecology of sea turtles. *Copeia* 1985: 736–751.
Bjorndal, K. A., and A. B. Bolten.
1988. Growth rates of immature green turtles, *Chelonia mydas*, on their feeding grounds in the southern Bahamas. *Copeia* 1988:555–564.
1995. Comparison of length-frequency analyses for estimation of growth parameters for a population of green turtles. *Herpetologica* 51:160–167.
Bjorndal, K. A.; A. B. Bolten; A. L. Coan, Jr.; and P. Kleiber.
1995. Estimation of green turtle (*Chelonia mydas*) growth rates from length-frequency analysis. *Copeia* 1995: 71–77.

Bolten, A. B.; K. A. Bjorndal; J. S. Grumbles; and D. W. Owens.
1992. Sex ratio and sex-specific growth rates in immature green turtles, *Chelonia mydas*, in the southern Bahamas. *Copeia* 1992:1098–1103.
Boulon, R. H., Jr.
1994. Growth rates of wild juvenile hawksbill turtles, *Eretmochelys imbricata*, in St. Thomas, United States Virgin Islands. *Copeia* 1994:811–814.
Boulon, R. H., Jr., and N. B. Frazer.
1990. Growth of wild juvenile Caribbean green turtles, *Chelonia mydas*. *Journal of Herpetology* 42:441–445.
Collazo, J. A.; R. Boulon, Jr.; and T. L. Tallevast.
1992. Abundance and growth patterns of *Chelonia mydas* in Culebra, Puerto Rico. *Journal of Herpetology* 26:293–300.
Frazer, N. B., and L. M. Ehrhart.
1985. Preliminary growth models for green, *Chelonia mydas*, and loggerhead, *Caretta caretta*, turtles in the wild. *Copeia* 1985:73–79.
Frazer, N. B.; J. W. Gibbons; and J. L. Greene.
1990. Exploring Faben's growth interval model with data on a long-lived vertebrate, *Trachemys scripta* (Reptilia: Testudinata). *Copeia* 1990:112–118.
Frazer, N. B., and R. C. Ladner.
1986. A growth curve for green sea turtles, *Chelonia mydas*, in the U.S. Virgin Islands, 1913–14. *Copeia* 1986:798–802.
Frazier, J.
1985. Tetracycline as an in vivo label in bones of green turtles, *Chelonia mydas* (L.). *Herpetologica* 41: 228–234.
Klinger, R. C., and J. A. Musick.
1992. Annular growth layers in juvenile loggerhead turtles (*Caretta caretta*). *Bulletin of Marine Science* 51: 224–230.
Limpus, C. J.
1992. The hawksbill turtle, *Eretmochelys imbricata*, in Queensland: Population structure within a southern Great Barrier Reef feeding ground. *Wildlife Research* 19:489–506.
Zug, G. R.
1990. Age determination of long-lived reptiles: Some techniques for seaturtles. *Annales des Sciences Naturelles, Zoologie*, Paris 11:219–222.
1991. Age determination in turtles. *SSAR Herpetological Circular* 20:1–28.
Zug, G. R.; A. H. Wynn; and C. Ruckdeschel.
1986. Age determination of loggerhead sea turtles, *Caretta caretta*, by incremental growth marks in the skeleton. *Smithsonian Contributions to Zoology* 427:1–34.

Deborah T. Crouse
Center for Marine Conservation
1725 DeSales Street, N.W.
Washington, D.C. 20036 USA

Nat B. Frazer
Savannah River Ecology Laboratory
P.O. Drawer E
Aiken, South Carolina 29802 USA

Population Models and Structure

The size and growth of any population depend on the annual number of births and deaths, and on the timing of maturation, reproduction, and death in each individual's life. Population modelling consists of using known or estimated birth, growth, and death rates to make projections about the future. The accuracy and usefulness of a model depend upon its design and the quality of input variables. To provide meaningful output for management decisions the model must reasonably mimic the life history of the population and the input values must be realistic. By varying each rate in a model and comparing its effects on population growth, the relative impact of natural or human-induced changes can be evaluated. This can help to prioritize management strategies and future research needs.

Since the first publication of this volume, estimates of classic demographic parameters (i.e., age-specific fecundity, survivorship, and age at maturity) have become available for two populations of loggerheads—one in the southeastern United States and one in Queensland, Australia. The data used by Richardson (1982) and by Richardson and Richardson (this volume) for an early model of loggerhead population dynamics have been reanalyzed (Frazer 1983a,b, 1984, 1986) and combined with additional data on growth (Frazer 1983a; Frazer and Schwartz 1984; Frazer and Ehrhart 1985) and survival (Frazer 1983a, 1987) to provide estimates suitable for an age-specific life table. Using these parameter estimates, Crouse et al. (1987) constructed a stage-based matrix model for this population and concluded that changes in survivorship of larger juveniles and adults would likely have a much greater effect on future population growth than changes in the egg/hatchling stage. In other words, effort put into nest protection might be wasted if large juvenile and adult survival is not also increased, by reducing mortality due to drowning in shrimp trawl nets.

The data gathered by Limpus (1985) for Australian loggerheads also have been recast to provide inputs for a stage-based matrix model (Heppell, Limpus, Crouse, Frazer, and Crowder, unpublished manuscript). Although there are important differences between the two populations, including a longer estimated maturation time for the Australian loggerheads (Frazer et al. 1994), the Heppell et al. model affirms many of the findings of the Crouse et al. (1987) model.

Interesting insights from the loggerhead models relate to the size (age) structure of populations. Delayed maturity requires that a very large number of eggs, hatchlings, juveniles and subadults be maintained in the population in order to sustain a relatively small but important number of reproductively active adults. For example, in the U.S. loggerhead model (Crouse et al. 1987), more than 498,000 female eggs and immatures would be necessary to maintain a stable adult female population of only 1277. Thus, a large number of observed immature individuals (for example 300,000) does not necessarily mean that a population is robust; nor does releasing large numbers of juveniles into the wild ensure that a population will grow. Likewise, because of the relatively small number of adults compared to other age classes, focusing protection only on adults may be less effective than increasing survivorship in the much larger juvenile (or subadult) population. Finally, the effects of management may take decades to materialize, particularly if monitoring focuses only on one stage, such as adults on the nesting beach. Additional modelling by Crowder et al. (1994) has shown that it may take up to 70 years or more before the deployment of turtle excluder devices (TEDs) on shrimp trawls results in any substantial observable increase in the numbers of nesting turtles. If juvenile and adult survival have been low, an increase in adult survival may result in more nesting for a decade or two, as current adults survive to return to nest more often. However, this increase may be followed by a transient decrease in nesting animals (Crowder et al. 1994), as the relatively fewer offspring produced by the smaller adult population decades earlier begin to mature.

Early computer programmers developed a concept known as GIGO (garbage in, garbage out) to help them remember that even the best models could not provide reliable output if input values were unrealistic or flawed. Providing good input values for sea turtle models is no small task. Reliable estimates of fecundity require almost constant beach patrols every night of the nesting season to determine clutch frequency (Frazer and Richardson 1985), which then must be adjusted for the renesting or remigration interval (Frazer 1984). Adult female survivorship estimates must take into account the confounding effects of both tag loss and remigration interval (Frazer

1983b). Because there are no reliable means of aging sea turtles, time to maturity must be estimated using capture/recapture data and growth trajectories (Frazer and Ehrhart 1985) or data of questionable value from captive studies (Frazer and Schwartz 1984).

Owing to the complex life cycle of sea turtles, with different life stages inhabiting distant habitats, data gathered in the wild may cover only a portion of the size class distribution and lead to erroneous conclusions. For example, small loggerheads (<50 cm) do not typically appear in Florida (USA) waters and were not included in the growth analysis by Frazer and Ehrhart (1985). The small loggerheads inhabiting the eastern Atlantic near the Azores probably come from western Atlantic rookeries. These turtles grow much slower than predicted by Frazer and Ehrhart's (1985) Florida model (A. Bolten, pers. comm.). Thus, Frazer and Ehrhart probably underestimated average age at maturity for the Florida adult females. Fortunately, the results of such errors can be assessed by conducting a sensitivity analysis to determine the effect of changing the values on model input, and the loggerhead models are fairly robust in terms of the effects of errors in estimating age at maturity (Crouse et al. 1987).

The problems of relative inaccessibility or unknown whereabouts of certain life stages will continue to plague those who seek dependable input values for sea turtle models. Fortunately, some insight can be gained from long-term studies of freshwater turtles. Small and large, young and old, these turtles inhabit the same pond in close proximity. Thus, they can be followed and recaptured much more easily to determine survival and growth rates. Furthermore, age at maturity often can be determined exactly, as reliable growth (age) annuli may be discernable on specimens that are first caught when relatively young. Unlike sea turtles, the shells of these turtles can be permanently marked with notches or by drilling holes, and individuals recaptured even decades later can then be reliably aged once again.

Population models of late-maturing freshwater species such as Blanding's turtles, *Emydoidea blandingii* (maturing at 14 to 17 years old: Congdon et al. 1993), and snapping turtles, *Chelydra serpentina* (maturing at 11 to 16 years old: Congdon et al. 1994), indicate that high survival rates are necessary at each life stage to ensure that enough individuals survive and reproduce to maintain a population. These models lead to the inescapable conclusions that: (a) successful management and conservation programs for turtles must protect all life stages, and (b) sustainable commercial harvests of long-lived, late-maturing turtle species may well be an impossibility (Congdon et al. 1993, 1994). Given the similarities in their life histories, the results for freshwater turtles add support to the implications provided by the sea turtle models (Crouse et al. 1987; Crowder et al. 1994).

For now, the best models we have indicate that successful conservation efforts for sea turtles will be those that do not neglect any life stage, and that commercial harvests of sea turtles probably are not sustainable. In the future, those who model sea turtle populations must remember the concept of GIGO and recognize that providing reliable input values for survivorship, fecundity, and age at maturity is not a simple task. Given the robustness of the current models, we suspect that the recent findings for sea turtles (Crouse et al. 1987; Crowder et al. 1994; Heppell et al., unpublished manuscript) and freshwater turtles (Congdon et al. 1993, 1994) will be further substantiated as even better estimates of demographic parameters emerge over the next decade.

Literature Cited

Congdon, J. D.; A. E. Dunham; and R. C. Van Loben Sels
1993. Delayed sexual maturity and demographics of Blanding's turtles (*Emydoidea blandingii*): Implications for conservation and management of long-lived organisms. *Conservation Biology* 7:826–833.
1994. Demographics of common snapping turtles (*Chelydra serpentina*): Implications for conservation and management of long-lived organisms. *American Zoologist* 34:397–408.

Crouse, D. T.; L. B. Crowder; and H. Caswell
1987. A stage-based population model for loggerhead sea turtles and implications for conservation. *Ecology* 68:1412–1423.

Crowder, L. B.; D. T. Crouse; S. S. Heppell; and T. H. Martin
1994. Predicting the impact of excluder devices on loggerhead sea turtle populations. *Ecological Applications* 4:437–445.

Frazer, N. B.
1983a. Demography and life history evolution of the Atlantic loggerhead sea turtle, *Caretta caretta*. Ph.D. dissertation, University of Georgia.
1983b. Survivorship of adult female loggerhead sea turtles, *Caretta caretta*, nesting on Little Cumberland Island, Georgia, USA. *Herpetologica* 39:436–447.
1984. A model for assessing mean age-specific fecundity in sea turtle populations. *Herpetologica* 40:281–291.
1986. Survival from egg to maturity in a declining population of loggerhead sea turtles, *Caretta caretta*. *Herpetologica* 42:47–55.
1987. Preliminary estimates of survivorship for juvenile loggerhead sea turtles (*Caretta caretta*) in the wild. *Journal of Herpetology* 21:232–235.

Frazer, N. B., and L. M. Ehrhart
1985. Preliminary growth models for green, *Chelonia mydas*, and loggerhead, *Caretta caretta*, turtles in the wild. *Copeia* 1985:73–79.

Frazer, N. B.; C. J. Limpus; and J. L. Greene
1994. Growth and estimated age at maturity of Queensland loggerheads. In *Proceedings of the 14th Annual Symposium on Sea Turtle Biology and Conservation*, compilers K. A. Bjorndal, A. B. Bolten, D. A. Johnson, and P. J. Eliazar, pp. 42–46. NOAA Technical Memorandum NMFS-SEFC-351.

Frazer, N. B., and J. I. Richardson
1985. Annual variation in clutch size and frequency for loggerhead sea turtles, *Caretta caretta*, nesting on Little Cumberland Island, Georgia, USA. *Herpetologica* 41:246–251.

Frazer, N. B., and F. J. Schwartz
1984. Growth curves for captive loggerhead turtles, *Caretta caretta*, in North Carolina, USA. *Bulletin of Marine Science* 34:485–489.

Limpus, C.
1985. A study of the loggerhead sea turtle, *Caretta caretta*, in eastern Australia. Ph.D. thesis, University of Queensland.

Richardson, J. I.
1982. A population model for adult female loggerhead sea turtles (*Caretta caretta*) nesting in Georgia. Ph.D. dissertation, University of Georgia.

Colin J. Limpus
Conservation Strategy Branch
Queensland Department of Environment and Heritage
P.O. Box 541
Capalaba 4157 Australia

Global Overview of the Status of Marine Turtles: A 1995 Viewpoint

Caretta caretta, loggerhead turtle

Many of the problems described by Ross (this volume) for this species remain. Major status changes include the 50–80% decline in nesting females at eastern Australian rookeries since the mid 1970s (Limpus and Reimer 1994) and the listing of the loggerhead as an endangered species under the Australian Endangered Species Protection Act 1992. Nesting populations of Georgia and South Carolina, USA, have undergone substantial declines while Florida rookeries appear to have changed little (National Research Council 1990). This latter conclusion may reflect the limited long-term census data. In contrast, the South African loggerhead population has more than doubled since the early 1960s when strong protective measures were introduced (Hughes 1989). Populations currently not identified as in decline may well be under threat from a diversity of impacts.

Chelonia mydas, green turtle

Most population declines identified in 1979 by King (this volume) are continuing, along with many additional ones. In Indonesia, green turtle populations subjected to long-term intense egg harvests (Berau Turtle Islands, Pangumbahan) have suffered order-of-magnitude declines in egg production since the 1940s (Limpus 1994); the Aru Island green turtle population has been decimated by harvest of nesting females, largely to supply turtles to Bali (J. P. Schulz, pers. comm.). In Japan, the decimation of the Ogasawara green turtle stock through overharvest of the nesting turtles during the late 1800s and early 1900s has been summarized (Horikoshi et al. 1994). The French Polynesian nesting population appears to have declined by approximately 90% (P. Siu, pers. comm.).

The largest slaughter of green turtles globally occurs within the Australasian region, including Indonesia, Papua New Guinea, Australia, Solomon Islands, Vanuatu, New Caledonia, and Fiji where, in order of magnitude, 100,000 green turtles, mostly big females, are harvested each year. This harvest threatens the stability of the large nesting populations in Australia and contributes to the decline of other regional stocks. Near-total egg harvest still characterises the green turtle nesting populations of Indonesia, Thailand, and Terengganu in Malaysia. The decline in the Sarawak Turtle Islands population remains the best documented case study of the long-term impact of intense harvest of green turtle eggs (Fig. 1). Strong conservation management operating for decades on depleted populations has resulted in some recoveries: Sabah (Fig. 1; Basintal and Lakim 1994), Hawaii (Wetherall and Balazs, in press), and Florida (Owen et al. 1994).

Eretmochelys imbricata, hawksbill turtle

Contrary to the 1979 summary (King, this volume), hawksbill turtles do/did nest in large aggregations (dispersed nesting today is probably the result of overharvest of previously large colonies, e.g. Caribbean Panama, U.S. Virgin Islands). Recent studies have highlighted the ongoing overharvest of this turtle (Bjorndal et al. 1993; Miranda and Frazier 1994). The last remaining large rookeries appear to be in Australia where thousands of hawksbills breed annually, but these have no long-term census data.

Although the Japanese Government banned the importation of turtle products in 1994, a substantial harvest for domestic consumption of meat and scale continues in Cuba, Indonesia, Solomon Islands, Fiji and elsewhere. A large proportion of eggs appear to be harvested in Malaysia (Terengganu), Thailand, Indonesia, and possibly Philippines.

Hawksbill turtles are apparently increasing in only one region (Trono 1994). After 25 years of strong protection, hawksbill nesting at the Sabah Turtle Islands, Malaysia, has increased more than tenfold since 1969 (Fig. 2).

Natator depressus, flatback turtle

The marine turtle endemic to the Australian continental shelf currently has an annual nesting population of the order of five to ten thousand females but most populations have never been monitored. Therefore, long-term population trends cannot be detected. Significant negative impacts include large egg losses through pig and varanid predation and potentially large losses in trawl and gillnet fisheries around Australia, Indonesia, and Papua New Guinea.

Lepidochelys kempii, Kemp's ridley turtle

Decades of intense conservation management effort appear to be succeeding. Clutch production at Rancho Nuevo, Mexico, has increased steadily in recent years (Fig. 3). Whether this is the result of increased hatchling production at the rookeries or increased survivorship of adults and near adults resulting from the use of turtle excluder devices (TEDs) in shrimp trawls remains to be seen.

Lepidochelys olivacea, olive ridley turtle

This species remains the most abundant of the world's marine turtles (Ross, this volume). Long-term population trends for the species at its arribada beaches are not well documented. In Mexico, although only the Escobilla arribada functions as a major population, the future for the depleted stocks is promising given the current protection regime. The Orissa, India, arribada has in recent times been under threat of incidental mortality in gillnet fisheries, trawling, coastal and port developments and substantial egg harvests.

There are no demonstrated recovering populations for this species. Declines have been demonstrated for some small populations. Terengganu (Malaysia) stock has declined from possibly thousands annually to approximately 20 per year in the early 1990s, and the stock in the Andaman Sea of Thailand has been decimated to only tens of females nesting annually as a result of long-term excessive egg harvest. The Surinam arribada has declined to a few hundred females annually in part due to trawling bycatch. This species is probably suffering from the assumption that, because it occurs in large numbers at some localities, it does not warrant special conservation effort. This species may not be as secure as it superficially appears.

Dermochelys coriacea, leatherback turtle

Since 1979, census data on most rookeries have improved and the global population estimate increased, especially because of reassessment of the Pacific Mexican population and the discovery of the large population in Irian Jaya, Indonesia (Pritchard 1982). For most populations, long-term census data from which to assess population stability are insufficient (Ross, this volume). In Malaysia, the dramatic decline to fewer than 20 nesting females in 1993 at the famous Terengganu rookery illustrates the consequences of harvesting most eggs for more than a generation (Fig. 4). Incidental mortality in fishing nets probably accelerated this decline (Chan et al. 1988). A similar decimation of the nesting population of the Andaman Sea area of western Thailand apparently occurred

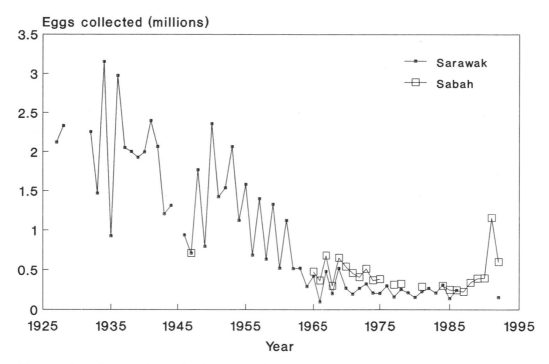

Figure 1. Annual egg production by the green turtle populations of Sarawak and Sabah in Malaysia. Data supplied by the Sarawak Museum and the Sabah National Parks staff.

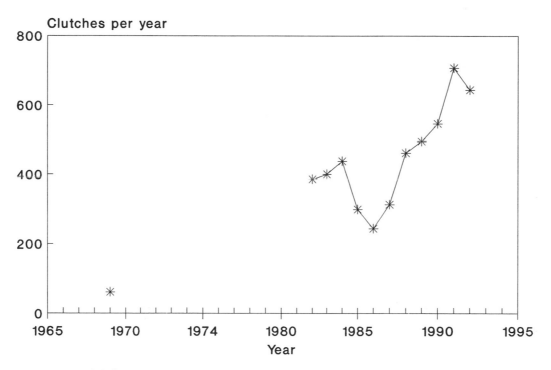

Figure 2. Hawksbill turtle nesting population (using egg production as an index) in the Sabah Turtle Islands National Park since the park was declared. Data supplied by the Sabah National Parks staff.

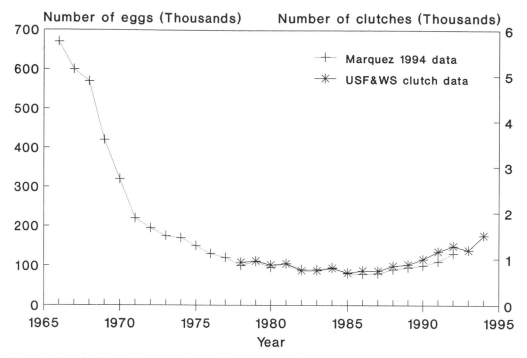

Figure 3. Rancho Nuevo Kemp's ridley turtle nesting population using egg production as an index.

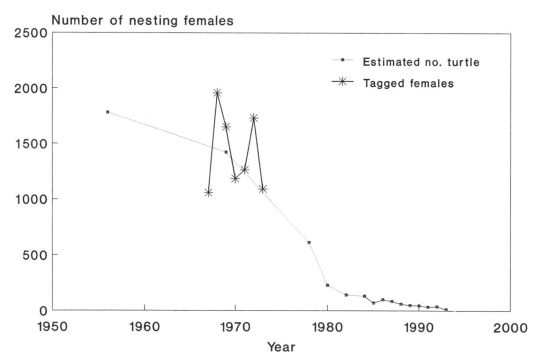

Figure 4. Leatherback nesting population in Terengganu, Malaysia, estimated from egg production data and from tagging data of nesting leatherbacks in Terengganu in the late 1960s and early 1970s. Data supplied by Malaysian Fisheries Department.

through near-total, long-term egg harvest. Similar levels of egg loss through harvest and pig predation characterize the nesting beaches of Irian Jaya, Indonesia, and northern Papua New Guinea, the last major leatherback nesting populations of the Australasian region.

In South Africa, thirty years of strong protection have been paralleled by an increase in annual nesting population from 20 to 90 leatherbacks (Hughes 1989).

Since the 1979 Washington World Conference, the losses have been extensive and the gains small in terms of numbers of turtles. Globally, marine turtles are still on the decline. No species is secure. The devastation caused by long-term near-total egg harvests is clear. However, major progress in understanding marine turtle biology and management should make it easier for turtle managers to make better headway in the next 15 years. Marine turtle population declines can be reversed, but decades of dedicated conservation effort are necessary to achieve just small progress. Marine turtles cannot cope with increased mortality substantially above natural levels. They can survive only if we provide effective long-term conservation management.

Literature Cited

Basintal, P., and M. Lakim
1994. Status and management of sea turtles at Turtle Island Park. In *Proceedings of the First ASEAN Symposium-Workshop on Marine Turtle Conservation, Manila, Philippines 1993*, pp. 139–149. Manila: World Wildlife Fund.

Bjorndal, K. A.; A. B. Bolten; and C. J. Lagueux
1993. Decline of the nesting population of hawksbill turtles at Tortuguero, Costa Rica. *Conservation Biology* 7:925–927.

Chan, E. H.; H. C. Liew; and A. G. Mazlin
1988. The incidental capture of sea turtles in fishing gear in Terengganu, Malaysia. *Biological Conservation* 43:1–7.

Horikoshi, K.; H. Suganuma; H. Tachikawa; F. Sato; and M. Yamaguchi
1994. Decline of Ogasawara green turtle population in Japan. In *Proceedings of the Fourteenth Annual Symposium on Sea Turtle Biology and Conservation*, compilers K. A. Bjorndal, A. B. Bolten, D. A. Johnson, and P. J. Eliazar, pp. 235–236. NOAA Technical Memorandum NMFS-SEFSC-351.

Hughes, G. R.
1989. Sea turtles. In *Oceans of Life off South Africa*, eds. A. I. L. Payne and R. J. M. Crawford, pp. 230–243. Cape Town: Vlaeberg Publishers.

Limpus, C. J.
1994. Current declines in south east Asian turtle populations. In *Proceedings of the Thirteenth Annual Symposium on Sea Turtle Biology and Conservation*, compilers B. A. Schroeder and B. E. Witherington, pp. 89–92. NOAA Technical Memorandum NMFS-SEFSC-341.

Limpus, C. J., and D. Reimer
1994. The loggerhead turtle, *Caretta caretta*, in Queensland: a population in decline. In *Proceedings of the Australian Marine Turtle Conservation Workshop*, Queensland Department of Environment and Heritage and Australian Nature Conservation Agency, pp. 39–59. Canberra: Australian Nature Conservation Agency.

Márquez, R.
1994. Sinopsis de datos biologicos sobre la tortuga lora, *Lepidochelys kempi* (Garman, 1880). *FAO Sinopsis sobre la Pesca* 152:1–141.

Miranda, E., and J. Frazier
1994. Isla Holbox, Mexico: an analysis of five nesting seasons of a major hawksbill nesting beach. In *Proceedings of the Fourteenth Annual Symposium on Sea Turtle Biology and Conservation*, compilers K. A. Bjorndal, A. B. Bolten, D. A. Johnson, and P. J. Eliazar, pp. 85–88. NOAA Technical Memorandum NMFS-SEFSC-351.

National Research Council
1990. *Decline of the Sea Turtles: Causes and Prevention*. Washington, D.C.: National Academy Press.

Owen, R. D.; S. A. Johnson; J. L. Prusak; W. E. Redfoot; and L. M. Ehrhart
1994. Marine turtle activity at the Archie Carr NWR in 1992: a third consecutive year of above-average loggerhead activity and the best season on record for the Florida green turtle. In *Proceedings of the Thirteenth Annual Symposium on Sea Turtle Biology and Conservation*, compilers B. A. Schroeder and B. E. Witherington, pp. 256–258. NOAA Technical Memorandum NMFS-SEFSC-341.

Pritchard, P. C. H.
1982. Nesting of the leatherback turtle, *Dermochelys coriacea* in Pacific Mexico, with a new estimate of the world population status. *Copeia* 1982:741–747.

Trono, R. B.
1994. The Philippine-Sabah Turtle Islands: a critical management area for sea turtles of the ASEAN region. In *Proceedings of the First ASEAN Symposium-Workshop on Marine Turtle Conservation, Manila, Philippines 1993*, pp. 167–180. Manila: World Wildlife Fund.

Wetherall, J. A., and G. H. Balazs
In press. Historical trends in the green turtle nesting colony at French Frigate Shoals, Northwestern Hawaiian Islands. *Marine Ecology Progress Series*.

Karen L. Eckert
Wider Caribbean Sea Turtle Conservation Network
(WIDECAST)
17218 Libertad Drive
San Diego, California 92127 USA

Anthropogenic Threats to Sea Turtles

Persistent overexploitation, especially of breeding adults, for meat, shell, oil and/or skin is largely responsible for depleted populations of sea turtles (Groombridge and Luxmoore 1989; Eckert 1993). Once-vast populations are on the verge of ruin (e.g., Indonesia: Greenpeace 1989). In some cases, egg collection alone has been implicated in population demise (e.g., Malaysia: Chan 1991). International trade has played a prominent role in the depletion of sea turtles, especially hawksbills (Milliken and Tokunaga 1987). Japanese sea turtle imports since 1970 represent a minimum of 2,250,000 turtles (Canin 1991). Japan recently banned the import of hawksbill shell (Donnelly 1991) and removed the last of its CITES reservations on sea turtles in July 1994.

Incidental capture in shrimp trawls has been implicated in population declines in the USA and Mexico (National Research Council 1990), Suriname (Reichart and Fretey 1993), and Australia (Limpus and Reimer 1994). Turtle excluder devices (TEDs) release trapped turtles, generally without compromising shrimp catch (Crouse et al. 1992), and are mandatory in U.S. waters. Longlines hook and kill an estimated tens of thousands of turtles every year (Nishemura and Nakahigashi 1990; Balazs and Pooley 1994) and driftnets kill many thousands more (Wetherall et al. 1993). Losses due to purse seines and tangle nets (e.g., Frazier and Brito 1990) have not been estimated.

Nesting beaches are degraded or lost to expanding urban development and a thriving global tourist industry. The literature is replete with specifics, including rising human activity, high-density coastal construction, beachfront lighting, erosion control structures, sand mining, port development, and indiscriminate waste disposal (Groombridge 1990; Eckert and Honebrink 1992; Meylan et al., in press; Witherington and Martin, in press). At sea, oil spills (Lutcavage et al., in press), persistent marine debris (Balazs 1985; Witzell and Teas 1994), pollution (Hutchinson and Simmonds 1992), and damage to

611

coral reefs and seagrass (e.g., anchoring, dredging) threaten dispersal and migratory routes, nursery habitats, and foraging grounds (e.g., Allen 1992).

Literature Cited

Allen, W. H.
1992. Increased dangers to Caribbean marine ecosystems. *BioScience* 42:330–335.

Balazs, G. H.
1985. Impact of ocean debris on marine turtles: entanglement and ingestion. In *Proceedings of the Workshop on the Fate and Impact of Marine Debris*, ed. R. S. Shomura and H. O. Yoshida, pp. 387–429. NOAA Technical Memorandum NMFS-SWFC-54.

Balazs, G. H.; S. G. Pooley; and 14 collaborators
1994. *Research Plan to Assess Marine Turtle Hooking Mortality: Results of an Expert Workshop Held in Honolulu, Hawaii, November 16–18, 1993*. NOAA Technical Memorandum NMFS-SWFSC-201.

Canin, J.
1991. International trade aspects of the Japanese hawksbill shell (bekko) industry. *Marine Turtle Newsletter* 54:17–21.

Chan, E. H.
1991. Sea Turtles. In *The State of Nature Conservation in Malaysia*, ed. R. Kiew, pp. 120–134. Kuala Lumpur: Malayan Nature Society.

Crouse, D. T.; M. Donnelly; M. J. Bean; A. Clark; W. R. Irvin; and C. E. Williams
1992. *The TED Experience: Claims and Reality*. Washington, D.C.: Center for Marine Conservation, Environmental Defense Fund, and the National Wildlife Federation.

Donnelly, M.
1991. Japan bans import of hawksbill shell effective December 1992. *Marine Turtle Newsletter* 54:1–3.

Eckert, K. L.
1993. *The Biology and Population Status of Marine Turtles in the North Pacific Ocean*. NOAA Technical Memorandum NMFS-SWFSC-186.

Eckert, K. L., and T. D. Honebrink
1992. *WIDECAST Sea Turtle Recovery Action Plan for St. Kitts and Nevis*. Caribbean Environment Programme Technical Report 17. Kingston, Jamaica: United Nations Environment Programme.

Frazier, J. G., and J. L. Brito
1990. Incidental capture of marine turtles by the swordfish fishery at San Antonio, Chile. *Marine Turtle Newsletter* 49:8–13.

Greenpeace
1989. *Sea Turtles and Indonesia*. Report prepared for the Seventh Conference of the Parties to CITES, Lausanne, 9–20 October 1989. Sussex, U.K.: The Beacon Press.

Groombridge, B.
1990. Marine Turtles in the Mediterranean: Distribution, Population Status, Conservation. Strasbourg, France: Council of Europe, *Nature and Environment Series*, No. 48.

Groombridge, B., and R. Luxmoore
1989. *The Green Turtle and Hawksbill (Reptilia: Cheloniidae): World Status, Exploitation and Trade*. Lausanne, Switzerland: CITES Secretariat.

Hutchinson, J., and M. Simmonds
1992. Escalation of threats to marine turtles. *Oryx* 26: 95–102.

Limpus, C., and D. Reimer
1994. The loggerhead turtle, *Caretta caretta*, in Queensland: a population in decline. In *Proceedings of the Australian Marine Turtle Conservation Workshop 14–17 November 1990*, compiler R. James, pp. 39–59. Queensland Department of Environment and Heritage and Australian Nature Conservation Agency.

Lutcavage, M. E.; P. L. Lutz; G. D. Bossart; and D. M. Hudson
In press. Physiologic and clinicopathologic effects of crude oil on loggerhead sea turtles. *Archives of Environmental Contamination and Toxicology*.

Meylan, A.; B. Schroeder; and A. Mosier
In press. Sea turtle nesting activity in the state of Florida, 1979–1992. *Florida Marine Research Publication*.

Milliken, T., and H. Tokunaga
1987. *The Japanese Sea Turtle Trade, 1970–1986*. Prepared by TRAFFIC (Japan) for the Center for Marine Conservation, Washington, D.C.

National Research Council
1990. *Decline of the Sea Turtles: Causes and Prevention*. Washington, D.C.: National Academy Press.

Nishemura, W., and S. Nakahigashi
1990. Incidental capture of sea turtles by Japanese research and training vessels: results of a questionnaire. *Marine Turtle Newsletter* 51:1–4.

Reichart, H. A., and J. Fretey
1993. *WIDECAST Sea Turtle Recovery Action Plan for Suriname*, ed. K. L. Eckert. Caribbean Environment Programme Technical Report 24. Kingston, Jamaica: United Nations Environment Programme.

Wetherall, J. A.; G. H. Balazs; R. A. Tokunaga; and M. Y. Y. Yong
1993. Bycatch of marine turtles in North Pacific high-seas driftnet fisheries and impacts on stocks. *North Pacific Commission Bulletin* 53(III):519–538.

Witherington, B. E., and R. E. Martin
In press. Understanding, assessing and resolving light-pollution problems on sea turtle nesting beaches. *Florida Marine Research Publication*.

Witzell, W. N., and W. G. Teas
1994. *The Impacts of Anthropogenic Debris on Marine Turtles in the Western North Atlantic Ocean*. NOAA Technical Memorandum NMFS-SEFSC-355.

Jeanne A. Mortimer
Caribbean Conservation Corporation
Mailing address:
Department of Zoology
P.O. Box 118525
University of Florida
Gainesville, Florida 32611-8525 USA

Headstarting as a Management Tool

Headstarting is the practice of growing hatchlings in captivity to a size that (theoretically) will protect them from the (presumably) high rates of natural predation that would have otherwise occurred in their early months. The rationale is that these turtles will continue to enjoy high rates of survival even after they are released to the wild. Headstarting has long been a subject of controversy among sea turtle biologists (note conflicting views expressed by Dodd, Ehrenfeld, Klima and McVey, and Reichart in this volume). Nevertheless, in virtually every country where sea turtles occur, people have repeatedly undertaken to headstart turtles. Donnelly (1994) reviews many of these cases and presents detailed evaluations of why three of the most prominent, well-funded and long-lived headstarting programs in the world—for green turtles in Florida (1959–1989), Kemp's ridleys in the United States (1978–1993), and hawksbills in Palau (1982–1991)—have all been terminated.

Critics argue that headstarting is not a proven management technique and may actually be harmful to turtles. They cite biological concerns that nutritional deficiencies and behavioral modifications associated with captivity (including insufficient exercise, lacking or inappropriate sensory stimuli, the unavailability of natural food, etc.) may interfere with the ability of headstarted turtles to survive in the open sea and with those imprinting mechanisms necessary to guide their breeding migrations (Mrosovsky 1983; Mortimer 1988; National Research Council 1990; Woody 1990, 1991; Taubes 1992; Donnelly 1994; Eckert et al. 1994). Disease is another problem, with some 27 sets of disease symptoms (many serious) common in captive turtles (Leong et al. 1989). In crowded conditions turtles bite each other causing injuries commonly invaded by secondary infections that can lead to loss of body parts (Mortimer 1988). Concern exists that captive-reared turtles might introduce or spread diseases among wild populations after their release (Woody 1981; Jacobson 1993; Donnelly 1994).

The most heavily subsidized headstarting programs, especially those at the Galveston Lab of NMFS and the Cayman Turtle Farm (both with multimillion dollar budgets), have produced useful information on sea turtle husbandry, behavior and physiology (Caillouet and Landry 1989; Caillouet 1993). Proponents claim that the emotional appeal commanded by headstarting enhances public concern for turtles (Allen 1990, 1992). Others argue that a few captive turtles would serve the same purpose and that the feel good appeal of headstarting siphons money from other programs which, though lacking popular appeal, are known to be effective (Mortimer 1988; Woody 1990, 1991; Donnelly 1994).

Ultimately, the success of headstarting as a management tool will be proven only by demonstrating that the proportion of nesting headstarted females has increased relative to the proportion of non-headstarted nesting females in the population (Mrosovsky 1983; Mortimer 1988; National Research Council 1990; Eckert et al. 1994). A critical point is that the headstarted turtles must nest at the appropriate beach in order to contribute effectively to the gene pool of the population (Bowen et al. 1994). Headstarting has always been considered experimental, but until recently, it has been an experiment lacking design and controls. In an effort to remedy this, based on recommendations by Wibbels et al. (1989) and Eckert et al. (1994), termination of the Kemp's ridley headstart program will be accompanied by intensive marking of wild hatchlings (as a control) and monitoring of those headstarted Kemp's turtles that have already been released (Byles 1993; Williams 1993; Donnelly 1994).

Modelling studies based on the analysis of reproductive value (Crouse et al. 1987) indicate that headstarting is unlikely to ever meet its goal of increased recruitment into the adult population without a simultaneous reduction in juvenile mortality in the wild (National Research Council 1990). Heppell and Crowder (1994) evaluated stage-based and age-based population models for Kemp's ridley and concluded that headstarting could not be a viable tool for species recovery because the addition of headstarted turtles is not sufficient to compensate for the annual loss of fecund adults. These models indicate that attempting to compensate for natural hatchling mortality without addressing the real causes of the decline of the species—i.e., overharvest, mortality in fishing gear, and habitat destruction—is not the best use of the limited resources available for conservation programs.

Literature Cited

Allen, C. H.
1990. Guest editorial: Give headstarting a chance. *Marine Turtle Newsletter* 51:12–16.

1992. It's time to give Kemp's ridley head-starting a fair and scientific evaluation! *Marine Turtle Newsletter* 56:21–24.

Bowen, B. W.; T. A. Conant; and S. R. Hopkins-Murphy
1994. Where are they now? The Kemp's ridley headstart project. *Conservation Biology* 8:853–856.

Byles, R.
1993. Head-start experiment no longer rearing Kemp's ridleys. *Marine Turtle Newsletter* 63:1–3.

Caillouet, C. W.
1993. *Publications and Reports on Sea Turtle Research by the NMFS Galveston Laboratory 1979–1992.* NOAA Technical Memorandum NMFS-SEFC-328.

Caillouet, C. W., and A. M. Landry, Jr., editors
1989. *Proceedings of the First International Symposium on Kemp's Ridley Sea Turtle Biology, Conservation and Management.* Texas A&M University, Sea Grant College Program, TAMU-SG-89-105.

Crouse, D. T.; L. B. Crowder; and H. Caswell
1987. A stage-based population model for loggerhead sea turtles and implications for conservation. *Ecology* 68:1412–1423.

Donnelly, M.
1994. *Sea Turtle Mariculture: A Review of Relevant Information for Conservation and Commerce.* Washington, D.C.: Center for Marine Conservation.

Eckert, S. A.; D. Crouse; L. B. Crowder; M. Maceina; and A. Shah
1994. *Review of the Kemp's Ridley Sea Turtle Headstart Program.* NOAA Technical Memorandum NMFS-OPR-3.

Heppell, S. S., and L. B. Crowder
1994. Is headstarting headed in the right direction? In *Proceedings of the Thirteenth Annual Symposium on Sea Turtle Biology and Conservation*, compilers B. A. Schroeder and B. E. Witherington, p. 278. NOAA Technical Memorandum NMFS-SEFSC-341.

Jacobson, E. R.
1993. Implications of infectious diseases for captive propagation and introduction programs of threatened/endangered reptiles. *Journal of Zoo and Wildlife Medicine* 24:245–255.

Leong, J. K.; D. L. Smith; D. B. Revera; Lt. J. C. Clary III; D. H. Lewis; J. L. Scott; and A. R. DiNuzzo
1989. Health care and diseases of captive-reared loggerhead and Kemp's ridley sea turtles. In *Proceedings of the First International Symposium on Kemp's Ridley Sea Turtle Biology, Conservation and Management*, eds. C. W. Caillouet, Jr., and A. M. Landry, Jr., pp. 178–201. Texas A&M University, Sea Grant College Program, TAMU-SG-89-105.

Mortimer, J. A.
1988. Management options for sea turtles: re-evaluating priorities. *Florida Defenders of the Environment Bulletin 25.* 4 pp.

Mrosovsky, N.
1983. *Conserving Sea Turtles.* London: British Herpetological Society.

JEANNE A. MORTIMER

National Research Council
1990. *Decline of the Sea Turtles: Causes and Prevention.*
       Washington, D.C.: National Academy Press.

Taubes, G.
1992. A dubious battle to save the Kemp's ridley sea tur-
       tle. *Science* 256:614–616.

Wibbels, T.; N. Frazer; M. Grassman; J. Hendrickson; and
P. Pritchard
1989. Blue Ribbon Panel review of the National Marine
       Fisheries Service Kemp's ridley headstart program.
       Report to the National Marine Fisheries Service,
       submitted to the Southeast Regional Office, August
       1989. 11 pp.

Williams, P.
1993. NMFS to concentrate on measuring survivorship,
       fecundity of head-started Kemp's ridleys in the wild.
       Marine Turtle Newsletter 63:3–4.

Woody, J. B.
1981. Head-starting of Kemp's ridley. *Marine Turtle
       Newsletter* 19:5–6.
1990. Guest editorial: Is "headstarting" a reasonable con-
       servation measure? "On the surface, yes; in reality,
       no." *Marine Turtle Newsletter* 50:8–11.
1991. Guest editorial: It's time to stop headstarting
       Kemp's ridley. *Marine Turtle Newsletter* 55:7–8.